Materials Horizons: From Nature to Nanomaterials

Series Editor

Vijay Kumar Thakur, School of Aerospace, Transport and Manufacturing,
Cranfield University, Cranfield, UK

Materials are an indispensable part of human civilization since the inception of life on earth. With the passage of time, innumerable new materials have been explored as well as developed and the search for new innovative materials continues briskly. Keeping in mind the immense perspectives of various classes of materials, this series aims at providing a comprehensive collection of works across the breadth of materials research at cutting-edge interface of materials science with physics, chemistry, biology and engineering.

This series covers a galaxy of materials ranging from natural materials to nanomaterials. Some of the topics include but not limited to: biological materials, biomimetic materials, ceramics, composites, coatings, functional materials, glasses, inorganic materials, inorganic-organic hybrids, metals, membranes, magnetic materials, manufacturing of materials, nanomaterials, organic materials and pigments to name a few. The series provides most timely and comprehensive information on advanced synthesis, processing, characterization, manufacturing and applications in a broad range of interdisciplinary fields in science, engineering and technology.

This series accepts both authored and edited works, including textbooks, monographs, reference works, and professional books. The books in this series will provide a deep insight into the state-of-art of Materials Horizons and serve students, academic, government and industrial scientists involved in all aspects of materials research.

Review Process

The proposal for each volume is reviewed by the following:

1. Responsible (in-house) editor
2. One external subject expert
3. One of the editorial board members.

The chapters in each volume are individually reviewed single blind by expert reviewers and the volume editor.

Satyaranjan Bairagi · Shakeel Ahmed ·
S. Wazed Ali
Editors

Nanotechnology in Textile Finishing

Advancements and Applications

Springer

Editors
Satyaranjan Bairagi
Materials and Manufacturing Research Group (MMRG)
School of Engineering
University of Glasgow
Glasgow, UK

Shakeel Ahmed
Postgraduate Department of Chemistry
Government Postgraduate College Rajouri
Rajouri, Jammu and Kashmir, India

S. Wazed Ali
Department of Textile and Fibre Engineering
Indian Institute of Technology Delhi, Hauz Khas
New Delhi, Delhi, India

ISSN 2524-5384 ISSN 2524-5392 (electronic)
Materials Horizons: From Nature to Nanomaterials
ISBN 978-981-97-2695-0 ISBN 978-981-97-2696-7 (eBook)
https://doi.org/10.1007/978-981-97-2696-7

This Springer imprint is published by the registered company Springer Nature Singapore Pte Ltd.
The registered company address is: 152 Beach Road, #21-01/04 Gateway East, Singapore 189721, Singapore

Preface

Nanotechnology has become a revolutionary force in the field of materials science, altering the limits of what is achievable in many industries. Nanotechnology has fundamentally transformed our understanding, creation, and application of textiles. The book *Nanotechnology in Textile Finishing* aims to thoroughly examine the convergence of nanotechnology and textile finishing, providing valuable insights into the latest advancements that are driving the textile industry towards a transformative era of opportunities.

Textile finishing, the ultimate stage in the textile production process, has conventionally prioritized the enhancement of the visual and practical characteristics of materials. Nanotechnology, by manipulating materials at the nanoscale, offers numerous possibilities to redefine and enhance the performance characteristics of textiles. The incorporation of nanoparticles into textile finishing techniques enables the realization of unparalleled functionality, longevity, and environmental friendliness, ranging from antibacterial treatments to water-repellent coatings.

This book aims to provide a helpful resource for researchers, scientists, engineers, and industry experts who are interested in gaining a comprehensive grasp of the complex relationship between nanotechnology and textile finishing. By conducting a methodical investigation of fundamental ideas, approaches, and practical uses, individuals will acquire a deep understanding of the most recent breakthroughs and developing patterns in this ever-changing domain. The chapters in this collection are written by professionals to present a combination of theoretical foundations and practical issues, providing a comprehensive view of the subject.

As we explore the contents of *Nanotechnology in Textile Finishing*, we start on a voyage across the domains of fundamental of nanotechnology, various nanomaterials, nanomaterials manufacturing methods, nanomaterials properties, and their utilization in textile finishing. The book is organized in a manner that leads readers through basic ideas, investigates particular nanomaterials and their characteristics, examines advanced manufacturing techniques, and presents practical examples that demonstrate the revolutionary capabilities of nanotechnology in textile finishing.

The authors of this collection are the experts in their specific areas of expertise, who add a substantial amount of knowledge and experience to the discussion on nanotechnology and textiles. Their combined knowledge sheds light on the continuous transformation in textile engineering, where nanotechnology acts as a driving force for creativity, environmental consciousness, and improved functionality.

This book aims to stimulate curiosity, ignite invention, and function as a helpful resource for individuals fascinated by the intersection of nanotechnology and textiles. As we explore the complex terrains of nanomaterials and textile finishing, let us welcome the boundless opportunities that arise from this thrilling intersection.

Ultimately, the editors of this book express their gratitude to the contributors for their exceptional contributions in finishing this exquisite book, as well as to Springer Nature for publishing this book.

New Delhi, India
Spring 2024

Satyaranjan Bairagi, Ph.D.
Shakeel Ahmed, Ph.D.
S. Wazed Ali, Ph.D.

About This book

This book is focused on the latest developments and practical applications of nanotechnology in textile finishing. It covers the fundamentals of nanotechnology, including the properties and behaviour of nanoparticles, and how they can be used to enhance the performance of textiles. The book also explores the various types of nanomaterials that are used in textile finishing, such as nanoparticles, nanocomposites, and nano-coatings and their properties, advantages, and limitations.

The book covers the different types of textile finishing techniques, including dyeing, printing, and coating, and how nanotechnology is used to improve their performance. It also covers the environmental, health, and safety aspects of using nanotechnology in textile finishing and the challenges and opportunities that lie ahead.

The book is targeted at textile scientists, engineers, and researchers working in the textile industry, as well as students and academics in textile science and engineering. It is also useful for those in related fields, such as materials science, chemistry, and chemical engineering.

Key Features of This Book

1. Provides a comprehensive overview of the latest developments and practical applications of nanotechnology in textile finishing.
2. Covers the fundamentals of nanotechnology, including the properties and behaviour of nanoparticles, and how they can be used to enhance the performance of textiles.
3. Explores the various types of nanomaterials that are used in textile finishing, such as nanoparticles, nanocomposites, and nano-coatings and their properties, advantages, and limitations.

4. Discusses the different types of textile finishing techniques, including dyeing, printing, and coating, and how nanotechnology is used to improve their performance.
5. Addresses the environmental, health, and safety aspects of using nanotechnology in textile finishing and the challenges and opportunities that lie ahead.

Contents

1 Introduction to Nanotechnology in Textile Finishing 1
Chintan R. Madhu

2 Types of Nanomaterials Used in Textile Finishing 23
Sudhanshu Singh, Zunjarrao Kamble, and Ghanshyam Neje

3 Metal Nanoparticles: Biosynthesis and Functional Application to Textiles 49
B. H. Patel and M. Z. Channiwala

4 Advanced Characterization Techniques of Nanoparticles for Textile Finishing 79
Sourav Banerjee, Mandira Mondal, Swagata Banerjee, S. Wazed Ali, and S. N. Naik

5 Modern Developments in Dyeing and Printing Using Nanotechnology 97
Anjali Agrawal and Kunal Singha

6 Nanopigments in Textiles 123
Kamaladin Gharanjig, Hamid Gharanjig, Hooman Imani, and Siavash Goudarzi

7 Nanotechnology in Coating and Finishing of Textiles 165
Lelona Pradhan, Saptarshi Maiti, and Ravindra V. Adivarekar

8 UV Protection and Antimicrobial Properties of Textiles Using Nanotechnology 191
Sharif Ahmed, A. K. M. Ashiqur Rahman Mazumdar, Md. Syduzzaman, and Md. Sultan Mahmud

9 Water-Repellent and Oil-Repellent Finishing on Textiles Using Nanotechnology 233
Saikat Ghosh and Subhadeep Paul

10 Nanotechnology in Flame Retardancy of Textile 271
Sayak Nandi, Ishmam Zahin Chowdhury, and Sumit Mandal

11 Nano-finishing of Natural Fibres 313
Subhadeep Paul and Saikat Ghosh

12 Nanofinishing of Synthetic Fibres 353
Manisha Yadav, Nagender Singh, Shelly Khanna, and Amit Madhu

13 Nanofinishing of Nonwoven Fabrics 381
Chintan R. Madhu

14 Nano-finishing in Apparel Textiles 415
Nagender Singh, Amit Madhu, and Manisha Yadav

15 Nanotechnology in Smart Textiles 441
Danmei Sun, Madiha Ahmad, Muhammad Owais Raza Siddiqui, and Kashif Iqbal

16 Biocompatibility and Toxicity of Nanoparticles in Textile Finishing 461
Mandira Mondal, Sourav Banerjee, Swagata Banerjee, S. N. Naik, and S. Wazed Ali

17 Challenges and Opportunities in Nanotechnology for Textile Finishing 487
Amit Madhu, Nagender Singh, and Manisha Yadav

18 Nanofinishing and Circular Economy 513
Monabbir Rafsan Fahim, Mahamu-Ul Islam, Dipa Saha, Humayra Akhter Himu, Shamima Akter Smriti, and Ummul Khair Fatema

About the Editors

Satyaranjan Bairagi, Ph.D. is a postdoctoral research associate under a project funded by Engineering and Physical Sciences Research Council (EPSRC), at the University of Glasgow, United Kingdom. He completed his Ph.D. in textile-based flexible piezoelectric materials from Indian Institute of Technology Delhi, India. His area of research interest includes development of nanotechnology-based piezoelectric and triboelectric materials for nanogenerators as well as textile-based flexible advanced materials for renewable energy harvesting. He has published numerous research articles (> 35) in various reputed peer-reviewed journals. In addition, Dr. Bairagi has published and written more than 40 book chapters which have been published and submitted in different books under various publishers such as Elsevier, Wiley, Springer, and so forth.

Shakeel Ahmed, Ph.D. is working as an assistant professor of chemistry at Higher Education Department, Government of Jammu and Kashmir, and assistant professor at Postgraduate Department of Chemistry, Government Postgraduate College Rajouri, Jammu and Kashmir, India. He worked as an assistant professor, Department of Chemistry, Government Degree College Mendhar. He obtained a first degree in general science from Government Postgraduate College Rajouri (University of Jammu) followed by master's degree and doctoral degree in chemistry from Jamia Millia Islamia, New Delhi. He gained postdoctoral experience in biocomposite materials at Indian Institute of Technology, Delhi. He has published several research publications in area of nanomaterials and composite materials for various applications including biomedical, packaging, and water treatment. He has published more than 30 books in the area of nanomaterials and green materials with publishers of international repute. He is the fellow of The Linnean Society of London and the recipient of many awards from different agencies.

S. Wazed Ali, Ph.D. is an associate professor in the Department of Textile and Fibre Engineering at Indian Institute of Technology (IIT) Delhi, India. After obtaining his Ph.D. from IIT Delhi, he joined General Electric (GE) India Technology Centre Pvt. Ltd. (JFWTC), Bengaluru, as a research scientist. Subsequently he moved to

SABIC Research and Technology Pt. Ltd., Bengaluru, and worked as a scientist before joining IIT Delhi in 2014. During his R&D service at GE India Technology, he was awarded with 'Bronze Medal' for above and beyond performance. He was the recipient of two International Fellowships during his Ph.D. studies. He was awarded with 'Gold Medal' for his highest score in M.Tech. in Fibre Science and Technology from IIT Delhi in 2006. He has been awarded prestigious 'Commonwealth Split-Site Doctoral Fellowship' 2008 from the Commonwealth Fellowship Commission, London, to pursue 1 year research work at UK during his Ph.D. studies. He was a recipient of 'Bolton Fellowship' from the University of Bolton, UK, for 3 year's Ph.D. program at IIT Delhi. His research areas broadly include nano fibrous web for energy, nanocomposite-based flexible piezoelectric materials, green chemistry in textiles, etc. He was also awarded with Young Scientist start-up Research Grant from Science and Engineering Research Board (SERB), Government of India. He has published 50 papers in different international peer-reviewed journals and presented research outcome in 39 different conferences. He has contributed and authored about 65 book chapters. He has applied 8 Patents (6 Indian, 1 European as well as 1 US Patent). He is a life member of the Indian Natural Fibre Society and regular reviewer of different peer review international journals. He is a member of Editorial Board of *American Journal of Polymer Science and Technology* and an associate editorial board member of *Current Applied Polymer Science.*

Chapter 1
Introduction to Nanotechnology in Textile Finishing

Chintan R. Madhu

1 Introduction

Nanotechnology has attracted researchers from the last 45 years. Nanotechnology is considered as the assemblies having at least one nanometer dimension while manufacturing the material which has been defined by National Nanotechnology Initiative (NNI). Also, nano-sized devices, materials, or assemblies characterized and observed significant enhancement in properties [80]. The range of nanoparticles are between 1 and 100 nm. Nanotechnology controls the material structure by producing very tiny assemblies, which shows the activities at the level of atom and can be apply for human service [69]. Mutually, nanotechnology and nano-science can be applied in the field of polymer and textiles. Performance of the textiles has enhanced by applying nano-based processing, which has been verified already. The new approach of nanotechnologies in textiles can introduced an improved and new product [44]. A new improved characteristics can be introduced with the help of nanotechnology. Science and technology have performed many experiments and proved that nanotechnology is growing at a great extent. In textile industries, nanotechnology is also performing greatly as a diverse applied field. The application in textiles has started with nanocomposites, nanofibers, and today it has been reached at smart coating using polymer for high performance. It also functionalized conventional finishing process by improving performance. Nanotechnology encroachments in textiles are advantageous with consistency, durability, and repeatability.

Coating of nanoparticles to various textiles improves the functionality during various textile processing operations. Nano-Tex performed the key work for nanotechnology application in textiles [80]. Then, the application was adopted and researched by other industries. One of those is application of nanoparticles with the combination of a carries, a surfactant and other ingredients, and the result amended

C. R. Madhu (✉)
Department of Textile Processing Technology, R. C. Technical Institute, Ahmedabad, India
e-mail: chintan.madhu@gujgov.edu.in

S. Bairagi et al. (eds.), *Nanotechnology in Textile Finishing*, Materials Horizons: From Nature to Nanomaterials, https://doi.org/10.1007/978-981-97-2696-7_1

the surface of textiles [34]. Various application methods are applied for application of nanoparticles such as spraying, transfer printing, washing, rinsing, and padding. Padding is most commonly used from all of the above methods [24, 33]. In padding, one can maintain the nip pressure, speed for the application of nanoparticles, and the padding is followed by drying and curing. Several properties like water repellency, soil resistance, wrinkle resistance, anti-bacteria, anti-static and UV protection, flame retardation, improvement of dye-ability can be improved or applied to textiles with nanotechnology.

2 Nanomaterials for Textile Finishing

The performance of textiles has been improved by improving the functionality with the nanomaterials such as nano-composite fibers, carbon nanofibers, nanoparticles, clay nanoparticles, carbon nanotubes (CNT), cellulosic and carbon nano-whiskers, nano-cellular foam structure, and inorganic color pigments like TiO_2, ZnO, Fe_2O_3, MgO, SiO_2, from which CNT is utilized as conductive fiber for smart textiles, and they can be classified as shown in Fig. 1 [83, 84]. Textiles can be produced by coating the cotton with polyelectrolyte-based coating of CNTs for imparting detecting abilities to the coated textiles. In medical textiles, albumin can be detected with the help of CNTs-coated cotton yarns. For transforming the solar energy into electrical energy, textiles can be modified with CNTs, which may be applied for improving the self-assembled power energy garments and their assemblies [49]. Mechanical strength, resistance against cracking at higher temperature, and better thermal insulation can be improved with the application of nano-cellular foam structures, which are polymeric materials with lighter in weight, and having the nano-size porosity [40, 83, 84]. Cellulosic materials can be overpowered with nano-composites especially for mechanical properties. Moreover, nano-composites with high strength are biodegradable as they are bio-based [82, 85, 86]. Various factors affect the selection of materials for preparing nanoscale materials which are utilized for finishing of textiles, and those factors are desirable size of nanoparticles, surface characteristics, degree of biodegradability, biocompatibility, toxicity levels and inherent properties like solubility, and stability.

3 Nanofinishing Application Methods

Various methods used for applying the nanomaterials on textiles viz. coating, electro deposition, sol–gel, etc.

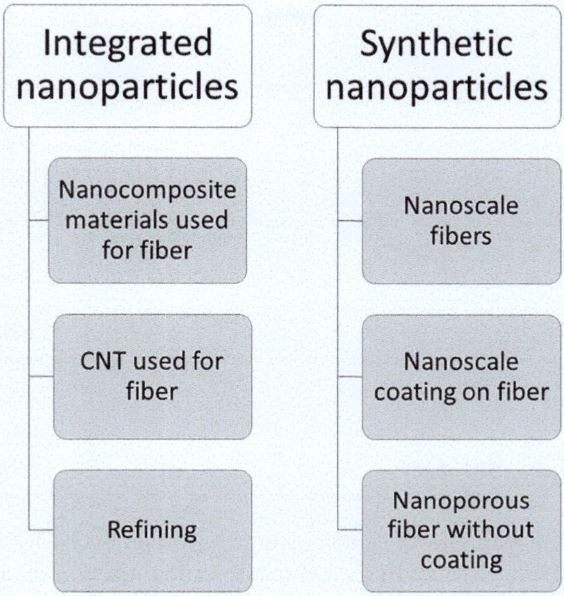

Fig. 1 Classification of nanoparticles based on applications

3.1 Coating

Dip coating is most commonly utilized for coating the textiles as it requires minimum effort. In this process, textile materials are dipped in fluid having low viscosity, and the dipped materials run with the persistent speed and followed by drying and curing. Foam coating is also used by textile industries, but issues like reproducibility hindered the application's popularity [13].

3.2 Electro Deposition

Electro deposition is performed with the principle of electroless deposition (plating), which can also be denoted by chemical plating. The metal ions are deposited on the surface of textiles in the presence of a reducing agent. This is the technique in which nanoparticles are deposited on the textiles, not coated. Commonly, researchers have performed this plaiting of silver, copper, nickel, or combinations of these nanomaterials in textiles. This plaiting is based on chemical reaction. These plated textiles are utilized in the EMI shielding, self-cleaning, antimicrobial, and electrical conductivity [57, 59, 60, 89].

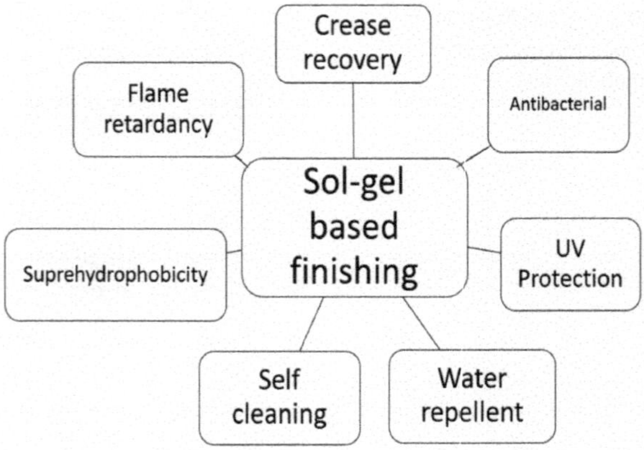

Fig. 2 Sol–gel-based nanofinishing

3.3 Sol–Gel

Nanoparticles or nano-coating can be produced by the sol–gel method which is ultimately hydrolysis and condensation reaction of silanes, the metal oxides [58]. For enhancing the performance of textile surface for achieving self-cleaning, flame retardant, water repellent, EMI shielding, antimicrobial finishes, done with pad-dry-cure, or dip coating methods. Wide investigations have been performed for sol–gel coating of TiO_2, ZnO, Cu_2O, and silica nanoparticles [89]. Various applications of sol–gel finishing of textiles are shown in Fig. 2.

4 Nanofinishing of Textiles

According to the application, nanofinishing can be classified into three major categories suggested in Fig. 3. and properties can be enhanced as suggested in Fig. 4.

4.1 UV Protection

As the atmospheric effect changes continuously, humans are aware about UV protection, and also want such a textile which can benefit with the same. UV protection can be imparted on textiles with inorganic or organic chemicals. Commonly used organic chemicals are phenyltriazines, phenyl salicylate, benzophenone, etc. Organic UV protection chemicals are toxic in nature and also have poor activity and durability. While inorganic chemicals used for UV protection possess a broad range of UV blocking effects. Biosynthesized ZnO nanoparticles have been applied on cotton, and the finished product suggested admirable UV protection and antibacterial characteristics. Also this ZnO decreases the toxicity of cotton finished with nanoparticles

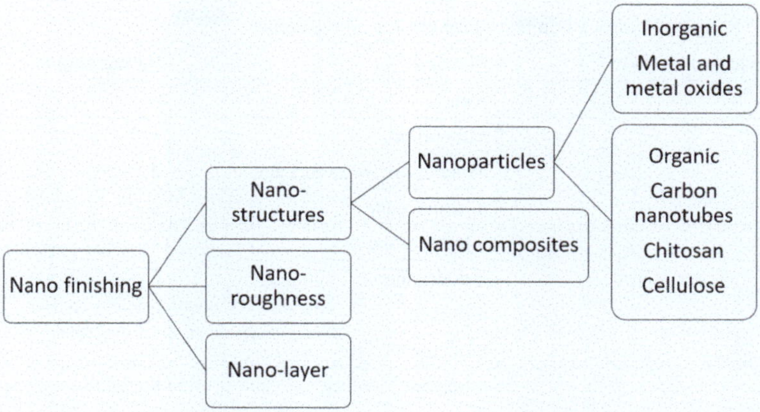

Fig. 3 Classification of nanofinishing

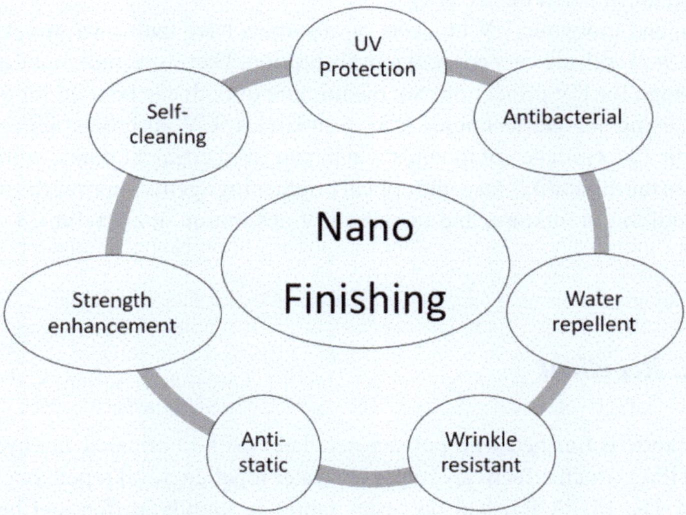

Fig. 4 Various properties enhanced by nanofinishing

[17]. Very small amount of ZnO nanoparticles coated on cotton, also maintains the antibacterial and UV protection properties with durability, i.e., after washing and abrasion, the finished cotton suggested outstanding result [47]. ZnO nanoparticles encapsulated with seaweed applied on cotton, and treated fabric shows excellent Ultraviolet Protection Factor (UPF), which ranges from 43 to 45 [54].

SiO_2 nanoparticles-chitosan-based inorganic UV blockers applied on polyester and viscose rayon, and the finished fabric resulted in an increment of 260% in the UPF (UV protection factor) compared to unfinished fabric [6]. Aramid fiber finished

Table 1 Rating of hydrocarbon resistance test for oil repellency

Surface tension (mN m^{-1})	Materials	Rating number
–	None	0
31.5	Mineral oil	1
29.6	Mineral oil: *n*-hexadecane	2
27.3	*n*-hexadecane	3
26.4	*n*-tetradecane	4
24.7	*n*-dodecane	5
23.5	*n*-decane	6
21.4	*n*-octane	7
19.8	*n*-heptane	8

with amorphous TiO_2 resulted in UV protection till 10 washing cycles, it also ensures better mechanical performance [11].

Organic and inorganic UV blockers or absorbers have their own wavelength for absorption, and both are very particular in absorption. Therefore, individual absorbers are not enough for UV protection. So, the mixture of both can be used for better UV protection using sol–gel technique [55]. Sometimes, UV absorbers are applied by mixing with alkoxysilane compounds which can form covalent bonds with textiles, and increase the durability. Also, the optical brightening agents improve the whiteness with the addition in mixture, and increase UV absorption, majorly in UV-A region [26].

4.2 *Soil Repellent*

Polyester fabric is finished with polyacrylate latex with fluorinated and hydrophilic groups (FHPA) which effectively improves water repellency, oil repellency, and soil repellency. The FHPA finished polyester fabric is slightly stiffen and bending is increased in less amount. The AATCC test method (AATCC test method—118) for oil repellency is shown in Table 1. As the number increases in rating, the resistance to oil also increases [75].

4.3 *Oil/Water Separation*

The technique in which oil can pass from the surface and transfer denoted as hydrophobic or oleophilic, resist the water, and if it's hydrophilic or oleophobic, then allow the passing of water and resist the oil [39]. Silica nanoparticles with the help of sono-chemical irradiation applied on the cotton fabric, and the result achieved was a

CA of 159 1.2° [71]. The effectiveness of separating the oil from water is increased to the maximum at 98.2%, and also it can be decreased to 94% after 30 cycles of separating. Eco-friendly antibacterial and flame retardant finishing obtained on cotton with the application of chitosan phosphate, BTCA, and TiO_2 by sol–gel technique [15]. The resulting fabric shows the improvement in surface area for hydrophobicity. Surface of cotton also modified with bifunctional fluorosilsesquioxane and superhydrophobic properties with a contact angle of 151° has been achieved with the resistance to 10 washing cycles [61]. Superhydrophobic cotton was also prepared by sol–gel technique with the use of TiO_2 particles, which also provides UV protection properties.

4.4 Self-cleaning

Plant leaves have the self-cleaning property, and Professor W. Barthlott of the University of Bonn, Germany, has observed the same phenomenon by bringing the concept of lotus effect for superhydrophobicity. The effect, called the Lotus effect, is patented by Barlett. Based on the lotus effect phenomenon, nanospheres were constructed which can provide stain repellency, oil and water repellency. Not only the plant leaves but insects also have the self-cleaning properties. Rice leaves, salvinia molesta, butterfly wings, fish scales, shark skin, and mosquito eyes are all suggest lotus effects which are bio-inspired [8, 20, 38, 41]. Cellulose nanofibers (CNFs) collected from hardwood and mixed with (2,2,6,6-tetramethylpiperidine-1-oxyl)oxidany for the application on cotton and polyester to achieve self-cleaning effect. Treated fabric suggests average WCA 162° and 152°. These angles clearly suggest the superhydrophobicity characteristics [66]. The mixture of anatase TiO_2 and polydimethylsiloxane (PDMS) gives the fluorine-free self-cleaning properties to cotton with the WCA of 153.8° [32], and the main attracted zone of this technique is fluorine-free product. Self-cleaning finish on garments having less stability with the exposure to UV light. Specially, the outer area of a garment is affected when it directly comes in the exposure of UV light, which one is the issue of durability while obtaining a self-cleaning surface. Many scholars suggested that the nature of self-cleaning is not affected by stabilizing them with the UV exposure. Major demerit of the self-cleaning textiles is the metal corrosion.

4.5 Self-cleaning by Photocatalytic Action

The fabric is treated with colloids of TiO_2 prepared through sol–gel method via 1-dip a-nip and curing process [87], and finished fabric resulted degradation with red wine stains under the photolytic exposure, also treated fabric self-cleaned itself till the 50 frequent washes. This application has also been performed on wool for imparting durable self-cleaning finishes. The TiO_2 nanoparticles are stable on cotton fabrics even after 50 washes, and this technique can be further applied to protein fibers like wool to develop durable self-cleaning coating layers [12]. The technique includes –OH, –SH, and –NH functional groups present in the structure of wool, especially on the surface with TiO_2 nanoparticles. Some scholars also worked for novel application techniques for TiO_2 nanoparticles application on fabric commercially [10]. To increase the stability of nanoparticles on the surface of textiles, various methods can be used like cross-linking, enzymatic pretreatment, sono-processing, and plasma [50]. Enhancement in photo efficiency of nanoparticles which also affects superhydrophobicity can be achieved by increasing the concentration of silica. Shorter exposure of illumination can inhibit the photodegradation of the fabric while it is under the exposure of incident light. Increment in concentration of silica blocks the pores of woven fabric and so, it affects negatively on air permeability and bending rigidity of the fabric. Also, nanoparticles on wool and silk show poor wash fastness, which is the demerit of such a functional finish. So, capacity for activeness can be increased for TiO_2 nanoparticles in the visible region, by doping them with noble metals, and then the mixture is applied for finishing the fabric. Scholars have worked for preparing the mixture of nanoparticle and noble metals, such as noble metals such as silver (Ag), gold (Au), and platinum (Pt) into the synthesis of TiO_2 nanoparticles to shift the threshold to visible region [52]. On the exposure of UV light and the visible light, finished cotton and wool fabrics with TiO_2/Au, TiO_2/Ag, TiO_2/Pt, and TiO_2/Au/ SiO_2 suggested higher efficiency. Mn (manganese) and Cu_2O (copper oxide) doping also increased the photocatalytic efficiency of TiO_2 nanoparticles while applied on cotton and polyester [18, 51, 91]. Dyes can also mix with metal doping for photocatalytic activity, which also increases the light activity of TiO_2 nanoparticles. The TiO_2-coated cotton fabric passed through the dye solution which contains meso-tetra (4-carboxyphenyl) porphyrin (TCPP) and Cu/TCPP, and the resulting fabric shows photostability, especially in later cases. Also, the coated fabric shows reduction of dye only 5% after the exposure of light for 30 h [3].

4.6 Superhydrophobic

Cotton, polyester, and their blends have been spray finished with nano-sol, which contained SiO_2 nanoparticle, 3-aminopropyltriethoxy silane (APTES) and hexadecyltrimethoxy silane (HDTMS), and kept overnight for drying at room temperature. WCA analysis of the finished fabric resulted by achieving superhydrophobicity with WCA higher than 150°. As the HDTMS or silica nanoparticles concentration increases, CA of untreated fabric is higher due to lower surface energy. Till the 5 cycles of washing, cotton and polyester showed superhydrophobicity, and drying was performed after washing, and the fabric retained its hydrophobicity till 600 [96]. For protein fiber fabric, like wool, fluoro-free-organosilicone prepared and applied for achieving durable superhydrophobicity. A nanocomposite prepared by using oligomer of methyltrimethoxysilane (HDTES) (HD-oligomer) and HDTES modified silica nanoparticle (HD-silica) via modified Stöber method, applied on wool for converting hydrophobic wool and superhydrophobic. For application of this nanocomposite on wool, a dip coating methods was used by utilizing catalyst such as toluene and triethylamine, and ultrasonic treatment was performed on wool for imparting superhydrophobicity with great environmental stability and abrasion resistance [93]. Water-repellent finish can impart on cotton by using ZnO, Al_2O_3, TiO_2, and ZrO_2 nanoparticles. In this process, primarily the cotton surface is made rough with the application of ZnO, Al_2O_3, TiO_2, and ZrO_2 nanoparticles, and then altered by HDMS for hydrophobicity. A mixture of Al_2O_3 nanoparticles and HDMS shows superhydrophobicity on cotton, even the properties maintained after washing also [48]. A mixture of PDMS, fluoroalkyl silane, and fluoroalkyl silane-modified SiO_2 dip coated on the fabric shows durable superhydrophobicity, and the finished fabric maintained the superhydrophobicity with WCA above 160° after 500 washing cycles, and also shows outstanding durability against acids and alkalis [98]. Anisotropic Ag nanoparticles applied on cotton using post-fluorination technique with color (red, blue, and yellow) for preparing superhydrophobic antibacterial fabric. For preparing such fabric, at the first stage, the fabric was dipped in the solution of polyethylenimine (PEI) for 20 min to impart positive charge on the fabric. In the second stage, PEI-treated fabric is immersed in the solution of Ag NPs with the weight ratio of 300:1 for 30 min, and finally the fabric is treated with fluorinated decyl polyhedral oligomeric silsesquioxane (F-POSS) ethanol dispersion (5 mg/mL). Due to surface plasmon resonance (SPR) of Ag NPs, various colors have been obtained on the fabric. Size, shape, and thickness of nanoparticles directly affect the color. This method prepares a protective layer on the fabric which increases the durability of Ag NPS. Finished fabric gives superhydrophobicity with WCA 169° and SA 3°. Also, prepared samples show antibacterial resistance against *Escherichia coli* (*E. coli*) and *Bacillus subtilis* (*B. subtilis*) with good washing fastness [81].

4.7 Electrospun Nanofibers

Superhydrophobicity can be achieved by creating the roughness with the electrospinning of nanofibers. Nanofibers with the several nanometers to hundreds of nanometers of diameter which is the base of polymer types in the spinning solution. Also, adding the nanomaterials in the nanofiber structure with the use of polymers having the low surface energy can increase the superhydrophobicity on nanofibers by increasing the roughness [42, 68]. Many researchers have also worked for applying nanofibers on natural and synthetic fibers as a coating material to increase the superhydrophobicity of electrospun mats, like various fibers such as wool, cotton, polyester, viscose, and acetate have been cultivated with silicone (polymethylsilsesquioxane) nanofilaments [99]. Also, results strongly support that size, distribution, and density of nanofilaments on the fibers affect the substrate. Different textiles were coated with above material, and from all the treated textiles polyester shows the great superhydrophobicity with SA 2°. As the layer of coating is damaged, SA will increase simultaneously.

4.8 Nanometals/Metal Oxides

Different nanomaterials like copper (Cu), silver (Ag), gold (Au), and iron (Fe) have been used to achieve excellent antimicrobial and electromagnetic shielding (EMI) and to get better antimicrobial and electromagnetic shielding (EMI) properties in textiles. Also, for achieving different function characteristics like electrical conductivity, antimicrobial, photocatalytic, UV absorption, and photo oxidations, nanometal oxides TiO_2, ZnO, Al_2O_3, and MgO were utilized. The details of metal and metal oxides used for textile finishing are tabulated in Table 2.

4.8.1 Copper

Amazing antibacterial and antifungal properties have been shown by silver nanoparticles when applied in various biomedical and antimicrobial textiles, and also on the everyday used products [53]. Medical textile consumes maximum silver nanoparticles, but the warning statements have been issued for limiting the usage of the said nanoparticles, especially cytotoxicity. Different cell activities are controversial as the application field of nanoparticles has limited source, and so the disturbing of nanoparticles happens at their size, shape, concentration, aggregation, or agglomeration level. In wound dressing, the presence of silver resisting bacteria has been confirmed, and to resolve this issue, metallic Cu, Cu_2O, CuO and their blends based nanoparticles had been studied by scholars, and resulting applications show excellent antimicrobial activity. Also, for producing Ag-based nanoparticles silver nitrate is used as precursor salt, which is expensive, while production of Cu nanoparticles is more economical than Ag nanoparticles, and so it is more viably used for imparting antimicrobial properties on textiles. Cellulosic textiles are effectively finished with

Table 2 Metals and nonmetals for textile finishing

Metal/metal oxides	Properties	Compounds
Metal oxides	Photocatalytic Self-cleaning UV protection Flame retardant antimicrobial	TiO_2
	Photocatalytic Self-cleaning UV protection Flame retardant antimicrobial	ZnO
	Antimicrobial	CuO and Cu_2O
	Antimicrobial	ZrO_2
Metal	Antimicrobial	Au
	Antimicrobial	Ag
	Antimicrobial	Cu

Cu nanoparticles, and those particles have exclusive tact [77]. Cellulose has a very lower ion exchange characteristic which can be incapacitating by adding carboxyl and amino groups in the fiber with the binding of Cu_2 ions. Many reducing agents have been utilized for Cu reduction by considering attendance of stabilizers, and salt precursors which affects the size, chemical structure, and aggregation of the nanoparticles [62, 65]. Main stage of synthesis is to take care of oxidation state transition of copper, also it must look at the time of storing and controlling exploitation. For core/shell structure concepts, characterization is multifaceted while mixing it with other compounds. Antimicrobial activity level and its mechanism is not common for all the types and states for copper nanoparticles, such as Cu, Cu_2O, CuO, and their mixtures, and also provides excellent antimicrobial properties against bacteria. For wound dressing, Cu-based nanoparticles application has been studied by various research scholars for their mechanism behind antimicrobial property, their action, and their boundaries. Commonly, for textiles, Cu nanoparticles have been researched for their antimicrobial activity and durability of the same up to certain washings. Moreover, Cu nanoparticles are majorly applied to textiles goods in recent era and must be studied for environmental concern [56, 79].

4.8.2 ZnO

About 15-nm diameter has been studied and observed for Ag-based nanoparticles, in the same concern, zinc oxide (ZnO) nanoparticles are most durable even after deposition with 2.5% doping quantity [19]. Also, ZnO nanoparticles finished cotton has shown outstanding hydrophobicity, including UV resistance, antibacterial activity, and anti-mildew. Above all, ZnO finished fabric shown 139° contact angle while for 3% Ag/ZnO suggested 103°. Also, UV resistivity has been observed in Ag/ZnO finished fabric. When the doping quantity of ZnO increased with 3% Ag/ZnO with 91%, 96%, and 98% for finishing the textile materials, antibacterial rates also increased for *S. aureus*, *E. coli*, and *C. albicans* correspondingly. Ag/ZnO nanoparticles also improve anti-mildew property, and in some cases, it is the highest on cotton, especially for *A. flavus*. Surface of the polyester fabric utilized for coating with ZnO, prepared by sol–gel and hydrothermal synthesis methods for spherical nanoparticles and nanorods [79]. Sol–gel solution with 0.08, 0.12, and 0.14 molar concentration ZnO particle prepared, and size was measured with particle size analyzer, and polydispersity index of ZnO in solution. Polyester was impregnated in 0.08 M sol–gel solution at the first stage for applying ZnO nanoparticles, and then later stage was allotted for development of nanorods. For hydrothermal synthesis to develop and grow nanorods of ZnO particles, the surface temperature was kept at 98 °C. Uniformity, orientation, and density can be studied for nanorods by considering their structure at center point. The presence of ZnO on the polyester surface and it has made a bond with the fiber has been confirmed with ATR-FTIR analysis by suggesting 404 cm^{-1} peak, which is the additional peak compared to polyester fabric ATR-FTIR data. Thermogravimetric analysis has also proved that ZnO nanorods developed on the polyester fabric by reducing the thermal stability. Finally, after growth of ZnO nanorods on polyester fabric, it turned into superhydrophobic fabric compared to hydrophobic polyester fabric.

4.8.3 TiO_2

TiO_2 nanoparticles are majorly used as a photocatalyst for self-sterilization and environment water splitting. Recent markets are majorly concerned on TiO_2-based application due to only the said property. Sol–gel, sputtering, and chemical vapor deposition (CVD) are the methods utilized for preparing TiO_2 which is present in the combination with films or fabrics [74, 95, 97]. Pad-dry-cure method, the bath dyeing method, is used for preparing composites of reduced graphene oxide (rGO)-TiO_2 and polyester [78]. This rgo-TiO_2 is responsible for improving the antimicrobial characteristic up to 98% for both *S. aureus* and *E. coli.* Cotton fiber finished with Polyaniline/ TiO_2 (PANI/ TiO_2) was prepared by doping with sulfosalicylic (SSA) and sodium dodecylbenzene sulfonate (SDBS), and the whole assembly was added to cotton with one-step in situ polymerization [90]. This combination with cotton suggests better conductivity, UV protection, sunlight absorption, and photocatalytic activity, till 10 washing cycles. For achieving superhydrophobicity on cotton with the contact angle of 151° till 10 washing cycles, the surface of cotton was treated with bifunctional fluorosilsesquioxane having different functional [61]. Sol–gel coating method was also utilized for TiO_2 application on cotton for imparting superhydrophobicity, also it improves the UV shielding characteristic, and similar to routine convention process, this method is not in the requirement of any expensive assemblies [88].

For enhancing the visible-light performance of TiO_2 nanoparticles, they are treated with noble and transition metals. These metals reduce the recombination rate of electrons with the holes, when they deposit on TiO_2 nanoparticles. So, this method increases the carriers which have been generated by photo application, and shows excellent photocatalytic activity [46]. Metal like Ag can also increase the photocatalytic activity while the surface has been imparted Plasmon resonance. Also, achieving the surface, in which the wavelength is higher than 450 nm, metal excites, and electrons flow from the metal surface to the TiO_2 nanoparticles conduction band by supporting interfacial charge transfer. Under visible light, as the number of negative electrons increases, the photocatalyst surface shows more efficiency than before it was observed under the same visible light. At higher deposition on metal, the metal reduces the photocatalytic activity by combining again, and so it deposits on the TiO_2 surface in smaller amounts. Also, the site which can combine with photocatalyst present on the fabric surface can be blocked by metals, and due to such reaction photocatalytic activity decreases. Therefore, TiO_2 can be doped with transition metals which can impart an innovative energy level, and also for nanoparticles, it reduces the band gaps in crystalline region [46]. So, the energy of light which can excite the photocatalyst decreases, and therefore, at the wide range of wavelengths, the response of photocatalyst will increase. The quantum proficiency decreases due to the presence of transition metals as they are behaving or acting like electron–hole recombination centers. Metal–nonmetal, metal–metal, and nonmetal–nonmetal types of co-doping can also advance the photocatalytic activity under the effect of visible light. Silica (SiO_2) can also be imparted with TiO_2 nanoparticles to improve the photocatalytic activity, where in TiO_2/SiO_2 composite—for the activity of silica—is related with two mechanisms. Likely or unlikely, hydrophilicity is increased with the introduction of SiO_2 in the system, and this hydrophilicity can withstand for higher duration compared to TiO_2 nanoparticles. As the TiO_2/SiO_2 system has maximum hydrophilicity, the surface of the fabric treated with the said system will absorb a higher number of hydroxyl groups. So, the hydroxyl group will reacts in the system only, i.e., with each other, and produce more and more hydroxyl radicals. So, charge will disproportionate with each other with the presence of silica, and it will increase the photocatalytic activity by producing Lewis acidic sites. So, the photocatalytic activity increases with increasing the number of hydroxyl group. Finally, this system increase the photo-efficiency under UV light and controlled photocatalytic activity on a fabric with increasing exposure of visible light.

Some of the applications of nanomaterials on textile substrates for enhancing their specific properties are listed in Table 3.

4.9 Nano-finishes for Sensing

The textile which can sense like humans by receiving the indications from the environment and generating the response for specific action is defined in the categories of smart textiles which sense and stimulate with the environment. One can combine the nanoparticles which can generate response with textile substrates. Light, temperature, electricity, mechanical force, moisture, and chemical stimuli can be imparted in textiles and such sensing textiles are suggested in Fig. 5.

5 Future Aspects

The chapter discussed about the application of nanotechnology in textile finishing for photocatalytic self-cleaning, superhydrophobic, oil and water separation, UV protection, and EMI shielding effects. Till nanotechnology for new functionalities, new application methods and new resources for textiles used in routine daily needs have the question of application. Exposure of human to the nanomaterials is hazardous/ risky or not, still is assumed condition. Scholars have been studied about the life cycles of nanomaterials, but still many questions related to their behavior and environmental aspects are there for textile field. But no a single researcher can suggest the negative remark of nano-finishing, as it imparts permanent multifunctional finishes on textiles, which is required by buyer, and also a buyer is concern about health and environment in recent era. Therefore, various governments are working for preparing the norms for eco-friendly production of nanomaterials, and also applications in safer

Table 3 Examples (application of nanomaterials)

Nanomaterial	Textile material/s	Properties	Application	References
Cu and CuO	Cotton	Antimicrobial	Wound dressing	[2, 63]
ZnO (glycidoxypropyl) trimethoxysilane (GPTS)	Cotton	Antiviral Antibacterial	Medical textile (bioactive textiles)	[2]
TiO_2 Ag NPs with polycarboxylic acid	Cotton/wool Viscose/wool	Antibacterial UV protection Self-cleaning Anti-wrinkle	Conventional uses	[28]
TiO_2 (NPs)-PVDF (poly(vinylidene fluoride))	Cotton	Antimicrobial UV protection Superhydrophobic	Multifunction fabric	[25]
ZnO/SiO_2 with DMDHEU	Cotton	UV protection Anti-wrinkle Self-cleaning	Multifunction fabric	[73]
ZnO nano-PU dispersion	Polyester	UV protection	Summer apparel	[1]
Ag NPs with chitosan	Viscose	Antibacterial	Medical textile	[63]
Iron oxide	Cotton	Electrical conductivity UV protection	Electric industry	[72]
SiO_2/TiO_2 nanosols	Cotton	Antibacterial UV protection Water repellent	Multifunction fabric	[70]
Ag and Cu NPs	Cotton Polyester	Antibacterial	Filtration	[23]
Ag and Zn NPs with citric acid	Cotton/wool	Antibacterial UV protection Self-cleaning Anti-wrinkle	Multifunctional fabric	[29]
TiO_2 NPs	Polyester/ wool	Self-cleaning	Hydrophobic	[22]
ZnO	Cotton	Antibacterial	Medical textile	[4]
Ag NPs	Denim	Antibacterial	Surgical gown	[21]
Cu NPs	Cotton Polyester/ Cotton	Antibacterial	Hygienic textiles	[35]
SiO_2 NPs	Denim	Flame retardant Hydrophilic Self-cleaning	Diverse environmental conditions	[76]
ZnO	Cotton	Antibacterial UV protection Self-cleaning	Multifunctional fabric	[30]

(continued)

Table 3 (continued)

Nanomaterial	Textile material/s	Properties	Application	References
ZnO	Denim	Self-cleaning	Diverse environmental conditions	[5]
Ceramic nanoparticles of cobalt ferrite	Polyester	Photocatalytic Antibacterial	Environmental	[92]
TiO_2 ZnO	Cotton	UV protection Antibacterial Superhydrophobic	Multifunctional fabric	[36]
Cationized TiO_2 NPs	Cotton	Antibacterial UV protection Self-cleaning	Multifunctional fabric	[64]
TiO_2 nano-colloidal solution	Polyester	Softness Wrinkle resistant UV protection	Fine silky PET	[37]
ZnO–TiO_2 hybrid nanocrystal	Cotton	Antibacterial UV protection Wash durable	Multifunctional fabric	[45]
SiO_2 NPs	Eri silk	Superhydrophobic	Water repellent	[9]
ZnO	Cotton	Antimicrobial	Medical textiles	[7]
ZnO	Jute	Fire retardant	Fireproof fabric	[67]
ZnO–TiO_2 with citric acid	Denim	Hydrophilic Self-cleaning	Multifunctional fabric	[27]
SiO_2 NPs and nanoclay	Polyester microfiber single jersey	Air permeability	Jersey	[14]
ZnO with PVA, PU and BA binder	Cotton	Antibacterial Wrinkle resistant	Health care apparel	[43]
Carbinol polydimethylsiloxane (CPDMS) 2,4-toluene diisocyanate (TDI) Polyethylene glycol (PEG)	Cotton	Antimicrobial Softness Anti-crease	Multifunctional fabric	[16]
SiO_2 NPs	Cotton	Superhydrophobic Transparency	Water-repellent coating	[31]
SiO_2 TiO_2	Wool	UV protection	Autumn apparel	[94]

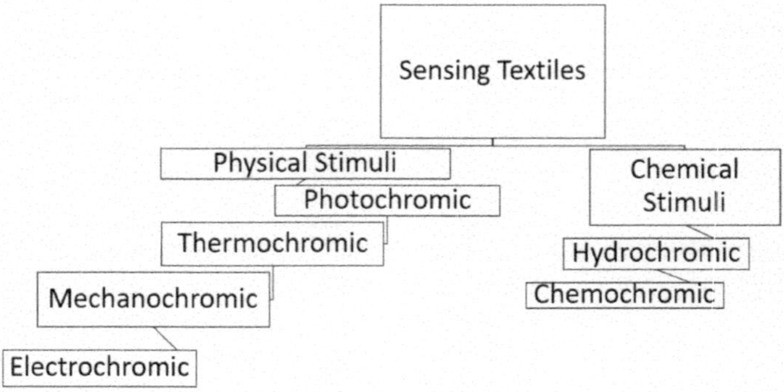

Fig. 5 Classification of sensing (stimuli)

way. This is the future research zone, development of nanomaterials by considering environment and human health at primary stage, and need of society.

References

1. Abo El-Ola SM, Elshakankery MH, Kotb RM (2022) Integration of nanocomposite finishing on polyester fabric for enhanced UV protection, performance, and comfort properties. J Eng Fibers Fabr 17:155892502211194. https://doi.org/10.1177/15589250221119447
2. Afzal F, Ashraf M, Manzoor S, Aziz H, Nosheen A, Riaz S (2023) Development of novel antiviral nanofinishes for bioactive textiles. Polym Bull 80(8):8447–8466. https://doi.org/10.1007/s00289-022-04461-2
3. Afzal S, Daoud WA, Langford SJ (2013) Photostable self-cleaning cotton by a Copper(II) Porphyrin/TiO_2 visible-light photocatalytic system. ACS Appl Mater Interfaces 5(11):4753–4759. https://doi.org/10.1021/am400002k
4. Ashar A, Bhutta ZA, Shoaib M, Alharbi NK, Fakhar-e-Alam M, Atif M, Kulyar MF-A, Mahfooz A, Boruah P, Eletmany MR, Al-Saeed FA, Ezzat Ahmed A (2023) Cotton fabric loaded with ZnO nanoflowers as a photocatalytic reactor with promising antibacterial activity against pathogenic *E. coli*. Arab J Chem 16(9):105084. https://doi.org/10.1016/j.arabjc.2023.105084
5. Asim F, Naeem F (2023) Investigation of self-cleaning attributes of denim fabric modified with naturally synthesized ZnO nanoparticles. Pigm Resin Technol 52(6):738–746. https://doi.org/10.1108/PRT-04-2022-0042
6. Attia NF, El Ebissy AA, Morsy MS, Sadak RA, Gamal H (2021) Influence of textile fabrics structures on thermal, UV shielding, and mechanical properties of textile fabrics coated with sustainable coating. J Nat Fibers 18(12):2189–2196. https://doi.org/10.1080/15440478.2020.1724233
7. Balasubramanian S, Subbiah DK, Nesakumar N, Kulandaisamy AJ, Das A, Lee JH, Balaguru Rayappan JB (2023) Synthesis and functional finishing of textile materials using zinc-based functional nanoparticles. In: Antiviral and antimicrobial coatings based on functionalized nanomaterials. Elsevier, pp 365–387. https://doi.org/10.1016/B978-0-323-91783-4.00001-2
8. Bixler GD, Bhushan B (2012) Bioinspired rice leaf and butterfly wing surface structures combining shark skin and lotus effects. Soft Matter 8(44):11271. https://doi.org/10.1039/c2sm26655e

9. Borah MP, Jose S, Kalita BB, Shakyawar D, Pandit P (2020) Water repellent finishing on eri silk fabric using nano silica. J Text Inst 111(5):701–708. https://doi.org/10.1080/00405000.2019.1659470
10. Bozzi A, Yuranova T, Kiwi J (2005) Self-cleaning of wool-polyamide and polyester textiles by TiO_2-rutile modification under daylight irradiation at ambient temperature. J Photochem Photobiol A 172(1):27–34. https://doi.org/10.1016/j.jphotochem.2004.11.010
11. Chen F, Zhai L, Huang Z, Luo Y, Yang H, Xing T, He A, Yu Z, Liu J, Zhang X, Xu W (2021) Decorating aramid fibers with chemically-bonded amorphous TiO_2 for improving UV resistance in the simulated extreme environment. SSRN Electron J. https://doi.org/10.2139/ssrn.3995390
12. Daoud WA, Leung SK, Tung WS, Xin JH, Cheuk K, Qi K (2008) Self-cleaning keratins. Chem Mater 20(4):1242–1244. https://doi.org/10.1021/cm702661k
13. Dastjerdi R, Montazer M (2010) A review on the application of inorganic nano-structured materials in the modification of textiles: focus on anti-microbial properties. Colloids Surf B 79(1):5–18. https://doi.org/10.1016/j.colsurfb.2010.03.029
14. Elgory Z, Seddik K, Yahia M, El-Gabry L (2019) The enhancement of the functional properties of polyester microfiber single jersey using some nano-materials. Egypt J Chem. https://doi.org/10.21608/ejchem.2019.2868.1804
15. El-Shafei A, ElShemy M, Abou-Okeil A (2015) Eco-friendly finishing agent for cotton fabrics to improve flame retardant and antibacterial properties. Carbohyd Polym 118:83–90. https://doi.org/10.1016/j.carbpol.2014.11.007
16. Fahmy HM, Okda HMY, Hassabo AG, El-Rafie MH, Youssef MA (2022) Synthesis of new silicone-based adducts to functionalize cotton fabric. Silicon 14(13):7613–7621. https://doi.org/10.1007/s12633-021-01451-3
17. Fouda A, EL-Din Hassan S, Salem SS, Shaheen TI (2018) In-vitro cytotoxicity, antibacterial, and UV protection properties of the biosynthesized Zinc oxide nanoparticles for medical textile applications. Microb Pathog 125:252–261. https://doi.org/10.1016/j.micpath.2018.09.030
18. Gaminian H, Montazer M (2015) Enhanced self-cleaning properties on polyester fabric under visible light through single-step synthesis of cuprous oxide doped nano-TiO_2. Photochem Photobiol 91(5):1078–1087. https://doi.org/10.1111/php.12478
19. Gao D, Liu J, Lyu L, Li Y, Ma J, Baig W (2020) Construct the multifunction of cotton fabric by synergism between nano ZnO and Ag. Fibers and Polymers 21(3):505–512. https://doi.org/10.1007/s12221-020-9347-4
20. Gao X, Yan X, Yao X, Xu L, Zhang K, Zhang J, Yang B, Jiang L (2007) The dry-style antifogging properties of mosquito compound eyes and artificial analogues prepared by soft lithography. Adv Mater 19(17):2213–2217. https://doi.org/10.1002/adma.200601946
21. Habib S, Raza ZA, Kishwar F, Abid S (2023) Citric-acid anchored silver nano-finishing of commercially dyed denim fabric for potential non toxic antibacterial surgical gowns. Pigm Resin Technol. https://doi.org/10.1108/PRT-03-2023-0021
22. Haji A, Mousavi Shoushtari A, Mazaheri F, Tabatabaeyan SE (2015) RSM optimized self-cleaning nano-finishing on polyester/wool fabric pretreated with oxygen plasma. J Text Inst 1–10. https://doi.org/10.1080/00405000.2015.1077023
23. Hontañón E, Meyer J, Blanes M, Cambra V, Guo X, Masuhr M, Muntean A, Santos L, Nirschl H, Kruis E (2016) A sustainable route for antibacterial nanofinishing of textiles. Int J Theoret Appl Nanotechnol. https://doi.org/10.11159/ijtan.2016.003
24. Hosne Asif AKMA, Hasan MZ (2018) Application of nanotechnology in modern textiles: a review. Int J Curr Eng Technol 8(02). https://doi.org/10.14741/ijcet/v.8.2.5
25. Hou J, Yang Y, Yu D-G, Chen Z, Wang K, Liu Y, Williams GR (2021) Multifunctional fabrics finished using electrosprayed hybrid Janus particles containing nanocatalysts. Chem Eng J 411:128474. https://doi.org/10.1016/j.cej.2021.128474
26. https://sdgs.un.org/2030agenda (2023) United Nations transforming our world: the 2030 Agenda for sustainable development. Available online
27. Ibrahim NA, Eid BM, Abdel Aziz MS, Hamdy SM, Abd Allah SE (2018) Green surface modification and nano-multifunctionalization of denim fabric. Cellulose 25(10):6207–6220. https://doi.org/10.1007/s10570-018-1985-y

28. Ibrahim NA, Eid BM, Emam E-AM (2022) An eco-friendly facile approach for imparting multi-functional protection properties to cellulose/wool blends. Polym Bull 79(11):10313–10331. https://doi.org/10.1007/s00289-021-04072-3
29. Ibrahim NA, Emam E-AM, Eid BM, Tawfik TM (2018) An eco-friendly multifunctional nano-finishing of cellulose/wool blends. Fibers Polym 19(4):797–804. https://doi.org/10.1007/s12221-018-7922-8
30. Irfan M, Naz MY, Saleem M, Tanawush M, Głowacz A, Glowacz W, Rahman S, Mahnashi MH, Alqahtani YS, Alyami BA, Alqarni AO, Alsaiari MA (2021) Statistical study of nonthermal plasma-assisted ZnO coating of cotton fabric through ultrasonic-assisted green synthesis for improved self-cleaning and antimicrobial properties. Materials 14(22):6998. https://doi.org/10.3390/ma14226998
31. Jeong SA, Kang TJ (2017) Superhydrophobic and transparent surfaces on cotton fabrics coated with silica nanoparticles for hierarchical roughness. Text Res J 87(5):552–560. https://doi.org/10.1177/0040517516632477
32. Jiang C, Liu W, Yang M, Liu C, He S, Xie Y, Wang Z (2018) Facile fabrication of robust fluorine-free self-cleaning cotton textiles with superhydrophobicity, photocatalytic activity, and UV durability. Colloids Surf A 559:235–242. https://doi.org/10.1016/j.colsurfa.2018.09.048
33. Joshi M (2008) The impact of nanotechnology on polyesters, polyamides and other textiles. In: Polyesters and polyamides. Elsevier, pp 354–415. https://doi.org/10.1533/9781845694609.2.354
34. Joshi M, Bhattacharyya A (2011) Nanotechnology—a new route to high-performance functional textiles. Text Prog 43(3):155–233. https://doi.org/10.1080/00405167.2011.570027
35. Kanade P, Patel BH (2017) Copper nano mediated hygienic textiles with improved aesthetic properties. Res J Text Appar 21(2):146–158. https://doi.org/10.1108/RJTA-01-2017-0004
36. Khan MZ, Baheti V, Ashraf M, Hussain T, Ali A, Javid A, Rehman A (2018) Development of UV protective, superhydrophobic and antibacterial textiles using ZnO and TiO_2 nanoparticles. Fibers Polym 19(8):1647–1654. https://doi.org/10.1007/s12221-018-7935-3
37. Khanjarpanah O, Dastjerdi R (2017) Multifunctional ultra-fine silky PET fabrics; nanofunctionalization via an ultrasound-assistant nanocolloidal preparation. J Eng Fibers Fabr 12(2):155892501701200. https://doi.org/10.1177/155892501701200210
38. Kwon DH, Huh HK, Lee SJ (2014) Wettability and impact dynamics of water droplets on rice (*Oryza sativa* L.) leaves. Exp Fluids 55(3):1691. https://doi.org/10.1007/s00348-014-1691-y
39. Lei S, Shi Z, Ou J, Wang F, Xue M, Li W, Qiao G, Guan X, Zhang J (2017) Durable superhydrophobic cotton fabric for oil/water separation. Colloids Surf A 533:249–254. https://doi.org/10.1016/j.colsurfa.2017.08.012
40. Mahmud R, Nabi F (2017) Application of nanotechnology in the field of textile. IOSR J Polym Text Eng 04(01):01–06. https://doi.org/10.9790/019X-0401010106
41. Mayser MJ, Bohn HF, Reker M, Barthlott W (2014) Measuring air layer volumes retained by submerged floating-ferns Salvinia and biomimetic superhydrophobic surfaces. Beilstein J Nanotechnol 5:812–821. https://doi.org/10.3762/bjnano.5.93
42. Meng L-Y, Han S, Jiang N, Meng W (2014) Study on superhydrophobic surfaces of octanol grafted electrospun silica nanofibers. Mater Chem Phys 148(3):798–802. https://doi.org/10.1016/j.matchemphys.2014.08.051
43. Mulchandani N, Karnad V (2022) Application of zinc oxide nano particles using polymeric binders on cotton fabric. Res J Text Appar 26(4):310–322. https://doi.org/10.1108/RJTA-02-2021-0018
44. Musante C, White JC (2012) Toxicity of silver and copper to Cucurbita pepo: differential effects of nano and bulk-size particles. Environ Toxicol 27(9):510–517. https://doi.org/10.1002/tox.20667
45. Nautiyal A, Shukla SR, Prasad V (2022) ZnO-TiO_2 hybrid nanocrystal-loaded, wash durable, multifunction cotton textiles. Cellulose 29(10):5923–5941. https://doi.org/10.1007/s10570-022-04595-6
46. Nolan NT, Seery MK, Hinder SJ, Healy LF, Pillai SC (2010) A systematic study of the effect of silver on the chelation of formic acid to a titanium precursor and the resulting effect on the

anatase to rutile transformation of TiO_2. J Phys Chem C 114(30):13026–13034. https://doi.org/10.1021/jp1016054
47. Noorian SA, Hemmatinejad N, Navarro JAR (2020) Ligand modified cellulose fabrics as support of zinc oxide nanoparticles for UV protection and antimicrobial activities. Int J Biol Macromol 154:1215–1226. https://doi.org/10.1016/j.ijbiomac.2019.10.276
48. Onar N, Mete G (2016) Development of water repellent cotton fabric with application of ZnO, Al_2O_3, TiO_2 and ZrO_2 nanoparticles modified with ormosils. Text Apparel 26(3):295–302
49. Ozyigit II, Dogan I, Hocaoglu-Ozyigit A, Yalcin B, Erdogan A, Yalcin IE, Cabi E, Kaya Y (2023) Production of secondary metabolites using tissue culture-based biotechnological applications. Front Plant Sci 14. https://doi.org/10.3389/fpls.2023.1132555
50. Pakdel E, Daoud WA (2013) Self-cleaning cotton functionalized with TiO_2/SiO_2: focus on the role of silica. J Colloid Interface Sci 401:1–7. https://doi.org/10.1016/j.jcis.2013.03.016
51. Pakdel E, Daoud WA, Sun L, Wang X (2014) Visible and UV functionality of TiO_2 ternary nanocomposites on cotton. Appl Surf Sci 321:447–456. https://doi.org/10.1016/j.apsusc.2014.10.018
52. Pakdel E, Daoud WA, Wang X (2015) Assimilating the photo-induced functions of TiO_2-based compounds in textiles: emphasis on the sol–gel process. Text Res J 85(13):1404–1428. https://doi.org/10.1177/0040517514551462
53. Panáček A, Kvítek L, Smékalová M, Večeřová R, Kolář M, Röderová M, Dyčka F, Šebela M, Prucek R, Tomanec O, Zbořil R (2018) Bacterial resistance to silver nanoparticles and how to overcome it. Nat Nanotechnol 13(1):65–71. https://doi.org/10.1038/s41565-017-0013-y
54. Pandimurugan R, Thambidurai S (2017) UV protection and antibacterial properties of seaweed capped ZnO nanoparticles coated cotton fabrics. Int J Biol Macromol 105:788–795. https://doi.org/10.1016/j.ijbiomac.2017.07.097
55. Park YR, Kim KJ (2005) Structural and optical properties of rutile and anatase TiO_2 thin films: effects of Co doping. Thin Solid Films 484(1–2):34–38. https://doi.org/10.1016/j.tsf.2005.01.039
56. Periyasamy AP, Militky J, Sachinandham A, Duraisamy G (2021) Nanotechnology in textile finishing: recent developments. In: Handbook of nanomaterials and nanocomposites for energy and environmental applications. Springer International Publishing, pp 1–31. https://doi.org/10.1007/978-3-030-11155-7_55-1
57. Periyasamy AP, Venkataraman M, Kremenakova D, Militky J, Zhou Y (2020) Progress in sol–gel technology for the coatings of fabrics. Materials 13(8):1838. https://doi.org/10.3390/ma13081838
58. Periyasamy AP, Vikova M, Vik M (2020) Spectral and physical properties organo-silica coated photochromic poly-ethylene terephthalate (PET) fabrics. J Text Inst 111(6):808–820. https://doi.org/10.1080/00405000.2019.1663633
59. Periyasamy AP, Yang K, Xiong X, Venkataraman M, Militky J, Mishra R, Kremenakova D (2020) Effect of silanization on copper coated milife fabric with improved EMI shielding effectiveness. Mater Chem Phys 239:122008. https://doi.org/10.1016/j.matchemphys.2019.122008
60. Periyasamy AP, Yang K, Xiong X, Venkatraman M, Militký J, Mishra R, Křemenáková D (2019) Influence of EMI shielding on Silane-coated conductive fabric. Textile Bioengineering & Informatics Society Ltd., p 12
61. Przybylak M, Maciejewski H, Dutkiewicz A (2016) Preparation of highly hydrophobic cotton fabrics by modification with bifunctional silsesquioxanes in the sol–gel process. Appl Surf Sci 387:163–174. https://doi.org/10.1016/j.apsusc.2016.06.094
62. Radetić M, Marković D (2019) Nano-finishing of cellulose textile materials with copper and copper oxide nanoparticles. Cellulose 26(17):8971–8991. https://doi.org/10.1007/s10570-019-02714-4
63. Raza ZA, Bilal U, Noreen U, Munim SA, Riaz S, Abdullah MU, Abid S (2019) Chitosan mediated formation and impregnation of silver nanoparticles on viscose fabric in single bath for antibacterial performance. Fibers Polym 20(7):1360–1367. https://doi.org/10.1007/s12221-019-1018-y

64. Riaz S, Ashraf M, Aziz H, Younus A, Umair M, Salam A, Iqbal K, Hussain MT, Hussain T (2022) Cationization of TiO_2 nanoparticles to develop highly durable multifunctional cotton fabric. Mater Chem Phys 278:125573. https://doi.org/10.1016/j.matchemphys.2021.125573
65. Román LE, Amézquita MJ, Uribe CL, Maurtua DJ, Costa SA, Costa SM, Keiski R, Solís JL, Gómez MM (2020) In situ growth of CuO nanoparticles onto cotton textiles. Adv Nat Sci Nanosci Nanotechnol 11(2):025009. https://doi.org/10.1088/2043-6254/ab8a2f
66. Roy S, Zhai L, Kim JW, Kim HC, Kim J (2020) A novel approach of developing sustainable cellulose coating for self-cleaning-healing fabric. Prog Org Coat 140:105500. https://doi.org/10.1016/j.porgcoat.2019.105500
67. Samanta AK, Bhattacharyya R, Jose S, Basu G, Chowdhury R (2017) Fire retardant finish of jute fabric with nano zinc oxide. Cellulose 24(2):1143–1157. https://doi.org/10.1007/s10570-016-1171-z
68. Sas I, Gorga RE, Joines JA, Thoney KA (2012) Literature review on superhydrophobic self-cleaning surfaces produced by electrospinning. J Polym Sci Part B Polym Phys 50(12):824–845. https://doi.org/10.1002/polb.23070
69. Sawhney APS, Condon B, Singh KV, Pang SS, Li G, Hui D (2008) Modern applications of nanotechnology in textiles. Text Res J 78(8):731–739. https://doi.org/10.1177/0040517508091066
70. Shaheen TI, Salem SS, Zaghloul S (2019) A new facile strategy for multifunctional textiles development through in situ deposition of SiO_2/TiO_2 nanosols hybrid. Ind Eng Chem Res 58(44):20203–20212. https://doi.org/10.1021/acs.iecr.9b04655
71. Shang Q, Liu C, Zhou Y (2018) One-pot fabrication of robust hydrophobia and superoleophilic cotton fabrics for effective oil-water separation. J Coat Technol Res 15(1):65–75. https://doi.org/10.1007/s11998-017-9947-0
72. Sharouf H, Saffour Z (2023) Synthesizing and using iron oxide nanoparticles as nanocomposite in cotton fabrics nanofinishing. Baghdad Sci J. https://doi.org/10.21123/bsj.2023.7743
73. Singh A, Chakraborty JN (2023) Multifunctional finishing of cotton using zinc oxide and silicon dioxide nanoparticles along with DMDHEU. Res J Text Appar 27(1):141–156. https://doi.org/10.1108/RJTA-11-2021-0130
74. Singh J, Khan SA, Shah J, Kotnala RK, Mohapatra S (2017) Nanostructured TiO_2 thin films prepared by RF magnetron sputtering for photocatalytic applications. Appl Surf Sci 422:953–961. https://doi.org/10.1016/j.apsusc.2017.06.068
75. Sun Y, Zhao X, Liu R, Chen G, Zhou X (2018) Synthesis and characterization of fluorinated polyacrylate as water and oil repellent and soil release finishing agent for polyester fabric. Prog Org Coat 123:306–313. https://doi.org/10.1016/j.porgcoat.2018.07.013
76. Talebi S, Montazer M (2020) Denim fabric with flame retardant, hydrophilic and self-cleaning properties conferring by in-situ synthesis of silica nanoparticles. Cellulose 27(11):6643–6661. https://doi.org/10.1007/s10570-020-03195-6
77. Tan X, Peng Q, Yang K, Saskova J, Periyasamy AP, Militký J, Wiener J (2021) Influence of UV light and ozonization on microbes state. In: Textiles and their use in microbial protection. CRC Press, pp 155–170. https://doi.org/10.1201/9781003140436-8
78. Touhid SSB, Shawon MdRK, Deb H, Khoso NA, Ahmed A, Fu F, Liu XD (2020) Nature inspired rGO-TiO_2 micro-flowers on polyester fabric using semi-continuous dyeing method: a binder-free approach towards durable antibacterial performance. Synth Met 261:116298. https://doi.org/10.1016/j.synthmet.2020.116298
79. Wang Y, Militký J, Periyasamy AP (2021) Disinfection mechanisms of UV light and ozonization. In: Textiles and their use in microbial protection. CRC Press, pp 135–154. https://doi.org/10.1201/9781003140436-7
80. Wong YWH, Yuen CWM, Leung MYS, Ku SKA, Lam HLI (2006) Selected applications of nanotechnology in textiles. AUTEX Res J 6(1):1–8. https://doi.org/10.1515/aut-2006-060101
81. Wu M, Ma B, Pan T, Chen S, Sun J (2016) Silver-nanoparticle-colored cotton fabrics with tunable colors and durable antibacterial and self-healing superhydrophobic properties. Adv Func Mater 26(4):569–576. https://doi.org/10.1002/adfm.201504197

82. www.imnovation-hub.com, 9 Oct 2023. https://www.imnovation-hub.com/science-and-technology/nanotechnology-textiles-t-shirts-body-olortemperature/ (http://nasi.nic.in/Nano/15%20-%20Mangla%20Joshi.pdf)
83. www.nanowerk.com, 9 Oct 2023. https://www.nanowerk.com/news2/newsid=27698.php
84. www.nanowerk.com, 9 Oct 2023. https://www.nanowerk.com/spotlight/spotid=19451.php
85. www.technicaltextile.net, 9 Oct 2023. https://www.technicaltextile.net/articles/self-cleaning-textile-an-overview-2646
86. www.thierry-corp.com, 9 Oct 2023. https://www.thierry-corp.com/plasma-knowledgebase/nanotechnology-fabrics
87. Xin JH, Daoud WA, Kong YY (2004) A new approach to UV-blocking treatment for cotton fabrics. Text Res J 74(2):97–100. https://doi.org/10.1177/004051750407400202
88. Xue C-H, Jia S-T, Chen H-Z, Wang M (2008) Superhydrophobic cotton fabrics prepared by sol–gel coating of TiO_2 and surface hydrophobization. Sci Technol Adv Mater 9(3):035001. https://doi.org/10.1088/1468-6996/9/3/035001
89. Yang K, Periyasamy AP, Venkataraman M, Militky J, Kremenakova D, Vecernik J, Pulíček R (2020) Resistance against penetration of electromagnetic radiation for ultra-light Cu/Ni-coated polyester fibrous materials. Polymers 12(9):2029. https://doi.org/10.3390/polym12092029
90. Yu J, Pang Z, Zheng C, Zhou T, Zhang J, Zhou H, Wei Q (2019) Cotton fabric finished by PANI/TiO_2 with multifunctions of conductivity, anti-ultraviolet and photocatalysis activity. Appl Surf Sci 470:84–90. https://doi.org/10.1016/j.apsusc.2018.11.112
91. Zahid M, Papadopoulou EL, Suarato G, Binas VD, Kiriakidis G, Gounaki I, Moira O, Venieri D, Bayer IS, Athanassiou A (2018) Fabrication of visible light-induced antibacterial and self-cleaning cotton fabrics using manganese doped TiO_2 nanoparticles. ACS Appl Bio Mater 1(4):1154–1164. https://doi.org/10.1021/acsabm.8b00357
92. Zarezadeh K, Sheibani S, Ataie A (2023) Photocatalytic and antibacterial characteristics of decorated polyester textile with ceramic nanoparticles of cobalt ferrite. Ceram Int 49(12):20104–20117. https://doi.org/10.1016/j.ceramint.2023.03.134
93. Zhang J, Li B, Wu L, Wang A (2013) Facile preparation of durable and robust superhydrophobic textiles by dip coating in nanocomposite solution of organosilanes. Chem Commun 49(98):11509. https://doi.org/10.1039/c3cc43238f
94. Zhang M, Xie W, Tang B, Sun L, Wang X (2017) Synthesis of TiO_2 & SiO_2 nanoparticles as efficient UV absorbers and their application on wool. Text Res J 87(14):1784–1792. https://doi.org/10.1177/0040517516659375
95. Zhao L, Lu J (2017) Fabrication and application of mesoporous TiO_2 film coated on Al wire by sol–gel method with EISA. Appl Surf Sci 402:369–371. https://doi.org/10.1016/j.apsusc.2017.01.023
96. Zhao Q, Wu LYL, Huang H, Liu Y (2016) Ambient-curable superhydrophobic fabric coating prepared by water-based non-fluorinated formulation. Mater Des 92:541–545. https://doi.org/10.1016/j.matdes.2015.12.054
97. Zheng Z, Gu Z, Huo R, Ye Y (2009) Superhydrophobicity of polyvinylidene fluoride membrane fabricated by chemical vapor deposition from solution. Appl Surf Sci 255(16):7263–7267. https://doi.org/10.1016/j.apsusc.2009.03.084
98. Zhou H, Wang H, Niu H, Gestos A, Wang X, Lin T (2012) Fluoroalkyl silane modified silicone rubber/nanoparticle composite: a super durable, robust superhydrophobic fabric coating. Adv Mater 24(18):2409–2412. https://doi.org/10.1002/adma.201200184
99. Zimmermann J, Reifler FA, Fortunato G, Gerhardt L, Seeger S (2008) A simple, one-step approach to durable and robust superhydrophobic textiles. Adv Func Mater 18(22):3662–3669. https://doi.org/10.1002/adfm.200800755

Chapter 2
Types of Nanomaterials Used in Textile Finishing

Sudhanshu Singh, Zunjarrao Kamble, and Ghanshyam Neje

1 Introduction: Nanofinishing

Nanotechnology is the design, manufacture, and use of materials that are between 1 and 100 nm in size, at least in one dimension. It enables the introduction of unusual chemical, biological, and physical qualities that differ from their bulk characteristics. The efficacy of nanostructures is attributed to size-related qualities and a large surface area-to-volume ratio that produce particles with improved attributes as compared to traditional macrometer-sized particles.

A novel type of textile finishing known as "Nanofinishing" has been introduced concurrently with the development of nanotechnology and nanoscience. Overcoming the limitations of traditional finishes, nanofinishing has created novel opportunities for enhancing current techniques or assisting in the achievement of brand-new functional features. Numerous novel and superior qualities, such as self-cleaning, electrical conductivity, antistain, antibacterial, UV blocking, and electromagnetic wave shielding have been added to textiles as a result of the incorporation of nanotechnology in textile finishing. Nanofinishing has made it feasible to create high-performance functional finishes that give technical textiles a variety of end uses. Nanofinishing has no negative effects on the hand or breathability of the material.

Dr. David Soane undoubtedly made the first attempts in textile nanofinishing when he sought to employ nanotechnology to provide natural and synthetic fabrics with remarkable qualities without altering the appearance or feel of the material. Nanofinishing can be generally categorized into two key areas: employing nanomaterials in traditional finishing compositions, and using a finishing composition that can produce nanoscale structures on the fabric's surface. Both display great efficiency while exhibiting utility or multifunctionality.

S. Singh · Z. Kamble (✉) · G. Neje
Department of Textile Technology, Dr B R Ambedkar National Institute of Technology Jalandhar, G.T Road, Amritsar Bypass, Jalandhar, Punjab 144008, India
e-mail: kamblezb@nitj.ac.in

S. Bairagi et al. (eds.), *Nanotechnology in Textile Finishing*, Materials Horizons: From Nature to Nanomaterials, https://doi.org/10.1007/978-981-97-2696-7_2

Using inorganic and organic nanomaterials, such as TiO_2, ZnO, Cu, Au, CNTs, and Ag to textile surfaces to add different versatile qualities is a large field of investigation. Research is also being done on chemical vapor, electrospraying, sol–gel, and nanocoating techniques. Surface activation and nanoscale roughness may both be produced using nanotechnology. The addition of several functionalities to textiles is possible through nanofinishing. It is difficult to manage the dimensions and even distribution of nanomaterials, especially when there are significant precursor levels and when synthesizing nanocomposites with many components.

2 Classification of Nanofinishing

Three primary types of nanotreatment (Fig. 1) for textile materials—nanostructures, nanolayers, and nanoroughness—are discussed in the below sections.

2.1 Nanostructure

The use of nanomaterials in textiles is one of the various fields of study in textile nanofinishing. This comprises nanoparticles having dimension in the nanoscale or nanocomposites made up of multiple constituents. Nanomaterials may be generated in a wide range of forms, including many more shapes with various physical and chemical characteristics, depending on many variables, including pH, temperature, synthesizing process, etc. As a result of their high surface-to-volume ratio and

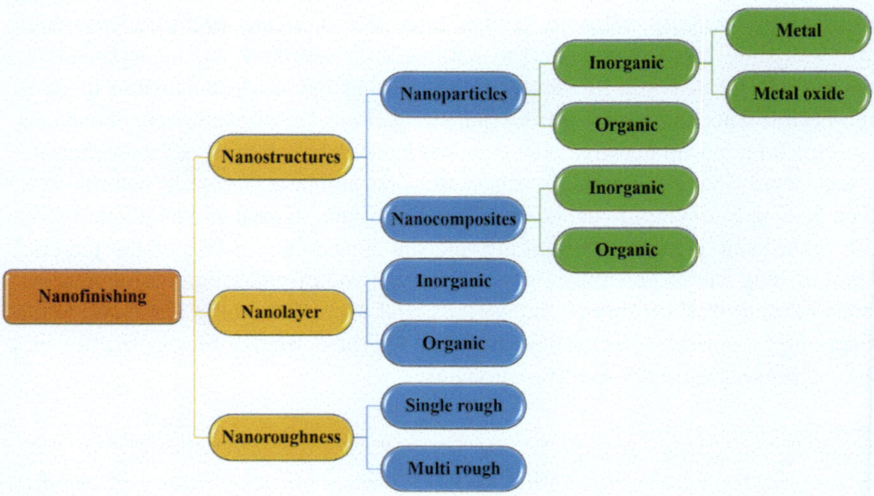

Fig. 1 Classification of textile nanofinishing

small size, nanoparticles exhibit distinct and potentially different properties from bulk materials. The manufacturing of nanoparticles typically involves one of two methods. A design created on a bigger size is reduced to a nanoscale using top-down approaches. Contrarily, bottom-up approaches start with molecules and work their way up to nanoscale structures.

In textile finishing, nanotechnology is also quite influential. It produced novel textile application techniques in addition to providing new finishes. Chemical finishes that are soft, variable, and highly functional have been developed. Technical and electrostatic methods can allow nanoparticle coatings to flow through certain textile regions [19]. Organic and inorganic nanoparticles can be used solely or together to create nanocomposites, which have the potential for several uses. Metals and metal oxides are additional classifications for inorganic nanoparticles. Semimetals have received attention recently as well. As an alternative, organic molecules with a diameter in nanometers to micrometers, such as nanocellulose and nanochitosan are also available [2].

A crucial stage in effective ex-situ nanofinishing is the creation of homogenous dispersion of nanoparticles. Agglomeration and aggregation are two concepts used to describe the clustering of nanoparticles. When weak physical connections hold the scattered particles together, agglomeration occurs, although it may be reversed. However, tightly bonded colloidal particles are a characteristic of aggregation, and the action is irreversible.

Therefore, aggregation is a specific arrangement of particles, whereas agglomeration occurs when the nano-elements are connected weakly, which may be easily disrupted by mechanical pressures. To achieve a homogeneous dispersion of nanomaterials and avoid their agglomeration, a variety of chemical, physical, and mechanical techniques can be used alone or in combination [30]. Surfactants are mostly used in chemical processes. The ability of nanoparticles to aggregate can be reduced by using stabilizers, resulting in particles in a suitable nanosize region [26].

TiO_2, Ag, and ZnO nanostructures, carbon nanotubes and nanotubes, SiO_2, Cu nanocrystals, Al and Fe compounds, metal oxides, and hydroxides, are examples of inorganic metal nanomaterials. Figure 2 shows the use of organic nanomaterials in various fields of textile nanofinishing.

2.1.1 Metal Nanoparticles

Metal nanomaterials play an important role to enhance and develop new functionality for textile materials. Metal nanoparticles facilitate showing textiles as smart futuristic materials by enhancing their properties as antimicrobial, electrical, thermal, etc. Figure 3 shows the use of metal nanomaterials in various fields of textile nanofinishing.

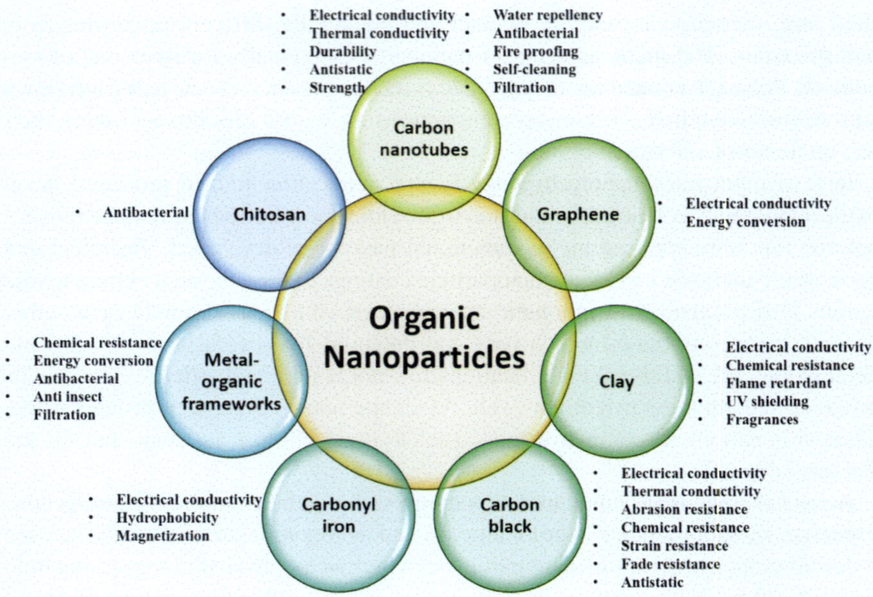

Fig. 2 Organic nanomaterials practiced in nanofinishing

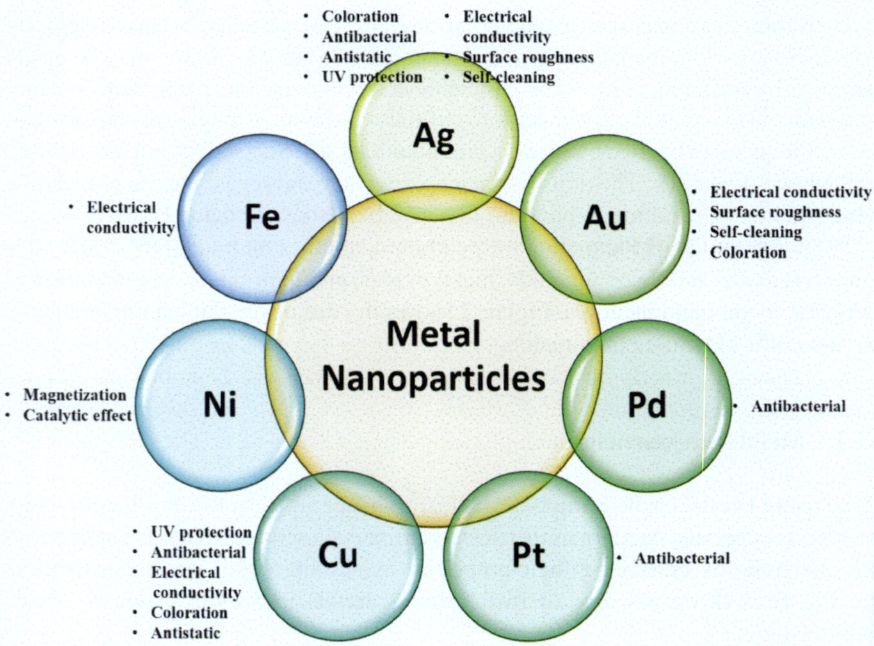

Fig. 3 Metal nanomaterials practiced in textile nanofinishing

Silver (Ag)

A wide variety of disease-causing microbes are resistant to the therapeutic properties of silver nanoparticles, which are known for their antibacterial capabilities. Numerous techniques can be used to synthesize them. The Ag ions' high specific surface area, nanoparticle size, and ability to interface closely with microbial membranes are often associated with the strengthening of the efficacy of Ag nanomaterials in fighting microbes. The major antibacterial mechanisms of silver nanoparticles include the destruction of the DNA, proteins, and lipids of the microbes as a result of electrostatic interactions with negatively charged microbial cell walls and attraction of Ag ions or nanomaterials to the microorganism [9].

Copper (Cu)

The heat transfer characteristics of copper nanomaterials, such as their high thermal conductivity, cheap production costs, effectiveness against germs, and optical and electrical characteristics, make them interesting. They are mostly created by a chemical reduction technique, just like other metal nanomaterials. However, when subjected to air, they are extremely susceptible to oxidation. An effective solution created to deal with this issue uses inert media like argon or nitrogen as well as capping or protective agents.

2.1.2 Metal Oxide Nanoparticles

Due to their capacity to produce charge carriers by absorbing energy, metal oxide nanomaterials are well known for their use in environmental remediation and electrical applications. Most metal oxide nanomaterials were suited as photocatalysts because of their electrical structure, ability to absorb light, and charge transport features. These nanomaterials have been used for their photocatalytic properties for the oxidation or reduction of materials to remove pollutants from the water and air, kill microorganisms, produce hydrogen, and convert carbon dioxide into hydrocarbons. Figure 4 shows the use of metal oxide nanomaterials in various fields of textile nanofinishing.

Zinc Oxide (ZnO)

ZnO is a semiconductor with a wide range of applications in antireflection coatings, UV protection, piezoelectric sensors, wastewater treatment, solar cells, sensors, sunscreens, and antibacterial materials. They additionally have the advantages of being inexpensive and white. Their photocatalytic excitation took place in the presence of light, just like it happened with TiO_2 nanomaterials. It has also been suggested to use noble metals, such as silver, to improve photocatalytic activity [5].

Titanium Dioxide (TiO_2)

TiO_2 nanomaterial comes in three different crystalline varieties: tetragonal anatase (more active), tetragonal rutile, and orthorhombic brookite which has a band gap of 3.2, 3.02, 2.96 eV. TiO_2 nanoparticles undergo redox reactions at the surface when

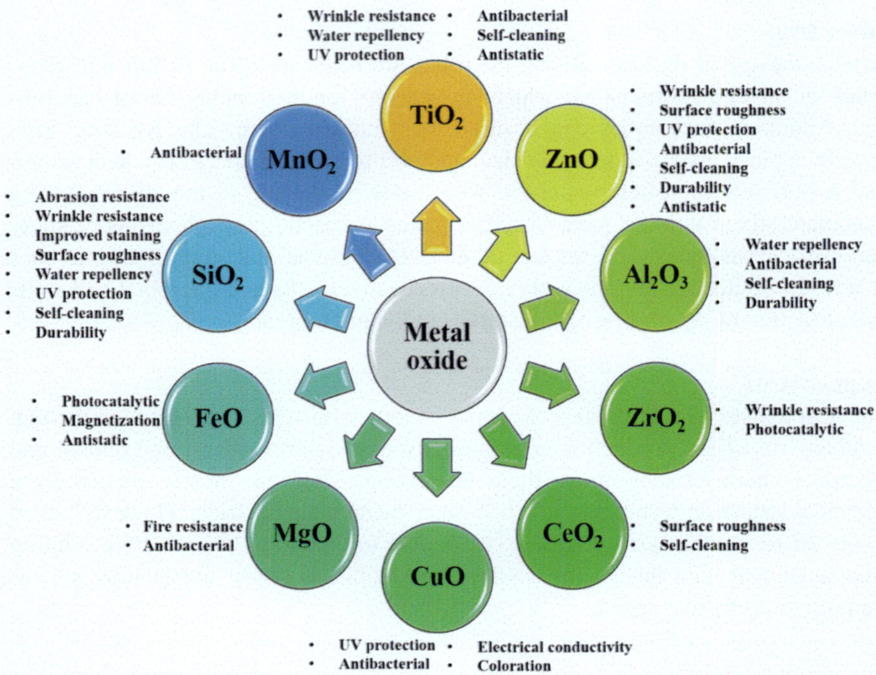

Fig. 4 Metal oxide nanomaterials practiced in textile nanofinishing

exposed to energy over their band gap. As hydroxyl radicals are produced by the positive holes and water, the newly formed oxygen and electrons react to form a superoxide anion. As a result, organic pollutants and pollutants from air and water, including microbes, can be broken down by TiO_2 [35].

2.2 Nanolayer

Another category of nanofinishing, known as nanocoating, is based on the production of one or more layers with a thickness in the nanorange on the surface of a textile substrate. Nanocoating has reduced the coat-to-weight ratio while addressing issues with traditional coating procedures such as poor adherence, weak abrasion resistance, strength loss, and decreased durability. The term "nanocoating" refers to the application of a thin layer or other substance with a nanometer-thickness to a surface. When applied to substrates, nanocoating significantly improves performance while also providing substantial strength, corrosion resistance, and durability with just the slightest modification in the substrates' basic characteristics. Chemical reduction, layer-by-layer coating, vapor deposition, and dip coating are a few techniques for applying nanocoating [16].

2.3 Nanoroughness

Creating the treated surface rougher on the nanoscale is dealt with in a different area of nanofinishing. This involves the dispersion of nanomaterials randomly on the surface of the material [16]. The dimension and amount of nanomaterials affect the surface roughness. Fluorochemicals, which migrate to the outermost layer and cover the nanoparticle-treated fiber surface, are materials with reduced surface energy that are occasionally used as well [36].

The stickiness and wettability of a surface are significantly influenced by the roughness of the material's surface. Atomic-force microscopy was used along with atmospheric pressure plasma to treat plain-woven polyester textiles made of spun and filament yarns. It indicated that after being exposed to plasma, the surface roughness intensified, significantly decreasing the water contact angle. The fiber surface's increased wettability led to improved water wicking and decreased static charging of the cloth.

Multiple-size nanoroughness has been the subject of intensive research by Dastjerdi et al. [10]. Their findings show that surface multiple-size nanoroughness was produced after colloidal Ag coupled with polysiloxane resin was applied to polyester textiles treated with nano-TiO_2. This resulted from the electrostatic interactions of differently sized, oppositely charged nanomaterials. Various hydrophobic or hydrophilic qualities were produced on the surface by changing the weight ratio of used materials. Figure 5 schematically depicts the development of surface nanoroughness of different dimensions.

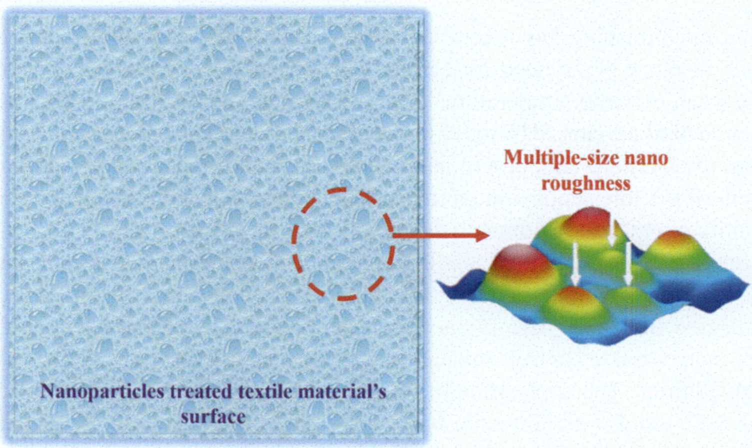

Fig. 5 Development of surface nanoroughness

3 Functionality of Various Nanomaterials in Textile Finishing

The textile industry has enormous future opportunities for nanotechnology. The main reason for this is that the methods currently used to give materials different properties usually do not produce effects that endure over time and might result in the utility of the cloth being lost after repeated washings or use. The development of nanotechnology in textile finishing processes has created a wealth of novel finishing and application options. Using different varieties of nanomaterials or developing structured surfaces based on nanotechnology is crucial for improving chemical finishing's usefulness, durability, and controllability [23].

In-situ techniques, which include the simultaneous synthesis and deposition of nanomaterials on textiles, have recently drawn more interest. In-situ procedures are therefore one-step processes without a final fixing stage. Numerous studies have been conducted recently to demonstrate the viability of in-situ nanofinishing for different textile-based products. The in-situ approach is quick, simple to execute, and economical since nanoparticle development and surface deposition happen together during the process.

Ex-situ techniques for applying nanomaterials often include several steps specifically synthesis, accumulation, and dispersion of nanomaterials on the outer layer in a distinct process. The operation duration will lengthen as a result, consuming more energy. Shorter duration, energy, and chemical intake, chemical and physical associations, excellent lifespan, improved physicomechanical characteristics, and minimal toxicity are the key advantages of in-situ nanomaterials manufacturing over ex-situ approaches.

In-situ nanofinishing has several advantages, but it also has certain drawbacks. One such instance is the need for certain nanoparticles, particularly metal oxides, to undergo an extreme temperatures final calcination step to fully crystallize. The degradation of fibers caused by the extreme pH of the synthetic bath is a difficult issue related to the in-situ production of nanomaterials on textile materials. Additionally, as a result of the low deposition on the surface, attempts need to be made to reduce the quantity of residual nanomaterials in the synthesis effluent [20].

A wide variety of technical textile applications, including sportswear, medical textiles, protective clothing, and automobile textiles, can benefit from the extraordinary degree of textile performances in nanofinishes, including hydrophobicity, antimicrobial, wrinkle control, antistatic, stain resistance, hydrophilicity, and shrink-proof capabilities. The application of finishing methods using nanomaterials in the form of nano-emulsions on textile bases also makes them smarter. Nanomaterials have a high surface energy and high surface area-to-volume ratio, thereby providing a higher affection for materials and increasing the function's longevity. More significantly, finishing based on nanostructures can offer textiles with remarkable levels of durability. As a result, a coating combined with nanomaterials on the outermost layer of textiles will not change their hand-feel qualities. Because of this, researchers and professionals are becoming more interested in applying nanotechnologies to the

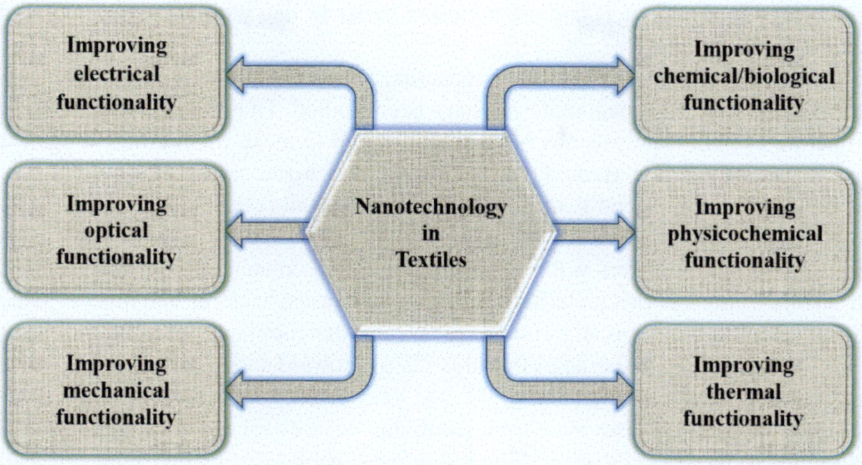

Fig. 6 Application of nanofinishing in various aspects of textiles

textile sector. Figure 6 shows the various applications of nanotechnology in the textile sector. The application of finishes using nanotechnology to enhance different textile properties is covered in these sections.

3.1 Biological and Chemical Functionality

3.1.1 Antibacterial

Consumer demand for antimicrobial fabrics has significantly grown in recent years due to greater awareness of health issues. Since the vulnerability of textiles to bacteria and fungus intensifies with moisture, nutrition, and temperature, it is crucial to prevent or at least minimize microbial development on them. The exterior, hard cell wall of the three types of bacteria—coccus (spheres), bacillus (rods), and spirillum (spirals)—gives these organisms their structure and support.

Antimicrobial, often known as antibacterial or antifungal based on the kind of microbe, is a substance that has a detrimental impact on the viability of microorganisms. Typically, two methods are used to carry out antimicrobial actions [4]. In the first method, antimicrobial drugs are released under controlled conditions and then dispersed to destroy the microorganisms. As a result, antibacterial agents are not very durable, and they seldom chemically connect with textile fibers. A significant aspect determining effectiveness in this respect is the antimicrobial agent's leaching rate. The second method is the direct contact method, where the antibacterial substance is firmly linked to the fibers of the textile and functions when it comes into contact with microbes. By using this mechanism, antimicrobial drugs might potentially have the fewest adverse impacts and are therefore better in line with environmental and toxicological standards.

For their durability, safety health benefits, antibacterial capabilities, and self-cleaning abilities, TiO_2 nanomaterials have been studied. TiO_2 nanoparticles' ability to operate as photocatalysts has been demonstrated to make them efficient against microorganisms. It is classified as a generally recognized safe product since they are environmentally friendly, inexpensive, and chemical resistant. Most permanent surfaces, whether hard or soft, can be coated with TiO_2. When exposed to UV radiation, TiO_2 coating lowers water vapor in the air and creates free oxygen radicals. On surfaces, attack germs including mold spores, viruses, bacteria, and other organic substances. Numerous investigations have revealed that they have powerful antibacterial and antifungal characteristics that are effective toward a variety of Gram-positive and Gram-negative bacteria [11].

Silver nanoparticle uses vary from sanitizing devices for healthcare and household items to water purification. Silver ions are capable of sticking to the cell wall because of the electrostatic interaction between silver ions and sulfur proteins. The attached silver ions will make the microbial cell wall easier to penetrate, eventually resulting in cell wall breakdown. Respiratory enzymes may become inactive once the free silver ions enter microbial cells, producing reactive oxygen species. The rupturing of the membrane of cells can be accelerated by reactive oxygen species [15]. To increase the antibacterial properties or give multifunctional features, Ag-based nanotechnology containing elements such as sodium carboxymethyl cellulose, SiO_2, hydroxyapatite, chitosan, and many more can be used. Additionally, several efforts have been undertaken to make fabrics with antibacterial silver-based finishes that are less harmful and more biodegradable. Additionally, several studies have looked into how silver-based nanomaterials, including silver chloride nanocrystals, can be used to make textile substrates more effective at resisting bacteria.

The formation of reactive oxygen species, one of the main mechanisms leading to antibiotic action, and the destruction of lipids and proteins in bacterial cell membranes are two ways that ZnO NPs can lead to cell death in bacteria. Reduction of bacterial growth is another result of H_2O_2 production from ZnO NP surfaces. When biomolecules and ZnO NPs contact, the microbial cell wall is destroyed, and the bacteria within it die. Figure 7 provides a schematic representation of a potential mechanism for the antibacterial effect of Cu-based NPs [12].

3.1.2 Water and Air Filtration

Pollution is the introduction of any sort of particles, organic molecule, or hazardous substance that is a gas, liquid, or solid into the environment and constitutes a threat to the system, damages or affects living things, and changes the ecology of a region. Metal oxide nanomaterials have shown a significant capacity for the harmful adsorption of hazardous substances, including air contaminants, weapons of mass destruction, acidic gases, and insecticides.

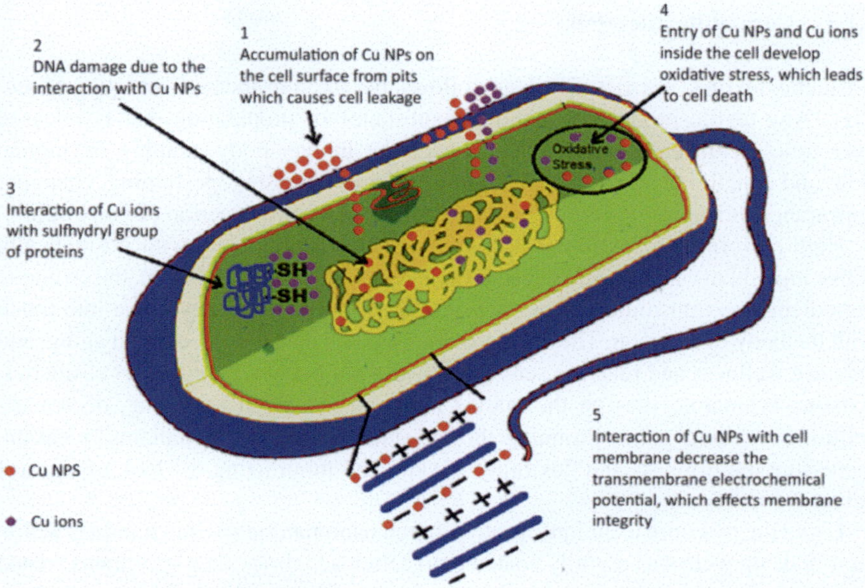

Fig. 7 Illustration of a potential Cu-based NP antibacterial activity mechanism ([12] open access)

The removal of numerous organic and inorganic impurities from the air and water using CNTs has been demonstrated to have excellent potential. The porous nature of CNTs and the existence of a variety of functional elements on their surfaces essentially contribute to their high adsorption capacity. These properties may be modified chemically or thermally to give CNTs the required efficiency.

Ceramic and polymer membranes for wastewater treatment are being made using nanomaterials. These nanomembranes include biomimetic membranes, organic–inorganic hybrid nanocomposite membranes, and ceramic membranes covered with catalytic and zeolite nanoparticles. Membranes made from biopolymers that consist of block copolymer membranes, CNT membranes, and protein molecules with comparable porosity are all examples of bio-mimic membranes. Water permeability, tolerance of contaminating molecules, and mechanical durability are used to enhance the efficiency of such membranes [1].

Particularly for the treatment of wastewater, particles with certain sizes and ranges have certain surface and physical–chemical characteristics that are helpful for multiple uses. The size of the particles might range from a few to several hundred nanometers. Nanotechnology is utilized to identify and remove these various contaminants from wastewater and remove them because conventional procedures like adsorption, oxidation, and extraction are efficient but sometimes quite expensive. There are several nanotechnologies, such as adsorption, nanofiltration, photocatalysis, and electrochemical oxidation employing metal or metal oxides like TiO_2 [7].

3.1.3 Cosmetotextiles

Cosmetic textiles are described as follows by the European Cosmetics Directive, "Any textile product containing a substance or preparation that is released over time on different superficial parts of the human body, notably on human skin, and containing special functionalities such as cleansing, perfuming, changing appearance, protection, keeping in good condition or the correction of body odor".

High-performance textiles that combine fabric components with cosmetically active ingredients are known as cosmetotextiles. To attain a certain aim, the cosmetic ingredients are transmitted from the textile to the skin when they come into touch with the body of a person. They can provide skincare advantages, combat aging, and enhance wellness and relaxing sensations while not being recognized as cosmetics. Existing cosmetotextiles on the market make claims about hydrating, decreasing cellulite, smelling good, slimming the physique, energetic, revitalizing, pleasant, increasing the firmness and flexibility of skin, and minimizing the look of signs of aging and wrinkles.

Using the pad-cure technique, protein-based nanomaterials with vitamin E within their makeup were successfully coated onto cellulose fabrics. The specimens coated with nanomaterials containing the highest levels of vitamin E during the oil phase showed the best antioxidant properties. This method demonstrated how the ejection of the active components from coated specimens was influenced by abrasion, perspiration, or enzymes. They will work together to facilitate the escape and transfer of the substance deposited to the skin for practical use in cosmetotextiles [17].

While the hydrophilic exterior of β-cyclodextrins (CDs) can graft or load onto the textile material, the hydrophobic interior of CDs enhances their bonding with aromatic compounds. Although the combination of these nanocapsules on textiles is often attached using adhesives, the physical finishing approach is not environmentally or human-friendly. In contrast, several new environmentally friendly physical finishing techniques may directly load these nanocapsules onto fabrics without the need for adhesives [25].

3.2 Physicochemical Functionality

3.2.1 Water and Oil Repellent

According to a surface's hydrophobic or hydrophilic properties, roughness significantly influences the extent to which it can absorb water. Roughness hence affects both the properties of the surfaces. Typically, there are two contact angle assumptions. Regardless of surface roughness, the Cassie–Baxter method depicts a water droplet remaining on the protuberances rather than occupying the holes, in contrast to the Wenzel approach (Fig. 8), in which it fills the cavities. Interfaces with low surface energy and high degrees of roughness are needed for developing superhydrophobic substrates [6].

Fig. 8 Effect of the surface structure on the wetting property (as per Wenzel and Cassie–Baxter state)

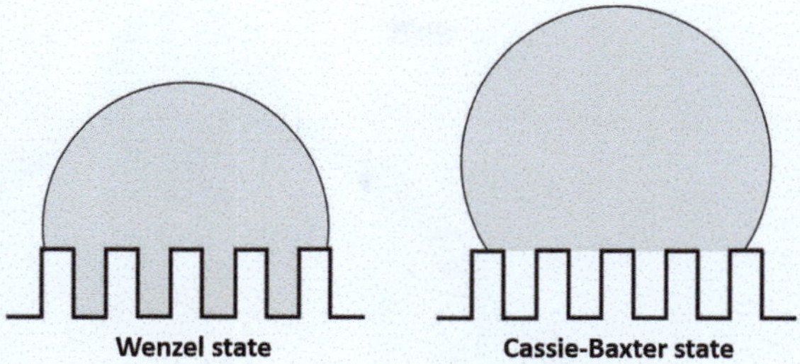

This process is known as the superhydrophobic physical self-cleaning effect, self-cleaning effect, or the lotus effect because water droplets readily roll off the low-adhesion superhydrophobic region, grab up polluting elements, and wipe down the outer layer. In a hydrophilic coating (Fig. 9), water flows across the surfaces and removes dirt and other contaminants, whereas the hydrophobic approach causes the water drops to roll and glide across the surfaces, wiping them [18].

The usage of organo-modified nanoclays for creating a heat, oil, and solvent-resistant finish is another intriguing finding that has been made. Fluoroelastomers that include nanoclays have a better rate and quality of cure [22]. By using nanoclays, it is possible to alter the oil- and water-repellent finish's curing conditions and increase the oil- and water-repellent finish's effectiveness while using less fluorocarbon.

Surfaces can become superhydrophobic with just the right amount of roughness and low surface energy. To create superhydrophobic materials, it has been essential to

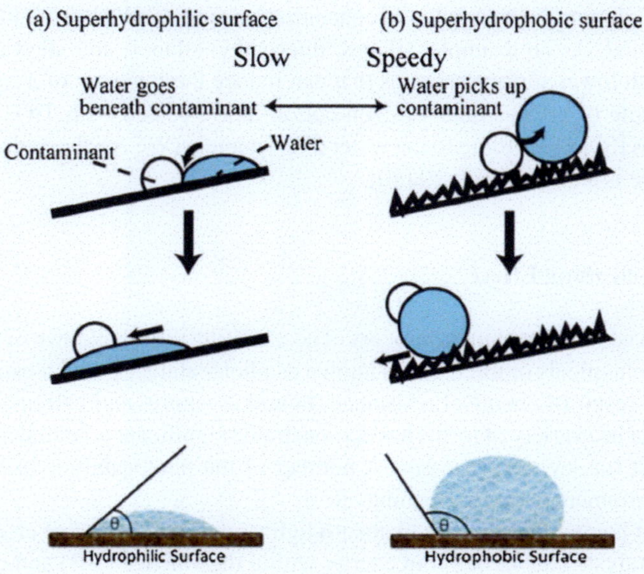

Fig. 9 Self-cleaning procedures are shown schematically on **a** a superhydrophilic and **b** a superhydrophobic surface ([18] open access)

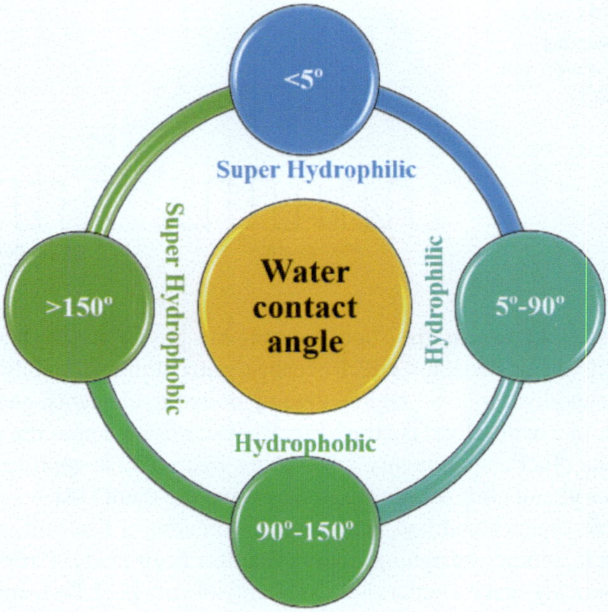

Fig. 10 Surface hydrophobicity based on measurement by WCA

improve the surface roughness of textiles and decrease surface energy by employing nanomaterials. To increase the surface roughness, the surfaces' morphological nature, the particles' dimension, formulation concentration, and the thickness of the coating can be adjusted. As an example, silanes, fluoroalkyl silanes, and alkyl amines are additives with lower surface energies that can reduce the intensity of a surface with a contact angle of greater than 120° on cotton fabric coated with TiO_2 and silver nanoparticles [32]. According to the water contact angle (WCA), the hydrophobicity of the surface is shown in Fig. 10 [40].

3.2.2 Self-cleaning Effect

A significant advancement in the process of accelerating the breakdown or reaction of undesirable chemicals or microbes is known as photocatalysis, which uses the sun's power. As a result of chemical resistance, nontoxic nature, and extremely reactive nature, titania in a variety of forms has demonstrated significant potential as a suitable photocatalyst for several processes. A number of the photocatalytic uses for TiO_2 include environment and self-cleaning, etc.

A catalyst having the capacity to absorb light and produce pairs of electron holes is a photocatalyst. The variation in energy within the valence (VB) and conduction (CB) bands of a semiconductor is known as the band gap. After being excited by photons, band gaps are transferred from VB to CB. Before excitement, both holes and electrons remain in the VB. The fundamental processes for producing hydrogen using photocatalysis and purifying water or air using photocatalysis are, respectively, reduction and oxidation processes.

The necessary band gap, appropriate shape, large surface area, stability, and repeatability are the key characteristics of the photocatalytic process. According to the photocatalyst's band gap energy, the light may be visible, ultraviolet, or a combination of the two. TiO_2 and ZnO nanomaterials are extensively explored with other metal oxide nanomaterials among the various metal oxide nanophotocatalysts [28].

The hydroxyl radicals are created when the photogenerated holes in the VB interact with adsorbed water molecules and diffuse to the surface. As a result, the adjacent organic molecules on the surface are oxidized by the hydroxyl radicals and photogenerated holes. The superoxide radical anion is created when reduction activities involving electrons in the conduction band interact with air molecules to form molecular oxygen. Contaminants can interact with radicals and anions to break down or change them into less hazardous compounds.

As shown in Fig. 11, with the aid of water, these compounds may be eliminated off the surface due to the photocatalytic activity of TiO_2 [18]. These characteristics derive from TiO_2's capacity to produce reactive oxygen species, which may break down organic molecules into carbon dioxide and water. The content of TiO_2 is one of the most significant elements affecting the self-cleaning characteristics, and these qualities often improve as TiO_2 levels rise.

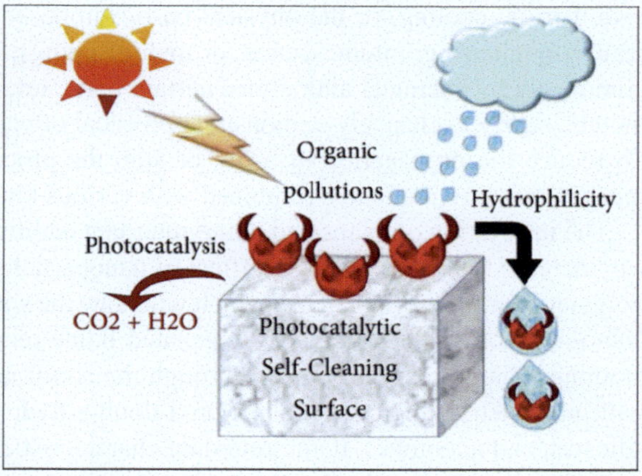

Fig. 11 Photocatalytic processes on the surface of a self-cleaning material ([18] open access)

3.3 Thermal Functionality

3.3.1 Thermal Insulation and Fire Resistance

Whenever fuel, heat, and sufficient oxygen are available, an ordinary fire situation happens. The temperature of the burning polymers keeps rising early in the fire incident to the point where it starts to cause the chemical bonds in the polymer's backbones to break. Char (also known as a carbonaceous layer) is produced by burning on the polymer surface during the stage of condensation. According to the amount of char present, the layer of char might either result in another decomposition process that produces additional volatiles or decrease the process of combustion. Degraded polymer also combines with atmospheric oxygen in the gas phase to produce free radical species, such as H and OH. To create more combustible combinations that can maintain combustion and produce additional energy, these radicals are then recombined with oxygen. The heat emitted by combusting polymeric materials reaches a certain level over time, and these materials are capable of burning despite the absence of an external heat source. The continual availability of heat, fuel, and oxidizers at this stage allows the combustion procedure to become self-sustaining.

Interacting with the chemical and physical mechanisms associated with polymeric combustion, flame retardant is a material applied to regulate the combustibility of polymeric materials. Depending on the working principle, flame retardants are frequently divided into several categories. A flame-retardant substance can prevent the propagation of a flame by either acting physically (by dilution of the fuel, endothermic cooling, or the physical barrier imposed by nanoparticles), or chemically (by charring, intumescence, or melt dripping). No flame retardant can work completely by chemical alterations; certain flame retardants, such as hydroxides of metals, nearly exclusively demonstrate physical effects while suppressing burning. To reduce the propagation of a fire or stop the process of combustion, chemical mechanisms are frequently combined with various kinds of physical activities [41].

The integration of nanosized flame-retardant additives into polymeric substances can increase the physical barrier effect of nanoparticles. These filters will cause the polymer matrix to form a maze, which will slow the spreading of combustible gases. Nano-antimony pentoxide and halogenated flame retardants work together to give clothing a flame-resistant finish. Through the addition of polymer nanocomposites with nanometric fillers such as layered double hydroxides, metal oxides, layered silicates, and nanotubes, flame-retardant characteristics may also be added [21].

TiO_2 may enhance the thermal behavior and flame retardancy of textile materials by developing a physical thermally protective char barrier, which is an inorganic substance with outstanding thermal stability. It serves as a layer of insulation and can lessen the amount of fuel, heat, and oxygen flow across the flame and fibers, thereby lowering the pace and level of burning. When anatase is thermally annealed at temperatures over 500 °C, the production of the TiO_2 barrier is accelerated, and the degree of protection is directly determined by the TiO_2 content and the dimension of the particles.

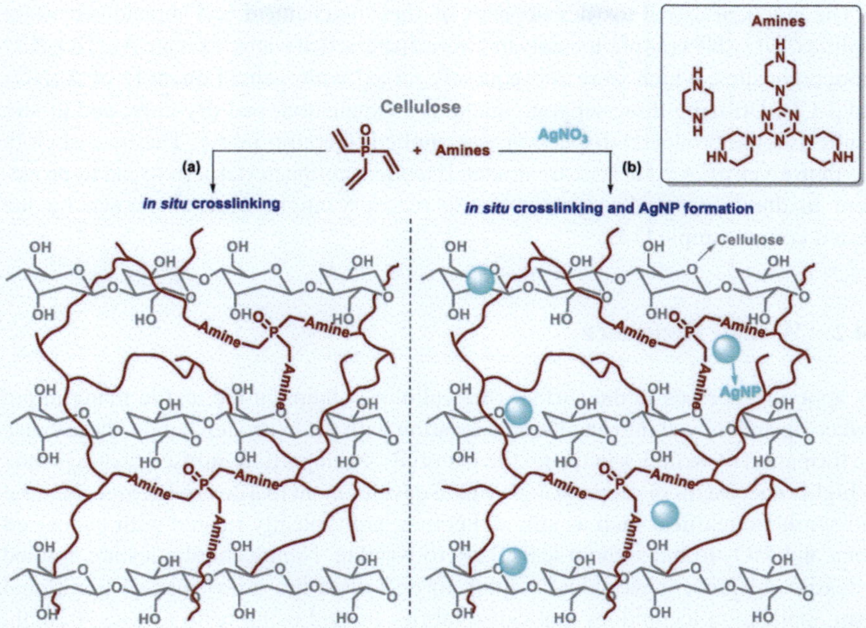

Fig. 12 **a** Trivinylphosphine oxide and amine in-situ crosslinking on cotton, **b** silver nanoparticle production on cotton while phosphine oxide and amine are simultaneously crosslinked in situ ([31] open access)

Utilizing the formation of silver nanoparticles and in-situ Michael addition reaction, cotton textiles were given flame resistance and antibacterial properties. Through Michael addition to the cotton fabric, trivinylphosphine oxide was able to interact with amine moieties to create the polymer network. The cellulose structure was able to contain the generated phosphine oxide macromolecules (Fig. 12) [31].

3.4 Mechanical Functionality

3.4.1 Abrasion Resistance and Durability

The abrasion resistance and durability of textile materials are some of the most significant improvements. A complicated combination of forces causes an impact on the abrasion and durability of a textile material that happens throughout maintenance and usage. The kind of activities engaged by the user, the areas of the fabric most exposed to the movement the design and fitting of the apparel, and consumer care methods are a few of the elements influencing the degree of abrasion. Nanomaterials can bear loads and boost durability and resistance to abrasion since these particles are equally dispersed across polymer matrices.

The practical use of textiles depends on their mechanical and abrasion-resistant qualities. TiO_2-NPs (as photocatalysts) were dispersed in water to create Ag/TiO_2/CD nanocomposites, which were subsequently mixed with a small quantity of $AgNO_3$ and β-CD. Utilizing three separate methods—exhaustion, pad-dry-cure, and in-situ synthesis—the produced dispersion was applied to cotton fabric. The treated cloth was then given a CA/SHP post-treatment. Excellent antibacterial activity, an improvement in durability, and a greater crease recovery angle were all shown by the treated cotton sample [14].

3.4.2 Wrinkle Resistance

By applying stresses to the surface, the cellulose chains in the textile material are broken up and moved into an alternate location with the hydrogen bonds, these cause the formation of wrinkles in the textile materials. Nanoparticle-infused coatings may be highly successful in suppressing wrinkle development to a desired degree. Natural and synthetic textiles, such as silk and cotton, are typically treated with nanosized silica and TiO_2 to make them less likely to wrinkle. Nanotechnologies are applied to textiles to prevent wrinkles, helping to overcome the drawbacks of traditional resin application techniques that cause fabrics treated using resin to lose strength, dyeability, resistance to abrasion, breathability, and moisture vapor transmission. In contrast with traditional crease-resistant finishes, nano-engineered crosslinking agents provide wrinkle resistance to textiles during the fabric finishing procedure without negatively affecting the handling and comfort qualities of treated cotton fabrics [21].

As catalysts for wrinkle-free finishes, nanoparticles of TiO_2, Ag, SiO_2, and ZrO_2 have been studied. Due to their photocatalytic properties, TiO_2NPs are widely employed because they make it easier for crosslinking agents to interact with substrates. For wrinkle resistance, these nanomaterials have not been utilized on their own. They have been utilized as cocatalysts in several wrinkle-free finishes. Using the pad-dry-cure process, a wrinkle-free treatment was created using TiO_2 nanoparticles distributed in self-crosslinking DMDHEU. By limiting the mobility of intermolecular chains, TiO_2 may have improved wrinkle resistance. Due to its small dimensions, it was able to penetrate fiber gaps and adhere to cellulose's hydroxyl chains. Additionally, the existence of TiO_2 improved the crosslinking process of DMDHEU [24].

3.5 *Optical Functionality*

3.5.1 Protection from UV

To increase UV shielding, textiles are treated with UV-blocking nanoparticles (UVB and UVA radiations) to create UV protection materials. The ultraviolet protection factor (UPF), which is dependent on the fabric's composition, is a measurement tool for UV protection effectiveness.

Textiles can be shielded against UV rays from the sun's harmful degradation with the application of UV-blocking compounds. Because they behave non-toxicly and are more stable than organic UV blockers, inorganic UV blockers are better-suitable for textile materials. ZnO, SiO_2, TiO_2, and Al_2O_3 nanomaterials can do an enhanced and more efficient job of absorbing and scattering UV radiation and shielding UV.

UV rays contain an extremely high energy level and a wavelength that varies between 290 and 400 nm. The three ultraviolet radiation spectrums are UV-A (320–400 nm), UV-B (290–320 nm), and UV-C (100–290 nm). However, UV-B exposure is the greatest offender since it adversely affects the possibility of cancer of the skin [29].

There is a reciprocal relation between the size of the nanoparticles and the wavelength of the fourth power of the wavelength for scattering at the nanoscale. TiO_2 and an acrylic binder produced on the cellulosic textile material using the pad-dry process were used by Sundaresan et al. [39] to create a UV-protective cotton textile material. Because of this, the chemically processed fabric had better UPF ratings and offered better UV resistance than raw material.

As per the band theory of solids, due to TiO_2's semi-conducting characteristics, it has been discovered that UV light may be absorbed regardless of the size of the particles. TiO_2's high refractive index has been suggested to be responsible for its UV-shielding capabilities, which are accomplished through UV scattering and reflection. However, it is widely accepted that these processes only matter in the region where TiO_2 is either not absorbing at all or just barely absorbing. Additionally, it is asserted that submicrometer particles have very low UV radiation absorption and that only TiO_2 nanoparticles may absorb UV light. UV radiation is barely transmitted through the textile surfaces as an outcome of UV radiation being absorbed, reflected, and scattered by TiO_2, shielding the consumer from UV radiation's adverse impacts (Fig. 13) [33].

3.5.2 Improve the Dyeability

The need for multifunctional fabrics has grown recently, necessitating a robust, interdisciplinary approach to reducing chemical consumption. The dyeability of synthetic and natural fibers has also been enhanced by pretreatment with nanomaterials. Dispersing agents like carriers and solvents are typically needed to improve the dyeing efficacy if nanoscale pigments are found to be employed in textile dyes.

Since it may eliminate the need for metal salt mordants when using basic dyes, this rise in color uptake by cotton acquired is very remarkable. The pretreatment of nanoclay also led to an enhancement in the fabric's light fastness and washing. Even with the lowest nanoclay levels applied, the fabric's drape ability was improved. However, the fabric's tensile characteristics were reduced as a result of the nanoclay preparation. In contrast, the introduction of Ag nanomaterials to an acid blue dye improved the dye uptake, washing fastness, tenacity, and other properties of wool

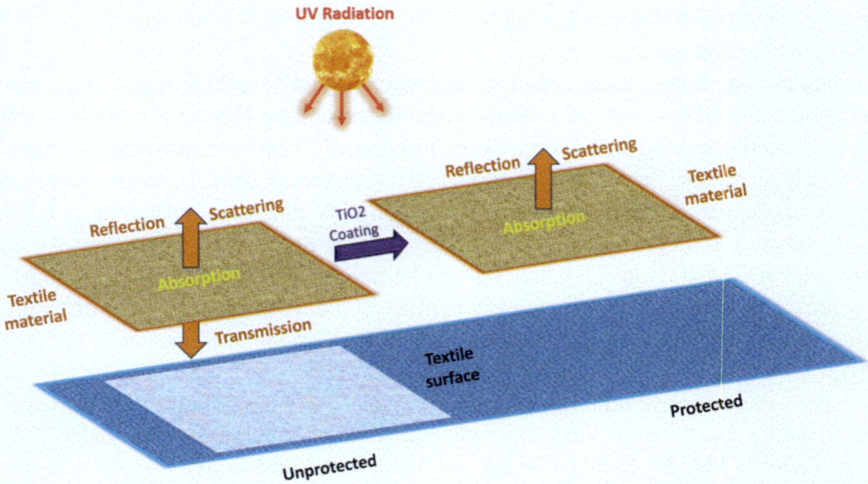

Fig. 13 Modified textile's UV transmission, reflection, and absorption (Redrawn from [33])

fabric. It was determined that the anionic groups in wool fibers, the acid dye, and the Ag nanoparticles formed a combination [13].

The viscose material changed colors of yellow while the acrylic material changed shades of beige according to the dimension and aggregation of the nanomaterials. Due to the chemical link between the silver nanomaterials and the fabric's substance, the color fastness against friction, laundering, and acidic and basic sweat was good to very good. However, the light-fastness scores were slightly worse. To enhance the photocatalytic ability of nano-TiO_2 on textiles made of cotton, the photonic qualities of Au nanomaterials were also used. The degradation developed in a Remazol Blue solution in water in visible light increased as a result of the nanostructured coating's improvement of the visible-light catalytic sensitivity of nano-TiO_2 [13].

In addition to being utilized as nanoscale dyes, nanomaterials can also be employed to increase dyeability. Chitosan is cationic in acidic conditions, making it simple for absorbing anionic compounds such as acid and reactive dyes. This has increased the dyeability of PP. Natural fibers may be dyed using this feature. Inkjet printing is a cutting-edge finishing technique that has just recently been invented and is continually being improved. In terms of process economy, management of waste, color brilliance, and storage stability, it enables the usage of dispersed dyes in nanoparticulate form. Plasma treatments are receiving extra attention among the finishing methods mentioned above because of their adaptability and reduced ecological impact [38].

3.6 Electrical Functionality

3.6.1 Electromagnetic (EM) Shielding

Shielding materials, especially textiles designed for this use, are created and studied at frequencies that are comparable to those in which EMI shielding is important, especially in the microwave region. The use of EMI-shielding fabrics for protecting from radio waves, radar, and other sources is also possible. Substances must react to the electrical or magnetic waves of radiation to be able to absorb or reflect EMR. Conductive and magnetic materials are the two groups of materials that are utilized to block electromagnetic radiation. Textile fabrics that have particular conductive coatings or additives integrated into them are electrically conductive, so they interact with the electrical portion of the EMR.

Due to their great electrical conductivity, metals like Ag, Cu, Au, and Al are the best options for EM shielding. It can be employed as particulates, fibers, or wire (Coated Cu, Ag, Au, Ni, or Zn Wires). Due to their low density, large specific surface area, flexibility in processing, and superior electrical conductivity, carbon compounds, such as carbon nanofiber, graphite, carbon black, CNT, and graphene, may also be utilized as EMI-shielding compounds in addition to metals. According to reports, conductive networks produced by conductive substances of various forms can be filled in by nanoscale elements, creating a denser and more complete network.

In the textile industry, nanocoating is a relatively recent method. Several techniques are used to create conductive nanocomposite yarns. Typically, CNT or metallic nanomaterials are deposited into a dielectric matrix during the formation of nanocomposite materials. Due to their generally inexpensive price and simple processability, polymeric matrices are of special importance. The shape of the incorporated metallic or carbon nanostructures, which relies on both metal concentration and layer thickness, directly affects the electric characteristics of these composite materials.

In a study by Bonaldi et al. [34], conductive fillers which include MWCNT, conductive polymer, and metal nanoparticles were applied using a knife-over-roll coating process to EM-shielding textiles comprised of knitted cotton and nonwoven polyester. The coating thickness was between 100 and 200 mm, while the measured EM range was between 200 MHz and 1 GHz.

3.6.2 Antistatic

Atoms and molecules are held together by the electrostatic force that results from the contact and separation of two materials. The air's breakdown strength regulates the maximum charge that may build up on a surface. The charge on the surface dissipates as the dielectric strength of the air is compromised. Two types of charging cause electrostatic discharge: conduction and induction. Contact and friction charging are two different categories of conduction charging. Simple contact charging is accomplished by simply bringing two materials into touch and separating them without friction. The method of friction charging is more complicated than simple contact charging since it involves moving the two components while they are in touch, which increases the static charge buildup on both surfaces after separation. There is no physical contact made between the two materials during induction charging. In this situation, bringing two objects close together might result in a charge [37].

Static electricity buildup in the industry may make fibers, yarns, and textiles more prone to attracting dirt and sticking to other materials or machine parts. Furthermore, the buildup of static charges during production can cause textiles to stick together, resulting in low-quality and inefficient cutting, finishing, and dyeing processes. Electrical charges may trigger parachutes to cling, which can result in fatally inadequate opening. Any non-conducting surface may be controlled for the static charge using two separate methods: the first method is to manage the amount of static generation when a substance contacts various surfaces and the second is to regulate the surface conductivity to enable quick transmission of any produced charge to ground [37]

To eliminate electrical charges on the textile material, antistatic chemicals are used. Since synthetic materials have weak antistatic qualities to begin with, the development of such properties is quite concerning. Due to its low water absorption nature, the majority of synthetic fibers are more likely to accumulate static charges. As a result, it has been reported in several papers that nanosized materials including nano-Sb-doped tin oxide with silane nanosols, ZnO whiskers, and TiO_2 are electrically conductive materials that may be used to discharge static electricity from fibers. As a result, ZnO and TiO_2 nanomaterials are well known for giving synthetic fibers antistatic qualities. The increased moisture in the cellulose fibers causes the internal accumulation of static charge. When compared to traditional approaches, nanomaterials are more resilient to cleaning and laundering processes than antistatic chemicals [40].

3.6.3 E-Textiles

The introduction of sensor-based materials using TiO_2 and MgO nanoparticles has transformed the design and development of wearable textiles. The addition of microcrystalline ceramic particles to textiles gives finished fabrics a particular functionality that makes them capable of translating applied mechanical stress into electrical impulses.

While wearing a garment made from completed cloth close to your skin, physiological factors like heart rate and pulse may be recorded. TiO_2 and ZnO nanostructures can transmit electrical power, dissipate static charges on fibers, repel static, and demonstrate a sterilizing effect, making them an ideal component of wearable smart technology. Figure 14 shows the development of the e-textiles from phase-one to the current fourth-generation phase [27].

According to certain studies, the inclusion of sensors and actuators into textiles is made possible by the electrically conductive properties of conducting materials and graphene, each of which is desirable for making textiles. Stretchable graphene

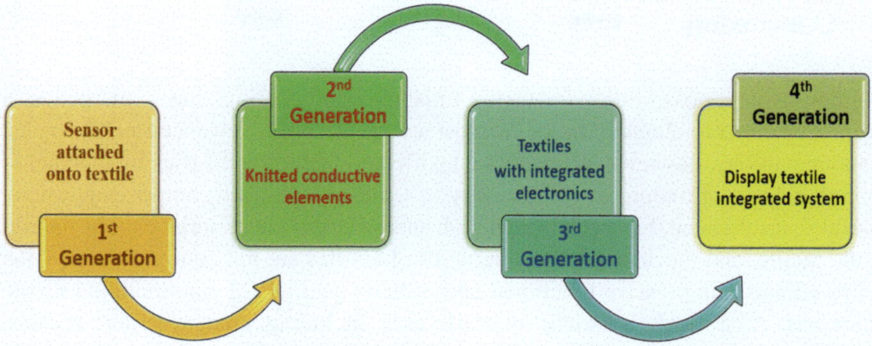

Fig. 14 Development of the e-textiles for different generations (Redrawn from [27])

and PPy (Polypyrrole)-based supercapacitors, triboelectric nanogenerators, flexible fiber and stripe batteries, and lightweight fabric carbon nanotube supercapacitor electrodes have all been proposed in academic publications. These e-textiles would become new types of practical clothing if electronic elements were added to them. By incorporating optical and optoelectronic devices into textiles, OLEDs in fiber form might result in ground-breaking innovations. [3]. Figure 15 shows the E-textiles for wearable applications [8].

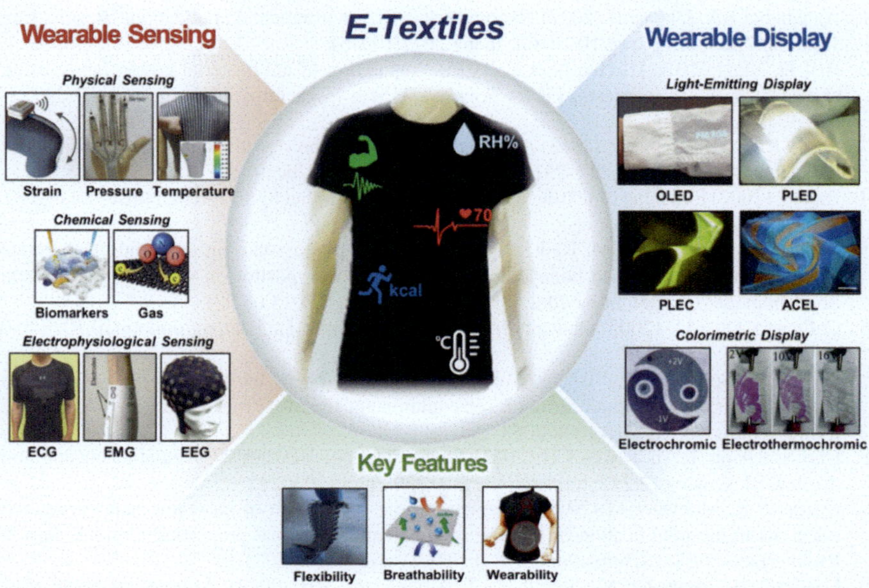

Fig. 15 E-textiles for wearable sensing and display ([8] open access)

4 Conclusion

Numerous new possibilities for textile finishings and coatings have emerged as a result of nanotechnologies. One of the most revolutionary and advanced technologies, nanotechnology, has revolutionized the textile and fashion industries by enhancing functionality and finding use in a variety of textile fields. This comprises physical, thermal, mechanical, biological, chemical, and electrical characteristics. The natural bulk, tactile, and mechanical characteristics of textiles are not compromised by the extraordinarily impressive functional and esthetic qualities of nanomodified textile materials. As a result, a variety of textile uses, including smart wearable textiles, healthcare and defensive textiles, athletics, antimicrobial, self-cleaning, and garments for extremely cold climates, have successfully used nanotechnology. The significant impact of nanotechnology on the textile industry is related to the radical transformation of material characteristics brought about by scaling down to the nanoscale level. The system has also outperformed traditional methods, which only provide short-term functioning. Consequently, the textile sector may benefit from nanotechnology, which can also be explored more thoroughly.

References

1. Aghababai BA, Jabbari H (2022) Nanomaterials for environmental applications. Results Eng 15:100467. https://doi.org/10.1016/j.rineng.2022.100467
2. Arif D, Niazi MBK, Ul-Haq N et al (2015) Preparation of antibacterial cotton fabric using chitosan-silver nanoparticles. Fibers Polym 16:1519–1526. https://doi.org/10.1007/s12221-015-5245-6
3. Asif AKMAH, Hasan MZ (2018) Application of nanotechnology in modern textiles: a review. Int J Curr Eng Technol 8:227–231. https://doi.org/10.14741/ijcet/v.8.2.5
4. Bajaj P (2002) Finishing of textile materials. J Appl Polym Sci 83:631–659. https://doi.org/10.1002/app.2262
5. Behzadnia A, Montazer M, Rad MM (2015) In situ photo sonosynthesis and characterize nonmetal/metal dual doped honeycomb-like ZnO nanocomposites on wool fabric. Ultrason Sonochem 27:200–209. https://doi.org/10.1016/j.ultsonch.2015.05.021
6. Callies M, Quéré D (2005) On water repellency. Soft Matter 1:55. https://doi.org/10.1039/b501657f
7. Chadha U, Selvaraj SK, Vishak TS et al (2022) A review of the function of using carbon nanomaterials in membrane filtration for contaminant removal from wastewater. Mater Res Express. https://doi.org/10.1088/2053-1591/ac48b8
8. Cho S, Chang T, Yu T, Lee CH (2022) Smart electronic textiles for wearable sensing and display. Biosensors 12:222. https://doi.org/10.3390/bios12040222
9. Dastjerdi R, Montazer M (2010) A review on the application of inorganic nano-structured materials in the modification of textiles: focus on anti-microbial properties. Colloids Surf B Biointerfaces 79:5–18. https://doi.org/10.1016/j.colsurfb.2010.03.029
10. Dastjerdi R, Montazer M, Stegmaier T, Moghadam MB (2012) A smart dynamic self-induced orientable multiple size nano-roughness with amphiphilic feature as a stain-repellent hydrophilic surface. Colloids Surf B Biointerfaces 91:280–290. https://doi.org/10.1016/j.colsurfb.2011.11.015
11. Dicastillo CLD, Correa MG, Martínez FB, et al (2021) Antimicrobial effect of titanium dioxide nanoparticles. In: Antimicrobial Resistance—A one health perspective. IntechOpen

12. Din MI, Arshad F, Hussain Z, Mukhtar M (2017) Green adeptness in the synthesis and stabilization of copper nanoparticles: catalytic, antibacterial, cytotoxicity, and antioxidant activities. Nanoscale Res Lett 12:638. https://doi.org/10.1186/s11671-017-2399-8
13. Dolez PI (2019) Application of nanomaterials in textile coatings and finishes. In: Nanomaterials-based coatings. Elsevier, pp 139–169
14. Eid BM, Ibrahim NA (2021) Recent developments in sustainable finishing of cellulosic textiles employing biotechnology. J Clean Prod 284:124701. https://doi.org/10.1016/j.jclepro.2020.124701
15. Galatage ST, Hebalkar AS, Dhobale SV, et al (2021) Silver nanoparticles: properties, synthesis, characterization, applications and future trends. In: Silver micro-nanoparticles—properties, synthesis, characterization, and applications. IntechOpen
16. Gashti MP, Pakdel E, Alimohammadi F (2016) Nanotechnology-based coating techniques for smart textiles. In: Active coatings for smart textiles. Elsevier, pp 243–268
17. Ghaheh FS, Khoddami A, Alihosseini F et al (2017) Antioxidant cosmetotextiles: cotton coating with nanoparticles containing vitamin E. Process Biochem 59:46–51. https://doi.org/10.1016/j.procbio.2017.04.020
18. Gholami Z, Jalilisadrabad S, Amrollahi R (2023) Investigating the impact of using modified cool materials by titanium dioxide (TiO_2)-based photocatalytic self-cleaning nanoparticles in urban facades on urban microclimate parameters. Case Stud Constr Mater 19:e02268. https://doi.org/10.1016/j.cscm.2023.e02268
19. Gulrajani ML (2013) The use of nanotechnology in the finishing of technical textiles. In: Advances in the dyeing and finishing of technical textiles. Elsevier, pp 280–308
20. Harifi T, Montazer M (2015) A review on textile sonoprocessing: a special focus on sonosynthesis of nanomaterials on textile substrates. Ultrason Sonochem 23:1–10. https://doi.org/10.1016/j.ultsonch.2014.08.022
21. Jhanji Y (2022) Nanomaterials for textile applications—a critical review. J Text Assoc 83:414–421
22. Kader MA, Nah C (2004) Influence of clay on the vulcanization kinetics of fluoroelastomer nanocomposites. Polymer (Guildf) 45:2237–2247. https://doi.org/10.1016/j.polymer.2004.01.052
23. Khan FR, Misra SK, Alonso JG et al (2012) Bioaccumulation dynamics and modeling in an estuarine invertebrate following aqueous exposure to nanosized and dissolved silver. Environ Sci Technol 46:7621–7628. https://doi.org/10.1021/es301253s
24. Lam YL, Kan CW, Yuen CW (2011) Wrinkle-resistant finishing with dimethyloldihydroxyethyleneurea (DMDHEU)—the effect of co-catalyst. Text Res J 81:1419–1426. https://doi.org/10.1177/0040517510387206
25. Ma J, Fan J, Xia Y et al (2023) Preparation of aromatic β-cyclodextrin nano/microcapsules and corresponding aromatic textiles: a review. Carbohydr Polym 308:120661. https://doi.org/10.1016/j.carbpol.2023.120661
26. Maryan AS, Montazer M, Harifi T (2015) Synthesis of nano silver on cellulosic denim fabric producing yellow colored garment with antibacterial properties. Carbohydr Polym 115:568–574. https://doi.org/10.1016/j.carbpol.2014.08.100
27. Meena JS, Bin CS, Jung SB, Kim JW (2023) Electronic textiles: new age of wearable technology for healthcare and fitness solutions. Mater Today Bio 19:100565. https://doi.org/10.1016/j.mtbio.2023.100565
28. Montazer M, Harifi T (2018) Nanofinishing: fundamental principles. In: Nanofinishing of textile materials. Elsevier, pp 19–34
29. Mousa M, Khairy M (2020) Synthesis of nano-zinc oxide with different morphologies and its application on fabrics for UV protection and microbe-resistant defense clothing. Text Res J 90:2492–2503. https://doi.org/10.1177/0040517520920952
30. Nafeie N, Montazer M, Nejad NH, Harifi T (2016) Electrical conductivity of different carbon nanotubes on wool fabric: an investigation on the effects of different dispersing agents and pretreatments. Colloids Surf A Physicochem Eng Asp 497:81–89. https://doi.org/10.1016/j.colsurfa.2016.02.029

31. Özer MS, Gaan S (2022) Recent developments in phosphorus based flame retardant coatings for textiles: synthesis, applications and performance. Prog Org Coatings 171:107027. https://doi.org/10.1016/j.porgcoat.2022.107027
32. Pakdel E, Wang J, Kashi S et al (2020) Advances in photocatalytic self-cleaning, superhydrophobic and electromagnetic interference shielding textile treatments. Adv Colloid Interface Sci 277:102116. https://doi.org/10.1016/j.cis.2020.102116
33. Rashid MM, Simončič B, Tomšič B (2021) Recent advances in TiO_2-functionalized textile surfaces. Surf Interfaces 22:100890. https://doi.org/10.1016/j.surfin.2020.100890
34. Rubežienė V, Žuravliova SV (2020) EMI shielding textile materials. In: Materials for potential EMI shielding applications. Elsevier, pp 357–378
35. Schneider J, Matsuoka M, Takeuchi M et al (2014) Understanding TiO_2 photocatalysis: mechanisms and materials. Chem Rev 114:9919–9986. https://doi.org/10.1021/cr5001892
36. Sebdani ZM, Khoddami A, Mallakpour S (2011) Improvement in hydrophobicity of polyester fabric finished with fluorochemicals via aminolysis and comparing with nano-silica particles. Colloid Polym Sci 289:1035–1044. https://doi.org/10.1007/s00396-011-2426-8
37. Seyam AM, Oxenham W, Theyson T (2015) Antistatic and electrically conductive finishes for textiles. In: Functional finishes for textiles. Elsevier, pp 513–553
38. Srinivas K (2016) The role of nanotechnology in modern textiles. J Chem Pharm Res 8:173–180. https://doi.org/10.9790/019X-04010106
39. Sundaresan K, Sivakumar A, Vigneswaran C, Ramachandran T (2012) Influence of nano titanium dioxide finish, prepared by sol–gel technique, on the ultraviolet protection, antimicrobial, and self-cleaning characteristics of cotton fabrics. J Ind Text 41:259–277. https://doi.org/10.1177/1528083711414962
40. Syduzzaman M, Hassan A, Anik HR et al (2023) Nanotechnology for high-performance textiles: a promising frontier for innovation. ChemNanoMat. https://doi.org/10.1002/cnma.202300205
41. Wen OY, Tohir MZM, Yeaw TCS et al (2023) Fire-resistant and flame-retardant surface finishing of polymers and textiles: a state-of-the-art review. Prog Org Coat 175:107330. https://doi.org/10.1016/j.porgcoat.2022.107330

Chapter 3
Metal Nanoparticles: Biosynthesis and Functional Application to Textiles

B. H. Patel and M. Z. Channiwala

1 Introduction

Nanotechnology, comprising nanoscale research, engineering, and technology, is the study of understanding, managing, and modifying matter at the nanoscale. It includes the planning, building, and use of functional structures having at least one nanometer-scale characteristic dimension. Because the constituent particles or molecules of materials and systems produced with nanotechnology have limited sizes, they display new and significantly better mechanical, chemical, biological, and structural properties, phenomena, and processes [1–3]. The unit of measurement known as a "nanometer" is 1 nm, or 10^{-9} m, a billionth of a meter.

1.1 Nanoparticles: Approach and Synthesis

Nanoparticles can be produced via pyrolysis or attrition, among other techniques. In attrition, ball mills, planetary ball mills, or other size-reducing devices are used to grind macro- or micro-scale particles. To recover nanoparticles, the resultant particles undergo air classification. A vaporous precursor (liquid or gas) is burned in pyrolysis by forcing it through an aperture under high pressure. The resultant solid, which is essentially soot, is air-categorized to extract oxide particles from gases that are byproducts. While sol–gel, precipitation, liquid and gas phase reaction methods, etc. are excellent examples of bottom-up approaches in the synthesis of nanoparticles,

B. H. Patel · M. Z. Channiwala (✉)
Department of Textile Chemistry, Faculty of Technology and Engineering, The Maharaja Sayajirao University of Baroda, Vadodara, Gujarat, India
e-mail: murtaza.1857@gmail.com

B. H. Patel
e-mail: b.h.patel-tchem@msubaroda.ac.in

S. Bairagi et al. (eds.), *Nanotechnology in Textile Finishing*, Materials Horizons: From Nature to Nanomaterials, https://doi.org/10.1007/978-981-97-2696-7_3

attrition is a typical top-down method [4]. The numerous popular physicochemical techniques for creating nano-sized material are microemulsion approach, solvent-gel synthesis, biosynthesis, plasma spraying, spray conversion, sputtering, chemical vapor deposition, inert gas condensation, laser ablation, sputtering, and liquid-phase reaction techniques.

In light of environmental concerns and the involvement of other methods in the manufacture of nanoparticles, the biosynthetic process has gained popularity in recent times. Enzyme-catalyzed biosynthesis is a multistep process that transforms substrates into more complex molecules. Simple chemicals undergo modifications, transformations into other compounds, or unions to produce macromolecules during the biosynthesis process. Metabolic pathways are commonly involved in this process. While some of these metabolic routes involve enzymes found in several cellular organelles, others are contained in a single organelle. The term "green process" also refers to the synthesis or integration of nanoparticles on polymeric substrates. The following are this process's benefits: (a) additional reducing agent(s) are not required, (b) the procedure can be carried out in an aqueous solution at room temperature and normal pressure, and (c) lastly, the nanoparticles produced by this approach have exceptional qualities, such as long-term dispersion stability.

It is necessary to create an environmentally acceptable method for synthesizing nanomaterials, one that avoids using hazardous chemicals in the synthesis process. The relationship between nanoparticles of inorganic material and biological structures is one of the most exciting fields of study in modern nanotechnology and science [5, 6]. It has been shown that form and size [7, 8] have a significant impact on the electromagnetic [9], optical [10], and catalytic [11] properties of noble metal nanocrystals. This has prompted the creation of synthesis pathways that enable more precise control over size and shape. Many techniques, including hardtemplate, bioreduction, and solution-phase syntheses, have been used to create noble metal nanomaterials [12, 13].

2 Biosynthesis of Nanoparticles

This section includes a list of the species that are employed in the biosynthesis of nanomaterials. The section also outlines the qualities that must be innate to produce nanoparticles with the appropriate features. Various species are utilized in the synthesis of nanoparticles.

- The use of bacteria,
- The use of actinomycetes,
- The use of yeast,
- The use of fungi,
- The use of plants.

2.1 Use of Bacteria

It has been demonstrated that Pseudomonas stutzeri isolated from silver mines can manufacture silver nanoparticles [14]. It has been discovered that magnetotactic bacteria can be used to synthesize magnetic nanoparticles [15]. Magnetospirillum magneticum is one type of magnetotactic bacteria. Make two different kinds of particles: some make graphite (Fe_3S_4) nanoparticles and some make magnetic (Fe_3O_4) nanoparticles in chains. Some individuals produce both types of particles. Similar to this, it has been demonstrated that the sulfate-reducing bacterium Desulfovibrio desulfuricans NCIMB 8307 is synthesizing palladium nanoparticles in the presence of an external electron donor [16, 17]. Furthermore, common Lactobacillus strains found in buttermilk have been reported to have facilitated the formation of well-defined morphologically distinct tiny crystals of precious metals such as silver, gold, and golden-silver alloys.

A filamentous cyanobacterium called Plectonema boryanum UTEX 485 has been employed to provide morphological influence on the shape of gold nanoparticles. When it was treated with aqueous Au $(S_2O_3)_2{}^3$ and $AuCl_4$ solutions at 25–100 °C for up to one month and 200 °C for one day, respectively, cubic gold nanoparticles and octahedral gold platelets precipitated [18]. Cyanobacteria-aqueous gold(III)-chloride interaction first facilitated the precipitation of amorphous gold(I)-sulfide nanoparticles at the cell walls and ultimately resulted in the deposition of metallic gold in the form of octahedral (III) platelets in solutions and close to cell surfaces [19]. These mechanisms of gold bioaccumulation by cyanobacteria [20] from gold (III)-chloride solutions have been documented. To further improve the procedure, the gold(I)-thiosulfate complex was converted into elemental gold using an enrichment of sulfate-reducing bacteria. This revealed that three possible mechanisms including iron sulfide, localized conditions for reduction, and metabolism might be involved in this [21].

2.2 Use of Actinomycetes

Thermomonospora sp., an extremophilic actinomycete, has been shown to decrease metal ions extracellularly in response to gold ions, producing gold nanoparticles with much increased polydispersity [22]. However, [23] conducted the reduction of $AuCl_4$-ions using an extremophilic *Thermomonospora* sp. biomass, which has led to the efficient synthesis of monodisperse gold nanoparticles, in an effort to elucidate the process or environmental factors that promote the production of nanoparticles with the required characteristics. The reduction of the metal ions and the stability of the gold nanoparticles were assumed to be caused by enzymatic activity [24]. Comparing this study to their earlier work of synthesizing polydisperse gold nanoparticles using *Fusarium oxysporum*, they concluded that the synthesis of monodisperse gold nanoparticles might have been caused by severe biological parameters, which include alkaline as well as slightly perched conditions of temperature. Based on this theory, alkalotolerant *Rhodococcus* sp. was used for the intracellular synthesis of superior monodisperse nanoparticles. They noticed that the cytoplasmic membrane had a higher quantity of nanoparticles than the cell wall. This may be the result of enzymes found in the cell wall and on the cytoplasmic membrane, but absent from the cytosol, reducing the metal ions. These metal ions multiplied even after the biogenesis of gold nanoparticles and were not harmful to the cells that were generating them.

2.3 Use of Yeast

This was the first time that *Schizosaccharomyces pombe* and Candida glabrata were used to produce cadmium sulfide (CdS) nanocrystals. These nanocrystals, which are now utilized in quantum semiconductor crystallites, were created using cadmium salts. [25]. In order to increase the amount of semiconductor CdS nanocrystals produced using *Schizosaccharomyces pombe* cells, more research has been done. When these cells were exposed to 1 mM Cd during their mid-log phase of growth, the highest amounts of nanocrystals were formed [26]. According to this study, the yeast growth phase affected the production of CdS nanocrystals. The addition of Cd during stationary phase resulted in either no CdS synthesis or a decrease in CdS nanocrystal production and absorption. CdS nanocrystals were formed when Cd was introduced to yeast during the early exponential phase of growth; however, this time, the yeast's cellular metabolism was affected, causing the Cd to exit the cells [27]. A plausible rationale for the possible reduction in the synthesis of CdS nanocrystals is that exposure to Cd as a stressor triggered several metabolic reactions intended to offset the negative effects of the metal. Initially, cytoplasmic Cd was chelated by phytochelatin synthase to form phytochelatins (PC), which in turn formed a low molecular weight PC-Cd complex. The PC-Cd compound was then transferred across the vacuolar membrane (HMT) via a vacuolar membrane protein of the ATP-binding cassette type. A large molecular weight PCCdS complex is formed in the membrane when sulfide is added to this complex in addition to Cd, allowing them to be sequestered into a vacuole [28].

Furthermore, the yeast strain MKY3, which is tolerant of silver, has been used to produce conditions for the synthesis of notable amounts of silver nanoparticles. It is also mentioned that freezing the samples at different times was necessary for the process of separating these silver particles [29]. It has recently been discovered that certain strains of yeast are able to produce gold nanoparticles; this was achieved by controlling the formation of the particles as well as other cellular processes to control their size and shape [30].

2.4 Use of Fungi

It is possible to create nanoparticles with high monodispersity and well-defined dimensions using fungi. This was demonstrated in an experiment utilizing the fungus Verticillium to bioreduce aqueous $AuCl_4$ ions, which produced gold nanoparticles with good monodispersity and very well-defined dimensions [31]. These results have shown that $AuCl_4$ ions may be trapped on the surface of fungal cells via electrostatic interaction with positively charged groups (like lysine residues) in enzymes present in the cell wall of the mycelia. Here, enzymes in the cell wall reduced the gold ions, causing the metal atoms to aggregate and create gold nanoparticles. They were unable to pinpoint the precise mechanism behind the creation of gold nanoparticles, though. According to their findings, fungus rather than bacteria could be a source for the industrial manufacture of nanoparticles. Fungi may create a great deal more nanoparticles using a biosynthetic approach because they are known to secrete a significant amount of proteins. An in vitro method was employed to successfully reduce $AuCl_4$ ions to gold nanoparticles using species-specific NADH dependent reductase, which was released by *Fusarium oxysporum*, in order to clarify the mechanism of nanoparticle creation. For the first time, this has made a revolutionary in vitro fungal/enzyme-based strategy for the production of nanomaterials possible [32, 33]. Additionally, it was used to create an extremely stable silver hydrosol based on *Fusarium oxysporum*'s properties [34]. When Ag^+ and $AuCl_4$ ions are added to the acidophilic fungus *Verticillium* sp., they can produce both gold and silver nanoparticles [35]. However, a unique biological technique has been reported for the manufacture of silver nanoparticles both extracellularly and intracellularly employing the fungus *Fusarium oxysporum* and Verticillium, respectively. This has created an intriguing option in which the nanoparticles might be generated in solution or entrapped in the biomass as a film, both of which have intriguing commercial potential [36]. *Aspergillus flavus* generated an accumulation of silver nanoparticles on the surface of its cell wall when it was cultured in a silver nitrate solution [37].

A key factor in the creation of nanoparticles utilizing fungal cultures is the growth conditions. Extracellular nanoparticles were formed when Trichothecium sp. biomass was incubated with gold ions in a stationary environment. Under shaking circumstances, intracellular gold nanoparticles were formed as a result of this. The proteins and enzymes involved in the creation of nanoparticles may be the cause of this. When the mixture remained stationary, the protein molecules flowed into the solution; but, when it was shaking, they were not released [38, 39]. *Fusarium oxysporum* and *Verticillium* sp. were used to create magnetic nanoparticles at room temperature. Both fungi secreted proteins that could hydrolyze the precursors to iron extracellularly and create iron oxides, primarily in the magnetite (Fe_3O_4) phase. Further data suggested that a shuttle quinone and a nitrate-dependent reductase from several *Fusarium oxysporum* strains were required for the extracellular synthesis of silver hydrosol or nanoparticles that was not the case with all *Fusarium* sp. When *Fusarium moniliforme* is incubated with silver ions, it produces the reductase enzyme but is unable to make silver nanoparticles [40]. However, by altering the quantity of cofactor NADH, highly stable Au–Ag alloy nanomaterials with varied compositions could have been created. This method can also be used to create several kinds of composite nanoparticles [41]. To shed light on the mechanism of nanoparticle synthesis, *Fusarium oxysporum*-derived phytochelatin and a NADPH-dependent nitrate reductase have been used to produce silver nanoparticles in vitro [42].

Utilizing the fungus *Fusarium oxysporum*, silica and titania nanoparticles have recently been produced from aqueous anionic complexes SiF_6^{2-} and TiF_6^{2-}, respectively. While silica must be calcined at 300 °C to crystallize, crystalline titania particles can be easily synthesized at room temperature by the hydrolysis of anionic complexes by extracellular proteins [43].

2.5 Use of Plants

There are several uses for quantum dots in nano-biotechnology, and plants are a useful source for making quantum dots. The possibility of living plants synthesizing quantum dots is documented in one of the intriguing studies involving alfalfa. Ag (0) from an agar media can be absorbed by alfalfa roots and transferred to the plant's shoots in the same oxidation state. Larger arrangements were formed by the joining of these Ag atoms, which allowed them to arrange themselves into nanoparticles within the shoot. The accumulation of Ag atoms in the plant tissue underwent nucleation, based on analysis with a transmission electron microscope (TEM) and a scanning electron microscope (SEM), which led to the production of nanoparticles [14]. Furthermore, shape and size variability can be produced in plant-based nanoparticles by varying the pH of the reaction conditions. The biomass of *Avena sativa* (oats) shows a pH-dependent pattern of Au (III) ion binding, which is followed by the creation of Au nanoparticles with varying sizes, as reported by [44]. The investigation discovered that rod-shaped nanostructures appeared common in all pH levels. At pH 3 and 4, significant amounts of smaller particles (5–20 nm) were produced, whereas minor amounts of bigger nanoparticles (25–85 nm) were synthesized at pH 2. In aqueous solutions, Au (III) is found as an anion. At low pH values, biomass may include more positive functional groups, which enhances the ability of Au (III) ions to bind more tightly. The pH-dependent variations in the size and form of nanoparticles may be explained by this mechanism. Soft ligands are the main means by which Au (III), a soft metal, interacts with biomass. Furthermore, at low pH, positive amino and sulfhydryl groups make Au (III) accessible for binding and permit Au (III) reduction to Au (0). Carboxylic groups, which are common in oat biomass and are known to be protonated at low pH values, may also aid in the binding of Au (III) ions [45]. At low pH (2), nucleation to form new nanoparticles was not preferred over the process of Au nanoparticles aggregating to form larger nanoparticles. At pH 3 and 4, additional functional groups might be available for gold binding, which could lead to more Au (III) complexes attaching to the biomass at once. This made it possible for more nanoparticles with lower diameters to develop later on. Thus, altering the pH of the reaction conditions allows for the control of nanoparticle size.

Emblica Officinalis fruit extract has been used as a reducing agent in the extracellular synthesis of very stable Ag and Au nanoparticles [46]. It is possible to regulate the form of nanoparticles generated by plants and their byproducts in addition to their stability. It has been demonstrated by the quick manufacture of high-concentration stable gold nano-triangles employing tamarind leaf extract as a reducing agent [47]. Similarly, by using Aloe vera leaf extract as a reducing agent, gold nano-triangles and silver nanoparticles were created in a single-crystalline triangular form. The Aloe vera plant extracts were combined with aqueous chloroaurate ions for this purpose. By varying the amount of extract in the reaction media, the proportion of gold nano-triangles to spherical particles and the size of the nano-triangles (50–350 nm) have been altered. The nanoparticle solution's optical properties were significantly controlled by the ratio of spherical to triangle nanoparticles [48]. The formation of gold nano-triangles was largely dependent on the slow reduction of aqueous gold ions ($HAuCl_4$) and the shape-directing characteristics of the carbonyl compounds present in the plant extract. Conversely, only spherical Ag nanoparticles are produced when Aloe vera extract is treated with aqueous silver ions ($AgNO_3$). The creation of silver and gold nanoparticles is indicated by the emergence of a faint yellow color and a brownish-red color, respectively, in the reactions.

It has recently been discovered that *Cinnamomum camphora* leaf extract may manufacture gold and silver nanoparticles, expanding the list of plants that exhibit nanoparticle production ability [49]. The notable differences in shape control between silver and gold nanoparticles were attributed to the respective advantages of reductive and protecting biomolecules. It was primarily found that the water-soluble heterocyclic components and the polyol components were responsible for the stabilization of the nanoparticles and the reduction of silver or chloroaurate ions, respectively.

3 Techniques for Adding Nanoparticles to Textiles

Nonetheless, there are numerous ways to incorporate nanoparticles into textile fabrics. Typically, textile materials are treated with nanoparticles at several phases, including:

(1) The preparation stage for filaments or fibers; (2) the finishing stage; and (3) the garment processing stage.

Nanoparticles can significantly change the surface characteristics of textile materials and provide them new functions in some common applications for textile finishing. When compared to larger particles, the nano-size particles have a greater surface area. Since the particles are clear and in the nanoscale range, they do not distort color or brightness on textile substrates.

There are two steps to applying nanoparticles to fabrics. First comes the creation of novel, stable nanomaterials, which must first be shielded from characteristics determined by particle size. In the best-case scenario, the production of foils, emulsions, and dispersions using traditional finishing techniques occurs in the second stage and can be added to the finished textile product [50].

The most popular method for applying nanoparticles to textiles is coating, which involves combining chemicals, a surfactant, a carrier medium, and nanoparticles. The following are some coating processes that alter the textile's surface:

- Spraying: With control over the depth and targeted application to particular locations, the coating formulation is applied by spraying onto fabrics.
- Soaking and dipping in an immersion container, subsequently letting it dry.
- Transfer techniques for printing like flexography, inkjet, and rotary.
- In order to wash, use a washing solution that contains nanoparticles throughout the machine's wash or rinse cycles.
- Padding: in this technique, fabrics are coated with nanoparticles by pressing a padder onto them.

3.1 Stage of Filament or Fiber Preparation

Sols can be utilized as a fundamental answer for fiber spinning. This process involves dry spinning silica sols to create basic glass fibers [52, 53]. Inorganic composite fibers based on other metal oxides or mixes of both, as well as pure SiO_2, can be produced by spinning straight from suitable sols. Inorganic/organic and pure inorganic composite fibers are the two main categories of sol–gel-based fibers [54]. In the realm of materials research, polymer-based organic and inorganic nanocomposites are receiving more and more attention. The benefits of an organic polymer and an inorganic material are combined in inorganic/organic nanocomposite composites. Polymer/inorganic nanocomposite composites have been created using a variety of inorganic nanoscale components, including semiconductors, metal nanoparticles, oxides, and silicates. Polymer/SiO_2 nanocomposites have garnered significant interest from academia and industry in recent times. Numerous potential and beneficial applications for this type of nanocomposites have been described, including optical devices, protective coatings, flame-retardant substances, and photo-resist materials. In general, four methods can be utilized to create polymer-silica nanocomposites [55]:

- The intercalation of melt,
- Approach to solution,
- In situ polymerization [56],
- The electrospun [57].

3.2 *Stage of Final Finishing*

3.2.1 Methods of Pad-Dry-Cure and Spray Finishing

The traditional methods used for textile finishing can be applied to the textile materials to apply nanosols. Simple dipping, cushioning, or spraying techniques, as well as foam finishing, are examples of these methods. Textiles can even be printed with nanosol-based methods. The simplest technique is probably to dip a textile sample into the nanosol. The cloth is submerged in the sol, left there for a predetermined amount of time, and then removed to either dry naturally or undergo heat curing in a stenter frame or oven. It is not required for the sol to completely penetrate the material for some applications or finishes. Repellency, for instance, might only be required on one side of the item. In certain situations, dipping or padding cannot be used to apply the sol. Applying a spray application to just one side of the textile fabric or casting the nanosol is two practical methods for this [58].

3.2.2 Method for Foam Finishing

Utilizing foam to apply chemical finishes to fabric is a crucial application technique. Water applied to the fabric can be greatly decreased by substituting air for a portion of the water in the chemical formulation. Surfactants are also added to the formulation to create foam. They may lessen the effect of repellent finishes, even if they are carefully chosen. Combining the chemical formulation with air in a foam generator produces large amounts of foam that can be applied to fabrics in different ways. The ratio of liquid to air in a foam, or the "blow ratio," is easily calculated using equation [59].

$$\text{Blow Ratio} = 1/\text{Foam Density}$$

3.2.3 Technology of Plasma

It is critical to comprehend what plasma is before delving into its applications. To put it simply, plasma is just a gas that has been ionized. Exposure of a surface to plasma results in a reciprocal contact between the gas and the substrate. Volatile surface materials contaminate and mix with the plasma, and the substrate's surface is attacked by electrons, ions, neutrals, radicals, and ultraviolet (UV) rays from the plasma [59].

Covalent bonds can be broken by the electrons in the plasma. Consequently, it is possible to alter the surfaces of different substrates chemically and physically. The wide range of results that can be obtained from plasma treatment is because of the usage of these in various combinations and on various substrates.

Layering, adhesion, or chain reaction: The processes of attaching films to textile substrates using gasses activated by plasma are referred to as coating and layering, respectively. The remarkably brief treatment period and low application temperatures of plasma treatment are its main advantages; also, drying procedures following plasma finishing can be omitted, as can the use of water or solvents [60].

3.3 Stage of Garment Finishing

Chemical finishes for clothes are usually applied with exhaustive finishing agents, like softening agents, antimicrobial agents, ultraviolet (UV)-absorbing materials, and so on. These finishes are added to the machine's bath after all other wet processing stages have been finished. Sometimes, nevertheless, one wants a non-exhaustible finish, like an easy-care finish. There are specific protocols in place for this kind of scenario. One method involves draining the clothes off as much water as possible before submerging them in the final solution either in a different trough or inside the garment machine itself and then repeating the extraction, drying, and curing processes. After water extraction, a second method is to precisely spray the finished solution into the garment processing equipment. Rotating the clothing for a sufficient amount of time to enable the finish to spread evenly throughout the item will achieve this. Curing and drying bring the procedure to a close. Both techniques have been proven in the marketplace; however, the immersion process requires more chemical finish but requires less capital investment than the spray method, which entails retrofitting current clothing machines with a precise spraying attachment [61].

4 Characterization Techniques for Polymer Nanocomposites

Numerous characterization techniques are employed in the examination of the morphology, microstructure, chemical structure, and physical characteristics of the nanocomposites. To comprehend structure–property correlations completely, many characterization techniques are frequently used. The composition, particle size, interfacial contact, and other factors of the nanocomposites all have a significant impact on their properties. The preparation procedure has a considerable impact on the surface interaction between the polymer and silica, which in turn affects the physical, thermal, and other properties of the nanocomposites.

Three effective microscopy methods for examining the morphology of nanocomposites are TEM, SEM, and AFM. DSC is typically used to study the crystallization characteristics of nano-silica particle loaded composites. The solid-state NMR and FTIR spectra are commonly employed for evaluation of polymer/silica nanocomposites chemical structure. Temperature affects a material's thermal characteristics.

The physical characteristics of polymer nanocomposites are of utmost importance, and as such, to improve the physical performance of polymers, inorganic fillers are frequently used. They are investigated using thermal analysis techniques such as dielectric thermal analysis, TGA, DTA, and DSC. As a result, it is typically required to characterize nanocomposites' mechanical properties from several angles. Many characteristics have been used to evaluate nanocomposites, including as tensile strength, flexural strength, fracture toughness, impact strength, hardness, and so on [51].

5 Nano-edge-Enhanced Functional Textiles

The textile sector has significant commercial potential with nanotechnology. This is mostly because traditional techniques for giving materials unique qualities frequently result in transient effects that disappear after washing or wearing [62].

Due to the huge surface area-to-volume ratio and high surface energy of nanoparticles, which present a greater affinity for fabrics and boost function durability, nanotechnology can give high durability for fabrics. Furthermore, a layer of nanoparticles on textiles will not change how breathable or tactile they are. Consequently, the textile sector is becoming more interested in utilizing nanotechnologies.

Nanoparticles can often be applied to textiles using a method called coating. Several methods can apply coating onto fabrics, including spraying, transfer printing, washing, rinsing, and padding. Of these methods, padding is the most commonly used. Nanotechnology is used to give textiles characteristics including soil resistance, improved dyeability, flame retardancy, water repellence, antibacterial, antistatic, wrinkle resistance, and UV protection. Although nanotechnology has many possible uses in the textile business, the following critically highlights just a few of the well-known qualities that are conferred by nano-treatment.

- Antibacteria,
- Antistatic,
- Easy care finish,
- Flameproof and retardant finish,
- Hydrophilic nano-finishes,
- Lotus effect,
- Odor control finish,
- Self-cleaning textiles,
- Thermal protective finish,
- UV protection,
- Water repellence,
- Wrinkle resistance.

5.1 Water Repellence Functionality

Water-repellent coatings alter the fiber's surface without obstructing the interstices. Fabric can therefore let air and water vapor through. Earlier water-repellent coatings were simple to take off for laundering or dry cleaning. Wax emulsions, pyridinium compounds, N-methylol compounds, silicons, and fluorochemicals are being employed to give a variety of synthetic and natural fibers water repellency. Recently, several products have been released to enhance the qualities of wrinkle resistance, water repellency, and stain resistance. One of the greatest options is the nano-whisker that Nanotex offers. By adding nano-whiskers, which are hydrocarbons that measure 1/1000 of the size of a regular cotton fiber, to the fabric to provide a peach fuzz effect without weakening the cotton, Nanotex enhances the fabric's ability to repel water. Water stays on top of the whiskers and above the fabric's surface because the spacing between the whiskers on the fabric are bigger than those of a typical drop of water but still smaller than those of individual water molecules. Unlike the bulky laminated fabrics or topical coatings that have been employed for this purpose in the past, they adhere to the cloth permanently. Hydrocarbons are added to the fibers in an aqueous solution to form the whiskers. The fabric's inherent breathability and hand feel are unaffected by the fiber modifications. There having been reports of a "peach fuzz" look, the cloth exhibits excellent wrinkle resistance, and the processing is imperceptible. Compared to conventional methods, a nanoscale emulsification process allows for a more thorough, even, and exact application of the finish to fabrics. Rolls of woven cotton fabric from textile mills are submerged in liquids that contain trillions of nano-whiskers, which make the fabric more dense and waterproof. Following treatment, the cotton is dried in ovens, joining the minuscule fibers to the significantly larger cotton strands. This makes the fabric's outer layer more surface tension-resistant, preventing liquid from penetrating through. The finished product offers a practically solid barrier against fluids or wrinkles, for example, even though its appearance remains identical. Unlike standard finishes, nano-whiskers offer an even application and do not alter the surface qualities. Because nanoparticles are so small, their addition to fabric cannot be seen with the naked eye [63].

Since nanoparticles added to textile materials are invisible to the unaided eye, the products' original color will not change. Rather of altering the chemical makeup of textile materials, nanoparticles create a protective coating on their surface; as a result, they do not release any harmful compounds or have any adverse consequences. Products made from textiles that have been treated with nanoparticles are more resilient than those with conventional finishes that require frequent cleaning. Fabrics with nano-finishes can have additional functionality added to them, such as the ability to repel stains and wrinkles simultaneously. On the other side, if appropriate precautions are not taken, nanotech clothing may perform poorly. The right way to take care of these clothes is to wash them in a delicate machine, use drip dying, refrain from dry cleaning, and avoid using chlorine bleach and wringing them out.

5.2 UV Protection Functionality

Textile materials can be treated with a variety of nanoparticles or nano-compounds to provide UV protection. The most often utilized nanoparticles are nano-sized zinc oxide and titanium dioxide. By diffusing, absorbing, or reflecting damaging UV rays, they offer a protective advantage [64].

Because inorganic UV blockers are non-toxic and chemically stable when exposed to UV and high temperatures, they are preferred over organic UV blockers. Inorganic UV blockers include several semiconductor oxides, including Al_2O_3, TiO_2, ZnO [65], and SiO_2. Zinc oxide (ZnO) and titanium dioxide (TiO_2) are two of these semiconductor oxides that are frequently utilized. It was shown that zinc oxide and titanium dioxide in nanoscale form were more effective in scattering and absorbing UV light than their normal-size counterparts, making them more capable of blocking UV light. This is because, in comparison with traditional materials, nanoparticles have a greater surface area per unit mass and volume, which increases their ability to block UV light [66].

5.3 Antibacterial Functionality

Quaternary ammonium compounds are widely employed as antibacterial agents. Fabrics that are resistant to bacterial development are also made of a variety of chlorinated organic compounds and organometallic compounds that contain zinc, copper, iron, manganese, or silver. The use of nano-silver particles offers durable antibacterial finishing to the textile [67].

Silver, copper, titanium dioxide, zinc oxide, and nano-sized silver are employed to impart antibacterial characteristics [68]. To some extent, metallic substances and ions have a sterilizing effect [69]. Part of the oxygen in the air or water is thought to be converted to active oxygen by catalysis with the metallic ion, which dissolves the organic material and has the sterilizing effect. The number of particles per unit area can be enhanced by using nanoparticles, maximizing the antibacterial properties. "Oxidation–reduction" is the name of this cascade reaction. The photocatalyst may break down typical organic compounds in the air, including bacteria, viruses, and odor molecules, through the reaction. When exposed to light, nanoparticles exhibit potent photocatalytic characteristics, which is why they are used to give textiles antimicrobial qualities [63].

Mahltig et al. [70] classified biocidal fabrics into three categories:

- Textiles with non-diffusible biocides,
- Textiles with photoactive characteristics,
- Textiles with an integrated biocide that releases under control.

One illustration for the first category would be the TiO_2-coated fabric. Daoud et al. [71]. By absorbing light, TiO_2's valence electrons can be elevated to a higher energy state, which produces excited charge carriers, an electron, and a positively charged electron–hole. According to Parkin and Palgrave [72], positively charged holes have the ability to oxidize organic molecules, while electrons can react with oxygen to produce hyperoxid radicals that attack and damage microorganisms' cell membranes. TiO_2 can also be utilized to produce textiles that clean themselves through the photocatalytic process that has been described. This process can also contribute to the destruction of stains.

The biocide in the second group's textiles is rendered inert either inside the coating's matrix or within the fiber matrix. According to Patel et al. [73], the bacterium is damaged and is unable to grow or reproduce due to the reaction between the positively charged biocide and the negatively charged cell membranes, which is what causes the antibacterial effect. The third class of coating is that in which biocides (such Cu^{2+}, Ag^+, and triclosan) permeate out of the matrix. The air causes the nanoparticles in the matrix of metallic antibacterial agents to oxidize. Following their formation, the cations diffuse out of the matrix and stop bacterial growth.

In the majority of the research publications that were examined, TiO_2, Ag, or a mixture of both NPs were used to produce a biocidal effect [74–77]. While chitosan core–shell-NP is used by Ye et al. [78] to create a bacterial-resistant coating for cotton fabrics, Vigneshwaran et al. [79] uses ZnO nanoparticles to get the desired effect in the textile. Mahltig et al. [70] also discuss the potential for adding other organic (like triclosan, chitosan) or inorganic (like Ag, Cu) biocides to the silica coating matrix.

TiO_2 is added to the cloth using sol–gel techniques in the majority of the reviewed research articles [80–83]. First, a so-called nanosol, or TiO_2 dispersion containing particles in the nano-range, is made as per Mahltig et al. [70]. After that, the fiber, yarn, or completed fabric is dipped into the nanosol, causing the particles to aggregate on the textile's surface and creating a lyogel layer that contains solvent. Mahltig et al. [70] state that a solid, porous xerogel layer is produced after heating the lyogel to eliminate the solvent. The antibacterial coatings are mentioned by Mahltig et al. [70]. The sol–gel methods also produce the biocides that are integrated into a Si-matrix. Different dip-pad-dry-cure techniques are used by Yuranova, [84] and Vigneshwaran et al. [79] and Lee et al. [85] to apply antibacterial ZnO, Ag or Ag/TiO_2 coatings to the cloth. To extract the solvent, the fabric is dipped into a nanosol, dried, and heated. Using a layer-by-layer deposition technique, Dubas et al. [86], achieve an antibacterial coating on the finished cloth by progressively dipping it into a polyelectrolyte solution and an Ag-nanosol. Another approach to creating an antimicrobial coating is to co-sputter Ag-NP with plasma polymerization by Hegemann et al. [87], and chemical vapor deposition of TiO_2 NP by Szymanowski et al. [88]. As opposed to methods where NP is applied as a coating, it is also feasible to incorporate NP at an earlier stage of the manufacturing process. For instance, Jeong et al. [89] describe direct melt compounding of Ag and the textile polymer, or Yeo et al. [90] describe producing Ag/PP master batches and then melting them.

Antibacterial textiles may find use in the following product categories: sports and leisure, protective clothing, protective clothing interiors, medical textiles, furniture, and apparel. These functional fabrics could be used to create furniture textiles, car interiors, or "self-cleaning" or "anti-odor" clothing in the fashion, interior trim, and automobile industries. Xin and Daoud [91]. Antibacterial fabrics could be used in towels, sponges, and dish cloths, among other home goods. Potential items in the medical industry include reusable surgical gloves and masks, patient clothing, bed linens, and antibacterial wound dressings. Sun [92] and Daoud and Xin [91]. Sun [92] suggests that protective face masks and suits against biohazards, as well as cosmetic items like antibacterial face masks and toothbrushes, are further potential uses.

For each of the uses listed above, there are commercially available items, with the exception of those pertaining to the automotive industry. Several businesses in the apparel industry sell antibacterial or self-cleaning shoe insoles, socks, even underwear Polo shirts as per Good Weaver Textile; United Textile Mills; Ag Active; Nano babies; Song Sing Nanotechnology; Nano-Infinity Nanotech [93–98]. According to Ag Active, green yarn technology found antimicrobial bedsheets and baby blankets as examples of products connected to interior trim and upholstery [99, 100]. According to Smith and Nephew [101], biocidal wound dressings are available in the medical product group, while face masks and toothbrushes are examples of cosmetic appliances according to Nanover, Song Sing Nanotechnology; Nano-Infinity Nanotech [98, 99, 102]. Emergency Filtration Products Inc. offers protective face masks against viral and bacterial diseases as well [103]. Additionally, a number of businesses provide textile treatment technologies, additives, or fabrics with antibacterial qualities to their industrial clientele. Thomson Research Associates [104] offers an antibacterial textile treatment that is used on a wide range of products including antibacterial underwear, kitchen sponges or clothes, bed sheets, towels, carpets, or sport helmet paddings. Nanopool [105] and Nanogate [106] both offer antibacterial coating technologies, while antibacterial polymer master batches are available from Murray [107]. Antibacterial fabrics that can be processed into outdoor clothing or household products were developed by HeiQ Materials Ltd. [108] and Polartec [109].

5.4 *Antistatic Functionality*

Because synthetic fibers like nylon and polyester do not absorb much water, static tends to accumulate in them. Traditionally, surfactants are employed to disperse the trace moisture on the fiber's surface, allowing the charge to escape. Silver is among the finest electrically conductive nanoparticles. The efficient dissipation of static charge is aided by silver nanoparticles.

Because they contain more moisture, cellulosic fibers are better at dissipating static charges and preventing them from building up. Because synthetic fibers do not have good antistatic qualities, research has been done on how to use nanotechnology to improve textiles' antistatic qualities. Synthetic fibers were found to possess antistatic capabilities when treated with nano-sized titanium dioxide, zinc oxide whiskers, nano-ATO (antimony-doped tin oxide), and silane nanosol. Since ATO, ZnO, and TiO_2 are electrically conductive materials, they have antistatic properties. This kind of substance aids in efficiently dispersing the static charge that builds up on the fabric.

However, silane nanosol enhances antistatic qualities because the fiber's silane gel particles bind water by absorbing moisture and water from the air through amino and hydroxyl groups [63].

5.5 Wrinkle Resistance Functionality

Wrinkles arise when a fiber is too crinkled. When a material is bent, the hydrogen bonds dividing the chains of molecules in the amorphous portions of the fiber or fabric weaken and permit the chains to pass one another. The fibers or fabric are kept in their crumpled forms, and the linkages regenerate in new locations. Reduced fiber strength, abrasion resistance, water absorbency, dyeability, and breathability are among the drawbacks of traditional resin treatments.

Conventional methods often utilize resin to give fabric resistance to wrinkles. Applying resin does have certain drawbacks, too, such as a reduction in the fiber's tensile strength, abrasion resistance, water absorbency, dyeability, and breathability. Several researchers have used nano-silica and nano-titanium dioxide to increase the wrinkle resistance of silk and cotton, respectively, to get around the drawbacks of employing resin. Under UV light, nano-titanium dioxide was used as a catalyst along with carboxylic acid to catalyze the process that links the acid and cellulose molecule. Conversely, when maleic anhydride was used as a catalyst together with nano-silica, the outcomes showed that the combination of the two may effectively increase silk's resistance to wrinkles [110].

5.6 Self-cleaning Functionality

In 1990, Wilhelm Barthlott, a German researcher at the Bonn Institute of Botany, discovered that the lotus plant, which is highly appreciated for the beauty of its blooms and leaves, possessed the ability to clean itself due to a high density of minute surface protrusions. Soil deposits are caught by these protrusions, which inhibits them from sticking. Zinc oxide, titanium oxide, and silver nanoparticles [111] are utilized to impart antibacterial and self-cleaning qualities. To some extent, metallic substances and ions have a sterilizing effect. The number of particles per unit area is greatly increased when nanoparticles are used. This is the most effective way to deal with the environmental issues that traditional finishing will cause. The ability of textile textiles to self-clean will help them overcome the negative effects of dirt and stains.

Products that repel stains reduce the fabric's crucial surface tension, preventing stains and dirt from sticking to the fabric. Water and oil roll off the fabric after beading up. Superior water and oil repellency, as well as stain and spill prevention, are offered by invisible repellent finishes when applied to textiles. Stain-release treatments allow spills and stains to seep into the fabric; when stains are applied, the cloth may bead somewhat and become slightly resistive to water and oil; however, the stain-release finish is undetectable, making it simple to remove ground stains during laundry. Stains that have been absorbed can be easily removed with home laundering thanks to this finish's water-loving component. The stain-release coating is the most recent finish in the market and works in dual-action repel and release. The coating combines the benefits of repellant and stain-release treatments into one. Together with an enhanced stain release, this dual protection provides a special blend of repellency that helps loosen even the most stubborn stains, especially those that are deeply embedded. Customers can get double the stain protection in an easily maintained fabric with dual-action repellents and release finishes.

5.7 Odor Control Functionality

A popular topic in the clothing and hosiery industries is odor management. The development of bacteria causes odors. It is possible to control odor by using an antimicrobial finish, which eliminates odor molecules as they arise, or by using a fragrance to mask the stench. To get rid of smell, cyclodextrins can be added to a cloth treatment. The distinctive molecular structure of cyclodextrins consists of a hydrophobic chamber with a hydrophilic outside. Because they are hydrophobic, the odor molecules become stuck in the cyclodextrins' cavities and are released after washing [112].

5.8 Hydrophilic Nano-functionality

The garment industry is limited in its use of synthetic fabrics like polyester and polyamides due to their low moisture absorption properties. The new line of hydrophilic nano-finishes from Nanotex, USA, called "Cotton Touch" and "Coolest Comfort" gives synthetic fabrics the appearance and feel of cotton. Durable cellulose wrapping is provided by Nanotouch for synthetic fibers like polyester and polyamides. The concentric structure formed by the synthetic core and cellulose shell addresses the negative aspects of synthetic materials, including blinding luster, rough handling, and static discharge. It is anticipated to stop the fall in demand for synthetic microfiber and expand the use of synthetics into new areas. It can also withstand up to 50 launderings. An innovation in moisture-wicking technology allows the nano-dry finish to quickly dry while drawing moisture away from the body. It enhances polyester and polyamides' ability to absorb moisture, making them more comfortable and hydrophilic. Sportswear and near-body garments that need to absorb sweat are the primary uses for it [112].

5.9 *Super Hydrophobic Functionality–Biomimatic Self-cleaning–Lotus Effect*

The contact angle is a measure of a surface's hydrophobicity. A surface is more hydrophobic, the higher the contact angles. Surfaces are classified as hydrophilic or hydrophobic depending on whether the contact angle is less than 90° or greater. Certain plants are referred to as super-hydrophobic because they exhibit contact angles of up to 160°, implying that just 2–3% of a drop's surface is in contact. A droplet's real contact area is about 0.6%; however, plants with a double-structured surface, like the lotus, can reach a contact angle of 170°. This all contributes to a self-cleaning outcome. Water droplets pick up dirt particles with a very small contact area, making it simple to remove them from the surface. Regardless of the chemistry of the dirt particle, there is more adhesion between the droplet and the surface when a water droplet rolls across a contaminated surface. Since the high surface tension of water is the basis for this self-cleaning effect, organic solvents will not operate with it.

Similar to a lotus leaf, some nanotechnologists have created treatments, coatings, paints, roof tiles, textiles, and other surfaces that can self-clean and remain dry. Usually, compositions comprising microscale particles or specific silicone or fluorochemical treatments on structured surfaces can accomplish this [111].

5.10 *Easy Care Functionality*

There are two basic methods to create a hydrophobic surface: either a rough surface can be modified using a material with low surface free energy, or a rough surface can be created on a hydrophobic surface. These two methods have been applied to textile substrates to impart a hydrophobic finish. One significant class of hydrophobic coatings is fluorocarbon finishes. A class of organic compounds known as fluorocarbons has a perfluoroalkyl residue in which fluorine has completely replaced all of the hydrogen atoms. Fluorine has taken the position of these chemicals because of their low reactivity and exceptional heat stability. They lower the surface tension significantly. For $-CF_3$, the critical surface tension (Ýc) is 6 Nm^{+1}. Dispersions of perfluorinated acrylates with co-monomers are known as fluorocarbon finishes. Reactive methanol or epoxy groups found in durable fluorocarbon finishes have the potential to react and create a cross-linked network that can also covalently attach to the fiber surface. The low-energy coatings created by these treatments shield the treated materials' fibers [111].

5.11 Flame-Retardant Functionality

A fire retardant is used to make materials more difficult to ignite by slowing down the pace of decomposition and increasing the temperature at which they burn. It functions in multiple ways, such as preventing oxygen from reaching the fuel or rerouting energy away from the flames. Flame-retardant polymer nanocomposites are attractive, and the nanoscale silica particle is a novel type of nanoparticle for these nanocomposites. The definition of a polymer's flammability behavior depends on several factors, including ignition properties, spread rates, and burning rates. The production of char, which forms an insulating barrier between the fuel and the heat and prevents oxygen from penetrating, is a crucial feature of the fire retardant. Char can be formed from phosphorus compounds, which are among the finest materials. It has been demonstrated that epoxy resins with sol–gel-produced nanoscale epoxy/silica hybrids containing phosphorus have better flame-retardant properties.

5.12 Thermal Protective Functionality

Thermal properties are the aspects of a material that change with temperature. Thermal analysis methods like as dielectric thermal analysis, TGA, DTA, and DSC are used to study them. In general, by acting as a better mass transport barrier and insulator for the volatile compounds created during the breakdown, adding inorganic particles smaller than nano-meters to the polymer matrix can increase thermal stability. Simultaneously, the addition of inorganic particles with nanoscale sizes, like silica, successfully reduces the polymer matrix's CTE. Research looking into how adding silica filler affects glass-transition temperature of the polymer has shown a variety of characteristics. Since the Tg of polymer nanocomposites varies due to filler loading, dispersion conditions, and filler size among other things, there is disagreement regarding the effect of nano-fillers in polymer composites on the glass-transition behavior of the polymer matrix [57].

6 Commercial Applications of Nano-engineered Textiles

The potential commercial areas of user listed in Fig. 1 are predicated on claims made in the research articles that were reviewed and on online searches for manufacturers who assert to use of nanotechnology within their goods. Online consumer inventory and publications about nanotechnology were reviewed as part of the research process. The manufacturer's claims serve as the basis for information regarding the employed nanoparticles or nanostructure in the commercially accessible products. Commercial goods for industrial customers, like textile additives or nanoengineered

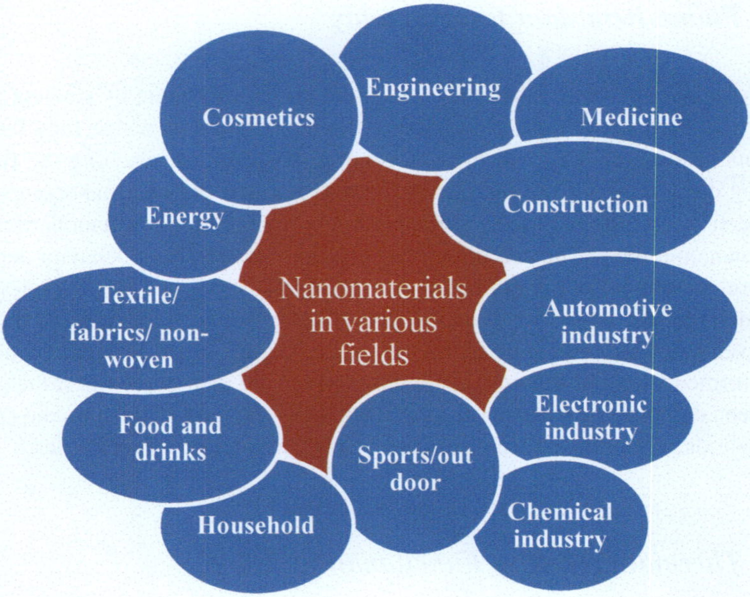

Fig. 1 Outline of the various domains in which goods based on nanomaterials are used

coating agents, are included in the list of applications; however, impregnation agents or textile coating for end users are not, as they are not considered. "textile products."

A range of textile tasks that could be accomplished through nano-engineering is illustrated by the research on nanotechnology on the Web of Science. It lists the various NP and production techniques that are now being researched. The literature describes the state in which the NP appears in the resulting fabric, the applied manufacturing process, the incorporated NP/nanostructure, and the mechanism that yields the intended effect for each textile function.

The theoretical frameworks presented in the reviewed research publications and online searches for consumer products with nano-engineered features form the basis for the commercial uses of the various functional textiles. The product groups to which the suggested products are assigned include clothing, clothing for protection, interior finishing and fabric upholstery, sports and recreation, homes, cosmetics, medical equipment, building materials, and auxiliary/intermediate products for industrial operations. Even though they are not sold directly to consumers, all goods acquired by industrial clients and used in the production of a nano-engineered finished product are listed in the final product category. These auxiliary or intermediate items include textile ingredients, fabrics that are made into apparel or furniture, and rights for the use of particular technology or production procedures. Many companies and research institutions already provide textile and nanotechnology applications. Here are a few of them [4]:

- Greensbara, North Carolina, USA; Nanotex, LLC,

- Sweden's Texcote Technology (International) Ltd,
- Located in Switzerland, Schoeller Textiles AG,
- Situated in China, Beijing Zhong-shong Century Nanotechnology Co. Ltd.

7 Environmental Factors and Safety for Health

Environmental and health safety issues are being raised by the textile industry's anticipated widespread use of nanotechnologies, as well as those of any other sector utilizing nanotechnology.

Businesses have become interested in nanomaterials due to their enormous economic potential, in addition scientists and researchers, because of their unique qualities. According to the National Science Foundation, by 2015, the market for products and services associated with nanotechnology will reach a value of USD 1 trillion. This sum exceeds the information technology and telecoms industries' total revenue. In the next ten years, nanotechnology is expected to generate several hundred billion euros in revenue. By 2007, the market for nanomaterials could reach $4 billion. To meet the 1 trillion dollar yearly output need of the world in 10–15 years, it was projected that 2 million new jobs would be generated [113].

The environment may benefit from nanotechnology as well. Nanotechnology has the potential to improve living standards and save raw materials by utilizing fewer resources without compromising performance.

7.1 Effect on the Environment

It is unavoidable that the extensive use of nanoparticles has led to a rise in emissions into the soil, air, and water in the ground. The unique characteristics of nanoparticles can have unfavorable consequences on the environment if they are released into it. Furthermore, in addition to their direct hazardous qualities, nanomaterials may also bind nutrients or interact with molecules in an undesirable way because of their unique structure, surface, or charge. This also applies to nanoparticles that are utilized in the textile sector.

Throughout their lifetime, nanomaterials may come into contact with the environment. The estimated concentrations of ambient nanoparticles in soil, water, and air could be in the mg/l range. How long they live there and in what form, i.e., for how long, is yet up for inquiry. When compared to known toxicity data for lethal and sublethal effects, these doses were substantially lower than those predicted to have biological impacts, suggesting a minimum degree of danger. It is crucial to realize that these predictions could change if new uses and particles are developed, as well as if more information about uptake routes, fate and behavior, and atmospheric entry becomes available. Furthermore, once incorporated into the environment, nanomaterials may accumulate in surrounding organisms. The fate of the beginning products of the nanoscale compounds and their transformation products must be monitored (life-cycle studies, exposure scenarios) and assessed in the target compartments under the exposure pathways resulting from manufacture, processing, and usage. Among the actions that must be taken is the identification of persistent nanoparticles that accumulate in the environment using suitable methods for measuring sediment, soil, and water; an analysis of the behavior of the nanomaterials during disposal, landfilling, incineration, or repurposing; and testing of ecotoxicity throughout the entirety of their lives. The stability of these nanoparticles is a critical aspect in determining the danger of exposure to nanomaterials; specifically, it should be investigated how long-lived and stable these forms are, as well as whether and under what circumstances they undergo alterations once entering the environment.

More and more information is becoming available on the possible outcomes of nanoparticles in the environment. Recent studies have shown that, in addition to the chemical and physical makeup of the nanomaterial and its amount present, the behavior of nanoparticles in their surroundings depends on the characteristics of the receiving environment. Due to the tiny size of nanoparticles, they can be widely disseminated by air, which allows for the partial application of study results about the behavior and impacts of natural ultrafine dust or ultrafine dust formed during incineration.

Because of their broad, active surfaces, which have the ability to bind and mobilize pollutants like organic compounds or heavy metals, soil-borne nanoparticles pose a risk to groundwater. Depending on the receiving environment, nanomaterials eventually tend to clump together and settle on the substrate if they do not dissolve or break down.

Wastes and products from industry typically wind up in rivers that eventually empty into the ocean. Dispersed nanoparticles are predicted to act in a way consistent with colloid science phenomena upon the release of water. To ascertain how these nanoparticles could interact with organic matter and perhaps be adsorbed, it is crucial to estimate concentrations and consider the surface characteristics of the nanomaterials as well as the physical and chemical properties of the aqueous phase.

There is a description of how cadmium nanocrystals are formed on phytoplankton; however, there is not much information available regarding how plants absorb or interact with these particles. It was found that there was a roughly linear relationship between toxicity and the escape of silver ions from the particles collected in the phytoplankton. It can be discovered that lipophilic nanoparticles, like carbon nanotubes, may be absorbed by root systems and microbial communities and end up building up in the tissues of plants. Similarly, plant tissues may act as scaffolding for the in situ accumulation of metallic nanoparticles.

Ultimately, our knowledge of the behavior and effects of nanomaterials in the environment as well as on living things has grown nearly exponentially as a result of the scientific community's keen interest and increased financing. Nevertheless, the field is still in its infancy. According to current projections, there is little chance of major ecotoxicological harm, and concentrations in the environment are anticipated to be far lower than those that have been shown to have biological impacts in laboratory settings. It appears that the quantity of nanoparticles utilized in food packaging contributes very little to the overall concentration in the environment.

Furthermore, there may be advantages to nanoparticles being present in the environment. The application of nanotechnology to the modification and detoxification of environmental contaminants is currently the subject of several investigations. Through the use of reactive nanoparticles, which facilitate both chemical reduction and catalysis to reduce the pollutants of concern, it is possible to do nano-remediation in situ without extracting soil to be treated or pumping groundwater out for elevated treatment. It is stated that nano-remediation can be done in situ and has the potential to lower some contaminant concentrations to almost zero, shorten cleanup times for large-scale contaminated sites, and do away with the need to treat and dispose of contaminated dredged soil. Naturally, in this instance as well, before this approach is widely applied, adequate assessment of these nanoparticles; including extensive studies spanning the entire ecosystem must be conducted to prevent any potential negative environmental effects.

Another interesting aspect of nanoparticles' impact on the environment is their use as "nano-additives" for two conflicting purposes: breaking down and stabilizing polymers under different environmental conditions and durability under different environmental conditions. A new utility of nanoparticles that has the potential to be widely exploited shortly is reviewed in a recent study that examines the state of research internationally in this area.

7.2 Effects on the Health of People

Nanoparticles can enter the body through three main routes: ingestion, skin penetration, and inhalation. There is mounting scientific evidence that certain free nanoparticles can penetrate cellular barriers and cause oxidative damage and inflammatory responses when they come into contact with them. Many individuals are concerned about the possibility of indirect exposure when it comes to nanomaterials used in food packaging because of the possibility of nanoparticle migration. Workers in the factories that produce the nanoparticles are nearly exclusively responsible for the inhalation and skin penetration of these compounds when it comes to food packaging nanomaterials. It is advised that these workers use masks with specific high-efficiency filters, gloves, and glasses for personal protection.

The consequences of these nanomaterials entering the body are not well understood. The risk evaluation of a tiny number of nanoparticles utilized in the packaging of food has been investigated after ingestion. Certain studies on TiO_2, nanoparticles of silver, and carbon nanomaterials/nanotubes suggest that the gastrointestinal tract may serve as a route for nanoparticles to enter the bloodstream. These processes are anticipated to be influenced by the chemical and physical characteristics of the nanoparticles, including size and the physiological state of the organs of entry [114]. Although the translocation fractions appear to be rather modest, there is currently a lot of study being done on this topic. Once the nanoparticles enter the circulation, the two primary organs for distribution are the liver and spleen. When the surface of the nanoparticles is positively charged and hydrophilic, the amount of time that the particles circulate increases dramatically. In the case of some nanoparticles, every organ studied thus far may be at risk since the distribution of the nanoparticles to various organs can be inferred from the chemical makeup of the particles or the particles themselves. The brain and the testicles/reproductive system are examples of these organs. Additionally, distribution to the fetus during gestation has been noted. Given the paucity of information regarding the long-term behavior of nanoparticles, it is prudent to presume that insoluble nanoparticles may accumulate during chronic exposure to secondary target organs, with as-yet-unstudied consequences. Particularly concerning is the potential for nanoparticles to enter the developing fetus's brain. To support or refute the theory between different brain illnesses and nanoparticles, research must be done in these two domains. The possible health impacts of other particles used in the packaging of food, like fullerenes and ZnO nanoparticles, are being investigated [65].

8 Summary

Particularly in the area of textile finishing, the development of nanoparticle, nanocomposites, and nanosphere applications has advanced extremely quickly in recent years. These nanoparticles can enhance the physical properties of traditional textiles, such as their strength, resistance to germs, anti-infrared, water repellency, antistatic, soil resistance, flame-retardant properties, and dyeability. These amazing nanoparticles' potential applications could lead to the creation of fabrics with wound-healing, self-cleaning, and self-repairing properties in the future.

Future developments of nanotechnologies in textiles will have a twofold focus:

(1) Improving textile materials' current capabilities and functions,
(2) Creating textiles with previously unheard-of functions that are clever and intelligent.

From the perspective of homeland security and technological advancement, the latter is more pressing. Wearable solar cells and energy storage, sensors information gathering and transfer, multiple and sophisticated protection and detection, healthcare and wound healing, and self-cleaning and repairing are some of the new textile functions that will be developed.

Indeed, textiles have a very bright future thanks to nanotechnology. Within a decade, it is predicted that nanotechnology will have a significant impact on the market for novel materials, potentially yielding hundreds of billions of dollars. Textiles play a significant role in this material industry.

References

1. Utracki LA (2004) Clay-containing polymeric nanocomposites, vols 1 and 2. RAPRA Technology Ltd., Shawbury, UK
2. Paul DR, Robeson LM (2008) Polymer nanotechnology: nanocomposites. Polymer 49:3187–3204
3. Sinha Ray S, Bousmina M (2008) Polymer nanocomposites and their applications. American Scientific Publishers, Valencia, CA, USA
4. Patel BH, Chattopadhyay DP (2007) Nano-particles & their uses in textiles. Indian Text J 118(3):23–31
5. Zhu J, Liu S, Palchik O, Koltypin Y, Gedanken A (2000) Shape-controlled synthesis of silver nanoparticles by pulse sonoelectrochemical methods. Langmuir 16:6396
6. Forster S, Antonietti M (1998) Amphiphilic block copolymers for templating application. Adv Mater 10:195
7. Xuan Y, Li Q (2000) Heat transfer enhancement of nanofluids. Int J Heat Fluid Flow 21:58
8. Eastman JA, Choi SUS, Li S, Yu W, Thompson LJ (2001) Anomalously increased effective thermal conductivities of ethylene glycol-based nanofluids containing copper nanoparticles. Appl Phys Lett 78:718
9. Rao CNR, Kulkarni GU, Thomas PJ, Edwards PP (2000) Metal nanoparticles and their assemblies. Chem Soc Rev 29:27
10. Anderson DJ (2006) A SERS-active system based on silver nanoparticles tethered to a deposited silver film. J Phys Chem B 110:13722
11. Tatasov S, Kolubaev A, Belyaev S, Lerner M, Tepper F (2002) Study of friction reduction by nanocopper additives to motor oil. Wear 252:63
12. Daniel M-C, Astruc D (2004) Gold nanoparticles: assembly, supramolecular chemistry, quantum size-related properties and application toward biology, catalysis and nanotechnology. Chem Rev 104:293
13. Rosi NL, Mirkin CA (2005) Nanostructures in biodiagnostics. Chem Rev 105:1547
14. Klaus T, Joerger R, Olsson E, Granqvist CG (1999) Silver-based crystalline nanoparticles, microbially fabricated. Proc Natl Acad Sci USA 96:13611–13614
15. Klaus T, Joerger R, Olsson E, Granqvist CG (2001) Bacteria as workers in the living factory: metal-accumulating bacteria and their potential for materials science. Trends Biotechnol 19:15–20
16. Lengke M, Southam G (2006) Bioaccumulation of gold by sulfate-reducing bacteria cultured in the presence of gold(I)-thiosulfate complex. Geochim Cosmochim Acta 70:3646–3661
17. Lengke M, Fleet ME, Southam G (2006) Morphology of gold nanoparticles synthesized by filamentous cyanobacteria from gold(I)-thiosulfate and gold(III)-chloride complexes. Langmuir 22:2780–2787
18. Lengke M, Ravel B, Fleet ME, Wanger G, Gordon RA, Southam G (2006) Mechanisms of gold bioaccumulation by filamentous cyanobacteria from gold(III)-chloride complex. Environ Sci Technol 40:6304–6309
19. Husseiny MI, El-Aziz MA, Badr Y, Mahmoud MA (2007) Biosynthesis of gold nanoparticles using *Pseudomonas aeruginosa.* Spectrochim Acta A Mol Biomol Spectrosc 67:1003–1006
20. Jianrong C, Yuqing M, Nongyue H, Xiaohua W, Sijiao L (2004) Nanotechnology and biosensors. Biotechnol Adv 22:505–518
21. Joerger R, Klaus T, Olsson E, Granqvist CG (1999) Spectrally selective solar absorber coatings prepared by a biomimetic technique. Proc Soc Photo-Opt Instrum Eng 3789:2–7
22. Ahmad A, Senapati S, Khan MI, Kumar R, Ramani R, Srinivas V, Sastry M (2003) Intracellular synthesis of gold nanoparticles by a novel alkalotolerant actinomycete, *Rhodococcus* species. Nanotechnology 14:824–828
23. Ahmad A, Senapati S, Khan MI, Kumar R, Sastry M (2003) Extracellular biosynthesis of monodisperse gold nanoparticles by a novel extremophilic actinomycete, *Thermomonospora* sp. Langmuir 19:3550–3553

24. Sastry M, Ahmad A, Khan MI, Kumar R (2003) Biosynthesis of metal nanoparticles using fungi and actinomycete. Curr Sci 85:162–170
25. Dameron CT, Reeser RN, Mehra RK, Kortan AR, Carroll PJ, Steigerwald ML, Brus LE, Winge DR (1989) Biosynthesis of cadmium sulphide quantum semiconductor crystallites. Nature 338:596–597
26. Kowshik M, Deshmukh N, Vogel W, Urban J, Kulkarni SK, Paknikar KM (2002) Microbial synthesis of semiconductor CdS nanoparticles, their characterization, and their use in the fabrication of an ideal diode. Biotechnol Bioeng 78:583–588
27. Kowshik M, Ashtaputre S, Kharrazi S, Vogel W, Urban J, Kulkarni SK, Paknikar KM (2003) Extracellular synthesis of silver nanoparticles by a silver-tolerant yeast strain MKY3. Nanotechnology 14:95–100
28. Ortiz DF, Ruscitti T, McCue KF, Ow DM (1995) Transport of metal-binding peptides by HMT-1, a fission yeast ABC type vacuolar membrane protein. J Biol Chem 270:4721–4728
29. Williams P, Keshavarz-Moore E, Dunnil P (1996) Production of cadmium sulfide microcrystallites in batch cultivation by *Saccharomyces pombe*. J Biotechnol 48:259–267
30. Gericke M, Pinches A (2006) Biological synthesis of metal nanoparticles. Hydrometallurgy 83:132–140
31. Ahmad A, Mukherjee P, Mandal D, Senapati S, Khan MI, Kumar R, Sastry M (2002) Enzyme-mediated extracellular synthesis of CdS nanoparticles by the fungus *Fusarium oxysporum*. J Am Chem Soc 124:12108–12109
32. Ahmad A, Mukherjee P, Senapati S, Mandal D, Khan MI, Kumar R, Sastry M (2003) Extracellular biosynthesis of silver nanoparticles using the fungus *Fusarium oxysporum*. Colloids Surf B Biointerf 28:313–318
33. Bansal V, Rautaray D, Ahmad A, Sastry M (2004) Biosynthesis of zirconia nanoparticles using the fungus *Fusarium oxysporum*. J Mater Chem 14:3303–3305
34. Bansal V, Rautaray D, Bharde A, Ahire K, Sanyal A, Ahmad A, Sastry M (2005) Fungus-mediated biosynthesis of silica and titania particles. J Mater Chem 15:2583–2589
35. Kumar SA, Ayoobul AA, Absar A, Khan MI (2007) Extracellular biosynthesis of CdSe quantum dots by the fungus, *Fusarium oxysporum*. J Biomed Nanotechnol 3:190–194
36. Mukherjee P, Ahmad A, Mandal D, Senapati S, Sainkar SR, Khan MI, Sastry M et al (2001) Fungus-mediated synthesis of silver nanoparticles and their immobilization in the mycelia matrix: a novel biological approach to nanoparticle synthesis. Nano Lett 1:515–519
37. Mukherjee P, Ahmad A, Mandal D, Senapati S, Sainkar SR, Khan MI, Kumar R et al (2001) Bioreduction of $AuCl_4^-$ ions by the fungus, *Verticillium* sp., and surface trapping of the gold nanoparticles formed. Angew Chem Int Ed 40:3585–3588
38. Mukherjee P, Senapati S, Mandal D, Ahmad A, Khan MI, Kumar R, Sastry M (2002) Extracellular synthesis of gold nanoparticles by the fungus *Fusarium oxysporum*. Chem Bio Chem 3:461–463
39. Durán N, Marcato PD, De Souza GIH, Gabriel IH, Alves OL, Esposito E (2007) Antibacterial effect of silver nanoparticles produced by fungal process on textile fabrics and their effluent treatment. J Biomed Nanotechnol 3:203–208
40. Vigneshwaran N, Ashtaputre NM, Varadarajan PV, Nachane RP, Paralikar KM, Balasubramanya RH (2007) Biological synthesis of silver nanoparticles using the fungus *Aspergillus flavus*. Mat Lett 61:1413–1418
41. Senapati S, Mandal D, Ahmad A, Khan MI, Sastry M, Kumar R (2004) Fungus-mediated synthesis of silver nanoparticles: a novel biological approach. Indian J Phys 78A:101–105
42. Bharde A, Rautaray D, Bansal V, Ahmad A, Sarkar I, Yusuf SM, Sastry M et al (2006) Extracellular biosynthesis of magnetite using fungi. Small 2:135–141
43. Durán N, Marcato PD, Alves OL, De Souza GIH, Esposito E (2005) Mechanistic aspects of biosynthesis of silver nanoparticles by several *Fusarium oxysporum* strains. J Nanobiotechnol 3:8
44. Armendariz V, Herrera I, Peralta-Videa JR, Jose-Yacaman M, Troiani H, Santiago P, Gardea-Torresdey JL (2004) Size-controlled gold nanoparticle formation by *Avena sativa* biomass: use of plants in nanobiotechnology. J Nanoparticle Res 6:377–382

45. Huang J, Li Q, Sun D, Lu Y, Su Y, Yang X, Chen C et al (2007) Biosynthesis of silver and gold nanoparticles by novel sundried *Cinnamomum camphora* leaf. Nanotechnology 18:105104–105114
46. Ankamwar B, Damle C, Absar A, Mural S (2005) Biosynthesis of gold and silver nanoparticles using Emblica officinalis fruit extract, their phase transfer and transmetallation in an organic solution. J Nanosci Nanotechnol 10:1665–1671
47. Ankamwar B, Chaudhary M, Mural S (2005) Gold nanotriangles biologically synthesized using tamarind leaf extract and potential application in vapor sensing. Synth React Inorg Metal-Org Nanometal Chem 35:19–26
48. Chandran SP, Chaudhary M, Pasricha R, Ahmad A, Sastry M (2006) Synthesis of gold nanotriangles and silver nanoparticles using Aloe vera plant extract. Biotechnol Prog 22:577–583
49. Shankar SS, Rai A, Ahmad A, Sastry M (2004) Rapid synthesis of Au, Ag, and bimetallic Au core–Ag shell nanoparticles using neem (*Azadirachta indica*) leaf broth. J Colloid Interf Sci 275:496–502
50. Patel BH, Channiwala MZ, Chaudhari SB, Mandot AA (2016) Biosynthesis of copper nanoparticles; its characterization and efficacy against human pathogenic bacterium. J Environ Chem Eng 4(2):2163–2169
51. Chattopadhyay DP, Patel BH (2014) Nano metal particles: synthesis, characterization and application to textiles. In: Manufacturing nanostructures. One Central Press (OCN) Manchester (UK), pp 184–215
52. Ahmadi SJ et al (2004) Review synthetic routes, properties and future applications of polymer-layered silicate nanocomposites. J Mater Sci 39:1919–1925
53. Information from the website: http://www.worldscibooks.com/materialsci/6961.html
54. Information from the website: http://217.218.200.220/c/document_library/get_file?p_l_id=18503&folderId=21410&name=DLFE-12410.pdf
55. Terlingen JGA (1993) Introduction of functional groups at polymer surfaces by glow discharge techniques. University of Twente, Enschede
56. Information from the website: aasimahmed.files.wordpress.com/…/a-review-of-plasma-treatment-for-application-on-textile-substrate-aasim-ahmed.pdf
57. Zou H et al (2008) Polymer/silica nanocomposites: preparation, characterization, properties, and applications. Chem Rev 108:3928–3947
58. Russell et al (2002) Nanotechnologies and the shrinking world of textiles. Text Horiz 9(10):7–9
59. Anonymous (2003) Small-scale technology with the promise of big rewards. Tech Text Int 3:13–15
60. Xin JH et al (2004) A new approach to UV-blocking treatment for cotton fabrics. Text Res J 74:97–100
61. Patel BH, Patel PN (2012) Synthesis and application of nano-sized SiO_2 to textiles: a review. Int Dyer 35–39
62. Chattopadhyay DP, Patel BH (2009) Improvement in physical and dyeing properties of natural fibers through pre-treatment with silver nanoparticles. Indian J Fibre Text Res 34:368–373
63. Wong YWH et al (2006) Selected applications of nanotechnology in textiles. AUTEX Res J 6(1):1–6
64. Chaudhari S, Shaikh T, Patel B, Pandey P (2014) Engineering polypropylene TiO_2 nanocomposite filament to improve UV resistance. Int J Adv Manag Technol Eng Sci 4(1):22–25
65. Chattopadhyay DP, Patel BH (2011) Modification of cotton textiles with nanostructural zinc particles. J Nat Fibers 8(1):39–47
66. Song XQ et al (2001) The effect of nano-particle concentration and heating time in the anti-crinkle treatment of silk. J Jilin Inst Technol 22:24–27
67. Chattopadhyay DP, Patel BH (2010) Effect of nanosized colloidal copper on cotton fabric. J Eng Fiber Fabrics 5(3):1–6
68. Mandot AA, Chaudhari SB, Patel BH (2012) Nanocomposite: manufacturing and applications in textiles. Melliand Int 18(3):188–189

69. Chaudhari SB, Mandot AA, Patel BH (2009) Functionalized nano-finishing to textiles using Ag nano-colloids. Melliand Int 15(5–6):214–216
70. Mahltig B, Haufe H, Bottcher H (2005) Functionalisation of textiles by inorganic sol–gel coatings. J Mater Chem 15(41):4385–4398
71. Daoud WA, Xin JH, Zhang YH (2005) Surface functionalization of cellulose fibers with titanium dioxide nanoparticles and their combined bactericidal activities. Surf Sci 599(1–3):69–75
72. Parkin I, Palgrave R (2004) Self-cleaning clothes. J Mater Chem 15:1689–1695
73. Patel BH, Channiwala MZ, Chaudhari SB, Mandot AA (2015) Green synthesis of silver nano-sols by leaf extract of Ocimum sanctum and their efficacy against human pathogenic bacterium. J Green Sci Technol 2(1):1–6. https://doi.org/10.1166/JGST.2015.1039
74. Chattopadhyay DP, Patel BH (2014) Functional properties of silver nano-sol treated wool and silk fabric. Man-Made Text India 42(9):340–344
75. Patel BH, Parmar SS, Dave BP (2016) Preparation and characterization of anti-bacterial knitted textile by novel green root. Int J Sci Res Dev 4(2):1929–1931
76. Chattopadhyay DP, Patel BH (2015) Studies on the effect of nano zinc treatment on jute fabric. J Text Sci Eng 5(189):2
77. Chaudhari SB, Mandot AA, Patel BH (2012) Effect of nano TiO_2 pretreatment on functional properties of cotton fabric. Int J Eng Res Dev 1(9):24–29
78. Ye WJ et al (2005) Novel core-shell particles with poly (*n*-butyl acrylate) cores and chitosan shells as an antibacterial coating for textiles. Polymer 46(23):10538–10543
79. Vigneshwaran N et al (2006) Functional finishing of cotton fabrics using zinc oxide-soluble starch nanocomposites. Nanotechnology 17(20):5087–5095
80. Daoud WA, Xin JH (2004) Nucleation and growth of anatase crystallites on cotton fabrics at low temperatures. J Am Ceram Soc 87(5):953–955
81. Qi KH et al (2006) Self-cleaning cotton. J Mater Chem 16(47):4567–4574
82. Bozzi A, Yuranova T, Kiwi J (2005) Self-cleaning of wool-polyamide and polyester textiles by TiO_2-rutile modification under daylight irradiation at ambient temperature. J Photochem Photobiol A Chem 172(1):27–34
83. Daoud WA, Xin JH (2004) Low temperature sol–gel processed photocatalytic titania coating. J Sol–Gel Sci Technol 29(1):25–29
84. Yuranova T et al (2006) Performance and characterization of Ag-cotton and Ag/TiO_2 loaded textiles during the abatement of E-coli. J Photochem Photobiol A Chem 181(2–3):363–369
85. Lee HJ, Yeo SY, Jeong SH (2003) Antibacterial effect of nanosized silver colloidal solution on textile fabrics. J Mater Sci 38(10):2199–2204
86. Dubas ST, Kumlangdudsana P, Potiyaraj P (2006) Layer-by-layer deposition of antimicrobial silver nanoparticles on textile fibers. Colloids Surf A Physicochem Eng Aspects 289(1–3):105–109
87. Hegemann D, Hossain MM, Balazs DJ (2007) Nanostructured plasma coatings to obtain multifunctional textile surfaces. Prog Org Coat 58(2–3):237–240
88. Szymanowski H et al (2005) Plasma-enhanced CVD deposition of titanium oxide for biomedical applications. Surf Coat Technol 200(1–4):1036–1040
89. Jeong SH, Yeo SY, Yi SC (2005) The effect of filler particle size on the antibacterial properties of compounded polymer/silver fibers. J Mater Sci 40(20):5407–5411
90. Yeo SY, Lee HJ, Jeong SH (2003) Preparation of nanocomposite fibers for a permanent antibacterial effect. J Mater Sci 38(10):2143–2147
91. Chattopadhyay DP, Patel BH (2017) In-situ nano silver-impregnated silk fabric. Man-Made Text India 45(11):369–372
92. Sun G, Halamine (2006) Chemistry and its applications in biological and chemical protective textiles. Presentation on the Nano Europe conference, St. Gallen. Division of Textiles and Clothing, University of California, USA
93. Goodweaver Textile Co. Ltd (nd) Goodweaver textile. Cited 9 Oct 2007. Available from: http://www.goodweaver.com

94. United Textile Mills Co. Ltd (nd) I-Tex "silver nano"—anti bacterial polo-shirt. Cited 9 Oct 2007. Available from: http://www.unitedtextilemills.com/index.asp
95. AgActive (nd) Sports anklet sock. Cited 9 Oct 2007. Available from: http://www.agactive.co.uk/index.cfm/fuseaction/produ
96. Nanobabies (nd) Nanobabies. Cited 9 Oct 2007. Available from: http://nanbabies.com/SupportProductsStore.htm
97. Song Sing Nanotechnology Co. Ltd (nd) Nano silver products. Cited 9 Oct 2007. Available from: http://www.ssnano.net/ehtml/0418b.php
98. Nanoinfinity Nanotech Co, Ltd (nd) Healthy deodorant series products. Cited 9 Oct 2007. Available from: http://www.nano-infinity.com.tw/product01.htm
99. AgActive (nd) 100% cotton sheet sets. Cited 9 Oct 2007. Available from: http://www.agactive.co.uk/index.cfm/fuseaction/product.display/ProductID8/.htm
100. Greenyarn Technology Ltd. (nd) Greenyarn technology: weaving together a healthy tomorrow. Cited 9 Oct 2007. Available from: http://www.Greenyarn.com/technology.htm
101. Smith and Nephew (nd) Acticoat product line. Cited 9 Oct 2007. Available from: http://wound.smith-nephew.com/us/node.asp?NodeId=2867
102. Nanover (nd) Nanover wet wipes. Cited 9 Oct 2007. Available from: http://www.nanogist.com/English/products/tissue.htm
103. Emergency Filtration Products Inc (nd) Emergency filtration products. Cited 9 Oct 2007. Available from: http://www.emergencyfiltration.com/Products/NanoMask.htm
104. Thomson Research Associates (nd) Using ultra fresh. Cited 10 Oct 2007. Available from: http://www.ultrafresh.com/tra/Content/Using_Specific/using_apparel1.cfm
105. Nanopool (nd) Nanopool. Cited 10 Oct 2007. Available from: http://www.nanoingermany.com/smr/content/company/company_id/88/exhibition_id/16/subexhibition/0/
106. Nanogate (nd) Ist keimfrei, bleibt keimfrei. Cited 9 Oct 2007. Available from: http://www.nanogate.de/de/produktveredelung/funktionen/nanoadd/antimikrobiell.php
107. Murray J (2006) Introducing a novel nanosilver additive for antimicrobial textile functionality. In: Presentation at the NanoEurope conference
108. HeiQ Materials Ltd., St. Gallen, Microban (nd) Home Products. Cited 10 Oct 2007. Available from: http://www.microban.com/Americas/products/category.html?lang=en&CategoryID=1&SubcategoryID=200
109. Polartec (nd) Forward fabric. Woodrow Wilson International Center for Scholars. Cited 10 Oct 2007. Available from: http://www.polartec.com/
110. Rathinamoorthy R et al (2009) Synthesis of nanoparticles and their application in textile. Melliand Int 1:31–32
111. Gopalakrishnan D, Tamil Alagan D (2009) Textile nano-finishing. Asian Dyer 6(6):41–44
112. Joshi M (2009) Nanotechnology-based functional finished and coating. Asian Dyer 2:30–34
113. Chattopadhyay DP, Patel BH (2009) Introduction to nanotechnology in textiles and effect on environment & human health. In: UGC sponsored national conference on environmental sustainability of textile industry. International College for Girls, Jaipur, India
114. Kanade P, Patel BH (2017) Copper nano-mediated hygienic textiles with improved aesthetic properties. Res J Text Appar 21(2):146–158. https://doi.org/10.1108/RJTA-01-2017-0004

Chapter 4
Advanced Characterization Techniques of Nanoparticles for Textile Finishing

Sourav Banerjee, Mandira Mondal, Swagata Banerjee, S. Wazed Ali, and S. N. Naik

1 Introduction

"*Necessity is the Mother of Invention*"- *Plato.*

'*What, why, and how*' are three key questions that drive scientific creativity. It has also been discovered that many instruments needed to assess basic visible stuff, such as a solid cube, also require a centimeter scale to give its measurements. As we discuss nanomaterials, which are essentially invisible to the human eye, we also need a ruler to determine their dimensions or shape. This is the fundamental idea behind the invention of different characterization tools [1].

However, what are nanomaterials?

Nanomaterials are characterized by having one or more external dimensions ranging between 1 and 100 nm [2]. Their minute size enables increased surface functionality within a specified volume and results in distinct physical properties that often diverge from those of larger-scale counterparts. These variances encompass electronic, optical, and magnetic features. Notably, nanoparticles exhibit significantly higher surface-to-mass ratios compared to bulk materials [3–5]. Consequently,

S. Banerjee · S. W. Ali (✉) · S. N. Naik
School of Interdisciplinary Research (SIRe), Indian Institute of Technology, Delhi, Hauz Khas, New Delhi 110016, India
e-mail: wazed@iitd.ac.in

M. Mondal · S. Banerjee · S. W. Ali
Department of Textile and Fibre Engineering, Indian Institute of Technology, Delhi, Hauz Khas, New Delhi 110016, India

S. N. Naik
Centre for Rural Development and Technology, Indian Institute of Technology, Delhi, Hauz Khas, New Delhi 110016, India

S. Bairagi et al. (eds.), *Nanotechnology in Textile Finishing*, Materials Horizons: From Nature to Nanomaterials, https://doi.org/10.1007/978-981-97-2696-7_4

surface atoms and their associated energy play a substantial role in determining material properties, contributing to effects such as reduced lattice constants and lower melting points [3, 6].

Nanoparticles exhibit exceptional properties that render them immensely attractive for diverse applications, encompassing catalysis, energy storage, photovoltaics, electronics, optics, biomedical technology, and beyond. They are increasingly predominant not only in research but also in various consumer products. When precisely engineered at the atomic level and integrated into textiles, these nanomaterials can significantly modify the functional properties of textiles. This integration holds substantial commercial promise for the textile industry. Unlike conventional methods that often result in temporary effects, nanotechnology offers durable functionalization to fabrics. The durability arises from the substantial surface area-to-volume ratio and heightened surface energy of nanoparticles. This enhances fabric adhesion, ensuring prolonged functionality following laundering or extended usage [7, 8].

In the textile industry, nanoparticles have enormous potential. They can replace conventional micron-sized particles in finishing processes to achieve various benefits, including improved dyeability, soil release, stain resistance, flame retardancy, wrinkle resistance, moisture management, and antibacterial and antistatic effects. Textiles' physical characteristics, including strength, electrical and thermal conductivity, and shrinking, are also affected by their integration. Traditional textile finishing methods often involve well-established characterization techniques like Fourier-transform infrared spectroscopy (FTIR) for analyzing new bond formations between the textile substrate and finishing agents. This is because the materials and their effects are commonly understood and predictable within conventional processes. However, the application of synthesized nanomaterials to textiles necessitates additional or specialized characterization methods. Nanomaterials exhibit unique properties at the nanoscale, including size, surface area, and reactivity, markedly different from conventional materials. These distinct features can substantially impact the properties of textiles in innovative ways. Therefore, specialized characterization methods are vital for accurately assessing and comprehending these nanoscale modifications and their influence on the performance and behavior of textiles [9–13].

Why is characterization knowledge necessary, particularly for nanomaterials?

This book is divided into several parts that explain the synthesis of nanoparticles, how they are used in textile industries, and the toxicity issues related to human health and the environment. Characterization tools are an essential component of working with nanomaterials. These instruments are necessary for assessing release evaluations, synthesis techniques, and functional characteristics [14, 15]. Therefore, for a thorough understanding and efficient application of nanomaterials in textiles, it is imperative to emphasize existing characterization methodologies and their underlying theories. Also, it is important to point out that no single method can completely reveal every aspect of a nanoparticle. Every approach adds a different piece to the puzzle; thus, multiple approaches are always necessary. The most precise description of the particle under investigation is produced by a sequence of well-selected characterization techniques that build upon earlier data sets to provide a complete knowledge of the particle sample. This chapter aims to provide an extensive set of instructions for analyzing and describing the four primary characteristics that identify a material as a nanoparticle: size, shape, surface charge, and surface morphology [16]. Our aim is to present a comprehensive survey of the state-of-the-art methods—both specialized and established—that are now utilized for nanoparticle assessment, exploring their relative merits and demerits [17].

2 Characterization Techniques Used for Nanoparticles Synthesis

The summarization (as given in Table 1) outlines diverse techniques used to analyze distinct properties of nanoparticles. These analyses aid in refining the nanoparticle synthesis process and ensuring repeatability. Some of the characterization techniques are extensively explained in the context of their underlying theories, alongside recent studies closely associated with them. We have compiled recent research on nanomaterial-based textile finishing and the corresponding characterization techniques in Table 2.

2.1 Scanning Electron Microscopy (SEM)

Scanning electron microscopy (SEM) diverges from optical microscopes by utilizing an electron beam instead of light. A focused and intensified stream of electrons is emitted from an electron gun at the device's apex within SEM machine. There are two different kinds of electron guns: thermionic guns, which release electrons from atoms by heating a filament, and field emission guns, which release electrons from atoms by producing a strong electric field. Three different types of electrons are released by the specimen when the incident electron beam comes into contact with it: Auger electrons, secondary electrons, and backscattered or primary electrons. In SEM, both primary and secondary electrons are harnessed for imaging purposes [18]. The SEM uses secondary electrons to produce high-resolution images that reveal features at about 1–5 nm. Energy-dispersive X-ray spectroscopy (EDX) utilizes characteristic X-rays to identify elemental compositions. Backscattered or secondary electrons are also employed to produce images in this technique [19, 20]. A schematic representation (Fig. 1) below explains the operational framework of SEM.

Typically, to increase their conductivity, test samples require a thin layer of conductive material to be applied. This coating is achieved by sputtering gold, utilizing argon gas atmosphere and a required electric field. Positively charged ions are created when one electron is removed from argon. These ions are then attracted to

Table 1 Different characterization tools and corresponding analytical information for nanomaterial analysis [17]

Characterization techniques	Primary data obtained
XRD (X-ray diffraction)	Arrangement of crystals, components, and size of crystalline grains
XAS (X-ray absorption spectroscopy)	X-ray absorption coefficient, chemical state, interatomic distances, and Debye–Waller factors
SAXS (small-angle X-ray scattering)	Particle size distribution, growth kinetics, and dimensions
XPS (X-ray photoelectron spectroscopy)	Electronic configuration, elemental constitution, stages of oxidation, and ligand binding
FTIR (fourier-transform infrared spectroscopy)	Functional group and binding
NMR (nuclear magnetic resonance)	Ligand density and arrangement, electronic structure, atomic composition, the influence of ligands on shape and size
BET (Brunauer–Emmett–Teller) analysis	Specific surface area and porosity
TGA (thermogravimetric analysis)	Mass and composition analysis under heat
LEIS (low-energy ion scattering spectroscopy)	Nanoparticles' self-assembled monolayers' thickness and chemical composition
UV–vis (ultraviolet–visible spectroscopy)	Nanomaterials shape suggestions, size, concentration, and optical characteristics
PL spectroscopy (photoluminescence spectroscopy)	The relationship between optical properties and structural factors, including size, composition, and defects
DLS (dynamic light scattering)	Hydrodynamic dimensions and agglomeration detection
NTA (nanoparticle tracking analysis)	The size and distribution of nanoparticle
DCS (differential centrifugal sedimentation)	Same as NTA
ICP-MS (inductively coupled plasma mass spectrometry)	Size and distribution of nanoparticles in addition to elemental composition
SIMS, and ToF–SIMS (time-of-flight secondary ion mass spectrometry)	Surface-sensitive chemical information, molecule orientation and conformation, surface topography, functional group, and size evaluation
SQUID-nanoSQUID (superconducting quantum interference device)	Blocking temperature, magnetization saturation, and magnetization remanence
VSM (vibrating sample magnetometer)	Similar to zero field cooling curves and magnetic hysteresis graphs used in SQUID
Mössbauer	Iron oxides and their nanoparticles can be distinguished by their oxidation state, symmetry, surface spins, magnetic ordering of the atoms, magnetic anisotropy energy, and thermal unblocking

(continued)

Table 1 (continued)

Characterization techniques	Primary data obtained
FMR (ferromagnetic resonance)	Size, distribution, shape, surface composition, crystallographic defect, magnetic anisotropy constant, and demagnetization field of a nanoparticle
XMCD (X-ray magnetic circular dichroism)	Magnetic moments in ferro- and ferri-magnetic materials and transition metal ion site symmetry
Superparamagnetic relaxometry	Hydrodynamic size distribution, core characteristics, and the ability to identify and locate superparamagnetic nanoparticles
TEM (transmission electron microscopy)	Shape, size, agglomeration, and kinetics of growth of nanoparticles in matrices
HRTEM (high-resolution transmission electron microscopy)	In addition to the data on the crystal structure of individual particles, it provides all of the information as TEM. It also provides a distinction between amorphous, polycrystalline, and monocrystalline nanoparticles. Abnormalities in structure
Liquid TEM	Show the real-time development of nanoparticles and investigate their mechanisms, single particle motion, and the production of superlattices
Cryo-TEM	Examine intricate development processes and aggregation routes for molecular biology
Electron diffraction	Order–disorder transformation, lattice parameters, and crystal structure
STEM (scanning transmission electron microscopy)	For the investigation of morphology, crystal structure, and elemental composition, combined with EDX. Information about the atomic configuration of hetero-interfaces
EELS (electron energy-loss spectroscopy)	Types of atoms, their chemical state, how they interact with one another collectively, and plasmon resonance
Electron tomography	3D particle visualization and quantitative data at the atomic level
EBSD (electron backscatter diffraction)	Analyze microstructures for details on grain shape, deformation, texture, and flaws
AFM (atomic force microscope)	Shape and size of nanoparticle in three dimensions; assessment of surface coverage with morphology; dispersion of nanoparticles in cells and other matrices/supports; accuracy of nanoparticle lateral dimensions; rapid element composition analysis

Table 2 Recent research on nanomaterial finished textile and use of different characterization techniques

Nanoparticle	Finished textile material	Properties	Characterization techniques used	References
Titanium dioxide	Polyamide 66 fabric	Antimicrobial	SEM	[36]
TiO_2 NPs	Cotton fabric	Antimicrobial, UV protection, and self-cleaning	SEM	[37]
Silver, zinc oxide, and titanium dioxide	Cotton and wool fabric	Mosquito-repellent, fragrance, and antibacterial	SEM, DLS, Zeta potential, and FTIR	[38, 39]
Zinc oxide and silicon dioxide	Cotton fabric	Self-cleaning, antibacterial, and ultraviolet (UV) protection	UV vis spectroscopy	[40]
Silver chloride-titanium dioxide ($AgCl–TiO_2$)	Polypropylene SMS nonwoven fabric	Superhydrophobic and antimicrobial	DLS, SEM, and FESEM	[41]
ZnO	Linen fabric	Ultraviolet protection, antibacterial, and hydrophobicity	SEM, TGA, and XRD	[42]
Casein and chitosan nanoparticles	Cotton and polyester fabric	Fire safety and antibacterial	SEM, DLS, and TGA	[43]
ZnO and chitosan nanoparticles	Cotton and viscose fabric	Anti-crease, anti-UV, and antibacterial	FTIR, TEM, SEM, and EDX	[44]

the negatively charged gold foil and settle onto the specimen. In a typical SEM analysis, it is crucial to eliminate water from the sample as water molecules tend to evaporate in a vacuum. Their presence can impede the path of electron beams, reducing image clarity. The high-resolution imaging capabilities of SEM enable the examination of surface topography, uniformity, and roughness of the nanocoating. Additionally, SEM provides visual insights into the distribution and texture of nanoparticles across the textile surface.

2.2 *Transmission Electron Microscopy (TEM)*

The electron beam in transmission electron microscopy (TEM) interacts with the material as it passes through the material. The electrons transmitted from electron source are utilized to create an image by expanding and focusing them through an

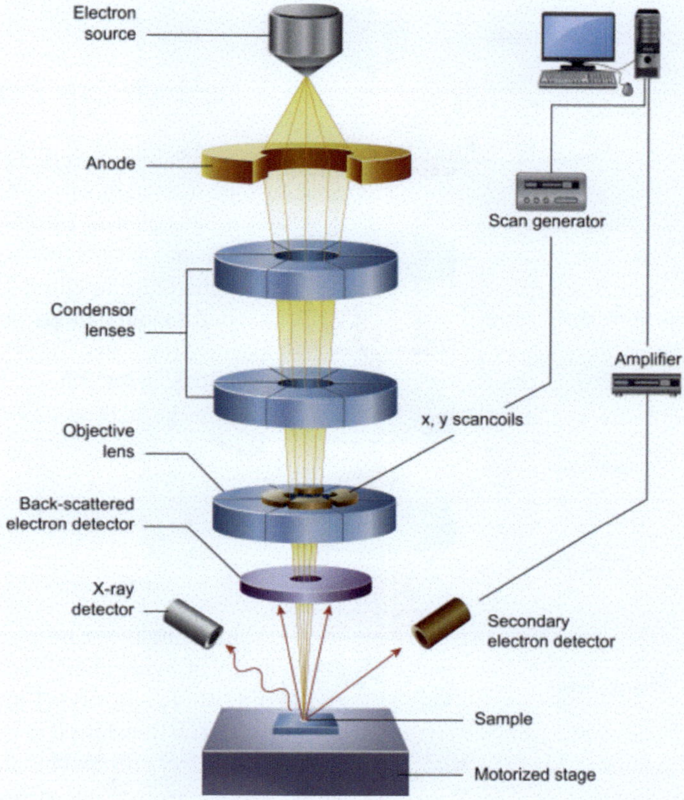

Fig. 1 Different components of scanning electron microscope (SEM) [19]. Reproduced with Permission Copyright © 2016 Elsevier Ltd

objective lens [21]. Figure 2 depicts the core elements utilized in transmission electron microscopy (TEM). The degree of diffraction intensity varies depending on the orientation of the plane concerning the electron beam. At specific angles, the electron beam experiences strong diffraction away from the axis, while at other angles, it transmits through. Specimen holders are crafted to tilt the sample, enabling specific diffraction conditions. An aperture is introduced to obstruct the deflected electrons, permitting only the unscattered electrons to pass. These unscattered electrons are then utilized to produce a contrast image termed as a light field mode image. Alternatively, the diverted electrons can also be controlled to generate an image known as a dark field mode image. This approach employs electrons deviating from their original trajectory to produce an image distinguished by unique contrast characteristics. Reduced-size materials are manufactured by applying a diluted sample that includes the specimen on support grids [19]. It offers insights into crystallinity, shape, and size distribution at nanoscale levels, crucial for understanding textile–nanoparticle interactions.

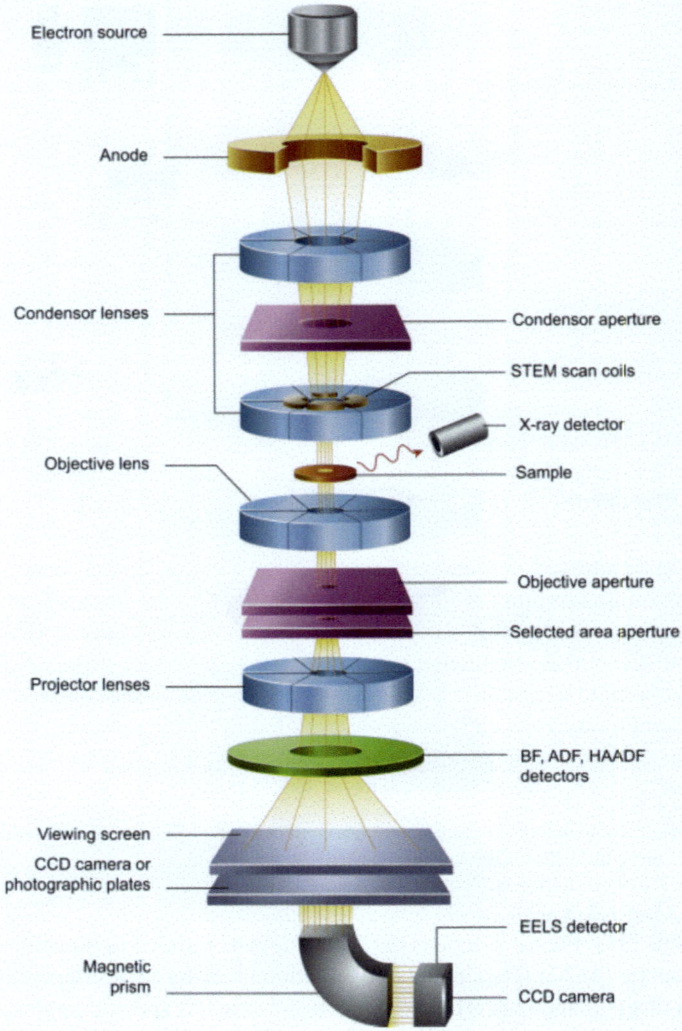

Fig. 2 Different components of transmission electron microscope (TEM) [19]. Reproduced with Permission Copyright © 2016 Elsevier Ltd

2.3 *Dynamic Light Scattering (DLS)*

Dynamic light scattering (DLS) determines particle size by observing the Brownian motion of particles in a solution, where, in comparison with smaller particles, larger particles scatter more light and move at a slower pace. Its benefits include a dynamic range covering nanometer scale to around 10-μm, fast analysis durations, and minimum sample requirements. DLS works well with a wide range of solvent compositions, that is why it is the perfect tool for quick routine characterization.

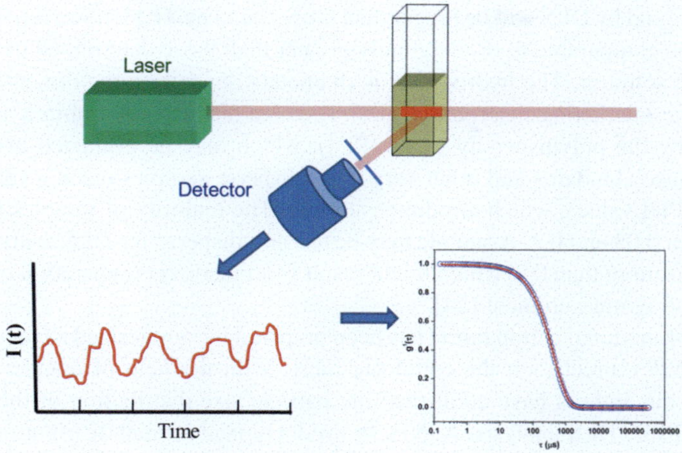

Fig. 3 Mechanism of dynamic light scattering [24]. Reproduced with Permission © 2014 American Chemical Society

The free diffusion coefficient of suspended particles is measured in order to evaluate particle size. Scattered light intensity fluctuates over time due to the random thermal movement of the particles inside a measuring cell subjected to laser pulses. Correlation analysis of these fluctuations provides insights into particle diffusion throughout the laser point. Particle diffusion coefficients proportional to the scattering correlation time are obtained by DLS through autocorrelation analysis of the time-domain signal (as shown in Fig. 3). The distribution of population sizes, which affects the autocorrelation curve's shape, is also investigated in this analysis [22, 23].

Particle size is deduced from the diffusion coefficient using the Stokes–Einstein equation (Eq. 1)

$$D = \frac{k_b T}{6\pi \eta R_h} \tag{1}$$

In this equation, D represents the diffusion coefficient, k_b is the Boltzmann constant, T signifies the temperature during analysis, and η denotes the solvent kinematic viscosity. R_h signifies the hydrodynamic radius of the particles, which refers to the radius of a spherical particle with a diffusivity equivalent to that of the sample being studied. It is important to note that R_h may significantly differ from the actual particle dimensions for nonspherical particles. Therefore, prior knowledge of the particle shape is essential to interpret this parameter for more representative size estimations accurately. When the solvent's kinematic viscosity is either known or can be determined using calibrated monodisperse particle suspensions, optically transparent particle suspensions can be used for dynamic light scattering. It provides information on agglomeration as well. In an aggregated solution, the hydrodynamic diameters found by DLS will be bigger than those found by TEM; however, the diameters will be comparable to or marginally greater than those determined by TEM in a dispersed solution. The hydrodynamic diameters depend on the dimensions and shape of the interacting macromolecules. The size distribution within a sample is measured by the polydispersity index (PDI), which may be evaluated using DLS analysis. Bead standards and other very monodisperse samples reach a value close to 0.05 in PDI values, which are dimensionless. The majority of samples typically fall between 0.05 and 0.7. A sample may be too polydisperse for DLS analysis if the PDI value is more than 0.7, which might result from incorrect preparation or sample changes during measurement [24].

A color-functional cotton fabric has been prepared by synthesized silver nanoparticles (AgNPs) directly on the cellulosic fabric with the help of madder [25]. In this study, the authors have confirmed the particle size distribution of the synthesized AgNPs by DLS analysis. AgNPs showed a typical normal distribution pattern in their particle size distribution curve, indicating a higher degree of stability in the system. In the mordant dyeing solution, the average particle size of AgNPs measured approximately 27.8 nm, while the average size of epoxy-modified AgNPs in the crosslinked dyeing solution was around 21.7 nm. It is important to note that these diameters represent the hydrodynamic size of the nanoparticles in the solution. Due to the lack of ion shielding, particle sizes in deionized water were 2–10 nm greater than the true size of the nanoparticles. In a separate study, silver nanoparticles have been synthesized through a hydrothermal process using chitosan as a bio-based reductant [26]. Subsequently, these nanoparticles have been functionalized with succinate to facilitate their impregnation onto the cellulose surface through esterification. Indeed, PDI analysis is valuable for assessing particle distribution. In this context, the lower PDI value (0.187) observed for the functionalized silver nanoparticles (SNPs) signifies a narrower size distribution in the aqueous medium compared to the plain SNPs (with a PDI of 2.266). In another study focused on bio-based nanoparticle synthesis, researchers endeavored to develop a sustainable and environmentally friendly method for producing sodium lignosulfonate nanoparticles (SLS NPs). This has been achieved through the utilization of chitosan, another biomolecule, in that synthesis process [27]. The hydrodynamic size of the nanoparticles displayed a range from 100 to 400 nm, as shown in Fig. 4. DLS along with SEM measurements confirmed the formation of SLS nanoparticles within the nanometer range.

2.4 X-Ray Diffraction (XRD)

X-ray diffraction is a flexible method used to examine a wide range of structural characteristics in crystalline materials. This technique provides information on the typical shape and size of crystals as well as macroscopic data such as the distribution of crystal constituents. Determining the degree of crystallinity, variations in an element's composition from its ideal state, and the structural condition is all made

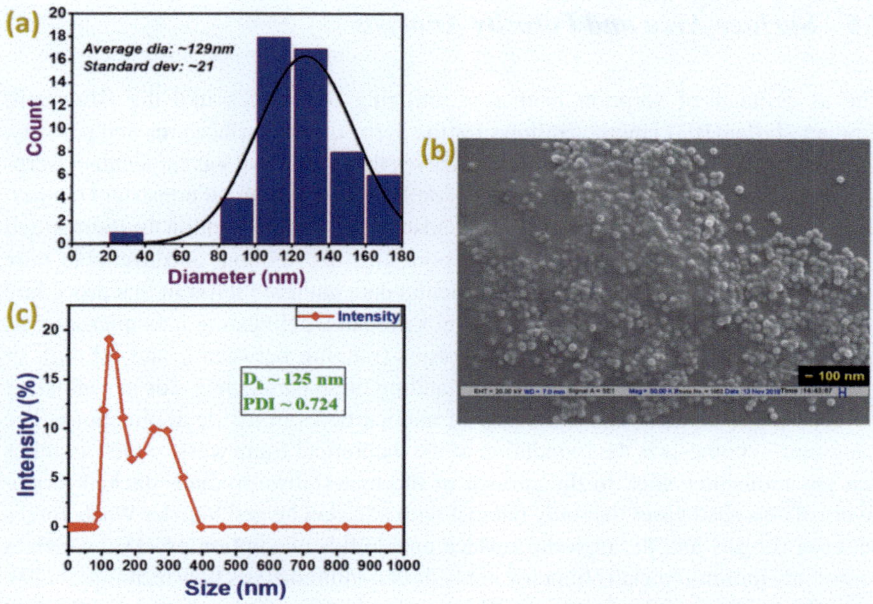

Fig. 4 **a** Dynamic light scattering (DLS) measurement curve displaying the synthesized SLS nanoparticles, **b** morphological images showcasing the synthesized SLS nanoparticles, and **c** diameter distribution chart illustrating the size distribution of the synthesized SLS nanoparticles [27]. Reproduced with Permission © 2022 Elsevier B.V

possible by analyzing the data. A portion of an X-ray beam is transmitted when it interacts with atomic planes in the sample; the remaining portion is absorbed, refracted, scattered, and diffracted. Because of their atomic configurations and the sorts of atoms they contain, each element in the sample diffracts X-rays in a different way, providing diffraction patterns.

2.5 *Fourier-Transform Infrared Spectroscopy (FTIR)*

Fourier-transform infrared spectroscopy (FTIR) involves measuring the absorption of electromagnetic radiation in the mid-infrared range (4000–400 cm^{-1}). When a molecule absorbs IR radiation, it undergoes modifications in its dipole moment, rendering the molecule IR active. The resulting spectrum from FTIR reveals bands corresponding to bond strengths, nature, and specific functional groups within the molecule. This spectral data provides valuable insights into molecular structures and interactions [28].

2.6 Surface Area and Porosity Analysis

The assessment of sorption isotherms, utilizing techniques like the Brunauer–Emmett–Teller (BET) method, allows for the derivation of surface area and porosity. Characterizing porous materials often involves the analysis of gas molecule adsorption and desorption, particularly nitrogen, on solid surfaces under consistent temperature conditions. This method captures adsorption and desorption isotherms and generates critical data on surface area, pore size distribution, and available pore volume within the material. The experimental data gathered through this procedure facilitates the classification of the material's porosity, delineating it as microporous (with pores smaller than 2 nm), mesoporous (ranging between 2 and 50 nm), or macroporous (exceeding 50 nm). Understanding these parameters aids in evaluating the adsorption capacity and efficiency of nanoparticles in textile applications. The Langmuir hypothesis is the foundation of the theoretical framework, which assumes that gas molecules stick to the surface of the material to produce stacked monolayers of absorbed gases that only interact with adjacent layers. Van der Waals forces between the gas and the material surface enable this adsorption process, which is commonly monitored via volumetric analysis or continuous gas flow techniques [29].

In the case of nonporous materials, the volume-specific surface area (VSSA) serves as a classification parameter to categorize it as a nanomaterial [as given in Eq. (2)].

$$\mathrm{VSSA} = \frac{S}{V} = \mathrm{SSA} \times \rho \tag{2}$$

Here, S denotes the external surface area of the sample, V represents the solid volume, SSA signifies the specific surface area (surface per mass), and ρ indicates the material density. For nonporous nanomaterials, determining the external surface area is feasible through gas adsorption measurements, with analysis conducted using the BET method. However, in the case of porous nanoparticles, the BET surface area comprises both external and internal surfaces. The adsorbed volume of nitrogen is graphed against the layer thickness of the adsorbed material, acquired from standard isotherms for the substance [15].

In a study, zinc oxide nanoflowers (ZnONFs) have been synthesized using the pod extract of Peltophorum pterocarpum as a green resource. These nanoflowers' specific surface area and pore volume were assessed using Brunauer–Emmett–Teller (BET) analysis for characterization [30]. The mesoporous nature was affirmed through BET analysis, revealing a substantial specific surface area of 19.61 m^2/g. Additionally, the calculated pore volume of 0.0826 cm^3/g supported the presence of a mesoporous structure, with pore sizes measuring approximately 16.85 nm. The presence of porous structures and a higher surface area like these zinc oxide nanoflower could be a probable solution for effectively removing harmful dyes from industrial wastewater. In a recent study, the surface area and pore size of CuO nanoparticles have been determined using isotherm measurements through BET analysis. This method enabled the characterization of the nanoparticles' surface area and total pore volume, which were 5.6 m^2/g and 0.0058 cm^3/g, respectively [31].

2.7 Zeta Potential

Colloidal or nanoparticles acquire a surface charge upon suspension. When an electric field is applied, these charged particles undergo motion due to the interplay between the electric field and the particle's charge. The velocity and direction of this movement rely on factors like the electric field's intensity, the particle's charge, and the characteristics of the suspending medium. The particle velocity measurement is facilitated by observing the Doppler shift in the scattered light. The velocity, which is proportionate to the electrical potential of the particle at the shear plane, is termed the zeta potential [22]. Hence, optical observation of particle movement in an electric field allows for determining the zeta potential. The zeta potential's magnitude serves as a marker for particle stability. Aggregation tendencies correspond to zeta potential values: particles tend to aggregate between 0 and 5 mV, while those that are minimally stable range from 5 to 20 mV. Moderately stable particles display values between 20 and 40 mV, whereas highly stable particles exhibit zeta potential values surpassing 40 mV [32, 33].

2.8 Atomic Force Microscopy (AFM)

Atomic force microscopy (AFM) is a robust and adaptable microscopy technology employed for studying samples at the nanoscale. It generates three-dimensional topographic images and offers diverse surface measurements, catering to the requirements of engineers and scientists. AFM excels at producing atomic-resolution images and providing height information at an Angstrom scale, requiring minimal sample preparation [34]. In the domain of polymer nanocomposites, it aids in measuring surface roughness and visualizing surface textures across various materials. It is nondestructive and boasts high three-dimensional spatial resolution. As the surface nears the AFM's cantilever with a sharp tip, a minor deflection is detected in the cantilever toward the surface due to the attractive force between the tip and the surface. Depending on the measurement mode, this deflection can manifest as vertical or lateral displacement of the cantilever, or alterations in the amplitude, frequency, or phase of the cantilever's oscillation. Generally, AFM operates in three distinct modes: contact, noncontact, and tapping. In contact mode, the AFM tip remains continuously with the sample surface. The repulsion between the tip and the surface atoms causes a vertical deflection of the cantilever, revealing the underlying topography. A piezo crystal drives the cantilever to oscillate at or close to its resonance frequency in noncontact mode. These oscillations occur slightly above the sample surface, ensuring the tip never directly contacts the surface. Tapping mode, also known as intermittent contact mode, shares similarities with noncontact method. An oscillating cantilever is utilized, but in this mode, instead of strictly oscillating above the sample surface, the tip intermittently taps the surface during its oscillations [35].

3 Conclusion and Future Goal

The advancement of nanoparticle synthesis processes and their application is crucial in addressing global issues. However, characterization serves as the mediator between them. Therefore, further research is needed to advance nanoparticle characterization for technology progression. The process of characterizing nanoparticles for textile finishing is multifaceted and requires a range of analytical approaches. When these techniques are combined, a thorough understanding of the characteristics of nanoparticles is produced, enabling the safe and efficient use of these particles in textile treatments. The usefulness and health and environmental consequences of nanoparticles are closely linked to their four properties: dimension, shape, surface charge, and porosity. In order to translate the potential advantages of nanomaterials into particular applications like textile coating, it is crucial to measure their characteristics depending on their final form. An analytical protocol that is conclusive, harmonized, and user-friendly is essential for characterizing the properties of nanomaterials. Such a protocol should be globally acceptable cum applicable. Several concerns should consistently be considered to gain a better understanding of the characterized data. For instance, it is highly advisable to utilize multiple techniques to analyze a specific property whenever feasible, serving as a confirmation process (such as employing both dynamic light scattering (DLS) and scanning electron microscopy (SEM) for analyzing a dispersion). Similarly, pharmaceutical formulations often have stringent statistical requirements for quality assurance compared to industrial coating materials. Given the present circumstances, the primary focus of future research should be the development of cost-effective analytical techniques. Simultaneously, enhancing existing techniques and broadening their boundaries and limitations is crucial. Novel approaches that promote cross-validation and allow for the assessment of physicochemical parameters that are difficult to evaluate at the moment are required. However, as new materials are continuously synthesized every day, there is a growing need for repeatability and quality control. We believe this drive will also stimulate advancements in the vital field of research and technology related to nanoparticle measurement.

References

1. Modena MM, Rühle B, Burg TP, Wuttke S, Modena MM, Zurich E, Rühle B, Burg TP, Wuttke S (2019) Nanoparticle characterization: what to measure? In: Rühle MB, Burg TP, Wuttke

S (eds) Advanced mater, vol 31(32). Wiley Online Library. https://doi.org/10.1002/adma.201901556
2. Wazed Ali S, Chowdhury A, Bairagi S, Banerjee S (2022) Green nanomaterials for multifunctional textile finishes. Green Funct Nanomater Environ Appl 343–364. https://doi.org/10.1016/B978-0-12-823137-1.00013-0
3. Heiligtag FJ, Niederberger M (2013) The fascinating world of nanoparticle research. Mater Today 16(7–8):262–271. https://doi.org/10.1016/J.MATTOD.2013.07.004
4. Zhao P, Li N, Astruc D (2013) State of the art in gold nanoparticle synthesis. Elsevier
5. Duan H, Wang D, Li Y (2015) Green chemistry for nanoparticle synthesis. pubs.rsc.org. https://doi.org/10.1039/c4cs00363b
6. Talapin DV, Shevchenko EV (2016) Introduction: nanoparticle chemistry. Chem Rev 116(18):10343–10345. https://doi.org/10.1021/ACS.CHEMREV.6B00566/ASSET/IMAGES/LARGE/CR-2016-00566V_0002.JPEG
7. Rezić I (2011) Determination of engineered nanoparticles on textiles and in textile wastewaters. TrAC Trends Anal Chem 30(7):1159–1167. https://doi.org/10.1016/J.TRAC.2011.02.017
8. Dastjerdi R, Montazer M, Shahsavan S (2009) A new method to stabilize nanoparticles on textile surfaces. Colloids Surf A Physicochem Eng Asp 345(1–3):202–210. https://doi.org/10.1016/J.COLSURFA.2009.05.007
9. Gao L-Z, Bao Y, Cai H-H, Zhang A-P, Ma Y, Tong X-L, Li Z, Dai F-Y (2020) Multifunctional silk fabric via surface modification of nano-SiO_2. Text Res J. https://doi.org/10.1177/0040517519897112
10. Ali SW, Chowdhury A, Bairagi S, Banerjee S (2022) Green nanomaterials for multifunctional textile finishes. Green Funct Nanomater Environ Appl 343–364. https://doi.org/10.1016/B978-0-12-823137-1.00013-0
11. Venkatraman PD, Sayed U, Parte S, Korgaonkar S (2021) Development of advanced textile finishes using nano-emulsions from herbal extracts for organic cotton fabrics. Coatings 11(8):939. https://doi.org/10.3390/COATINGS11080939
12. Yasmin R, Shahadat M, Ali SW, Ahammad SZ, Adnan R (2022) Chitosan-based green nanomaterials for treatment of textile industry dyes. Green Nanomater Ind Appl 153–176. https://doi.org/10.1016/B978-0-12-823296-5.00003-4
13. Ali SW, Banerjee S, Bairagi S (2013) Sustainable textile finishing using natural materials. Sustain Text Chem Process, 153–182. https://doi.org/10.1201/9781032629933-7
14. Mishra R, Militky J (2018) Nanoparticles and textile technology. Elsevier Ltd. https://doi.org/10.1016/B978-0-08-102609-0.00004-3
15. Titus D, James Jebaseelan Samuel E, Roopan SM (2018) Nanoparticle characterization techniques. Elsevier Inc. https://doi.org/10.1016/B978-0-08-102579-6.00012-5
16. Mishra R, Militky J, Arumugam V (2018) Nanorisks and nanohazards. Elsevier Ltd. https://doi.org/10.1016/B978-0-08-102609-0.00008-0
17. Mourdikoudis S, Pallares RM, Thanh NTK (2018) Characterization techniques for nanoparticles: comparison and complementarity upon studying nanoparticle properties. Nanoscale 10(27):12871–12934. https://doi.org/10.1039/c8nr02278j
18. Crouzier L, Delvallee A, Ducourtieux S, Ultramicroscopy LD (2019) A new method for measuring nanoparticle diameter from a set of SEM images using a remarkable point. Elsevier
19. Inkson BJ (2016) Scanning electron microscopy (SEM) and transmission electron microscopy (TEM) for materials characterization. Mater Charact Using Nondestruct Eval Methods 17–43. https://doi.org/10.1016/B978-0-08-100040-3.00002-X
20. Crouzier L, Delvallée A et al (2019) Development of a new hybrid approach combining AFM and SEM for the nanoparticle dimensional metrology, beilstein-journals.org
21. Pyrz WD, Buttrey DJ (2008) Particle size determination using TEM: a discussion of image acquisition and analysis for the novice microscopist. Langmuir 24(20):11350–11360. https://doi.org/10.1021/LA801367J/ASSET/IMAGES/MEDIUM/LA-2008-01367J_0003.GIF
22. Bhattacharjee S (2016) DLS and zeta potential—what they are and what they are not? J Control Release 235:337–351. https://doi.org/10.1016/J.JCONREL.2016.06.017

23. Chan MY, Dowling QM, Sivananthan SJ, Kramer RM (2017) Particle sizing of nanoparticle adjuvant formulations by dynamic light scattering (DLS) and nanoparticle tracking analysis (NTA). Methods Mol Biol 1494:239–252. https://doi.org/10.1007/978-1-4939-6445-1_17/COVER
24. Hassan PA, Rana S, Verma G (2015) Making sense of brownian motion: colloid characterization by dynamic light scattering. Langmuir 31(1):3–12. https://doi.org/10.1021/LA501789Z/ASSET/IMAGES/LARGE/LA-2014-01789Z_0004.JPEG
25. Dai Y, Li H, Wan J, Liang L, Yan J (2024) Green in-situ synthesis of silver nanoparticles from natural madder dye for the preparation of coloured functional cotton fabric. Ind Crops Prod 208:117871. https://doi.org/10.1016/J.INDCROP.2023.117871
26. Khan A, Raza ZA, Bhatti HN, Sarwar T (2023) Synthesis of succinate-functionalized silver nanoparticles and their application in cellulose fabric coating. Cellulose 30(17):11231–11247. https://doi.org/10.1007/S10570-023-05538-5/FIGURES/11
27. Yadav V, Banerjee S, Bairagi S, Baisoya S, Ali SW (2022) Green synthesis of sodium lignosulfonate nanoparticles using Chitosan for significantly enhanced multifunctional characteristics. Int J Biol Macromol 211:380–389. https://doi.org/10.1016/j.ijbiomac.2022.05.069
28. Petreanu I, Niculescu VC, Enache S, Iacob C, Teodorescu M (2023) Structural characterization of silica and amino-silica nanoparticles by fourier transform infrared (FTIR) and Raman spectroscopy. Anal Lett 56(2):390–403. https://doi.org/10.1080/00032719.2022.2083144
29. Hwang N, Barron AR (2011) BET surface area analysis of nanoparticles. researchgate.net
30. Vinayagam R, Pai S, Murugesan G, Varadavenkatesan T, Selvaraj R (2023) Synthesis of photocatalytic zinc oxide nanoflowers using peltophorum pterocarpum pod extract and their characterization. Appl Nanosci 13(1):847–857. https://doi.org/10.1007/S13204-021-01919-Z/FIGURES/7
31. Javid-Naderi MJ, Sabouri Z, Jalili A, Zarrinfar H, Samarghandian S, Darroudi M (2023) Green synthesis of copper oxide nanoparticles using okra (*Abelmoschus esculentus*) fruit extract and assessment of their cytotoxicity and photocatalytic applications. Environ Technol Innov 32:103300. https://doi.org/10.1016/J.ETI.2023.103300
32. El-Borady OM, Fawzy M, Hosny M (2023) Antioxidant, anticancer and enhanced photocatalytic potentials of gold nanoparticles biosynthesized by common reed leaf extract. Appl Nanosci 13(5):3149–3160. https://doi.org/10.1007/S13204-021-01776-W/FIGURES/8
33. Serrano-Lotina A, Portela R, Baeza P, Alcolea-Rodriguez V, Villarroel M, Ávila P (2023) Zeta potential as a tool for functional materials development. Catal Today 423:113862. https://doi.org/10.1016/J.CATTOD.2022.08.004
34. Ganjeh-Anzabi P, Jahandideh H, Kedzior SA, Trifkovic M (2023) Precise quantification of nanoparticle surface free energy via colloidal probe atomic force microscopy. J Colloid Interface Sci 641:404–413. https://doi.org/10.1016/J.JCIS.2023.03.061
35. Hoo CM, Starostin N, West P, Mecartney ML (2008) A comparison of atomic force microscopy (AFM) and dynamic light scattering (DLS) methods to characterize nanoparticle size distributions. J. Nanoparticle Res. 10(Suppl. 1):89–96. https://doi.org/10.1007/S11051-008-9435-7/TABLES/1
36. Abdel Salam KA, Ibrahim NA, Maamoun D, Abdel Salam SH, Fathallah AI, Abdelrahman MS, Mashaly H, Hassabo AG, Khattab TA (2023) Anti-microbial finishing of polyamide fabric using titanium dioxide nanoparticles. J Text Color Polym Sci 20(2):171–174. https://doi.org/10.21608/JTCPS.2023.195787.1174
37. Mahmoud M, Sherif N, Fathallah AI, Maamoun D, Abdelrahman MS, Hassabo AG, Khattab TA (2023) Antimicrobial and self-cleaning finishing of cotton fabric using titanium dioxide nanoparticles. J Text Color Polym Sci 20(2):197–202. https://doi.org/10.21608/JTCPS.2023.188833.1165
38. Mohamed AL, Hassabo AG (2022) Core–shell Titanium@ silica nanoparticles impregnating in poly(itaconic acid)/poly(*N*-isopropylacrylamide) microgel for multifunctional cellulosic fabrics. J Polym Res 1(2):3. https://doi.org/10.1007/s10965-022-02921-x
39. Mohamed AL, El-Zawahry M, Hassabo AG, Abd El-Aziz E (2023) Encapsulated lemon oil and metal nanoparticles in biopolymer for multifunctional finishing of cotton and wool fabrics. Ind Crops Prod 204:117373. https://doi.org/10.1016/J.INDCROP.2023.117373

40. Singh A, Chakraborty JN (2023) Multifunctional finishing of cotton using zinc oxide and silicon dioxide nanoparticles along with DMDHEU. Res J Text Appar 27(1):141–156. https://doi.org/10.1108/RJTA-11-2021-0130/FULL/XML
41. İbili H, Daşdemir M (2023) AgCl–TiO_2/dendrimer-based nanoparticles for superhydrophobic and antibacterial multifunctional textiles. J Text Inst 114(5):861–873. https://doi.org/10.1080/00405000.2022.2089836
42. Sui Z, Guo Z, Li Y, Zhang Q, Zu B, Zhao X (2022) Study on preparation and performance of multifunctional linen fabric finishing agent. Text Res J 93(7–8):1686–1698. https://doi.org/10.1177/00405175221134974
43. Attia NF, Mohamed A, Hussein A, El-Demerdash AGM, Kandil SH (2023) Greener bio-based spherical nanoparticles for efficient multilayer textile fabrics nanocoating with outstanding fire retardancy, toxic gases suppression, reinforcement and antibacterial properties. Surf Interfaces 36:102595. https://doi.org/10.1016/J.SURFIN.2022.102595
44. Ibrahim NA, Ameen HA, Eid BM (2023) Green synthesized Chitosan and ZnO nanoparticles for sustainable use in multifunctionalization of cellulosic fabrics. Polym Bull 1–20. https://doi.org/10.1007/S00289-023-04887-2/FIGURES/6

Chapter 5
Modern Developments in Dyeing and Printing Using Nanotechnology

Anjali Agrawal and Kunal Singha

1 Nanotechnology in Dyeing

1.1 Introduction

The term "colouring" refers to the process of applying colour to fibre, yarn or fabric by the use of dyes, which are often synthetic organic dyes. During this process, the textile is treated with dye and extra processing chemicals in order to provide a consistent level of hue and colourfastness that is suitable for the application that is going to be carried out. Different fastness requirements might be relevant to the textile, depending on the purpose for which it is intended to be used. Two examples of this would be swimwear that is resistant to water and does not bleed when it is submerged, as well as automotive textiles that must be resistant to fading after prolonged exposure to the sun. There are a number of dyes and chemical compounds that are utilised in order to accomplish these properties [1]. This process is becoming more extensive and employing new technology on an annual basis. The demand for textiles that are cleaner, more cost-effective and have a higher value-added is the impetus behind these innovations.

With the passage of time, the production and utilisation of synthetic dyes for the purpose of dyeing fabric has developed into a considerable industry. The practice of artificial colouring has a detrimental effect on the natural environment. Because it contains sulphur, naphthol, vat dyes, nitrates, acetic acid, soaps, enzymes, chromium compounds, heavy metals such as lead, cadmium, copper, mercury, nickel, sodium, and cobalt, and a variety of auxiliary chemicals, the effluent from the textile industry is extremely hazardous to human health. Chlorinated stain removers, colour-fixing

A. Agrawal (✉) · K. Singha
Department of Textile Design, National Institute of Fashion Technology, Kolkata, West Bengal 700098, India
e-mail: anjali.agrawal@nift.ac.in

S. Bairagi et al. (eds.), *Nanotechnology in Textile Finishing*, Materials Horizons: From Nature to Nanomaterials, https://doi.org/10.1007/978-981-97-2696-7_5

agents based on formaldehyde, softeners based on hydrocarbons and dyeing chemicals that are not biodegradable are all potential extra hazardous substances that could be found in the water. By combining these organic components, undesirable and potentially hazardous by-products are produced. There are also a variety of techniques that can be utilised in the process of dyeing textiles, including continuous pad applications and batch processing equipment exhaust dyeing. It is normal practice to colour woven materials in a continuous manner, but the dyeing process for knit fabrics is often carried out in batch equipment employing exhaust techniques. There is a surge in the chemical wet processing of textiles on a yearly basis as a consequence of the proliferation of cutting-edge technologies and the introduction of new products that take the place of obsolete ones. The quest for textile products that are more functional, more affordable, and less harmful to the environment is the fundamental motivator [2].

The textile industry is now working to develop and implement nanotechnology in order to create textiles in a manner that is beneficial to the environment. This will result in a reduction in the utilisation of hazardous and toxic chemicals that are detrimental to the environment.

1.2 Nanotechnology

The study and application of materials with dimensions ranging from one to one hundred nanometres comes under the umbrella of the field of nanotechnology. Because of the considerable breakthroughs that have been made in nanotechnology over the course of the last few decades, the textile sector has reaped significant benefits. The traditional methods of dyeing textiles, which include the use of water as a carrier to transfer dyes to polymeric fibres, result in a significant amount of wastewater that is either colourful or hazardous, as well as high expenses associated with the treatment of wastewater. For the purpose of developing a variety of dyeing techniques, it is unquestionably essential to make use of upcoming technologies, such as nanotechnology.

The goal of nanotechnology is to develop and apply an understanding of how objects operate at the nanometre (nm) range to the construction of intricate structures that may be used in a variety of applications that are based in the actual world. A vast market might be created with the help of the smallest materials, which have the ability to change the business. To assist nanotechnology, which makes use of particles of matter with a size measured in billions of metres, businesses are doing research and creating goods in the domains of chemicals, cosmetics, medications, technology and textiles. This is done in order to promote nanotechnology. The labels of nanotechnology are currently being included on a wide range of products sold by well-known clothing companies. These products include Coolest Comfort, Resists Static and Repels and Releases Stains for example. Different metal additives, polymer additives, inorganic compounds and organic compounds are being observed by researchers from all over the world in order to determine whether or not they have the potential to add a variety of fascinating characteristics to textiles at the nanoscale.

In comparison with the qualities that nanomaterials exhibit at the bulk scale, the properties that nanomaterials exhibit at the nanoscale are completely different [3, 4]. The surface effect is a phenomenon that describes the observation that nanoparticles have a higher ratio of surface area to volume in comparison with bulk materials (Fig. 1). Because of this, they have a greater number of reactive sites per unit mass, which results in an increase in the material's reactivity as well as an improvement in its physical, chemical and mechanical properties.

In addition, the quantum effect, which is a phenomenon that grows more evident as the size of the nanomaterial decreases, is the second explanation. It is possible for dynamic changes in the configurations of electronic energy levels to result in the generation of a variety of properties, such as optical, electrical and magnetic properties, as demonstrated in Fig. 2. Nanomaterials of smaller sizes are able to interact with other molecules in a much more straightforward manner [6] as also mentioned by, National Research Council Committee (US) [7]. This is because of the increased reactivity that these nanomaterials possess.

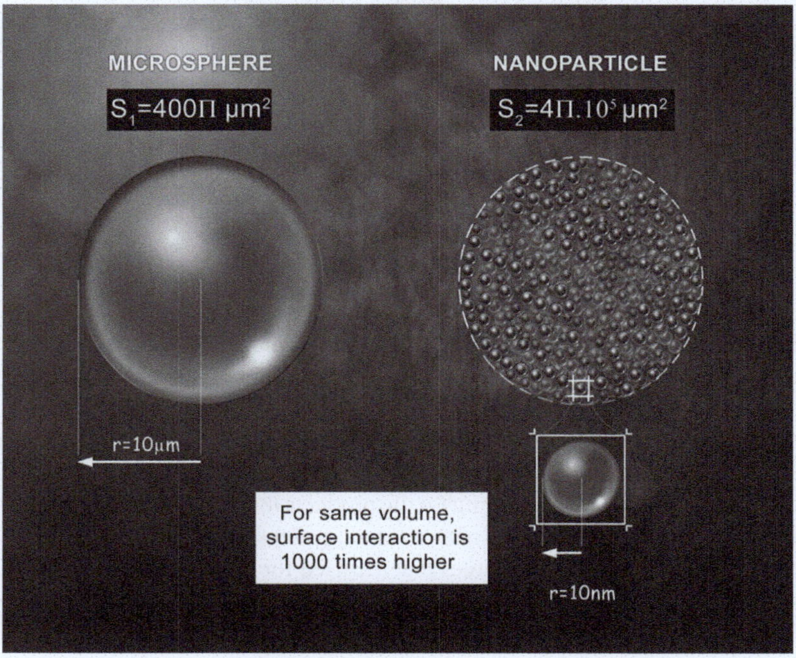

Fig. 1 Diagrammatic representation of surface-to-volume ratio of a microparticle ($S1 = 1$ mm) versus a nanoparticle ($S2 = 100$ nm) [5]

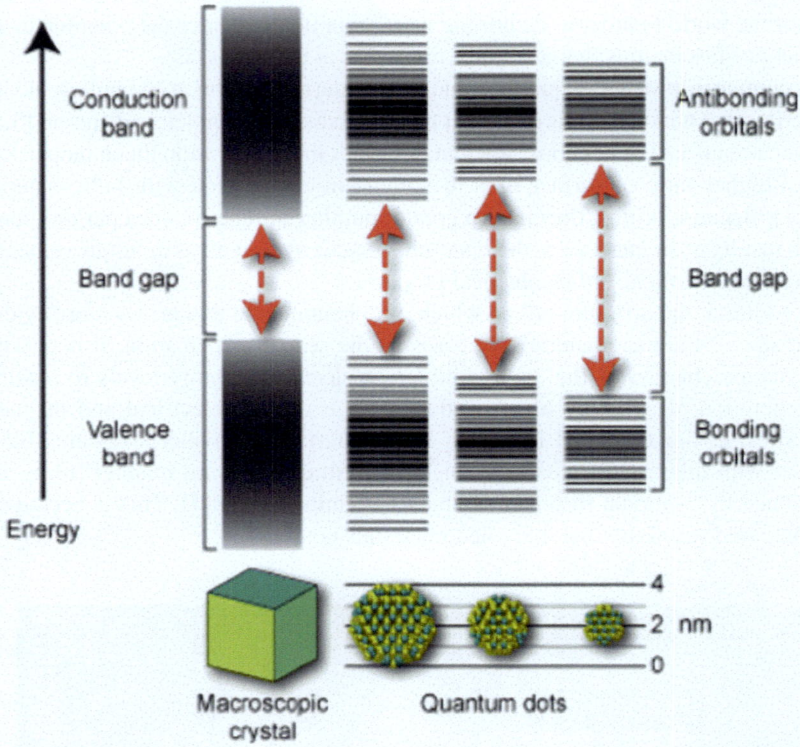

Fig. 2 Quantum effect demonstrated by nanomaterials [8]

1.3 Nanotechnology in Dyeing

The textile industry is one of the most important industrial sectors that is making advantage of the growth of nanotechnology. When it comes to functionalizing textiles with nanomaterials, the textile substrates offer a surface area that is notably larger than the volume or weight of the material. This is a quality that is useful. Researchers have demonstrated a significant amount of interest in the application of nanoparticles in textiles, with the goal of developing textile products that are capable of performing novel functionalities. The usage of nanotechnology does not affect performance, but it does reduce the amount of resources that are used for raw materials.

The following are some of the reasons why nanodyeing is possible, in addition to the use of nanotechnology for the functionalization and finishing of textiles:

(a) To reduce the amount of time spent on heating, power and processing during the dyeing process for textiles [9].
(b) An effort to reduce the amount of water it takes to dye.
(c) To eliminate colour from wastewater agents that have been released by utilising effluent with a lower total dissolved solids.

(d) To refrain from the use of hazardous substances or to discharge compounds that are less hazardous into the ambient environment.
(e) From the beginning of the process, begin to generate large earnings.

1.4 *Techniques for Colouration of Textiles using Nanotechnology*

1.4.1 Pigment Particles Consisting of Nanoscale Size Used in Textile Colouring

Several research have been conducted on the application of pigment particles with nanoscale diameters in the textile processing industry as a result of the development of nanotechnology.

Nanopigments, distinguished by their minuscule size, possess unique properties that set them apart from conventional pigments. Pigments are an excellent option for colouring textiles in an environmentally friendly manner since they offer a multitude of advantages, such as decreased efficiency, decreased effluent load, decreased energy and water consumption. Pigments are a type of textile colourant that can be classed as either organic or inorganic, soluble in water and non-interacting with polymeric fabrics. They are used to colour textiles.

A potential application would be the direct application of pigment nanoparticles in the colouring of textiles. It is possible that such an approach will be successful (Fig. 3), provided that the particles can be spread uniformly and reduced to a size that is tiny enough to prevent them from aggregating out in dye baths.

More stable suspension can be made with ultrasmall sizes of particles as the force of gravity is negligible on these particles. Physical properties improve extraordinarily, and optical and electro-optical attributes improve impressively due to particles in the range of 50–500 nm, which otherwise are impossible to attain [11, 12]. To attain comparable colour value in comparison with conventional dispersion, a small amount

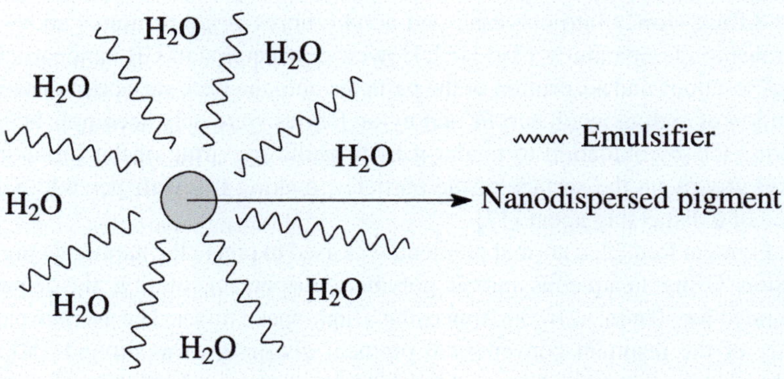

Fig. 3 Nanodispersion of an organic pigment [10]

of nanopigment dispersion is required along with a lesser amount of binder, to bind pigments with textiles. Consequently, the effect of binder concentration on fabric hand feel will be minimal.

The vividness and chromaticity of pigments may be enhanced by tailoring their particles at nanoscale. The combination of structural features due to reduced particle size (like structural colours which exist in nature) and chromophore aspects of pigments may produce textiles with extraordinary brightness and chromaticity.

Pigment printing, pigment padding and batch exhaustion are some of the methods employed when pigments are used to colour textiles. Using pigment nanoparticles directly for textile colouring is one potential use for nanotechnology. To prevent aggregation in dye baths, pigment particles can be effectively dispersed if they can be made smaller.

Fang et al. [13] have dyed cotton modified with a cationic reagent by an exhaustion method utilising nanoscale pigment dispersions. Measurements were made of cotton's pigment uptake, colour production and fastness characteristics. It was discovered that the cationic reagent concentration, pretreatment temperature, pretreatment bath pH and cationization period were all strongly correlated with pigment uptake. These elements also had an impact on the fabric's fastness qualities. When applied to cotton, nanoscale pigment dispersions produced a significantly greater colour yield than traditional pigment dispersions. For wash-down effects, fastness qualities were deemed appropriate. It is evident that employing nanoscale pigment dispersions for exhaust dyeing has several benefits, including reduced pigment requirements, enhanced handling and appearance and environmental protection.

According to a recent work by Langhals, it is possible to achieve lipophilic perylene bisimide pigment nanodispersion by employing UV/Vis spectroscopy to analyse the details of the pigment molecules' immediate surroundings in the nanoparticles. The pigment's nanodispersion's UV/Vis spectra matched those of a homogenous lipophilic solution, such as chloroform, demonstrating the dispersant's full coverage of the chromophore at the nanoscale [14, 15].

Additionally, studies have demonstrated that in a thermal colouring process, nominally 8 nm-sized carbon black (CB) nanoparticles dispersed in surfactants were able to diffuse slowly into polyester and acrylic fibres at temperatures above their glass transition temperatures (Tg) [16]. However, dispersion of CB nanoparticles in aqueous solutions and adsorption of the particles onto the fibre surface in exhaustion colouration of cotton, wool, acrylic and nylon fabrics were only accomplished after modifying CB by oxidation to render it hydrophilic by virtue of the formation of carboxyl groups on the surface of the particles, making it self-dispersible without the need of a dispersing agent [17].

C.I. Pigment Red 22, a normal pigment, was used to create the nanoscale pigment dispersion. Using an aqueous anionic polymeric dispersant solution, the dispersion was agitated for 30 min. at 10,000 rpm using a high-speed mixer. The average particle diameter of the resultant conventional pigment dispersion was around 1500 nm. With the use of a microfluidizer and 2.5 h of processing at 22,000 Pa of pressure,

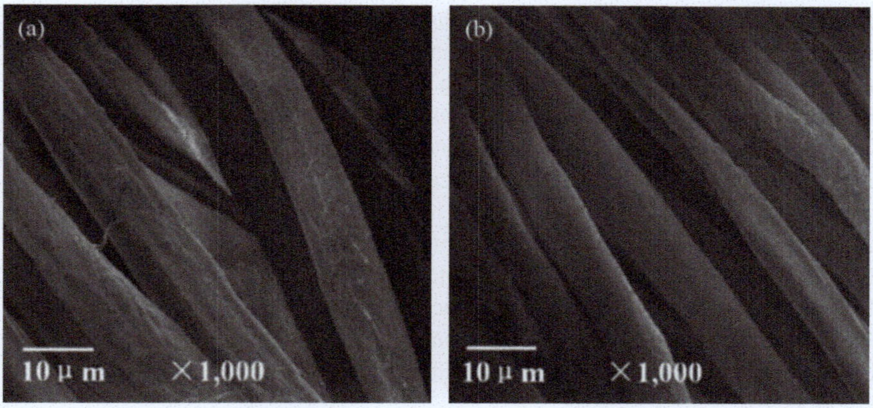

Fig. 4 Scanning electron micrographs of cotton fibre **a** conventional dyeing method, **b** ultrasonic dyeing method [19]

the dispersion was transformed into a nanoscale dispersion with an average particle diameter of 128 nm and a final pigment dispersion with 40% solid content [18].

Hao et al. [19] utilized ultrasound to make a stable nanoscale pigment dispersion. The purpose was to investigate the effect of ultrasound on dyeing cationized cotton with nanopigments. The effect of sonic cavitation on the dispersion of nanopigments, which make the pigment particle deposition on the cotton surface, or diffusion solely into the big pores between the yarns, is another way that ultrasound can be used to promote pigment adsorption for insoluble pigments. They discovered that using ultrasonic is a useful method for enhancing the nanopigments' equilibrium adsorption. SEM pictures (Fig. 4) demonstrated that applying ultrasonic can help spread and deposit nanoscale pigments on cotton surfaces more evenly.

However, natural silk fibres can also be dyed using plant colourants as a substitute for synthetic colouring [14]. Natural colours are ground into a powder at the nanoscale using a ball mill grinder. A variety of artificial and natural mordants are used to dye silk fabrics, using them as a substratum. The application of various extraction solvents and mordants during the dyeing process led to a variation in colour hues on silk textiles. Tests for colourfastness to washing revealed colour changes from 3/4 to 4/5 and colourfastness to wash from 3 to 4/5. The colourfastness of dyed cotton yarn to washing was evaluated using a launder-o-meter in accordance with the procedure outlined in IS 105-C10 (2006).

By using natural colours, the nanoscale dye powder would benefit the local fashion industry as well as the dyeing and printing sectors such as batik producers. One of the main types of synthetic fibres are synthetics like polyester, which are frequently dyed with a variety of colours. The majority of the powder was created in the size range of μm (10–6), but some was also created as nanosized powder in the 200–800 nm range, which allowed polyester fabrics to be dyed at a lower temperature.

Multiple metals, including copper, gold, silver, platinum and palladium, exhibit colouration when their particle size is in the nanometre range. Copper is one of the metals that exhibits colouration. The hues of the particles shift in proportion to the size of the particles. The phenomenon of surface plasmon resonance (SPR) is the main cause of this occurrence. When light strikes nanoparticles (NPs) of the aforementioned metals (Fig. 5) that are smaller than the wavelength of light, the electromagnetic field that is oscillating interacts with free metal electrons to form their coherent oscillation. The consequence of this is that a charge separation in the shape of a dipole takes place with respect to the ionic lattice. There is an oscillation of this dipole that occurs in the direction of the electric field. The highest amplitude of the oscillation that takes place at a particular frequency is referred to as surface plasmon resonance [20–22]. Due to the fact that the resonance frequency is located in the visible area of the spectrum, a strong hue that shifts in accordance with the change in particle size that is brought about by the change in resonance frequency is exhibited [23].

The use of gold and silver nanoparticles is quite new, having only recently begun. Sreekumar et al. published one of the first studies on the colouring of textiles with NPs [24]. They created fibres that ranged in colour from red to blue by incorporating silver nanoparticles (20–120 nm) into a polyacrylonitrile copolymer. There have also been reports on the surface deposition of nanoparticles (NPs) and their capacity to impart colour on cotton, wool and silk.

Tang et al. [26] were able to synthesise anisotropic silver nanoparticles (NPs) with a variety of sizes and morphologies. These nanoparticles, when applied to wool fabrics, generated colours that were extremely vivid. In another research, distinct coloured silver nanoparticles were generated directly on wool fabric. Trisodium citrate was used to decrease silver nitrate, and it also functioned as a linker, which allowed the silver nanoparticles to be quickly attached to the wool fibres and

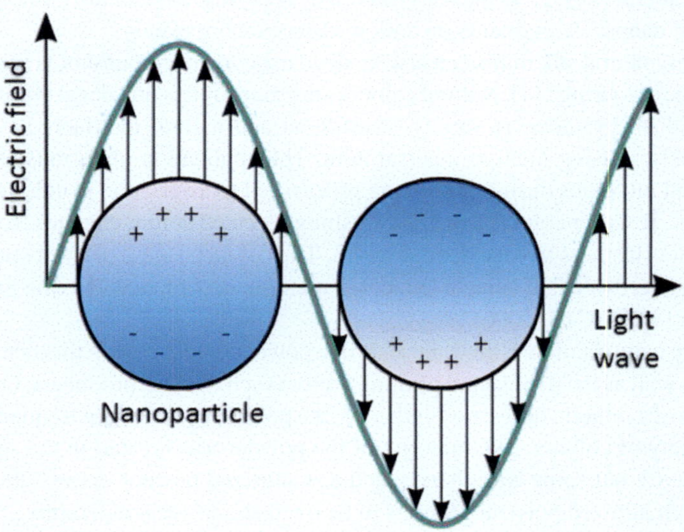

Fig. 5 Schematic representation of surface plasmon resonance on nanoparticles [25]

stabilised. The result of this was the production of a one-of-a-kind colourant for wool fibres that is resistant to repeated washing and rubbing [27].

NPs made of gold and silver have been utilised in the process of colouring cotton textiles. Before depositing red, yellow and blue silver nanoparticles on cotton fabric, [28] researchers treated the cloth with poly (diallyl dimethyl ammonium chloride) to make it more resistant to water. The colourfastness of the printed cotton fabric that had been treated was excellent. It was possible to make a multifunctional coloured cotton fabric by employing in-situ reduction and deposition of silver nanoparticles. The amount of silver nitrate was reduced as a result of the terminal aldehyde groups that were present on the remaining starch (sizing material) on the cloth. In addition, several coloured gold nanorods (NRs) have been synthesised by adjusting the length-to-diameter aspect ratio. These nanorods have been applied to cotton and silk textiles in order to impart colour [29].

As an exciting additional dimension of colour, the structural hues of nature are extremely interesting. Because surface micro- and nanostructures are present, structural colours, which are vibrant and eye-catching, are the outcome of fundamental optical phenomena such as single and multilayer interference, photonic crystals, diffraction grating effect, light scattering, and other similar processes. These hues are used to colour a variety of items, including sea creatures, jewel beetles, and the wings of morpho butterflies, amongst other things. Structural hues are fascinating for a variety of reasons, including the fact that they could be utilised in paint, textiles, automobiles, cosmetics and other industries [30–33].

1.4.2 Use of Plasma Technology in Nanodyeing

The interaction of an electromagnetic field with a gas at the right pressure results in the formation of plasma, which is a partially ionised gas that contains ions, electrons and neutral particles. Modern plasma-chemical procedures are highly superior to chemical modification, which involves the use of corrosive reagents such as acids, hydroxides, alkaline-earth metals and their compounds. These techniques are related to ecologically clean technologies and are a significant improvement over chemical modification. In this regard, the use of plasma technologies for the pretreatment and finishing of textiles is gradually replacing the use of wet chemical processes.

Low-temperature plasma treatment is one of the most promising and advanced approaches for polymer modification. This technology enables the surface properties to be modified over a large range, and it also allows for the area of application of polymeric materials to be significantly extended. An increase in the hydrophilicity of the treated fibre is brought about by this surface modification. It is a significant characteristic of plasma treatment because it only affects the surface of the material that is being treated, as well as a very thin layer that is located near the surface. The thickness of this layer can range anywhere from 100 Å to several micrometres, depending on the estimates that are given. Under these conditions, the majority of the polymer is preserved to its original state, keeping the mechanical, physicochemical and electrophysical properties that were present in the material in the first place

[34, 35]. The ultraviolet photons that are released by the plasma possess sufficient energy to break chemical bonds (such as C–C and C–H) and to generate radicals that are able to move up the chain and react with other radicals there. The activity of the plasma can result in activation (the generation of radicals), unsaturation (the synthesis of double bonds), chain scission, and cross-linking [36]. These outcomes are dependent on the circumstances of the plasma as well as the types of polymers that are present.

By using a pad–dry–bake technique, also known as the thermosol process, or by subjecting polyester textiles to high temperatures and pressure, dispersion dyes are typically used to colour the fabric. Alternatively, carriers based on phenol are required in order to swell the fibre during the process of dyeing under atmospheric pressure, which has the potential to harm the environment. The treatment of PET fibres was therefore accomplished through the utilisation of low-temperature plasma rather than the chemical technique. It was found that treating PET fabrics with both argon and air plasma resulted in an increase in the colour depth immediately following the colouring process. The increase in surface roughness and surface area that was created by the plasma was responsible for this phenomenon. Additionally, the introduction of hydrophilic groups, which can be generated by both reactive and chemically inert plasmas, has the potential to enhance the water swelling capability of PET fibres as well as their affinity for dyes that include polar groups. This, in turn, adds to the increase in the K/S values of dyed PET specimens [37].

Because of the high surface area that they offer, nanostructured surfaces are a subject of significant attention. Consequently, ultrathin coatings are capable of achieving a high level of functionality. In a regime that included both deposition and etching operations, plasma polymerization of acetylene mixed with ammonia (C_2H_2/NH_3) onto PET fibre was utilised. This resulted in the formation of a nanoporous, cross-linked network that contained accessible functional groups [38, 39] (Fig. 6).

In contrast to other natural fibres, wool possesses a surface structure that is more complex than that of other fibres, and it is one of the most important fibres used in the textile industry. On the other hand, it has a few technical issues, such as wettability, which has an effect on the dyeability characteristics. Because of the presence of

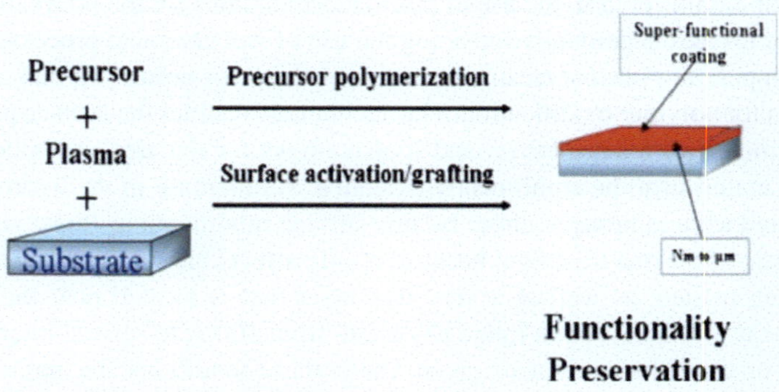

Fig. 6 Plasma coating on nanostructures [10]

a high number of disulphide cystine cross-linkages (–S–S–) in the A-layer of the exocuticle and of fatty acids on the surface of the fibre, the surface of the wool fibre is hydrophobic. This is because of the nature of the fibre. It is believed that the surface shape of wool fibre contributes to the determination of the diffusion [40].

As a result, the chemical approach of treating wool was replaced by the use of low-temperature plasma, also known as LTP. LTP treatments have been shown to improve the colouration behaviour of wool fibres in a variety of colouration systems, as has been confirmed. It is possible that this is because of plasma-induced cystine oxidation, which leads to a decreased number of cross-linkages in the fibre surface. This, in turn, makes it easier for a transcellular dye diffusion to occur in addition to the intercellular dye diffusion [38, 40–43].

Reactive dyes are frequently utilised for the application of cotton fibres due to the fact that they offer a comprehensive colour range and are simple to apply, particularly in circumstances when exhaust colouration is concerned. Reactive dyes, on the other hand, have only a moderate affinity for cotton fibre. Over the course of the past, there have been multiple attempts made to circumvent this limitation. One of the most practical approaches is to improve the interaction between the dye and the fibre by utilising cationized cotton [44]. In this particular area of interest, plasma-treated cotton fibre in the presence of amine compounds demonstrated better dyeability with reactive dyes in comparison with the fabric that had not been treated [45, 46].

1.4.3 Microbial Nanopigments

Recently, there has been a resurgence in interest in textile dyeing due to the biocompatibility, sustainability, natural functioning and cleaner synthesis of microbial pigments, which are the most promising biomass pigments. Because their resources are not constrained by climatic, seasonal or geographic limits, microbial pigments hold a great deal of promise for industrial application. This is in contrast to natural pigments that originate from plants and insects.

Prodigiosin is a pyrrole structure microbial pigment that has lately attracted the attention of researchers due to the fact that it possesses antibacterial properties and bright colours due to its composition. Gong et al. found that bacteria produce red pigments within cells, which had pyrrole structures resembling nanoscale particles when extracted from extracellular culture systems. During the course of this research, an in vitro platform called a mimic phospholipid bilayer membrane was developed in order to investigate the behaviour of transmembrane transport. In order to generate a dye suspension, biomimetic membranes were utilised for the very first time in order to evaluate the transmembrane transport of microbial nanopigments.

This research, which identified the successful synthesis of microbial colour liquor based on fermentation broth [47, 48], has had a significant impact on the use and development of microbial pigments.

1.4.4 Dyeable Polypropylene Using Nanotechnology

Nanocomposites are produced by incorporating nanoparticles, also known as filler, into a macroscopic sample material called the matrix (Fig. 7). The ever-expanding field of nanotechnology encompasses this particular element. Following the incorporation of nanoparticles into the matrix material, the nanocomposite that is produced may exhibit greatly enhanced properties.

The dyeability of polypropylene (PP) is a problem, despite the fact that it is more affordable than other materials like as polyester and nylon. The dyeability of polypropylene (PP) can be improved through the use of a few conventional procedures, including as copolymerization, poly blending and grafting. However, these techniques are more expensive than others, which results in an increase in the overall cost of PP fibre. Now, owing to nanotechnology, there is a solution to this problem. Nanocomposite polypropylene, also known as nano-PP, was successfully manufactured by utilising nanoclays that had been modified with quaternary ammonium salt [50].

This novel polypropylene can be coloured with either dispersion dyes or acid dyes, depending on the desired effect. The nano-PP's dyeability with acid is attributed to the ionic interaction between the positively charged quaternary ammonium ions in the nanoclay and the negatively charged acid dye. It is possible that hydrogen bonding and van der Waals forces are also considered to be key contributors to the acidic colouring of the nano-PP. It was hypothesised that the disperse dye's attraction to nanoclay in the context of colouration is mostly caused by van der Waals forces and hydrogen bonding. This was the conclusion reached by the researchers. Furthermore, it has been demonstrated in a recent study [51] that nanoclay/PP nanocomposites with four primary colours—C.I. Acid Blue 80, C.I. Acid Orange 74, C.I. Acid Red 266, and C.I. Acid Yellow 135, can be dyed with acid by employing three distinct chemical structures: anthraquinone, premetallized and monoazo molecules.

Additionally, it was shown that nanoclay/polyamide 6 (PA6) nanocomposite yarn colours itself more rapidly when dyes are disseminated than when PA6 yarn is empty; however, this is not the case when dyes are acidic or 2:1 metal complex. This was

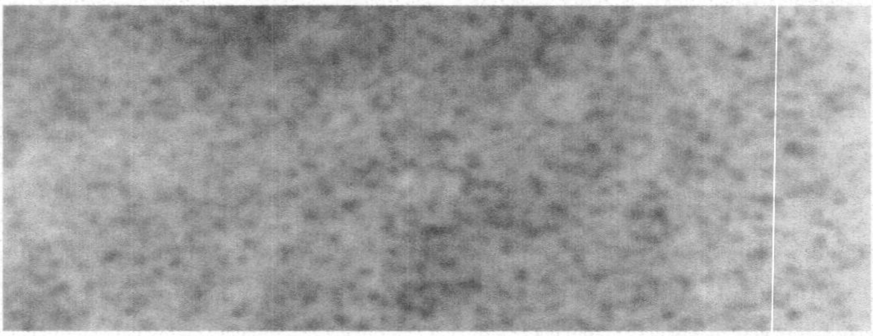

Fig. 7 Nanocomposite fibre [49]

a significant finding. When it comes to the latter two circumstances, the nanoclay is able to cling to the amino sites, which prevents the acid or metal complex dyes from joining together [52].

Recently, polyester and silica (SiO_2) nanoparticles were mixed to generate polyester nanocomposite fibre. This was done with the intention of providing polyester (PET) with a new added value and improving its dyeability with disperse dye. It has been demonstrated through published research that when PET/SiO_2 nanocomposite fibres were treated to alkaline hydrolysis, they experienced a greater reduction in weight compared to pure PET fibres. The addition of more lasting superfine features, such as craters, cavities and fractures, was done in order to produce a specific application-like deep colour [53–55].

Conversely, nanotechnology can also be utilised to improve textiles' capacity to absorb dyes, especially synthetic fabrics. Because the majority of polypropylenes have so far resisted dyeing, they have not been considered acceptable for use in consumer items such as tablecloths, floor coverings, or window dressings. An innovative method in development involves incorporating dye-friendly clay nanoparticles into raw polypropylene stock prior to its extrusion into fibres. Colours can be absorbed by the resulting composite material without degrading the cloth.

1.4.5 Nanodisperse Dye

The most processed dyeing and synthetic printing is typically the consequence of the application of the largest consumable volumes of disperse colours. The creation of stable suspensions requires the use of commercial disperse dyes, which are unable to dissolve in water due to their exceptionally low solubility. When grinding or milling dyes, it is necessary to use dispersants that have a desirable concentration of 150% depending on the weight of the dye in order to achieve the ideal dye particle size [56]. The excessive use of dye dispersants, on the other hand, results in the degradation of dyes and the staining of fibres, in addition to contributing to environmental pollution [57, 58]. As a result of the development of polyester microfibres and nanofibres [59], the specific surface area of these fibres has greatly grown in comparison with the usual surface area and a greater quantity of dyes is required to achieve a particular level of shade [56, 60–62].

As a result of this, the production of micro- and nanodye particles is required in order to colour microfibre [59]. During the nanodyeing process, the dye particles are reduced to a size of approximately 100 nm. When the dye particles are smaller, they have a greater surface area, which allows them to contact with the dispersing agent more forcefully. This interaction results in a dispersion that is more uniform and stable. On the other hand, this can also slow down the rate at which the dye runs out on the fibre. This is because the nanodispersions have a greater surface area and a longer durability than traditional dispersions.

Choi and Kang [63] dyed regular and ultramicrofibre polyester using nanodispersions of disperse dyes. They created a nanodispersion dye (Fig. 8) by first creating a mixed nanoemulsion using caprylic triglyceride and sodium lauryl sulphate as an emulsifier, continuously precipitating in the oil phase and then using ultrasonic treatment to break up the dye particles. Exhaustion yields were less than 40% of the applied dye when the nanodye emulsions were applied to standard polyester and ultramicrofibre polyester; these were lower than those of commercial disperse dyes. The solubilisation of dye particles into surfactant micelles may have contributed to the low exhaustion yields of nanodispersion dye emulsions that were reported, delaying the uptake of dyes by polyester fibre. The low exhaustion yields may possibly be due to the excellent stability of the nanoemulsions. However, after dyeing with nanodyes generated on ultramicrofibres, it was found that the microfibre site displayed greater K/S values (in the range of 1.5–22.4) than the standard polyester site, suggesting a potential for higher efficiency on ultramicrofibre.

An ultrasonic homogenizer was used in a study to reduce C.I. Disperse Red 60 and C.I. Disperse Blue 5 to nanosize [64]. Due to the presence of zinc oxide NPs,

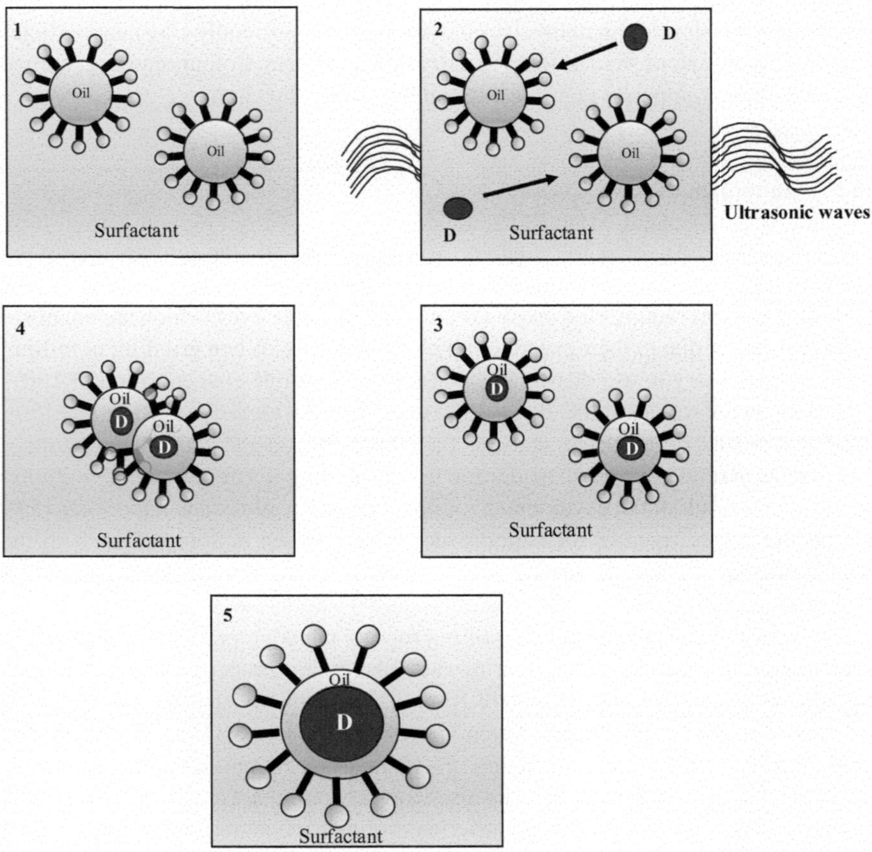

Fig. 8 Nanodisperse dye formation (TO BE DELETED AND REPLACED BY ONE SENT TO THE EDITOR IN MAIL). Adapted from Choi and Kang [63]

coloured cloth exhibited increased colour fading and UV protection when applied in conjunction with zinc oxide NPs dispersion [65].

Using an ultrasonic homogenizer, Disperse Blue 56 and Disperse Red 60 were brought to the nanoscale and transmission electron microscopy was used to characterise the results. Depending on how long the dyes are sonicated for, there is an increase in fastness. The use of nano-ZnO powder in the fabrics during pre-, simultaneous- and post-treatment has increased the sonicated dyes' light fastness more than the non-sonicated ones. Better outcomes were also seen in the post-treatment scenario with regard to light fastness, which raised UV protection and slowed the rate of fading. The polyester samples that were dyed at 130 °C showed a significant increase in both the fastness qualities and the value of K/S after employing two dispersion dyes. This could be the result of the sonicated dyes being evenly distributed throughout the dyeing bath [66].

Using hydroxypropyl sulfonated alkali lignin (HSAL) dispersion, a new nanodisperse dye was effectively synthesised by Qin et al. [67] Half of the milling dye was the dispersant dosage in the manufactured nanodisperse dye. Additionally, it showed minimal reducing impact, strong stability at high temperature and good dispersion. The lowering of the phenolic hydroxyl content in the HSAL molecule prevented the reducing effect on the nanodye (azo structure). Quartz Crystal Microbalance with Dissipation (QCM-D) and Atomic Force Microscopy (AFM) were used to thoroughly examine the production mechanism of the nanodispersion dye preparation as well as the interaction between dye particles and hydroxypropyl sulfonated alkali lignin. Strong electrostatic repulsion can be produced by a high concentration of sulfonic group that is linked to the long alkyl chain in HSAL molecules, allowing it to extend into the aqueous phase and scatter dye particles to generate nanodisperse dye. This work expanded the use of biomass lignin in the dyeing and printing industries and created a new, very effective nanodispersion dye.

In this work conducted by Kale and his team [68], the disperse dyeing of nylon fabric using nanoemulsion assistance exhibits colour strength comparable to commercial dyeing samples. By lowering the size of the dye particles, nanoemulsions provide extremely fine dispersion and serve as a transport medium for the dye into the fibre core without the need for additional dye. When compared to the usual process, this results in significant savings on dyeing and dye manufacturing costs as well as a reduction in the load on effluent. All dyeings using nanoemulsion had similar fastness ratings. The employment of nanodisperse dyes reduces the time-consuming step of milling the disperse dye using the dispersing agent. Because of this, employing nanodisperse dyes instead of finished commercial dispersion dyes to dye nylon can save a significant amount of energy, time and money during the manufacturing process.

1.4.6 Nanobubble Dyeing

The innovative method known as e-flow, which is based on nanobubbles, was developed and patented by Jeanologia. Using e-flow results in a more comfortable feel,

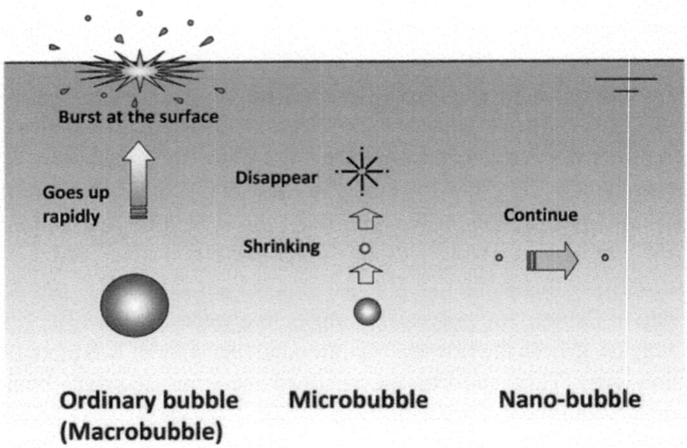

Fig. 9 Schematic diagram showing macro, micro and nanobubbles [47]

breaks up the surface of the garment and reduces the amount of shrinking that occurs. This process does not result in any waste, and it only requires a small amount of water to be carried out. Within the context of this technique, ambient air is introduced into an e-flow reactor, which is then subjected to electromechanical shock in order to generate a flow of moist air and nanobubbles [47, 69].

Nanobubbles, which are very small bubbles with a diameter of less than 200 nm, have been the subject of researchers' attention for a variety of applications. As can be seen in Fig. 9, microbubbles gradually decrease in size and eventually collapse as a result of prolonged stagnation and the dissolution of internal gas into the waters that surround them. On the other hand, nanobubbles continue to exist in their current state for months and do not suddenly burst out [47, 70].

1.5 Conclusion

This chapter basically deals with application of nanotechnology in dyeing and printing. The first section of the chapter describes the various advancements in dyeing by employing nanotechnology. Despite the fact that the application of nanotechnology in the process of textile colouring is still in its infancy, it has the potential to significantly alter the colour of fibres, particularly high-performance fibres such as aramids and ultrahigh molecular weight polyethylenes. By requiring only one "drop-in" pretreatment step for the greige products straight from knitting and a modification of the exhaust dye jet process, the nanodye technique enhances performance and lowers total dyeing costs. This is accomplished by reducing the amount of time and money spent on dyeing. The dyeing industry need novel ways and best practices in order to prevent more damage to our quantum climate. This is because the dyeing industry is significantly concentrated in developing nations. The nanodye technology, which has been the subject of extensive research and development over the course of several years, is the future of environmentally friendly exhaust dyeing. The completed fabric has a more pleasant colour and aroma, a handle that is more comfortable and a colour tone that is more distinct.

2 Nanotechnology in Printing

The textile industry experienced a period of prosperity in the twenty-first century as a result of the tremendous spread of social media platforms. The majority of the market is currently controlled by large brands on a global scale and the textile business generates millions of dollars annually. As time goes on, nanotechnology is gradually being included into the textile fibre industry. This is due to the fact that nanotechnology possesses unique and beneficial properties. The short-term effects of quality and durability were typically constrained in the past by materials that were created using traditional procedures [71–73]. However, the development of nanotechnology has made it possible for these fabric companies to use nanotechnology to make clothing that is of superior quality, flexible, and durable and that continues to be wearable and durable even after being washed and worn. Nanomaterials have a high surface-to-volume ratio, which allows them to maintain a high surface energy and improve their affinity to their textiles [56, 74]. This allows them to ensure that they are durable over an extended period of time. Because of the nanoparticle coating and the thin layering, the fabric is not only comfortable to the touch but also allows air to circulate through it. This stacking is accomplished through the use of processes like as printing, washing, padding, rinsing, drying and curing. The goal is to affix nanoparticles to the surface of the fabric. It is possible to achieve water repellence, soil resistance, flame resistance, hydrophobicity, wrinkle resistance, antibacterial and antistatic characteristics and increased dyeability by the utilisation of these processes [75].

Through the use of nanotechnology, the capacity of fabrics to absorb ultraviolet light and add dye pigmentation is much improved. This technology deflects the sun's damaging rays, which sets the cloth apart from others [76]. It is preferable to other materials because it provides excellent protection.

As a result of the financial benefits that are linked with their application, large-scale organisations have been lured to the unique properties that nanoparticles possess in the textile industry. This is the reason why the nanotextile industry is experiencing a rapid expansion in the number of competitors, which may soon force the old textile industry to become obsolete [74, 77]. The ability to remove soil and malodour, abrasion resistance, a modified form of surface friction and colour enhancement through nanomaterials are some of the advantages that are associated with nanotextile engineering and industry [78]. Other advantages include improved cleaning surfaces, resistance to soil, wrinkles, stains and colour damage, higher wettability and strike-through qualities and the ability to remove soil and malodour.

2.1 Recent Developments in Nanotechnology-Based Printing Process

2.1.1 Recent Advancements in Printing Electrode Devices for Electrochemical Sensors that are Based on Nanotechnology

A comprehensive review of the current state of the art in screen, inkjet and three-dimensional (3D) printing electrode technologies, production methods and applications is described in this work (Fig. 10). This is the first time that such an analysis has been presented. Emerging printing electrode-based technologies have been the impetus behind the development of printed electrode devices and systems. These technologies aim to improve the creation of printed electrode materials utilising printing electrodes. As a result, a wide range of electrochemical applications now have access to previously unavailable opportunities. In addition, the features of nanostructured materials can be modified for specific electrochemical applications by the process of embedding and/or immobilising the materials. There are a number of enticing qualities of nano-based printed or modified electrodes, including high selectivity towards a variety of targeted electroactive analytes, cost-effectiveness, scalability and better performance. The fact that the nanoscale-printed electrodes are portable and flexible makes them suitable for use in on-site testing facilities. However, the performance of printed electrodes is affected by the type of printed electrode materials that are utilised as well as the fabrication procedures that are employed. Consequently, the purpose of this research is to investigate the many kinds of electrode materials, printing methods, and applications for biosensors, as well as the identification of organic and inorganic compounds. This article provides an overview of the printed electrode materials that focus on qualities that have been discovered in the research literature. These characteristics include selectivity, sensitivity and limit of detection. An examination of the possibilities, challenges and opportunities presented by these cutting-edge printing electrode technologies is presented in the last section [78].

2.1.2 Advances in Screen Printing of Conductive Nanomaterials for Stretchable Electronics

Stretchable electronics have demonstrated that they have the potential to give significant promise in a variety of applications, including but not limited to soft robotics, persistent human–machine interfaces, wearable health care, and better diagnostic testing. However, their usefulness is limited since they are dependent on fabrication procedures that are both expensive and difficult to produce in large quantities. The typical Micro-Electromechanical System/Nano-Electromechanical System (MEMS/NEMS) metallization process and off-contact direct printing techniques are not suited for application on a large scale or in big quantities. Screen printing, on the other hand, is a well-established printing method that is capable of producing a great volume of

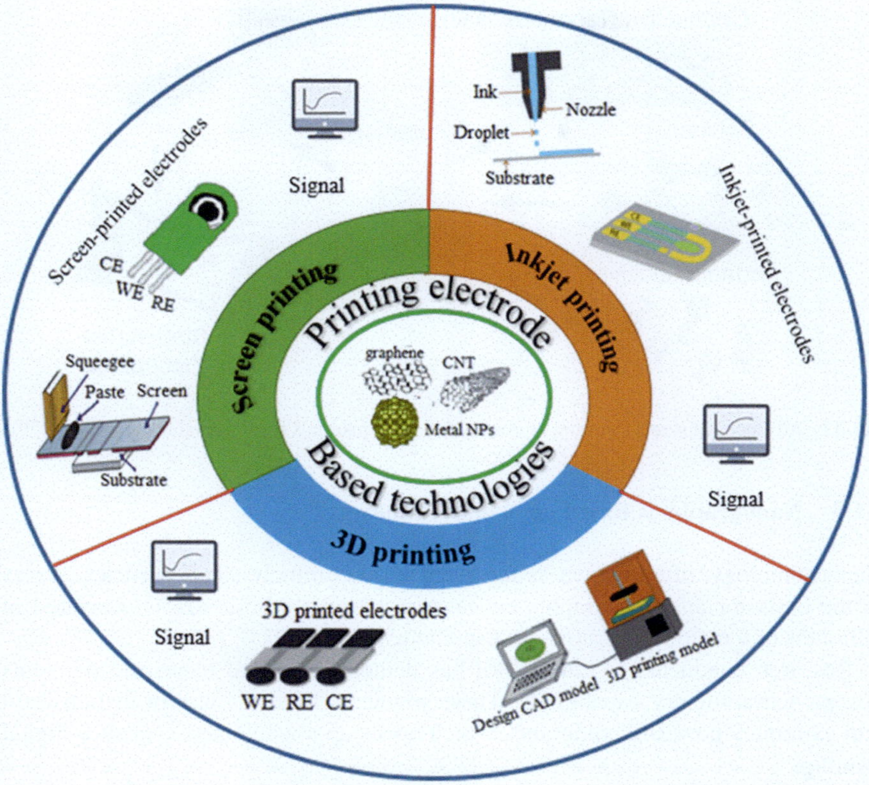

Fig. 10 Recent developments in nanotechnology-based printing electrode systems for electrochemical sensors [78]

prints. Because of the recent discovery of conductive nanomaterial inks that are intrinsically stretchy, it is now possible to manufacture stretchable electronics on a wide scale. The design of screen-printed inks is restricted due to the fact that it must adhere to stringent rheological standards throughout the printing process, substrate–ink attraction and nanomaterial properties that dictate dispersibility and percolation threshold. In this study, a concise summary of these significant constraints is provided, along with an explanation of the current efforts being made to resolve them. It begins with an explanation of the fluid dynamics that drive screen printing, then on to infer the ideal rheological properties of the ink based on these features, and finally presents an explanation of the combining of nanomaterials, solvents, binders and rheological agents in order to produce high-performing inks. Although the primary focus of this investigation is on conductive linkages, materials that are sensing, insulating, photovoltaic, and semiconducting can also reap significant benefits from the application of these approaches. Therefore, it can be inferred that by examining the prospects and obstacles that lie ahead for screen-printing stretchy electronics and their broader applications (Fig. 11), [79] this conclusion can be reached.

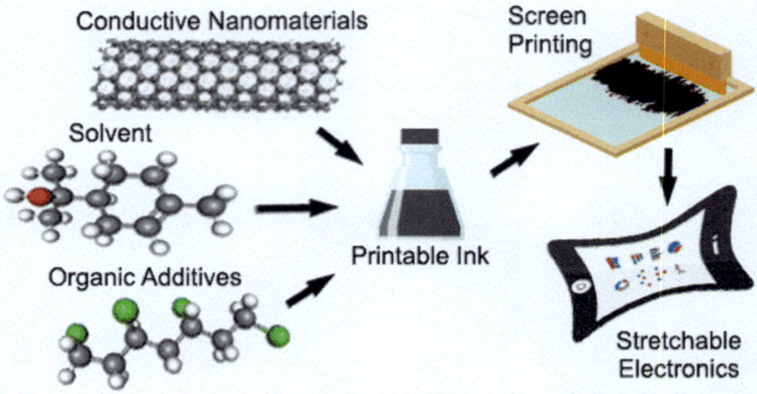

Fig. 11 Advances in screen printing of conductive nanomaterials for stretchable electronics [79]

2.1.3 Nanography is Based on Nanotechnology

Nanotechnology, often known as the science of extremely small particles, serves as the basis for any nanographic endeavour: particles whose size is measured in billionths of a metre, also known as nanometres.

Research conducted at Landa [80] has demonstrated that nanosized pigments possess extraordinary capabilities. These pigments have the ability to transform into extremely powerful colourants, which opens up a whole new space for digital printing.

The broadest CMYK colour gamut, ultrasharp dots with exceptionally high uniformity, the ability to print on any common untreated paper stock, coated or uncoated, the capacity to print on almost any plastic packaging film or label stock, and remarkable abrasion- and scratch-resistant images are some of the benefits of nanography (Fig. 12) [81].

2.1.4 Conductive (Metal-Based) Fibres in Nanoprinting

As can be seen in Fig. 12, conductive fibres are constructed by combining yarns or threads with various metals (such as nickel, titanium, copper, zinc, and silver, amongst others), carbon, and stainless steel. Metallic yarns and threads can be utilised in knitted or woven textiles for a number of smart electronic applications [82–84]. This is due to the fact that they enable electrical connections to be made between different components.

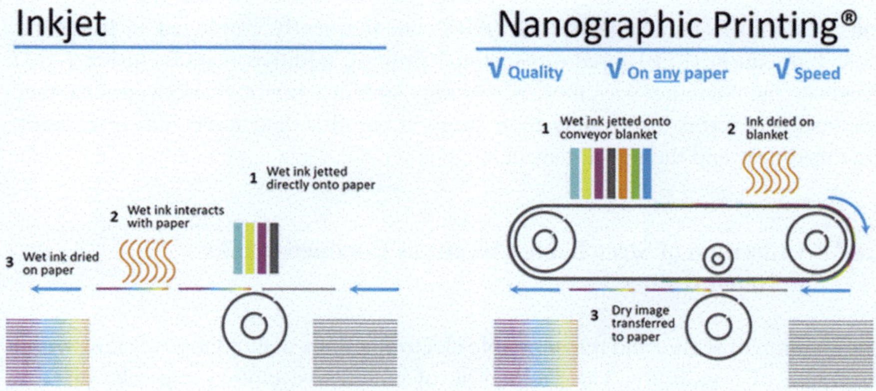

Fig. 12 Advantages of modern nanographic printing over normal inkjet printing process [81]

2.1.5 Conductive Ink in Nanoprinting

Recent years have seen the development of a wide variety of conductive fibres that are made of a variety of semiconducting and conductive materials. These fibres have been utilised to create electrical circuits in addition to smart fabrics. It is interesting to note that these conductive fibres, due to their rigidity, do not always possess sufficient flexibility to be utilised in smart textile domains. Using a unique process known as conductive inks, printing methods have recently made it easier to manufacture electrical circuits. This has been made possible by the use of printing methods. The increased attention that conductive inks have received can be attributed to the fact that they are simple to produce and employ. The application of conductive inks as a stretchy printing medium can be accomplished with elastic polymers (elastomers) that contain conductive metallic additives, in contrast to the usage of conductive fibres [82, 83].

2.1.6 3D Nanoprinting in Nanocomposite, Thermoplastic and Biology or Medical Applications

Furthermore, the research conducted by Dos Santos et al. [85] has showed how 3D printing and nanotechnology can be applied in the field of personalised medicine in a number of medical treatment scenarios and multiscale alliances.

Following a similar line of thought, Yamamoto et al.'s most recent study, which was published in 2019 [86], proved that nanocomposite materials may be created by the use of additive manufacturing, three-dimensional printing and minimal graphene loading. Nanocomposites are widely used in the field of three-dimensional printing, sometimes known as 3D printing. This printing technique involves the quick prototyping of parts through the use of geometric modelling. A number of thermoplastic polymers, including polylactic acid (PLA), polybutylene terephthalate (PBT), and acrylonitrile butadiene styrene (ABS) are frequently employed in fused filament fabrication (FFF) three-dimensional printing additive manufacturing (AM). However, put it another way, these nanocomposites are superior to conventional thermoplastics in terms of not only their strength but also their molecular orientation, their usability, and their cost margin.

2.1.7 Utilisation of Stretchable Circuits as Conductive Inks in Nanoprinting

The creation of secure and transportable electronic links within fabrics is now within the realm of possibility. The production of conductive clothes can take place in any location. For the production of textiles with conductive regions, screen printing and inkjet printing with conductive ink are the methods of choice. It is possible to enhance the conductivity of standard printing inks by including metals such as nickel, carbon, gold, silver, and copper into the formulation. The regions of a textile that have been printed can be used as pressure pads or electronic switches. Through its usage as an active wearable and a monitoring tool, conductive ink has proven to be beneficial in the field of medicine. When compared to alternative conductive materials such as conductive yarn and polymers, these inks are much more favourable in terms of affordability, flexibility, and comfort [87, 88]. At this time, the emphasis is being placed on dielectric encapsulating inks and conductive inks based on silver. Conductive inks that are based on silver are made out of polyurethane film that has been heated to 120 °C in order to manage silver particles. A negligible amount of electrical resistance can be found there. It is also possible to employ insulated inks to cover or wrap wires that include conductive inks at the same time. For the purpose of constructing stretchy circuits, insulating and conductive inks are used. Biological signals, which are easily detectable by conductive inks, are employed for the purpose of monitoring the physiological functioning of humans. A number of companies have developed electronic inks that are based on textiles. Once these inks have been printed on fabrics, they are connected to a number of sensory components that monitor the rate of the heart, the pace of respiration and the amount of stress. It is possible for a battery module that is in wireless connection with a smart device to detect impulses that are emanating from the body [89]. This is accomplished by the application of embedded sensing elements on shirts. Conductive ink was first introduced by DuPont, as can be seen in Fig. 11.

2.2 Conclusions

This work provides a comprehensive review of the latest advancements in screen, inkjet and 3D printing electrode technologies, production methods and applications. These technologies aim to enhance the creation of printed electrode materials, providing new opportunities for electrochemical applications. Nanostructured materials can be modified for specific applications, offering high selectivity, cost-effectiveness, scalability and better performance. However, the performance of printed electrodes depends on the materials used and fabrication procedures. The research focuses on electrode materials, printing methods, and applications for biosensors and organic and inorganic compound identification.

References

1. Schlaeppi F (1998) Optimizing textile wet processes to reduce environmental impact. Text Chem Colorists 30:19–26
2. Moore SB, Ausley LW (2004) Systems thinking and green chemistry in the textile industry: concepts, technologies and benefits. J Clean Prod 12:585–601
3. Roduner E (2006) Nanoscopic materials: size-dependent phenomena. In: Roduner E (ed) Nanoscience and nanotechnology. Royal Society of Chemistry Publishing, UK
4. Roduner E (2006) Size matters: why nanoparticles are different? Chem Soc Rev 35:583
5. Laroui H, Wilson DS, Dalmasso G, Salaita K, Murthy N, Sitaraman SV, Merlin D (2011) Nanomedicine in GI. In: AJP gastrointestinal and liver physiology, pp 371–383
6. Joshi M, Bhattacharyya A (2011) Nanotechnology—A new route to high performance functional textiles. Text Prog 43(3):155–233
7. National Research Council (US) (2002) Committee for the review of the national nanotechnology initiative. In: Small wonders, endless frontiers: a review of the national nanotechnology initiative, National Academies Press, Washington, US
8. Li J, Mao H, Kawazoe N, Chen G (2017) Insights into the interactions between nanoparticles and cells. Biomater Sci 5:173–189
9. Maryan AS, Montazer M, Rashidi A (2013) Introducing old-look, soft handle, flame-retardant and antibacterial properties to denim garments using nano clay. J Eng Fibres Fabr 8(4):68–77
10. Ahmed NSE, El Shishtawy Reda M (2010) The use of new technologies in coloration of textile fibers. J Mater Sci 45:1143–1153
11. Horn D, Rieger J (2001) Organic nanoparticles in the aqueous phase—Theory, experiment and use. Angew Chem Int Ed 40:4330
12. Piri N, Shams-Nateri A, Mokhtari J (2017) Solar spectral performance of nanopigments. Sol Energy Mater Sol Cells 162:72–82
13. Fang K, Wang C, Zhang X, Xu Y (2005) Dyeing of cationised cotton using nanoscale pigment dispersions. Color Technol 121:325
14. Langhals H (2007) In 4th International conference of textile research division NCR. Egypt, Cairo, pp 543–546
15. Langhals H (2008) Brightly shining nano particles: lipophilic perylene bisimides in aqueous phase. New J Chem 32:21–23
16. Li DP, Sun G (2003) Coloration of polyester and acrylic fabrics with carbon black nanoparticles. AATCC Rev 3(12):19–22
17. Li D, Sun G (2007) Coloration of textiles with self-dispersible carbon black nanoparticles. Dyes Pigm 72(2):144–149
18. Wang C, Zhang Y (2007) Effect of cationic pretreatment on modified pigment printing of cotton. Mater Res Innovations 1112730
19. Hao L, Wang R, Liu J, Liu R (2012) Ultrasound assisted adsorption of anionic nanoscale pigments on cationised cotton fabrics. Carbohyd Polym 90(4):1420–1427
20. Huang X, El-Sayed MA (2010) Gold Nanoparticles: optical properties and implementations in cancer diagnosis and photothermal therapy. J Adv Res 1:13–28
21. Lu X, Rycenga M, Skrabalak SE, Wiley B, Xia Y (2009) Chemical synthesis of novel plasmonic nanoparticles. Annu Rev Phys Chem 60:167–192

22. Willets KA, Van Duyne RP (2007) Localized surface plasmon resonance spectroscopy and sensing. In: Leone S, Groves J, Ismagilov R, Richmond G (eds) Annual review of physical chemistry, vol 58, pp 267–297
23. Link S, EL-Sayed MA (2000) Shape and size dependence of radiative, non-radiative and photothermal properties of gold nano-crystals. Int Rev Phys Chem 19:409–453
24. Sreekumar TV, Das A, Chandra L, Srivastava A, Rao BKU (2009) Inherently coloured antimicrobial fibres employing silver nanoparticles. J Biomed Nanotechnol 5(1):115–120
25. Unser S, Bruzas I, He J, Sagle L (2015) Localized surface plasmon resonance biosensing: current challenges and approaches. Sensors 15:15684–15716
26. Tang B, Wang J, Tafrin S, Lsunand W, Wang X (2011) Application of anisotropic silver nanoparticles: multifunctionalization of wool fabric. J Colloid Interface Sci 356:513–518
27. Tang B, Li J, Hou X, Afrin T, Sun L, Wang X (2013) Colorful and antibacterial silk fiber from anisotropic silver nanoparticles. Ind Eng Chem Res 52:4556–4563
28. Tang B, Zhang M, Hou X, Li J, Sun L, Wang X (2012) Coloration of cotton fibres with anisotropic silver nanoparticles. Ind Eng Chem Res 51:12807–12813
29. Zheng Y, Xiao M, Jiang S, Ding F, Wang J (2013) Coating fabrics with gold nanorods for colouring, UV-protection, and antibacterial functions. Nanoscale 5:788–795
30. Fu Y, Tippets CA, Donev EU, Lopez R (2016) Structural colors: from natural to artificial systems. Wiley Interdisc Rev Nanomed Nanobiotechnol 8(5):758–775. https://doi.org/10.1002/wnan.1396
31. Kinoshita S, Yoshioka S (2005) Structural colors in nature: the role of regularity and irregularity in the structure. Chem Phys Chem 6:1443–1459
32. Sun J, Bhushan B, Tong J (2013) Structural coloration in nature. RSC Adv 3:14862–14889
33. Zhao Y, Xie Z, Gu H, Zhu C, Gu Z (2012) Bio-inspired variable structural colour materials. Chem Soc Rev 41:3297–3317
34. Gil'Man AB (2003) Low-temperature plasma treatment as an effective method for surface modification of polymeric materials. High Energy Chem 37:17–23
35. Knittel D, Schollmeyer E (2000) Technologies for a new century—Surface modification of fibres. J Text Inst 3:151–165
36. Vallon S, Hofrichter A, Drdvillon B, Klemberg-Sapieha JE, Martinu L, Poncin-Epaillard F (1996) Improvement of the adhesion of silica layers to polypropylene induced by nitrogen plasma treatment. Thin Solid Films 290–291:68–73
37. Raffaele-Addamo A, Selli E, Barni R, Riccardi C, Orsini F, Poletti G, Meda L, Massafra MR, Marcandalli B (2006) Cold-plasma induced modification of the dyeing properties of poly (ethylene terephthalate) fibres. Appl Surf Sci 252:2265–2275
38. Hegemann D (2006) Plasma polymerization and its application in textiles. Indian J Fibre Text Res 31:99–115
39. Hegemann D, Hossain MM, Balazs DJ (2007) Nanostructured plasma coatings to obtain multifunctional textile surfaces. Prog Org Coat 58:237–240
40. Kan CW, Yuen CWM (2006) Dyeing behaviour of low temperature plasma treated wool. Plasma Processes Polym 3(8):627–635
41. Kan CW, Chan K, Yuen CWM, Miao MH (1998) Plasma modification on wool fibre: effect on the dyeing properties. J Soc Dyers Colorists 114(2):61–77
42. Kan CW, Chan K, Yuen CWM, Miao MH (1998) Surface properties of low-temperature plasma treated wool fabrics. J Mater Process Technol 83:180–184
43. Kan CW, Chan K, Yuen CWM, Miao MH (1998) The effect of low temperature plasma on the chrome dyeing of wool fibre. J Mater Process Technol 82:122–126
44. El-Zawahry MM, Ibrahim NA, Eid MA (2006) The impact of nitrogen plasma treatment upon the physical-chemical and dyeing properties of wool fabric. Polym Plastics Technol Eng 45(10):1123–1132
45. Hauser PJ (2000) Reducing pollution and energy requirements in cotton dyeing. Text Chem Color Am Dyest Rep 32(6):44–48
46. O¨zdogan E, Saber R, Ayhan H, Seventekin N (2002) A new approach for dyeability of cotton fabrics by different plasma polymerization methods. Color Technol 118(3):100–103

47. Bashari A, Shakeri M (2022) Nanotechnology enabled approaches for textile applications, In: Yusuf M, Shahid M (eds) Emerging technologies for textile coloration, CRC Press
48. Gong J, Liu J, Tan X, Li Z, Li Q, Zhang J (2019) Bio-preparation and regulation of pyrrole structure nano-pigment based on biomimetic membrane. Nanomaterials 9:114
49. Maity S, Roy S, Banerjee AN (2022) Nanotechnology in textiles. In: Ahmed W, Nourafkan E (eds) Science and applications of nanoparticles. Jenny Stanford Publishing Pte. Ltd., USA, ISBN 978-981-4267-34-2
50. Fan Q, Ugbolue SC, Wilson AR, Dar YS, Yang Y (2003) Nanoclay modified polypropylene dyeable with acid and disperse dyes. AATCC Rev 3:25–28
51. Dar YS, Fan D, Ugbolue SC, Wilson AR, Yang Y (2006) NSTI nanotechnology conference and trade show, In: NSTI nanotech 2006 technical proceedings, vol 2, pp 795–798
52. Razafimahefa L, Chlebicki S, Vroman I, Devaux E (2005) Effect of nanoclay on the dyeing ability of PA6 nanocomposite fibres. Dyes Pigm 66:55–60
53. Roessler A, Jin X (2003) State of the art technologies and new electrochemical methods for the reduction of vat dyes. Dyes Pigm 59(3):223–235
54. Yang Y, Gu H (2006) Superfine structure, physical properties and dyeability of alkaline hydrolyzed soly (ethylene terephthalate)/silica nanocomposite fibers prepared by *in situ* polymerization. J Appl Polym Sci 102:3691–3697
55. Zollinger H (1987) Color chemistry: synthesis, properties and applications of organic dyes and pigments. Verlag GmbH & Co, Weinheim
56. Kissa E (1990) Partitioning and stability of aqueous dispersions: particle size of dye dispersions. Langmuir 6(2):478–481
57. Dilling P (1988) Effect of cation type on lignosulfonate dispersant performance in disperse dye systems. Text Chem Colorists 20(5):17–22
58. Yeap KL, Teng TT, Poh BT, Morad N, Lee KE (2014) Preparation and characterization of coagulation/flocculation behavior of a novel inorganic–organic hybrid polymer for reactive and disperse dyes removal. Chem Eng J 243(5):305–314
59. Uddin MZ, Watanabe M, Shirai H, Hirai T (2002) Dyeing conventional and microfiber polyester with disperse dyes. Text Res J 72(1):77
60. Bian H, Luo J, Wang R, Zhou X, Ni S, Shi R, Fang G, Dai H (2019) Recyclable and reusable maleic acid for efficient production of cellulose nanofibrils with stable performance. ACS Sustain Chem Eng 7(24):20022–20031
61. Koh JS, Kim YG, Kim JP (2001) Dyebath reuse in dyeing of nylon microfiber non-woven fabric with 1: 2 metal complex dyes. Fiber Polym 2(1):35–40
62. Lee KS, Lee BS, Park YH, Park YC, Kim YM, Jeong SH, Kim SD (2005) Dyeing properties of nylon 66 nano fiber with high molecular mass acid dyes. Fiber and Polymer 6(1):35–41
63. Choi JH, Kang MJ (2006) Preparation of nano disperse dyes from nanoemulsions and their dyeing properties on ultramicrofiber polyester. Fiber Polym 7(2):169–173
64. Epps HH (2003) Basic principles of textile coloration. Color Res Appl 28:230–231
65. Kamel MM, Mashaly HM (2013) Coloration of fibrin fabric with berberine natural dye with ultrasonic technique using ZnO nano powder as light fastness improving. World Appl Sci J 26(8):994–1001
66. Mashaly HM, Abdelghaffar RA, Kamel MM, Youssef BM (2014) Dyeing of polyester fabric using nano disperse dyes and improving their light fastness using ZnO nano powder. Indian J Sci Technol 7(7):960–967
67. Qin Y, Yuan M, Hu Y, Lu Y, Lin W, Ma Y, Lin X, Wang T (2020) Preparation and interaction mechanism of nano disperse dye using hydroxypropyl sulfonated lignin. Int J Biol Macromol 1(152):280–287
68. Kale RD, Pratap A, Kane PB, Gorade VG (2017) Dyeing of nylon fabric using nanoemulsion. Indian J Sci Res 14(2):257–261
69. Riaz S, Ashraf M, Hussain T, Hussain MT, Rehman A, Javid A, Iqbal K, Basit A, Humera A (2018) Functional finishing and coloration of textiles with nanomaterials. Color Technol 1–20
70. Shah MA, Pirzada BM, Price G, Shibiru AL, Qurashi A (2022) Applications of nanotechnology in smart textile industry: a critical review. J Adv Res 38:55–75

71. https://fashinza.com/brands-and-retail/technology/nanotechnology-in-the-textile-industry/, Nanotechnology in The Textile Industry, Fashinza platform, downloaded on 27.10.2023, at 12.12 pm IST
72. Malik S, Muhammad K, Waheed Y (2023) Nanotechnology: a revolution in modern industry. Molecules 28(2):661
73. Temesgen AG, Turşucular ÖF, Eren R, Ulcay Y (2018) Novel applications of nanotechnology in modification of textile fabrics, properties and apparel. Int J Adv Multi Res 5(12):49–58
74. Krifa M, Prichard C (2020) Nanotechnology in textile and apparel research—An overview of technologies and processes. J Text Inst 111(12):1778–1793
75. Yetisen AK, Qu H, Manbachi A, Butt H, Dokmeci MR, Hinestroza JP, Skorobogatiy M, Khademhosseini A, Yun SH (2016) Nanotechnology in textiles. ACS Nano 10(3):3042–3068
76. Ambaye AD, Kefeni KK, Mishra SB, Nxumalo EN, Ntsendwana B (2021) Recent developments in nanotechnology based printing electrode systems for electrochemical sensors. Talanta 225:121951
77. Zavanelli N, Yeo WH (2021) Advances in screen printing of conductive nanomaterials for stretchable electronics. ACS Omega 6(14):9344–9351
78. https://landanano.com/technology/nanographic-printing/, Nanographic printing, An entirely new category, Landa Digital printing, downloaded on 27.10.2023, 12.11 pm IST
79. Shaheen TI (2022) Nanotechnology for modern textiles: highlights on smart applications. J Text Inst 113(10):2274–2284
80. Gao X, Xiao X, Dong B, Hu J, Tang S, Qi W, Zou D (2019) Converting commercial sewing threads into high-performance and flexible/wearable fiber-shaped supercapacitors via facile vapor deposition polymerization. Chem Sel 4(18):5355–5365
81. Shahzad A, Ali Z, Ali U, Khaliq Z, Zubair M, Kim IS, Hussain T, Khan MQ, Rasheed A, Qadir MB (2019) Development and characterization of conductive ring spun hybrid yarns. J Text Inst 110(1):141–150
82. Liu F, Qiu X, Xu J, Huang J, Chen D, Chen G (2019) High conductivity and transparency of graphene-based conductive ink: Prepared from a multi-component synergistic stabilization method. Prog Org Coat 133:125–130
83. Rosen Y, Marrach R, Gutkin V, Magdassi S (2019) Thin copper flakes for conductive inks prepared by decomposition of copper formate and ultrafine wet milling. Adv Mater Technol 4(1):1800426
84. Zhu C, Guan X, Wang X, Li Y, Chalmers E, Liu X (2019) Mussel inspired flexible, durable & conductive fibers manufacturing for finger-monitoring sensors. Adv Mater Interfaces 6(1):1801547
85. dos Santos J, de Oliveira RS, de Oliveira TV, Velho MC, Konrad MV, da Silva GS, Deon M, Beck RC (2021) 3D printing and nanotechnology: a multiscale alliance in personalized medicine. Adv Func Mater 31(16):2009691
86. Yamamoto BE, Trimble AZ, Minei B, Ghasemi Nejhad MN (2019) Development of multifunctional nanocomposites with 3-D printing additive manufacturing and low graphene loading. J Thermoplast Compos Mater 32(3):383–408
87. Hassan BS, Islam GMN, Haque ANMA (2019) Applications of nanotechnology in textiles: a review. Adv Res Text Eng 4(2):1038
88. Rius J, Manich S, Rodríguez R, Ridao M (2007) Electrical characterization of conductive ink layers on textile fabrics: model and experimental results. In: Proceedings of XXII conference on circuits and integrated systems, Session D (7)
89. Vafaei S, Tuck C, Ashcroft I, Wildman R (2016) Surface microstructuring to modify wettability for 3D printing of nano-filled inks. Chem Eng Res Des 109:414–420

Chapter 6
Nanopigments in Textiles

Kamaladin Gharanjig, Hamid Gharanjig, Hooman Imani, and Siavash Goudarzi

1 Introduction

One of the wonders which Mother Nature gave us humans is to see different shades of color. It gives us an opportunity to see the beauty of our surroundings or to make any object beautiful. Colorants including dyes and pigments are materials which gives a color to a given substrate when added. Plants, creatures, and minerals have been utilized as essential hotspots for colorants since old times. Pigments are the most important substances used to add or change the color of any substance [1, 2].

Pigments are white, black, or colorful chemical compounds. They are insoluble in water, organic solvents or used media and are used as a dispersed particle. Pigments can be categorized into two primary types: synthetic and natural. They are widely employed in various industries such as food, pharmaceuticals, textiles, cosmetics, furniture, fluorescent polymers, and other consumer goods [3, 4]. Although there are several classifications of pigments, these compounds are also divided to two large groups of organic and inorganic pigments.

The nanomaterials that have similar chemistry as in mass structure might have different physio-chemical properties. Materials diminished to the nanoscale can unexpectedly show totally different properties contrasted with the properties they have on a macroscale. As an example, certain substances that are not transparent,

K. Gharanjig (✉) · H. Gharanjig · H. Imani · S. Goudarzi
Departments of Organic Colorants, Institute for Color Science and Technology, PO Box 16765-654, Tehran, Iran
e-mail: gharanjig@icrc.ac.ir

H. Gharanjig
e-mail: gharanjikhamid@gmail.com

H. Imani
e-mail: hooman.imani69@gmail.com

S. Goudarzi
e-mail: goudarzi-s@icrc.ac.ir

S. Bairagi et al. (eds.), *Nanotechnology in Textile Finishing*, Materials Horizons: From Nature to Nanomaterials, https://doi.org/10.1007/978-981-97-2696-7_6

such as copper, can undergo a transformation and become transparent. Similarly, materials that are normally inactive, like platinum, can acquire the ability to catalyze chemical reactions. Stable materials, such as aluminum, can change their properties and become flammable. Solid elements like gold can undergo a change and become fluid at normal room temperature. Additionally, insulating materials, like silicon, can be transformed into conductors [2].

Nanomaterials are often created by top-down techniques, creating extremely small structures from larger particles, such as silicon microchips etched to provide circuits. Bottom-up methods are also possible, which involve building up structures atom by atom or molecule by molecule to obtain nanoparticles. One of the techniques for doing so is self-assembly, where molecules arrange themselves into structures based on their properties. The syntheses of large molecules, as well as the growth of crystals for the semiconductor industry, are examples of self- assembling compounds [2].

Nanopigments are a class of pigments that have been engineered at the nanoscale, typically ranging from 1 to 100 nm in size [5]. Nanopigments have gained a wide range of industrial applications recently, despite the fact that current manufacturing practices demand particle sizes in the 100–200 nm range [6, 7]. For instance, mica-based pigments with a pearlescent effect and a particle size of 20 nm are utilized in polymers, automotive coatings, and beauty products [8]. They offer unique properties and advantages compared to conventional pigments owing to their tiny size and great surface-to-volume ratio. These properties make them highly desirable for a variety of requests in various industries, including paints and coatings, polymers, textiles, cosmetics, and printing inks [7].

The traditional pigment industry has relied on micro-sized particles for coloration and other functionalities. However, the emergence of nanotechnology has opened up new possibilities in pigment design and performance. Nanopigments exhibit enhanced color strength, improved lightfastness, increased stability, and unique optical properties compared to their larger counterparts [6]. Also, some groups of nanopigments enable some special properties such as antimicrobial, self-cleaning, UV protection, and flame retardant. These compounds are known as functional nanopigments. These attributes arise from the quantum confinement and surface effects that occur at the nanoscale, where the behavior of materials deviates from their bulk properties [9, 10].

The synthesis of nanopigments involves the use of different methods and techniques [11]. Physical methods include processes such as precipitation, milling, and grinding, which reduce the particle size of pigments to nanoscale [11, 12]. Chemical procedures, such as sol–gel processes, hydrothermal production, and chemical vapor deposition, offer precise control over the composition and morphology of nanopigments [13]. Biological methods utilize microorganisms or natural extracts to produce bio-inspired pigments, adding a sustainable and environmentally friendly dimension to the field [13–15]. Here are some common types of nanopigments.

Carbon-based Nanopigments: These nanopigments include carbon black, carbon nanotubes, and graphene. These materials can possess excellent stability, high tinting strength, and suitable electrical conductivity. Carbon black is widely used as a black pigment, while carbon nanotubes and graphene offer interesting properties for applications in conductive inks, coatings, and electronics [16–18].

Titanium Dioxide Nanoparticles (TiO_2): TiO_2 is commonly used as white pigments due to their high refractive index and light-scattering properties. It provides excellent opacity to coatings, paints, and cosmetics [19]. TiO_2 nanoparticles can possess photocatalytic properties, enabling self-cleaning surfaces and air purification applications [19, 20].

Iron Oxide Nanoparticles: These nanopigments including magnetite (Fe_3O_4) and hematite (Fe_2O_3) are used as pigments in various colors, including red, yellow, and black [21]. They exhibit strong color saturation and have high stability. Iron oxide nanoparticles find applications in paints, inks, cosmetics, and magnetic storage media [22, 23].

Other Inorganic Nanopigments: There are several other inorganic nanopigments used in different applications. For example, zinc oxide (ZnO) nanoparticles can provide UV protection and are used in sunscreens and coatings [24]. Silica nanoparticles can be used as transparent pigments in paints and coatings, providing scratch resistance and enhancing durability [26].

Organic Nanopigments: These pigments are made from organic molecules or colorants that have been engineered at the nanoscale. They provide intense and vibrant colors and can exhibit improved stability, lightfastness, and durability [26]. Organic nanopigments are used in various applications, including paints, inks, textiles, and plastics [26, 27].

Characterization approaches are essential for comprehending the structure, morphology, and characteristics of nanopigments. Microscopy techniques, such as scanning electron microscopy (SEM) and transmission electron microscopy (TEM), provide visual insights into the size, shape, and distribution of nanoparticles. Analysis instruments, such as UV–V is spectrophotometer and Fourier-transform infrared spectroscopy (FT-IR), help analyze the optical and chemical properties of nanopigments. X-ray diffraction (XRD) allows the evaluation of crystal structure and phase composition, while thermal analysis techniques such as thermogravimetric analysis (TGA) and differential scanning calorimetry (DSC) assess the thermal stability and behavior of nanopigments [28–31].

The applications of nanopigments are vast and diverse. In the paints and coatings industry, nanopigments offer improved color intensity, UV protection, and resistance to fading and weathering. In plastics, they provide vibrant colors, transparency, and enhanced mechanical properties. Nanopigments find use in textiles for coloration and functional finishes, such as antimicrobial, UV protection, fire retardancy, and self-cleaning properties. Additionally, nanopigments are utilized in printing inks to enhance print quality and provide special effects [7, 32–34].

While nanopigments bring numerous advantages, it is essential to consider environmental and safety considerations. Proper handling, storage, and disposal of nanopigments are crucial to prevent potential harm to human health and the environment. Regulations and guidelines are in place to ensure the safe use and implementation of nanopigments [2].

This chapter explains various methods of syntheses of nanopigments including hydro/solvothermal, sol–gel, layer-by-layer deposition, sonosynthesis, microwave aided, biosynthesis, precipitation, emulsion polymerization, microfluidic synthesis, and self-assembly methods. Nanopigment dispersion is an innovative breakthrough in the field of materials science that carries substantial ramifications for multiple sectors and academic areas. Therefore, dispersion of nanopigments in applied media is illustrated, and its effects on properties are shown. Finally, applications of nanopigments in textile industry and their properties are described.

2 Syntheses of Nanopigments

2.1 Synthesis of Inorganic Nanopigments

Nanoparticles are synthesized and applied on textiles using various methodologies, including in situ and ex situ production [35]. Ex situ techniques involve sequential processes, such as synthesis, dispersion, and deposition, which can be time-consuming and energy-intensive. In situ synthesis, on the other hand, involves simultaneous production and deposition, making it faster, more facile, and cost-efficient. This method offers advantages such as reduced time, energy, and chemical consumption, improved durability, and low toxicity. However, limitations include high temperature calcination, extended calcination duration, inadequate deposition, and uncontrolled shape and size in the presence of high precursor content [35]. There exist various techniques for synthesizing inorganic nanopigments, which can be classified as follows.

2.1.1 Hydro/Solvothermal

The solvothermal and hydrothermal processes exhibit considerable similarities. The hydrothermal method refers to a technique employed for the crystallization of a substance under conditions of elevated temperature and vapor pressure, utilizing an aqueous solution containing the material. In contrast, the solvothermal process distinguishes itself by employing an organic solvent in lieu of water. The hydrothermal method capitalizes on the phenomenon that altering temperature and pressure beyond the boiling point of water which can induce significant modifications on its fundamental properties and, consequently, its efficacy as a solvent. Key features, including ionic product density, thermal conductivity, viscosity, heat capacity, and dielectric constant, exhibit remarkable dependence on pressure and temperature [36]. By adjusting the synthesis settings, it is possible to achieve desired solvent properties.

The solvothermal approach often offers superior capabilities in terms of form manipulation, size control, morphology, and crystallinity of nanoparticles compared to the hydrothermal process [37]. The hydro/solvothermal reaction occurs within a pressurized vessel known as autoclave. These are steel vessels that are hermetically sealed, designed to endure prolonged periods of high heat and pressure cycles. Additionally, it is imperative that the vessel exhibits chemical inertness and refrains from participating in the hydrothermal process. Various nanoscale structures of TiO_2 can be synthesized using the hydro/solvothermal technique [36, 37]. In addition to the ex situ synthesis of nanopigments using this method, it is also possible to achieve in situ synthesis. A study has been conducted on the use of silver nanoparticles (AgNP) for the antibacterial finishing of cotton fabric. The findings indicate that the deposition of AgNP and color strength K/S values are higher, and the coloration of the fabric is more uniform and durable after washing, compared to in situ synthesis at room temperature [38].

2.1.2 Sol–Gel

Sol–gel has emerged as a significant technology in the fabric industry due to several reasons, including enhanced effectiveness, reduced costs, decreased chemical consumption, improved functionality and durability of essential qualities, and notably, reduced environmental implications [39]. Sol–gel technique is an aqueous treatment procedure commonly applied for nanomaterial synthesis. In this process, the first step involves dissolving the precursor in a solvent reaction, which can be either water or an organic solvent. This step is crucial and remains consistent regardless of whether the starting material is a metal alkoxide or an inorganic salt. The sol–gel processes encompass five distinct phases, namely hydrolysis, condensation (gelation), gel aging, applications, and curing [40]. This methodology yields materials of exceptional purity, taking on many forms such as powders, thin-film coatings, fibers, monoliths, and self-supported bulk structures. The hydroxyl group, which is formed as a result of the hydrolysis stage, has the ability to form hydrogen and covalent bonds with the functional groups present in fibers [39, 40].

Textile fabrics can typically be treated with sol–gel generated solutions by the utilization of a straightforward pad-dry-cure or dip coating technique [35]. In a study, researchers used a sol–gel synthesis approach to deposit TiO_2 and Au/TiO_2 coatings onto cotton fibers. The findings indicate that titanium dioxide and gold particles exhibit nanoscale dimensions and have the ability to manifest a purple hue when applied to cotton fabric [41]. Additionally, the nanopigments exhibit a high degree of bonding with the fiber and demonstrate exceptional resistance to multiple washing cycles. The cotton fibers coated with Au/TiO_2 exhibit a notable efficacy in the photodegradation of methylene blue dye, hence indicating a substantial potential for photocatalytic self-cleaning attributes when exposed to solar-like light conditions [41]. Another study has investigated the manufacture of ZnO nanoparticles by the sol–gel process and their application on cotton fabric to promote antibacterial finishing. The proposed methodology enables the efficient coating of fibers with

ZnO nanoparticles in a single step, requiring a significantly low temperature of 50 °C. This strategy effectively minimizes the consumption of chemicals, energy, and time, hence enhancing overall use of resources. Following the treatment, the fabric retains its white color. However, reflectance tests conducted on the fabric before and after the use of ZnO nanoparticles indicate a marginal rise in reflectance within the visible spectrum region. This increase can be attributed to the presence of ZnO nanoparticles in the treated fabric. Furthermore, it is worth noting that a reduction in reflectance within the UV region is exclusively observed in the case of ZnO nanoparticles treatment. This outcome is anticipated as a result of the UV absorption characteristics inherent to ZnO material [42].

2.1.3 Layer-By-Layer (LbL) Deposition

The layer-by-Layer (LbL) deposition method is a technique for creating thin composite films on solid surfaces. It involves stepwise deposition of opposite-charge polycations and polyanions, forming polyelectrolyte multilayer films. The process of creation one layer starts with charging a substrate, immersing it in a solution, and then rinsing, and then, the process will be continued to creation of more layer on the surface. This method allows for the creation of bilayers, trilayers, or quadlayers with diverse functionalities, allowing for up to 20 ultrathin layers [43].

The primary driving force behind the assembly of multilayer films is the interaction between charges of opposing polarity, which is predominantly governed by electrostatic forces. However, contemporary investigations have expanded the scope of understanding to include several other interactions such as donor/acceptor interactions, covalent bonds, hydrogen bonding, stereo complexes production, and π–π interactions [44].

Traditional methods for immobilizing nanoparticles onto textile substrates often involve multiple stages and chemical substances, leading to thick, unstable coatings. However, recent research suggests LBL deposition as a chemically gentle alternative for creating ultrathin, transparent, and durable coatings. This method can be applied to fabrics through immersion or vertical or horizontal spraying techniques, enhancing fabric breathability and durability [43].

The technique of LbL deposition has promise in the field of textile finishing since it involves the use of polyelectrolyte solutions with extremely low concentrations of active chemicals. A minimal concentration of a few milligrams per milliliter is sufficient to provide the required qualities. [44].

The application of UV protection and self-cleaning properties on cotton fabric has been effectively achieved by the process of LBL deposition of titanium dioxide. The initial stage of this process involves surface preparation. The cotton surface can be modified by introducing cationic sites through the use of 2,3-epoxypropyltrimethylammonium chloride [45]. This modification prepares the fabric for immersion in a TiO_2 anionic dispersion, facilitating the deposition of the initial layer onto the surface. Following the rinsing process, the subsequent layer will be deposited through the dipping of the fabric in a solution containing TiO_2 cationic dispersion. The deposition process was iterated until several layers of nano-TiO_2 films were successfully coated onto the cotton substrate. Subsequently, the fabric was dried and cured to ensure the stability of the deposited layers. The fabric that has undergone treatment exhibits a notable capacity for shielding against ultraviolet radiation, maintaining this ability even after undergoing 10 cycles of washing. Despite a drop in UV protection factor (UPF) values after 20 washing cycles, the garments still demonstrated exceptional UV protection. The results indicate that the nano-TiO_2 layers, which were created using the described procedure, exhibited a high level of durability even after undergoing multiple washing cycles [45].

In another investigation, researchers utilized zinc oxide nanoparticles that were stabilized within the polymer poly-(diallyldimethylammonium chloride) (PDDA) for the purpose of achieving a self-cleaning property on cotton fabric. This was accomplished by the application of an LbL deposition technique. The methodology employed for surface treatment was the oxidation of the carboxylate group present in cellulose, thereby rendering the cotton fabric suitable for the subsequent deposition of several layers of ZnO-PDDA and polystyrene sulfonate (PSS). The findings indicate that as the number of bilayers increases, there is a corresponding increase in the rate of methylene blue photodegradation. This can be attributed to the larger concentration of ZnO within the cotton fibers. The application of a single layer of AgNP on the treated cotton surface resulted in a rise in photodegradation efficiency, indicating the presence of a synergistic impact between these nanopigments [46].

2.1.4 Sonosynthesis

Ultrasound is a type of mechanical energy that necessitates an intermediate for its transmission and is not readily absorbed by molecules. The phenomenon of acoustic cavitation in liquid media, which involves the development, growth, and subsequent collapse of bubbles, is widely acknowledged to be responsible for the effects of ultrasound. The rapid and forceful implosion of bubbles within a timeframe of less than one microsecond results in the liberation of substantial thermal energy, hence generating transient regions of elevated temperature [47].

Ultrasound irradiation has the potential to induce the degradation of organic molecules, as well as the oxidation and/or reduction of inorganic compounds. Additionally, it can facilitate the sonolysis of water, resulting in the formation of reactive radicals such as hydroxyl radicals ($^{\bullet}OH$) and hydrogen radicals ($H^{\bullet}$). Some of the impacts of ultrasound include the acceleration of chemical processes, enhancement of mass transfer, reduction of reaction cycles, improvement of reaction yield, modification of reaction pathway, increase in surface area between reactants, and acceleration of dissolution [47, 48].

Ultrasound acoustic cavitation, a sonochemistry technique, generates intense, temporary conditions for nanomaterial creation. This generates hot spots, allowing for gentler synthesis under lower temperatures and pressures. This approach offers enhanced yield and reduced time requirements, while reducing nanoparticle size and maintaining durability on textiles through strong adsorption. This method also reduces the need for additional solvents and preparation steps [48].

Multiple investigations have been carried out regarding the utilization of different nanoparticles, such as TiO_2, ZnO, Ag, and CuO, in textile sonosynthesis [48]. In a detailed view, the sonosynthesis of ZnO on dopamine-treated nylon/cotton fabric will be considered. Zinc acetate as precursor would be hydrolyzed in alkali solution to zinc hydroxide. For instance, research has been done on the sonosynthesis of ZnO on nylon/cotton fabric treated with dopamine. In an alkali solution, zinc acetate as a precursor would hydrolyze to produce zinc hydroxide. The decomposition of zinc hydroxide into ZnO nanoparticles can be observed by Eqs. 1 and 2 when subjected to the effects of transient cavitation bubbles.

$$Zn(COOCH_3)_2 \rightarrow 2H_2O + 2Na_2CO_3 \rightarrow Zn(OH)_2 + 2CH_3COONa + 2H_2O \quad (1)$$

$$Zn(OH)_2 \rightarrow^{\text{ultrasound}} ZnO \downarrow + H_2O \quad (2)$$

During the process of ultrasonic irradiation, the momentum carried by sound waves was absorbed by the water, resulting in the formation of circulatory flows within the mixture. The migration of ZnO nanoparticles and $Zn(OH)_2$ toward the nylon/cotton fabric was induced by the fluxes. In addition, various functional groups, including hydroxyl, ketone, and amide, were seen to form robust connections with ZnO nanoparticles when subjected to ultrasound at a temperature of 70 °C [49].

2.1.5 Microwave Assisted

Traditional heating techniques, such as oil baths and furnaces, facilitate heat transfer by means of convective thermal gradients, which enable the movement of heat from the surface to the interior of the material or solution. Consequently, the elevated temperature of the container's surface relative to other components within the system results in nonuniform heating. This non-homogeneous heating can lead to the decomposition of products or reagents, as well as the formation of undesirable by-products [50]. These effects ultimately reduce the overall reaction yield and necessitate additional purification procedures. Microwaves, classified as electromagnetic waves with wavelengths ranging from 1 to 1 mm, induce heating by two primary mechanisms: dipolar polarization and conduction [50, 51]. The electric component of the electromagnetic field is responsible for this heating phenomenon. The utilization of microwave-assisted synthesis in nanoparticle synthesis has numerous benefits, including increased reaction rate, improved yield, reduced reaction time, decreased particle size, narrowed particle size dispersion, better purity of materials, and improved physicochemical attributes [51].

Using of microwaves and ultrasounds for synergistic synthesis is possible. Microwaves and ultrasounds exhibit distinct characteristics, although their combined utilization enables the substantial optimization of chemical reaction parameters. The integration of microwaves and ultrasound has emerged as a viable opportunity for efficient, cost-effective, and environmentally friendly synthetic processes. The positive outcomes can be attributed to the amalgamation of two variables, namely the distinctive process of microwave heating and the occurrence of cavitation. The utilization of microwave technology is commonly employed in conjunction with efficient heating methods, while the incorporation of sonification enhances the efficacy of temperature-induced stimulation in chemical processes [52].

Based on what was mentioned, a study has been conducted on the immobilization of ZnO nanoparticles onto cotton fabric using a microwave-ultrasonic synergistic technique. The antibacterial activity and durability of zinc oxide coated cotton fabrics were found to be higher in products prepared using a combined microwave-ultrasonic method, as compared to those prepared using individual in situ methods of ultrasonic or microwave irradiation. This can be attributed to the superior characteristics of the ZnO nanoparticles in the former, including small size, excellent purity, strong adhesive force, identical morphology and great loading capacity [53].

2.1.6 Biosynthesis

Chemical methods employed in nanoparticle synthesis exhibit several restrictions that hinder their potential for biological applications [54]. These limitations include the utilization of substantial quantities of toxic and costly precursors, as well as the generation of undesirable by-products. The advancement of clean, non-toxic, and environmentally sustainable processes is crucial for enhancing the production and fabrication of nanoparticles. This may involve the application of various species, including bacteria, fungi, and plants, to build eco-friendly methods. Therefore, it is well-established that both unicellular and multicellular organisms have the capability to generate inorganic materials either within their cells or outside of them [55].

The fundamental process of nanoparticle formation is essentially consistent for both microbes and plants. Metal salts, which consist of metal ions, are initially converted into individual atoms through the use of a reducing agent. The atoms acquired thereafter undergo nucleation, forming tiny clusters that subsequently develop into particles [55, 56].

Several investigations have been conducted on the creation of nanoparticles using the in situ approach. Silver nanoparticles were generated on cotton by use of ashes from the burning of leaves and stems of the Seidlitzia rosmarinus woody, Keliab in Iran [57]. Keliab has a significant concentration of sodium and potassium carbonate, allowing for the generation of an alkaline environment. The biogenesis of silver nanoparticles occurs when a solution containing $AgNO_3$ and cotton is introduced into an alkaline solution with Keliab. Cellulose, a polyhydric alcohol molecule, can undergo oxidation in an alkaline environment, resulting in the development of oxycellulose, which in turn leads to the generation of ketone, aldehyde, or carboxylic groups [58].

Silver nitrate undergoes oxidation to form silver oxide (Ag_2O), which dissociates into Ag^+ ions as the temperature rises. The generation of electrons resulting from the oxidation of cellulose. The redox processes under consideration are complex, involving simultaneous oxidation reactions and reduction events. In an alternative process, cellulose undergoes a transformation in an alkaline environment, resulting in the formation of a cellulosate anion. The presence of positively charged silver ions facilitates their adsorption onto cellulose fibers by electrostatic attraction, resulting in the formation of Ag nanoparticles. The process of nucleation for silver nanoparticles commences at this stage, and their growth occurs without any steric hindrance caused by polymeric chains [57].

2.2 *Synthesis of Organic Nanopigments*

The process of synthesizing organic nanopigments entails the fabrication of nanoscale particles utilizing organic chemicals as the fundamental constituents. Various techniques are employed to accomplish the synthesis of these pigments, each presenting distinct benefits and enabling precise manipulation of the ultimate particle dimensions and structure [11]. Below are many frequently employed techniques.

2.2.1 Precipitation Method

The precipitation method is a frequently employed technique in the synthesis of organic nanopigments [59]. This process entails the deliberate induction of precipitation of organic molecules or polymers from a solvent. In general, the process involves the combination of a solution comprising organic precursors with a non-solvent or antisolvent, resulting in the generation of nanoparticles with dimensions on the nanoscale. The formed precipitates are subsequently gathered, rinsed, and subjected to a drying process in order to acquire the organic nanopigments. The manipulation of particle size and morphology can be achieved through the adjustment of many parameters, including the choice of solvent, temperature, and reaction conditions [59].

2.2.2 Emulsion Polymerization

Emulsion polymerization is a widely employed methodology applied in the production of organic nanopigments derived from polymer components. The process entails the dispersion of droplets of monomers within an aqueous phase that contains surfactants and initiators. Subsequent to the monomer units, polymerization occurs, resulting in the generation of polymer particles at the nanoscale. The organic nanopigments obtained can undergo additional modifications through the inclusion of dyes or pigments into the polymer matrix during the polymerization procedure, thereby imparting them with distinct color characteristics [60].

2.2.3 Microfluidic Synthesis

The application of microfluidic synthesis represents a very adaptable approach in the production of organic nanopigments, offering meticulous regulation of both particle dimensions and morphology [61]. The methodology encompasses the utilization of microfluidic devices, comprising of diminutive channels and chambers, for the execution of reactions at the microscale magnitude. The organic precursors are delivered into the microfluidic channels, where they undergo controlled mixing and reaction to generate nanosized particles. Microfluidic synthesis offers several advantages, including precise regulation of particle size, a restricted range of size distribution, and the capability to fabricate intricate nanostructures [62, 63].

2.2.4 Self-Assembly Method

The employment of molecular self-organization is an approach used in the production of organic nanopigments, sometimes referred to as self-assembly. The approach entails the conceptualization and fabrication of organic molecules or polymers possessing distinct functional groups, which possess the capability to interact and self-assemble into nanostructures in a spontaneous manner. The operation of self-assembly can be initiated by a range of parameters, including temperature, solvent composition, and pH levels. Nanosized pigments with regulated size, shape, and arrangement can be achieved through the process of self-assembly [64].

2.2.5 Sol–Gel Method

The sol–gel technique encompasses the process of hydrolyzing and condensing precursor molecules in order to generate a gel. Subsequently, the gel undergoes a processing procedure aimed at eliminating the solvent and acquiring particles with dimensions on the nanoscale. The approach frequently employs organic precursors, specifically organosilanes, to synthesize nanopigments that exhibit a hybrid organic–inorganic nature [65, 66].

3 Nanopigment Dispersion and Stabilization

With broad ramifications for numerous sectors and academic fields, nanopigment dispersion is a ground-breaking development in materials science. Its significance stems primarily from the exceptional capacity of nanoparticles to dramatically improve the optical and mechanical characteristics of textiles, notably in the context of color. A stabilized nanopigment dispersion makes it possible to produce pigment particles at the nanoscale that are evenly dispersed throughout a matrix, producing materials with improved color intensity and durability. This level of precision and control in pigment size distribution is invaluable in fields such as textile printing and coloration, where the application of nanopigment dispersions leads to exceptional color, brilliance, longevity, and possible functional performance. For example, incorporating nanopigments with tailored particle size distribution into textiles provides additional qualities like improved UV resistance, antibacterial properties, and increased mechanical strength. Gaining a desirable nanopigment dispersion and dispersion stability is highly reliant on pigment surface chemistry and the employment of a suitable dispersing agent [67].

The surface chemistry of nanopigments has a significant impact on the dispersion stability of nanopigments in a variety of environments [68]. The nature of the surface functional groups or chemical alterations that nanopigments exhibit is an important determinant. The affinity of nanopigments for the dispersion medium, which might include solvents, polymers, or other carrier molecules, is determined by surface chemistry. Functional groups, such as hydrophilic or hydrophobic moieties, can either promote or inhibit nanopigment dispersibility in a given medium. Because of their affinity for water, hydrophilic nanopigments have improved dispersibility in aqueous solutions, but hydrophobic nanopigments may require surface changes to improve dispersion stability in such settings [68, 69]. Furthermore, surface chemistry influences pigment and dispersant interaction, which can stimulate or prevent aggregation, which is a critical factor impacting dispersion stability. Understanding and tailoring the surface chemistry of nanopigments is therefore critical for achieving the desired dispersion characteristics.

On the other side, to achieve finely dispersed and stable colloidal dispersion of nanopigments, the employment of an appropriate dispersant is essential. Dispersants are able to create stabilization through various mechanisms primarily electrostatic, steric hindrance, and electrosteric interactions [68, 70]. A desirable dispersion stability can be produced using a polymeric dispersant where electrostatic interactions can be formed through anionic functional groups on the polymer chain and steric hindrance which is provided by the thickness of adsorbed polymer onto the pigment surface. Contrary to surfactants, a polymeric dispersion can be constructed with anchor chemical groups that efficiently link to the pigment surface. Additionally, polymeric dispersants containing higher number of anchor groups has a considerable benefit over traditional wetting agents in terms of extending and effectively stabilizing dispersion [70].

With increasing particle size, the thickness of the adsorbed layer required for efficient dispersion stability increases [71]. In other words, to achieve adequate stability, dispersions containing organic or inorganic pigment particles with sizes ranging from 200 to 1000 nm are predicted to require an adsorbed layer of 10 to 20 nm. As a result, stabilization of nanoscale pigments with polymeric dispersants results in relatively large diameters, requiring only a lower concentration colloid; as a result, lower molecular weight dispersants are preferred for nanopigment dispersion and stabilization [71]. The importance of effective pigment dispersion as well as the incorporation of sustainable dispersants has recently been conducted for nanopigments application in textiles which will be covered in the rest of this part.

To understand the importance of particle size distribution and suitable dispersion process on the application of some novel nanopigments, examples for a hybrid nanopigment and two organic nanopigments are presented. Rhodamine B/montmorillonite hybrid nanopigments were dispersed in a polypropylene matrix, and the dispersion quality was investigated [72]. The nanopigment was synthesized in a way that the cationic rhodamine B dye was intercalated into sodium ions of montmorillonite, resulting in enhanced UV stability of colorants. The influence of nanopigment dispersion on photostability was investigated. The light stability results showed that adequate mixing improved the photostability of the nanopigment even more. Following the proper mixing process, the color difference was reduced from 6 to 2.6 [72]. The increased light stability is primarily due to the achievement of finer particle size distribution of nanopigments, as well as further intercalation by polymeric media and increased specific surface area of pigments. Consequently, it resulted in further blocking UV rays and providing additional protection to the dye molecules within the pigment particles present in the inner layers.

In another investigation, nanosized benzopyran coloring agents exhibited excellent color fastness, strong ecological performance, enhanced color performance, and non-selectivity to various fibers. The nanopigments of benzopyran derivatives were produced in an environmentally friendly manner, with average particle sizes ranging from 25 to 57 nm [73]. It permits a broad market for textile products and offers strong technical support in order to revolutionize the traditional printing and dyeing sector with cutting-edge technology. To highlight the impact of pigment size on color strength and color fastness, the color strength of pigment-printed polyester, linen, and polyester/cotton fabrics utilizing the synthesized pigment was tested. The results show that lowering particle size across all pigments leads to a considerable rise in K/S values. Furthermore, all printed textiles have demonstrated good to very good fastness to washing, rubbing, light, and perspiration [73]. The size control and decrement was only achievable after efficient dispersion process.

Moreover, the pH impact on nanopigment dispersion was explored as an applied example of the environmental implications of the dispersion process. Nanopigments were synthesized from vat dyes and utilized in printing cotton garments in this study [74]. To do this, the vat dye was first transformed into the soluble leuco form by utilizing various reducing agents such as β-cyclodextrin, glucose, and so on. The leuco form was then subjected to a rapid oxidation process to produce nanosized pigment particles. A homogeneous morphology with a fairly uniform particle size distribution in the region of 6–16 nm was attained under these conditions. The pH of the nanopigment dispersion, according to the authors, has a considerable effect on the color strength values of the printed materials. The highest K/S values have been obtained for printed cotton fabrics using nanopigment dispersion of pH 7 with or without binder in the printing ink, regardless of the type of reducing agent utilized or the manner of fixation used. The nanopigment dispersion's increased color strength at pH 7 may be related to the nanopigment's lower particle size distribution. At pH 7, the particle size diameters of the nanopigment were in the range of 6–27 nm.

As was already noted, dispersing agents with lower molecular weights are more suited for stable nanopigment dispersion and have the least effect on nanopigment size increment. Polyoxyethylene lauryl ether (PLE) was utilized as dispersant for C.I. Pigment Red 112 and C.I. Pigment Blue 15:3 [75]. PLE is a nonionic, high HLB, ethoxylated ether of lauryl alcohol that has a broad range of pH stability. Excellent solubility, low foaming, and chemical stability are the hallmarks of this nonionic surfactants. It is widely utilized in many sectors and has a good biodegradable quality. It was employed in the textile industry for sizing and as an emulsion stabilizer. Authors reported that pigment dispersions with particle diameters between 150 and 200 nm could be achieved using this dispersant. For the obtained dispersion, zeta potential ranged from −24 to −32 mV, indicating a significant surface charge interaction between dispersant and pigments. The stability of the dispersion over high-speed centrifugation and freeze–thaw treatment was then thoroughly examined which demonstrated equal performance of the dispersant for both pigments. A weakly stabilized dispersion normally flocculates throughout the heat cycles, increasing the particle size [75].

For printing ink applications in aqueous systems, organic nanopigments can work just as well as dyes in terms of color stability and permanence. However, the limited stability of pigment suspensions for printing applications is the fundamental problem. They have a tendency to agglomerate or aggregate into larger particles with time due to their hydrophobic nature or high surface energy. The dispersion stability of Laponite RD-methylene blue hybrid nanopigments was studied in presence of poly (sodium-4-styrene sulfonate) and poly (diallyldimethylammoniumchloride) as anionic and cationic dispersants, respectively [76]. Laponite RD is a trade name for a synthetic layered silicate, tri-octahedral smectites-type mineral with a cation exchange capacity. Zeta potential evaluations confirmed the adsorption of polymeric dispersant on to the nanopigments where the reported values for zeta potential were around −48 and + 78 mV for nanopigment dispersion containing anionic and cationic dispersants, respectively.

The reported particle size distribution results also demonstrate that the polymeric dispersants should be used in an optimized concentration. In terms of particle size, when the colloidal dispersion is prepared without any polymeric dispersant, the reported average particle size is 99.6 nm which is ideal for a colloidal nanopigment dispersion. However, over a short period of time, the particles will gradually agglomerate, and the dispersion will eventually sediment. When an inadequate quantity of dispersant is used, it was unable to coat all nanoparticles. This can potentially end up an attraction between the oppositely charged particles which has led to agglomeration into larger particles with an average particle size above 1000 nm. On the other hand, when over concentrated dispersant is added, even though the dispersion is stable, the excess of polymer tends to work as a flocculent and force nanopigment agglomeration, resulting a particle size diameter above 1000 nm. Nevertheless, results indicate that at optimized dispersant concentration, less agglomeration has occurred where average particle size is 42.8 nm, with 97% of the particles within 40–50 nm [76].

The utilization of eco-friendly or sustainable dispersing agents in textile applications holds significant importance in the context of contemporary textile industry practices. Traditional dispersing agents, often based on non-renewable resources and containing environmentally harmful components, raise concerns about their ecological impact. Eco-friendly dispersing agents, on the other hand, are derived from renewable sources and formulated to minimize adverse environmental effects. They not only aid in achieving vibrant and consistent coloration but also align with the principles of sustainability and environmental responsibility. These agents reduce the ecological footprint of textile processes by mitigating water pollution, minimizing energy consumption, and decreasing chemical waste.

Among the variety of biopolymers including proteins and polysaccharids, cellulase was suggested as an effective dispersing agent for indigo nanopigments and coloration of cotton fabrics [77]. Cellulase is one of the most widely used protein in the textile industry because it may be employed in a sustainable manner to enhance the characteristics of cellulosic materials without causing significant damages [77]. Due to the hydrophobic and hydrophilic regions of cellulase molecules being completely distinct, it has been claimed that the protein exhibits a special and inherent colloidal behavior akin to dispersants [78]. Therefore, nanocolloidal dispersion of indigo with an average particle size of 210 nm was prepared for cotton coloration. The zeta potential tests demonstrated that the amphoteric nature of cellulase proteins allowed for the adjustment of pigment surface charges as well as the pH characteristics of the surrounding media. The result of dispersion stability in high-speed centrifuge, freeze–thaw treatment, and ion intensity showed great application of cellulase as an effective sustainable dispersing agent. TGA graphs evidenced that cellulase proteins coated the pigment particles, and XRD analysis confirmed that the combined effects of cellulase and ultrasonic power did not result in the change of the pigment's crystal structures. The practical test of textile coloration indicated that the color strength of cotton fabrics is higher at pHs lower than 4 due to the potential established charge attractions between cotton surface and the well-dispersed nanopigment at this condition.

In another investigation to diminish the environmental impacts of textile production especially during the finishing process, a more sustainable polymeric dispersant is recommended for nanopigments [79]. The new dispersant was synthesized by using alkali lignin as a bio-based raw material. To this end, sulfomethylated lignin is initially synthesized, and it is then grafted with pyrrolidone. Pyrrolidone is mainly responsible for effective dispersant adsorption on to the pigment surface. Both pyrrolidone-free and pyrrolidone-containing polymers were used as dispersants to understand the importance of pigment affinitive functional group of this polymeric dispersant. The average particle size reported for cobalt blue nanopigment pastes were 52 and 82 nm for grafted and non-grafted dispersants. After 7 days of thermal storage, the average mean diameter of pigments almost remained unaffected for the grafted polymer, whereas for the initial polymer with no pigment affinity pyrrolidone, this value increased up to 285 nm along with pigment sedimentation. The greater efficiency of the grafted polymer as a dispersant is also understandable according to the dispersant adsorption isotherms where higher concentrations of dispersants are available on the nanopigment surface (Fig. 1) [79].

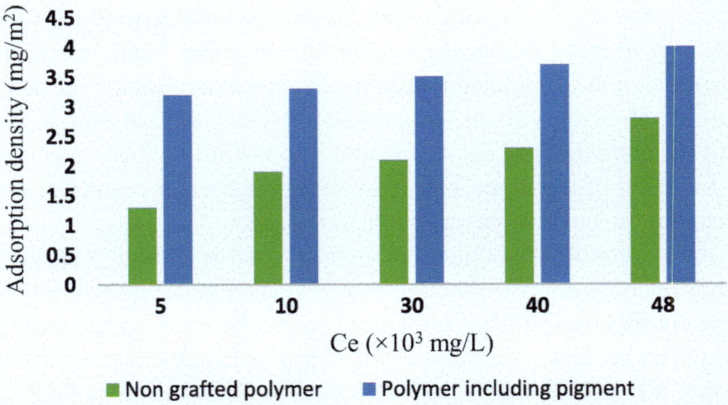

Fig. 1 Polymer adsorption curves of grafted and non-grafted dispersants on nanopigment surface [79]

4 Optical Properties of Nanopigments

Broadly speaking, pigments can be categorized into two primary groups: inorganic pigments and organic pigments. In general, it can be observed that inorganic pigments possess high refractive indices, resulting in their characteristic high opacity when applied. Conversely, organic pigments are characterized by low refractive indices, hence imparting transparency to the materials in which they are utilized [80]. Nevertheless, these substances exhibit a notable tinctorial strength and possess commendable brightness. Changing the particle size and shape can modify the opacity grades of organic and inorganic pigments. For example, it is possible to produce transparency in specific inorganic pigments, such as iron oxides, through the manipulation of their particle size and shape, hence enhancing this particular characteristic [80].

The optical characteristics of nanopigments are influenced by various parameters, including but not limited to their size, chromophore composition, crystal structure, refractive index, concentration, and organic or inorganic nature. The optical properties of nanopigments can be influenced significantly by their size, making it a crucial component for their classification. This criterion can be succinctly summarized as follows [81]:

- Nanoparticles within the size range of 30–100 nm exhibit notable characteristics such as exceptional transparency, minimal scattering, and significant absorption capabilities.
- Particles within the size range of 100–800 nm exhibit an escalating level of scattering, reaching a peak within this range.
- Nanoparticles within the size range of 300–700 nm exhibit moderate absorption characteristics and possess a high degree of opaqueness.
- Particles within the wavelength range of 800 nm to 2 μm exhibit characteristics such as reduced scattering, minimal absorption, and moderate transparency.

A mathematical theory exists that can expect the spectral characteristics of monodispersed nanopigments. These pigments have varying sizes, refractive indices, and volume fractions and are incorporated in a non-absorbing resin with a refractive index of 1.54. The theory applies to the ultraviolet to near-infrared areas of spectrum. Furthermore, it demonstrates how these features can influence the optical performance of a layer containing these pigments. The results of the calculations indicate that nanopigments characterized by their small size and index of refraction and exhibit a high degree of isotropic light scattering [82]. Anisotropic scattering is detected when particle size and refractive index increase. The dependence of reflectance, transmittance, and absorption of a nanopigmented layer on particle size, refractive index, and concentration has been demonstrated [82]. Furthermore, the modification of these parameters has the potential to influence both the color properties and the positioning of the pigmented layer inside the CIEL*a*b* color gamut. Additionally, it can also impact the levels of dullness and brightness in various manners [82].

Inorganic nanopigments can be categorized into two distinct types based on their optical properties: semiconductor and metallic nanoparticles. The significance of the nanoscale in semiconductors arises from the phenomenon of quantum confinement, which gives rise to quantum dot materials. When the size of the particle reductions below the Bohr radius of the semiconductor, which can range from 1 nm to over 100 nm depending on the specific material, the electron experiences increased confinement within the particle [83, 84]. Confinement refers to the deliberate limitation of the movement of electrons, which are inherently characterized by random mobility, to discrete energy levels. When the size of particles is reduced to the nanoscale, the confinement of dimensions results in the discretization of energy levels and a rise in the bandgap of the material [85]. For example, spherical nanoparticles of CdSe with varying sizes have been synthesized and reported to exhibit a spectral shift across the whole visible region. This shift ranges from red emission for the biggest particles to blue emission for the tiniest clusters [84].

Conversely, surface plasmon resonance (SPR) is the mechanism for the optical characteristics of metallic nanoparticles. SPR is a phenomenon where the free electrons in metal nanoparticles, such as in silver and gold, move unrestricted across the material. The mean free path in these materials is around 50 nm, preventing scattering phenomena from the bulk. When the wavelength of light exceeds the nanoparticle's size, standing resonance conditions can be established. This oscillation pattern, known as a standing wave, is determined by absorption and scattering spectroscopy, which depends on some parameters of the nanoparticle such as shape, size, and dielectric constants [84].

The phenomenon of surface plasmon resonance is not observed in isolated atoms or in bulk materials. According to the proposed theory, it is generally seen that the absorption peak of surface plasmon resonance undergoes a red shift toward longer wavelengths as the particle size increases [86].

The color of an organic nanopigment is primarily determined by its molecular composition, although two additional characteristics, namely size and shape distribution, also contribute to their color. The optical characteristics of organic nanomaterials were discovered to exhibit fundamental distinctions from their inorganic counterparts. This disparity arises due to the inherent nature of intermolecular interactions in organic compounds, which mostly consist of weak forces such as hydrogen bonding, π-π stacking, and van der Waals interactions. Several reports have been published regarding the impact of the optical characteristics of pyrazoline compounds [87–89]. In a previous study, the synthesis of 1-phenyl-3-((dimethylamino)-styryl)-5-((dimethylamino) phenyl)-2-pyrazoline (PDDP) was carried out to produce an organic nanomaterial [88]. The observed bathochromic effect was postulated to arise from the enhanced overlap of the π-orbitals of the pyrazoline ring and the intermolecular contacts between PDDP molecules, which occurred as a consequence of the increase in nanoparticle size. The absorption transitions of a different pyrazoline molecule, namely 1,3-diphenyl-5-(2-anthryl)-2-pyrazoline (DAP) nanoparticles, were seen to undergo a bathochromic shift toward lower energy levels as the particle size increased [89]. This shift can be attributed to the enhanced intermolecular interactions occurring within the nanoparticles. The cause of the hypsochromic effect observed in the emission of DAP nanoparticles was determined to be the significant reduction in the Stokes shift resulting from the inhibition of vibrionic relaxation and the rearrangement of molecular configuration caused by enhanced intermolecular contact [89].

5 Nanopigments in Development of Functional and Smart Textiles

Functional textiles are specialized textile materials or fabrics that are developed and manufactured to carry out specific functions in addition to their standard function of offering protection or coverage. To provide the fabric particular qualities or capabilities, these textiles are made by integrating various additives, coatings, or innovative production procedures. Numerous industries, including but not limited to healthcare, sports, defense, and environmental protection, use functional textiles for a variety of purposes [90]. Functional textiles can exhibit a variety of functional attributes, which may include antimicrobial properties, breathability, thermal regulation, UV protection, fire resistance, water repellent/proofing, enhanced durability, and environmental sustainability [90, 91].

On the other side, smart textiles are cutting-edge components and textile structures that integrate electronic parts, sensors, actuators, or other functional aspects to offer distinctive and dynamic capabilities beyond the conventional functions of textiles. The ability to sense, evaluate information, and react to environmental changes or external stimuli enables smart textiles to adapt, communicate, or carry out specific tasks [92]. Key characteristics and attributes of smart textiles include integration of electronics, sensing capabilities, responsive activity, and communication. Smart textiles find applications in diverse fields, including healthcare (e.g., wearable health monitors), sports and fitness (e.g., performance tracking garments), fashion (e.g.,

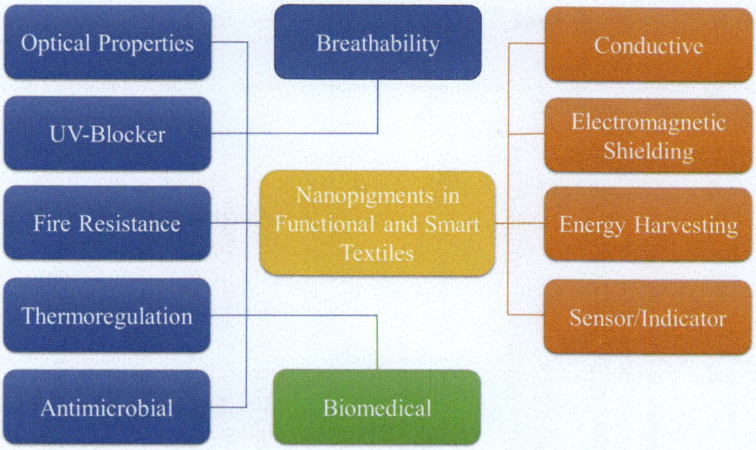

Fig. 2 Application of nanopigments in the functional and smart textiles

color-changing fabrics), military and defense (e.g., camouflage textiles), and automotive (e.g., sensor-integrated car seats) [92, 93]. Some of the major application of nanopigments in development of functional and smart textiles is shown in Fig. 2.

Nanotechnology has been probably one of the most significant drivers in the advancement of recently developed functional and smart textiles. Additionally, nanotechnology enabled the development of novel nanopigments which could potentially offer functional properties for polymers and textiles. The development of functional textiles using nanopigments has had a number of important effects and benefits, advancing textile technology and improving the performance and value of textiles in a variety of applications. Some of the most significant and recent impacts of nanopigments on the development of functional textiles is reported in Table 1. This section will provide an extensive review of the utilization, importance, and application of nanopigments in the development of smart and functional textiles.

5.1 *Nanopigments as an Effective UV-Blocker for Textiles*

Textile fabric materials are susceptible to harmful ultraviolet rays, making them essential for protecting against their harmful effects. UV radiation can be categorized into three zones: UV-C (200–280 nm), UV-B (280–315 nm), and UV-A (315–400 nm). The ozone layer effectively absorbs UV-C, limiting its harmful effects on human health. The stratospheric layer acts as a barrier, limiting UV-C's impact. However, UV-B rays penetrate the Earth's surface due to ozone depletion. Currently, 90–95% of UV-A rays traverse the Earth's atmosphere without any hindrance due to their longer wavelength. The Ultraviolet protection factor (UPF) is a metric used

Table 1 Recent novel application of nanopigments in development of functional and smart textiles

Nanopigment	Textile	Functional properties	Key points	References
Nano-encapsulated dye	Leather	UV blocking	• Nano-encapsulated bis-benzoxazole based fluorescent dyes in polymethylmethacrylate was used	[94]
Titanium dioxide nanoparticles (TiO_2 NPs)	Polyacrylonitrile/polyurethane membranes	UV blocking waterproof breathability	• It can be employed in development of protective clothing, self-cleaning materials, and medical application • TiO_2 NPs were effective in reflecting UV rays, and no UV absorption was reported	[95]
TiO_2 NPs ZnO NPs CuO NPs	Cotton and wool fabrics	UV blocking Antimicrobial activity Self-cleaning	• UV blocking performance CuO > TiO_2 > ZnO • TiO_2 NPs were the most effective antibacterial agent on cotton and wool	[96]
TiO_2 NPs ZnO NPs TiO_2 NPs	Cotton/Polyester fabrics	UV blocking Antimicrobial activity Self-cleaning	• Antibacterial performance TiO_2 > ZnO > SiO_2 • UV blocking performance TiO_2 > ZnO > SiO_2 • Self-cleaning capacity TiO_2 > ZnO > SiO_2	[97]
TiO_2 NPs ZnO NPs	Polyester fabrics	Antibacterial activity UV blocking Self-cleaning	• ZnO NPs have shown greater reduction of bacterial strain cells compared to TiO_2 NPs • Higher UPF was achieved by TiO_2 • Color strength was improved in presence of nanopigments	[98]
ZnO NPs	PET fibers Cotton fabric	Antibacterial, antifungal, and antiviral properties UV blocking IR reflectance	• ZnO NPs were effective against both Gram-negative and Gram-positive bacteria • Nanopigments showed antifungal activity against toxigenic fungi • Improved UPF was recorded on cotton • Fixation of ZnO NP on cotton elevated all functional properties	[99–101]

(continued)

Table 1 (continued)

Nanopigment	Textile	Functional properties	Key points	References
TiO_2 NPs	Polyester Cotton	IR reflectance Thermoregulating	• TiO_2 nanopigments were effective in reflection both visible light and near-infrared rays • Size and concentration of TiO_2 nanopigments are key effective parameters on cooling behavior	[102, 103]
Lanthanide-doped aluminum	Cotton fibers	Superhydrophicity Fluorescent Antibacterial activity	• Coated fabrics showed an unfatigued, reversible, and quick fluorescent photochromism • Can be useful for specialist fields like brand protection, security encoding, and military camouflage	[104]
Anthocyanin hybrid	Biopolyester	pH-sensitive sensor	• Natural anthocyanins were encapsulated in different nanoclays • Thermal and chemical stability of anthocyanin was enhanced after encapsulation	[105]
Organo-clay	Polyester fabrics	Flame retardant Electromagnetic reflection	• Reactive dye covalently linked to nanoclay particles • Dye to nanoclay ratio is a key factor to boost flame-retardant and electromagnetic reflection properties	[106]
Polyethyleneimine–Polypyrrole	Polyethylene fabrics	NIR absorption Antipathogen	• The coated mask fabric with nanopigments is capable of performing infrared red photothermal conversion efficacy • The produced face masks were washable, reusable, breathable, biocompatible, and photothermally conversable for the active eradication of pathogenic bacteria	[107]

to assess protection against UV rays, primarily quantified within the 280–400 nm wavelength range [108, 109].

There exist two primary mechanisms for UV protection. The initial step involves the absorption of ultraviolet irradiation by semiconductor materials, such as TiO_2 and ZnO, which results in the excitation of electrons from the valence band to the conduction band. The second mechanism pertains to the prevention of the transmission of irradiated ultraviolet rays through the surface of the fabric [35]. Further studies have been conducted on employment of TiO_2 nanoparticles and its impact on UV-blocking textiles [95]. To this end, a colloidal dispersion of TiO_2 nanoparticles and polyacrylonitrile/polyurethane polymers were electrospun to fabricate nanofiber membranes. The achieved enhanced UV resistance properties were shown to be more due to the UV reflectance properties of TiO_2 nanoparticles rather than UV absorption. UV reflectance of TiO_2 nanoparticles was claimed due to its high reflective index and no UV-A and UV-B absorption of the TiO_2 loaded membranes.

The multifunctional properties of inorganic nanopigments have also been resulted in another comparative investigation where TiO_2 nanoparticles performance on cotton/polyester mixed fabric was compared to ZnO and SiO_2 nanoparticles [97]. Results indicated that the treatment of cotton/polyester fabrics with these nanopigments effectively caused improved UV blocking, wettability, antibacterial properties, self-cleaning, and durability. In terms of wettability, all the nanopigments showed similar performance, whereas other functional properties significantly varied by changing metal oxide. In this case, the best functional properties were achieved by incorporation of TiO_2 while ZnO outperformed SiO_2 in all functional properties. It has also been suggested that substrate surface modification in order to create carboxylic groups can boost all the mentioned functional properties by providing extra adsorption sites for nanopigments [97].

One issue associated with the application of photocatalysts, such as ZnO and TiO_2 nanoparticles, pertains to their reliance on UV irradiation for activation, which is often limited during daylight hours. In order to address this issue, the introduction of doping in these nanoparticles presents a viable solution. The alteration of the energy band gap through doping results in an expansion of the adsorption spectrum, hence enhancing the range of light absorption. In order to achieve this objective, the synthesis of nano-cerium dioxide/zinc oxide and nano-cerium dioxide/titanium dioxide will be conducted on the surface of cotton fabric. Based on the obtained results, it can be observed that the treated samples exhibit a lower spectrum compared to the raw samples [110]. In alternative terms, the samples containing nanomaterials exhibit a greater degree of UV protection compared to the unprocessed samples. According to the findings, it can be inferred that the sample treated with CeO_2 nanoparticles exhibits a comparatively reduced level of UV irradiation protection when compared to samples treated with CeO_2/ZnO and CeO_2/TiO_2. The reason for this phenomenon can be attributed to the ultraviolet adsorption capacity of TiO_2 and ZnO [110]. The potential combination of metallic and semiconductor nanomaterials could potentially have a synergistic impact in terms of UV protection for cotton fabric. This hypothesis is supported by a study that utilized Ag-deposited ZnO nanomaterials, which demonstrated an improvement in UV protection compared to undeposited ZnO [110].

5.2 Antimicrobial Textiles Fabricated by Nanopigments

Textile fabrics and garments play a crucial role in facilitating the growth of microbes, particularly bacteria [111]. The application of antimicrobial finishes on textiles has become essential in manufacturing various products for protection, decoration, and technical purposes [111, 112]. The effectiveness of antibacterial agents depends on the specific bacterial strain being targeted. Antibacterial agents are classified into two classes: bacteriostatic agents, which inhibit microorganism growth and dissemination, and bactericides, which effectively eradicate microbes by killing them.

Carboxylic peripheral groups in bacterial cell walls create a negative charge, which dissociates when exposed to aqueous solutions. When cationic substances like silver, zinc, and copper are near bacteria, the electrostatic attraction initiates an initial engagement, leading to strong adhesion and robust binding with antibacterial agents. Antibacterial drugs can induce bactericidal effects or imped bacterial growth.

One of the key attributes of the nanopigments is their relatively high antimicrobial activity, with some of them exhibiting antibacterial, antiviral, and antifungal capabilities in textiles [98–100]. Photocatalytic properties of nanomaterials are linked to their antimicrobial efficacy. In the presence of light, photocatalytic nanopigments in water and air undergo a catalytic process, separating into electrons and holes. This results in highly chemically active hydrogen–oxygen free radicals (HO●) and superoxide anion radicals (O_2-), which effectively target and degrade organic compounds in microorganisms. The primary process of photocatalytic antimicrobial reactions involves the chemical oxidation of cell walls and membranes, primarily through reactive oxygen species (ROS) reactions with key cellular constituents like lipids and proteins [113].

Recent research has demonstrated the enduring antimicrobial characteristics of TiO_2/ZnO nanoparticles. The antibacterial properties of towels treated with these nanoparticles was evaluated by testing, revealing a prolonged duration of effectiveness that persisted for up to 30 cycles of washing. The efficacy of microbial growth suppression on textiles under investigation is contingent upon the intensity of UV light, the specific microbe species, and the number of wash cycles performed (15 and 30 iterations). The significance of light intensity in the initiation of photocatalysis in textiles that have been treated with metal nanoparticles is significant [113].

As mentioned earlier, TiO_2 nanoparticles have generally shown superior properties in the development of functional properties in textiles compared to zinc oxide nanoparticles. Although TiO_2 nanoparticles have shown effective antimicrobial properties [96, 97], ZnO nanoparticles have been extensively introduced as a superior antimicrobial nanopigment for functional textiles [98]. In order to examine the antimicrobial effectiveness of these two nanopigments, pathogenic bacterial strains and fungal yeast *Candida albicans* were located in diverse types of polyester fabrics. Gram-positive, gram-negative, and pathogenic yeast are all significantly reduced by the ZnO treated fabrics, with reductions up to 93%. Additionally, the TiO_2 treated sample has shown excellent antimicrobial activity against all pathogenic bacterial strains and yeast, with reductions ranging within 35 to 67% [98].

Zinc oxide nanopigment has gained the attentions as biocides due to their biocompatibility and lack of toxicity for humans as well as their ability to be manufactured in a variety of morphologies, including nanoparticles, nanobuds, nanowires, and nanorods. Despite not being as effective as silver nanoparticles, they can be a preferred nanopigment for textile applications due to other important factors including cost and availability. Antimicrobial tests including *Bacillus subtilis* and *Escherichia coli* and the *Fusarium graminearum* and *Penicillium s.p* were conducted on ZnO-loaded electrospun PET fibers [99]. Reported results demonstrates that the inhibitory zone against *Bacillus subtilis* and *Escherichia coli* have increased with ZnO concentration. Release of cationic zinc, creation of reactive oxygen species, and cumulating of nanoparticles in the cell membrane or cytoplasmic area are among the different proposed mechanism for ZnO antibacterial effect [99].

The ZnO nanopigment was introduced as a strong antibacterial agent against both Gram-positive and Gram-negative bacteria. The findings demonstrate that ZnO nanoparticles were liberated from the textile matrix and diffused significantly into the culture medium, where they formed bacterial inhibitory halos that were within the desired range. Similar antibacterial effects for ZnO have been reported by other studies [100, 101]. In addition to antibacterial activity, ZnO nanopigments were able to create antifungal activity against both toxigenic fungi including *Fusarium graminearum* and *Penicillium s.p.* As the concentration of nanoparticles rose, a potent fungicidal property was granted. It is reasonable that ZnO nanoparticles cause the membrane lipid bilayer of fungi to be disturbed [99].

One of the most important steps in reaping the benefits of nanopigments is to successfully adsorb them onto fibers. Enhanced nanoparticle adsorption on to the fibers can be achieved by either increasing adsorption sites on the fiber surface [97], or by surface modification of the nanopigments and formation of suitable anchor groups on the nanoparticles [101]. The later approach was employed to enhance the adsorption and stabilization of ZnO nanoparticles on the cotton fibers. To this end, a bifunctional regent glycidoxypropyl trimethoxysilane (GPTS) was used where the silane part can react with ZnO and the epoxy end can react with cotton under alkaline condition. For samples treated using this technique, excellent antibacterial activity and antiviral characteristics were reported [101].

To evaluate the antiviral properties of cotton treated with ZnO nanoparticles, ISO 18184 method can be used to perform quantitative antiviral testing. To briefly explain the method, the effectiveness of the treated fabric's antiviral performance was evaluated by injecting virus suspension (dengue virus and the hepatitis C virus) onto the product and then rinsing it clean. The virus's infectious titer in the washed-out suspension was calculated, and the test suspension's viral infectivity titer was assessed by quantifying the number of plaque creating units [114]. The antiviral mechanism of action of nanoparticles, such as ZnO, is not fully understood. However, several theories suggest that the antiviral activity is attributed to the release of Zn^{2+} ions and reactive oxygen species. These substances can cause damage to proteins, lipid membranes, and nucleic acids, and can also penetrate host cells. As a result, viruses are inactivated and the host cells undergo cell death. Dengue and hepatitis C virus resistance was demonstrated for fabric treated with zinc oxide nanoparticles [114]. After being exposed to dengue and hepatitis C for 60 min, cotton fibers treated with ZnO showed a viral decrease of over 70%.

Furthermore, the utilization of metal nanoparticles, such as silver, can demonstrate antimicrobial properties, in addition to the employment of metal oxides. The incorporation of Ag nanoparticles onto polyester can effectively mitigate issues associated with the growth of microorganisms [115]. This is attributed to the hydrophobic nature of the polyester surface, which discourages proliferation, as well as the substantial adsorption of metabolic by-products originating from the skin's perspiration and sebaceous glands. Nevertheless, the techniques employed for the deposition of Ag nanoparticles often fail to provide satisfactory adherence of the nanoparticles. The absence of functional groups on the polyester framework is the cause of this.

In a recent publication [115], the authors employed a multi-layer approach involving the application of Chitosan or HMDSO (hexamethyldisiloxane) onto a textile substrate. These layers were strategically applied both prior to and subsequent to the deposition of Ag nanoparticles. The primary objectives of this approach were to enhance the adhesion of nanoparticles onto the textile, safeguard the Ag nanoparticles against oxidation, regulate their release, and subsequently investigate their antimicrobial efficacy. The findings indicate that the use of a binder in conjunction with nanoparticles leads to increased microbial eradication compared to the exclusive usage of nanoparticles. This highlights the significance of adhesion in enhancing the antimicrobial efficacy. However, the application of a sole layer of HMDSO did not enhance the attachment of Ag nanoparticles. Conversely, when chitosan is employed, a single layer is sufficient to significantly enhance the adhesion. Furthermore, the introduction of both an initial and final layer of chitosan resulted in the achievement of a regulated antibacterial action. Additionally, the combination of chitosan and Ag nanoparticles demonstrated synergistic antimicrobial effects, which can be attributed to the antimicrobial activity of this biopolymer [115, 116].

5.3 Thermoregulating Performance Achieved by Nanopigments

Due to their unique structure and capacity to control heat and moisture transfer, textile materials can be remarkably functional in human thermal comfort. An extensive analysis has been published on advanced functional photonic fabric designed to improve thermoregulating efficiency. The review highlights the methods employed to create these textiles, which are aimed at enhancing personal thermal management and energy-saving initiatives. It has been shown that flexible and lightweight textiles with enhanced thermal properties can be fabricated by using nanopigments. Moreover, NIR reflecting systems containing nanopigments and mixture of nanopigments with organic coloring agents are effective in reflective electromagnetic light within 700–1500 nm [117].

One of the most sought-after nanopigments for the development of thermoregulating textiles is titanium dioxide. They offer additional functional qualities including antibacterial and UV-blocking capabilities and are affordable, accessible, and biocompatible. Additionally, the Rutile phase of titanium dioxide (TiO_2) has a high reflectivity (~ 85.0%) in both the visible and near-infrared spectrums and a very high refractive index (~ 2.7) desired for cooling textile technologies [102, 103]. To better understanding the key variables impacting IR reflectance properties of nanopigments, TiO_2 pigments with various concentration and sizes (30, 250, and 500 nm) have been applied to polyester and cotton, respectively [102, 103]. Light reflectance of adsorbed titanium dioxide on the polyester fibers was increased as pigment particle size increased. This is not unexpected according to Fresnel's low describing the achievable optimal reflection for particles with sizes equal to half the wavelength of the exposed light. In case of heat reflection which is attributed to wavelength within 700–1100 nm, the average half wavelength would be around 450 nm. Furthermore, increasing nanopigment over weight of fiber has resulted in enhanced light and NIR reflectance [102].

In another similar study, it was confirmed that increasing nanopigment size of TiO_2 from 35 to 250 nm increases IR reflectance from less than 88% to over 91% resulting in 5 °C cooler temperatures compared to the untreated cotton under NIR [103]. Likewise, by increasing the TiO_2 content from 1–3 g/l, the temperature difference between control and coated textiles increased, and eventually hit a plateau. It was also advised against using higher concentrations of TiO_2 nanopigments because the excessive quantity can impact on fabrics' reflective properties and color appearance, while the change in temperature difference becomes negligible [103].

5.4 Sensor Textiles Developed by Nanopigments

As defined in the beginning of this chapter, smart clothing refers to textiles that alter their qualities in answer to external inputs. A subset of smart textiles are fluorescent materials, which change color in reaction to light. In an innovative approach to develop smart and functional materials for textiles lanthanide-doped aluminum strontium oxide pigment phosphor nanoparticles (LDASOPP) transparent nanocomposite film is cured onto the cotton fabrics. The resulted cotton fabrics are proven to show super hydrophobic activity, UV protection, and antibacterial properties. Moreover, the coated fabric showed an unfatigued, reversible, and quick fluorescent photochromism [104].

Based on the colorimetric and spectrophotometric analysis reported by authors, the addition of novel LDASOPP nanoparticles did not alter fiber whiteness.

However, photoluminescence data shows that all the treated fibers have instantaneous, reversible photochromic, and fluorescent properties under UV light. The coated cotton fibers emit a greenish-yellow fluorescence when exposed to UV light. The absorbance values gradually rise as the UV exposure time is increased, causing a color shift to green-yellow that can be reversed to come back to its original white when the effect is eliminated. It is further reported that rapid reversibility can only be obtained for low pigment concentrations, whereas afterglow effect with slow and gradual discoloration was observed for samples with high pigment concentrations. Long-lasting emission is produced in part by the lanthanide ions Dy^{3+} and Eu^{2+}. These cations, often referred to as trapping cations, can greatly delay photon fatigue [104]. Such technologies can be useful for specialist fields like brand protection, security encoding, and military camouflage due to outstanding thermal and photostability, reversibility, and fatigue resistance.

The pH-responsive textiles, which have found use in medical apparel, environmental monitoring equipment, food quality control, chemical, and agricultural applications, are another intriguing use of functional smart materials. In one of the recent attempts to develop a sustainable and natural nanopigments as a potential pH-responsive material, an organic–inorganic hybrid nanopigment was synthesized from natural anthocyanin and different nanoclays [105]. The anthocyanin usage by itself as a pH-responsive sensor is very limited due to poor environmental stability of anthocyanins. However, in this work, it has been reported that the encapsulation of anthocyanins into nanoclays have improved its thermal and chemical stability. Also, the resulted hybrid nanopigments has shown a wide color gamut ranging from red to blue hues indicating its potential pH-responsive sensor performance.

5.5 *Flame-Retardant Nanopigments*

To develop flame-retardant polyester, a new class of hybrid nanopigments was designed and synthesized by reacting C.I. Reactive Blue 19 and nanoclay particles of montmorillonite [106]. Unlike earlier example of entrapment of anthocyanin in nanoclay particles, the reactive dye has been covalently linked to the hydroxyl groups of nanoclay plates. Anthocyanin pigments exhibit vibrant coloration and are highly responsive to variations in environmental circumstances. There exist multiple techniques for the stabilization of anthocyanins. [118, 119]. The newly developed organo-clay nanopigments were used in coloration of polyester where color strength was raised by increasing dye-to-clay ratio. Reported results indicate that increasing pigment concentration on fiber surface shows a decrease in transition in the electromagnetic wave transmission. This has been claimed to be due to the layered structure of the nanopigments [106]. The findings of the horizontal flammability analysis also showed that the nanopigments reduced the burning time and length of burning. It is suggested that coloring polyester fabrics with the new nanopigments might create multifunctional textiles for use in electronic and military applications.

Metal and metal oxide nanoparticles have been employed in textile polymers to enhance their flame-retardant properties. Several mechanisms have been proposed within this domain. Metal nanoparticles have the ability to impede the passage of heat and mass, alter the degradation pathway of polymers, limit the movement of polymer chains, and absorb active species like free radicals. In addition, it is possible that they can enhance the thermal conductivity within the material, so impeding the movement of bubbles, diminishing the heat dissipation, and lowering the local oxygen content as a result of the oxidation–reduction reactions involving oxides [120]. Nevertheless, several studies indicate that these materials, when used in isolation, may not effectively mitigate combustion. However, they do exhibit synergistic effects, enhancing their effectiveness when combined with other flame retardant compounds [10].

The investigation focused on the utilization of ZnO nanoparticles as a component of a flame-retardant system for cotton fabric, with the intention of gaining a more detailed perception. Zinc oxide nanoparticles are recognized for their catalytic properties, making them a highly efficient material in promoting char formation during the combustion of textiles. It has been established that ZnO nanoparticles, when consumed as co-catalysts, boost the flame retardancy of cotton fabric by strengthening its thermal stability, as well as increasing the durability of the flame retardant [121].

A recent study has demonstrated the remarkable flame retardancy of cotton fabric achieved with the use of zinc oxide nanoparticles. Nevertheless, the duration of the after-glow period is too protracted, rendering it unsuitable for application as a flame-retardant material. The duration of the after-glow of the cloth was seen to reduce significantly after the application of boric acid [122].

5.6 Photothermal Nanopigments

The urgent need for photothermal textiles may be addressed by the employment of engineering-designed nanopigments. To this end, nano-based polypyrrole-polyethyleneimine was synthesized which showed remarkable photothermal convertible capability as well as significant NIR absorption [107]. By itself, polypyrrole is a conductive polymer and dye that can absorb NIR, increasing the efficiency of photothermal conversion. In this example, the PE-fiber-constructed membrane's mask fabric was coated by the cationic polyethyleneimine polypyrrole nanopigments' binding affinity. Face masks currently on the market are made of cloth, which when exposed to water loses its ability to function as an air filter. The disposable face masks would be very difficult to reuse because their filtering efficiency would be considerably decreased. The coated mask fabric with polyethyleneimine polypyrrole nanopigments, however, has been found to be capable of performing infrared red photothermal conversion efficacy to give biocompatible warming, remotely photothermally captured antipathogen, and antivasoconstriction. This pioneering study showcased the feasibility of creating washable, reusable masks with a coating that allows for breathability, biocompatibility, and the ability to actively eliminate dangerous germs by photothermal conversion. It also has heat retention and antivasoconstriction properties [107].

5.7 Application of Nanopigments in Self-cleaning Textiles

The demand for hygienic surfaces has led to a surge in interest in self-cleaning protective materials and surfaces. Two main methods for achieving self-cleaning qualities are hydrophobicity and hydrophilicity. Hydrophobic coatings help water droplets roll and eliminate debris, while hydrophilic coatings promote water sheeting and dirt removal [9]. Absorbed dirt by hydrophilic materials can also undergo chemical degradation when exposed to sunlight, using a hydrophilic photocatalytic coating. Photocatalysis involves semiconductors like TiO_2 and ZnO being exposed to light that matches or surpasses the band gap, causing electrons to migrate toward the semiconductor's surface. This process leads to the formation of radicals that break down pollutants. This process also promotes soil degradation through solar energy [9, 35].

The utilization of nano-TiO_2 as a self-cleaning agent for textiles has generated considerable attention. In a recent investigation, scholars employed rutile nanotitania as a coating substance on cotton textile. The improvement of the self-cleaning characteristic can be achieved by increasing the concentration and size of nanotitanium dioxide. However, the changes in color observed in stained samples treated with methylene blue after being exposed to daylight irradiation are considered insignificant, suggesting that the rutile structure does not operate satisfactorily as a photocatalytic agent [123].

Nanotitanium dioxide has demonstrated many limits despite its extensive utilization as a photocatalyst in textile finishing. The sole observation of the photocatalytic capabilities of nano-TiO_2 occurs when ultraviolet energy is present. Furthermore, the increased rate of electron–hole recombination leads to a reduction in the efficiency of titania. Efforts aimed at tackling these problems have predominantly focused on the utilization of an additional component in combination with titanium dioxide. A research study was undertaken to examine the concurrent application of coloring and functional finishing on cotton by modifying nanotitane with extracted dye from walnut husk. The results suggest that the efficiency of methylene blue degradation is enhanced when titania is combined with a natural dye. Moreover, the photocatalytic activity is intensified by increasing the concentration of nano-TiO_2 [124].

Although TiO_2 has received much attention for its photocatalytic characteristics, which are comparable to those of ZnO nanoparticles, its implementation in large-scale industrial settings is currently constrained by its high cost. The comparative analysis of titanium dioxide and zinc oxide in relation to their photodegradation capacities for Reactive Blue 19 demonstrates that ZnO demonstrates a higher level of efficiency. However, the application of ZnO nanoparticles is accompanied by two notable obstacles that require resolution. There are two key issues that are inherently linked to this particular situation. The initial concern is to the occurrence of photo-corrosion, which results in a notable decline in the photocatalytic capabilities. The second obstacle is to the issue of inadequate resistance to acid [125].

In the present scenario, the application of ZnO catalysts enhanced by active metal particles is a promising approach to tackle these difficulties. Furthermore, this methodology facilitates the control of particle size, hence augmenting the stability and characterization of the catalysts. The findings indicate that the incorporation of nano-ZnO onto a silica support confers advantageous properties, including enhanced thermal stability, outstanding mechanical properties, and a high degree of safety and biocompatibility. Furthermore, it is possible to augment the flame-retardant and self-cleaning properties of many substrates, including polyester fabric. The application of SiO_2 as a coating material on ZnO nanoparticles led to enhanced photocatalytic efficacy and a heightened rate of methylene blue degradation on polyester fabric [125].

5.8 Nanopigments as Antistatic Materials

Textile fibers are classified as insulators due to their poor electrical conductivity, making them dielectric materials. Conductive materials facilitate the equalization of electron flow when surfaces are separated, while insulating materials maintain the difference in charge over time. Static electricity can negatively impact human health and textile processes. Antistatic compounds are chemical substances used to reduce static charges in synthetic fabrics by minimizing electrostatic charge creation and enhancing conductivity. Most of these agents operate through conductivity, with advancements in nanotechnology leading to improved antistatic finishing of fabrics [126].

Metal and metal oxide nanoparticles possess a higher level of electroconductivity in contrast to textile fibers, making them well-suited for the purpose of dissipating static charge in these fibers [126, 127]. A paper has been published that provides confirmation of the antistatic property exhibited by zinc oxide in the context of textile finishing. The use of F-doped ZnO as a coating on the surface of fabric, serving as a textured inner layer, has the potential to enhance the fabric's conductivity. The ease of removing static charges is attributed to the significantly low reported value of surface resistivity. Furthermore, the antistatic feature is assessed using the gold-leaf electroscopic technique, wherein the absence of any displacement of the gold leaf indicates the presence of antistatic characteristics in the treated fabric [128]. A supplementary study has been conducted about the antistatic characteristics of wool fabric that has been dyed using silver nanoparticles. Wool fabrics that underwent treatment with Ag nanoparticles produced in situ exhibited significantly reduced surface charges in comparison with fabrics that were left untreated. The surface resistivity of wool exhibited a reduction as the concentration of applied silver increased [129].

5.9 Application of Nanopigments in Printing of Textile Fabrics

The application of pigment-printed textile fabric has gained significant recognition globally in recent years due to its numerous advantages in comparison with alternative printing methods. The advantages of this method encompass its ease of implementation, versatility in its ability to be applied to various combinations of natural and synthetic fibers, and the ability to circumvent the need for a final wash off, so resolving any concerns related to the disposal of waste materials [130, 131].

Nanotechnology is a scientific discipline that involves the utilization and production of materials, devices, and systems at the nanometer measure. It encompasses a perspective that views all aspects of the world through the lens of atomic or molecular constituents. This field has already exerted a significant impact on a wide array of human technological endeavors. When a bulk material is subjected to size reduction resulting in particles with diameters in the nanometer range, the material exhibits distinct features that meaningfully deviate from those of the bulk material [132–134].

Nanoparticles can modify the surface characteristics and confer new functions on textile materials in certain applications involving textile finishing [135]. Matter undergoes changes in its properties as it gets smaller, including color and chemical activity—the ability to interact with other matter. The alteration of the electrical properties is what leads to the change in features. Reducing the particle size increases the material's surface area. As a result, more atoms have the ability to interact with other materials, such as the resin matrix.

Because they are transparent and have a greater surface area than larger particles, nanoparticles do not significant change the original color or brightness of the product they are contained in or the textile substrates on which they are used. Furthermore, because of their huge surface area to volume ratio and easy attachment to fabrics, nanoparticles boost the functional endurance of the materials they come into contact with. Additionally, the breathability and tactile feel of the fabric are unaffected by the usage of nanoparticles [135]. Not only is pigment printing the most traditional printing technique, but it is also the most straightforward to use. Pigment printing accounts for almost 80% of printed items because of its many benefits, including ease of use and adaptability in producing nearly final prints during the printing stage itself [136].

Pigment printing is distinct from other printing techniques in that color is created using finely powdered insoluble pigments that are indifferent to fiber [137]. In contrast to printing with fiber affinitive dyes, a binder clings to the fiber and creates a continuous coating that encloses the pigment particles on the fabric, fixing the pigment to the fabric [138].

Nowadays, the use of nanotechnology in textile printing and the utilization of nanopigments are increasing due to their advantageous properties. The use of nanopigments in textile printing can have an impact on various other characteristics in addition to adding new features such as UV protection, self-cleaning properties, antimicrobial properties, and more [139].

Pigments used in traditional textile printing generally have larger particle sizes, usually varying in size from a few micrometers to 10 of micrometers. In contrast, nanopigments have significantly smaller particle sizes, typically in the nanometer range. The smaller particle size of nanopigments allows for better dispersion and more uniform color distribution in the printing process, resulting in improved color vibrancy and resolution. Nanopigments also exhibit enhanced color intensity and saturation compared to conventional pigments. The smaller particle size of nanopigments results in a larger surface area, allowing for better light absorption and reflection. This leads to more vibrant and vivid colors on printed textiles, providing a greater visual impact [140, 141].

Nanopigments often offer improved color fastness properties, such as resistance to fading and color bleeding, compared to traditional pigments. The smaller particle size and improved dispersion of nanopigments allow for better adhesion to textile fibers, leading to enhanced wash fastness and rub resistance. This ensures that the printed designs maintain their original colors and appearance even after repeated washing or exposure to external factors. The smaller particle size of nanopigments enables higher printing resolution and finer details in textile printing [142]. Nanopigments can achieve sharper and more precise prints, allowing for intricate patterns, small text, and complex designs to be accurately reproduced on fabrics. This finer printing resolution enhances the overall quality and visual appeal of the printed textiles. Nanopigments can be engineered to be more environmentally friendly compared to traditional pigments. They can be formulated using non-toxic and sustainable materials, reducing the environmental impact of textile printing processes. This aligns with the growing demand for sustainable and eco-friendly practices in the textile industry [143].

Nanopigments exhibit improved dispersion stability compared to conventional pigments. The smaller particle size of nanopigments allows for better dispersion in ink or coating formulations, resulting in a more stable and homogeneous mixture. This ensures consistent color performance and prevents particle settling or agglomeration during storage or printing. Traditional pigments used in textile printing often have an opaque or semi-opaque nature, which can limit their application in certain designs or fabric types. In contrast, nanopigments can be engineered to be more transparent, allowing for more versatile application on different fabric colors and textures. Transparent nanopigments enable the creation of layered or blended effects, expanding the design possibilities in textile printing [143, 144].

Nanopigments can be formulated into various types of inks used in textile printing, including water-based, solvent-based, and UV-curable inks. They offer compatibility with different ink formulations and printing processes, providing flexibility for textile printers to choose the most suitable ink system for their specific requirements [144].

The use of nanopigments in textile printing can contribute to energy efficiency in the printing process. Due to their smaller particle size, nanopigments require lower energy input for dispersion, mixing, and printing compared to larger-sized pigments. This can result in reduced energy consumption and improved printing efficiency. Nanopigments generally have a higher cost compared to traditional pigments. The manufacturing processes involved in producing nanopigments with controlled particle sizes and properties can be more complex and expensive. However, advancements in nanotechnology and increased demand for nanopigments are leading to improved cost-effectiveness over time [142, 144].

Nanopigments are still an area of active research and development in the field of textile printing. Ongoing efforts are focused on optimizing their properties, such as color range, stability, and environmental impact, to meet the specific requirements of textile applications. This includes exploring new synthesis methods, surface modifications, and pigment formulations to further enhance the performance and versatility of nanopigments in textile printing [144]. It is important to consider the specific requirements of each textile printing application when choosing between traditional pigments and nanopigments. Factors such as desired color intensity, resolution, fastness properties, transparency, and environmental considerations should be taken into account to determine the most suitable pigment type for a particular printing project.

6 Conclusion

This chapter deals with an overview of the importance, benefits, and application of nanopigments especially in textile finishing. Nanopigments have emerged as a revolutionary technology with the potential to bring remarkable changes in multiple applications, such as coloration, printing, and eventual textile finishing. Nanopigment reduced to the nanoscale may exhibit entirely distinct characteristics from those observed at the macroscale. The reduced dimensions of nanopigments lead to nanostructured materials, which result in enhanced chromatic intensity, heightened resistance to fading caused by light exposure, and superior dispersibility in different mediums. Various methods of nanopigments syntheses were discussed in this chapter which mainly included hydro/solvothermal, sol–gel, layer-by-layer deposition, sonosynthesis, microwave aided, biosynthesis, precipitation, emulsion polymerization, microfluidic synthesis, and self-assembly.

Furthermore, the importance of nanopigment dispersion on finishing quality is disclosed. The main reason for the importance of nanopigment dispersion lies in the remarkable ability of nanoparticles to greatly enhance the optical and mechanical properties of textiles, particularly in relation to color. The utilization of a stabilized nanopigment dispersion enables the production of pigment particles at the nanoscale, which are uniformly distributed within a matrix. This results in materials that exhibit enhanced color intensity and durability. The high level of accuracy and regulation in the distribution of pigment sizes is extremely valuable in applications like textile printing and coloration [145]. The use of nanopigment dispersions in these fields results in outstanding color, brightness, durability, enhanced resistance to fading, and potential functional performance.

Nanopigments have found diverse applications in the textile industry, particularly in the development of functional and smart textiles. The additional optical and functional benefits of nanopigments applied to textiles were disclosed. Nanopigments can offer functional properties such as antibacterial, temperature regulation, UV protection, fire resistance, photothermal, antistatic, and self-cleaning properties. Further investigation should be conducted to develop novel organic, inorganic, or hybrid nanopigments to create colorants with improved fastness properties, unique color characteristics or desired functional effects. Also, optimized methods to disperse such nanomaterials prior and during application on textiles need to be investigated to step up advancements in textile finishing.

References

1. Gürses A, Açıkyıldız M, Güneş K, Gürses MS (2016) Dyes and pigments: their structure and properties. Springer, Berlin. https://doi.org/10.1007/978-3-319-33892-7_2
2. Srivastava R, Agrawal M, Bano S (2022) Nano-pigments: applications and ecological impact: a review 6:92–99. https://doi.org/10.47001/IRJIET/2022.605012
3. Hari RK, Patel TR, Martin A M (2009) An overview of pigment production in biological systems: functions, biosynthesis, and applications in food industry 10:49–70. https://doi.org/10.1080/87559129409540985
4. Soltani Panah H, Khosravi A, Gharanjig K, Khorassani M, Khatib Zadeh M, Afshar Taromi F (2010) Synthesis and characterization of new fluorescent polymerizable dyes based on naphthalimide 491–500
5. Ahmed NSE, El-Shishtawy RM (2009) The use of new technologies in coloration of textile fibers 45:1143–1153. https://doi.org/10.1007/s10853-009-4111-6
6. Hu Z, Xue M, Zhang Q, Sheng Q, Liu Y (2008) Nanocolorants: a novel class of colorants, the preparation and performance characterization 76:173–178. https://doi.org/10.1016/j.dyepig.2006.08.026
7. Cavalcante PMT, Dondi M, Guarini G, Raimondo M, Baldi G (2009) Colour performance of ceramic nano-pigments 80:226–232. https://doi.org/10.1016/j.dyepig.2008.07.004
8. Maile FJ, Pfaff G, Reynders P (2005) Effect pigments—past, present and future 54:150–163. https://doi.org/10.1016/j.porgcoat.2005.07.003
9. Saad SR, Mahmed N, Abdullah MMAB, Sandu AV (2016) Self-cleaning technology in fabric: a review. https://doi.org/10.1088/1757-899X/133/1/012028
10. Silva de Oliveira CR, Batistella MA, Guelli U, de Souza SMA, Ulson de Souza AA (2021) Functionalization of cellulosic fibers with a kaolinite-TiO_2 nano-hybrid composite via a solvothermal process for flame retardant applications 266. https://doi.org/10.1016/J.CARBPOL.2021.118108
11. Mohan C, Kumari N, Jeandet P, Kumari P, Negi A (2023) Synthesis of nano pigments using clay minerals and organic dyes and their application as colorants in polymer matrix 14:1087. https://doi.org/10.3390/mi14051087
12. Harish V, Ansari MM, Tewari D, Manish, Yadav AB, García-Betancourt M-L, Abdel-Haleem FM, Bechelany M, Barhoum A (2022) Nanoparticle and nanostructure synthesis and controlled growth methods 12:3226. https://doi.org/10.3390/nano12183226
13. Huston M, DeBella M, DiBella M, Gupta A (2021) Green synthesis of nanomaterials 11:2130. https://doi.org/10.3390/nano11082130
14. Sharma D, Kanchi S, Bisetty K (2019) Biogenic synthesis of nanoparticles: a review 12:3576–3600. https://doi.org/10.1016/j.arabjc.2015.11.002
15. Sarmiento-Tovar AA, Silva L, Sánchez-Suárez J, Diaz L (2022) Streptomyces-derived bioactive pigments: ecofriendly source of bioactive compounds 12:1858. https://doi.org/10.3390/coatings12121858
16. Smith T (1983) A review of carbon black pigments 12:14–16. https://doi.org/10.1108/eb041897

17. Silva TA, Moraes FC, Janegitz BC, Fatibello-Filho O (2017) Electrochemical biosensors based on nanostructured carbon black: a review 1–14. https://doi.org/10.1155/2017/4571614
18. Kamyshny A, Magdassi S (2014) Conductive nanomaterials for printed electronics 10:3515–3535. https://doi.org/10.1002/smll.201303000
19. Thirugnanasambandan T (2017) Review on Titania Nanopowder-processing and applications. https://doi.org/10.48550/arXiv.1704.00981
20. Januszkiewicz K (2019) Air purification in highly-urbanized areas with use TiO_2: new approach to design the urban public space to benefit human condition. https://doi.org/10.1088/1757-899X/603/3/032100
21. Mufti N, Atma T, Fuad A, Sutadji E (2014) Synthesis and characterization of black, red and yellow nanoparticles pigments from the iron sand. 1617:165–169. https://doi.org/10.1063/1.4897129
22. Rahman MM, Khan SB, Jamal A, Faisal M, Aisiri AM (2011) Iron oxide nanoparticles. https://doi.org/10.5772/27698
23. Bhavani P, Rajababu CH, Arif MD, Subba Reddy IV, Ramamanohar Reddy N (2017) Synthesis of high saturation magnetic iron oxide nanomaterials via low temperature hydrothermal method 426:459–466. https://doi.org/10.1016/j.jmmm.2016.09.049
24. Abdou Y, Elhussiny F, Morsy R (2023) Novel green synthesis of UV-sunscreen ZnO nanoparticles using solanum lycopersicum fruit extract and evaluation of their antibacterial and anticancer activity. https://doi.org/10.1007/s10904-023-02744-3
25. Pacheco-Blandino R, Vanner C (2012) Buzea: toxicity of nanoparticles 427–475. https://doi.org/10.1533/9780857096357.427
26. Yuan J, Yuan J, Xing W, Gu G, Wu L (2008) The properties of organic pigment encapsulated with nano-silica via layer-by-layer assembly technique 76. https://doi.org/10.1016/J.DYEPIG.2006.10.002
27. Pandian B, Ramalingam S, Sreeram KJ, Rao JR (2021) Natural pigment: preparation of brown pigment from lignin biomass for coloring application 195. https://doi.org/10.1016/J.DYEPIG.2021.109704
28. Wang C (2023) Precise analysis of nanoparticle size distribution in TEM image 6. https://doi.org/10.3390/mps6040063
29. Uncu O, Ozen B, Tokatli F (2019) Use of FTIR and UV-visible spectroscopy in determination of chemical characteristics of olive oils 201. https://doi.org/10.1016/J.TALANTA.2019.03.116
30. Raja PB, Munusamy KR, Perumal V, Ibrahim MNM (2022) Characterization of nanomaterial used in nanobioremediation 57–83. https://doi.org/10.1016/B978-0-12-823962-9.00037-4
31. Blanco I, Siracusa V (2021) The use of thermal techniques in the characterization of bio-sourced polymers 14. https://doi.org/10.3390/MA14071686
32. Mahović Poljaček S, Tomašegović T, Leskovšek M, Stankovič Elesini U (2021) Effect of SiO_2 and TiO_2 nanoparticles on the performance of UV visible fluorescent 11. https://doi.org/10.3390/COATINGS11080928
33. Rashid MM, Simončič B, Tomšič B (2021) Recent advances in TiO_2-functionalized textile surfaces 22. https://doi.org/10.1016/J.SURFIN.2020.100890
34. Elshamy SM, Khalil MMH, Hanafi AMG (2017) Preparation and characterization of nano pigments by wet milling technique and their applications in printing inks industry 13–24. https://doi.org/10.21608/EJAPS.2017.183768
35. Montazer M, Harifi T (2018) Nanofinishing of textile materials. Woodhead Publishing
36. Nyamukamba P, Okoh O, Mungondori H, Taziwa R, Zinya S (2018) Synthetic methods for titanium dioxide nanoparticles: a review. Titanium dioxide-material for a sustainable environment. Presented at the October 12
37. Malekshahi Byranvand M, Nemati Kharat A, Fatholahi L, Malekshahi Beiranvand Z (2013) A review on synthesis of nano-TiO_2 via different methods 3:1–9. https://doi.org/10.7508/jns.2013.01.001
38. Jain A, Kongkham B, Puttaswamy H, Butola B, Malik H, Malik A (2022) Development of wash-durable antimicrobial cotton fabrics by in situ green synthesis of silver nanoparticles

and investigation of their antimicrobial efficacy against drug-resistant bacteria 11:864. https://doi.org/10.3390/antibiotics11070864
39. Ismail W (2016) Sol–gel technology for innovative fabric finishing—a review 78:698–707. https://doi.org/10.1007/s10971-016-4027-y
40. Periyasamy AP, Venkataraman M, Kremenakova D, Militky J, Zhou Y (2020) Progress in Sol-Gel technology for the coatings of fabrics 13:1838. https://doi.org/10.3390/ma13081838
41. Uddin MJ, Cesano F, Scarano D, Bonino F, Agostini G, Spoto G, Bordiga S, Zecchina A (2008) Cotton textile fibres coated by au/tio$_2$ films: synthesis, characterization and self cleaning properties 199:64–72. https://doi.org/10.1016/j.jphotochem.2008.05.004
42. d'Água R, Branquinho R, Duarte M, Maurício E, Fernando A, Martins R, Fortunato E (2018) Efficient coverage of ZnO nanoparticles on cotton fibres for antibacterial finishing using a rapid and low cost: in situ synthesis 42:1052–1060. https://doi.org/10.1039/C7NJ03418K
43. Gulrajani M, Deepti G (2011) Emerging techniques for functional finishing of textiles. 388–397
44. Magovac E, Vončina B, Jordanov I, Grunlan J, Bischof S (2022) Layer-by-layer deposition: a promising environmentally benign flame-retardant treatment for cotton, polyester. Polyamide Blended Text 15:432. https://doi.org/10.3390/ma15020432
45. Ugur Ş, Sariişik M, Aktaş A (2010) The fabrication of nanocomposite thin films with TiO_2 nanoparticles by the layer-by-layer deposition method for multifunctional cotton fabrics 21:325603. https://doi.org/10.1088/0957-4484/21/32/325603
46. Matsushita AFY, Nascimento JRV, Leite JMA, Pessoa CA, Garcia JR (2023) Incorporation of ZnO/Ag-ZnO nanoparticles into a cotton fiber by the LbL method for Self-Cleaning applications. 375:121336. https://doi.org/10.1016/j.molliq.2023.121336
47. Li Z, Dong J, Wang L, Zhang Y, Zhuang T, Wang H, Cui X, Wang Z (2021) A power-triggered preparation strategy of nano-structured inorganics: sonosynthesis. 3:2423–2447. https://doi.org/10.1039/D1NA00038A
48. Harifi T, Montazer M (2015) A review on textile sonoprocessing: a special focus on sonosynthesis of nanomaterials on textile substrates 23:1–10. https://doi.org/10.1016/j.ultsonch.2014.08.022
49. Aminloo E, Montazer M (2021) Clean Sono-synthesis of ZnO on Cotton/Nylon fabric using dopamine: photocatalytic. Hydrophilic Antibacterial Features 22:97–108. https://doi.org/10.1007/s12221-021-9237-4
50. Diaz de Grenu B, de Los Reyes R, Costero AM, Amorós P, Ros-Lis JV (2020) Recent progress of microwave-assisted synthesis of silica materials 10:1092. https://doi.org/10.3390/nano10061092
51. Motshekga S, Pillai S, Ray S, Jalama K, Krause R (2012) Recent trends in the microwave-assisted synthesis of metal oxide nanoparticles supported on carbon nanotubes and their applications. e691503. https://doi.org/10.1155/2012/691503
52. Pawełczyk A, Sowa-Kasprzak K, Olender D, Zaprutko L (2018) Microwave (MW), Ultrasound (US) and combined synergic MW-US strategies for rapid functionalization of pharmaceutical use phenols 23:2360. https://doi.org/10.3390/molecules23092360
53. Yang H, Zhang Q, Chen Y, He Y, Yang F, Lu Z (2018) Microwave–ultrasonic synergistically assisted synthesis of ZnO coated cotton fabrics with an enhanced antibacterial activity and stability 1:340–346. https://doi.org/10.1021/acsabm.8b00086
54. Kundu D, Hazra C, Chatterjee A, Chaudhari A, Mishra S (2014) Extracellular biosynthesis of zinc oxide nanoparticles using Rhodococcus pyridinivorans NT2: multifunctional textile finishing, biosafety evaluation and in vitro drug delivery in colon carcinoma 140:194–204. https://doi.org/10.1016/j.jphotobiol.2014.08.001
55. Punjabi K, Choudhary P, Samant L, Mukherjee S, Vaidya S, Chowdhary A (2015) Biosynthesis of nanoparticles: a review 30:219–226
56. Mohanpuria P, Rana N, Yadav S (2008) Biosynthesis of nanoparticles: technological concepts and future applications 10:507–517. https://doi.org/10.1007/s11051-007-9275-x
57. Aladpoosh R, Montazer M, Samadi N (2014) In situ green synthesis of silver nanoparticles on cotton fabric using Seidlitzia rosmarinus ashes. 21:3755–3766. https://doi.org/10.1007/s10570-014-0369-1

58. Lewin M (1997) Oxidation and aging of cellulose. 118:715–724. https://doi.org/10.1002/masy.19971180192
59. Zhao S, Guo J, Li W, Guo H, You B (2018) Fabrication of cobalt aluminate nanopigments by coprecipitation method in threonine waterborne solution 151. https://doi.org/10.1016/J.DYEPIG.2017.12.062
60. Sahoo S, Gopalan A, Ramesh S, Nirmala PV, Ramkumar G, Shifani SA, Subbiah R, JoshuaRamesh JI, Lalvani R (2021) Preparation of polymeric nanomaterials using emulsion polymerization. https://doi.org/10.1155/2021/1539230
61. Thomée E (2017) Microfluidic nanoparticle synthesis: a short review
62. McMillan A (2016) Microreactors and microfluidics in chemistry—A review
63. Gimondi S, Reis RL, Ferreira H, Neves N (2022) Microfluidic-driven mixing of high molecular weight polymeric complexes for precise nanoparticle downsizing 43. https://doi.org/10.1016/j.nano.2022.102560
64. Synthesis and applications of optically active nanomaterials 14. https://doi.org/10.1142/9789813222991_0001
65. Yulianto A, Fitriawan M, Sulhadi K, Sumpono I (2019) Synthesis of iron sand into yellow nano-pigment using sol-gel method 1170. https://doi.org/10.1088/1742-6596/1170/1/012049
66. Ohishi T, Kawamura T (2017) Synthesis and properties of latent pigment nano- and micro-encapsulated in high-durability inorganic capsules 3. https://doi.org/10.15761/FNN.1000161
67. Agbo C, Jakpa W, Sarkodie B, Boakye A, Fu S (2018) A review on the mechanism of pigment dispersion 39:874–889. https://doi.org/10.1080/01932691.2017.1406367
68. McKeen L (2015) Fluorinated coatings and finishes handbook: the definitive user's guide. William Andrew
69. Baez E (2011) Nanopigment development and dispersion in aqueous systems
70. Farrokhpay S (2009) A review of polymeric dispersant stabilisation of Titania pigment 151:24–32. https://doi.org/10.1016/j.cis.2009.07.004
71. Solomon D, Hawthorne D (1983) Chemistry of pigments and fillers. Wiley, New York
72. Raha S, Ivanov I, Quazi N, Bhattacharya S (2009) Photo-stability of rhodamine-B/montmorillonite nanopigments in polypropylene matrix 42:661–666. https://doi.org/10.1016/j.clay.2008.06.008
73. Ahmed K, El-Hennawi H, Shahin A, Ragheb A (2019) Preparation of nano disperse dye based on benzopyran in one pot reaction using microwave irradiation and its appliance in textile printing 1:1–9. https://doi.org/10.1007/s42452-019-0926-1
74. Hebeish A, Rekaby M, Shahin A, Ragheb A (2016) Novel nanopigment derived from vat dyes for printing cotton fabrics 59:99–114. https://doi.org/10.21608/ejchem.2016.339
75. Agbo C, Acheampong C, Zhang L, Li M, Fu S (2019) Preparing stable pigment dispersion utilizing polyoxyethylene lauryl ether as dispersant 48:1–8. https://doi.org/10.1108/PRT-10-2017-0081
76. Baez E, Quazi N, Ivanov I, Bhattacharya S (2009) Stability study of nanopigment dispersions 20:267–272. https://doi.org/10.1016/j.apt.2009.02.005
77. Hao L, Wang R, Fang K, Liu R, Men Y (2016) Utilizing cellulase as dispersant to prepare stable nano-pigment suspension and investigating its coloring performance on cotton fabrics 27:164–170. https://doi.org/10.1016/j.apt.2015.12.001
78. Gusakov AV, Sinitsyn AP, Berlin AG, Markov AV, Ankudimova NV (2000) Surface hydrophobic amino acid residues in cellulase molecules as a structural factor responsible for their high denim-washing performance 27:664–671. https://doi.org/10.1016/S0141-0229(00)00264-7
79. Yang R, Tang Q, Qian Y, Pang Y, Yang D, Zheng D, Zhou M (2023) Preparation of sulfomethylated lignin grafted by pyrrolidone for utilization as a dispersant in nano pigment paste 49:16578–16586. https://doi.org/10.1016/j.ceramint.2023.02.015
80. Christie R, Abel A (2021) Organic pigments: general principles 6:807–834. https://doi.org/10.1515/psr-2020-0187
81. Pfaff G (2021) Transparent pigments 6:549–555. https://doi.org/10.1515/psr-2020-0200

82. Piri N, Shams-nateri A, Mokhtari J (2017) Solar spectral performance of nanopigments 162:72–82. https://doi.org/10.1016/j.solmat.2016.12.036
83. Smith A, Nie S (2010) Semiconductor nanocrystals: structure, properties, and band gap engineering 43:190–200. https://doi.org/10.1021/ar9001069
84. Eustis S, El-Sayed M (2006) Why gold nanoparticles are more precious than pretty gold: noble metal surface plasmon resonance and its enhancement of the radiative and nonradiative properties of nanocrystals of different shapes 35:209–217. https://doi.org/10.1039/B514191E
85. Ramalingam G, Kathirgamanathan P, Ravi G, Elangovan T, Kumar BA, Manivannan N, Kasinathan K (2020) Quantum Confinement effect of 2D nanomaterials. Presented at the October 14
86. Zeng S, Yong K, Roy I, Dinh X, Yu X, Luan F (2011) A review on functionalized gold nanoparticles for biosensing applications. 6:491–506. https://doi.org/10.1007/s11468-011-9228-
87. Fu H, Loo B, Xiao D, Xie R, Ji X, Yao J, Zhang B, Zhang L (2002) Multiple emissions from 1,3-Diphenyl-5-pyrenyl-2-pyrazoline nanoparticles: evolution from molecular to nanoscale to bulk materials 114:1004–1007. https://doi.org/10.1002/1521-3757(20020315)114:63.0.CO;2-R
88. Fu H-B, Yao J-N (2002) Size effects on the optical properties of organic nanoparticles 123:1434–1439. https://doi.org/10.1021/ja0026298
89. Xiao D, Xi L, Yang W, Fu H, Shuai Z, Fang Y, Yao J (2003) Size-tunable emission from 1,3-diphenyl-5-(2-anthryl)-2-pyrazoline nanoparticles 125:6740–6745. https://doi.org/10.1021/ja028674s
90. Riaz S, Ashraf M, Hussain T, Hussain M, Rehman A, Javid A, Iqbal K, Basit A, Aziz H (2018) Functional finishing and coloration of textiles with nanomaterials 134:327–346. https://doi.org/10.1111/cote.12344
91. Rivero P, Urrutia A, Goicoechea J, Arregui F (2015) Nanomaterials for functional textiles and fibers 10:501–501. https://doi.org/10.1186/s11671-015-1195-6
92. Syduzzaman M, Patwary S, Farhana K, Ahmed S (2015) Smart textiles and nano-technology: a general overview 5:1–7. https://doi.org/10.4172/2165-8064.1000181
93. Jatoi AS, Khan FSA, Mazari SA, Mubarak NM, Abro R, Ahmed J, Ahmed M, Baloch H, Sabzoi N (2021) Current applications of smart nanotextiles and future trends. Presented at the October 14 (2021)
94. Alvarez J, Zwierzyński K, New barrier Nanopigment form against UV radiation that improves the functionality and aesthetical properties of selected products of light industry
95. Xu Y, Sheng J, Yin X, Yu J, Ding B (2017) Functional modification of breathable polyacrylonitrile/polyurethane/TiO_2 nanofibrous membranes with robust ultraviolet resistant and waterproof performance. 508:508–516. https://doi.org/10.1016/j.jcis.2017.08.055
96. Kafafy H, Helmy H, Zaher A (2021) Treatment of cotton and wool fabrics with different nanoparticles for multifunctional properties 64:5255–5267. https://doi.org/10.21608/ejchem.2021.72780.3609
97. Ibrahim NA, Eid BM, El-Aziz EA, Elmaaty TMA, Ramadan SM (2017) Loading of chitosan—Nano metal oxide hybrids onto cotton/polyester fabrics to impart permanent and effective multifunctions 105:769–776. https://doi.org/10.1016/j.ijbiomac.2017.07.099
98. Abdelghaffar R, Hamoda D, Elgohary D (2022) Surface coatings of polyester fabrics using titanium dioxide and zinc oxide for multifunctional medical applications 113:1–13. https://doi.org/10.1080/00405000.2021.2001892
99. Vázquez K, Vanegas P, Cruzat C, Novoa N, Arrué R, Vanegas E (2021) Antibacterial and antifungal properties of electrospun recycled PET polymeric fibers functionalized with zinc oxide nanoparticles 13:3763. https://doi.org/10.3390/polym13213763
100. Hu R, Yang J, Yang P, Wu Z, Xiao H, Liu Y, Lu M (2020) Fabrication of ZnO@Cotton fabric with anti-bacterial and radiation barrier properties using an economical and environmentally friendly method. 27:2901–2911. https://doi.org/10.1007/s10570-019-02965-1
101. Afzal F, Ashraf M, Manzoor S, Aziz H, Nosheen A, Riaz S (2023) Development of novel antiviral nanofinishes for bioactive textiles. 80:8447–8466. https://doi.org/10.1007/s00289-022-04461-2

102. Asadi L, Shams Nateri A (2023) Relationship between cooling and optical properties of polyester fabric treated with TiO_2 nanopigments 1–15. https://doi.org/10.1080/00405000.2023.2225719
103. Shams-Nateri A, Kazemian S, Piri N (2020) Nano-TiO_2 coated cotton fabrics with temperature regulating properties 111:1223–1230. https://doi.org/10.1080/00405000.2019.1690916
104. Subaihi A, Al-Qahtani S, Attar R, Alkhamis K, Alzahrani H, Alhasani M, El-Metwaly N (2022) Preparation of fluorescent cotton fibers with antimicrobial activity using lanthanide-doped pigments 29:6393–6404. https://doi.org/10.1007/s10570-022-04665-9
105. Micó-Vicent B, Ramos M, Viqueira V, Luzi F, Dominici F, Terenzi A, Maron E, Hamzaoui M, Kohnen S, Torre L, Jiménez A, Puglia D, Garrigós MC (2021) Anthocyanin hybrid nanopigments from pomegranate waste: colour, thermomechanical stability and environmental impact of polyester-based bionanocomposites 13:1966. https://doi.org/10.3390/polym13121966
106. Almasian A, Mirjalili M, Maleknia L, Giahi M (2019) Preparation of multi-functional polyester fabrics with the use of a new organo-clay nanopigment in high temperature/high pressure condition. 20:1387–1395. https://doi.org/10.1007/s12221-019-8524-9
107. Hsiao Y-C, Jheng P-R, Nguyen HT, Chen Y-H, Manga YB, Lu L-S, Rethi L, Chen C-H, Huang T-W, Lin J-D, Chang T-K, Ho Y-C, Chuang E-Y (2021) Photothermal-irradiated polyethyleneimine-polypyrrole nanopigment film-coated polyethylene fabrics for infrared-inspired with pathogenic evaluation 13:2483–2495. https://doi.org/10.1021/acsami.0c17169
108. Attia N, Osama R, Elashery S, Kalam A, Al-Sehemi A, Algarni H (2022) Recent advances of sustainable textile fabric coatings for UV protection properties 12:1597. https://doi.org/10.3390/coatings12101597
109. Osterwalder U, Rohwer H (2002) Improving UV protection by clothing—recent developments. In: Recent results in cancer research. Springer, Berlin
110. Bekrani M, Zohoori S, Davodiroknabadi A (2020) Producing multifunctional cotton fabrics using nano CeO_2 doped with nano TiO_2 and ZnO. 20:78–84. https://doi.org/10.2478/aut-2019-0057
111. Sadeghi-Kiakhani M, Arami M, Gharanjig K (2013) Application of a biopolymer chitosan-poly(propylene)imine dendrimer hybrid as an antimicrobial agent on the wool fabrics 22. https://doi.org/10.1007/S13726-013-0193-8
112. Shaki H, Gharanjig K, Khosravi A (2015) Synthesis and investigation of antimicrobial activity and spectrophotometric and dyeing properties of some novel azo disperse dyes based on naphthalimides 31. https://doi.org/10.1002/BTPR.2107
113. Gutarowska B, Matyjas-Zgondek E, Kulpiński P, Mroczyńska-Florczak M, Rutkowski E (2021) Long-lasting photocatalytic and antimicrobial activity of cotton towels modified with TiO_2 and ZnO nanoparticles 11:952. https://doi.org/10.3390/catal11080952
114. Imoto Y, Seino S, Nakagawa T, Yamamoto TA (2017) Quantitative methods for testing antiviral activities of textile fabrics 3. https://doi.org/10.4172/2472-1212.1000146
115. Ribeiro AI, Shvalya V, Cvelbar U, Silva R, Marques-Oliveira R, Remião F, Felgueiras HP, Padrão J, Zille A (2022) Stabilization of silver nanoparticles on polyester fabric using organo-matrices for controlled antimicrobial performance 14:1138. https://doi.org/10.3390/polym14061138
116. Sadeghi-Kiakhani M, Hashemi E, Gharanjig K (2020) Treating wool fibers with chitosan-based nano-composites for enhancing the antimicrobial properties 10. https://doi.org/10.1007/S13204-019-01203-1
117. Mazhar M, Abdouss M, Gharanjig K, Teimuri-Mofrad R (2016) Synthesis, characterization and near infra-red properties of perylenebisimide derivatives 101. https://doi.org/10.1016/J.PORGCOAT.2016.08.018
118. Gharanjig H, Gharanjig K, Farzi G, Hosseinnezhad M, Jafari SM (2020) Novel complex coacervates based on Zedo gum, cress seed gum and gelatin for loading of natural anthocyanins. 164:3349–3360. https://doi.org/10.1016/j.ijbiomac.2020.08.218
119. Gharanjig H, Gharanjig K, Hosseinnezhad M, Jafari SM (2020) Development and optimization of complex coacervates based on zedo gum, cress seed gum and gelatin 148:31–40. https://doi.org/10.1016/j.ijbiomac.2020.01.115

120. Norouzi M, Zare Y, Kiany P (2015) Nanoparticles as effective flame retardants for natural and synthetic textile polymers: application. Mech Optim 55:531–560. https://doi.org/10.1080/15583724.2014.980427
121. Javed A, Wiener J, Saskova J, Müllerová J (2022) Zinc oxide nanoparticles (ZnO NPs) and N-methylol dimethyl phosphonopropion amide (MDPA) system for flame retardant cotton fabrics 14:3414. https://doi.org/10.3390/polym14163414
122. Chao L, Lamei G (2019) Fabrication of high Nano-ZnO doped with boric acid assembled on cotton fabric with flame retardant properties 136:65–70. https://doi.org/10.5604/01.3001.0013.1820
123. Asadnajafi S, Shahidi S, Dorranian D (2019) In situ synthesis and exhaustion of nano TiO_2 on fabric samples using laser ablation method 111:122–128. https://doi.org/10.1080/00405000.2019.1624035
124. Nazari A (2019) Superior self-cleaning and antimicrobial properties on cotton fabrics using nano titanium dioxide along with green walnut Shell Dye 20:2503–2509. https://doi.org/10.1007/s12221-019-1135-7
125. Nozari B, Montazer M, Rad M (2021) Stable ZnO/SiO_2 nano coating on polyester for anti-bacterial, self-cleaning and flame retardant applications 267:124674. https://doi.org/10.1016/j.matchemphys.2021.124674
126. Abdel-Halim ES, Abdel-Mohdy FA, Al-Deyab SS, El-Newehy MH (2010) Chitosan and monochlorotriazinyl-β-cyclodextrin finishes improve antistatic properties of cotton/polyester blend and polyester fabrics 82:202–208. https://doi.org/10.1016/j.carbpol.2010.04.077
127. Choudhury A (2017) Principles of textile finishing. Woodhead Publishing
128. Thennakoon C, Rajapakshe R, Rajapakse R, Rajapakse S (2022) Anti-stain and durable super-hydrophobic/antistatic dual functionality surface for fabric materials based on F-ZnO/TiO_2 composite 101:529–538. https://doi.org/10.1007/s10971-022-05734-y
129. Hassan M (2019) Enhanced colour, hydrophobicity, UV radiation absorption and antistatic properties of wool fabric multi-functionalised with silver nanoparticles 581:123819. https://doi.org/10.1016/j.colsurfa.2019.123819
130. Miles LWC (1994) Textile printing. Second Edition. Edited by Leslie W C Miles Society of Dyers and Colourists
131. Wang C, Yin Y, Wang X, Bu G (2008) Improving the color yield of ultra-fine pigment printing on cotton fabric
132. Franks A (1987) Review article: nanotechnology. https://doi.org/10.1088/0022-3735/20/12/001
133. Gilman JJ (2016) Nanotechnology 5:12–14. https://doi.org/10.1007/s100190100124
134. Tutak M, Bilget Ö (2015) Development of bio-active pigment printed fabric loaded with nano-sized silver/zinc particles 44. https://doi.org/10.1108/PRT-01-2014-0001
135. Patel BH, Chattopadhyay D (2014) Nano metal particles: synthesis, characterization and application to textiles
136. Iqbal M, Mughal J, Sohail M, Moiz A, Ahmed K, Ahmed K (2012) Comparison between pigment printing systems with acrylate and butadiene based binders. https://doi.org/10.4236/jasmi.2012.22016
137. Bishnoi K, Balakrishnaiah B (1999) Effect of chemical pretreatments on the color yield of pigment printed fabrics
138. Shaabani A, Dabiri M, Bazgir A, Gharanjig K (2006) Microwave-assisted rapid synthesis of 1,4-diketo-pyrrolo(3,4-c)-pyrroles' derivatives under solvent-free conditions 71. https://doi.org/10.1016/J.DYEPIG.2005.06.002
139. Patra JK, Gouda S (2013) Application of nanotechnology in textile engineering: an overview 5. https://doi.org/10.5897/JETR2013.0309
140. Lin CH, Lin CM (2013) The study in the application of nanotechnology digital printing in clothing pattern design. https://doi.org/10.4028/WWW.SCIENTIFIC.NET/KEM.562-565.674
141. Hakeim OA, Arafa AA, Zahran MK, Abdou LAW (2014) UV-curable encapsulation of surface—Modified organic pigments for inkjet printing of textiles 447. https://doi.org/10.1016/J.COLSURFA.2014.01.075

142. Rizk HF, El-Borai MA, Hemeda OM, El Sadek MM (2023) Novel nanopigments with a Thiazole moiety for printing paper, carton, and polyester fabrics: synthesis, characterization, and color strength with comparative study 24. https://doi.org/10.1007/s12221-023-00056-4
143. Mahmoodi A, Jiryaei Z, Dadras A, Khorasani M, Shi X (2021) A hybrid yellow nanopigment as an environmentally sound alternative to lead chromate pigment for pavement markings 319. https://doi.org/10.1016/J.JCLEPRO.2021.128733
144. Zavodchikova AA, Safonov VV, Solina EV, Ivanov VB (2014) Ultraviolet paints based on nanopigments for printing on textile materials 84. https://doi.org/10.1177/0040517514523177
145. Gharanjig H, Gharanjig K, Sarli MA, Ozguney AT, Jalili M, Gharanjik A, Jahankaran S (2022) Effect of molecular composition of comb-like polycarboxylate dispersants on hydrophobic dye dispersion properties 350. https://doi.org/10.1016/j.molliq.2022.118615

Chapter 7
Nanotechnology in Coating and Finishing of Textiles

Lelona Pradhan, Saptarshi Maiti, and Ravindra V. Adivarekar

1 Introduction

The terms "nanomaterials" and "nanotechnology" have garnered significant attention in the contemporary world. "Nanomaterials" refer to the materials with at least one dimension falling within the nanoscale range, typically ranging from 1 to 100 nm [10, 41]. Meanwhile, "nanotechnology" encompasses a broad spectrum of activities related to these materials, encompassing their synthesis, characterization, and diverse applications. Nanotechnology is not a recent concept, it dates back to 1950s when Nobel Prize laureate Richard Feynman discussed these topics during the annual meeting of the American Physical Society in his presentation titled "There's Plenty of Room at the Bottom". While the term "nanotechnology" was not explicitly defined at that time, the discussion primarily revolved around exploring diverse application possibilities while manipulating materials at the atomic level. The term "nanotechnology" was officially introduced by Norio Taniguchi in 1974 through one of his published papers [7].

The inception of tangible progress began when IBM successfully developed the scanning tunneling microscope (STM) in 1981, revolutionizing the field by enabling the detailed imaging of materials at the atomic level. This momentous achievement marked the commencement of an ongoing evolution, which later witnessed the advent of the atomic force microscope (AFM) in 1986, among other advancements. Concurrently, fullerenes emerged as some of the earliest nanomaterials studied [97]. Since that point, nanotechnology has been steadily advancing, encompassing all facets of materials, starting from atoms to macro molecules. The continuous refinement of tools and techniques has further strengthened the research interest in this domain [75].

L. Pradhan · S. Maiti (✉) · R. V. Adivarekar
Department of Fibres and Textile Processing Technology, Institute of Chemical Technology, Matunga (E), Mumbai 400019, India
e-mail: sa.maiti@ictmumbai.edu.in

S. Bairagi et al. (eds.), *Nanotechnology in Textile Finishing*, Materials Horizons: From Nature to Nanomaterials, https://doi.org/10.1007/978-981-97-2696-7_7

Nanomaterials have found their way into a multitude of industries, including paint, functional coatings, cosmetics, sensors, and various biomedical applications such as biosensors and drug delivery [7, 41]. Moreover, the textile industry has become a specialized field where nanomaterials have achieved considerable success, with ample opportunities for further advancements. This success can be attributed to the customizable properties and additional functionalities offered by nanotechnology, without compromising the overall feel and comfort of the materials. Applications range from conventional uses such as imparting flame retardancy and antimicrobial properties to fabrics, creating superhydrophobic textiles, and extending to the realm of smart textiles, including nanomaterial-embedded wearables, sensor materials, color-changing fabrics, etc. [109]. Notably, these innovations have transitioned beyond research phases and have been commercialized with significant acceptance in the market. As the textile market continues to expand, so too does the potential for leveraging nanomaterials to achieve desired outcomes.

This chapter deals with the basic concepts regarding nanotechnology and nanomaterials and then proceeds to introduce the readers about various multifunctional coating and finishing using these nanomaterials on textile substrates. A short discussion on the characterization techniques of these nanoranged materials has also been included to deliver a complete idea about their synthesis, functionality, and applications in textile field.

2 Types

Nanomaterials have been classified into several categories, each distinguished not only by their sources but also by their characteristics, which encompass factors such as size, structural arrangement, pore dimensions, and levels of toxicity [47, 67]. Furthermore certain studies have provided a broader classification, classifying these materials into five main groups: (i) carbon-based NPs, (ii) metal NPs, (iii) ceramic NPs, (iv) semiconductor NPs, and (v) polymeric NPs [55], [56] (Fig. 1).

Moreover, nanomaterials can be sorted into categories based on their shapes, although comprehensive information on this aspect is somewhat limited. In this conversation, we will explore the categorization of nanomaterials by their shapes. These materials, based on their shapes, can be generally organized into six types, and we will provide a brief description of each type below.

2.1 Nanoparticles

Nanoparticles are materials categorized as nanostructures with dimensions spanning from 1 to 100 nm. If the size falls below 1 nm, they are more precisely referred to as atom clusters rather than nanoparticles. They exhibit diverse shapes including spherical, cylindrical, spiral, and others while maintaining relatively consistent internal

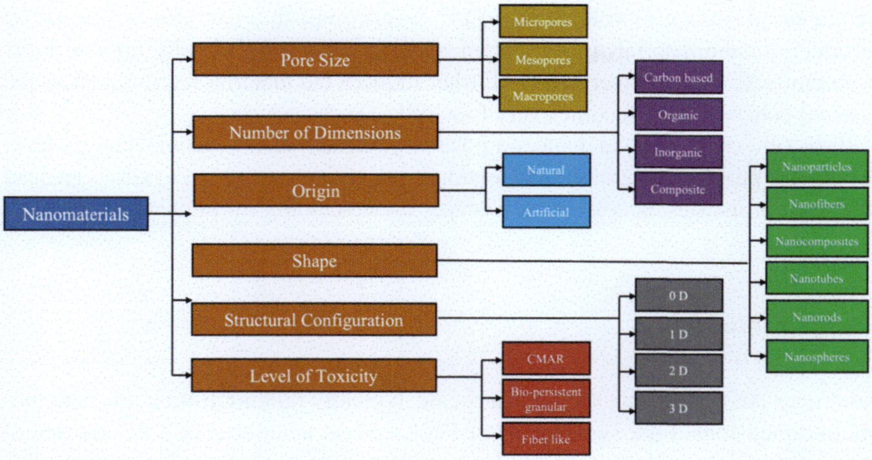

Fig. 1 Classification of nanomaterials

axis dimensions. These nanoparticles can further be classified into organic, inorganic, and carbon-based varieties [52].

2.2 *Nanofibers*

Fibers having diameters in the range of 1 nm to 1 μm are considered to be nanofibers. These can be produced from different polymers by techniques such as template synthesis, electrospinning, self-assembly, and more. Nanofibers have found successful applications in diverse biomedical fields, including tissue engineering and wound dressings. [26].

These materials also exhibit intelligent properties, such as their utilization in direct electron transfer, where polyaniline nanofibers serve as electrodes for immobilizing glucose oxidase, enabling electron transfer without structural deformations [113]. Conductive nanofibers are another intriguing research area, involving the synthesis of nanofibers with intrinsically conducting polymers or the addition of conductive additives for various applications [5].

2.3 *Nanocomposites*

Nanocomposites, akin to traditional composites, are materials formed when phase mixing occurs at the nanolevel. This results in materials that exhibit superior properties compared to their macroscopic or conventional counterparts [54]. The reduction in particle size has been proven to deliver unique and superior properties owing to the increased surface-to-volume ratio [15]. Nanocomposites can also be formed by introducing nanomaterials in the polymer matrix. The exceptionally high surface-to-volume ratio and number density further reduces the material requirement as the required concentration becomes very low [68].

However, challenges in nanocomposite synthesis, such as achieving uniform dispersion, proper orientation, and maintaining the correct aspect ratio, can lead to inferior properties, necessitating a deeper understanding for improvement [40].

2.4 Nanotubes

Nanotubes are hollow tubes at the nanoscale, typically ranging from a few nanometers in dimensions. First synthesized in 1991, carbon nanotubes (CNTs) are one of the most promising allotropes of carbon family having fine structure–property relationships which can again be single-walled nanotubes (SWNTs) and multi-walled nanotubes (MWNTs) depending on the concentric cylindrical layers [3, 36]. CNTs find extensive applications in energy storage, flat displays, X-rays to microwave generators, water purification, as charge carriers and drug delivery in biomedicals, and many more [25]. Recent advances revolve around finding sustainable sources for CNT synthesis like agro-waste, bio-char, industrial waste, plastics, and others [33].

Beyond carbon based, nanotubes have also been synthesized from other functional materials like boron nitride and titanium dioxide. Boron nitride nanotubes (BNNTs) possess unique properties and are used in mechanical reinforcement, filtration, hydrogen storage, sensing, etc. [57]. Similarly, TiO_2 nanotubes being non-toxic, environment friendly, and corrosion resistant have found wide applications in catalytic activities, solar cells, electrochromic devices, drug delivery, biomedical coatings, etc. [91]. High price and limited bulk production are some of the challenges faced by these materials.

2.5 Nanorods

Nanorods are materials with dimensions ranging from 1 to 100 nm and can be composed of metals, nonmetals, or semiconductors. They have been synthesized from various materials, including silver nitrate, carbon, tin, manganese, and gold, each exhibiting unique physical and chemical characteristics. Nanorods can be fabricated through methods such as electrospinning and hydrothermal techniques [71]. Nanorods have demonstrated success in applications such as batteries, electronics, water purification, and solar cells, often offering superior properties compared to conventional materials [76]. Gold nanorods have found suitable applications in photo-acoustic molecular imaging where the nanoscale range of these materials greatly influenced the properties [19], they have also been utilized in plasmonic spectroscopies, optoelectronic devices, etc. [18].

2.6 Nanospheres

Nanomedicine is an emerging branch where the diagnosis and treatment of diseases takes place in the scale range of 1–500 nm, thus paving a path of noninvasive procedures [53]. This requires nanocarriers or nanomaterials which brings nanospheres into picture. The size of these homogeneous nanospheres range from 100 to 1000 nm and have been utilized for targeted drug delivery. Various polymers can be used to produce these nanospheres like polycaprolactam, polylactic acid, polyethylene, ethyl cellulose, etc., where the active material like drug, antioxidants are embedded for applications in biomedicals, bone tissue regeneration, cosmetics, etc. [37, 104].

3 Synthesis Techniques

Nanoparticles can be synthesized from various methods or techniques depending on the origin materials and the properties expected. These include physical, chemical, and biological methods [81, 92]. Accessing all these methods, the idea to the synthesis of nanoparticles mainly revolves around two ideas, i.e., bottom-up approach and top-down approach which are briefly discussed below.

3.1 Bottom–Up Approach

As the name suggests, this approach deals with the synthesis of nanoparticles when the interaction takes place by combining sub- or nanoscale (atoms or molecules) under different conditions to produce the nanoparticle [27]. This may include various chemical reactions where the raw materials are mostly liquids or gases, which are ionized, dissociated, evaporated, etc., to form the final product whose dimensions are always less than 100 nm [38]. This approach provides more control over the process conditions, thus resulting in a well-defined shape, size, and characteristics of synthesized nanoparticles [1].

3.1.1 Sol Gel

In this technique, the precursor is dissolved in a medium like water, alcohol, etc., and is heated under stirring conditions to form gel. This gel is then dried by evaporating or burning the medium used and is then powdered followed by calcination to form the nanoparticles. This method is usually used for the synthesis of metal oxide nanoparticles [11].

3.1.2 Wet Chemical Synthesis

This method is gaining huge interest due to the controlled synthesis of nanoparticles having required shape, size, and crystallinity and is mostly used for the production of monocrystalline metal oxides. This is not a single method, rather includes four different techniques like microwave-assisted synthesis, thermal decomposition, solvothermal synthesis, and continuous flow synthesis [73]. Shape control of amorphous nanoparticles faces many difficulties, however, wet chemical synthesis has proven to be successful in this regard [112].

3.1.3 Self-assembly

It is a process where the molecules arrange themselves to form ordered structures without any external interference resulting in rationally structured matters with dimensions less than 100 nm [106]. It can be driven by various interactions like hydrogen bonding, electrostatic interaction, van der Waals forces, etc., to achieve required morphology [29].

3.1.4 Hydrothermal

This method of synthesis requires the solubility of the precursor in water which may be in room temperature or elevated temperature either in low-pressure or high-pressure conditions to produce the nanoparticles [34].

3.1.5 Template

This method involves synthesizing nanomaterials by using membranes to obtain desired morphologies ranging from nanorods to nanofibers to many more. Two types of membranes have generally been used for template synthesis, i.e., track-etch and porous alumina depending on the required properties and processing conditions [103]. The precursor is deposited in the templates by various chemical strategies like electrochemical deposition, electroless deposition, chemical polymerization, etc. [30, 45].

3.1.6 Chemical Vapor Deposition (CVD)

This method involves the layering of gaseous reactant having high vapor pressure, on top of a substrate which is then subjected to chemical reactions inside a reaction chamber. The process could involve temperature and pressure variations depending on the processing conditions. Nanoparticles are synthesized on the substrate in the form of a film, as a result of these reactions. This film is then recovered and used.

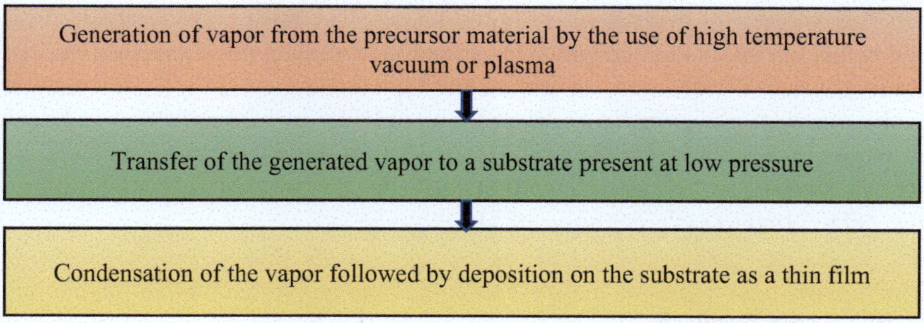

Fig. 2 Steps involved in PVD

Advantages include synthesis of strong, pure nanoparticles while the upside of this process involves toxic by-products and costly apparatus [47]. This method can be divided into various types depending on the requirement of pressure, characteristics of vapor, and use of plasma [31].

3.1.7 Physical Vapor Deposition (PVD)

This technique involves mainly three steps [28], (Fig. 2).

This process is capable of depositing a uniform layer ranging from nano- to visible-scale and can be used to utilize various kinds of inorganic and organic materials [86].

3.1.8 Electrospinning

Electrospinning involves the stretching of a viscous polymer solution under high voltage condition to generate nanofibers [94]. Various morphologies can be tuned by controlling the machine parameters like voltage, feed distance, feed rate, etc., and polymer parameters like concentration, viscosity, etc. Surface roughness as well as different kinds of porous to nonporous materials can be synthesized in the nanorange using this technique [69]. However, high cost and low production are some of the obstacles faced by this technique. New advancements are being carried on increasing the production opening scope for commercialization, among which multi-needle electrospinning is one of them, where instead of single needle multiple needles simultaneously work to produce the nanofibers [43, 83].

3.1.9 Supercritical Fluid Synthesis

Supercritical fluid is defined as a substance having temperature and pressure above its critical point, where it possesses the viscosity of a gas and density of a liquid, with nonexistence of distinct gas or liquid phases. The most popular supercritical fluid is that of carbon dioxide which has found successful applications in areas like cleaning, material extraction, environmental remediation, etc. [107]. This method involves growing the particles in desired morphologies by size reduction and does not require any further processing [14]. Apart from these, supercritical fluids are great alternatives to toxic organic solvents and can be recovered to 100% for reuse, thus, are referred to as "green solvent" [100].

3.1.10 Laser Pyrolysis

In this method, the precursor material in the form of a gas is introduced to a chamber with the help of a carrier gas, typically argon. The precursor then comes in contact with a high-power laser which initiates nucleation and thus the synthesis of nanoparticles. These nanoparticles are then collected by suitable methods [105]. This process is generally used for the synthesis of carbide, nitrite, oxides, metals, and composites nanoparticles [89]. Advantages of this process include highly pure materials without reactor corrosion and single-step procedure [64].

3.1.11 Molecular Condensation

A stream of atoms is generated from a target material by the use of heat inside a vacuum chamber. The vapor is then reacted and cooled to undergo condensation under a controlled environment. The cooled vapor keep on growing up to a certain point which are collected as nanoparticles with high reactivity [22].

3.2 *Top–Down Approach*

This approach is completely opposite to bottom–up approach and deals with the synthesis of nanomaterials where the starting bulk material is broken down to its nanoform to finally form the nanoparticles with the use of energy. Less control over the processing conditions result in complex particle morphologies, uneven structures, and wide size distribution. These methods are used for nanomaterials having dimensions over 100 nm [21, 38].

3.2.1 Mechanical Grinding

Mechanical grinding, or more commonly known as ball milling, is one of the low-cost and high-yield technique for the synthesis of nanomaterials. The bulk raw material is introduced in the milling chamber along with balls made up of different materials like ceramic, glass, stainless steel, etc. During processing, the high impact from the balls breaks down the precursor materials into its nanoform having same constituent. Processing parameters like number of balls, milling speed, temperature, medium used (if any), etc., affect the properties of the nanoparticles [84].

3.2.2 Laser Ablation

In this technique, the bulk material is irradiated with laser causing it to absorb heat which then evaporates or sublimates. This evaporation followed by deposition on a collector produces nanoparticles of the initial bulk material [88].

3.2.3 Sputtering

Sputtering causes the ejection of particles from the target species. These particles in the range of nanoscale are then deposited as a thin film. Various factors affect the size and shape of these synthesized nanoparticles like temperature, thickness of the layer, substrate, etc [47]. This technique removes the drawbacks of using precursor chemicals and the decomposition products [66].

3.2.4 Lithography

Lithography can be associated with the techniques of nano-architecture formation using a beam of light or electrons. This can be of two types, (i) masked lithography and (ii) mask-less lithography. In masked lithography, specifically designed masks or templates are used to transfer the nanodesigns onto a substrate, while in mask-less lithography, no additional masks or templates are used, instead the patterns are carried out directly with the help of light or electron beams [7, 35]. This technique has the advantage of making it possible to tune the nanostructure size, period, and materials while making it possible to fabricate a larger area with ease and lost cost [12].

3.2.5 Coprecipitation

When a material is soluble in a medium, after a certain point, it will reach its supersaturation. This supersaturation then results in the nucleation process followed by diffusion of the material onto the surface producing nanoparticles. The process needs to be controlled carefully to control the nanoparticle characteristics [72].

3.2.6 Thermal Decomposition

This method is typically used for the synthesis of monodispersed nanoparticles. Here, precise temperature conditions are introduced to the principal material to initiate bond breakage, thus triggering chemical reactions and formation of new by-products. Important process parameters include processing time, temperature, reactant concentration, etc., which play a greater role in controlling the shape and size of the nanoparticles thus produced [56].

3.3 Biological Synthesis

This technique is often referred to as "green synthesis" and involves non-toxic, environment friendly, sustainable media for the synthesis of nanomaterials. The disadvantages faced by other synthesis techniques like the physical and chemical methods have forced researchers to find suitable alternatives, and green synthesis has adequate potential. However, the issues like extraction of raw materials, quality of final product, and homogeneity of particle size still pose problems in the commercialization of such techniques [110]. The conversion of precursors to nanomaterials is usually mediated by a naturally availably organism or substance, some of which are discussed below [77], [96].

3.3.1 Bacteria

Bacteria have various inorganic compounds present in their intra or extracellular matrices. These compounds help in the breakdown of materials and reducing their dimensions to nanoscale, thus producing nanomaterials. The most widely used bacteria for such applications include the prokaryotic and actinomycetes.

3.3.2 Fungi

Monodispersed nanoparticles with well-defined structures are the main advantages of using fungi-based synthesis techniques. They act on the principle of enzymatic reduction and deliver better properties along with the scope for larger production. The enzymes are secreted from the extracellular matrix of the fungi and take part in the synthesis process. This technique was first utilized in the synthesis of silver nanoparticles.

3.3.3 Yeast

These organisms are present in the eukaryotic cells as unicellular organisms. Yeasts like *Candida glabrata, Schizosaccharomyces pombe, and Torulopsis sp.* have been successfully utilized to synthesize various types of nanoparticles.

3.3.4 Plant

The wide availability of plants along with their functional properties have evoked interest in their use for the synthesis of nanoparticles. Various plant species like Aloe vera, lemon, coriander, neem, etc., have been utilized to synthesize silver and gold nanoparticles and other bio-nanocomposites.

4 Characterization Techniques

For better utilization of nanomaterials, a detailed knowledge of their properties is must. Nanomaterials are characterized by various techniques depending on material properties like size, surface area, composition, net charge, morphology, crystallinity, etc. The techniques also vary depending on the state of matter like solid, liquid, gas, etc. Important characterization techniques for nanomaterials are listed in Table 1, [4], [9, 51, 111].

5 Methods of Application on Textile Substrate

Coatings and finishing techniques for a variety of substrates have been employed for centuries in various industries. However, over time, the desired outcomes of these methods have evolved, necessitating more advanced coating procedures and specific material requirements. Initially, coatings were primarily used to protect and decorate substrates, but as time has progressed, the requirements have expanded to include functionalization to impart additional properties that were not originally inherent to the substrate [63].

Textile industry is one of the oldest industries and has been constantly growing with the inception of advanced technologies. Coatings and finishing of textiles mainly proceed through various application techniques including the physical and chemical processes using different chemical finishes. The industry demands and the regulations to control the toxicity of add-ons on the textile substrates have forced the industry to look for new materials which could deliver comparable properties with low add-on. Due to the exceptional properties possessed by nanomaterials, these materials garnered significant interest for their utilization in coating and finishing industries.

Table 1 Characterization methods for nanomaterials

Sr. no	Parameter	Characterization technique
1.	Particle size	• Atomic force microscopy (AFM) • Laser diffractometry • Photon correlation spectroscopy (PCS) • Scanning mobility particle sizer (SMPS) • Optical particle counter • Scanning electron microscopy (SEM) • Transmission electron microscopy (TEM) • Dynamic light scattering (DLS) • Localized surface plasmon resonance (LSPR)
2.	Surface area	• BET isotherm • Simple titration • Nuclear magnetic resonance (NMR) • Scanning mobility particle sizer (SMPS) • Differential mobility analyzer (DMA)
3.	Synthesis stability	• UV–Vis spectroscopy
4.	Dispersibility	• Atomic force microscopy (AFM) • Critical flocculation temperature (CFT)
5.	Morphology	• Scanning electron microscopy (SEM) • Transmission electron microscopy (TEM) • Scanning tunneling microscopy (STM)
6.	Crystallography	• X-ray diffraction (XRD) • Neutron diffraction • High-resolution transmission electron microscopy (HR-TEM)
7.	Concentration	• Condensation particle counter
8.	Surface hydrophobicity	• Rose Bengal (dye) binding • Water contact angle measurement • X-ray photoelectron spectroscopy
9.	Charge	• Laser Doppler anemometry • Zeta potentiometer
10.	Surface chemical analysis	• Static secondary ion mass spectrometry • Sorptometer • Energy-dispersive X-ray (EDX)

The initial use of nanomaterials in textiles dates back to 2600 BC, where nanoscale pigments were used in the dyeing of fabrics. Nanotechnology can be introduced in the coating and finishing of textiles by three ways [24].

i. Crosslinking the substrate with nanoparticle,
ii. Use of nanoparticle dispersion,
iii. Direct deposition of nanomaterials like nanofibers or nanoparticles on the substrate by electrospinning or spray coating.

Although the terms “nanocoating” and “nanofinishing” are often used interchangeably, they have distinct meanings. While nanocoating refers to the deposition

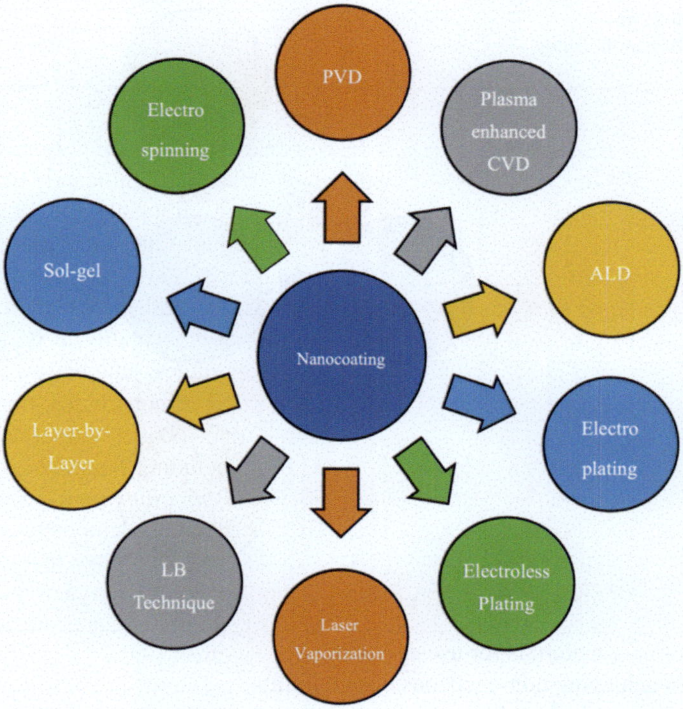

Fig. 3 Application techniques for nanocoatings in textiles (*PVD: Physical vapor deposition, CVD: Chemical vapor deposition, ALD: Atomic layer deposition, LB: Langmuir–Blodget*)

of nanorange film on the textile substrate, nanofinishing refers to the functionalization of textile substrate with nanomaterials by covalent bonding or other methods. Both these processes use different techniques. The methods of application of nanotechnology on textiles through coating and finishing are illustrated in Figs. 3 and 4, [24], [58].

Nanotechnology coatings and finishes to functionalize textile have found significant commercial applications. Some of which are listed below [58].

- Ag fresh product (Nisshinbo, Japan): antibacterial and anti-odor finish,
- Resist spill, release stains (Nanotex, USA): Antistain finish,
- Nano Matrix (Toray, Japan): Antistain finish,
- Perma-Freshy (Teijin, Japan): Antistain finish,
- Percenta Nano Textile & Leather Sealant" (Percenta AG, Germany): Water- and oil-resistant finish,
- Perma-Freshy (Teijin, Japan): Sweat Deodorizing, etc.

Nanotechnology still faces some major issues in their applications, some of which are listed below [79],

- Nonuniform size and shape,

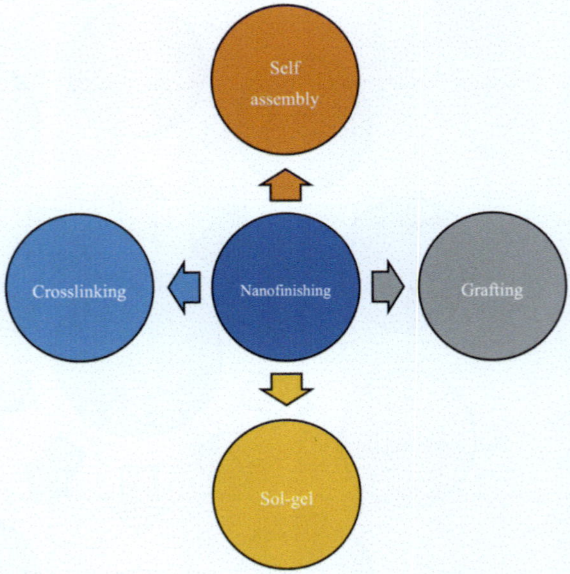

Fig. 4 Application techniques for nanofinishing in textiles

- Purity of nanomaterials for use as coatings,
- Issues in achieving homogeneous dispersibility,
- Regulations toward health and safety,
- Aggregation issues,
- Durability problems.

These issues needed to be addressed to increase the applicability of these materials and wider acceptance. Nonetheless, various application areas have been explored using these nanomaterials which will be discussed in brief in the next section.

6 Application Areas

Applications of nanotechnology is not confined to a single area, rather covers a wide range of industries. The applications can be mainly divided into three sectors, namely nanobiotechnology, nano-electronics, and nanomaterials. These fields are not exclusive and often overlap with each other depending on the final applicability [99]. Limiting our focus to only textile industries, nanotechnology in the form of nanocoatings and nanofinishing has found a number of applications excelling in properties like water repellency, conductivity, UV protection, easy care, antibacterial, self-cleaning, antistatic finish, and many other ones [44].

Most commonly used nanomaterials in textiles include various metal nanoparticles like silver, titanium, zinc, copper; various inorganic materials like carbon nanotubes, graphene; various bio-based products [16, 90]. Nanoparticles does not significantly affect the feel and flexibility of textile substrates as compared to the conventional coatings and finishes, and thus, these materials have an upper hand in imparting the same functionalities along with low add-on [102]. Some of the important application areas relating nanotechnology to the textiles are discussed below in brief.

6.1 Flame Retardancy

Conventional flame retardants face a number of compatibility issues with the textiles ranging from reduced tensile properties to reduced dyeability to reduced flexibility along with the requirement of large concentrations or add-on to achieve desired results. Commercial products like PYROVATEX and PROBAN face similar issues affecting the properties of the material. Thus, the search for alternatives began which could deliver flame retardancy without adversely affecting the textile properties, and nanomaterials are the most promising candidates. Nanoparticles like zinc oxide, titanium dioxide, aluminum trioxide, nanoclay, nanosilica, etc., have been successfully studied for flame retardancy. Zinc nanoparticles aided a better flame retardancy with a reduced add-on by almost 12 times than the bulk zinc along with 25% increase in char formation thus increased LOI. Similarly, just 0.5% of titanium dioxide deliver flame retardancy with other additional functionalities like hydrophilicity, self-cleaning, etc. [2].

Nanoparticles treated with naturally derived materials like phytic acid and chitosan, further enhance the properties like hydrophilicity, increasing tensile properties, improving antibacterial properties, etc. [59, 114]. Alternatively, nanoparticles can be functionalized using the phosphorous or halogens to induce flame retardancy during their synthesis which can then be coated onto the textiles.

Although these nanofinishes provide higher LOI, better thermal stability with comparatively reduced add-on, still they lack the properties to self-extinguish and thus fail many regulatory tests. Use of polymer nanocomposites in conjunction to conventional flame retardants could synergistically utilize the properties of these nanomaterials [74].

6.2 Nano/Ultrafiltration

Filtration primarily revolves around three areas, namely effluent treatment, water filtration, and gas filtration. Use of nanotechnology in these areas is of great interest relating to their exceptional properties. Textile effluents containing larger quantity of dyes are a greater cause of headache considering various environmental and aqua toxicity factors. A study reports a polyester fabric coated with polystyrene-acrylic/ZnO-based nanocomposite delivered better dye removal properties up to 60% [80]. Another study used carbon fibers derived from PAN and functionalized with silver nanoparticles to successfully eliminate various microbes like *E. coli, C. albicans, A. niger*, etc., and thus can be helpful in reducing water toxicity mainly in rural regions [65].

Considering gas filtration, studies have used electrospun nanofibers to produce hydrophobic textile substrates which are capable of gas filtration along with successful water filtration that is achieved [49]. Similarly, treatment with nanoparticles further increased the filtration properties of single-layered nanowovens while improving the physical and mechanical properties [46].

6.3 Anti-odor

Many reasons including bacterial growth causes odor in textiles which is a major health concern. Materials used for odor elimination usually have specific selectivity toward the odor, like carboxyl groups for ammonia, amine group for formaldehyde, etc. Other processes including oxidation and reduction can also be used in the elimination of odor. Studies have reported the use of nanocomposite emulsions containing titanium dioxide, silver, etc., successfully eliminated the odor from textiles more efficiently as compared to conventional formulations [17].

6.4 Antibacterial/Antimicrobial

Due to the inherent properties of textiles like moisture absorption and temperature, these materials are very much prone to bacterial and microbial growth which can result in many health concerns. Antimicrobial finishes are categorized into different types depending on their efficiency, mode of action, mechanism of application, durability, etc. On the basic of activity, the agents can be biocidal or biostatic, and on the basis of durability they can be leaching or bound antimicrobials. Biocides act by killing the bacterial while biostatic acts by inhibiting the growth. Similarly leaching antimicrobials include agents having controlled release and are not chemically bound to textile substrate and thus are prone to wear out with reduced efficacy during gradual use. On the other hand, bound antimicrobials are chemically bound to the textiles and thus are more durable in their action. Conventional products include quaternary ammonium compounds, N-halamines, chitosan, silver, polybiguanides, etc. [32].

Various nanoparticles like silica, silver, zinc, chitosan, etc., have successfully being utilized in the coating and finishing of textiles by techniques like one-step dip coating, spray coating [108], sonochemical coating [82], layer-by-layer self-assembly [48], pad-dry-curing [95], etc. Apart from synthetic or metal nanoparticles, herbal nanoparticles are gaining a major interest in this application area. Aluminum-based nanoparticles obtained from floral base [95], carnauba wax nanoparticles [8], nanoparticles from biopolymers like chitosan, etc., are some of the examples.

6.5 Water Repellent

Many end applications require complete water repellency like rainwear, sportswear, protective clothing, automobile interiors, and more. To achieve these, industries are looking forward to superhydrophobic coatings having a water contact angle > 150°. Wetting properties of textiles are influenced by two parameters, i.e., surface free energy and surface roughness. These two properties are intrinsic to materials and, can be controlled depending on the end uses by the application of various chemical formulations or techniques [87].

To achieve superhydrophobic textiles, hydrophobic materials are used along with nanoparticles to achieve low surface energy with increased surface roughness. Nanomaterials including zinc oxide NPs (nanoparticles), zinc oxide nanorods, silica NPs, titanium dioxide NPs, silver NPs, titanium dioxide NPs, carbon nanotubes, etc., have successfully being utilized in this regard. Techniques like layer-by-layer, spray coating, and immersion coating are used to functionalize textiles with these properties. Along with water repellency, these finishes also provide additional benefits like transparent surfaces with self-cleaning, oil – water separation, antistain, antiwater absorption, and antibacterial properties [20, 50, 101].

6.6 Oil Repellent

Similar to water repellency finishing, oil repellency requires textiles to have extremely low surface energy to mimic the lotus leaf effect. Silica nanoparticles have been widely used in a pad-dry-cure method to achieve oil repellency properties when coated onto textile substrates. Good mechanical strength is also an additional benefit of such coating [60, 78]. Apart from nanoparticles, silicon nanofilaments could also be used to achieve the oleophobic properties [61].

6.7 UV Protection

One of the significant carcinogens in our atmosphere is the UV radiation from the sun. Prolonged exposure could result in aging, skin cancer, etc. UV protection has become a must and textiles are the most basic protective materials used by humans. UV radiation can be divided into three types, transmission, absorption, and reflection when they come in contact with the fabric. Blocking the transmission of UV rays serves the main principle in achieving UV protective clothing [70].

Most commonly used nanoparticle is titanium dioxide nanoparticles as they provide high UPF factors along with good wrinkle resistant or crease recovery properties [62, 85]. Similar to titanium dioxide nanoparticles, zinc nanoparticles are gaining quite success which can be padded or micro-encapsulated on textile substrates to obtain UV protection [98]. Zinc nanoparticles can also be biosynthesized to reduce the environmental load and in fact biosynthesized zinc particles show better efficiency than the ones obtained by chemical methods of synthesis [6]. Research is also moving toward finding alternative natural-based nanoparticles for such applications. Iron titanate nanoparticles derived from natural resources (like ilmenite sand) have shown comparable results in blocking the UV rays [23].

6.8 Smart Textiles

Smart textiles are classified into three types,

(a) *Passive smart textiles*: Only sense the changes,
(b) *Active smart textiles*: Sense the changes and act accordingly,
(c) *Ultra-smart/Super smart textiles*: Sense the change, react according, and further adapt their functions according to the external environment.

First generation of smart textiles, i.e., passive one, includes textiles coated with nanoparticles to achieve various properties like antimicrobial, UV protection, etc. Active smart textiles include shape changing textiles, thermochromic fibers, etc., while the ultra-smart textiles include the nanogenerators, wearable electronics, sensors, actuators, energy storage textiles, etc. [42].

Use of nanotechnology is evolving, and smart textiles is one of the potential fields of application of these multifunctional materials. Micro- and nanofibers of various polymers are being utilized for energy storage applications. Flexible fiber batteries, organic light-emitting diodes, and use of carbon nanotubes onto soft textile fabrics are being extensively researched [44, 93]. Some of the conducting polymers include polyaniline, polyacetylene, polythiophene, polypyrrol, etc., which are used as sensing and actuating materials when applied on textile substrates [13].

7 Future Trends

Although the incorporation of nanotechnology in textiles is not a new concept, this field continues to expand into new horizons day by day. The demand for fashionable and multifunctional textiles is on the rise, necessitating advanced materials and streamlined manufacturing techniques. The future of nanotechnology in textiles hinges primarily on four key factors: (i) the development of synthesis techniques that yield homogeneous and uniformly distributed nanoparticles, (ii) cost-effective production methods, (iii) scalability for commercialization, and (iv) compliance with laws and regulations regarding use and safety.

Simplifying synthesis techniques involves not only the introduction of advanced methods but also the adaptation of conventional approaches to make the most of existing resources. Streamlining processes to one-step methods is preferred over multi-step procedures. Consumer demands and market responses significantly shape the future of nanotextiles. Currently, many synthesis techniques require sophisticated instruments, driving up production costs, which should be addressed to enhance affordability. Additionally, nanotechnology-based textiles should be amenable to large-scale production to reduce costs and expand their range of applications. Concerns about the toxicity and health hazards associated with nanomaterials have led to the formulation of regulations aimed at safeguarding human health by imposing limits on the use of these materials. The textile industry must adhere to these regulations rigorously to ensure product quality for the consumer market [24].

Along with the conventional finishing properties like water repellency, stain repellency, antibacterial, antimicrobial, etc., nanofinishes and nanocoatings are capable of delivering various other added benefits like drug release, sensor properties, conductivity, etc., and have the potential to convert conventional textiles to smart textiles [16].

Organizations worldwide are keenly interested in harnessing the potential of nanotechnology in the textile industry, recognizing the need for further exploration, and understanding. Key areas of focus include smart fabrics, self-healing fabrics, advanced materials like luminescent polymers, absorbent textiles, innovative active wear and sportswear, temperature-sensitive fabrics, anti-odor solutions, and flame-retardant textiles [39].

8 Conclusion

Nanotechnology, characterized by materials with nanoscale dimensions, represents an advanced field of research that is inherently technology-driven, serving multiple purposes. Introducing this cutting-edge science into one of the oldest industries, namely textiles, presents a multitude of challenges. This technology has found successful applications in various aspects of textile production, particularly in coating and finishing processes. Coating and finishing mainly focuses on the functionalization of textile substrates by delivering additional properties or increasing the appeal. These processes mainly involve the use of various chemicals to achieve desired results. Introduction of nanotechnology to this field is quite interesting as one of the most significant advantages lies in the ability of nanomaterials to provide multifunctional properties in a simple, one-step procedure. While nanotechnology offers significant advantages, it also comes with certain drawbacks such as concerns about durability, toxicity, and release profiles. Nevertheless, embracing this modern scientific discipline in the textile industry holds the promise of enhanced performance and numerous advancements. Therefore, a thoughtful and deliberate approach should be taken when considering the incorporation of nanotechnology for textile coatings and finishing. The potential of nanotechnology is remarkably high, especially in sectors like technical textiles and smart textiles, and warrants thorough exploration for optimal utilization.

References

1. Abid N, Khan AM, Shujait S, Chaudhary K, Ikram M, Imran M, Haider J, Khan M, Khan Q, Maqbool M (2022) Synthesis of nanomaterials using various top-down and bottom-up approaches, influencing factors, advantages, and disadvantages: a review. Adv Colloid Interface Sci 300:102597. https://doi.org/10.1016/j.cis.2021.102597
2. Ali SW, Basak S, Shukla A (2020) Nanoparticles: a potential alternative to classical fire retardants for textile substrates. In: Handbook of nanomaterials for manufacturing applications. Elsevier, pp 265–278. https://doi.org/10.1016/B978-0-12-821381-0.00011-9
3. Andrews R, Jacques D, Qian D, Rantell T (2002) Multiwall carbon nanotubes: synthesis and application. Acc Chem Res 35(12):1008–1017. https://doi.org/10.1021/ar010151m
4. Ealia SAM, Saravanakumar MP (2017) A review on the classification, characterisation, synthesis of nanoparticles and their application. IOP Conf Ser: Mater Sci Eng 263:032019. https://doi.org/10.1088/1757-899X/263/3/032019
5. Arrigoni A, Brambilla L, Castiglioni C, Bertarelli C (2022) Conducting electrospun nanofibres: monitoring of iodine doping of P3HT through infrared (IRAV) and Raman (RaAV) Polaron spectroscopic features. Nanomaterials 12(23):4308. https://doi.org/10.3390/nano12234308
6. Asmat-Campos D, Delfín-Narciso D, Juárez-Cortijo L (2021) Textiles functionalized with ZnO nanoparticles obtained by chemical and green synthesis protocols: evaluation of the type of textile and resistance to UV radiation. Fibers 9(2):10. https://doi.org/10.3390/fib9020010
7. Baig N, Kammakakam I, Falath W (2021) Nanomaterials: a review of synthesis methods, properties, recent progress, and challenges. Mater Adv 2(6):1821–1871. https://doi.org/10.1039/D0MA00807A
8. Bashari A, Koohestani AHS, Salamatipour N (2020) Eco-friendly dual-functional textiles: green water-repellent and anti-bacterial. Fibers Polym 21(2):317–323. https://doi.org/10.1007/s12221-020-9568-6
9. Bhatia S (2016) Nanoparticles types, classification, characterization, fabrication methods and drug delivery applications. In: Bhatia S (ed) Natural polymer drug delivery systems. Springer International Publishing, pp 33–93. https://doi.org/10.1007/978-3-319-41129-3_2
10. Bhushan B (2016) Introduction to nanotechnology: history, status, and importance of nanoscience and nanotechnology education. In: Winkelmann K, Bhushan B (eds) Global Perspectives of Nanoscience and Engineering Education. Springer International Publishing, pp 1–31. https://doi.org/10.1007/978-3-319-31833-2_1
11. Bokov D, Turki Jalil A, Chupradit S, Suksatan W, Javed Ansari M, Shewael IH, Valiev GH, Kianfar E (2021) Nanomaterial by sol-gel method: synthesis and application. Adv Mater Sci Eng 2021:1–21. https://doi.org/10.1155/2021/5102014
12. Brady B, Wang PH, Steenhoff V, Brolo AG (2019) Nanostructuring solar cells using metallic nanoparticles. In: Metal nanostructures for photonics. Elsevier, pp 197–221. https://doi.org/10.1016/B978-0-08-102378-5.00009-X
13. Bs H, Gmn I, Anma H (2019) Applications of nanotechnology in textiles: a review. Adv Res Text Eng 4(2):1038
14. Byrappa K, Ohara S, Adschiri T (2008) Nanoparticles synthesis using supercritical fluid technology—towards biomedical applications. Adv Drug Deliv Rev 60(3):299–327. https://doi.org/10.1016/j.addr.2007.09.001
15. Camargo PHC, Satyanarayana KG, Wypych F (2009) Nanocomposites: synthesis, structure, properties and new application opportunities. Mater Res 12(1):1–39. https://doi.org/10.1590/S1516-14392009000100002
16. Chakrabarty S, Jasuja K (2022) Applications of nanomaterials in the textile industry. In: Pandey LM, Hasan A (eds) Nanoscale engineering of biomaterials: properties and applications. Springer Nature Singapore, pp 567–587. https://doi.org/10.1007/978-981-16-3667-7_20

17. Chen C-C, Wang C-C, Yeh J-T (2010) Improvement of odor elimination and anti-bacterial activity of polyester fabrics finished with composite emulsions of nanometer titanium dioxide-silver particles-water-borne polyurethane. Text Res J 80(4):291–300. https://doi.org/10.1177/0040517508100626
18. Chen H, Shao L, Li Q, Wang J (2013) Gold nanorods and their plasmonic properties. Chem Soc Rev 42(7):2679–2724. https://doi.org/10.1039/C2CS35367A
19. Chen Y-S, Zhao Y, Yoon SJ, Gambhir SS, Emelianov S (2019) Miniature gold nanorods for photoacoustic molecular imaging in the second near-infrared optical window. Nat Nanotechnol 14(5):465–472. https://doi.org/10.1038/s41565-019-0392-3
20. Chirila L, Cinteza LO, Tanase M, Radulescu DE, Radulescu DM, Stanculescu IR (2020) Hybrid materials based on ZnO and SiO_2 nanoparticles as hydrophobiccoatings for textiles. Ind Text 71(04):297–301. https://doi.org/10.35530/IT.071.04.1814
21. Cook AB, Clemons TD (2022) Bottom-up versus top-down strategies for morphology control in polymer-based biomedical materials. Adv NanoBiomed Res 2(1):2100087. https://doi.org/10.1002/anbr.202100087
22. Daraio C, Jin S (2012) Synthesis and patterning methods for nanostructures useful for biological applications. In: Silva GA, Parpura V (eds) Nanotechnology for biology and medicine. Springer New York, pp 27–44. https://doi.org/10.1007/978-0-387-31296-5_2
23. Dhineshbabu NR, Bose S (2018) Smart textiles coated with eco-friendly UV-blocking nanoparticles derived from natural resources. ACS Omega 3(7):7454–7465. https://doi.org/10.1021/acsomega.8b00822
24. Dolez PI (2019) Application of nanomaterials in textile coatings and finishes. In: Nanomaterials-based coatings. Elsevier, pp 139–169. https://doi.org/10.1016/B978-0-12-815884-5.00006-5
25. Dubey R, Dutta D, Sarkar A, Chattopadhyay P (2021) Functionalized carbon nanotubes: synthesis, properties and applications in water purification, drug delivery, and material and biomedical sciences. Nanoscale Adv 3(20):5722–5744. https://doi.org/10.1039/D1NA00293G
26. Eatemadi A, Daraee H, Zarghami N, Melat Yar H, Akbarzadeh A (2016) Nanofiber: synthesis and biomedical applications. Artif Cells, Nanomed Biotechnol 44(1):111–121. https://doi.org/10.3109/21691401.2014.922568
27. Escudero A, Carrillo-Carrión C, Romero-Ben E, Franco A, Rosales-Barrios C, Castillejos MC, Khiar N (2021) Molecular bottom-up approaches for the synthesis of inorganic and hybrid nanostructures. Inorganics 9(7):58. https://doi.org/10.3390/inorganics9070058
28. Faraji G, Kim HS, Kashi HT (2018) Introduction. In: Severe plastic deformation. Elsevier, pp 1–17. https://doi.org/10.1016/B978-0-12-813518-1.00020-5
29. Feng Y, Li H (2017) Self-assembly of fullerenes. In: Comprehensive supramolecular chemistry II. Elsevier, pp 3–25. https://doi.org/10.1016/B978-0-12-409547-2.12636-8
30. Foss CA, Rahman AZMS, Eldressi KA, Elsheikhi SA, Salem WOA (2019) Optical properties of nanoparticle pair structures. In: Reference module in materials science and materials engineering. Elsevier, p B9780128035818024103. https://doi.org/10.1016/B978-0-12-803581-8.02410-3
31. Frishberg IV (2019) Gas-phase method of metal powder production. In: Handbook of non-ferrous metal powders. Elsevier, pp 187–200. https://doi.org/10.1016/B978-0-08-100543-9.00005-1
32. Fouda MMG (2012) Antibacterial modification of textiles using nanotechnology. In: Bobbarala V (ed) A search for antibacterial agents. InTech. https://doi.org/10.5772/45653
33. Gacem A, Modi S, Yadav VK, Islam S, Patel A, Dawane V, Jameel M, Inwati GK, Piplode S, Solanki VS, Basnet A (2022) Recent advances in methods for synthesis of carbon nanotubes and carbon nanocomposite and their emerging applications: a descriptive review. J Nanomater 2022:1–16. https://doi.org/10.1155/2022/7238602
34. Gan YX, Jayatissa AH, Yu Z, Chen X, Li M (2020) Hydrothermal synthesis of nanomaterials. J Nanomater 2020:1–3. https://doi.org/10.1155/2020/8917013

35. Grabowska E, Marchelek M, Paszkiewicz-Gawron M, Zaleska-Medynska A (2018) Metal oxide photocatalysts. In: Metal oxide-based photocatalysis. Elsevier, pp 51–209. https://doi.org/10.1016/B978-0-12-811634-0.00003-2
36. Gupta N, Gupta SM, Sharma SK (2019) Carbon nanotubes: synthesis, properties and engineering applications. Carbon Lett 29(5):419–447. https://doi.org/10.1007/s42823-019-00068-2
37. Guterres SS, Alves MP, Pohlmann AR (2007) Polymeric nanoparticles, nanospheres and nanocapsules, for cutaneous applications. Drug target insights, pp 147–157
38. Habiba K, Makarov VI, Weiner BR, Morell G (2014) Fabrication of nanomaterials by pulsed laser synthesis. In: Waqar A, Ali N (ed) Manufacturing nanostructures. UK: one central press (OCN), pp 263–92. ISBN: 9781910086070
39. Haque M (2019) Nano fabrics in the 21st century: a review. Asian J Nanosci Mater 2(2):131–148
40. Hári J, Pukánszky B (2011) Nanocomposites. In: Applied plastics engineering handbook. Elsevier, pp 109–142. https://doi.org/10.1016/B978-1-4377-3514-7.10008-X
41. Harish V, Tewari D, Gaur M, Yadav AB, Swaroop S, Bechelany M, Barhoum A (2022) Review on nanoparticles and nanostructured materials: bioimaging, biosensing, drug delivery, tissue engineering, antimicrobial, and agro-food applications. Nanomaterials 12(3):457. https://doi.org/10.3390/nano12030457
42. Hassabo A, Elmorsy H, Gamal N, Sediek A, Saad F, Hegazy B, Othman H (2022) Applications of nanotechnology in the creation of smart sportswear for enhanced sports performance: efficiency and comfort. J Text Colorat Polym Sci. https://doi.org/10.21608/jtcps.2022.181608.1147
43. He J, Zhou Y (2019) Multineedle electrospinning. In: Electrospinning: nanofabrication and applications. Elsevier, pp 201–218. https://doi.org/10.1016/B978-0-323-51270-1.00006-6
44. Hosne Asif AKMA, Hasan MdZ (2018) Application of nanotechnology in modern textiles: a review. Int J Curr Eng Technol. https://doi.org/10.14741/ijcet/v.8.2.5
45. Hulteen JC, Martin CR (1997) A general template-based method for the preparation of nanomaterials. J Mater Chem 7(7):1075–1087. https://doi.org/10.1039/a700027h
46. Ibrahim HM, Saad MM, Aly NM (2016) Preparation of single layer nonwoven fabric treated with chitosan nanoparticles and its utilization in gas filtration. Int J ChemTech Res
47. Ijaz I, Gilani E, Nazir A, Bukhari A (2020) Detail review on chemical, physical and green synthesis, classification, characterizations and applications of nanoparticles. Green Chem Lett Rev 13(3):223–245. https://doi.org/10.1080/17518253.2020.1802517
48. Javaid S, Mahmood A, Nasir H, Iqbal M, Ahmed N, Ahmad NM (2022) Layer-by-layer self-assembled dip coating for antifouling functionalized finishing of cotton textile. Polymers 14(13):2540. https://doi.org/10.3390/polym14132540
49. Jawad KH, Mohammed DB, Alsaffar MR (2019) Performed of filter nano textile for purification of aerosol water by electrospinning technique. J Phys Conf Ser 1298(1):012023. https://doi.org/10.1088/1742-6596/1298/1/012023
50. Jeong SA, Kang TJ (2017) Superhydrophobic and transparent surfaces on cotton fabrics coated with silica nanoparticles for hierarchical roughness. Text Res J 87(5):552–560. https://doi.org/10.1177/0040517516632477
51. Joshi M, Bhattacharyya A, Ali SW (2008) Characterization techniques for nanotechnology applications in textiles. Indian J Fibre Text Res
52. Joudeh N, Linke D (2022) Nanoparticle classification, physicochemical properties, characterization, and applications: a comprehensive review for biologists. J Nanobiotechnol 20(1):262. https://doi.org/10.1186/s12951-022-01477-8
53. Juan J, Marlen I, Luisa C, Diaz R, Luisa A, Casas N (2012) Nanocarrier systems for transdermal drug delivery. In: Sezer AD (ed) Recent advances in novel drug carrier systems. InTech. https://doi.org/10.5772/50314
54. Kandola BK (2001) Nanocomposites. In: Fire retardant materials. Elsevier, pp 204–219. https://doi.org/10.1533/9781855737464.204

55. Khan I, Saeed K, Khan I (2019) Nanoparticles: properties, applications and toxicities. Arab J Chem 12(7):908–931. https://doi.org/10.1016/j.arabjc.2017.05.011
56. Khan Y, Sadia H, Ali Shah SZ, Khan MN, Shah AA, Ullah N, Ullah MF, Bibi H, Bafakeeh OT, Khedher NB, Eldin SM, Fadhl BM, Khan MI (2022) Classification, synthetic, and characterization approaches to nanoparticles, and their applications in various fields of nanotechnology: a review. Catalysts 12(11):1386. https://doi.org/10.3390/catal12111386
57. Kim JH, Pham TV, Hwang JH, Kim CS, Kim MJ (2018) Boron nitride nanotubes: synthesis and applications. Nano Converg 5(1):17. https://doi.org/10.1186/s40580-018-0149-y
58. Kim YK (2019) Nanotechnology-based advanced coatings and functional finishes for textiles. In: Smart textile coatings and laminates. Elsevier, pp 189–203. https://doi.org/10.1016/B978-0-08-102428-7.00009-2
59. Kundu CK, Song L, Hu Y (2020) Nanoparticles based coatings for multifunctional polyamide 66 textiles with improved flame retardancy and hydrophilicity. J Taiwan Inst Chem Eng 112:15–19. https://doi.org/10.1016/j.jtice.2020.07.013
60. Kwon J, Jung H, Jung H, Lee J (2020) Micro/nanostructured coating for cotton textiles that repel oil, water, and chemical warfare agents. Polymers 12(8):1826. https://doi.org/10.3390/polym12081826
61. Lab D, Yat-sen S (2020) Fabrication and functions of nanomaterials for polymeric textile fibers. MedDocs Publishers LLC
62. Madhu A, Singh N, Kaur A, Prakash Sahu O (2022) UV protective fabric for face covering utility article using TiO_2 nanoparticles. Mater Today Proc 68:1022–1029. https://doi.org/10.1016/j.matpr.2022.08.252
63. Makhlouf ASH (2011) Current and advanced coating technologies for industrial applications. In: Nanocoatings and ultra-thin films. Elsevier, pp 3–23. https://doi.org/10.1533/9780857094902.1.3
64. Malekzadeh M, Mane VV, Xuan Z, Ohulchanskyy TY, Swihart MT (2023) Laser pyrolysis synthesis of upconverting lanthanide-doped NaYF 4 nanocrystals for anticounterfeiting applications. ACS Appl Nano Mater 6(5):3704–3717. https://doi.org/10.1021/acsanm.2c05434
65. Martins IM, Sampaio AG, Lima GMG, Oliveira E Campos MAC, Rodgher S, Rodrigues-Siqueli AC, Baldan MR, Marcuzzo JS, Koga-Ito CY (2023) Application of textile (PAN-based) activated carbon fibers decorated with silver nanoparticles in water treatment. Front Environ Sci 10:1100583
66. Meischein M, Ludwig A (2021) Upscaling nanoparticle synthesis by sputter deposition in ionic liquids. J Nanopart Res 23(6):129. https://doi.org/10.1007/s11051-021-05248-8
67. Mekuye B, Abera B (2023) Nanomaterials: an overview of synthesis, classification, characterization, and applications. Nano Select 4(8):486–501. https://doi.org/10.1002/nano.202300038
68. Mishra R, Militky J (2019) Nanocomposites. In: Nanotechnology in textiles. Elsevier, pp 263–310. https://doi.org/10.1016/B978-0-08-102609-0.00006-7
69. Mishra R, Militky J, Venkataraman M (2019) Electrospun nanofibers. In: Nanotechnology in textiles. Elsevier, pp 35–161. https://doi.org/10.1016/B978-0-08-102609-0.00002-X
70. Mondal S (2022) Nanomaterials for UV protective textiles. J Ind Text 51(4_suppl):5592S-5621S. https://doi.org/10.1177/1528083721988949
71. Naidu KCB, Kumar NS, Banerjee P, Reddy BVS (2021) A review on the origin of nanofibers/nanorods structures and applications. J Mater Sci Mater Med 32(6):68. https://doi.org/10.1007/s10856-021-06541-7
72. Nam NH, Luong NH (2019) Nanoparticles: synthesis and applications. In: Materials for biomedical engineering. Elsevier, pp 211–240. https://doi.org/10.1016/B978-0-08-102814-8.00008-1
73. Nikam AV, Prasad BLV, Kulkarni AA (2018) Wet chemical synthesis of metal oxide nanoparticles: a review. CrystEngComm 20(35):5091–5107. https://doi.org/10.1039/C8CE00487K

74. Norouzi M, Zare Y, Kiany P (2015) Nanoparticles as effective flame retardants for natural and synthetic textile polymers: application, mechanism, and optimization. Polym Rev 55(3):531–560. https://doi.org/10.1080/15583724.2014.980427
75. Nouailhat A (2006) An introduction to nanoscience and nanotechnology
76. Obeas LK, Ghalib AK, Kareem-Alsultan GA, Mijan NA, Yunus R (2021) A-review on nanorods-an overview from synthesis to emerging, device applications and toxicity (a-review). Orient J Chem 37(2):256–268. https://doi.org/10.13005/ojc/370201
77. Pal G, Rai P, Pandey A (2019) Green synthesis of nanoparticles: a greener approach for a cleaner future. In: Green synthesis, characterization and applications of nanoparticles. Elsevier, pp 1–26. https://doi.org/10.1016/B978-0-08-102579-6.00001-0
78. Parsamanesh M, Shekarriz S (2023) Water/oil-repellent cotton fabrics prepared using fluorocarbon compounds, polyurethane cross-linker, and silica nanoparticles. Fibers Polym. https://doi.org/10.1007/s12221-023-00290-w
79. Parvinzadeh Gashti M, Pakdel E, Alimohammadi F (2016) Nanotechnology-based coating techniques for smart textiles. In: Active coatings for smart textiles. Elsevier, pp 243–268. https://doi.org/10.1016/B978-0-08-100263-6.00011-3
80. Pasichnyk M, Gaálová J, Minarik P, Václavíková M, Melnyk I (2022) Development of polyester filters with polymer nanocomposite active layer for effective dye filtration. Sci Rep 12(1):973. https://doi.org/10.1038/s41598-022-04829-4
81. Patil N, Bhaskar R, Vyavhare V, Dhadge R, Khaire V, Patil Y (2021) Overview on methods of synthesis of nanoparticles. Int J Curr Pharm Res. https://doi.org/10.22159/ijcpr.2021v13i2.41556
82. Petkova P, Francesko A, Fernandes MM, Mendoza E, Perelshtein I, Gedanken A, Tzanov T (2014) Sonochemical coating of textiles with hybrid ZnO/chitosan antimicrobial nanoparticles. ACS Appl Mater Interfaces 6(2):1164–1172. https://doi.org/10.1021/am404852d
83. Pradhan L, Maiti S, Mallick A, Shahid M, More SP, Adivarekar RV (2023) 3 Coating- and lamination-based smart textiles: techniques, features, and challenges. In: Adak B, Mukhopadhyay S (eds) Smart and functional textiles. De Gruyter, pp 97–150. https://doi.org/10.1515/9783110759747-003
84. Prasad Yadav T, Manohar Yadav R, Pratap Singh D (2012) Mechanical milling: a top down approach for the synthesis of nanomaterials and nanocomposites. Nanosci Nanotechnol 2(3):22–48. https://doi.org/10.5923/j.nn.20120203.01
85. Rabiei H, Farhang Dehghan S, Montazer M, Khaloo SS, Koozekonan AG (2022) UV protection properties of workwear fabrics coated with TiO_2 nanoparticles. Front Public Health 10:929095. https://doi.org/10.3389/fpubh.2022.929095
86. Rafique MS, Rafique M, Tahir MB, Hajra S, Nawaz T, Shafiq F (2020) Synthesis methods of nanostructures. In: Nanotechnology and photocatalysis for environmental applications. Elsevier, pp 45–56. https://doi.org/10.1016/B978-0-12-821192-2.00003-6
87. Ramaratnam K, Iyer SK, Kinnan MK, Chumanov G, Brown PJ, Luzinov I (2008) Ultrahydrophobic textiles using nanoparticles: lotus approach. J Eng Fibers Fabr 3(4):155892500800300. https://doi.org/10.1177/155892500800300402
88. Rane AV, Kanny K, Abitha VK, Thomas S (2018) Methods for synthesis of nanoparticles and fabrication of nanocomposites. In: Synthesis of inorganic nanomaterials. Elsevier, pp 121–139. https://doi.org/10.1016/B978-0-08-101975-7.00005-1
89. Reau A, Guizard B, Mengeot C, Boulanger L, Ténégal F (2007) Large scale production of nanoparticles by laser pyrolysis. Mater Sci Forum 534–536:85–88. https://doi.org/10.4028/www.scientific.net/MSF.534-536.85
90. Riaz S, Ashraf M, Hussain T, Hussain MT, Younus A (2019) Fabrication of robust multifaceted textiles by application of functionalized TiO_2 nanoparticles. Colloids Surf A 581:123799. https://doi.org/10.1016/j.colsurfa.2019.123799
91. Roy P, Berger S, Schmuki P (2011) TiO_2 nanotubes: synthesis and applications. Angew Chem Int Ed 50(13):2904–2939. https://doi.org/10.1002/anie.201001374
92. Satyanarayana T (2018) A review on chemical and physical synthesis methods of nanomaterials. Int J Res Appl Sci Eng Technol 6(1):2885–2889. https://doi.org/10.22214/ijraset.2018.1396

93. Shah MA, Pirzada BM, Price G, Shibiru AL, Qurashi A (2022) Applications of nanotechnology in smart textile industry: a critical review. J Adv Res 38:55–75. https://doi.org/10.1016/j.jare.2022.01.008
94. Sharpe JM, Lee H, Hall AR, Bonin K, Guthold M (2020) Mechanical properties of electrospun, blended fibrinogen: PCL nanofibers. Nanomaterials 10(9):1843. https://doi.org/10.3390/nano10091843
95. Shikder A (2021) Effect of herbal based aluminium nanoparticles on anti-microbial property of cotton textiles, p 11
96. Singh J, Dutta T, Kim K-H, Rawat M, Samddar P, Kumar P (2018) 'Green' synthesis of metals and their oxide nanoparticles: applications for environmental remediation. J Nanobiotechnol 16(1):84. https://doi.org/10.1186/s12951-018-0408-4
97. Singh T (2020) Introduction to nanotechnology system. https://doi.org/10.13140/RG.2.2.12770.15049/1
98. Singha K, Maity S, Pandit P (2020) UV Protection via nanomaterials. In: Shabbir M, Ahmed S, Sheikh JN (eds) Frontiers of textile materials, 1st edn. Wiley, pp 153–166. https://doi.org/10.1002/9781119620396.ch7
99. Tile VG, Suraj HS, Uday BM, Sahana SG (2016) Recent trends in nanotechnology and its future scope: a review. Int J Emerg Technol (Special Issue on ICRIET-2016) 7(2): 377–385
100. Tsang MP, Philippot G, Aymonier C, Sonnemann G (2018) Supercritical fluid flow synthesis to support sustainable production of engineered nanomaterials: case study of titanium dioxide. ACS Sustain Chem Eng 6(4):5142–5151. https://doi.org/10.1021/acssuschemeng.7b04800
101. Tudu BK, Sinhamahapatra A, Kumar A (2020) Surface modification of cotton fabric using TiO_2 nanoparticles for self-cleaning, oil-water separation, antistain, anti-water absorption, and antibacterial properties. ACS Omega 5(14):7850–7860. https://doi.org/10.1021/acsomega.9b04067
102. Vigneshwaran N (2009) Modification of textile surfaces using nanoparticles. In: Surface modification of textiles. Elsevier, pp 164–184. https://doi.org/10.1533/9781845696689.164
103. Wade TL, Wegrowe J-E (2005) Template synthesis of nanomaterials. Eur Phys J Appl Phys 29(1):3–22. https://doi.org/10.1051/epjap:2005001
104. Wang H, Leeuwenburgh SCG, Li Y, Jansen JA (2012) The use of micro- and nanospheres as functional components for bone tissue regeneration. Tissue Eng Part B Rev 18(1):24–39. https://doi.org/10.1089/ten.teb.2011.0184
105. Wang S, Gao L (2019) Laser-driven nanomaterials and laser-enabled nanofabrication for industrial applications. In: Industrial applications of nanomaterials. Elsevier, pp 181–203. https://doi.org/10.1016/B978-0-12-815749-7.00007-4
106. Whitesides GM, Kriebel JK, Mayers BT (2005) Self-assembly and nanostructured materials. In: Huck WTS (ed) Nanoscale assembly. Springer US, pp 217–239. https://doi.org/10.1007/0-387-25656-3_9
107. Ye X, Wai CM (2003) Making nanomaterials in supercritical fluids: a review science and education. J Chem Educ 80: 198–204. https://doi.org/10.1021/ed080p198
108. Ye Z, Li S, Zhao S, Deng L, Zhang J, Dong A (2021) Textile coatings configured by double-nanoparticles to optimally couple superhydrophobic and antibacterial properties. Chem Eng J 420:127680. https://doi.org/10.1016/j.cej.2020.127680
109. Yetisen AK, Qu H, Manbachi A, Butt H, Dokmeci MR, Hinestroza JP, Skorobogatiy M, Khademhosseini A, Yun SH (2016) Nanotechnology in textiles. ACS Nano 10(3):3042–3068. https://doi.org/10.1021/acsnano.5b08176
110. Ying S, Guan Z, Ofoegbu PC, Clubb P, Rico C, He F, Hong J (2022) Green synthesis of nanoparticles: current developments and limitations. Environ Technol Innov 26:102336. https://doi.org/10.1016/j.eti.2022.102336
111. Zhang X-F, Liu Z-G, Shen W, Gurunathan S (2016) Silver nanoparticles: synthesis, characterization, properties, applications, and therapeutic approaches. Int J Mol Sci 17(9):1534. https://doi.org/10.3390/ijms17091534
112. Zhao H, Li F, Wang S, Guo L (2021) Wet chemical synthesis of amorphous nanomaterials with well-defined morphologies. Acc Mater Res 2(9):804–815. https://doi.org/10.1021/accountsmr.1c00104

113. Zhao M, Wu X, Cai C (2009) Polyaniline nanofibers: synthesis, characterization, and application to direct electron transfer of glucose oxidase. J Phys Chem C 113(12):4987–4996. https://doi.org/10.1021/jp807621y
114. Zhou Q, Chen J, Lu Z, Tian Q, Shao J (2022) In situ synthesis of silver nanoparticles on flame-retardant cotton textiles treated with biological phytic acid and antibacterial activity. Materials 15(7):2537. https://doi.org/10.3390/ma15072537

Chapter 8
UV Protection and Antimicrobial Properties of Textiles Using Nanotechnology

Sharif Ahmed, A. K. M. Ashiqur Rahman Mazumdar, Md. Syduzzaman, and Md. Sultan Mahmud

1 Introduction

In today's rapidly changing global environment, textiles have evolved far beyond their traditional uses, emerging as significant barriers against persistent concerns, including ultraviolet (UV) radiation and microbial attacks, which pose serious threats to human health and resources [1, 2]. UV and microbiological protection are vital issues because protecting against UV radiation defends human health by preventing skin damage and lowering the risk of skin cancer. In addition, antimicrobial protection in textiles maintains a healthy atmosphere, preventing the spreading of illnesses and enhancing general health. Although UV ray only makes up a tiny percentage of the sun's rays, it significantly impacts Earth and its inhabitants. The effect of UV radiation is all-encompassing, affecting everything from environmental processes to directly endangering human health. UV-A, UV-B, and UV-C are the most significant ultraviolet (UV) radiation types. UV-B rays are the leading cause of sunburn [3]. Although UV-C is the most energetic and possibly dangerous kind of UV radiation, it rarely reaches the surface as it is mostly absorbed by the Earth's atmosphere [4].

Microbes growing on textile surfaces can cause allergies, and skin irritation, and are even more troublesome when textiles act as a source of pollution that threatens

S. Ahmed · Md. S. Mahmud (✉)
Department of Yarn Engineering, Bangladesh University of Textiles, Tejgaon Industrial Area, Dhaka 1208, Bangladesh
e-mail: sultan.mahmud@ye.butex.edu.bd

A. K. M. A. R. Mazumdar
Department of Fabric Engineering, Bangladesh University of Textiles, Tejgaon Industrial Area, Dhaka 1208, Bangladesh

Md. Syduzzaman
Department of Textile Engineering Management, Bangladesh University of Textiles, Tejgaon Industrial Area, Dhaka 1208, Bangladesh

S. Bairagi et al. (eds.), *Nanotechnology in Textile Finishing*, Materials Horizons: From Nature to Nanomaterials, https://doi.org/10.1007/978-981-97-2696-7_8

a community's public health safety. Textile nanofinish emerges as a compelling choice for the protection of UV and bacterial infestation due to its versatility and efficacy. To accomplish these required capabilities, a variety of materials are used in the nanofinishing process. By absorbing or reflecting adversely impacting rays, nanoparticles of zinc oxide, titanium dioxide, and silver, among other materials, serve as crucial for UV and antimicrobial protection. Furthermore, the incorporation of these materials into textiles is being made more accessible by developments in nanotechnology, all without losing the textiles' general comfort, and other features [5].

This chapter describes different materials employed in nanofinishing to enhance the antibacterial qualities of textiles and shield them from harmful UV radiation. Diverse strategies for incorporating these elements into fabrics are addressed and investigated. The discussion also includes unique factors to consider, advancements in protecting against harmful ultraviolet (UV) rays, and improvements in the antimicrobial properties of fabrics. Prospects in this developing field are considered, along with an investigation into the environmental effects of these materials. The chapter distinguishes antibacterial properties and UV protection as distinct but related topics.

2 UV Radiation and Its Effect

2.1 UV Radiation

The electromagnetic spectrum is a range of waves, including the longest radio waves and the shortest gamma waves. Ultraviolet (UV) radiation is located between visible light and X-rays in this spectrum [6]. UV rays are invisible to the naked human eye, and they significantly impact ecosystems, the environment, and, most significantly, human health.

Figure 1 shows the electromagnetic spectrum while emphasizing various ray types. There are three distinctive categories within the ultraviolet ray segment: UV-A (315 nm to 400 nm), UV-B (280 nm to 315 nm), and UV-C (200 nm to 280 nm). The purple and blue tones illustrating these ultraviolet categories represent the corresponding wavelengths.

2.1.1 UV-A (Ultraviolet A)

About 95% of UV light that reaches the surface of the Earth has wavelengths between 315 and 400 nm (nm), most of which are UV-A rays. Despite having the lowest energy in the UV spectrum, these rays have significant effect on our health because of their widespread presence throughout daytime hours. Compared to UV-B rays, UV-A rays penetrate deeper into the skin's layers. It is often linked to DNA damage that causes long-term skin aging and can even cause the onset of skin cancer [7].

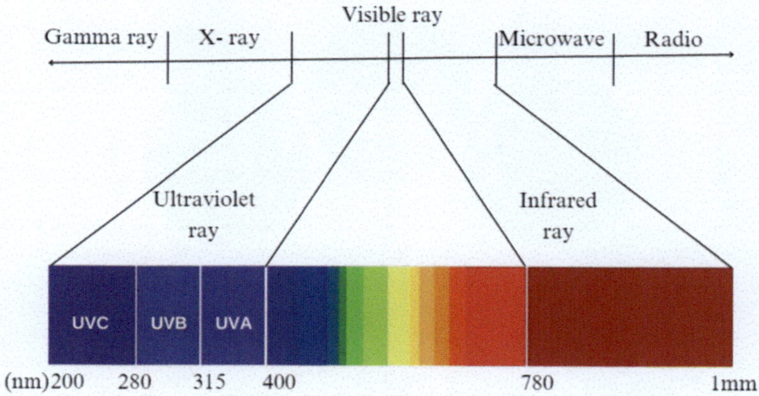

Fig. 1 UV-A, UV-B, and UV-C on the electromagnetic spectrum

2.1.2 UV-B (Ultraviolet B)

UV-B rays have shorter wavelengths than UV-A rays, ranging from 280 to 315 nm. Their intensity fluctuates based on factors including height, latitude, and season. Sunburns are primarily caused by UV-B rays, which directly harm the skin's outermost layers. It is commonly recognized that UV-B rays contribute to the rise of skin cancer, especially the deadly kind called malignant melanoma [8, 9]. Additionally, UV-B rays play a critical role in the skin's production of vitamin D, a vital component of human health [10].

2.1.3 UV-C (Ultraviolet C)

UV-C rays are the most damaging type of all UV rays since they have the most intensity and the shortest wavelengths (200 to 280 nm). Fortunately, most of the UV-C radiation is absorbed by the ozone layer in the Earth's atmosphere, preventing it from reaching the surface. UV-C rays have the potential to cause serious biological effects, which would drastically change life as we know it [4, 11]. But the ongoing deterioration of the ozone layer could lead to an increase in the levels of UV-C radiation, posing a significant risk to human health.

2.2 *Effects of UV Radiation on Human Health*

UV radiation has both negative and positive effects on human health. UV radiation has the ability to elevate human mood, heal certain diseases of the skin, and serve as a disinfectant. Additionally, they are necessary for the synthesis of vitamin D. On the other hand, UV radiation can age the skin, create sunburns, harm the eyes, and

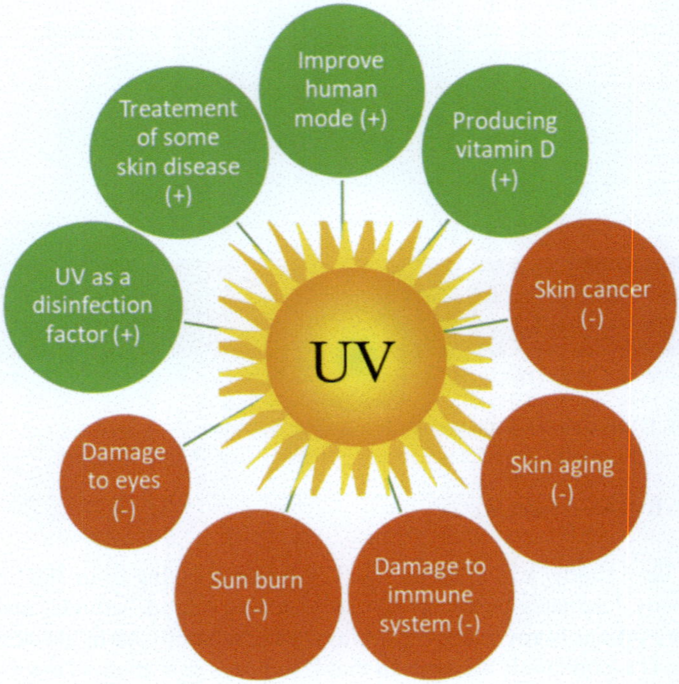

Fig. 2 Positive and negative effects of UV rays on the human body

trigger skin cancer. Additionally, they may weaken the immunological system [12]. Figure 2 shows both negative and positive effects of UV radiation on human health.

2.2.1 Skin Composition and UV Radiation

When delving into the effects of UV radiation, the forefront of consideration often turns to the human skin, as it is one of the most severely affected areas. Different skin layers interact differently with different types of UV rays. In general, there are three main layers of the human skin: epidermis, dermis, and hypodermis. Epidermis is the skin's outermost layer, and it is a barrier and defense against the external elements whereas dermis is the layer of skin underneath, which contains sweat glands, hair follicles, and connective tissues. Inner most part of the skin is hypodermis which comprises connective tissues and fat. Different UV rays penetrate the skin at different depths depending on the type. UV-A rays have longer wavelengths compared to UV-B rays, but their longer wavelengths actually allow them to penetrate the skin more superficially. UV-A rays primarily affect the epidermis, which is the outermost layer of the skin. On the other hand, UV-B rays, with shorter wavelengths, can penetrate more deeply into the skin, reaching the dermis [8]. Figure 3 illustrates different UV

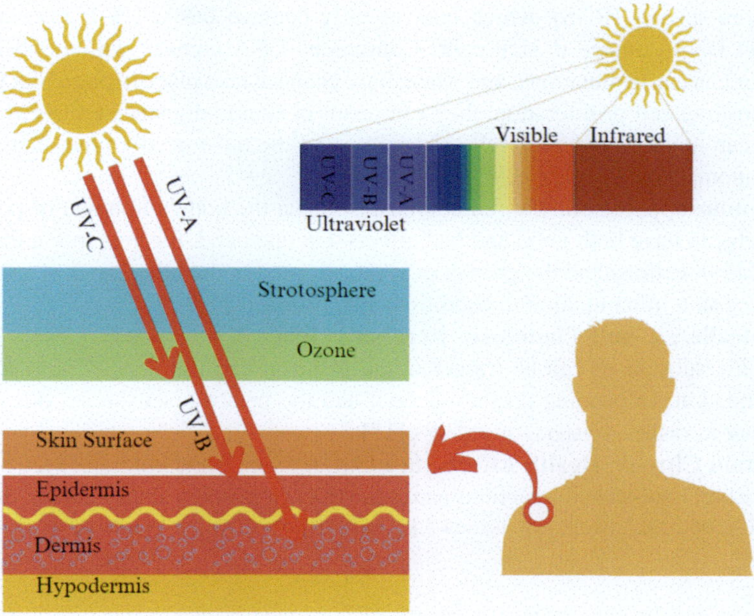

Fig. 3 Effect of different UV rays on human skin

rays coming from the sun and their penetration to the different layers of the skin. About all the UV-C energy is absorbed by the stratospheric ozone layer shielding humans from its damaging effects of it.

2.2.2 Detrimental Effects of UV Radiation on Skin

- **Sunburn**: Sunburn, which is predominantly brought on by UV-B rays, is one of the effects of UV radiation that is most obvious and quick. When skin gets too many rays, it makes extra melanin to protect itself. Melanin is the stuff that gives the skin its color. However, when the exposure exceeds the skin's capacity for protection, it causes inflammation, which causes the redness and agony of a sunburn [13].
- **Photoaging**: Exposure in UV-A rays for a long time can make the skin age faster, which dermatologists call "photoaging." The skin's premature aging characterizes this process. Some symptoms of photoaging include wrinkles, a leathery texture, black patches, and decreased skin suppleness. UV-A rays cause abnormal elastin formation in the dermis because of their deeper penetration and damage to the collagen fibers. To repair the damage, the skin unintentionally creates disorganized collagen, which cause wrinkles [14, 15].
- **DNA Damage and Skin Cancer**: UV-A and UV-B radiation can harm DNA in skin cells through various processes. While UV-A photons generate reactive

oxygen species (ROS), which inadvertently damage DNA, UV-B rays induce direct DNA damage to skin cells. Continuous DNA damage without adequate repair can cause mutations, and when these changes take place in particular genes controlling cell proliferation, they can result in skin malignancies such as basal cell carcinoma, squamous cell carcinoma, and the most deadly type, malignant melanoma [3, 8].

- **Immunosuppression**: UV radiation can weaken the body's immune responses, which can have both good and bad effects. On one hand, it can increase the risk of skin infections and the growth of skin cancers. On the other hand, it can help reduce skin inflammation in conditions like psoriasis [16].
- **Photoallergic and Phototoxic Reactions**: When UV rays come to the skin, specific substances can be triggered and result in harmful or allergic reactions on the skin. Depending on the material and the person's sensitivity, this could appear as rashes, blisters, or redness [17].
- **Certain Chronic Health Effects**: In addition to the acute consequences of UV exposure, long-term interactions can cause thicker, leathery skin, freckles, discolored spots, and even eye issues like cataracts and some types of eye cancer [18].

2.2.3 Beneficial Effects of UV Radiation on Skin

Vitamin D production is one of the most well-known advantages of UV ray for human skin [19]. UVB rays interact with 7-dehydrocholesterol when they come into contact with the skin, turning it into vitamin D3. The active form of vitamin D3 is then created in the liver and kidneys, where it is vital for calcium absorption, bone health, and several other physiological processes [20]. It also improves mood and promotes well-being by inducing the release of endorphins. Moderate UV radiation exposure promotes the healing of vitiligo and psoriasis, among other skin diseases. Moreover, UV light sterilizes surfaces and water by acting as a natural disinfectant [21, 22].

2.3 The UV Protection Factor (UPF) in Textiles

Textiles and ultraviolet (UV) light have a long history of interaction. Textiles' importance in protecting against UV radiation has increased as our knowledge of its impacts on human health has expanded. Here comes the UV protection factor (UPF), a measure used to assess how well materials can protect ultraviolet (UV) light [23–25]. The sun protection factor (SPF) used in sunscreens and the UV protection factor (UPF) in textiles and other materials are similar. SPF measures the percentage of UV rays that cause sunburns that reach the skin, whereas UPF measures the rate of UV radiation (UV-A and UV-B) penetrates through the fabric [25–27]. For instance, a fabric with a UPF rating of 50 allows just 1/50th of the sun's UV radiation to pass through, effectively blocking 98% of it. The international scientific community established the UPF to rate the UV protection of sunscreens, textiles, or colored glasses [28, 29]. Equation (1) shows how UFP is calculated, and its value depends on the UVR spectral irradiance, erythemal effectiveness, wavelength, and wavelength intervals.

$$\mathrm{UFP} = \frac{\sum_{290}^{400} E_\lambda . S_\lambda . \Delta_\lambda}{\sum_{290}^{400} E_\lambda . S_\lambda . T_\lambda . \Delta_\lambda} \tag{1}$$

where,

E_λ is the solar UVR spectral irradiance in W $m^{-2} \cdot nm^{-1}$,
S_λ is the relative erythemal effectiveness according to the CIE,
Δ_λ is the wavelength interval of the measurements,
T_λ is the spectral transmittance at wavelength λ, and
λ is the wavelength (in nm).

2.3.1 Factors Affecting the UPF of Textiles

There are several factors play a vital role in determining the UPF. These factors encompass the type of materials used, fabric thickness, colors and types, moisture content, fabric stretch, and the condition and age of the fabric. Understanding how these elements affect UPF is essential for ensuring effective UV protection through clothing.

3 Textile Properties Influencing UV Protection

3.1 Fiber Type and UV Protection

Natural and synthetic fibers are the two primary categories of textile fibers. UV protection properties differ across various fibers. For instance, polyester and nylon often offer greater UV protection than the majority of natural fibers because of their chemical composition, which naturally interferes with UV light. Protein fibers like wool and silk, which have molecular configurations, allow for better UV absorption than other natural fibers [24, 26, 28].

3.1.1 Natural Fibers

Untreated cotton fabrics typically have a lower UV protection factor (UPF). This is mainly caused by its flimsy molecular structure, which allows UV radiation to pass through [24]. Nevertheless, the UV-absorbing properties of cotton can be improved through particular treatments and particular dyeing. Wool contains lanolin, a natural wax that can absorb UV rays and provide some protection. However, just like cotton, the degree of protection depends on the weave and color of the cloth [30]. Silk has certain amino acids that absorb UV ray and offer some level of protection.

3.1.2 Synthetic Fibers

It has compact molecular structure and has the capacity to absorb, scatter, and reflect UV radiation [24–26]. Through particular treatments and fabric designs, polyester's UV resistance can be further increased. Like polyester, nylon has an innate capacity to withstand UV light for its chemical structure. However, its effectiveness may decline when moist, necessitating further treatments for reliable defense in a variety of conditions. Acrylic, rayon, and spandex fibers generally provide limited protection against UV radiation. Unless modified or treated, their structures allow a significant fraction of UV rays to pass through [24].

3.2 Fabric Construction and UV Protection

The ability of a fabric to block UV rays depends on how it is made, whether it is woven, knitted, or non-woven. Fewer holes allow UV rays to pass through tightly woven or knitted materials, improving protection. The weight and thickness of the fabric are important factors as well; heavier and thicker materials often provide higher UV protection than lighter, thinner alternatives [24, 26]. Fabric construction and UV protection are described shortly here.

3.2.1 Woven Structure

Three popular weave patterns—plain, twill, and satin—have varying degrees of UV protection because there are fewer interstitial spaces in plain weave, which is characterized by closely interwoven threads providing excellent UV protection. Distinguished by its diagonal patterns, twill weave is heavier than plain weave and generally offers superior UV protection. UV protection in satin weave is dependent on the type and color of yarn used [5].

3.2.2 Knit Structure

The degree of UV protection offered by knitted materials with interlocking loops varies. Stretching single jersey makes it porous and reduces UV protection. Double knit is denser and frequently provides better UV protection since it uses two yarn sets [31].

UV resistance can also be affected by finishing techniques, including calendaring, which results in a glossy finish, and brushing and raising, which provide a fuzzy surface. Creating a fabric that is both UV-protective and breathable sometimes may be challenging, because dense construction might provide superior UV protection but could be less breathable, making them less suitable for warmer climates. The longevity of UV protection, especially with fabric wear and tear, may be but another concern [32, 33].

3.3 Dye and UV Protection

The UV protection of a fabric can be greatly affected by dyeing. Darker colors typically absorb more UV radiation than lighter ones. Additionally, some pigments and dyes can absorb and diffuse UV rays, improving the fabric's ability to block sunlight. Some colors' chemical composition makes them more effective at absorbing UV radiation. A brief overview is given in the following sections.

3.3.1 Different Colors and UV Protection

Different colors have differing capacities to absorb UV rays. The darker colors, such as black, navy, and deep red, absorb more UV radiation. Conversely, lighter hues like white and pastels could reflect more light but also let more UV rays through, particularly in thin textiles [26, 34].

3.3.2 Dye Concentration and UV Protection

UV protection in cloth can be greatly impacted by the dye concentration utilized. Since chromophores are responsible for absorbing UV light and turning it into harmless heat, higher dye concentrations result in greater UV protection [34]. Additionally, the durability and uniformity of UV protection might be affected by the dyes' ability to bond with the cloth. Strong covalent connections between dye and fibers are typically responsible for the protection's extended duration.

3.3.3 Specialty Dyes and UV Protection

In order to improve textile UV protection, advanced textile technology has introduced a variety of specialty dyes. Among these, dyes are UV-absorbing dyes, which have substances added to them that effectively absorb large amounts of UV light and convert it into a less harmful type of radiation, typically heat. In addition to making cloth brighter in the sun, fluorescent dyes could absorb UV radiation and re-emit it in the visible spectrum, thereby lowering the quantity of UV radiation that may pass through textiles. Nano-enhanced dyes, which integrate nanoparticles like titanium dioxide or zinc oxide to scatter and absorb UV rays, are a growing interest in this field [35].

4 Nanomaterials for UV-Protective Textiles

4.1 Types of Nanomaterials for UV Protection

Different metal oxide nanoparticles such as titanium dioxide (TiO_2) and zinc oxide (ZnO), as well as carbon nanotubes and nanocomposites, play an important role in enhancing the UV protection of textiles. These nanoparticles absorb and scatter UV rays efficiently, strengthening the textile's ability to shield against radiation while retaining its visual appeal [36]. They may come in a variety of forms, including nanoparticles, nanofibers, and nanocoating, each having special qualities that can considerably enhance the UV protection of textiles [37–39]. A brief discussion about nanomaterials is given in the following sections.

4.1.1 Metal Oxides

The most common metal oxides used for UV protection are titanium dioxide (TiO_2) and zinc oxide (ZnO). When they are incorporated into textiles, they scatter and absorb UV rays effectively. As they are nanoparticles, they are virtually invisible, and they retain the esthetic properties of fabric as well as provide augmented UV protection [1, 38, 40]. A brief explanation of metal oxides is mentioned here.

i. **Titanium Dioxide** (TiO_2) **Nanoparticles**:

For having UV protection properties, it has been using in sunscreen formulations as well as pigments in paints due to its high refractive index and excellent photocatalytic capabilities [41, 42]. To protect human body from UV radiation, TiO_2 nanoparticles are now incorporated with textiles. These nanoparticles scatter and absorb ultraviolet rays (Fig. 4) offering strong UV radiation protection [43, 44]. It is a versatile compound employed in various applications, including sunscreen formulations, where it serves as an effective UV filter. Additionally, TiO_2 is utilized for its self-cleaning properties and antimicrobial attributes in diverse industrial and consumer products.

ii. **Zinc Oxide (ZnO) Nanoparticles**:

ZnO nanoparticles are widely used in sunscreen formulations because they offer effective protection against ultraviolet (UV) radiation due to their broad UV absorption spectrum and lack of white residue on the skin [45,46]. In addition, ZnO nanoparticles are photostable, which means they do not break down when exposed to sunlight [47, 48]. These nanoparticles scatter and absorb ultraviolet rays offering strong UV radiation protection.

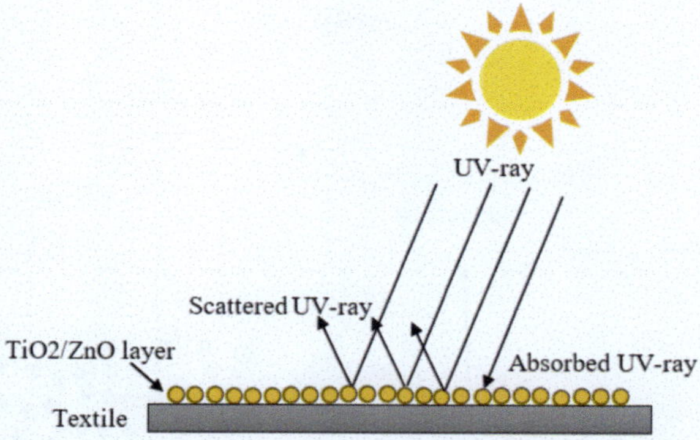

Fig. 4 Photocatalytic and self-cleaning mechanisms of TiO_2 nanocomposites [39]

4.1.2 Carbon-Based Nanomaterials

Carbon nanotubes (CNT) have demonstrated considerable promise in the realm of UV protection, showcasing their potential to create highly effective and durable barriers against UV radiation. When embedded into fabrics, they form a barrier that absorbs and reflects harmful UV radiation [49]. UV absorption through CNTs is attributed to the remarkable strength of C–C bonds, the extremely small diameter of carbon atoms, and the availability of free electrons of p bonds in graffiti-like form [50]. Although it is not yet extensively used for UV protection of textiles, but some researchers claimed that UV protection capability of CNT is same as TiO_2 and ZnO [49]. On the other hand, graphene, a transparent thin film of carbon, is also investigated for the protection of UV rays as it has good absorption capacity. For being transparent, it does not change the transparency of the base materials [51].

4.1.3 Polymer Nanocomposites

Polymers can be combined with nanoparticles like zinc oxide (ZnO) and titanium dioxide (TiO_2) to create a dense yet lightweight barrier that protects against UV radiation and also provides good microbial protection [52]. For a good UV-protective property, these are used as sunscreen for long times. To increase UV protection, textiles are also treated with polymer-based nanocomposites. This use is particularly pertinent to the textile sector, where it is crucial to combine UV protection with durability. The UV-blocking barrier of nanocomposite is dense yet lightweight because of the nanoscale structure [53, 54]. Nanoparticles embedded in the polymer matrix of fabrics have the ability to scatter and absorb UV rays, offering strong protection without changing the weight, appearance, or feel of the fabric [55]. Figure 5 illustrates UV absorption and scattering of nanoparticles.

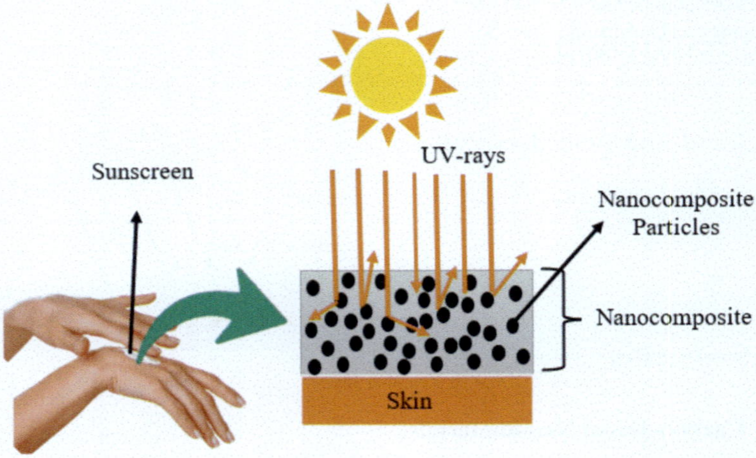

Fig. 5 UV protection of polymer nanocomposite [55]

4.2 *Mechanisms of UV Protection Using Nanotechnology*

4.2.1 Absorption

The most important process through which nanotechnology improves the UV-protective capabilities of textiles is absorption. Well-known UV blockers include nanoparticles like ZnO and TiO_2, which absorb UV radiation and transform it into less hazardous forms of energy like heat. These nanoparticles can be used to efficiently coat or embed textile fibers, ensuring the absorption of a wide range of UV rays. As a result of the use of such nanoparticles, treated fabrics' UPF values significantly increase [56, 57].

4.2.2 Scattering

Nanofinishing enhances UV radiation (multidirectional) scattering in fabrics and other fibrous materials and thus effectively reduces the detrimental impact of these radiations. For instance, when integrated into textiles, zinc oxide (ZnO) nanoparticles help to scatter UV rays and provide excellent UV protection [58]. The scattering is caused by the uneven surface of nanoparticles (Fig. 12), which scatters the incident in various directions and lessens the amount of UV transmission through the fabric. It has been demonstrated that this scattering mechanism enhances the UV-blocking capabilities of fabrics, particularly lightweight summer clothing [59, 60].

4.2.3 Reflection

UV reflection from the textile surface occurs when nanofinishing develops a reflective barrier on the textiles (Fig. 12). Utilizing metallic nanoparticles, especially silver (Ag), zinc oxide (ZnO), and titanium dioxide (TiO_2), have good properties of UV ray reflection [61]. It happens because nanoparticles act as tiny mirrors and stop UV rays from passing through the fabric and reaching the skin. The uniform coating of the textile, which maximizes the reflection of UV rays, increases the fabric's UV protection factor [62, 63].

4.3 Methods to Assess UV Protection in Textiles

Advancements in the textile sector have been sparked by the need to protect oneself from the damaging effects of UV radiation. However, this has also highlighted how important it is to have precise and regulated methodologies to evaluate the UV protection effectiveness of textiles [26, 32].

4.3.1 Spectrophotometric Method

A common way to evaluate fabrics' UV protection is the spectrophotometric method. A fabric sample is placed inside a spectrophotometer, a complex tool designed to precisely measure the amount of UV radiation passed through the fabric. The ultraviolet protection factor (UPF) of the material is calculated using this quantifiable information on UV transmission [64, 65].

4.3.2 Standardized Testing Procedures

Ensuring uniformity and reliability in the UV protection assessment necessitates standardized procedures. Some of the prominent standards include:

i. ***AS/NZS 4399*:** The method of evaluating UPF has been developed by the Australian and New Zealand standards committees. The method involves spectrophotometric measurements and provides guidelines for labeling UV-protective garments [28]. Table 1 shows the proposed UVR rating scheme.
ii. ***AATCC 183*:** The American Association of Textile Chemists and Colorists developed this standard of UPF measurement. It guides determining UPF, post-conditioning the textile (like wetting or stretching), and a framework for fabric classification based on UPF values [29].
iii. ***EN 13758*:** Originating from Europe, this standard deals with testing methods and labeling guidelines for UV-protective clothing. It emphasizes the aspects affecting UPF, such as fabric type, color, and moisture content [29].

Table 1 Proposed UVR rating scheme for textiles [28]

UVR protection	UV protection	Mean %UVR
Moderate	UPF 10 to 19	10 to 5.1
High	UPF 20 to 29	5.0 to 3.4
Very high	UPF 30 to 49	3.3 to 2.0
Maximum	UPF 50	2.0

5 Antimicrobial Properties

5.1 *The Need for Antimicrobial Textiles*

Textile materials used in everyday life provide a perfect habitat for microorganisms, particularly bacteria, to proliferate because of their huge surface area and profound capacity to absorb moisture. Being made of biomaterials, fabrics are naturally susceptible to microbial attack that can promote their degradation, loss of value, and functionality. Microorganisms have the capability to produce a wide range of enzymes that break down textile components like cellulose, hemicellulose, and keratin. They may also degrade textile chemicals such as varnishes, dyes, adhesives, and fillers and use them as a source of nutrition for their development. These pathogens can also accelerate the deterioration of synthetic fabrics such as polyester and nylon by targeting the additives in the fibers. When microorganisms are hosted to fabrics, they can lead to skin infections and irritations. Moreover, fungal respiratory infections and allergies might be exacerbated by the presence of spore-producing fungi on textiles. Textiles used in healthcare settings, such as staff uniforms, nose masks, curtains, and bedding, can engender healthcare-associated infections because they can harbor and spread pathogens such as methicillin-resistant *Staphylococcus aureus* (MRSA) and vancomycin-resistant enterococci (VRE) [66].

Antimicrobial textiles inhibit bacterial growth and prevent pathogen transmission in medical textiles [67]. With rising antibiotic resistance, nanomaterials, especially natural polymers like chitosan and alginic acid, are increasingly used for their biocompatibility and antibacterial properties, often enhanced with silver nanoparticles as nanocomposite (NC) forms [68].

5.2 *Mechanism of Bacterial Growth in Textile Fabrics*

Bacterial growth mechanisms in textile materials must be understood in order to successfully prevent their growth. Figure 6 shows the schematic representation of bacterial growth mechanisms in textile fabrics. Textile fabrics are typically made from either synthetic or natural fibers, consisting of natural or synthetic polymers. These polymers can be created from either chemical or natural monomers, depending on the type of fiber used. They provide a surface for airborne bacteria to attach to and

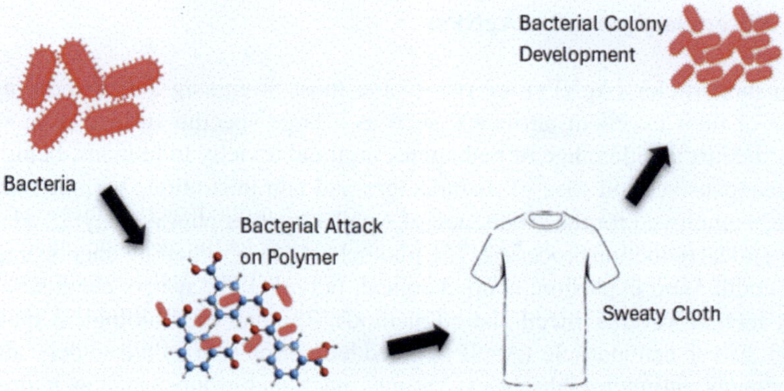

Fig. 6 Schematic representation of bacterial growth mechanism in textile fabrics [72]

serve as a nutrient source for bacterial growth and colony development [69]. Moisture from sweat also creates ideal conditions and atmosphere for bacterial growth [70]. As bacteria eat polymers, they help the polymer slowly break down into monomers inside the fabric. This process also causes metabolites to be released, which further breaks down the fiber. This collaborative effect of bacteria and their metabolites leads to fabric damage and the proliferation of bacterial colonies [71]. However, natural fibers are more susceptible to bacterial attacks compared to synthetic fibers, as the polymers in natural fibers provide a significant carbohydrate source for bacterial growth. In contrast, some chemical-based polymers in synthetic fibers are toxic to bacterial cells.

5.3 Types of Antimicrobial Agents Used in Textiles

Antimicrobial agents are used in textiles to prevent microorganism growth, with various chemicals like inorganic salts, antibiotics, and phenols. Natural bioactive agents from medicinal plants like neem, aloe vera, tulsi (basil), pomegranate, eucalyptus, and tea tree oil offer eco-friendly alternatives with fewer side effects and lower costs. Inorganic nanostructured materials like nanoparticles of Ag, Au, Cu, TiO_2, ZnO, carbon nanotubes, nanoclay, and modified species of nanoclay, including silver-chitosan/clay, clay-polyvinyl pyridinium, and silver-modified montmorillonites, are also utilized for antimicrobial textile finishes. Various parameters, like synthesis techniques, temperature, dispersion agent, surfactant, and so on, all have a significant impact on the effectiveness of the nanoparticles [73].

5.3.1 Silver Nanoparticles (AgNPs)

Silver nanoparticles (AgNPs) are one of the most promising antibacterial agents because of their excellent attributes, such as a large specific surface area, strong activity against a wide range of pathogens, minimal toxicity to humans, being relatively inexpensive, and ease of manufacture and administration. Various methods have been employed for the production of AgNPs, such as photo-catalytic reduction [74], chemical reduction processes [75], photochemical or radiation–chemical reduction, metallic wire explosion, sono-chemical, polyols [76], matrix chemistry [77], photoreduction, reverse micelle-based methods [78], and even biological synthesis [79–82]. Silver nanoparticle (AgNP)-embedded cotton fabrics are widely utilized for producing antimicrobial fabrics. Various functional groups, such as hydroxyls, amines, thioether [83], carboxylate [84], phosphonyl dichloride [85], and methyl [86], have been reported to bind AgNPs to cotton for increasing antibacterial properties. These functionalized textiles exhibit antibacterial activity against *E. coli*, *S. aureus*, *P. aeruginosa*, *K. pneumoniae*, *K. oxytoca*, and *P. mirabilis*, as well as antifungal properties against *Aspergillus niger* [87, 88]. It has found that cotton fabric treated with AgNPs had significant antibacterial, antifungal, and antiviral activity [89]. Cotton fabrics treated with AgNPs finish have demonstrated 100% antibacterial and good antifungal activity against various microorganisms, such as *E. coli, S. aureus, C. albicans,* and *A. niger* [90].

5.3.2 Gold Nanoparticles (AuNPs)

AuNPs (gold nanoparticles) are sufficiently inert and biocompatible with microbes. To develop antimicrobial properties, researchers have modified their surfaces with functional groups or integrated antibiotics. Gold nanoparticles coated with antibiotics have been used in textiles, providing efficient antibacterial coatings [91].

5.3.3 Metal Oxide Nanoparticles (MONPs)

Apart from metal nanoparticles, nanosized oxides of metal particles such as titanium, zinc, magnesium, copper, and iron, which are more stable in reaction contexts compared to metal nanoparticles, are also utilized to coat textiles to improve their antibacterial activity [92]. The solgel process is a well-established synthesis strategy for producing high-quality metal oxide nanoparticles (MONPs) and mixed oxide composites.

Zinc oxide (ZnO) in its various nanostructures form like nanotubes, nanobelts, and nanowires, making it valuable for applications in the fields of nanotechnology. Studies have highlighted its antibacterial properties against both gram-positive and gram-negative bacteria, including *S. aureus* and *E. coli.* ZnO nanoparticles have been used to create antimicrobial finishes on textile fabrics and cotton bandages, inhibiting the growth of pathogenic bacteria [93–95]. Titanium dioxide nanoparticles (TiO_2) have received significant interest due to their photocatalytic properties [96], self-cleaning [97], antibacterial agents by the solgel technique [98], and UV-protective agents. Reactive oxygen species (ROS) such as superoxide and hydroxyl radicals can be produced in the presence of TiO_2. ROS can cause bacteria to die by degrading their cell walls [36]. Cu-based NPs (CuO and their blends) have become popular nowadays in wound dressing fabrication due to the emergence of potentially Ag-resistant bacteria [99]. Overcoming the low ion-exchange capacity of cellulose fibers by introducing specific functional groups provides exciting possibilities for binding Cu^{2+} ions. These textiles maintain their antibacterial traits even through multiple washing cycles, [100, 101]. Magnesium oxide (MgO) nanoparticles have gained significant attention due to their applications in treating thermal burns, fighting cancer, and acting as antimicrobial agents [102, 103]. MgO nanoparticles synthesized from natural plant extracts exhibit strong antibacterial activity, particularly against *S. aureus*. This advancement opens up new possibilities for utilizing MgO nanoparticles to improve textile hygienes [103, 104].

5.3.4 Silica Nanoparticles (SiO_2 NPs)

Silica nanoparticles (SiO_2 NPs) are commonly employed to enhance antimicrobial properties in cotton fibers, where they form covalent bonds to augment mechanical robustness and longevity [105]. Cotton fabrics integrated with SiO_2 NPs showed considerable antibacterial efficacy against *S. aureus* and *E. coli* [106].

5.3.5 Polymer Nanocomposites

Polymer nanocomposites have become increasingly significant across various domains due to their ability to combine the advantages of multiple materials while mitigating their toxicity and other harmful impacts. Nanocomposite from mixture of silicone quaternary ammonium salts with zinc oxide and silver can exhibit strong antibacterial performance against *S. aureus* and *E. coli* with improved fabric properties like hydrophilicity, air permeability, softness, and durability even after multiple washes [107]. A novel nanocomposite named silver–titanium-dioxide-beta cyclodextrin-coated cellulose fabric showed enhanced antibacterial properties against *S. aureus* with self-healing abilities, dye degradation, increased strength, and an improved crease recovery angle [108]. Other nanocomposites like copper oxide–polyethylene glycol, copper oxide–titanium dioxide, and hydroxyapatite–silver nanoparticles have also been employed in textiles to combat *S. aureus* and *E. coli* [109].

5.3.6 Carbon-Based Nanoparticles

Carbon nanotubes (CNTs) modified with nanosized antibacterial agents are used to impart antibacterial effectiveness against bacteria like *S. aureus* in cotton fabrics [110, 111]. CNTs can be combined with other nanoparticles, like silver, to create nanocomposites that bond well with textiles and exhibit strong antibacterial properties even after multiple wash cycles [112]. Single-walled carbon nanotubes (SWNT) can damage bacterial cell membranes through direct contact which paved the way for developing SWNT-based antimicrobial coatings [113]. Carbon dots and fullerenes, which are zero-dimensional carbon nanostructures, show great potential to package antibiotics, improving their precision delivery to harmful bacteria. However, their capacity to confer antibacterial properties on textiles has not yet been fully explored [114, 115].

5.3.7 Chitosan Nanoparticles

Beside this, due to excellent biocompatibility, biodegradability, non-toxicity, high surface area and zeta potential stability, chitosan nanoparticles are widely used in cellulosic textiles along with metal and metal oxide nanoparticles. Chitosan nanoparticles can be prepared based on the ionotropic gelation of chitosan and tri-polyphosphate [116–118].

5.4 Antibacterial Mechanism

Bacteria are distinguished by an exterior cell wall structure that is important in categorizing them. They are classified as gram-positive or gram-negative. When compared to gram-negative bacteria, gram-positive bacteria have a thicker encapsulation of peptidoglycan in their cell walls (Fig. 7). They do not, however, contain the outer membrane of lipopolysaccharide that exists in gram-negative bacteria. As a result, gram-negative bacteria have an extra layer between the outer layer and the plasma membrane known as the periplasmic layer. Nonetheless, because this layer is thinner than that of gram-positive bacteria, it is more susceptible to penetration and eventual demise when exposed to antimicrobial nanoparticles [119].

Antibacterial activity can be achieved in metal and metal oxide nanoparticle-incorporated textiles through the generation of reactive oxygen species (ROS), which involves in the production of highly reactive molecules, such as superoxide radicals ($O_2{\cdot}{-}$), hydrogen peroxide (H_2O_2), and hydroxyl radicals ($\cdot OH$). ROS are highly reactive and can damage various biomolecules, including proteins, lipids, and DNA by peroxidation of the polyunsaturated phospholipid in the bacterial cell membrane. Once generated, ROS can interfere with bacterial replication and lead to mutations, potentially rendering the bacteria non-viable [120]. Figure 8 represents schematic drawings of antimicrobial mechanism of metal and metal oxide nanoparticles.

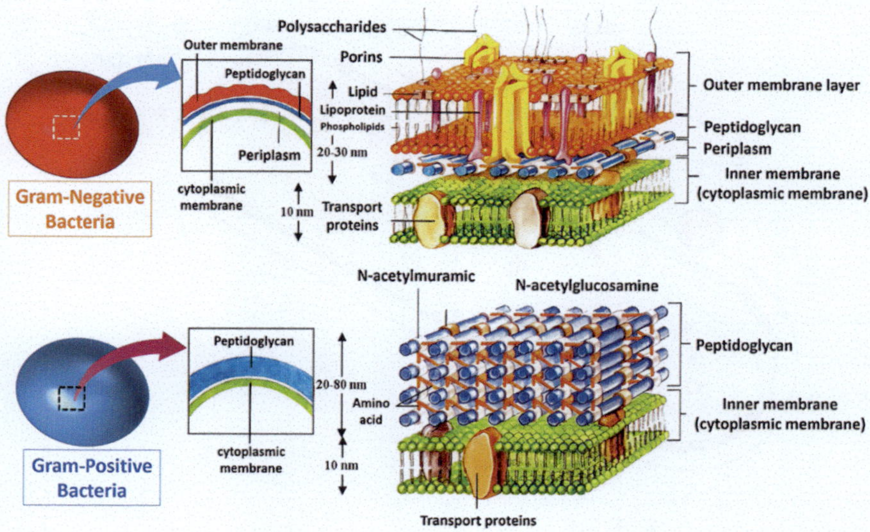

Fig. 7 Cell wall structures of gram-positive and gram-negative bacteria reprinted from [119] with permission

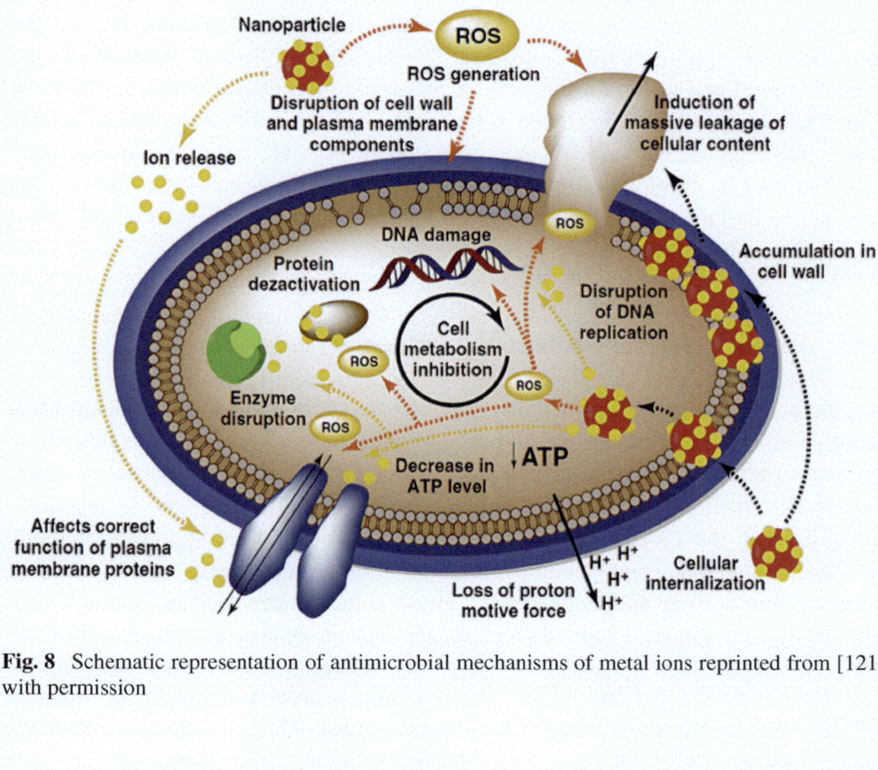

Fig. 8 Schematic representation of antimicrobial mechanisms of metal ions reprinted from [121] with permission

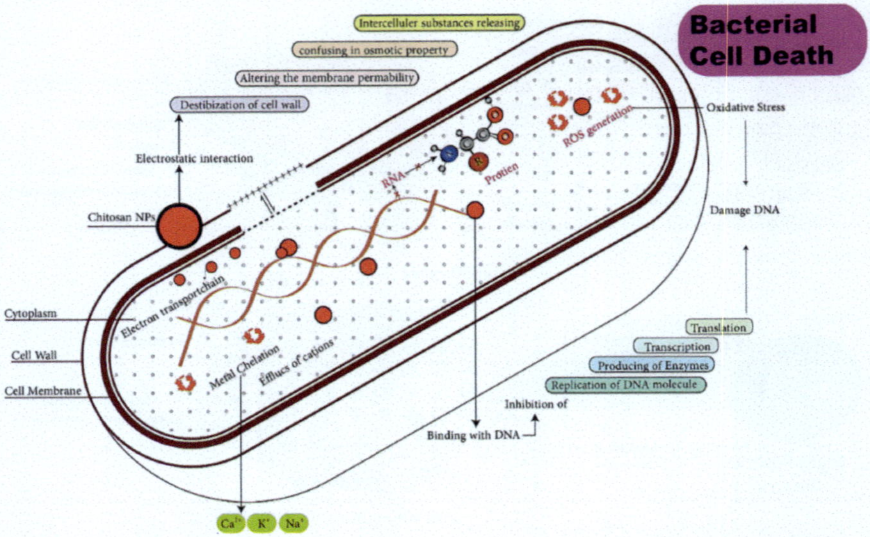

Fig. 9 Antibacterial mechanism of chitosan nanoparticles reprinted from [117] with permission

Chitosan (CS) shows cationic characteristics in acidic conditions due to the conversion of NH_2 to NH^{3+}. Chitosan's cationic nature allows it to infiltrate and break bacterial cell walls. Furthermore, CS has a similar effect on fungi. The most significant drawback of CS is that it is more effective against gram-positive bacteria than gram-negative bacteria. As demonstrated in Fig. 16, chitosan nanoparticles (CSNPs) may be employed to solve this issue because their nanostructure enables their access into microbial (bacterial and fungal) cells. The antibacterial mechanism of CSNPs has been represented in Fig. 9.

5.4.1 Factors Affecting Antibacterial Properties

In general, the uniqueness of nanoparticles for specific applications is determined significantly by the synthesis methods employed. The synthesis methods determine the size, structure, configuration, and electrostatic properties of nanoparticles, which impact their attributes. Three main methods—physical, chemical, and biological—are utilized in the production of nanoparticles. Each of these synthesis methods has benefits and disadvantages in demonstrating biological features, particularly in deciding their activity and fastness [72]. Physical techniques such as laser ablation, ball milling, and physical vapor deposition generate nanoparticles that are uniformly sized but are plagued by cost and complexity issues. On the other hand, chemical methods like solgel and hydrothermal synthesis provide control over size and stability but may introduce potential toxicity concerns [122]. In contrast, eco-friendly approaches, known as green or biosynthesis methods, utilize organisms or plant extracts (Table 2) to produce nanoparticles that are environmentally sound. However,

Table 2 Synthesis of metal nanoparticles from plant origin with their size and antimicrobial activity [125]

Nanoparticle	Plant origin	Size (nm)	Application
Au	*Croton sparsiflorus*	16.6–17	UV protection, antibacterial
Ag	*Elaeagnus umbellata*	40	Antimicrobial
Ag	*Cestrum nocturnum*	20	Antioxidant and antibacterial
Ag	*Malus domestica*	16	Antimicrobial
Ag	*Nauclea latifolia*	12	Antimicrobial and antioxidant
Cu	*Walnut shells*	15–22	Antibacterial, antioxidant
Cu	*Hagenia abyssinica*	34.76	Antimicrobial
Pd	*Rosmarinus officinalis*	15–90	Antibacterial and antifungal
Pt	*Nigella sativa L*	1–6	Antimicrobial and anticancer
ZnO	*Acalypha fruticosa*	50	Antimicrobial
TiO_2	*Mentha arvensis*	20–70	Antimicrobial
TiO_2	*Ledebouria revoluta*	47	Larvicidal, antibacterial, and anticancer

these methods often result in nanoparticles with limited stability and weak adhesion to textiles. To address this, additional processes like cationization can be employed to enhance the bond between the fabric and nanoparticles, thereby improving antibacterial effectiveness even after multiple washes. Selecting the appropriate synthesis approach is pivotal in achieving effective bactericidal finishes for textiles [119, 123, 124].

5.5 *Effectiveness and Durability of Antimicrobial Treatments*

The efficacy and resilience of antimicrobial treatments can exhibit substantial variations, contingent on multiple factors. These factors encompass the specific microorganism under consideration, the nature and dimensions (Table 3) of the nanomaterial employed as an antimicrobial agent, the application method, the quantity of loaded nanoparticles, potential cross-contamination, and more. Notably, the quantity of AgNPs integrated into the fabric escalates with rising concentrations of $AgNO_3$ in the synthesis process of AgNP-loaded fabrics. Consequently, these fabrics demonstrate durability by withstanding up to 20 repeated washing cycles [126]. Impressive antimicrobial efficacy was observed against *Alternaria sp.* and *Rhizopus stolonifera* when testing active nano-laminated films. These films comprised five layers alternating between polyelectrolyte solutions, including alginate, zein–carvacrol nanocapsules, chitosan, and chitosan–carvacrol emulsions [127]. Carboxymethyl chitosan (CMCTS) and silver nanoparticles (AgNPs) covalently linked to cotton fabric exhibit enduring antibacterial properties (BR > 95%) over 50 laundering cycles [128]. Cotton textiles treated with nanosized zinc oxide

(ZnO) colloids exhibit moderate antibacterial effects against both gram-positive and gram-negative microorganisms. At a concentration of 20 ppm nano-ZnO colloid solution, bacterial growth decreased by approximately 82.2% for *Staphylococcus aureus* and 81.9% for *Bacillus subtilis* (gram-positive bacteria), while gram-negative bacteria, represented by *Pseudomonas aeruginosa* and *Escherichia coli*, showed reductions of approximately 74.8% and 75.4%, respectively [129]. Another study revealed that copper oxide (CuO) nanoparticle-treated fabrics initially showed strong antibacterial activity against *E. coli* and *S. aureus*. However, this effectiveness gradually declined with successive wash cycles, and by the 25th wash, no reduction in bacterial growth was observed. The diminishing antibacterial efficacy was attributed to the leaching of physically adsorbed CuO nanoparticles from the fabric surface during washing [130].

6 Integration Methods for Nanoparticles in Textiles

Sometimes nanoparticles are integrated with textiles aiming to prevent UV radiation, sometimes they are used to improve antimicrobial properties, and sometimes both the properties are achieved at a time using multifunctional nanoparticles. Whether the nanoparticles are integrated with the textiles for UV protection or antimicrobial properties, the methods are generally same and the methods are with the consideration of maintaining the fabric's esthetic and tactile properties. While these nanoparticles serve this purpose well, they frequently encounter difficulties in establishing robust connections with the fabric's surface. To resolve this problem, procedures such as nanoparticle customization or alterations to the textile surface are necessary to firmly integrate the nanoparticles into the fabric. Many researchers have been trying to develop textile materials that can block microbial growth using three main approaches [67]:

(i) Addition of antimicrobial compounds in the textile fibers;
(ii) Grafting of antimicrobial compounds on the fiber surface; and
(iii) Coating of the fiber surface with antimicrobial metal and metal oxide nanoparticles (NPs).

The most prevalent procedures for integrating nanoparticles with textiles are as follows:

(i) Dip coating
(ii) Pad–dry–cure
(iii) Sonication
(iv) Layer-by-layer
(v) Spraying and coating
(vi) Electrospinning
(vii) Solgel technique
(viii) In situ polymerization technique.

Table 3 Antibacterial effectiveness of different sizes of nanoparticles with their mode of action [119]

NPs	Size (nm)	Fiber	Mode of action	Microorganism	Effectiveness
Ag	2.3	Cotton	Bacterial cell membrane destruction	*S. aureus* and *E. coli*	92% and 99% reduction for *E. coli* and *S. aureus,* respectively
Ag	20	Cotton	Controlled release of silver to the surface	*S. aureus* and *E. coli*	Significant activity after 20 washes for *S. aureus* and 5 washes for *E. coli*
Ag	60–100	Rayon and Cotton	Binding of silver to bacterial enzymes resulting in deactivation of cells and cell membrane destruction	*E. coli, S. aureus* and *E. faecali*	Bacterial growth inhibition for *E. coli-100%, S. aureus-98%, E. faecalis-97%*
Ag	11	Cotton	Bacterial cell membrane destruction	*E. coli* and *S. aureus*	Antimicrobial activity against *E. coli*-99.91%, *S. aureus*-99.31%
CuO	83	Cotton	Attachment of CuO NPs to the bacteria cell walls by electrostatic attraction, followed by the release of Cu^{2+} which causes damage to cell wall, and death to bacteria	*S. aureus, E. coli* and *K. pneumoniae*	Excellent antimicrobial activity with 93% antimicrobial durability after 50 washes
ZnO	100	Cotton	Generation of reactive oxygen species (ROS) that disrupts the normal metabolism of bacteria cells	*E. coli* and *S. aureus*	97% inhibition rate at the concentration of 1 g ZnO and 94% reduction with 10-g ZnO and 84% inhibition after 20 washes
TiO_2	9	100% Cotton and 45/55% knit polyester cotton	Generation of reactive oxygen species (ROS) that disrupts the normal metabolism of bacteria cells	*S. aureus* and *K. pneumonia*	All treated sample showed > 75% antibacterial efficiency. The knitted fabric revealed better bacterial reduction than pure cotton due to the resistance property of polyester

(continued)

Table 3 (continued)

NPs	Size (nm)	Fiber	Mode of action	Microorganism	Effectiveness
TiO_2	50	Cotton	Creation of free radical oxygen which breaks down bacterial cell wall by oxidation–reduction reaction	*E. coli* and *S. aureus*	> 95% bacterial reduction which is sustained after 20 washes
TiO_2-GO	< 100	Cellulose acetate	Generation of ROS under UV illumination	*B. subtilis*, *B. cereus*	> 95% bacterial reduction

Dip Coating

Dip coating, due to its increased controllability and large-scale manufacturing capability, is a simple and adaptable process for coating nanoparticles on cotton textiles [101, 131, 132]. It involves the deposition of a nanoparticle film by immersion of the substrate into a solution containing nanoparticles and its withdrawal at constant speed, and homogeneous liquid film containing nanoparticles is formed on the substrate's surface (Fig. 10). In recent studies, various methods [66, 72] for incorporating nanoparticles into fabrics for antibacterial properties have been explored. When employing dip coating, it had issues with thickness control and uniformity.

Pad–dry–cure

The pad–dry–cure technique is frequently used to integrate nanoparticles with fabric or other materials [133–135]. Sometimes the method is also called the dip–pad–dry–cure method. In this method (Fig. 11), the cloth is submerged in a nanoparticle-containing solution or dispersion. To ensure a uniform distribution of nanoparticles,

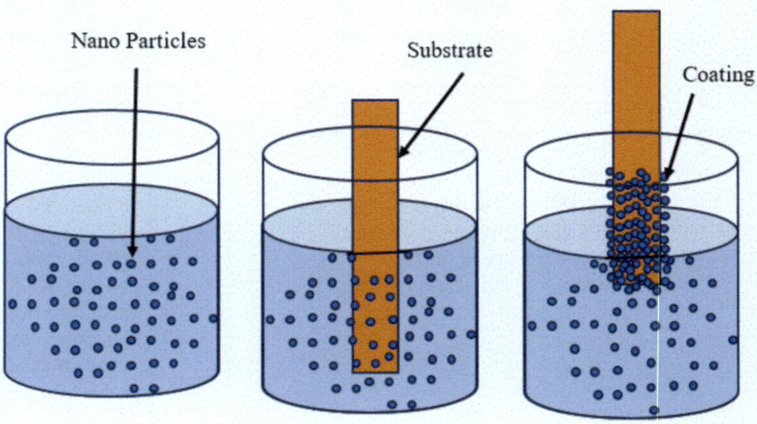

Fig. 10 A schematic view of the dip-coating method

extra solution is pushed out using padding rollers. The nanoparticles can be fixed or cured to the fabric by passing through heat chambers. This method is helpful since it is quick and efficient at coating textiles with a thin layer of nanoparticles. The concentration of the nanoparticle solution, the padding, and other variables all affect how well these techniques work [5, 136]. Sometimes binders (or cross-linkers) are required to fix the particles with textiles. The pad–dry–cure method is promising but incur increased production costs due to excess binder [87, 137].

Sonication

Sonication is an efficient nanofinishing technology for the deposition and insertion of nanoparticles on the fabric surface [138–140]. It ensures improved adhesion of the deposited nanoparticles at ultrasound (20 kHz to 1 MHz) [66], since the nanoparticles

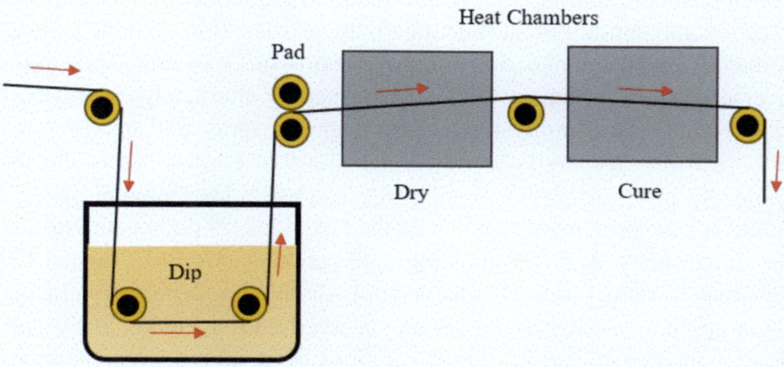

Fig. 11 Schematic representation of the pad–dry–cure method

Fig. 12 Schematic representation of the sonication method

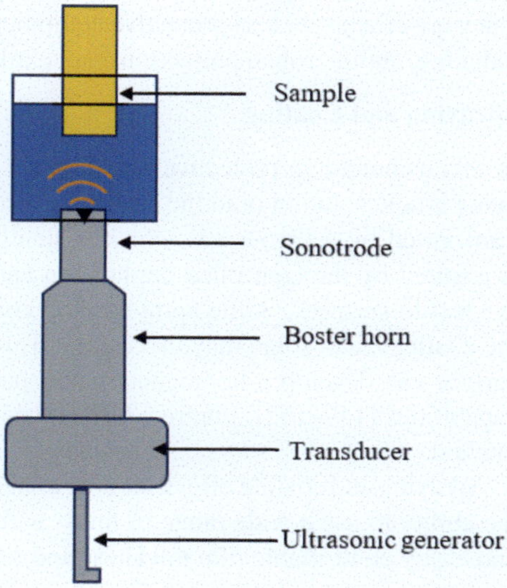

have been introduced into the substrate by the microjets through sonic bubble collapse near the surface. As a result of the ultrasonic waves spreading the nanoparticles over the fabric surface, a uniform and thick coating of nanoparticles is formed. Effectively coated silver nanoparticles in nylon, polyester, and cotton textiles using a sonochemical process demonstrated good antibacterial activity against both gram-positive and gram-negative bacteria [141]. Nuruzzaman et al. used sonochemical coating of textiles with ZnO for the production of UV-protective textiles. With this technique, the coated substrates exhibit a near-complete suppression of UV-A and UV-B radiation and consistently produce high UV protection factor values (>UPF50) for the illustrative application of UV-blocking capacity, even after 51 washing cycles [142].

Layer-by-Layer

The layer-by-layer approach has been reported as a unique deposition mechanism for UV protection and antimicrobial functionalizing textiles. It is a simple, convertible, and significant approach for coating non-polar textile surfaces with nanocomponents from diverse chemical groups, which is performed by alternately overlaying oppositely charged materials. Immersion, electromagnetic, spray, and spin processes are utilized to perform layer-by-layer functionalization of textiles where electrostatic forces hold the layers together. Metal oxide nanoparticles including TiO_2, SiO_2, and ZnO/SiO_2 have been investigated with the technique for photocatalytic, antibacterial, hydrophobicity, and UV-shielding applications [143–146]. Figure 13 illustrates the process where positively charged poly diallyldimethylammonium chloride (PDDA) is applied on slightly negatively charged cotton fabric. In this process, fabric is sequentially dipped in diluted solutions of metal oxide nanocomposite at different concentration ratios with PDDA dilute solution. The nanocomposite's Zn/Si ratio and layer count have a big influence on the efficacy of blocking UV rays. This approach is especially well suited for applications where precision and consistency are crucial such as protective equipment for employees exposed to strong UV radiation, antimicrobial protection, and textiles used in space exploration [147, 148].

Spraying and Coating

In this technique, a protective nanoparticle layer is created by directly spraying a nanoparticle solution onto the textile surface. The droplets of the nanoparticles are transported by a carrier gas which is finally evaporated. Effective UV protection is ensured by the controlled application and even distribution of nanoparticles on the textile surfaces. Some spraying and coating methods include hot plate to dry the coatings, and some may use hot air. Through this coating procedure, the textile surface can develop a homogenous nanoparticle layer, improving its UV defense capabilities [149–152]. Figure 14 illustrates a simple form of spraying and coating method.

Spraying and coating offer a number of benefits over other methods. These include the ability to use a wide range of fluids with various thicknesses, low liquid waste, and easily controllable film thickness and roughness.

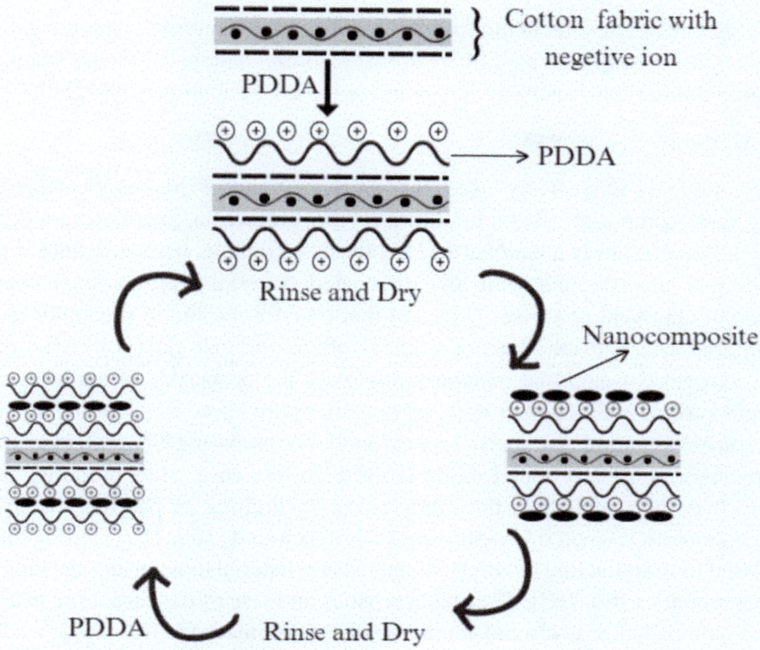

Fig. 13 Layer-by-layer (LbL) method

Fig. 14 Simple spraying and coating techniques

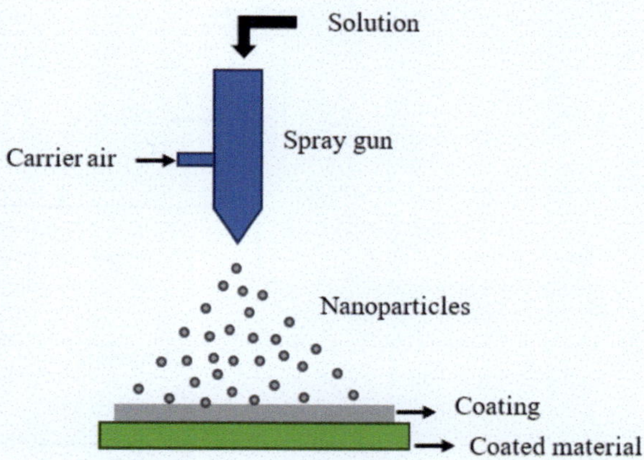

Electrospinning

For having capacity to produce textiles with improved UV protection, antimicrobial characteristics, and other desirable functionalities, electrospinning has attracted considerable attention. A polymer melt or solution containing nanoparticles is extruded through a small nozzle during the electrospinning process (Fig. 15). The solution experiences a phase shift as a result of the simultaneous application of a high voltage, which results in the formation of ultrafine nanofibers that are gathered on a substrate. When these nanofibers are formed, the nanoparticles are seamlessly incorporated into them, resulting in a homogeneous distribution [153–156].

Solgel Method

This technique is extensively used for applying metal oxide nanoparticles and creating nanocomposite fibers to develop UV protection properties in textiles [61, 157]. It is originally a nanoparticle synthesis technique, but sometimes dipping or padding is incorporated with this for coated material production. It entails a sequence of chemical processes (Fig. 16) that transform a liquid precursor (sol—a stable colloidal suspension of nanoparticles) into a solid material (gel). After drying the gel, xerogel is found that generally possesses the properties of higher porosity and larger surface area together with very smaller pore sizes. By grinding and high-temperature calcination, nanoparticles are found. For preparing UV-protected textiles as a precursor, generally metal oxide is used. In this case, after preparing solgel solution, it is then applied to the fabric using by dipping or padding. Then after drying, nanoparticle-coated UV-protected fabric is found [158, 159]. This technique is also used to manufacture a variety of nanoscale materials, including ceramics and other compounds [160–162]. The characteristics and size of the particles can also be precisely controlled. It is also an economically viable method [162, 163].

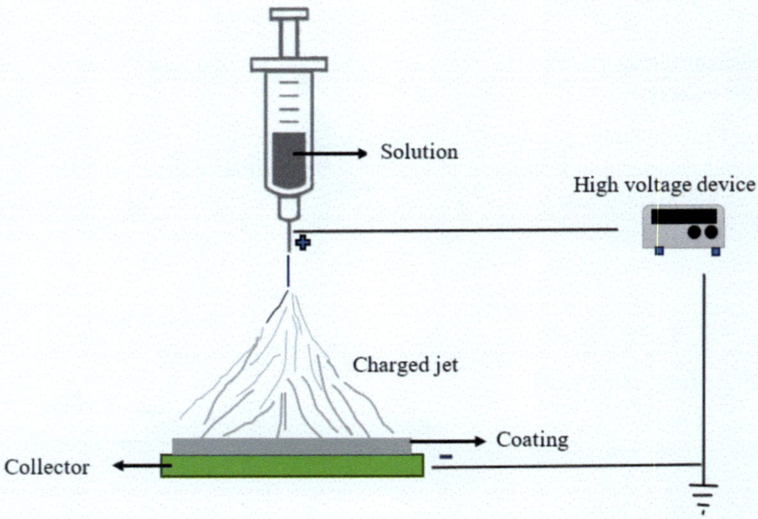

Fig. 15 Electrospinning method

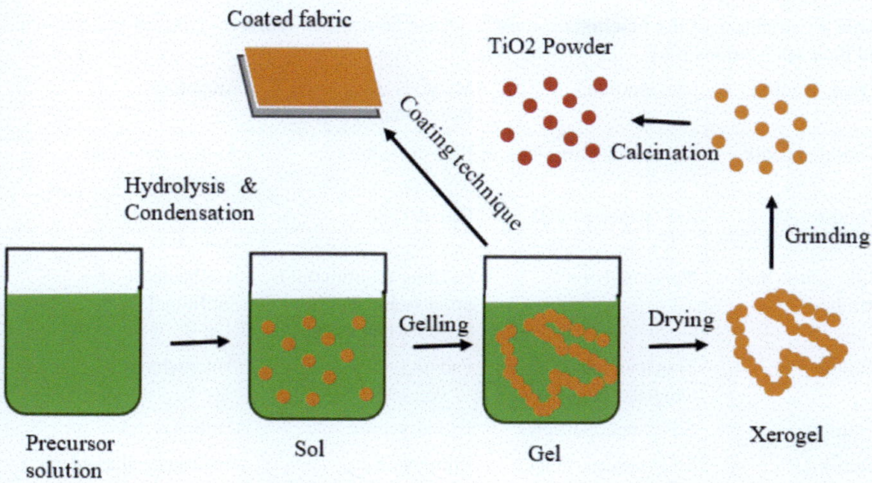

Fig. 16 Solgel method

In Situ Technique

Under the technique, the nanoparticles can be grown on/inside the textile fibrous structures. In this technique, first, appropriate nanomaterial is chosen and then, nanomaterials are mixed with monomer along with a compatible solvent or binder solution (Fig. 17). After polymerization, the polymer is formed with uniformly dispersed nanoparticles into it. In order to affix the nanomaterials securely onto the fabric, a drying or curing procedure is employed, the exact methodology of which is contingent on the type of nanomaterial utilized and the chosen binder or solvent system [164–166]. Table 4 represents functionalization of different textile fabrics with different types of nanoparticles with their applications.

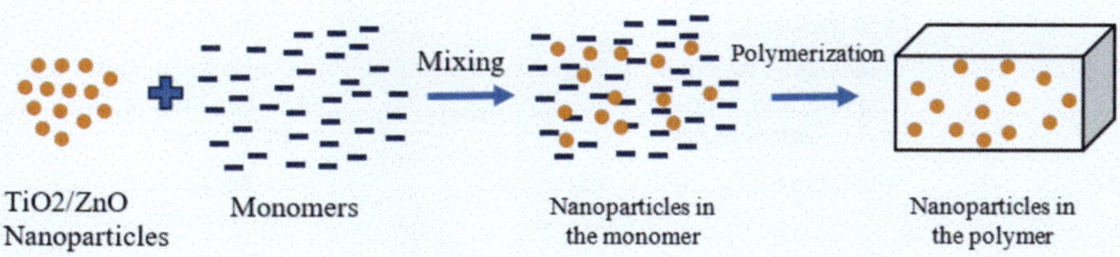

Fig. 17 In situ polymerization

Table 4 Summary of the functionalization of textile fabrics with different types of nanoparticles and their applications [66]

Types of fabric	Nanoparticles	Methods of Functionalization	Attributes
Cotton and silk	Silver nanoparticles	Pad–dry–cure	Antibacterial and antifungal
Cotton	ZnO nanoparticles	Pad–dry–cure	Antibacterial, UV protection
Silk, cotton and bamboo	Nanoemulsion	Padding (continuous) process and batch (exhaust) process	Antibacterial and antifungal
Cotton	Tragacanthin gum/ nanosilver hydrogel	Pad–dry–cure	Antibacterial
Cotton	Silver nanoparticles	Pad–dry–cure	Antibacterial
Cotton	SiO_2 nanoparticles	Immersion	Antibacterial and UV protection
Cotton	ZnO/TiO_2 nanoparticles	Pad–dry–cure	Antimicrobial activity, UV protection, and self-cleaning antibacterial
Cotton	Ag/Cu nanoparticles	Immersion	Antibacterial
Cotton	Ag/TiO_2 nanoparticles	Dip coating	Antimicrobial activity and self-cleaning
Cotton	Ag/ZnO/Cu nanoparticles	Immersion	Antibacterial activity, UV protection
Cotton	Ag/ZnO nanoparticles	Immersion	Antibacterial activity, UV protection
Wool	Silica, titanium, and silver nanoparticles	Electrostatic self-assembly	Antibacterial, hydrophilicity and self-cleaning
Wool	Selenium nanoparticles	Immersion	Antibacterial and UV-blocking properties
Wool, Cotton	Zinc oxide, titanium dioxide, and copper oxide nanoparticles	Pad–dry–cure	Antibacterial, antifungal, UV blocking, and self-cleaning
Silk	Platinum nanoparticles	Immersion	Antibacterial, catalysis and dyeability
Silk	Gold nanoparticles	Electrospinning	Antibacterial
Nylon	Titanium dioxide and zinc oxide nanoparticles	Layer-by-layer	Antibacterial, UV protection, and stain proof

7 Integration of UV Protection and Antimicrobial Properties

7.1 Possibilities of Combining UV Protection and Antimicrobial Functionalities

UV protection and antimicrobial functionalities together can be combined in the textiles using various techniques. Cotton fabrics treated with TiO_2NPs with *Moringa oleifera* extract can improve antibacterial and ultraviolet properties [167]. Cheng et al. [168] used layer-by-layer self-assembly to apply TiO_2NPs on cotton fabric. The UPF values demonstrated that the nanocotton fabrics provided excellent UV protection and had a good affinity between the nanoparticles and the fabric surface against laundering. AgNP synthesized by in situ and thermal reduction process imparted antimicrobial and UV-blocking properties of plasma-treated cotton fabrics [169].

7.2 Potential Applications

UV protection can also be provided by the HAp (Hydroxyapatite)-AgNPs coating. Cotton gauze textiles were cationized with chitosan or partially carboxymethylated (anionic alteration) to increase HAp and AgNP dispersion and adherence on the surface. The effects of cationic and anionic alterations, as well as HAp and AgNPs deposition, on the characteristics of cotton gauze (coloration, UV protection, antibacterial activities, and water absorption) were carefully investigated. Overall, the results show that chemical (anionic and cationic) modification of cotton gauze increases the deposition of HAp and AgNPs. Chitosan improves the biocompatibility and wound-healing qualities of cotton gauze. Deposition of AgNP onto cotton gauze textiles resulted in strong antibacterial activity against *Candida albicans* and gram-positive and gram-negative bacteria, as well as increased UV protection [67].

7.3 Challenges and Considerations in Developing Dual-Functional Textiles

Use of UV-protective and antibacterial particles to the textiles may influence the comfortability of textiles. Use of nanoparticles may be a potential solution of this. However, some studies have raised apprehensions about the safety of the nanoparticles, as they exhibit a tendency to be absorbed by the skin, thereby posing potential health hazards [170]. The release of AgNPs and Ag+ ions during textile use is a concern for diminishing antibacterial effects and safety, coupled with insufficiently studied cytotoxicity; however, green synthesis from fenugreek leaves yielded stable, eco-friendly AgNPs ranging from 20 to 30 nm [171]. Some particular concerns involve the toxic consequences of inhaling carbon nanotubes (CNT) and their similarity to asbestos fibers, as well as the inflammatory properties of inhaled high-dose nano-TiO_2. While there is a chance that nanoparticles could enter the bloodstream and affect other organs, it is generally believed that skin contact with dermal titanium and zinc oxides poses minimal risk because they cannot penetrate the skin's outermost layer. Lastly, it is advisable to exercise caution regarding the potential toxicities of nanostructured materials, including the pro-inflammatory effects of silver and inhaled TiO_2, the toxicity associated with inhaled silver, and the potential for allergic reactions to TiO_2. Additional research and investigation are necessary to gain a comprehensive understanding of the risks linked to these materials [172].

These tiny particles may accumulate in many areas of the environment and endanger plants and animals [173]. To determine how to lessen any potential detrimental effects, we must investigate how these particles behave in the environment. It is essential to consider these nanoparticles' entire life cycle, from production to disposal, to comprehend where they might end up and how to reduce potential concerns [174].

Sustainability and the secure application of nanotechnology are significant concerns in the field of textile finishing. This includes finding a balance between technological advancement and safeguarding the environment and people's health. To achieve this balance, we rely on using sustainable and biodegradable nanomaterials, which are eco-friendly and naturally degrade [175]. Research is constantly being done in this area, and new techniques are being developed, mainly for using nanomaterials made from renewable sources. In addition to safety measures, monitoring and adhering to established safety standards are essential [176].

8 Future Trends and Innovations

The demand for antimicrobial textiles has surged due to COVID-19, promising substantial growth in the industry. Enhancing durability in functional finishes by incorporating agents into the fiber structure is crucial. A potential future development is the micro- and nanoencapsulation of natural biocides, offering durability and safety benefits. Substances like neem extract, aloe vera extract, and chitosan show promise but need more research for commercial use. Efforts to integrate sustainable biopolymers into the market face economic feasibility issues. However, using natural agents as antimicrobials and UV protectors is cost-effective when sourced naturally, though they may have a shorter life span compared to synthetics. Despite challenges, the use of natural agents in antimicrobial textiles holds promise [2].

9 Conclusion

The need for protective measures from UV rays and textiles' microbial contamination is now a well-known concern to consumers, and nanofinishing has emerged to deal with both with promising outcomes. Because of the special properties of nanoparticles, both functionalities can be achieved with the same finishing, which means steps and cost eliminate. The chapter has elaborately described the updated details about the nanotechnologies linked with UV and microbial protection using textiles though investigations with newer approaches continue to appear. Functionality, durability, safety, and sustainability are probably the paramount questions in this domain to solve in future research.

References

1. Lee S (2009) Developing UV-protective textiles based on electrospun zinc oxide nanocomposite fibers. Fibers Polym. https://doi.org/10.1007/s12221-009-0295-2
2. Shah JN, Padhye R, Pachauri RD (2022) Studies on UV protection and antimicrobial functionality of textiles. J Nat Fibers 19:6810–6821. https://doi.org/10.1080/15440478.2021.1932678
3. Cadet J, Douki T, Ravanat JL (2015) Oxidatively generated damage to cellular DNA by UVB and UVA radiation. Photochem. Photobiol
4. Hessling M, Haag R, Sieber N, Vatter P (2021) The impact of far-UVC radiation (200–230 nm) on pathogens, cells, skin, and eyes—a collection and analysis of a hundred years of data. GMS Hyg Infect Control. https://doi.org/10.3205/dgkh000378
5. Mondal S (2022) Nanomaterials for UV protective textiles. J Ind Text
6. Diffey BL (2002) Sources and measurement of ultraviolet radiation. Methods. https://doi.org/10.1016/S1046-2023(02)00204-9
7. D'Orazio J, Jarrett S, Amaro-Ortiz A, Scott T (2013) UV radiation and the skin. Int J Mol Sci
8. De Gruijl FR (1999) Skin cancer and solar UV radiation. Eur J Cancer. https://doi.org/10.1016/S0959-8049(99)00283-X
9. Cross SE, Innes B, Roberts MS et al (2007) Human skin penetration of sunscreen nanoparticles: in vitro assessment of a novel micronized zinc oxide formulation. Skin Pharmacol Physiol. https://doi.org/10.1159/000098701
10. Wacker M, Holick MF (2013) Sunlight and vitamin D. Dermatoendocrinol 5:51–108. https://doi.org/10.4161/derm.24494
11. Polvani LM, Previdi M, England MR et al (2020) Substantial twentieth-century arctic warming caused by ozone-depleting substances. Nat Clim Chang
12. Bashari A, Shakeri M, Shirvan AR (2018) UV-protective textiles. In: The impact and prospects of green chemistry for textile technology
13. Gambetta JM, Holzapfel BP, Stoll M, Friedel M (2021) Sunburn in grapes: a review. Front Plant Sci
14. Fisher GJ, Kang S, Varani J et al (2002) Mechanisms of photoaging and chronological skin aging. Arch Dermatol. https://doi.org/10.1001/archderm.138.11.1462
15. Afaq F, Mukhtar H (2006) Botanical antioxidants in the prevention of photocarcinogenesis and photoaging. Exp Dermatol
16. Schwarz T (2005) Mechanisms of UV-induced immunosuppression. Keio J Med
17. Collaris EJH, Frank J (2008) Photoallergic contact dermatitis caused by ultraviolet filters in different sunscreens. Int J Dermatol 47:35–37. https://doi.org/10.1111/j.1365-4632.2008.03957.x

18. Hartono SP, Bedell VM, Alam SK et al (2022) Vascular endothelial growth factor as an immediate-early activator of ultraviolet-induced skin injury. Mayo Clin Proc
19. McKenna MJ, Kilbane M (2022) Vitamin D deficiency. In: Endocrinology and diabetes: a problem oriented approach, 2nd edn
20. Vitamin D and your health: breaking old rules, raising new hopes. Harv Mens Health Watch (2007)
21. De Castro GSG, Sant'Anna WC, Da Silva CVA et al (2020) Importância do ultravioleta no vitiligo e na psoríase. Life Style 7:69–76. https://doi.org/10.19141/2237-3756.lifestyle.v7.n1.p69-76
22. Vieyra-Garcia PA, Wolf P (2021) A deep dive into UV-based phototherapy: mechanisms of action and emerging molecular targets in inflammation and cancer. Pharmacol Ther 222:107784. https://doi.org/10.1016/j.pharmthera.2020.107784
23. Harrison SL, Buettner PG, MacLennan R (2005) The north queensland sun-safe clothing study: design and baseline results of a randomized trial to determine the effectiveness of sun-protective clothing in preventing melanocytic nevi. Am J Epidemiol. https://doi.org/10.1093/aje/kwi081
24. Aguilera J, De Gálvez MV, Sánchez-Roldán C, Herrera-Ceballos E (2014) New advances in protection against solar ultraviolet radiation in textiles for summer clothing. Photochem Photobiol. https://doi.org/10.1111/php.12292
25. Davis S, Capjack L, Kerr N, Fedosejevs R (1997) Clothing as protection from ultraviolet radiation: which fabric is most effective? Int J Dermatol. https://doi.org/10.1046/j.1365-4362.1997.00046.x
26. Lu JT, Ilyas E (2022) An overview of ultraviolet-protective clothing. Cureus. https://doi.org/10.7759/cureus.27333
27. Berry EG, Bezecny J, Acton M et al (2022) Slip versus slop: a head-to-head comparison of UV-protective clothing to sunscreen. Cancers (Basel). https://doi.org/10.3390/cancers14030542
28. Gies PH, Roy CR, Toomey S, McLennan A (1998) Protection against solar ultraviolet radiation. In: Mutation research—fundamental and molecular mechanisms of mutagenesis
29. Louris E, Sfiroera E, Priniotakis G et al (2018) Evaluating the ultraviolet protection factor (UPF) of various knit fabric structures. In: IOP conference series: materials science and engineering
30. UV PROTECTION. In: International wool textile organization
31. Reinert G, Fuso F, Hilfiker R, Schmidt E (1997) UV-protecting properties of textile fabrics and their improvement. Textile chemistry coloration
32. Osterwalder U, Herzog B (2009) Sun protection factors: world wide confusion. Br J Dermatol. https://doi.org/10.1111/j.1365-2133.2009.09506.x
33. Gambichler T, Hatch KL, Avermaete A et al (2002) Influence of wetness on the ultraviolet protection factor (UPF) of textiles: In vitro and in vivo measurements. Photodermatol Photoimmunol Photomed. https://doi.org/10.1034/j.1600-0781.2002.180105.x
34. Wong WY, Lam JK, Kan C, Postle R (2016) Influence of reactive dyes on ultraviolet protection of cotton knitted fabrics with different fabric constructions. Text Res J. https://doi.org/10.1177/0040517515591776
35. Emam HE, Ahmed HB (2023) Ultraviolet protection finishing agents in textile functionalization. In: Advances in healthcare and protective textiles. Elsevier, pp 423–446
36. Shah MA, Pirzada BM, Price G et al (2022) Applications of nanotechnology in smart textile industry: a critical review. J Adv Res 38:55–75. https://doi.org/10.1016/j.jare.2022.01.008
37. Tung WS, Daoud WA (2010) Nanoparticle coatings for UV protective textiles Takuya. Res J Text Appar 14:83–88. https://doi.org/10.1108/RJTA-14-02-2010-B010
38. Mankodi DH, Agarwal DB (2011) Studies on nano—UV protective finish on apparel fabrics for health protection. Res J Text Appar. https://doi.org/10.1108/RJTA-15-03-2011-B002
39. Constantinides G (2013) Nanoscience and nanoengineering of cement-based materials. In: Nanotechnology in eco-efficient construction. Elsevier, pp 9–37a

40. Yang D, Ramu AG, Choi D (2022) Synthesis of transparent ZnO–TiO_2 and its nanocomposites for ultraviolet protection of a polyethylene terephthalate (PET) film. Catalysts 12:1590. https://doi.org/10.3390/catal12121590
41. Mont FW, Kim JK, Schubert MF et al (2008) High-refractive-index TiO_2-nanoparticle-loaded encapsulants for light-emitting diodes. J Appl Phys 103. https://doi.org/10.1063/1.2903484
42. Ullah S, Ferreira-Neto EP, Pasa AA et al (2015) Enhanced photocatalytic properties of core@shell SiO_2@TiO_2 nanoparticles. Appl Catal B Environ. https://doi.org/10.1016/j.apcatb.2015.05.036
43. Montazer M, Seifollahzadeh S (2011) Enhanced self-cleaning, antibacterial and UV protection properties of nano TiO_2 treated textile through enzymatic pretreatment. Photochem Photobiol 87:877–883. https://doi.org/10.1111/j.1751-1097.2011.00917.x
44. Rashid MM, Tomšič B, Simončič B et al (2023) In situ tailoring of Ag-doped-TiO_2/TPMP/cotton nanocomposite with UV-protective, self-sterilizing and flame-retardant performance for advanced technical textiles. Polym Degrad Stab 216:110504. https://doi.org/10.1016/j.polymdegradstab.2023.110504
45. Battistin M, Pascalicchio P, Tabaro B et al (2022) A safe-by-design approach to reef safe sunscreens based on ZnO and organic UV filters. Antioxidants. https://doi.org/10.3390/antiox11112209
46. Schneider SL, Lim HW (2019) A review of inorganic UV filters zinc oxide and titanium dioxide. Photodermatol. Photoimmunol. Photomed.
47. Archana B, Manjunath K, Nagaraju G et al (2017) Enhanced photocatalytic hydrogen generation and photostability of ZnO nanoparticles obtained via green synthesis. Int J Hydrogen Energy. https://doi.org/10.1016/j.ijhydene.2016.11.099
48. Nisansala HMD, Rajapaksha GKM, Dikella DGN V et al (2021) Zinc oxide nanostructures in the textile industry. Indian J Sci Technol. https://doi.org/10.17485/ijst/v14i46.1052
49. Mahmoudifard M, Safi M (2012) Novel study of carbon nanotubes as UV absorbers for the modification of cotton fabric. J Text Inst 103:893–899. https://doi.org/10.1080/00405000.2011.622461
50. Mondal S, Hu JL (2007) A novel approach to excellent UV protecting cotton fabric with functionalized MWNT containing water vapor permeable PU coating. J Appl Polym Sci 103:3370–3376. https://doi.org/10.1002/app.25437
51. Nuraje N, Khan SI, Misak H, Asmatulu R (2013) The addition of graphene to polymer coatings for improved weathering. ISRN Polym Sci 2013:1–8. https://doi.org/10.1155/2013/514617
52. Samb-Joshi KM, Sethi YA, Ambalkar AA et al (2019) Lignin-mediated biosynthesis of Zno and TiO_2 nanocomposites for enhanced antimicrobial activity. J Compos Sci. https://doi.org/10.3390/jcs3030090
53. Nouh SA, Barakat MME, El-Nabarawy HA et al (2021) Color changes in some UV irradiated polymer nanocomposite materials for the application in textile industry. Fibers Polym. https://doi.org/10.1007/s12221-021-0871-7
54. Kaur H, Kalia A (2021) Low-cost nano-TiO_2 composites for remediation of textile dyes: appraisal on the effect of solar and ultraviolet irradiations. Microsc Res Tech. https://doi.org/10.1002/jemt.23775
55. Torbati TV, Javanbakht V (2020) Fabrication of TiO_2/Zn_2TiO_4/Ag nanocomposite for synergic effects of UV radiation protection and antibacterial activity in sunscreen. Colloids Surfaces B Biointerfaces 187:110652. https://doi.org/10.1016/j.colsurfb.2019.110652
56. Diffey BL (1994) A method for broad spectrum classification of sunscreens. Int J Cosmet Sci. https://doi.org/10.1111/j.1467-2494.1994.tb00082.x
57. Elmaaty T, Mandour B (2018) ZnO and TiO_2 nanoparticles as textile protecting agents against UV radiation: a review. Asian J Chem Sci 4:1–14. https://doi.org/10.9734/AJOCS/2018/40329
58. Yang RD, Tripathy S, Li Y, Sue H-J (2005) Photoluminescence and micro-Raman scattering in ZnO nanoparticles: the influence of acetate adsorption. Chem Phys Lett 411:150–154. https://doi.org/10.1016/j.cplett.2005.05.125
59. Khan A, Nazir A, Rehman A et al (2020) A review of UV radiation protection on humans by textiles and clothing. Int J Cloth Sci Technol

60. Silva MRF, Alves MFRP, Cunha JPGQ et al (2023) Nanostructured transparent solutions for UV-shielding: recent developments and future challenges. Mater Today Phys
61. Rabiei H, Farhang Dehghan S, Montazer M et al (2022) UV protection properties of workwear fabrics coated with TiO_2 nanoparticles. Front Public Heal 10. https://doi.org/10.3389/fpubh.2022.929095
62. Kocić A, Bizjak M, Popović D et al (2019) UV protection afforded by textile fabrics made of natural and regenerated cellulose fibres. J Clean Prod. https://doi.org/10.1016/j.jclepro.2019.04.355
63. Huang C, Cai Y, Chen X, Ke Y (2022) Silver-based nanocomposite for fabricating high performance value-added cotton. Cellulose 29:723–750. https://doi.org/10.1007/s10570-021-04257-z
64. Gambichler T, Altmeyer P, Hoffmann K (2002) Comparison of methods: determination of UV protection of clothing. Recent Results Cancer Res. https://doi.org/10.1007/978-3-642-59410-6_8
65. Yang SI, Liu S, Brooks GJ et al (2018) Reliable and simple spectrophotometric determination of sun protection factor: a case study using organic UV filter-based sunscreen products. J Cosmet Dermatol. https://doi.org/10.1111/jocd.12390
66. Aguda ON, Lateef A (2022) Recent advances in functionalization of nanotextiles: a strategy to combat harmful microorganisms and emerging pathogens in the 21st century. Heliyon 8:e09761. https://doi.org/10.1016/j.heliyon.2022.e09761
67. Said MM, Rehan M, El-Sheikh SM et al (2021) Multifunctional hydroxyapatite/silver nanoparticles/cotton gauze for antimicrobial and biomedical applications. Nanomaterials 11:1–25. https://doi.org/10.3390/nano11020429
68. Alavi M, Rai M (2019) Recent progress in nanoformulations of silver nanoparticles with cellulose, chitosan, and alginic acid biopolymers for antibacterial applications. Appl Microbiol Biotechnol 103:8669–8676. https://doi.org/10.1007/s00253-019-10126-4
69. Roy S, Ghosh S, Bhowmick N, Roy Choudhury PK (2019) Role of physicochemical factors on bacterial attachment in textile fibrous media. Water Environ J 33:21–30. https://doi.org/10.1111/wej.12365
70. Varshney S, Pandey P, Gupta D, Sharma S (2020) Role of fabric properties, moisture and friction in transfer of bacteria from fabric to fabric. Text Res J 90:478–485. https://doi.org/10.1177/0040517519866956
71. Gupta P, Bairagi N, Priyadarshini R et al (2017) Bacterial contamination of nurses white coats after first and second shift. Am J Infect Control 45:86–88. https://doi.org/10.1016/j.ajic.2016.07.014
72. Andra S, Balu SK, Jeevanandam J, Muthalagu M (2021) Emerging nanomaterials for antibacterial textile fabrication. Naunyn Schmiedebergs Arch Pharmacol 394:1355–1382. https://doi.org/10.1007/s00210-021-02064-8
73. Yaqoob AA, Umar K, Ibrahim MNM (2020) Silver nanoparticles: various methods of synthesis, size affecting factors and their potential applications—a review. Appl Nanosci 10:1369–1378. https://doi.org/10.1007/s13204-020-01318-w
74. Chang C, Lin C, Chan C et al (2006) Photocatalytic properties of nanocrystalline TiO_2 thin film with Ag additions 494:274–278. https://doi.org/10.1016/j.tsf.2005.08.152
75. Yu D (2007) Formation of colloidal silver nanoparticles stabilized by Na+—poly (γ-glutamic acid)—silver nitrate complex via chemical reduction process 59:171–178. https://doi.org/10.1016/j.colsurfb.2007.05.007
76. Nersisyan HH, Lee JH, Son HT et al (2003) A new and effective chemical reduction method for preparation of nanosized silver powder and colloid dispersion 38:949–956. https://doi.org/10.1016/S0025-5408(03)00078-3
77. Ayyad O (2009) From silver nanoparticles to nanostructures through matrix chemistry from silver nanoparticles to nanostructures through matrix chemistry. https://doi.org/10.1007/s11051-009-9620-3
78. Xie Y, Ye R, Liu H (2006) Synthesis of silver nanoparticles in reverse micelles stabilized by natural biosurfactant 279:175–178. https://doi.org/10.1016/j.colsurfa.2005.12.056

79. Durán N, Marcato PD, De Souza GIH et al (2007) Antibacterial effect of silver nanoparticles produced by fungal antibacterial effect of silver nanoparticles produced by fungal process on textile fabrics and their effluent treatment. https://doi.org/10.1166/jbn.2007.022
80. Kalishwaralal K, Deepak V, Ram S et al (2010) Colloids and surfaces B: biointerfaces biosynthesis of silver and gold nanoparticles using brevibacterium casei. Colloids Surf, B 77:257–262. https://doi.org/10.1016/j.colsurfb.2010.02.007
81. Sathishkumar M, Sneha K, Won SW et al (2009) Colloids and surfaces b: biointerfaces cinnamon zeylanicum bark extract and powder mediated green synthesis of nano-crystalline silver particles and its bactericidal activity 73:332–338. https://doi.org/10.1016/j.colsurfb.2009.06.005
82. Shahverdi AR, Minaeian S, Reza H et al (2007) Rapid synthesis of silver nanoparticles using culture supernatants of enterobacteria: a novel biological approach 42:919–923. https://doi.org/10.1016/j.procbio.2007.02.005
83. Pan Y, Yang P, Moloney MG et al (2019) Ag NP-loaded cotton fiber materials: preparation, surface deposition, and antibacterial activity with different chemical structures. ACS Appl Bio Mater 2:510–517. https://doi.org/10.1021/acsabm.8b00696
84. Sambalova O, Thorwarth K, Heeb NV et al (2017) Carboxylate functional groups mediate interaction with silver nanoparticles in biofilm matrix. ACS Omega 3:724–733. https://doi.org/10.1021/acsomega.7b00982
85. Alosmanov R, Buniyat-zadeh I, Soylak M et al (2022) Design, structural characteristic and antibacterial performance of silver-containing cotton fiber nanocomposite. Bioengineering 9. https://doi.org/10.3390/bioengineering9120770
86. Zhou J, Cai D, Xu Q et al (2018) Excellent binding effect of l-methionine for immobilizing silver nanoparticles onto cotton fabrics to improve the antibacterial durability against washing. RSC Adv 8:24458–24463. https://doi.org/10.1039/c8ra04401e
87. Aguda ON, Lateef A (2021) Novel biosynthesis of silver nanoparticles through valorization of Parkia biglobosa fermented-seed wastewater: antimicrobial properties and nanotextile application. Environ Technol Innov 24:102077. https://doi.org/10.1016/j.eti.2021.102077
88. Deeksha B, Sadanand V, Hariram N, Rajulu AV (2021) Preparation and properties of cellulose nanocomposite fabrics with in situ generated silver nanoparticles by bioreduction method. J Bioresour Bioprod 6:75–81. https://doi.org/10.1016/j.jobab.2021.01.003
89. Hamouda T, Ibrahim HM, Kafafy HH et al (2021) Preparation of cellulose-based wipes treated with antimicrobial and antiviral silver nanoparticles as novel effective high-performance coronavirus fighter. Int J Biol Macromol 181:990–1002. https://doi.org/10.1016/j.ijbiomac.2021.04.071
90. Quispe-Quispe LG, Limpe-Ramos P, Arenas-Chávez CA et al (2022) Physical and mechanical characterization of a functionalized cotton fabric with nanocomposite based on silver nanoparticles and carboxymethyl chitosan using green chemistry. Processes 10:1–12. https://doi.org/10.3390/pr10061207
91. Rai A, Prabhune A, Perry CC (2010) Antibiotic mediated synthesis of gold nanoparticles with potent antimicrobial activity and their application in antimicrobial coatings. J Mater Chem 20:6789. https://doi.org/10.1039/c0jm00817f
92. Jeevanandam J, Chan YS, Danquah MK (2016) Biosynthesis of metal and metal oxide nanoparticles. ChemBioEng Rev 3:55–67. https://doi.org/10.1002/cben.201500018
93. Akbar A, Sadiq MB, Ali I et al (2019) Synthesis and antimicrobial activity of zinc oxide nanoparticles against foodborne pathogens salmonella typhimurium and staphylococcus aureus. Biocatal Agric Biotechnol 17:36–42. https://doi.org/10.1016/j.bcab.2018.11.005
94. Emami-Karvani Z (2012) Antibacterial activity of ZnO nanoparticle on gram-positive and gram-negative bacteria. African J Microbiol Res 5:1368–1373. https://doi.org/10.5897/ajmr10.159
95. Singh G, Joyce EM, Beddow J, Mason TJ (2012) Evaluation of antibacterial activity of ZnO nanoparticles coated sonochemically onto textile fabrics. J Microbiol Biotechnol Food Sci 2:106–120

96. Wang H, Niu J, Long X, He Y (2008) Sonophotocatalytic degradation of methyl orange by nano-sized Ag/TiO_2 particles in aqueous solutions. Ultrason Sonochem 15:386–392. https://doi.org/10.1016/j.ultsonch.2007.09.011
97. Benedix R, Dehn F, Quaas J, Orgass M (2000) Application of titanium dioxide photocatalysis to create self-cleaning building materials. Lacer 157–168
98. Desai VS, Kowshik M (2009) Antimicrobial activity of titanium dioxide nanoparticles synthesized by solgel technique. Res J Microbiol 4:97–103. https://doi.org/10.3923/jm.2009.97.103
99. Marcus EL, Yosef H, Borkow G et al (2017) Reduction of health care–associated infection indicators by copper oxide–impregnated textiles: crossover, double-blind controlled study in chronic ventilator-dependent patients. Am J Infect Control 45:401–403. https://doi.org/10.1016/j.ajic.2016.11.022
100. Radetić M, Marković D (2019) Nano-finishing of cellulose textile materials with copper and copper oxide nanoparticles. Cellulose 26:8971–8991. https://doi.org/10.1007/s10570-019-02714-4
101. Román LE, Gomez ED, Solís JL, Gómez MM (2020) Antibacterial cotton fabric functionalized with copper oxide nanoparticles. Molecules 25:1–21. https://doi.org/10.3390/MOLECULES25245802
102. Jeevanandam J, Chan YS, Ku YH (2018) Aqueous Eucalyptus globulus leaf extract-mediated biosynthesis of MgO nanorods. Appl Biol Chem 61:197–208. https://doi.org/10.1007/s13765-018-0347-7
103. Suresh J, Gandhi RGR, Selvam S, Sundrarajan M (2013) Synthesis of magnesium oxide nanoparticles by wet chemical method and its antibacterial activity. Adv Mater Res 678:297–300. https://doi.org/10.4028/www.scientific.net/AMR.678.297
104. Cai L, Chen J, Liu Z et al (2018) Magnesium oxide nanoparticles: effective agricultural antibacterial agent against ralstonia solanacearum. Front Microbiol 9. https://doi.org/10.3389/fmicb.2018.00790
105. Abou Elmaaty T, Elsisi HG, Elsayad GM et al (2021) Fabrication of new multifunctional cotton/lycra composites protective textiles through deposition of nano silica coating. Polymers (Basel) 13:2888. https://doi.org/10.3390/polym13172888
106. Rethinam S, Ramamoorthy R, Robert B, Nallathambi G (2018) Production of silica nanoparticles bound fabrics and evaluation of its antibacterial/ultraviolet protection properties. Micro Nano Lett 13:1404–1407. https://doi.org/10.1049/mnl.2018.5033
107. Gao D, Li Y, Lyu B et al (2020) Silicone quaternary ammonium salt based nanocomposite: a long-acting antibacterial cotton fabric finishing agent with good softness and air permeability. Cellulose 27:1055–1069. https://doi.org/10.1007/s10570-019-02832-z
108. Attarchi N, Montazer M, Toliyat T et al (2018) Novel cellulose fabric with multifunctional properties through diverse methods of Ag/TiO_2/β-cyclodextrin nanocomposites synthesis. Cellulose 25:1449–1462. https://doi.org/10.1007/s10570-017-1638-6
109. Khani A, Talebian N (2017) In vitro bactericidal effect of ultrasonically solgel-coated novel CuO/TiO_2/PEG/cotton nanocomposite for wound care. J Coatings Technol Res 14:651–663. https://doi.org/10.1007/s11998-016-9870-9
110. Mojtahed F, Shahidi S, Hezavehi E (2016) Influence of plasma treatment on CNT absorption of cotton fabric and its electrical conductivity and antibacterial activity. J Exp Nanosci 11:215–225. https://doi.org/10.1080/17458080.2015.1044479
111. Chen X, Fang F, Zhang X et al (2016) Flame-retardant, electrically conductive and antimicrobial multifunctional coating on cotton fabric via layer-by-layer assembly technique. RSC Adv 6:27669–27676. https://doi.org/10.1039/C5RA26914H
112. Kumar A, Dalal J, Dahiya S et al (2019) In situ decoration of silver nanoparticles on single-walled carbon nanotubes by microwave irradiation for enhanced and durable anti-bacterial finishing on cotton fabric. Ceram Int 45:1011–1019. https://doi.org/10.1016/j.ceramint.2018.09.280
113. Kang S, Pinault M, Pfefferle LD, Elimelech M (2007) Single-walled carbon nanotubes exhibit strong antimicrobial activity. Langmuir 23:8670–8673. https://doi.org/10.1021/la701067r

114. Li H, Huang J, Song Y et al (2018) Degradable carbon dots with broad-spectrum antibacterial activity. ACS Appl Mater Interfaces 10:26936–26946. https://doi.org/10.1021/acsami.8b08832
115. Zhang Y, Zhang H, Zou Q et al (2018) An injectable dipeptide–fullerene supramolecular hydrogel for photodynamic antibacterial therapy. J Mater Chem B 6:7335–7342. https://doi.org/10.1039/C8TB01487F
116. Mujeeb Rahman P, Abdul Mujeeb VM, Muraleedharan K, Thomas SK (2018) Chitosan/nano ZnO composite films: enhanced mechanical, antimicrobial and dielectric properties. Arab J Chem. https://doi.org/10.1016/j.arabjc.2016.09.008
117. Mosaad RM, Alhalafi MH, Emam EAM, et al (2022) Enhancement of antimicrobial and dyeing properties of cellulosic fabrics via chitosan nanoparticles. Polymers (Basel) 14. https://doi.org/10.3390/polym14194211
118. Navlani-García M, Arias JL, Fernandes M et al (2022) Polysaccharides and metal nanoparticles for functional textiles: a review. https://doi.org/10.3390/nano12061006
119. Ogunsona EO, Muthuraj R, Ojogbo E et al (2020) Engineered nanomaterials for antimicrobial applications: a review. Appl Mater Today 18:100473. https://doi.org/10.1016/j.apmt.2019.100473
120. Choi O, Hu Z (2008) Size dependent and reactive oxygen species related nanosilver toxicity to nitrifying bacteria. Environ Sci Technol 42:4583–4588. https://doi.org/10.1021/es703238h
121. Wyszogrodzka G, Marszałek B, Gil B, Dorożyński P (2016) Metal-organic frameworks: mechanisms of antibacterial action and potential applications. Drug Discov Today 21:1009–1018. https://doi.org/10.1016/j.drudis.2016.04.009
122. Parashar M, Shukla VK, Singh R (2020) Metal oxides nanoparticles via solgel method: a review on synthesis, characterization and applications. J Mater Sci Mater Electron 31:3729–3749. https://doi.org/10.1007/s10854-020-02994-8
123. Benelli G (2015) Plant-mediated biosynthesis of nanoparticles as an emerging tool against mosquitoes of medical and veterinary importance: a review. https://doi.org/10.1007/s00436-015-4800-9
124. Siddiqi KS, Husen A (2016) Fabrication of metal nanoparticles from fungi and metal salts: scope and application. Nanoscale Res Lett 22–24. https://doi.org/10.1186/s11671-016-1311-2
125. Jadoun S, Arif R, Jangid NK, Meena RK (2021) Green synthesis of nanoparticles using plant extracts: a review. Environ Chem Lett 19:355–374. https://doi.org/10.1007/s10311-020-01074-x
126. El-Naggar ME, Shaarawy S, Hebeish AA (2018) Bactericidal finishing of loomstate, scoured and bleached cotton fibres via sustainable in-situ synthesis of silver nanoparticles. Int J Biol Macromol 106:1192–1202. https://doi.org/10.1016/j.ijbiomac.2017.08.127
127. Fabra MJ, Flores-López ML, Cerqueira MA et al (2016) Layer-by-layer technique to developing functional nanolaminate films with antifungal activity. Food Bioprocess Technol 9:471–480. https://doi.org/10.1007/s11947-015-1646-1
128. Xu QB, Xie LJ, Diao H et al (2017) Antibacterial cotton fabric with enhanced durability prepared using silver nanoparticles and carboxymethyl chitosan. Carbohydr Polym 177:187–193. https://doi.org/10.1016/j.carbpol.2017.08.129
129. Fouda A, EL-Din Hassan S, Salem SS, Shaheen TI (2018) In vitro cytotoxicity, antibacterial, and UV protection properties of the biosynthesized zinc oxide nanoparticles for medical textile applications. Microb Pathog 125:252–261. https://doi.org/10.1016/j.micpath.2018.09.030
130. Hasan R (2018) Production of antimicrobial textiles by using copper oxide nanoparticles. Int J Contemp Res Rev 9:20195–20202. https://doi.org/10.15520/ijcrr/2018/9/08/564
131. Balamurugan M, Saravanan S, Soga T (2017) Coating of green-synthesized silver nanoparticles on cotton fabric. J Coatings Technol Res 14:735–745. https://doi.org/10.1007/s11998-016-9894-1
132. Kumar S (2016) Study on antimicrobial effectiveness of sliver nano coating over cotton fabric through green approach 2:1–16
133. Ortelli S, Costa A, Dondi M (2015) TiO_2 nanosols applied directly on textiles using different purification treatments. Materials (Basel) 8:7988–7996. https://doi.org/10.3390/ma8115437

134. Saad SR, Mahmed N, Abdullah MMAB, Sandu AV (2016) Self-cleaning technology in fabric: a review. In: IOP conference series: materials science and engineering
135. Atwah AA, Khan MA (2023) Influence of microscopic features on the self-cleaning ability of textile fabrics. Text Res J 93:450–467. https://doi.org/10.1177/00405175211069881
136. Abbas M, Iftikhar H, Malik M, Nazir A (2018) Surface coatings of TiO_2 nanoparticles onto the designed fabrics for enhanced self-cleaning properties. Coatings 8:35. https://doi.org/10.3390/coatings8010035
137. Gokarneshan N (2018) A review of some significant insights on nano finishing of protein fibres
138. Behzadnia A, Montazer M, Rashidi A, Mahmoudi Rad M (2014) Rapid sonosynthesis of N-doped nano TiO_2 on wool fabric at low temperature: introducing self-cleaning, hydrophilicity, antibacterial/antifungal properties with low alkali solubility, yellowness and cytotoxicity. Photochem Photobiol 90:1224–1233. https://doi.org/10.1111/php.12324
139. Perelshtein I, Perkas N, Gedanken A (2019) The sonochemical functionalization of textiles. In: The impact and prospects of green chemistry for textile technology. Elsevier, pp 161–198
140. Shahidi S, Zarei L, Elahi SM (2014) Fabrication of ZnO nano particles using sonochemical method and applying on cotton fabric using in situ and pad-dry-cure methods. Fibers Polym 15:2472–2479. https://doi.org/10.1007/s12221-014-2472-1
141. Perelshtein I, Applerot G, Perkas N et al (2008) Sonochemical coating of silver nanoparticles on textile fabrics (nylon, polyester and cotton) and their antibacterial activity. Nanotechnology 19:245705. https://doi.org/10.1088/0957-4484/19/24/245705
142. Noor N, Mutalik S, Younas MW et al (2023) Sonochemical coating of textiles with zinc oxide: robust, silver-seeded growth on cotton for effective UV shielding. ACS Appl Eng Mater. https://doi.org/10.1021/acsaenm.3c00578
143. Abd El-Hady MM, Sharaf S, Farouk A (2020) Highly hydrophobic and UV protective properties of cotton fabric using layer by layer self-assembly technique. Cellulose. https://doi.org/10.1007/s10570-019-02815-0
144. Wågberg L, Erlandsson J (2021) The use of layer-by-layer self-assembly and nanocellulose to prepare advanced functional materials. Adv Mater 33. https://doi.org/10.1002/adma.202001474
145. de Villiers MM, Otto DP, Strydom SJ, Lvov YM (2011) Introduction to nanocoatings produced by layer-by-layer (LbL) self-assembly. Adv Drug Deliv Rev 63:701–715. https://doi.org/10.1016/j.addr.2011.05.011
146. Mamedov AA, Kotov NA (2000) Free-standing layer-by-layer assembled films of magnetite nanoparticles. Langmuir. https://doi.org/10.1021/la000560b
147. Srivastava S, Kotov NA (2008) Composite layer-by-layer (LBL) assembly with Inorganic nanoparticles and nanowires. Acc Chem Res 41:1831–1841. https://doi.org/10.1021/ar8001377
148. Ariga K, Hill JP, Ji Q (2007) Layer-by-layer assembly as a versatile bottom-up nanofabrication technique for exploratory research and realistic application. Phys Chem Chem Phys. https://doi.org/10.1039/b700410a
149. Del Secco B, Trabucco S, Ravegnani F et al (2022) Particles emission from an industrial spray coating process using nano-materials. Nanomaterials 12:313. https://doi.org/10.3390/nano12030313
150. Nørgaard AW, Jensen KA, Janfelt C et al (2009) Release of VOCs and particles during use of nanofilm spray products. Environ Sci Technol 43:7824–7830. https://doi.org/10.1021/es9019468
151. Göhler D, Stintz M (2014) Granulometric characterization of airborne particulate release during spray application of nanoparticle-doped coatings. J Nanoparticle Res 16:2520. https://doi.org/10.1007/s11051-014-2520-1
152. Gondal MA, Sadullah MS, Dastageer MA et al (2014) Study of factors governing oil-water separation process using TiO_2 films prepared by spray deposition of nanoparticle dispersions. ACS Appl Mater Interfaces 6:13422–13429. https://doi.org/10.1021/am501867b

153. Dadvar S, Tavanai H, Dadvar H et al (2011) UV-protection and photocatalytic properties of electrospun polyacrylonitrile nanofibrous mats coated with TiO_2 nanofilm via solgel. J Solgel Sci Technol 59:269–275. https://doi.org/10.1007/s10971-011-2495-7
154. Koozekonan AG, Esmaeilpour MRM, Kalantary S et al (2021) Fabrication and characterization of TiO_2 and MWCNT coated electrospinning nanofibers for UV protection properties. MethodsX 8:101354. https://doi.org/10.1016/j.mex.2021.101354
155. Koozekonan AG, Esmaeilpour MRM, Kalantary S et al (2021) Fabrication and characterization of PAN/CNT, PAN/TiO_2, and PAN/CNT/TiO_2 nanofibers for UV protection properties. J Text Inst. https://doi.org/10.1080/00405000.2020.1813408
156. Sun Y, Cheng S, Lu W et al (2019) Electrospun fibers and their application in drug controlled release, biological dressings, tissue repair, and enzyme immobilization. RSC Adv 9:25712–25729. https://doi.org/10.1039/C9RA05012D
157. Abidi N, Hequet E, Tarimala S, Dai LL (2007) Cotton fabric surface modification for improved UV radiation protection using solgel process. J Appl Polym Sci 104:111–117. https://doi.org/10.1002/app.24572
158. Ugemuge N, Parauha YR, Dhoble SJ (2021) Synthesis and luminescence study of silicate-based phosphors for energy-saving light-emitting diodes. In: Energy materials. Elsevier, pp 445–480
159. Periyasamy AP, Venkataraman M, Kremenakova D et al (2020) Progress in solgel technology for the coatings of fabrics. Materials (Basel) 13:1838. https://doi.org/10.3390/ma13081838
160. Yu M, Gu G, Meng W-D, Qing F-L (2007) Superhydrophobic cotton fabric coating based on a complex layer of silica nanoparticles and perfluorooctylated quaternary ammonium silane coupling agent. Appl Surf Sci 253:3669–3673. https://doi.org/10.1016/j.apsusc.2006.07.086
161. Mahltig B, Fiedler D, Böttcher H (2004) Antimicrobial solgel coatings. J Solgel Sci Technol 32:219–222. https://doi.org/10.1007/s10971-004-5791-7
162. Navas D, Fuentes S, Castro-Alvarez A, Chavez-Angel E (2021) Review on solgel synthesis of perovskite and oxide nanomaterials. Gels 7:275. https://doi.org/10.3390/gels7040275
163. Tranquillo E, Bollino F (2020) Surface modifications for implants lifetime extension: an overview of solgel coatings. Coatings 10:589. https://doi.org/10.3390/coatings10060589
164. Sahile K, Alemayehu E, Worku A et al (2023) Fabrication of enhanced UV protective cotton fabric using activated nano-biocarbon derived from teff hay grafted by polyaniline: RSM-based optimization and characterization. Molecules 28:5142. https://doi.org/10.3390/molecules28135142
165. Zohdy MH, El Hossamy MB, El-Naggar AWM et al (2009) Novel UV-protective formulations for cotton, PET fabrics and their blend utilizing irradiation technique. Eur Polym J 45:2926–2934. https://doi.org/10.1016/j.eurpolymj.2009.06.018
166. Zhang S, Fang K, Liu X et al (2023) Enhancement UV-resistance and hydrophobic of cotton fabrics through membrane formation by cationic colored nanospheres. Surf Interfaces 37:102701. https://doi.org/10.1016/j.surfin.2023.102701
167. El-Bisi MK, Othman R, Yassin FA (2020) Improving antibacterial and ultraviolet properties of cotton fabrics via dual effect of nano-metal oxide and Moringa oleifera extract. Egypt J Chem 63:3441–3451. https://doi.org/10.21608/ejchem.2020.39534.2805
168. Cheng D, He M, Cai G et al (2018) Durable UV-protective cotton fabric by deposition of multilayer TiO_2 nanoparticles films on the surface. J Coatings Technol Res 15:603–610. https://doi.org/10.1007/s11998-017-0021-8
169. El-Naggar ME, Khattab TA, Abdelrahman MS et al (2021) Development of antimicrobial, UV blocked and photocatalytic self-cleanable cotton fibers decorated with silver nanoparticles using silver carbamate and plasma activation. Cellulose 28:1105–1121. https://doi.org/10.1007/s10570-020-03537-4
170. Gulson B, McCall M, Korsch M et al (2010) Small amounts of zinc from zinc oxide particles in sunscreens applied outdoors are absorbed through human skin. Toxicol Sci 118:140–149. https://doi.org/10.1093/toxsci/kfq243
171. Senthil B, Devasena T, Prakash B, Rajasekar A (2017) Non-cytotoxic effect of green synthesized silver nanoparticles and its antibacterial activity. J Photochem Photobiol B Biol 177:1–7. https://doi.org/10.1016/j.jphotobiol.2017.10.010

172. Dastjerdi R, Montazer M (2010) A review on the application of inorganic nano-structured materials in the modification of textiles: focus on anti-microbial properties. Colloids Surf, B 79:5–18. https://doi.org/10.1016/j.colsurfb.2010.03.029
173. Zhu Y, Xu F, Liu Q et al (2019) Nanomaterials and plants: positive effects, toxicity and the remediation of metal and metalloid pollution in soil. Sci Total Environ 662:414–421. https://doi.org/10.1016/j.scitotenv.2019.01.234
174. Attia NF, Osama R, Elashery SEA et al (2022) Recent advances of sustainable textile fabric coatings for UV protection properties. Coatings 12:1597. https://doi.org/10.3390/coatings12101597
175. Dhineshbabu NR, Bose S (2018) Smart textiles coated with eco-friendly UV-blocking nanoparticles derived from natural resources. ACS Omega. https://doi.org/10.1021/acsomega.8b00822
176. Nosheen A, Hussain MT, Khalid M et al (2022) Development of protective cotton textiles against biohazards and harmful UV radiation using eco-friendly novel fiber-reactive bioactive agent. Process Saf Environ Prot 165:431–444. https://doi.org/10.1016/j.psep.2022.07.035

Chapter 9
Water-Repellent and Oil-Repellent Finishing on Textiles Using Nanotechnology

Saikat Ghosh and Subhadeep Paul

1 Introduction

Textile finishing plays a crucial role in enhancing the aesthetic appearance, comfort, durability, and performance characteristics of fabrics. Traditionally, various chemical and mechanical processes have been employed to achieve desired finishing effects. However, with the advancements in nanotechnology, a new paradigm in textile finishing has emerged, known as nanofinishing. Nanofinishing involves the application of nanomaterials or nanoparticles onto textile surfaces to impart specific properties or functionalities. Amongst the various functionalities achievable through nanofinishing, water and oil repellency are critical properties in textiles that have gained significant attention due to their impact on fabric performance and consumer demand for functional and high-quality textiles [1]. Water repellency is the ability of a textile surface to resist the penetration of water, preventing moisture absorption and ensuring dryness. Oil repellency, on the other hand, refers to the resistance of a fabric to the penetration of oils or oily substances. Both water- and oil-repellent finishes are highly desirable in textiles, as they offer numerous advantages in various applications, including outdoor apparel, protective clothing, upholstery, and technical textiles. Understanding the scientific principles behind water and oil repellency is essential for developing advanced textile technologies and meeting the evolving needs of consumers [2]. There are several approaches to achieve water and oil repellency through nanofinishing. One common method is the application of hydrophobic

S. Ghosh (✉)
Technical Textiles Research Centre, School of Arts and Humanities, University of Huddersfield, Huddersfield, UK
e-mail: S.Ghosh3@hud.ac.uk

S. Paul
Living Construction Group, Hub for Biotechnology in the Built in Environment, Applied Sciences Department, Northumbria University, Newcastle-upon-Tyne, United Kingdom
e-mail: subhadeep.paul@northumbria.ac.uk

S. Bairagi et al. (eds.), *Nanotechnology in Textile Finishing*, Materials Horizons: From Nature to Nanomaterials, https://doi.org/10.1007/978-981-97-2696-7_9

or oleophobic nanoparticles onto the textile surface. These nanoparticles, such as fluorinated compounds or silicones, exhibit inherent water- and oil-repellent properties. When applied to the textile, they form a thin layer that lowers the surface energy, causing water droplets or oil droplets to bead up and roll off the fabric surface. Another approach involves the modification of textile fibres themselves with nanomaterials. Nanoparticles can be incorporated into the fibre matrix during the spinning or extrusion process, resulting in inherent water and oil repellency throughout the fabric. This approach ensures long-lasting repellency, as the nanoparticles are not merely surface coatings that can wear off over time.

Furthermore, nanofinishing techniques can be combined with other functional finishes to achieve multifunctional textiles. For instance, nanoparticles can be incorporated into fabrics that already possess properties like antimicrobial or UV-protective functionalities. This integration of multiple functionalities into a single fabric is a significant advantage of nanofinishing, as it allows for the development of textiles with superior performance characteristics. In addition to their inherent water- and oil-repellent properties, nanomaterials also offer environmental benefits. Compared to conventional finishing methods, nanofinishing requires fewer chemicals and water, minimising waste generation and reducing the environmental impact. Nanoparticles can be applied in lower quantities whilst achieving comparable or even superior repellent effects, leading to more sustainable textile production practices. Despite the numerous advantages of nanofinishing for water and oil repellency, challenges remain. The durability of the repellent finishes, especially during repeated washing or exposure to harsh conditions, is an important consideration. Researchers and manufacturers are continuously working on improving the longevity of nanofinished textiles to ensure that their repellent properties remain intact over extended periods of use.

1.1 Background on Water and Oil Repellency

Water and oil repellency in textiles refers to the ability of a fabric's surface to resist wetting and penetration by water or oil droplets [1]. This behaviour is determined by the interplay between the surface tension of the liquid and the surface energy of the textile material. Surface tension governs the interaction between the liquid droplet and the fabric surface, whilst surface energy describes the affinity of the textile for the liquid [2, 3]. Hydrophobicity refers to the resistance of a fabric to water wetting, whilst oleophobicity refers to the resistance to oil wetting [4].

1.2 Significance in Textile Applications

Water and oil repellency hold significant significance in textile applications for several reasons [5]. Firstly, the ability of textiles to repel water and oil protects against liquid stains and spills [6]. Fabrics treated with water- and oil-repellent finishes are less prone to absorbing liquids, preventing stains from setting in [5, 7]. Fabrics with water- and oil-repellent properties are easier to clean and maintain, reducing the need for frequent laundering or dry cleaning [8, 9]. This is particularly valuable in applications such as upholstery, workwear, and outdoor apparel, where fabrics are exposed to potential staining agents [10–13]. Secondly, water- and oil-repellent textiles exhibit enhanced durability and longevity. When liquids are unable to penetrate the fabric surface, the risk of damage, such as fabric degradation, discolouration, or staining, caused by liquid absorption is significantly reduced [5, 14]. This prolongs the lifespan of the material, ensuring that it can withstand repeated use and maintain its original appearance and functionality and improves overall performance. Additionally, water- and oil-repellent fabrics offer improved comfort by preventing moisture absorption [15]. Fabrics that repel water and oil do not become saturated, maintaining their breathability and allowing moisture vapour to escape. This reduces the clammy feeling associated with wet fabrics, enhancing comfort, particularly in outdoor apparel and active sportswear [10, 11]. Water- and oil-repellent properties provide protection from environmental elements such as rain, snow, and wind. Fabrics that repel water keep the wearer dry and comfortable, whilst oil-repellent properties can protect against oil-based contaminants and pollutants. This is especially important in applications such as outdoor gear, protective clothing, and medical textiles. Water- and oil-repellent properties can be combined with other functionalities to create multifunctional materials [16, 17]. For example, fabrics can be treated to be both water-repellent and antimicrobial, or oil-repellent and flame-resistant. This versatility allows for the customisation of materials to meet specific application requirements. Fabrics that repel water and oil are less prone to discolouration, fading, or distortion caused by liquid contact [18]. This improves the appearance and maintains the desired properties of the fabric over time, resulting in increased customer satisfaction.

1.3 Current Challenges and Limitations

Whilst water- and oil-repellent textiles offer numerous advantages, certain challenges and limitations need to be addressed [19]. One significant concern is the environmental impact of conventional repellent treatments. Many of these treatments contain per- and polyfluoroalkyl substances (PFAS), which are persistent and potentially harmful to human health and the environment [20]. As a result, there is a growing demand for more sustainable and eco-friendly alternatives that can provide effective water and oil repellency without compromising safety or performance [21]. Another challenge lies in ensuring the durability and wash resistance of repellent finishes [21]. Frequent laundering or exposure to various cleaning agents can gradually diminish the effectiveness of repellent treatments [22]. Manufacturers and researchers are actively working on developing durable repellent finishes that can withstand repeated washing cycles and maintain their performance over an extended period [23]. Additionally, compatibility with other desirable textile properties is a consideration [24]. For example, incorporating water and oil repellency without compromising the softness [25], breathability, or colourfastness of the fabric can be challenging [26]. Balancing these properties requires careful formulation and optimisation to achieve the desired balance of functionality and aesthetics [27, 28]. However, there are challenges related to environmental concerns, durability, and compatibility with other textile properties. Ongoing research and development efforts are focused on addressing these challenges, seeking sustainable alternatives, and advancing repellent technologies to create high-performance textiles that meet consumer demands. Superhydrophobic finishing for both natural and synthetic materials are done through different process steps. The most common approach is to create micro- or nanolevel textures on the textiles through grafting low surface energy materials. The surface roughness can be created by deposition or corrosion or itching. For deposition mechanism, gas-phase coating or polymer deposition are usually preferred whereas in case of itching, plasma treatment or surface itching through chemical reaction are the way to achieve such micro- or nanolevel texture on textile surfaces. Durability, moisture management, and breathability are the challenges of these treatments. Fluorocarbons are well known for their dual hydrophobic and oleophobic behaviour; however, there are concerns about unwanted emission of PFOA and PFOS. To mitigate this problem, fluorine-free chemicals should be developed for applying on to the textile materials, which will impart hydrophobic property keeping its pristine hand feel intact.

2 Nanotechnology

Nanotechnology has emerged as a promising field with vast potential in various industries, including textiles [29]. In textile finishing, nanotechnology offers new opportunities to enhance fabric properties and introduce novel functionalities [30]. By utilising nanoscale materials and processes, textile manufacturers can achieve improved performance, durability, and functionality in their products [31]. This overview provides a glimpse into the applications, benefits, and challenges of nanotechnology in textile finishing.

2.1 Nanoparticles

Nanoparticles, which are particles with dimensions on the nanoscale range, are commonly employed in textile finishing [32]. At nanoscale, materials exhibit unique and enhanced properties due to their high surface area-to-volume ratio, quantum effects, and altered physical and chemical behaviour. By harnessing these characteristics, nanofinishing has opened up exciting possibilities for textile manufacturers to create fabrics with exceptional attributes. These nanoparticles can be made from various materials such as metals, metal oxides, carbon-based materials, and polymers.

They are typically incorporated into finishing formulations or coatings, allowing for precise control over their distribution on the fabric surface [33].

The integration of nanomaterials in textile finishing has revolutionised the industry by offering numerous advantages. One of the significant benefits is the ability to impart multifunctionality to textiles. Nanoparticles can be engineered to provide a wide range of functionalities such as water repellency, stain resistance, antimicrobial properties, UV protection, flame retardancy, self-cleaning capabilities, and even electronic conductivity. These functionalities can be incorporated into textiles without compromising their comfort, breathability, or aesthetic appeal. The application of nanofinishing techniques in textile manufacturing has also resulted in improved durability and longevity of finished fabrics. The nanoscale particles form a protective layer on the textile surface, enhancing its resistance to wear, tear, and fading. Additionally, the nanoparticles can enhance the mechanical strength of the fabrics, contributing to their overall durability. Moreover, nanofinishing processes often lead to reduced environmental impact compared to conventional finishing methods. By using nanomaterials, the amount of chemicals and water required for finishing can be significantly reduced. This not only minimises waste generation but also reduces energy consumption and the release of harmful substances into the environment. Nanofinishing techniques have also paved the way for innovative and futuristic applications in the textile industry. For instance, the integration of nanoparticles with textile fibres or coatings can enable the development of smart textiles capable of sensing, actuation, and energy harvesting. These advancements hold tremendous potential in various sectors, including health care, sports, fashion, and industrial textiles. Nanoparticles, such as silica [34, 35], titanium dioxide [35, 36], or clay nanoparticles [37, 38], can be applied to the fabric surface. These nanoparticles create a rough or structured surface, reducing the contact area between the liquid droplets and the fabric [39]. They can also be functionalised with hydrophobic or oleophobic materials to enhance repellency. Discussion on different nanoparticles has been presented later in this chapter.

2.2 *Nanofibres*

Electrospinning, a technique that uses an electric field to produce ultrafine fibres, can be employed to create a layer of nanofibres on the fabric surface [40]. The nanofibre layer acts as a physical barrier, preventing the penetration of water and oil droplets into the fabric [41]. The nanofibres themselves can be hydrophobic or oleophobic, further enhancing the repellency [42]. Nanofibres have emerged as a promising solution for enhancing both hydrophobicity (water repellency) and oleophobicity (oil repellency) properties. These unique characteristics make nanofibres highly effective in repelling both water and oil. Due to their high surface area-to-volume ratio and inherent hydrophobic nature, nanofibres create surfaces that repel water. The interconnected structure and small size of the fibres generate numerous air pockets, reducing the contact area between water droplets and the surface. This hydrophobicity can be further enhanced through the application of hydrophobic coatings or surface modifications. As a result, nanofibres find applications in waterproof coatings, water-resistant membranes, self-cleaning surfaces, and moisture barriers.

Nanofibres also demonstrate excellent oleophobicity, preventing the spreading or penetration of oil. Their low surface energy and unique structure create non-wetting surfaces for oil droplets. This repellency can be further improved by applying hydrophobic and oleophobic coatings or surface treatments. As a result, nanofibres find applications in oil spill clean-up materials, oil-resistant filters, protective clothing for oil-based environments, and antifouling coatings. The versatile nature of nanofibres, with their high surface area, lightweight nature, and tunable surface properties, makes them an ideal choice for both hydrophobic and oleophobic applications. Advances in nanofibre production techniques, such as electrospinning, have further facilitated the fabrication of nanofibres with controlled morphology and properties. However, comprehensive characterisation and testing are necessary to evaluate and optimise the repellency performance of nanofibre materials for specific applications.

2.3 Nanostructured Materials

Coatings made from nanostructured materials, such as nanoemulsions or solgel coatings, can be applied to the fabric surface. These coatings form a thin, uniform layer with desirable water- and oil-repellent properties. They can be customised to provide additional functionalities like self-cleaning or antimicrobial effects.

2.3.1 Nanoemulsions

Nanoemulsions offer a promising approach for achieving water and oil repellency in various applications. These nano-sized emulsions, composed of water, oil, and emulsifying agents, can be formulated to create coatings or treatments that repel both water and oil. By carefully selecting the emulsifying agents and optimising the formulation, nanoemulsions can create a hydrophobic and oleophobic barrier on surfaces, preventing water from wetting or oil from penetrating the treated material. The small droplet size of nanoemulsions allows for uniform coverage and enhanced surface area, leading to effective repellency properties. Nanoemulsions provide a versatile solution for developing water- and oil-repellent surfaces in applications such as textiles, coatings, packaging materials, and personal care products.

2.3.2 Nanocoatings

Nanocoatings offer a versatile solution for achieving both water and oil repellency in various applications. These thin films or layers, containing nanoscale particles or structures, can be applied onto surfaces to provide enhanced repellent properties. By incorporating hydrophobic and oleophobic nanoparticles or nanostructures in the nanocoating formulation, surfaces can become superhydrophobic and superoleophobic, repelling both water and oil. The reduced surface energy of the coated material causes water droplets to form spherical shapes and easily roll off the surface, whilst oil droplets are prevented from spreading or penetrating the coating. This dual functionality makes nanocoatings valuable for applications such as protective coatings, automotive surfaces, electronic devices, and medical equipment, where both water and oil repellency are desired.

2.3.3 Nanocomposites

Nanocomposites, which combine nanoscale fillers with a matrix material, offer a versatile solution for achieving enhanced water and oil repellency in various applications. By incorporating hydrophobic and oleophobic nanoparticles into the matrix, nanocomposites can repel both water and oil. The nanoparticles create a surface with reduced surface energy, preventing the wetting or spreading of water and oil droplets. This dual functionality makes nanocomposites ideal for applications such as coatings, membranes, packaging materials, and textiles, where resistance to both water and oil is desired. Additionally, the nanoscale fillers enhance the mechanical, thermal, and barrier properties of the composites, further expanding their potential applications in industries such as automotive, aerospace, and electronics.

2.4 Applications and Benefits

Nanotechnology offers a wide range of applications and benefits in textile finishing [43]. Nanotechnology has made significant contributions to the field of water and oil repellency, offering numerous applications and benefits in various industries. By harnessing the unique properties of nanoparticles, nanotechnology has revolutionised the achievement of repellent finishes, surpassing the limitations of conventional methods. The applications of nanotechnology in water and oil repellency have been discussed in detail.

- Outdoor Apparel: Nanotechnology has transformed the performance of outdoor apparel by providing durable water- and oil-repellent finishes. Fabrics treated with nanomaterials exhibit exceptional water repellency, preventing rain or snow from penetrating the fabric and keeping the wearer dry. This is particularly beneficial for outdoor enthusiasts, athletes, and workers who face challenging weather conditions. The repellent finishes also repel oils and stains, ensuring that the fabric remains clean and free from unsightly marks. Kong et al. developed a superhydrophobic surface for outdoor wear by employing fluoride-free hydrophobic fumed SiO_2 nanoparticles (NPs), which were spray-coated onto a polyester fabric

[44]. Zimmermann et al. conducted a study where they applied a one-step gas-phase coating of polymethylsilsesquioxane nanofilaments to 11 different fabrics composed of various textile fibres. Amongst these fabrics, PET fibres exhibited superior results, primarily attributed to the superhydrophobic effect. The nanoscopic, fibrous coating generated a unique surface geometry on the microscopic, fibrous textiles, contributing to the enhanced performance observed in PET fibres [45].

- Protective Clothing: In industries such as construction, chemical handling, and health care, protective clothing plays a vital role in safeguarding workers from hazardous liquids and substances. Nanotechnology offers advanced water- and oil-repellent finishes for protective clothing, enhancing their performance and effectiveness. The repellent properties of the fabric prevent liquids from permeating the clothing, reducing the risk of contact with harmful substances and enhancing the overall safety of the wearer. Yeerkeen et al. developed a novel self-cleaning and chemically stable superhydrophobic aramid fabric using a two-step coating technique. The fabric was coated with a readily available material system comprising a polytetrafluoroethylene (PTFE) membrane, along with particles of PTFE and silica (SiO_2). This innovative approach holds potential for the production of individual protective clothing for emergency rescue applications [46].
- Upholstery and Home Textiles: Nanotechnology has found applications in the realm of upholstery and home textiles, providing water- and oil-repellent finishes to furniture, carpets, curtains, and other household fabrics. These finishes protect the textiles from spills, stains, and liquid penetration, making them easier to clean and maintain. The repellent properties ensure that the fabrics remain fresh, clean, and visually appealing for an extended period. Kuruppu et al. developed a stain-repellent fabric using TiO_2 nanoparticles. The fabric maintained its stability after 25 washes and demonstrated hydrophobicity against tea, coffee, and water stains. This suggests its potential for use in aircraft upholstery [47].
- Technical Textiles: Technical textiles encompass a wide range of applications, including filtration systems, medical textiles, geotextiles, and automotive textiles. Nanotechnology has enabled the development of water- and oil-repellent finishes for technical textiles, enhancing their functionality and performance. For example, in filtration systems, nanocoatings can prevent water or oil from clogging the filters, ensuring efficient and continuous operation. In medical textiles, nanofinishing can provide repellency to bodily fluids, reducing the risk of contamination and enhancing hygiene. Raeisi et al. successfully fabricated superhydrophobic cotton fabrics through the utilisation of chitosan/titanium dioxide (TiO_2) nanocomposites. The cotton fabric's surface was completely coated with these nanoparticles using a dip-coating method. The resulting nanocomposite-coated fabrics demonstrated great potential for application in hospital-related settings, specifically as antibacterial and self-cleaning garments, showcasing their relevance in the field of technical textiles [48].

- Multifunctionality: Nanotechnology enables the integration of multiple functionalities into a single textile. With water and oil repellency as a base, nanofinishing can be combined with other functional finishes, such as antimicrobial, UV-protective, or self-cleaning properties. This multifunctional approach allows for the development of textiles with superior performance characteristics, catering to a wide range of applications and industry requirements. Ye et al. demonstrated the significant potential of textile coatings (dip and spray) combining the antifouling properties of superhydrophobic surfaces with the contact-killing antibacterial activity of surface-bound bactericides. This innovative approach offers promising applications in both medical and daily protection, particularly for protective clothing in moist and unsanitary environments. The strategy presented by the researchers enables the development of multifunctional textiles with enhanced functionality [49]. In their study, the researcher utilised various plain weave fabrics, including cotton (200 GSM), polyester (180 GSM), silk (160 GSM), nylon (100 GSM), cotton canvas (240 GSM), and cotton fibres.

2.5 Challenges and Considerations of Using Nanoparticles

Whilst nanotechnology offers exciting possibilities, there are challenges and considerations that have been addressed. The potential risks associated with the release of nanoparticles into the environment and their impact on human health are important considerations. Extensive research is being conducted to ensure the safe use and disposal of nanomaterials in textile applications. Maintaining the desired properties of nanocoatings or finishes over multiple wash cycles can be a challenge. Adequate adhesion and stability of nanoparticles on the fabric surface are crucial to ensure long-term performance. The integration of nanotechnology into textile processes may involve additional costs, including the synthesis of nanoparticles, equipment investment, and process optimisation. Ensuring cost-effectiveness and scalability of nanotechnology-based textile finishing processes is essential for commercial viability.

2.6 Measurement of Hydrophobicity of a Material

Hydrophobicity of a material could be characterised by different methods such as measuring water contact angle (WCA) and tilt angle, and the multiresonance thickness-shear mode sensors (MTSM). WCA is the most used method, where the contact angle between a water droplet and the material is characterised. If the contact angle (θ) is acute ($0° \leq \theta < 90°$), the surface is identified as hydrophilic, and if it is obtuse angle ($90° < \theta \leq 180°$), the surface is identified as hydrophobic [50]. A surface or material is termed superhydrophobic if they are extremely water repellent having a WCA of 150° or above and a lower contact angle hysteresis (CAH), the energy

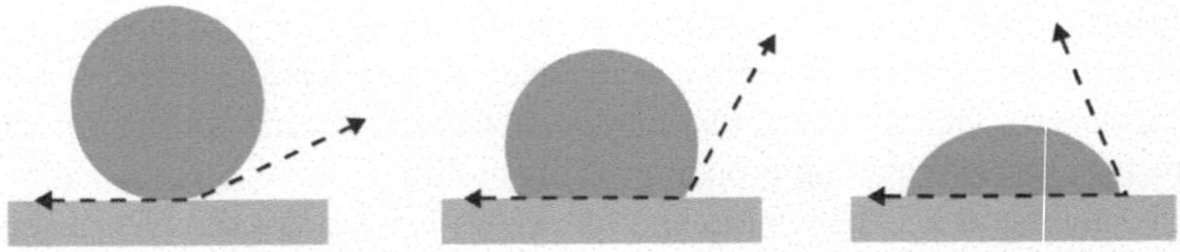

Fig. 1 Different contact angles: first one describes the nature of superhydrophobic material, where WCA is more than 150°; middle one has more than 90° contact angle hydrophobic-type material; third one describes hydrophilic material having contact angle less than 90° [53] Reprinted with permission from Willey

Table 1 Hydrophilicity or hydrophobicity nature based on contact angle

Water contact angle	Property
$\theta < 5°$	Superhydrophilic
$5° < \theta < 90°$,	Hydrophilic
$90° < \theta < 150°$	Hydrophobic
$\theta > 150°$	Superhydrophobic

requires for the movement of a droplet from one metastable state to another state on a particular surface. Such high contact angle of superhydrophobic material helps the droplets to attain spherical shape, and thus, it has a tendency to roll over the material surfaces carrying the dirt particles with them [51]. Roach et al. further added another classification; if the WCA is higher than 120°, they termed it as "ultrahydrophobic" [52, 53] (Fig. 1 and Table 1).

3 Different Models for Hydrophobicity

3.1 *The Wetting Behaviour of Liquid Droplets on a Surface*

The wettability of a material can be explained by Young's equation [54], where θ is the contact angle of a droplet, γSL is the surface tensions or interfacial energies acting between the solid and liquid, γSV between the solid and vapour, and γLV between the liquid and vapour [50].

$$\cos\theta = \frac{\gamma SV - \gamma SL}{\gamma LV} \tag{1}$$

As per the equation, the highest contact angle would be achieved at the lowest surface energy of material, γSV [51]. However, naturally occurring hydrophobic surfaces exhibit a much higher WCA than 120° which is attributed to the surface roughness [52].

3.2 Wenzel Model

The Wenzel model is a fundamental concept in understanding the behaviour of liquids on rough hydrophobic surfaces. It was proposed by Robert N. Wenzel in 1936 and provides insights into the wetting characteristics of a liquid droplet on a textured or rough surface [55]. According to the Wenzel model, when a liquid droplet comes into contact with a rough surface, it will conform to the surface's topography [56]. If the droplet is in a wetting state, it spreads extensively across the surface, resulting in increased contact area between the liquid and the solid. The Wenzel equation describes this phenomenon:

$$\cos\theta_w = r \times \cos\theta \tag{2}$$

where,

- θ_w is the apparent contact angle of the droplet on the rough surface.
- r is the roughness factor, defined as the ratio of the actual surface area to the projected surface area.
- θ is the intrinsic contact angle of the droplet on a flat, smooth surface.

$$\cos\theta_w = \frac{r(\gamma SV - \gamma SL)}{\gamma LV} = r\cos\theta \tag{3}$$

The roughness factor, r, quantifies the surface roughness by considering the ratio of the actual surface area to the projected surface area. A rough surface will have a higher roughness factor, indicating increased surface roughness and a larger contact area for the liquid droplet. This results in a decrease in the apparent contact angle (θ_w) compared to the intrinsic contact angle (θ) on a smooth surface.

The Wenzel model suggests that roughness amplifies the hydrophobic or hydrophilic nature of a surface [55]. For hydrophobic surfaces, the Wenzel model predicts that increasing surface roughness will enhance the hydrophobic behaviour, leading to a decrease in the apparent contact angle [56]. This is because the roughness provides more hydrophobic surface area for the droplet to make contact with, enhancing the overall hydrophobicity.

On the other hand, for hydrophilic surfaces, the Wenzel model predicts that increasing surface roughness will enhance the hydrophilic behaviour, leading to an increase in the apparent contact angle. The roughness exposes more hydrophilic surface area, promoting wetting and spreading of the liquid droplet.

It is important to note that the Wenzel model assumes complete wetting, meaning that the liquid droplet fully wets the surface without any air pockets or trapped gas. Additionally, the Wenzel model does not consider the effects of surface energy or intermolecular interactions, which can influence the wetting behaviour (Fig. 2) [57].

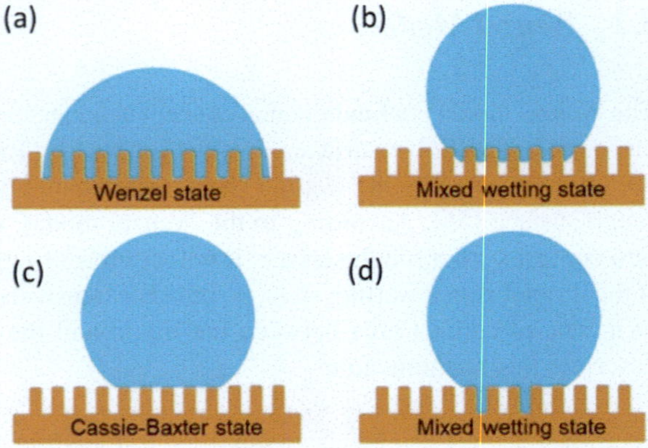

Fig. 2 **a** Droplet on a rough substrate under Wenzel wetting mode. **b** Droplet on a rough substrate under a mixed-wetting state. **c** Droplet on a rough substrate under Cassie–Baxter wetting mode. **d** Droplet on a rough substrate under a mixed-wetting state [57]. CC by 4.0

3.3 Cassie–Baxter Model

The Cassie–Baxter model is a concept that extends the understanding of wetting behaviour on rough surfaces beyond the assumptions of the Wenzel model. Proposed by Wenzel Cassie and Stuart Baxter in 1944, this model takes into account the presence of air pockets or trapped gas beneath the liquid droplet on a rough surface [55]. According to the Cassie–Baxter model, when a liquid droplet comes into contact with a rough surface, it can either completely wet the surface (wetting state) or partially suspend on top of the surface features (non-wetting state) [56]. The model assumes that the liquid droplet sits on a composite interface consisting of solid surface structures and air pockets. The Cassie–Baxter equation describes the wetting behaviour in this model:

$$\cos\theta c = \varphi \times \cos\theta s + (1 - \varphi) \tag{4}$$

where,

- θ_c is the apparent contact angle of the droplet on the rough surface.
- φ is the fractional area of the solid–liquid contact.
- θ_s is the intrinsic contact angle of the droplet on a smooth surface.

The Cassie–Baxter model introduces the concept of the fractional area of solid–liquid contact (φ), which represents the portion of the droplet's base that directly contacts the solid surface structures. The remaining fraction $(1 - \varphi)$ corresponds to the portion of the droplet suspended on air pockets [56]. In the Cassie–Baxter model, the wetting behaviour is governed by the balance between the surface energies of the solid and liquid, as well as the energy associated with the air–liquid and solid–liquid interfaces. When the liquid droplet partially suspends on air pockets, the apparent contact angle (θ_c) is determined by the combination of the intrinsic contact angle (θ_s) and the solid–liquid contact angle (θ_1) at the composite interface.

In the case of hydrophobic surfaces, the Cassie–Baxter model predicts that increasing surface roughness can promote the non-wetting state. The air pockets trapped beneath the liquid droplet create a composite interface with lower surface energy, reducing the contact area between the liquid and the solid. This leads to an increase in the apparent contact angle, as compared to the intrinsic contact angle on a smooth surface. The Cassie–Baxter model offers a more nuanced understanding of wetting behaviour on rough surfaces by accounting for the presence of air pockets. It acknowledges that surface roughness can influence the wetting state of a liquid droplet, leading to either enhanced hydrophobicity or superhydrophobicity [55]. This model has important implications in various fields, including self-cleaning surfaces, anti-icing coatings, and microfluidic devices, where precise control over wetting behaviour is desired. It is worth noting that the Cassie–Baxter model assumes that the liquid droplet remains in a metastable state on the rough surface, with the air pockets being stable and not easily displaced. However, in certain conditions, such as high pressures or vigorous agitation, the trapped air pockets may collapse, transitioning the droplet into the wetting state predicted by the Wenzel model.

3.4 *Transition State Between Wenzel and Cassie–Baxter States*

The transition state between the Wenzel and Cassie–Baxter states, often referred to as the Wenzel–Cassie transition, occurs when the wetting behaviour of a liquid droplet on a rough surface changes from one state to the other. This transition is influenced by factors such as surface roughness, surface energy, and the presence of air pockets or trapped gas. In the Wenzel state, the liquid droplet fully wets the rough surface, conforming to the surface topography. The apparent contact angle (θ_w) in the Wenzel state is reduced compared to the intrinsic contact angle (θ) on a smooth surface due to the increased contact area resulting from the roughness. The Wenzel state is characterised by a high degree of contact between the liquid and the solid, without the presence of air pockets. On the other hand, in the Cassie–Baxter state, the liquid droplet partially suspends on air pockets or trapped gas present in the rough surface. This state is characterised by a composite interface consisting of solid surface structures and air pockets. The apparent contact angle (θ_c) in the Cassie–Baxter state is higher than the intrinsic contact angle (θ) due to reduced contact area between the liquid and the solid. The transition between the Wenzel and Cassie–Baxter states occurs when the wetting behaviour shifts from complete wetting to partial wetting or vice versa. This transition is influenced by the roughness of the surface and the ability of the liquid to penetrate the gaps between the surface structures. Several factors can trigger this transition:

- Surface roughness: Increasing surface roughness promotes the transition from the Wenzel state to the Cassie–Baxter state. As the surface becomes rougher, the air pockets or trapped gas beneath the liquid droplet become more prominent, leading to a non-wetting state.

- Surface energy: The surface energy of the solid and the liquid also play a role in the transition. If the solid surface is highly hydrophobic (low surface energy), the transition to the Cassie–Baxter state is favoured. Lower surface energy reduces the adhesive forces between the liquid and the solid, making it easier for the droplet to suspend on air pockets.
- Surface chemistry: Surface treatments or coatings can modify the surface chemistry and alter the wetting behaviour. For example, introducing a hydrophobic coating can facilitate the transition to the Cassie–Baxter state by reducing the solid–liquid contact area and promoting the formation of air pockets.
- Capillary forces: Capillary forces, such as those caused by the geometry of the surface structures or the presence of surface tension gradients, can influence the transition. These forces can either enhance or hinder the penetration of the liquid into the gaps, affecting the wetting state.

It is important to note that the Wenzel–Cassie transition may not be a sharp transition but rather a gradual change in wetting behaviour as the surface roughness or other factors are varied. The transition can be influenced by the interplay of multiple factors and is not solely determined by a single parameter. Understanding and controlling the Wenzel–Cassie transition is of significant interest in various fields, including surface engineering, microfluidics, and materials science. By tailoring the surface roughness, chemistry, and other parameters, researchers can manipulate the wetting behaviour to achieve desired functionalities such as enhanced hydrophobicity, superhydrophobicity, or liquid repellency.

4 Characterisation of Hydrophobicity and Oleophobicity

4.1 Contact Angle Measurement

Static contact angle determines the surface wettability of a liquid droplet on a surface by measuring the extent of its spreading. The sessile drop technique is used for the characterisation of surface energies; this method involves placing a droplet of liquid of known surface tension on a solid surface and then by measuring the contact angle [58]. The theoretical description of contact angle arises from the consideration of a thermodynamic equilibrium between the three phases. Those are the liquid, the solid (*S*), and the gas phase (*G*) [54]. The shape of a liquid/vapour interface is determined by the Young–Laplace equation, with the contact angle playing the role of a boundary condition following the Young's equation:

$$\gamma SG - \gamma SL - \gamma LG \times \cos\theta = 0$$

where, γSG corresponds to the solid–vapour interface energy, γSL to the solid–liquid interfacial energy, γLG to the liquid–vapour interfacial energy (i.e. the surface tension), and θ the contact angle [54].

Water can be used as a test liquid to establish whether a surface is hydrophilic (angle < 45°) or hydrophobic (angle > 90°) (ASTM Standard D7334). The lower the contact angle, the more hydrophilic the surface [54, 58]. As the surface becomes more oxidised, or has more ionisable groups introduced to it, hydrogen bonding with water becomes easier and the droplet spreads along the hydrophilic surface, resulting in a lower contact angle. This method measures the contact angle formed by a liquid droplet on the material's surface. It provides information about the material's wetting behaviour. The standards used for this testing are ASTM D7334 and ASTM D5946. ASTM D7334 is a standard developed by the American Society for Testing and Materials (ASTM). It provides a test method for determining the contact angle of a liquid droplet on a solid surface to assess the material's wetting behaviour. The test involves placing a liquid droplet on the material's surface and measuring the contact angle formed between the droplet and the surface. The contact angle provides information about the material's hydrophobic or oleophobic properties. A higher contact angle indicates greater hydrophobicity or oleophobicity. ASTM D5946 is a standard developed by the American Society for Testing and Materials (ASTM). It provides guidelines for calculating the surface energy of solid materials using contact angle measurements. The test involves measuring the contact angles formed by different probe liquids with known surface tensions on the material's surface. Based on these measurements, the surface energy of the material can be calculated using mathematical models, such as the Owens–Wendt–Rabel–Kaelble (OWRK) method.

4.1.1 Drop Test

Hydrophilicity of wool fabrics are evaluated according to the AATCC Test Method 39–1980 by determining the wetting time in seconds using the drop test. The test schematically consists of laying down a droplet of water of constant volume on the surface of the fabric and measuring the time required for its complete absorption [59].

4.1.2 Roll-Off Angle Measurement

This method determines the angle at which a liquid droplet starts to roll off the material's surface. It indicates the material's self-cleaning ability and resistance to liquid adhesion. The standards used for this testing are ASTM D7335 and ISO 27447. ASTM D7335 and ISO 27447 are standards that specifically address roll-off angle measurement, which is a method used to evaluate the self-cleaning ability and liquid repellency of materials. ASTM D7335 provides a test method for determining the roll-off angle of a liquid droplet on a surface to assess the material's self-cleaning properties. The test involves tilting the material at various angles and placing a liquid droplet on its surface. The roll-off angle is the inclination angle at which the droplet starts to roll off the surface due to its interaction with the material. A lower roll-off angle indicates better self-cleaning ability and increased resistance to liquid adhesion. ISO 27447 provides guidelines for the measurement of roll-off angle to assess the liquid repellency and self-cleaning characteristics of surfaces. The standard outlines the procedure for determining the roll-off angle by tilting the surface and placing a liquid droplet on it. The roll-off angle is the inclination angle at which the droplet starts to roll off the surface. Lower roll-off angles indicate better liquid repellency and self-cleaning ability.

4.1.3 Surface Energy Calculation

Surface energy calculation involves measuring the contact angles formed by different probe liquids with known surface tensions. It provides a quantitative measure of the material's hydrophobic or oleophobic properties. The standards used for this testing are ASTM D5946 and ISO 27448. ASTM D5946 provides a test method for calculating the surface energy of solid materials using contact angle measurements. The test involves measuring the contact angles formed by different probe liquids with known surface tensions on the material's surface. Based on these measurements, the surface energy of the material can be calculated using mathematical models such as the Owens–Wendt–Rabel–Kaelble (OWRK) method. ISO 27448 provides guidelines for the calculation of surface free energy and contact angle measurements. The standard outlines the procedures for measuring the contact angles of different probe liquids on the material's surface and calculating the surface free energy using mathematical models such as the Fowkes method or the acid–base approach.

4.1.4 Oil Repellency Test

The oil repellency test evaluates the material's ability to resist oil penetration. It typically involves applying a droplet of oil to the material's surface and assessing its behaviour, such as spreading or beading up. The standards used for this testing are AATCC 118 and ASTM D1308. AATCC 118 and ASTM D1308 are standards that specifically address the oil repellency test, which is used to evaluate the resistance of materials to oil penetration. AATCC 118 provides guidelines for conducting the oil repellency test on fabrics. The test involves applying a specified amount of oil to the fabric's surface and observing its behaviour. The material is evaluated based on a numerical rating scale that indicates the degree of oil repellency. Higher ratings represent better resistance to oil penetration. ASTM D1308 provides a test method for evaluating the resistance of fabrics to penetration by liquids. The test involves applying a specified liquid, such as oil, to the fabric's surface. The material is evaluated based on the visual observation of how the liquid spreads or beads up on the fabric. The test method does not provide a numerical rating system.

4.1.5 Spray Rating Test

The spray rating test assesses the material's resistance to liquid penetration under a defined spray condition. It assigns a numerical rating indicating the material's ability to repel liquids. AATCC 22 and ISO 4920 are standards that specifically pertain to the spray rating test, which is a method used to evaluate the resistance of materials to liquid penetration under a defined spray condition. AATCC 22 is a standard developed by the American Association of Textile Chemists and Colourists (AATCC). It provides guidelines for conducting the spray rating test to assess the resistance of fabrics to liquid penetration. The test involves subjecting the fabric to a specified amount of liquid spray under controlled conditions. The material is evaluated based on a numerical rating scale ranging from 0 to 100, with higher values indicating greater resistance to liquid penetration. The test method outlines the apparatus, test procedure, and criteria for determining the spray rating of fabrics. It ensures consistency and reproducibility in evaluating the hydrophobic or oleophobic properties of materials. ISO 4920 is an international standard developed by the International Organization for Standardization (ISO). It specifies a test method for determining the resistance of fabrics to penetration by water-repellent sprays. The test involves exposing the fabric to a defined amount of water spray using a calibrated spray nozzle. The material is evaluated based on a numerical rating system, which indicates the level of water penetration. Higher ratings indicate better resistance to liquid penetration. ISO 4920 provides detailed instructions for the test set-up, sample preparation, spraying parameters, and evaluation criteria. It ensures consistency and comparability of results across different laboratories and testing facilities.

5 Mechanisms of Water and Oil Repellency in Textiles

Water and oil repellency in textiles relies on specific mechanisms that prevent the wetting and penetration of liquid droplets into the fabric [1, 5, 60, 61]. These mechanisms can be broadly categorised into two main types: physical barriers and surface modifications [62–64]. Plausible mechanisms of water and oil repellency in textiles have been discussed below.

5.1 Physical Barriers

(a) Microscopic Surface Roughness:

This mechanism relies on creating a fabric surface with microscopic roughness. The surface irregularities or protrusions on a microscopic scale reduce the contact area between the liquid droplet and the fabric. As a result, the liquid droplet rests on the top of the rough surface, forming a spherical shape and minimising contact with the

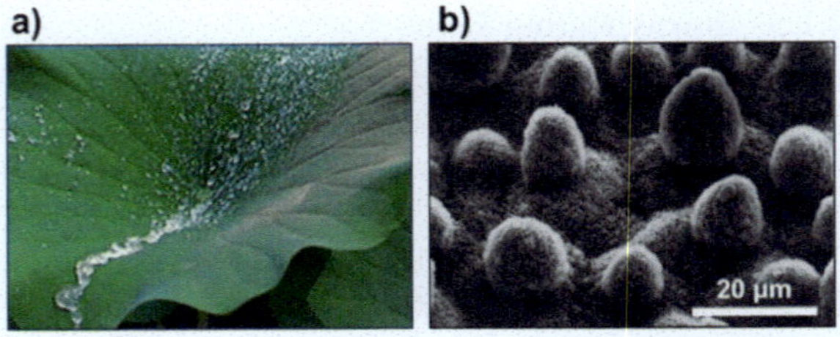

Fig. 3 Lotus leaf **a** [66], SEM of the lotus leaf **b** [65] (Copyright 2020, Elsevier)

fabric. This phenomenon is known as the lotus effect, inspired by the self-cleaning properties of lotus leaves. The surface roughness can be achieved through various techniques such as chemical etching, mechanical abrasion, or the use of textured fibres (Fig. 3) [65, 66].

(b) Air Trapping:

Air trapping is another physical barrier mechanism for water and oil repellency. Fabrics with a structure that can trap air within their fibres or between fibre layers create a barrier for liquid droplets. The trapped air prevents the liquid from directly contacting the fabric surface, causing the droplets to bead up and roll off the fabric. Fabrics with high porosity or open structures, such as certain types of non-wovens or meshes, often exhibit air-trapping mechanisms.

5.2 *Surface Modifications*

Surface modifications play a crucial role in achieving water and oil repellency by altering the surface energy and chemistry of the fabric. These modifications make the fabric less attractive to water and oil molecules. Hydrophobic surface modifications reduce the surface energy of the fabric, making it less likely to attract water molecules. This can be achieved through the application of hydrophobic coatings or finishes containing materials such as fluoropolymers (e.g. perfluorinated compounds), silicone, or wax. These coatings create a low surface energy barrier, causing water droplets to form high contact angles and roll off the fabric. Oleophobic surface modifications repel oil molecules in a similar manner to hydrophobic treatments. They reduce the surface energy and increase the contact angle of oil droplets, preventing their wetting and penetration into the fabric. Oleophobic coatings often incorporate fluorinated compounds or other antiwetting agents to achieve oil repellency.

Modifying the surface charge of the fabric can also contribute to water and oil repellency. By imparting a positive or negative charge to the fabric surface, the attraction between the fabric and the liquid molecules can be minimised. This can be achieved through the application of cationic or anionic finishes or by incorporating charged nanoparticles into the fabric structure. The repulsive forces between the charged fabric surface and the liquid droplets prevent wetting and penetration.

Advanced surface modifications involve the application of micro- or nanocoatings to the fabric surface. These coatings can be composed of nanoparticles, nanofibres, or nanostructured materials, which cannot react with the material to form any strong bonds. They create a thin layer on the fabric surface, altering its properties and providing water and oil repellency.

5.3 Others

5.3.1 Photocatalytic Effects

Photocatalytic materials, such as titanium dioxide (TiO_2), can be incorporated into textile finishes or coatings to achieve self-cleaning and water-repellent properties [67, 68]. When exposed to light, photocatalytic materials generate reactive oxygen species, such as hydroxyl radicals, through a photochemical reaction [69]. These reactive species have a strong oxidising effect and can break down organic compounds, including dirt and stains that may accumulate on the fabric surface [70]. By keeping the fabric clean, the photocatalytic effect enhances its water repellency by preventing the accumulation of contaminants that could reduce repellency [71].

5.3.2 Self-Cleaning

The combination of photocatalytic effects and water repellency creates a self-cleaning mechanism [72–74]. When water droplets come into contact with the fabric surface, they spread and carry away any dirt or stains due to the low surface energy and the hydrophilic properties of the photocatalytic materials [75]. This self-cleaning action helps to maintain the water-repellent properties of the fabric over time [76].

5.3.3 Surface Energy Gradient

Creating a surface energy gradient on the fabric can contribute to water and oil repellency [77]. This mechanism involves modifying the fabric surface with materials of varying surface energies [63, 78]. The fabric surface can be treated with different materials or coatings, each having a distinct surface energy [79]. For example, a hydrophobic coating with low surface energy can be applied to one area of the fabric, whilst another area is left untreated or treated with a higher surface energy material [80]. This creates a gradient of surface energy across the fabric surface [26, 81]. When water or oil droplets come into contact with the fabric, they tend to move towards a region of lower surface energy [82, 83]. This movement is driven by the difference in surface energies between the droplet and the fabric [84]. As a result, the droplets bead up and roll off the fabric, following the surface energy gradient [85].

5.3.4 Surface Texturing

Fabric surface texturing involves creating specific patterns or structures on the fabric surface to enhance water and oil repellency [86, 87]. Laser ablation is a technique that uses laser beams to remove or modify the fabric surface, creating patterns or textures [88]. By selectively removing material, the fabric surface can be roughened or structured, reducing the contact area between liquid droplets and the fabric [78]. This can enhance the water and oil repellency. Embossing is a process where heat and pressure are used to create patterns or textures on the fabric surface [89]. The embossed patterns can create microscale roughness or structures that promote repellency by reducing the contact area and allowing droplets to bead up and roll off the fabric [90]. Chemical etching involves treating the fabric surface with specific chemicals to create patterns or textures [91, 92]. The etching process selectively removes or modifies the fabric surface, resulting in a roughened or structured surface. This surface texture enhances the water [52, 93, 94].

6 Superhydrophobic, Superoleophobic, and Omniphobic

Superhydrophobic and superoleophobic surfaces exhibit extremely high contact angles and low contact angle hysteresis, resulting in exceptional water and oil repellency [95]. Superhydrophobic and superoleophobic surfaces often possess a combination of microscale and nanoscale roughness [96]. The roughness can be achieved through various techniques, such as etching, chemical vapour deposition, or self-assembly of particles [97]. The hierarchical roughness enhances the surface area and reduces the contact area between the liquid droplets and the fabric, promoting repellency. Superhydrophobic and superoleophobic surfaces are typically coated with materials that have low surface energy, such as fluoropolymers (e.g. perfluorinated compounds) or silicone [98]. These coatings further reduce the surface energy of the fabric, making it even less attractive to water and oil molecules. The combination of roughness and low surface energy creates a dual effect, causing water and oil droplets to bead up and roll off the fabric effortlessly [99].

Omniphobic refers to a surface or material that exhibits resistance to various types of liquids, including both polar and non-polar substances. An omniphobic surface repels a wide range of liquids, such as water, oils, solvents, and even viscous fluids like syrups. This property is achieved by reducing the surface energy of the material, causing the liquids to bead up and roll off without wetting or spreading on the surface. Omniphobic surfaces are highly non-stick and typically possess self-cleaning properties, making them resistant to staining, fouling, and contamination. These surfaces find applications in various fields, including textiles, coatings, biomedical devices, and environmental protection, where the ability to repel different liquids is desirable. Huang et al. developed a superomniphobic coating by spraying a hybrid material system consisting of fluorinated poly[dimethylsiloxane-co-(3-aminopropyl) methylsiloxane (PDMS-F), fluorinated silica nanoparticles (FSiO2-NPs), and perfluoroalkyl polymer (F-Polymer) onto different substrates, including cotton textile. The coating exhibited minimal contact angle hysteresis (CAH) ($<15°$) for various liquids, such as water, hexadecane, glycol, ethanol, and diiodomethane [100]. The cotton textile showed a static water contact angle (CA) of 163° and a hexadecane (oil) CA of 153°, demonstrating the effectiveness of the superomniphobic coating.

7 Different Nanoparticles

Water repellency of nanoparticles is primarily achieved through surface chemistry and structural features. Different types of nanoparticles exhibit varying degrees of water repellency based on their chemical composition and surface properties. Here, we will discuss the chemistry-based functions of three commonly used nanoparticles: silicon dioxide (SiO_2), titanium dioxide (TiO_2), and fluoropolymers.

7.1 *Silicon Dioxide (SiO_2)*

Silicon dioxide nanoparticles can exhibit hydrophobic properties, making them useful in creating water-repellent surfaces [101]. The water repellency of SiO_2 nanoparticles arises from their surface chemistry, which can be modified to reduce the surface energy and increase hydrophobicity [102] [103]. Common surface modifications include the introduction of hydrophobic functional groups, such as alkyl or fluoroalkyl groups [104], onto the nanoparticle surface [105]. The hydrophobic functionalisation of SiO_2 nanoparticles involves several steps. First, the nanoparticle surface is typically functionalised with a silane coupling agent, such as octyltrimethoxysilane (OTMS) [106] [107] or perfluorooctyltriethoxysilane (PFOTES) [108, 109]. These coupling agents possess hydrophobic alkyl or fluoroalkyl chains that attach to the surface of the SiO_2 nanoparticles. This functionalisation alters the surface chemistry, reducing the surface energy and making the nanoparticles water repellent.

The hydrophobic SiO_2 nanoparticles can be incorporated into coatings or films to create water-repellent surfaces. The roughness of the surface, achieved through the use of hierarchical micro–nanostructures, can further enhance the water repellency by reducing the contact area between water droplets and the surface. In a study conducted by Yin et al., a hybrid SiO_2/HTEOS/CPTS sol was prepared through acid-catalysed hydrolytic co-polycondensation of tetraethoxysilane (TEOS), γ-chloropropyltriethoxysilane (CPTS), and hexadecyltrimethoxysilane (HTEOS). This sol was used to deposit a water-repellent functional hybrid coating onto cellulose substrates. The fabric was submerged in the hybrid SiO_2/HTEOS/CPTS sol for 5 min and then padded twice using a padding mangle. The combination of low surface energy chemical compositions (HTEOS and CPTS) and a rough surface geometrical structure contributed to the improvement in water repellency. The contact angles of water on the coated fabric were significantly enhanced, reaching

139.8° [110]. Fu et al. conducted a study where they developed a facile coating for cotton fabric (CF) to achieve superhydrophobicity and flame retardancy. The coating consisted of chitosan (CS), phytic acid (PA), and silicon dioxide (SiO_2) modified with hexamethyldisilylamine (HMDS). Nanoparticles were synthesised via the solgel method. The coating was applied onto the CF using a simple repeated coating technique. Initially, the coating demonstrated a high contact angle of 150 degrees, indicating excellent water repellency. After undergoing 50 cycles of testing, the contact angle reduced to approximately 120 degrees, suggesting the durability of the coating under repeated use. This study highlights the potential of the CS/PA/SiO_2-HMDS coating for achieving both superhydrophobicity and flame retardancy on cotton fabric surfaces [111]. Chirila et al. employed the solgel method to synthesise flower-like ZnO nanoparticles and organically modified SiO_2 nanoparticles (ORMOSIL) for their study. These nanoparticles were then incorporated into a hybrid coating to achieve hydrophobic and oleophobic properties on textile materials. The dispersions of ZnO and SiO_2 nanoparticles exhibited stability, with SiO_2 nanoparticles showing a unimodal distribution of smaller quasi-spherical particles measuring approximately 158 nm in average size. The flower-like ZnO nanoparticles displayed a bimodal distribution, with an average size of 144 nm and 435 nm. The ZnO/SiO_2 nanoparticle dispersions were subsequently applied to 100% cotton fabrics using a laboratory-scale padding method in two phases. The resulting functionalised cotton fabrics experienced changes in surface morphology whilst maintaining the overall integrity of the cotton fibres. These treated fabrics demonstrated a low wetting capacity, characterised by a high water absorption time (>15 min), and exhibited a water-repellent degree of 75 (AATCC photographic scale) and 2.5 (ISO scale), indicating a notable hydrophobic effect. Moreover, the hybrid coating based on the ZnO/ SiO_2 nanoparticles effectively provided the cotton substrate with protection against oily substances, achieving an oil repellency degree of 6 [112]. In their study, Borah et al. utilised nanosilica extracted from rice husk to enhance the water repellency of eri silk fabric. To further improve the water-repellent properties, a silicone-based polymer was applied. The results demonstrated that the application of nanosilica conferred hydrophobic properties to the eri silk fabric. The hydrophobicity was further enhanced to a superhydrophobic level by incorporating the silica polymer. The spray test ratings indicated a significant improvement in water repellency, with a rating of 0 for the control fabric and 50–70 for the modified fabric treated with silica extracted from rice husk. When the polymer alone was used, the spray test rating reached 70–80. However, the combination of silica and the polymer resulted in a further increase to 80–90 in the spray test rating [113]. Jhatial et al. prepared a durable hydrophobic bamboo knitted fabric with multifunctional properties using a solgel coating method. They utilised titania and silica nanoparticles, combined with citric acid as a cross-linking agent. The bamboo fabric was functionalised through a dip–pad–dry–pad–cure process [114].

7.2 Titanium Dioxide (TiO_2)

Titanium dioxide nanoparticles are commonly used for their photocatalytic properties, but they can also exhibit water-repellent behaviour. The water repellency of TiO_2 nanoparticles is attributed to their superhydrophilic nature, which arises from their photocatalytic activity and self-cleaning properties [72]. In the presence of ultraviolet (UV) light, TiO_2 nanoparticles generate reactive oxygen species (ROS), such as hydroxyl radicals, on their surface. These ROS have strong oxidative properties and can break down organic compounds, including contaminants or organic pollutants on the surface. The photocatalytic activity of TiO_2 nanoparticles leads to self-cleaning surfaces as organic materials are degraded, preventing the accumulation of dirt or contaminants [115].

The superhydrophilic nature of TiO_2 nanoparticles is a result of the formation of a thin layer of water molecules on their surface [116]. This water layer, known as the "photoinduced hydrophilic layer", arises from the interaction between UV light, TiO_2 nanoparticles, and water vapour. The photoinduced hydrophilic layer reduces the contact angle of water droplets, causing them to spread and adhere to the surface rather than forming droplets. Tudu et al. created a superhydrophobic coating on cotton fabric using TiO_2 nanoparticles and perfluorodecyltriethoxysilane. The coated fabric shows excellent self-cleaning, stain resistance, rust stain resistance, water repellency, and antibacterial properties. It is also highly effective for oil–water separation, with excellent reusability [75]. However, in their study, Pakdel et al. utilised a low-temperature solgel method to functionalise wool fabrics with TiO_2 and TiO_2/SiO_2 nanocomposites. The nanocomposite sols were formulated with varying molar ratios of TiO_2/SiO_2 (70:30, 50:50, and 30:70). The sols were then applied to the wool fabric surfaces using the dip–pad–dry–cure process. The study revealed that the incorporation of silica in the TiO_2 nanocomposite played a crucial role in enhancing the functionality of the self-cleaning coating layer on wool fabric. By providing a higher surface area in the vicinity of TiO_2 and increasing the surface acidity of the photocatalyst, silica contributed to the improved performance of the coating. Specifically, the wool fabrics treated with TiO_2/SiO_2 50:50 and 30:70 exhibited superhydrophilic properties. Amongst them, the fabrics functionalised with TiO_2/SiO_2 30:70 demonstrated the highest efficiency in stain removal, followed by those treated with TiO_2/SiO_2 50:50 [117]. In their study, Xu et al. utilised SiO_2 and TiO_2 composite nanoparticles to create superhydrophobic and UV-blocking cotton fabrics. The fabrication process involved incorporating TiO_2/SiO_2 composite nanoparticles onto the cotton fabric surface, followed by hydrophobization using hexadecyltrimethoxysilane (HDTMS). The TiO_2/SiO_2 composite nanoparticles were synthesised through a solgel reaction. The resulting cotton fabric exhibited a water contact angle of 159.3°, indicating a high level of hydrophobicity. Additionally, the water shedding angle was measured at 8°, further demonstrating the fabric's superhydrophobic nature.

These findings highlight the potential variations in the performance of TiO_2-based coatings when combined with different chemical components. Based on the previously mentioned contradictory findings regarding the behaviour of TiO_2 nanoparticles, it can be inferred that the hydrophilicity and other properties of TiO_2 are highly dependent on various factors such as crystal phase, surface composition, and experimental conditions. Therefore, it is important to consider these factors when studying and evaluating the behaviour of TiO_2 nanoparticles. The contradictory results observed in different studies highlight the complexity of TiO_2 behaviour and emphasise the need for further research and a comprehensive understanding of the underlying mechanisms.

7.3 Fluoropolymers

Fluoropolymers, such as polytetrafluoroethylene (PTFE) or perfluorinated compounds, are known for their exceptional water repellency owing to the presence of fluorine atoms in their chemical structure [118, 119]. Fluorine is highly electronegative, creating strong covalent bonds with carbon and rendering the fluoropolymers highly hydrophobic and oleophobic. The water repellency of fluoropolymers is a result of the low surface energy and the non-polar nature of the carbon–fluorine bonds [120]. These bonds create a highly stable and inert surface that repels water molecules due to the weak van der Waals forces between the polar water molecules and the non-polar fluorinated surface. The low surface energy prevents water from wetting the surface, resulting in a high contact angle and water droplets easily rolling off. Fluoropolymers can be synthesised or applied as coatings to surfaces to impart water repellency [121]. The fluorinated surface forms a barrier that prevents water penetration and adhesion, making it useful for applications requiring water resistance, such as in waterproof textiles, non-stick coatings, or water-repellent paints. Ekanayake et al. conducted a study to develop superomniphobic polyester fabric with a water contact angle (WCA) of 152° by reducing the surface free energy through the polymerization of 3,3,4,4,5,5,6,6,7,7,8,8,8-tridecafluorooctyl methacrylate (TDM) on a ZnO seed layer grown on polyester. Two types of fabrics, one with and the other without the seed layer, were analysed using various techniques. The fabric with the seed layer exhibited a WCA of 153° compared to 142° for the fabric without the seed layer. The modified omniphobic fabric demonstrated oleophobicity, as indicated by a contact angle of 132° with SAE 40 motor oil, whilst the non-treated fabric did not exhibit a measurable contact angle [122]. The modified fabric displayed superior durability, withstanding 50 cycles of abrasion and over 1.5 h of washing (three washing cycles). Additionally, the modified fabric demonstrated omniphobic behaviour towards milk tea, coffee, ethanol, and coconut oil, whilst these liquids immediately penetrated the non-treated fabric without forming a contact angle. Jiang et al. successfully developed superhydrophobic coatings on cotton textiles using a simple one-step solgel process. A fluorinated copolymer, PHFA-MPS, was synthesised by copolymerizing hexafluorobutyl methacrylate (HFA) and

3-methacryloxypropyltrimethoxysilane (MPS) as monomers. The fluoro-containing silica sols were then formed through alkaline hydrolysis of tetraethyl orthosilicate (TEOS) and the PHFA-MPS solution in ethanol. These sols were applied to cotton textiles using a convenient dip-coating method, resulting in excellent superhydrophobicity. The cotton textiles exhibited a water contact angle (WCA) of 153.4°, indicating their remarkable water-repellent properties [123]. Furthermore, the superhydrophobic coatings demonstrated satisfactory stability, providing a simple and effective approach for achieving superhydrophobicity on cotton textiles.

7.4 Zinc Oxide (ZnO)

Zinc oxide nanoparticles possess unique properties that contribute to their water-repellent behaviour [124]. The surface chemistry and structure of ZnO nanoparticles play crucial roles in their water repellency. ZnO nanoparticles often exhibit a rough surface morphology, which can be achieved through various synthesis methods. The roughness increases the surface area and creates air pockets, reducing the contact area between water droplets and the nanoparticle surface. The surface of ZnO nanoparticles can be modified to introduce hydrophobic properties [125]. This can be achieved by functionalising the surface with hydrophobic organic molecules or polymers. For example, alkyl chains or fluorinated compounds can be attached to the surface of ZnO nanoparticles, reducing the surface energy and enhancing water repellency [126, 127].

The functionalisation process typically involves the use of coupling agents, such as alkylsilanes or perfluorosilanes, which form covalent bonds with the ZnO nanoparticle surface [128]. These hydrophobic functional groups create a barrier that repels water molecules, resulting in water droplets rolling off the surface. Tania et al. aimed to enhance cotton fabric with antibacterial, stain-release, and UV protection properties by depositing ZnO nanoparticles onto its surface. The nanoparticles were synthesised using a sonochemical strategy, and a mechanical thermo-fixation technique was employed to firmly affix them onto the fabric [129]. Kathirvelu et al. synthesised and characterised nano-sized zinc oxide (ZnO) particles. These particles were applied to cotton and polyester/cotton (P/C) fabrics to impart stain-release properties [130]. Yadav et al. employed a wet chemical method to prepare zinc oxide nanoparticles, which were subsequently coated onto bleached cotton fabrics. The coated fabrics exhibited desirable features, including 75% UV blocking, improved breathability, reduced friction, also investigated wash fastness, antimicrobial properties, abrasion resistance, and fabric handle properties [131]. Mohamed et al. proposed a simple and effective method for fabricating cotton fabric with superhydrophobicity and flame retardancy. The process involved coating the fabric with amino-functionalised silica nanoparticles in combination with ZnO nanoparticles. To achieve this, cotton was first functionalised with epoxy and carboxyl groups through grafting with a nanoemulsion containing glycidyl methacrylate (GMA) and acrylic acid (AA).

Subsequently, the fabric underwent conventional pad–dry–cure treatment for functional finishing. The resulting treated fabrics exhibited good water repellency and excellent flame-retardant properties [17].

7.5 Aluminium Oxide (Al_2O_3)

Aluminium oxide nanoparticles, also known as alumina, generally exhibit hydrophilic behaviour due to the presence of hydroxyl groups on their surface [132]. However, the water-repellent properties of Al_2O_3 nanoparticles can be achieved through surface modifications and coatings [133]. The hydrophilicity of Al_2O_3 nanoparticles can be reduced by modifying the surface chemistry. This can be achieved by functionalising the surface with hydrophobic molecules or polymers. For example, organic silanes with hydrophobic functional groups can be used to treat the surface of Al_2O_3 nanoparticles. The functionalisation process involves the reaction between the hydroxyl groups on the nanoparticle surface and the functional groups of the silane molecules. This results in the formation of covalent bonds, creating a hydrophobic coating that repels water. Surface coatings can also involve the deposition of hydrophobic materials, such as polymers or fluorinated compounds, onto the Al_2O_3 nanoparticle surface. These coatings reduce the surface energy and enhance water repellency. In the study conducted by Abidi et al., cotton fabric was subjected to a two-step functionalisation process. First, nanoparticles vapour deposition (NVD) of Al_2O_3 was performed on the fabric surface, followed by molecular vapour deposition (MVD) of (tridecafluoro-1,1,2,2,-tetrahydrooctyl)-trichlorosilane (FOTS). This dual treatment resulted in the modification of the fabric's surface properties. The deposition of nanoparticles increased the surface roughness of the cotton fabric, leading to an enhanced contact angle. Dynamic water contact angles were measured, and contact angles exceeding 150° were achieved, indicating the fabric's highly hydrophobic nature [134]. In the study by Xiao et al., a superhydrophobic surface was achieved on wool fabrics through the deposition of aluminium oxide using atomic layer deposition (ALD). The process involved exposing the fabrics to alternating pulses of trimethylaluminium and water at 80 °C. As a result, Al_2O_3 nanoparticles formed around the surface of the ALD-coated wool fibres, leading to increased surface roughness compared to control wool fibres. The functionalised wool fabrics exhibited significant improvements in hydrophobicity. The static water contact angles of the ALD-coated wool fabrics increased from 130° to around 160°, indicating a highly superhydrophobic surface. Furthermore, the hydrophobicity of the coated fabrics demonstrated higher durability compared to that of the control wool fabric [135].

7.6 *Gold (Au) and Silver (Ag)*

Gold [136] and silver [137, 138] nanoparticles can exhibit water-repellent properties based on their unique surface chemistry and plasmonic effects. Gold and silver nanoparticles can have a hydrophobic surface chemistry, depending on the ligands or capping agents used during their synthesis. These ligands can introduce hydrophobic functional groups, such as thiol (-SH) or alkyl chains, onto the nanoparticle surface. The hydrophobic surface chemistry reduces the surface energy and enhances water repellency. Gold and silver nanoparticles exhibit plasmonic effects, which arise from the collective oscillation of their conduction electrons when excited by light. These plasmonic effects can induce local heating, resulting in the formation of vapour nanobubbles on the nanoparticle surface when irradiated with laser light. Hassan et al. treated wool fabrics with silver nanoparticles using trisodium citrate as a reducing and capping agent. The treated fibres showed excellent antibacterial activity, UV radiation absorption, antistatic properties, and colourfastness to washing. The treatment proved highly durable, with minimal loss in colour and antibacterial activity even after 20 washes [16]. Xue et al. achieved superhydrophobic cotton textiles by in situ coating the fibres with silver nanoparticles (Ag NPs) produced through the reduction of $[Ag(NH_3)_2]^+$ complex with glucose. The coated fibres were further modified with hexadecyltrimethoxysilane, resulting in the creation of dual-size roughness on the textiles and the formation of superhydrophobic surfaces. The contact angle measured was approximately 157 degrees, indicating the highly water-repellent nature of the textiles [137]. Bu et al. developed a durable antibacterial and superhydrophobic viscose textile by in situ growing silver nanoparticles (NPs) on tannic acid (TA)-coated viscose textiles, followed by a hydrophobic treatment. The Ag/TA-coated viscose textile was immersed in a PFDT solution, resulting in the formation of PFDT/Ag/TA-coated viscose textile. The superhydrophobic textile exhibited a water contact angle more than 150°. These findings highlight the significance of in situ-generated Ag NPs in achieving superhydrophobicity [139].

7.7 *Iron Oxide (Fe_3O_4) Nanoparticles*

Iron oxide nanoparticles, specifically magnetite (Fe_3O_4), can exhibit both hydrophilic and hydrophobic behaviours depending on their surface chemistry. The water-repellent properties of Fe_3O_4 nanoparticles can be achieved through surface modifications and coatings. The surface of iron oxide nanoparticles is typically composed of iron and oxygen atoms. The presence of hydroxyl groups (OH) on the surface contributes to their hydrophilic behaviour. However, the surface chemistry can be altered to introduce hydrophobicity. Fe_3O_4 nanoparticles can be coated with hydrophobic materials to achieve water repellency. The coating process involves the deposition of hydrophobic polymers or organic molecules onto the nanoparticle surface. Zheng et al. developed a layer-by-layer assembling technique to create multilayer sandwiched hollow magnetic Fe_3O_4 nanospheres (HFO)/MXene on flexible cellulose non-woven fabrics. By modifying the surface of HFO and MXene nanosheets, the hydrophobicity (superhydrophobicity) of the composite fabrics was significantly improved, as indicated by an increase in the water contact angle from 94.7° to 135.3° [140]. The superhydrophobicity of the composite fabric (HFM11) was attributed to the micro–nano rough structure and low surface energy of the HFO nanospheres. Notably, HFM11 demonstrated not only excellent superhydrophobicity with water but also effective repellency against common household liquids like coffee and milk. The fabrics enabled spherical-shaped droplets of these liquids to stand freely on their surface. For example, polymers such as polyethylene glycol (PEG) or polydimethylsiloxane (PDMS) can be coated onto the Fe3O4 nanoparticles. These coatings reduce the surface energy and create a hydrophobic barrier, preventing water from wetting the surface.

7.8 Carbon Nanotubes (CNTs)

Carbon nanotubes are cylindrical structures composed of rolled-up graphene sheets. They can exhibit hydrophobic behaviour due to the presence of non-polar sp^2 carbon bonds in their structure. The surface of CNTs consists of carbon atoms arranged in a hexagonal lattice. This surface can be functionalised to introduce hydrophobicity or enhance water repellency. Shim et al. developed a superhydrophobic textile material with a self-cleaning function by incorporating carbon nanotubes (CNTs) and water-repellent agents. Hydrophobic fabrics were created on 100% polyester woven fabrics using either fluorocarbon or siloxane treatments. Fabrics treated with fluorocarbon exhibited higher contact angles (CA) and lower shedding angles compared to those treated with siloxane, attributed to the lower surface energy of the fluorocarbon repellent. Amongst the specimens, the ones made of 50 denier DTY and treated with CNT-Teflon AF® demonstrated the most superhydrophobic characteristics [141]. These fabrics displayed a static contact angle greater than 150° and a shedding angle less than 15°. The presence of CNTs on the fabric's surface contributed to nanoscale surface roughness, resembling the papillae on a lotus leaf, thereby enhancing the superhydrophobic properties by trapping air pockets.

7.9 Graphene

Graphene is a two-dimensional sheet of carbon atoms arranged in a hexagonal lattice. Its unique structure results in excellent mechanical, electrical, and thermal properties. The surface of graphene consists of carbon atoms bonded together by sp^2 hybridization, forming strong covalent bonds. The sp^2 carbon bonds contribute to the hydrophobic nature of graphene. Graphene can be modified to enhance its water-repellent properties through various methods. The surface of graphene can be chemically modified by attaching hydrophobic groups or polymers. This can be achieved through covalent functionalisation, where the hydrophobic functional groups are introduced onto the graphene surface through chemical reactions. Non-covalent functionalisation involves the adsorption of hydrophobic molecules or polymers onto the graphene surface. Physical interactions, such as $\pi-\pi$ stacking or hydrophobic interactions, can occur between the hydrophobic species and the graphene surface, enhancing its water repellency. Ouadil et al. developed a simple and cost-effective method for enhancing the properties of polyester fabric by coating it with graphene/silver nanoparticles (PET-G/Ag°) using a dip-coating technique. Graphene nanosheets were coated onto the polyester fabric, serving as a support for the deposition of silver nanoparticles (Ag°). The high specific surface area of graphene facilitated the distribution and stability of the silver nanoparticles, acting as a bifunctional agent for linking and stabilising them [142]. Compared to pristine polyester fabric, the coated polyester fabrics exhibited improved hydrophobic behaviour, as evidenced by water contact angles of 122°, 126°, and 124° for PET-GO, PET-G, and PET-G/Ag°, respectively. This indicates that the graphene coating enhances the fabric's ability to repel water. Tissera et al. accomplished hydrophobic properties on cotton fabric by grafting graphene oxide (GO) onto the fabric surface using a dyeing method. Graphite oxide was synthesised by oxidising natural flake graphite through an improved Hummer's method. The synthesised graphite oxide was then exfoliated in water using ultrasound energy to obtain graphene oxide (GO). By loading varying amounts of GO onto the cotton fibre surface, the fabric demonstrated different levels of hydrophobicity [143]. The highest water contact angle observed was 143°, achieved with the highest loading of graphene oxide. Additionally, the fabric surfaces grafted with GO exhibited adhesive-type hydrophobicity.

7.10 Carbon Dots

Carbon dots are small carbon-based nanoparticles with sizes typically less than 10 nm. They exhibit unique optical properties, making them suitable for various applications. Carbon dots possess a surface consisting of sp^2 carbon atoms, similar to graphene. However, the surface chemistry can vary depending on the synthesis method and surface passivation. To enhance the water-repellent properties of carbon dots, surface passivation can be performed. This involves the introduction of hydrophobic ligands or coatings onto the surface. Carbon dots can be functionalised with hydrophobic molecules or polymers to reduce their surface energy and enhance water repellency. The functional groups can include alkyl chains, fluorinated compounds, or hydrophobic polymers. Almufarij et al. conducted a study on the synthesis of carbon quantum dots (CQDs) derived from banana leaves using a hydrothermal process. These CQDs were then utilised to create a durable superhydrophobic coating on textile fabric for oil/water separation purposes [144]. The resulting CQD-based superhydrophobic membrane exhibited exceptional superhydrophobic properties, with water contact angles reaching 163° and low water sliding angles of only 1°. This demonstrates the effectiveness of the developed coating in repelling water and its potential for efficient oil/water separation applications.

7.11 Fullerenes

Fullerenes are carbon-based nanoparticles with spherical or cage-like structures, such as C60 and C70. They have unique electronic and optical properties. Fullerenes possess a surface composed of sp^2 carbon atoms, similar to other carbon-based nanoparticles. The hydrophobic nature of fullerenes is attributed to the non-polar carbon–carbon bonds. Fullerenes can be modified to enhance their water-repellent properties through covalent or non-covalent functionalisation. Covalent functionalisation involves the introduction of hydrophobic functional groups onto the fullerene surface through chemical reactions. This can be achieved by attaching hydrophobic moieties, such as alkyl chains or fluorinated compounds, to the surface. Non-covalent functionalisation involves the adsorption of hydrophobic molecules or polymers onto the fullerene surface through physical interactions. These interactions can include π–π stacking or hydrophobic interactions, resulting in improved water repellency. Partheeban et al. conducted a study on the interfacial precipitation of flower- and octahedron-like microcrystals of fullerene using an anisole/IPA liquid–liquid interfacial precipitation method. Both the flower- and octahedron-like fullerene microcrystals exhibited a superhydrophobic nature, displaying a high water contact angle of 158.8° [145]. In contrast, other C60 nanostructures such as nanowhiskers, nanosheets, and nanorods did not demonstrate a significant water contact angle.

Thin films composed of the flower-like fullerene crystals and octahedrons exhibited even better superhydrophobic properties, with a water contact angle of 158.8°. By comparison, the thin film of pristine fullerene (C60) showed a much lower value of 87.1° [145]. Fullerene and its derivatives are known for their inherent hydrophobic nature, making them suitable for various applications, including superhydrophobic coatings for vehicle windshields, anticorrosive/icing coatings, surgical tools, medical equipment, and textiles.

Efforts have been made to enhance the hydrophobicity of fullerene-based nano/microcrystals, leading to significant success. Zheng et al. reported hierarchical fullerene architectures using C60/C70 microstructures with a water contact angle of 154.12° [146]. Wei et al. developed C60 hollow microspheres with a water contact angle of 156.3° [147], whilst Nakanishi et al. reported globular objects of fullerene derivatives with a water contact angle value of 152.0° [148]. These findings highlight the potential of fullerene-based nanostructures in achieving superhydrophobic surfaces for a wide range of applications.

7.12 *Other Nanoparticles*

Apart from the nanoparticles mentioned above, many other nanoparticles can exhibit water-repellent properties based on their surface chemistry and modifications. Whilst it is not feasible to cover all of them in detail, a table has been provided below to summarise the key information.

Fabrication method	Nanoparticles	Other Materials	Contact angle	References
Solgel process	Fluorinated SiO_2	Tetraethyl orthosilicate (TEOS)-modified PET filter material	144.6 ± 1.6°	[149]
	Multiwalled carbon nanotube	Methyltrimethoxysilane	146	[150]
	SiO_2	Organo-silane without fluorine	153.1° ± 0.8°	[151]
	Silica	Trimethylsilane	156°	[152]
	Silica	Tetraethyl orthosilicate and methyltriethoxysilane	160°	[153]
Chemical vapour deposition	SiO_2	Poly(hexafluorobutyl acrylate) (PHFBA)	> 165°	[154]
	Cu	–	154.5°	[155]
Hydrothermal synthesis	ZnO nanorods	Non-fluorinated silane	170.2°	[156]
	TiO_2	Octadecyltrimethoxysilane (OTMS)	158.6°	[157]
	ZnO	Polydimethylsiloxane (PDMS)	> 150°	[156]
Polymer film generation	ZnO	Hexadecyltrimethoxysilane (HDTMS)	150°	[158]
	SiO_2	Siloxane	> 150°	[159]
	Nanoparticle spray	Polydimethylsiloxane (PDMS)	172°	[160]
	SiO_2	Ethyl-α-cyanoacrylate superglue	154°	[161]
Dip coating	Ag	Poly(ethylenimine)	> 150°	[162]
	Ag	Polydopamine	154°	[163]

8 Conclusion

Researchers have explored different nanomaterials and nanostructures to achieve superhydrophobic and superoleophobic surfaces. This chapter provides an overview on the different chemistry and type of chemicals involved in achieving superhydrophobic and superoleophobic finishes. The introduction of nanofinishing has created a huge impact for scaling up these nanoparticle-based finishing agents in commercialised textile products. The future prospects in water- and oil-repellent finishing using nanotechnology are promising as advances in various nanocomposites, metal organic frameworks, metal oxides, carbon-based NPs, fluoropolymers, etc. can be incorporated in their structure to impart or improve the performance of the finished textiles. Technologies like plasma etching and supercritical carbon dioxide-based finishing can be used to obtain superhydrophobic finishes keeping the aspects of circularity and net zero into consideration in textile finishing. Researchers can further explore the synthesis methods and application of novel nanomaterials with enhanced hydrophobic and oleophobic properties. Implementation of nanofinishes on natural fibre-based textiles using different coating methods can affect the inherent properties of the fibre. Thus, the compatibility of the nanoparticles with the textile fibres is still at a development stage keeping in mind the biodegradability and toxicity level. Additionally, the development of scalable and cost-effective methods for coating large areas and integrating nanomaterials into various products can drive the commercialization and widespread adoption of these technologies.

References

1. Li S et al (2017) A review on special wettability textiles: theoretical models, fabrication technologies and multifunctional applications 5(1):31–55
2. Schwartz AM (2002) Surface tension and surface energy. Textile science and technology. Elsevier, pp 57–91
3. Chatterjee P, Gupta B (2002) Surface tension and surface energy, p 57
4. Mahltig B (2015) Performance and protection, hydrophobic and oleophobic finishes for textiles, pp 387–428
5. Holmquist H et al (2016) Properties, performance and associated hazards of state-of-the-art durable water repellent (DWR) chemistry for textile finishing 91:251–264
6. Simončič B et al (2012) Multifunctional water and oil repellent and antimicrobial properties of finished cotton: influence of sol–gel finishing procedure 61:340–354
7. Williams JT (2017) Waterproof and water repellent textiles and clothing. Woodhead Publishing
8. Saha PK et al (2021) Functionalization of hydrophobic nonwoven cotton fabric for oil and water repellency 3(5):586
9. Galante AJ et al (2020) Superhemophobic and antivirofouling coating for mechanically durable and wash-stable medical textiles 12(19):22120–22128
10. Truong QT, Pomerantz N (2018) Military applications: development of superomniphobic coatings, textiles and surfaces. Waterproof and water repellent textiles and clothing. Elsevier, pp 473–531
11. Holmes D (2000) Textiles for survival 12:461

12. Choudhury AKR (2017) Principles of textile finishing. Woodhead Publishing
13. Goy R, Jenkins J (1970) Industrial applications of textiles 2(1):1–60
14. Khattab TA et al (2020) Textile dyeing industry: environmental impacts and remediation 27:3803–3818
15. Cloud R, Cao W, Song G (2013) Functional finishes to improve the comfort and protection of apparel. Advances in the dyeing and finishing of technical textiles. Elsevier, pp 258–279
16. Hassan MM, Physicochemical SA, Aspects E (2019) Enhanced colour, hydrophobicity, UV radiation absorption and antistatic properties of wool fabric multi-functionalised with silver nanoparticles 581:123819
17. Mohamed AL, El-Sheikh MA, Waly AI (2014) Enhancement of flame retardancy and water repellency properties of cotton fabrics using silanol based nano composites. Carbohyd Polym 102:727–737
18. Paul R (2015) Improving comfort, performance and protection. Functional finishes for textiles: an overview, pp 1–14
19. Gupta RK et al (2017) Oil/water separation techniques: a review of recent progresses and future directions 5(31):16025–16058
20. Holmquist H (2016) Environmental assessment of alternative durable water repellent chemicals. Chalmers University of Technology
21. Gargoubi S et al (2020) Water repellent treatment for cotton fabrics with long-chain fluoropolymer and its short-chain eco-friendly alternative 111(6):835–845
22. Dalawai SP et al (2020) Recent advances in durability of superhydrophobic self-cleaning technology: a critical review 138:105381
23. Nelson G (2002) Application of microencapsulation in textiles 242(1–2):55–62
24. Schindler WD, Hauser PJ (2004) Chemical finishing of textiles. Elsevier
25. Slater K (1977) Comfort properties of textiles 9(4):1–70
26. Ferrero F, Periolatto M, Tempestini L (2017) Water and oil repellent finishing of textiles by UV curing: evaluation of the influence of scaled-up process parameters 7(5):60
27. Garetti M, Taisch M (2012) Sustainable manufacturing: trends and research challenges. Control 23(2–3):83–104
28. Zahid M et al (2019) Environmentally benign non-wettable textile treatments: a review of recent state-of-the-art 270:216–250
29. Shah MA et al (2022) Applications of nanotechnology in smart textile industry: a critical review 38:55–75
30. Krifa M, Prichard C (2020) Nanotechnology in textile and apparel research–an overview of technologies and processes 111(12):1778–1793
31. Joshi M, Bhattacharyya A (2011) Nanotechnology—a new route to high-performance functional textiles 43(3):155–233
32. Mishra R et al (2014) The production, characterization and applications of nanoparticles in the textile industry 46(2):133–226
33. Rivero PJ et al (2015) Nanomaterials for functional textiles and fibers **10**:1–22
34. Yuce I et al (2022) Titanium and silicon dioxide-coated fabrics for management and tuning of infrared radiation 22(10):3918
35. Pakdel E, Daoud W (2013) Self-cleaning cotton functionalized with TiO_2/SiO_2: focus on the role of silica. Science 401:1–7
36. Montazer M, Pakdel E (2011) Functionality of nano titanium dioxide on textiles with future aspects: focus on wool. Reviews 12(4):293–303
37. Ghosh A (2011) Nano-clay particle as textile coating 11(5):34–36
38. Rehman ZU et al (2021) LBL generated fire retardant nanocomposites on cotton fabric using cationized starch-clay-nanoparticles matrix 274:118626
39. Jeong SA, Kang T (2017) Superhydrophobic and transparent surfaces on cotton fabrics coated with silica nanoparticles for hierarchical roughness 87(5):552–560
40. Bhardwaj N, Kundu SC (2010) Electrospinning: a fascinating fiber fabrication technique 28(3):325–347

41. Pervez MN et al (2022) Nanomaterial-based smart and sustainable protective textiles. Protective textiles from natural resources. Elsevier, pp 75–111
42. Sheng J et al (2017) Robust fluorine-free superhydrophobic amino-silicone oil/SiO_2 modification of electrospun polyacrylonitrile membranes for waterproof-breathable application 9(17):15139–15147
43. Mahmud R, Nabi F (2017) Application of nanotechnology in the field of textile 4(1)
44. Kong X et al (2020) Robust fluorine-free superhydrophobic coating on polyester fabrics by spraying commercial adhesive and hydrophobic fumed SiO_2 nanoparticles 138:105342
45. Zimmermann J et al (2008) A simple, one-step approach to durable and robust superhydrophobic textiles 18(22):3662–3669
46. Yeerken T et al (2019) Chemical stable, superhydrophobic and self-cleaning fabrics prepared by two-step coating of a polytetrafluoroethylene membrane and silica nanoparticles 89(23–24):4827–4841
47. Kuruppu KADD (2021) Surface treated cotton fabric with stain repellent property for the use in aircraft upholstery. Fabrics 16:15589250211005122
48. Raeisi M et al (2021) Superhydrophobic cotton fabrics coated by chitosan and titanium dioxide nanoparticles with enhanced antibacterial and UV-protecting properties. Int J Biol Macromol 171:158–165
49. Ye Z et al (2021) Textile coatings configured by double-nanoparticles to optimally couple superhydrophobic and antibacterial properties 420:127680
50. Cheng Q-Y et al (2017) Sustainable and biodegradable superhydrophobic coating from epoxidized soybean oil and ZnO nanoparticles on cellulosic substrates for efficient oil/water separation 5(12):11440–11450
51. Agrawal N et al (2017) Superhydrophobic palmitic acid modified ZnO nanoparticles 43(16):14271–14276
52. Roach P, Shirtcliffe NJ, Newton MI (2008) Progess in superhydrophobic surface development 4(2):224–240
53. Chieng BW et al (2019) Functionalization of graphene oxide via gamma-ray irradiation for hydrophobic materials. Synthesis, technology and applications of carbon nanomaterials. Elsevier, pp 177–203
54. Kwok DY, Neumann AW (1999) Contact angle measurement and contact angle interpretation. Science 81(3):167–249
55. Erbil HY, Cansoy CE (2009) Range of applicability of the wenzel and cassie−baxter equations for superhydrophobic surfaces. Langmuir 25(24):14135–14145
56. Wang B et al (2012) Advances in the theory of superhydrophobic surfaces 22(38):20112–20127
57. Wang X, Xu B, Chen Z, Del Col D, Li D, Zhang L, Mou X, Liu Q, Yang Y, Cao Q (2022) Review of droplet dynamics and dropwise condensation enhancement: theory, experiments and applications. Adv Colloid Interface Sci 305:102684
58. Yuan Y, Lee TR (2013) Contact angle and wetting properties. In: Surface science techniques. Springer, pp 3–34
59. Chen Q et al (2020) Wicking and drying behavior of knitted cotton fabrics under modified water drop test 1–16
60. Martin S et al (2017) Fabrication techniques for bioinspired, mechanically-durable, superliquiphobic surfaces for water, oil, and surfactant repellency 241:1–23
61. Loghin C et al (2018) Introduction to waterproof and water repellent textiles. Waterproof and water repellent textiles and clothing. Elsevier, pp 3–24
62. Bellanger H et al (2014) Chemical and physical pathways for the preparation of superoleophobic surfaces and related wetting theories 114(5):2694–2716
63. Nadi A et al (2018) Evolution in the surface modification of textiles: a review 50(2):67–108
64. Fabbri P, Messori M (2017) Surface modification of polymers: chemical, physical, and biological routes. Modification of polymer properties. Elsevier, pp 109–130
65. Barthlott W, Neinhuis C (1997) Purity of the sacred lotus, or escape from contamination in biological surfaces. Planta 202:1–8

66. Ivanović L, Vencl A, Stojanovic B, Markovic B (2018) Biomimetics design for tribological applications.
67. Padmanabhan NT, John H (2020) Titanium dioxide based self-cleaning smart surfaces: a short review 8(5):104211
68. Diaa M, Hassabo AG (2022) Self-cleaning properties of cellulosic fabrics (a review) 12(2):1847–1855
69. Nosaka Y, Nosaka AY (2017) Generation and detection of reactive oxygen species in photocatalysis 117(17):11302–11336
70. Tung WS, Daoud WA (2011) Self-cleaning fibers via nanotechnology: a virtual reality 21(22):7858–7869
71. Wang W et al (2019) Durable superamphiphobic and photocatalytic fabrics: tackling the loss of super-non-wettability due to surface organic contamination 11(38):35327–35332
72. Banerjee S, Dionysiou DD, Pillai SC (2015) Self-cleaning applications of TiO_2 by photo-induced hydrophilicity and photocatalysis 176:396–428
73. Pakdel E et al (2020) Advances in photocatalytic self-cleaning, superhydrophobic and electromagnetic interference shielding textile treatments 277:102116
74. Zong L et al (2021) The preparation of superhydrophobic photocatalytic fluorosilicone/SiO_2–TiO_2 coating and its self-cleaning performance 18(5):1245–1259
75. Tudu BK, Sinhamahapatra A, Kumar A (2020) Surface modification of cotton fabric using TiO_2 nanoparticles for self-cleaning, oil–water separation, antistain, anti-water absorption, and antibacterial properties 5(14):7850–7860
76. Forbes P (2008) Self-cleaning materials 299(2):88–95
77. Singh AK, Singh JK (2017) Fabrication of durable super-repellent surfaces on cotton fabric with liquids of varying surface tension: low surface energy and high roughness 416:639–648
78. Zhang X et al (2008) Superhydrophobic surfaces: from structural control to functional application 18(6):621–633
79. Song K et al (2019) Interaction of surface energy components between solid and liquid on wettability, and its application to textile anti-wetting finish 11(3):498
80. Oyola-Reynoso S et al (2015) Revisiting the challenges in fabricating uniform coatings with polyfunctional molecules on high surface energy materials 5(4):1002–1018
81. Shim MH, Kim J, Park CH (2014) The effects of surface energy and roughness on the hydrophobicity of woven fabrics 84(12):1268–1278
82. Zhang S et al (2019) Liquid mobility on superwettable surfaces for applications in energy and the environment 7(1):38–63
83. Wenzel RN (1936) Resistance of solid surfaces to wetting by water. Chemistry 28(8):988–994
84. Bahners T, Gutmann JS (2020) Procedures for the characterization of wettability and surface free energy of textiles-use, abuse, misuse and proper use: a critical review. Adhesives 5:259–293
85. Pan S et al (2018) Coatings super-repellent to ultralow surface tension liquids 17(11):1040–1047
86. Milionis A, Bayer I, Loth E (2016) Recent advances in oil-repellent surfaces 61(2):101–126
87. Naik VM et al (2009) Super functional materials: creation and control of wettability, adhesion and optical effects by meso-texturing of surfaces, pp 129–148
88. Toosi SF et al (2017) Fabrication of micro/nano patterns on polymeric substrates using laser ablation methods to control wettability behaviour: a critical review 5(1):55–78
89. Jucius D et al (2011) Hot embossing of PTFE: towards superhydrophobic surfaces 257(6):2353–2360
90. Kong T et al (2019) Bioinspired superwettability micro/nanoarchitectures: fabrications and applications 29(11):1808012
91. Coblas DG et al (2015) Manufacturing textured surfaces: state of art and recent developments 229(1):3–29
92. Ou J et al (2013) Superoleophobic textured copper surfaces fabricated by chemical etching/oxidation and surface fluorination 5(20):10035–10041

93. Wen G, Guo Z, Liu W (2017) Biomimetic polymeric superhydrophobic surfaces and nanostructures: from fabrication to applications 9(10):3338–3366
94. Liu H et al (2018) Bioinspired surfaces with superamphiphobic properties: concepts, synthesis, and applications 28(19):1707415
95. Peng J et al (2019) Durable self-cleaning surfaces with superhydrophobic and highly oleophobic properties 35(25):8404–8412
96. Chu Z, Seeger S (2014) Superamphiphobic surfaces 43(8):2784–2798
97. Rahmawan Y, Xu L, Yang S (2013) Self-assembly of nanostructures towards transparent, superhydrophobic surfaces 1(9):2955–2969
98. Bayer IS (2020) Superhydrophobic coatings from ecofriendly materials and processes: a review 7(13):2000095
99. Brown PS, Bhushan B (2015) Bioinspired, roughness-induced, water and oil super-philic and super-phobic coatings prepared by adaptable layer-by-layer technique 5(1):14030
100. Huang K-W et al (2022) Super-omniphobic surface prepared from a multicomponent coating of fluoro-containing polymer and silica nanoparticles 173:107174
101. Sharma K et al (2022) A review on challenges, recent progress and applications of silica nanoparticles based superhydrophobic coatings 48(5):5922–5938
102. Liu H-D et al (2019) Superhydrophobic property of epoxy resin coating modified with octadecylamine and SiO_2 nanoparticles 247:204–207
103. Hsieh C-T et al (2009) Super water-and oil-repellencies from silica-based nanocoatings 203(22):3377–3384
104. Ambrozewicz D et al (2013) Fluoroalkylsilane versus alkylsilane as hydrophobic agents for silica and silicates 2013:18–18
105. Belyakova L et al (1999) Surfaces properties of silica gels modified with hydrophobic groups 154(3):285–294
106. Ramezani M, Vaezi MR, Kazemzadeh A (2014) Preparation of silane-functionalized silica films via two-step dip coating sol–gel and evaluation of their superhydrophobic properties 317:147–153
107. Roe B, Zhang X (2009) Durable hydrophobic textile fabric finishing using silica nanoparticles and mixed silanes 79(12):1115–1122
108. Li L et al (2016) Roles of silanes and silicones in forming superhydrophobic and superoleophobic materials 4(36):13677–13725
109. Wu Y et al (2019) Preparation and characterization of perfluorine-SiO_2 nanoparticles and superhydrophobic fluorosilicone/silica hybrid composite coating 125(4):250
110. Yin Y et al (2013) Water-repellent functional coatings through hybrid SiO_2/HTEOS/CPTS sol on the surfaces of cellulose fibers 417:120–125
111. Fu J et al (2019) A facile coating with water-repellent and flame-retardant properties on cotton fabric 43(25):10183–10189
112. Chirila L et al (2020) Hybrid materials based on ZnO and SiO_2 nanoparticles as hydrophobic coatings for textiles 71(4):297–301
113. Borah MP et al (2020) Water repellent finishing on eri silk fabric using nano silica 111(5):701–708
114. Jhatial AK et al (2019) Sol–gel finishing of bamboo fabric with nanoparticles for water repellency, soil release and UV resistant characteristics 26:6365–6378
115. Diamanti MV et al (2013) Photocatalytic and self-cleaning activity of colored mortars containing TiO_2 46:167–174
116. Huang W et al (2009) Superhydrophilic nano-TiO_2 film with porous surface structure 13(4):459–463
117. Pakdel E, Daoud WA, Wang X (2013) Self-cleaning and superhydrophilic wool by TiO_2/SiO_2 nanocomposite 275:397–402
118. Améduri B (2020) The promising future of fluoropolymers. Physics 221(8):1900573
119. Puts GJ, Crouse P, Ameduri BM (2019) Polytetrafluoroethylene: synthesis and characterization of the original extreme polymer 119(3):1763–1805

120. Demir T (2015) Synthesis and characterization of oleophobic fluorinated polyester films. Clemson University
121. Sha M et al (2015) Synthesis and surface properties study of novel fluorine-containing homopolymer and copolymers for coating applications 349:496–502
122. Ekanayake UM et al (2020) Facile fabrication of fluoro-polymer self-assembled ZnO nanoparticles mediated, durable and robust omniphobic surfaces on polyester fabrics 235:109565
123. Jiang C et al (2018) Synthesis of superhydrophobic fluoro-containing silica sol coatings for cotton textile by one-step sol–gel process 87:455–463
124. Verbič A, Gorjanc M, Simončič BJC (2019) Zinc oxide for functional textile coatings: recent advances 9(9):550
125. Kwak G et al (2009) Superhydrophobic ZnO nanowire surface: chemical modification and effects of UV irradiation 113(28):12085–12089
126. Qing Y et al (2015) A facile method to prepare superhydrophobic fluorinated polysiloxane/ZnO nanocomposite coatings with corrosion resistance 326:48–54
127. Textor T, Mahltig B (2010) A sol–gel based surface treatment for preparation of water repellent antistatic textiles 256(6):1668–1674
128. Cohen N et al (2016) Superhydrophobic coatings and their durability 31(9):1143–1155
129. Tania IS et al (2022) Fabrication, characterization, and utilization of ZnO nanoparticles for stain release, bacterial resistance, and UV protection on cotton fabric 17:15589250221136378
130. Kathirvelu S, D'Souza L, Dhurai B (2010) Study of stain-eliminating textiles using ZnO nanoparticles 101(6):520–526
131. Yadav A et al (2006) Functional finishing in cotton fabrics using zinc oxide nanoparticles 29:641–645
132. Portilla L, Halik M (2014) Smoothly tunable surface properties of aluminum oxide core–shell nanoparticles by a mixed-ligand approach. Interfaces 6(8):5977–5982
133. Zhang B et al (2021) Scalable, fluorine free and hot water repelling superhydrophobic and superoleophobic coating based on functionalized Al_2O_3 nanoparticles 66:74–81
134. Abidi N et al (2012) Super-hydrophobic cotton fabric prepared using nanoparticles and molecular vapor deposition methods. In: Functional materials from renewable sources. ACS Publications, pp 149–165
135. Xiao X et al (2015) Durable superhydrophobic wool fabrics coating with nanoscale Al_2O_3 layer by atomic layer deposition 349:876–879
136. Liu S et al (2021) Gold nanoparticles modified graphene foam with superhydrophobicity and superoleophilicity for oil-water separation 758:143660
137. Xue C-H et al (2012) Superhydrophobic conductive textiles with antibacterial property by coating fibers with silver nanoparticles 258(7):2468–2472
138. Bayer IS et al (2011) Water-repellent cellulose fiber networks with multifunctional properties 3(10):4024–4031
139. Bu Y et al (2019) Fabrication of durable antibacterial and superhydrophobic textiles via in situ synthesis of silver nanoparticle on tannic acid-coated viscose textiles 26:2109–2122
140. Zheng X et al (2023) Superhydrophobic hollow magnetized Fe_3O_4 nanospheres/MXene fabrics for electromagnetic interference shielding 934:167964
141. Shim MH, Kim J, Park CH (2015) Development of superhydrophobic fabrics by surface fluorination and formation of CNT-induced roughness 21(1):68–73
142. Ouadil B et al (2019) A new route for the preparation of hydrophobic and antibacterial textiles fabrics using Ag-loaded graphene nanocomposite 579:123713
143. Tissera ND et al (2015) Hydrophobic cotton textile surfaces using an amphiphilic graphene oxide (GO) coating 324:455–463
144. Almufarij RS, Mohamed ME (2023) Green synthesis of a carbon quantum dots-based superhydrophobic membrane for efficient oil/water separation 16(15):5456
145. Partheeban T, Sathish M (2016) Selective growth of fullerene octahedra and flower-like particles by a liquid–liquid interfacial precipitation method for super-hydrophobic applications 6(82):78791–78794

146. Zheng S et al (2015) Facile method toward hierarchical fullerene architectures with enhanced hydrophobicity and photoluminescence 7(36):20285–20291
147. Wei L et al (2012) Fullerene hollow microspheres prepared by bubble-templates as sensitive and selective electrocatalytic sensor for biomolecules 4(3):1594–1600
148. Wang J et al (2009) Self-assembly made durable: water-repellent materials formed by cross-linking fullerene derivatives 48(12):2166–2170
149. Dong W et al (2021) Fabrication of superhydrophobic PET filter material with fluorinated SiO_2 nanoparticles via simple sol–gel process 98:224–237
150. Nasirizadeh N et al (2015) Preparation of hydrophobic and conductive cotton fabrics using multi-wall carbon nanotubes by the sol–gel method 73:14–21
151. Zhang XF, Zhao JP, Hu JM (2017) Abrasion-resistant, hot water-repellent and self-cleaning superhydrophobic surfaces fabricated by electrophoresis of nanoparticles in electrodeposited sol–gel films 4(13):1700177
152. Tao C et al (2016) Sol-gel based antireflective coatings with superhydrophobicity and exceptionally low refractive indices built from trimethylsilanized hollow silica nanoparticles 509:307–313
153. Hashjin RN et al (2022) Tuning up sol-gel process to achieve highly durable superhydrophobic coating 33:e102282
154. Şimşek B, Karaman MJ (2020) Initiated chemical vapor deposition of poly (hexafluorobutyl acrylate) thin films for superhydrophobic surface modification of nanostructured textile surfaces. Research 17(2):381–391
155. Wang J, Wang HJMPB (2017) Robust and durable superhydrophobic fabrics fabricated via simple Cu nanoparticles deposition route and its application in oil/water separation 119(1):64–71
156. Jeong S-W et al (2020) Fabrication of superhydrophobic surfaces based on PDMS coated hydrothermal grown ZnO on PET fabrics 34(1):102–113
157. Khan MZ et al (2021) Development of durable superhydrophobic and UV protective cotton fabric via TiO_2/trimethoxy (octadecyl) silane nanocomposite coating. 112(10):1639–1650
158. Rilda Y et al (2019) Hexamethyldisiloxane-modified ZnO-SiO_2-coated superhydrophobic textiles for antibacterial application 66(6):594–599
159. Aslanidou D, Karapanagiotis I, Panayiotou C (2016) Superhydrophobic, superoleophobic coatings for the protection of silk textiles 97:44–52
160. Celik N et al (2021) Robust superhydrophobic fabrics by infusing structured polydimethylsiloxane films 138(41):51358
161. Sasaki K et al (2016) Asymmetric superhydrophobic/superhydrophilic cotton fabrics designed by spraying polymer and nanoparticles 8(1):651–659
162. Wu M et al (2016) Silver-nanoparticle-colored cotton fabrics with tunable colors and durable antibacterial and self-healing superhydrophobic properties 26(4):569–576
163. Gao Y-N et al (2021) Multifunctional cotton non-woven fabrics coated with silver nanoparticles and polymers for antibacterial, superhydrophobic and high performance microwave shielding 582:112–123

Chapter 10 Nanotechnology in Flame Retardancy of Textile

Sayak Nandi, Ishmam Zahin Chowdhury, and Sumit Mandal

1 Introduction

In the intricate tapestry of our modern lives, the specter of fire hazards looms as a persistent and potentially devastating threat. Fire not only endangers human life, but it also damages material property. According to the NFPA (National Fire Protection Association) in the United States, fire killed 3790 civilians, injured over 13,000 people, and caused over \$18 billion in property damage in 2022 [1]. Furthermore, the impact of fire dangers goes beyond property and life loss. According to recent research, fire also produces several carcinogens and toxins [2, 3]. These toxins and carcinogens enhance the risk of several health problems and even lead to cancer [4]. Thus, fire poses a long-term harm to human health in addition to an immediate threat to human ecology. Furthermore, the ever-changing technological and industrial scene provides new aspects to fire dangers. Thus, there is a necessity for new avenues for flame-retardant solutions to meet the new changing landscape of the fire.

Flame-retardant textiles have been the widely accepted solutions for prevention of fire and its spreading. The technology of flame retardancy has also evolved, and currently there are numerous technologies available to obtain flame-retardant finishes. Natural cellulosic materials, protein materials, and other synthetic materials make up the textile market. These polymeric materials are excellent for its end use as a textile; however, most of textile materials in everyday usage are flammable. As a result, flame retardancy must be added in some form to make it more suitable for end-use applications that include any type of fire threat. From imparting fire-retardant coating [5–8], to developing intrinsically flame-resistant fiber materials [9], to the creation of new innovative materials such as aerogel [10], the technique of imparting flame retardancy has seen its fair share of advancement. This advancements in the flame retardancy have witnessed new and innovative mechanisms for achieving

S. Nandi · I. Z. Chowdhury · S. Mandal (✉)
Department of Design and Merchandising, Oklahoma State University, Stillwater, OK, USA
e-mail: sumit.mandal@okstate.edu

S. Bairagi et al. (eds.), *Nanotechnology in Textile Finishing*, Materials Horizons: From Nature to Nanomaterials, https://doi.org/10.1007/978-981-97-2696-7_10

flame retardancy and have also witnessed a surge in the application of nanotechnology owing to its higher specific surface area for achieving flame retardancy. In the domain of flame retardancy, nanotechnology provides superior fire protection [11, 12] with minimal impact on comfort properties [13] and lower weight [8]. These attributes are particularly advantageous for the development of protective gear for firefighters or for workers in the oil and gas industry [14, 15].

2 Basic Mechanism of Flame Retardancy

2.1 Polymer Combustion Process

In the process of polymer combustion, complex interactions of chemical reactions and physical phenomena occur, resulting in heat and mass transfer. The combustion and spread of fire rely on the availability of heat, fuel, and oxygen. The combustion process can be outlined into three distinct segments: the pyrolysis zone, the charring layer, and the burning zone [16].

In the pyrolysis zone, the polymer is subjected to flame radiation, conduction, and convection. The temperature of the burning polymer substrate rises to a certain threshold, initiating the breakdown of polymer bonds. This leads to surface regression and the endothermal decomposition of the polymer, releasing flammable volatiles [17, 18]. Here, gaseous products react with atmospheric oxygen, producing free radicals such as H· and OH· [19]:

$$\text{Initiation} : \text{RH} \rightarrow^{\circ} \text{R} +^{\circ} \text{H} \tag{1}$$

$$\text{Propagation} :^{\circ} \text{H} + \text{O}_2 \rightarrow^{\circ} \text{OH} +^{\circ} \text{O} \tag{2}$$

$$\text{Propagation} :^{\circ} \text{OH} + \text{CO} \rightarrow \text{CO}_2 +^{\circ} \text{H} \tag{3}$$

$$\text{Propagation} :^{\circ} \text{H} + \text{O}_2 \rightarrow^{\circ} \text{HO}_2 \tag{4}$$

$$\text{Termination} :^{\circ} \text{R} +^{\circ} \text{HO}_2 \rightarrow \text{R} - \text{HO}_2 \tag{5}$$

$$\text{Termination} :^{\circ} \text{R} +^{\circ} \text{R} \rightarrow \text{R} - \text{R} \tag{6}$$

The decomposition also results in the formation of a condensed charred layer, which can either release volatiles generated in the burning zone or suppress them. From the burning zone, free radicals reacting with oxygen release more combustible mixtures, acting as an additional fuel supply for the pyrolysis process. Consequently, the heat released from the burning polymer gradually increases to a level where an

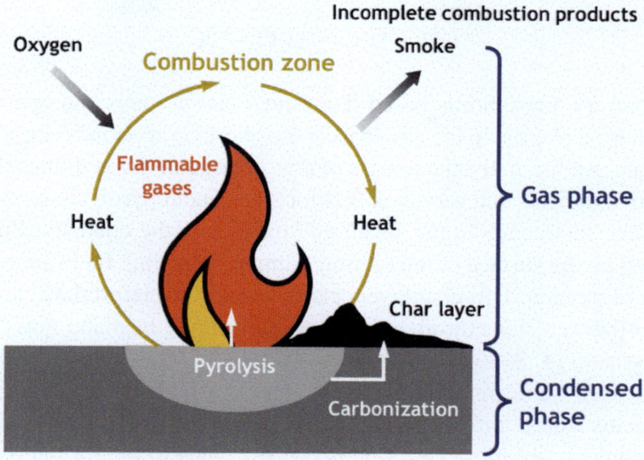

Fig. 1 Combustion process; taken from [23]

external heat source is no longer required for further combustion. This establishes a self-sustaining combustion process for the polymers [20–23]. Figure 1 depicts a schematic diagram of the combustion process. To limit the polymer combustion process, flame-retardant systems are being integrated into the polymer substrate.

2.2 *Conventional Flame-Retardant Mechanism*

Flame retardants serve as additives aimed at restricting the combustion process of polymers. These systems exert their influence by the byproducts generated in ongoing combustion reactions. These byproducts can function through physical means like fuel dilution, endothermic cooling, or forming a physical barrier to the reactive species, or through chemical actions such as flame inhibition, charring, intumescence, or melt dripping [24]. The mechanisms of flame retardant (FR) actions can be explained into gas and condensed phases. Based on the principle of intumescence, FR systems may further be categorized into non-intumescent and intumescent. Non-intumescent systems often incorporate halogen, phosphorus, nitrogen, silicone, and/or inorganic metal compounds [5, 8, 25, 26].

Halogen-based flame-retardant systems primarily act as scavengers of free radicals in the gas phase. During pyrolysis, radicals like Br· and Cl· are released from C–X bond breakage (X refers to the halogen atoms), absorbing highly reactive species like H· and OH· from the chain oxidation reactions [16, 19], as described by the following general reactions:

$$^{\circ}H + HX \rightarrow H_2 + ^{\circ}X \tag{7}$$

$$^{\circ}OH + HX \rightarrow H_2O + ^{\circ}X \tag{8}$$

On the contrary, phosphorus-based flame-retardant systems can operate in both gas and condensed phases. In the gas phase, phosphorus compounds can scavenge H· and O· radicals. Additionally, the release of non-flammable gases dilutes the concentration of active radicals and serves as fuel for subsequent pyrolysis, contributing to imparting flame retardancy to the substrate [16, 27]. In the condensed phase, char layers develop on the surface of the burning sample, impeding fuel vaporization and the diffusion of oxygen. This char layer acts as insulation against the generated heat, restricting further reactions through pyrolysis by limiting heat and mass transfer in the burning system [5, 8, 26].

Nitrogen-based flame retardants also function in both the condensed and gas phases. Systems based on compounds like melamine release ammonia during burning, diluting combustible gases. This process leads to the formation of a thermally stable layer due to product species condensation. Similarly, silicon-based FR systems create thermally stable layers that act as coatings, hindering heat and species flow. Reactions with pyrolyzed product species generally follow endothermic pathways, minimizing further heat generation [25, 28].

Intumescent flame retardants primarily operate in the condensed phase. The highly foamed carbonaceous layer (char) serves as insulation for the substrate. They typically consist of three components: a char-forming carbonizing agent like pentaerythritol or PA6, an acid source functioning as a dehydrating catalyst for the char-forming agents (such as a polyphosphate system), and a blowing agent like melamine or guanidine, which aids in expanding and swelling the char layer during thermal degradation [20, 25, 29]. Here, the quality and yield of the char significantly influence the FR performance. Higher char yield indicates greater accumulation of non-flammable residues, while improved char quality refers to a more continuous, compact, and uniform distribution of char over the polymer surface below, providing higher efficiency insulative attributes [16, 30–32].

2.3 General Mechanism of Nano-Based Flame-Retardant System

The limitations of conventional flame-retardant systems, such as their tendency to produce toxic gases and high levels of smoke, have spurred the search for innovative materials to serve as flame-retardant systems in polymers. Concerns have arisen regarding halogen-based systems due to their carcinogenic and greenhouse properties, and phosphorus-based compounds due to the link to the generation of formaldehyde, raising significant health and environmental issues [28, 33, 34].

In response, flame-retardant nanoparticles have emerged as an alternative to traditional FR systems. Nanoparticles have become a heavily researched area due to their potential to mitigate environmental problems while improving mechanical and thermal insulative properties [34–36]. Nanoscale flame retardants (size $\leq$ 100 nm) exhibit enhanced thermal stability, reduced flammability, and decreased smoke release, largely due to their superior surface-to-volume ratio.

During combustion process, nanoparticles exert FR influence through both physical means, like ceramization, and chemical actions, including catalyzed charring. Polymer ablation leads to the creation of networked floccules of nanoparticles. These networks serve as thermally stable ceramic surface layers, acting as shields against surface re-irradiation and as barriers to heat and oxygen transfer from the flame to the material, as well as the degradation products from the material to the flame. For example, nanosized fillers such as carbon nanotubes, graphene, graphene nanosheets, and nanoclays, when used as flame retardants in the polymer matrix, generate a physical barrier effect. They create a maze-like microstructure during burning, limiting the diffusion of volatile gases in the polymer matrix. Moreover, networks of floccules formed from these systems further restrict heat and mass transfer [37, 38].

The applications of nanoparticles over polymer substrates have also shown enhanced char quality and yield. This improvement arises from superior interfacial contact and interaction between the polymer and nanoparticles [11, 16, 39, 40]. The presence of nanoparticles contributes to the partial charring of the polymer, which in turn limits the production of combustible volatiles. For instance, polymer matrices like PA6 burn without charring, but the incorporation of 5 wt.% nanoclay induces extensive charring at the polymer surface [41].

Current research focuses heavily on the compatibility between nanoparticles and textile polymer substrates in flame-retardant applications. The dispersion of nanoparticles over textile polymer substrates is a key area of investigation, with modifications targeting optimal FR solutions through interactions such as van der Waals forces, hydrogen bonding, and electrostatic forces [5, 42]. Subsequent sections discuss the application nanoparticles in the FR applications of textile-based polymer substrate.

3 Nanoparticles

Nanoparticles constitute diminutive entities with dimensions on the order of 10^{-9} m. Characterized by their exceedingly minute size, these particles afford a higher specific surface area. This higher specific surface area is linked to superior physical properties, furthermore, offering superior efficiency in comparison with larger particulate counterparts [43]. The higher efficiency of nanoparticles compared to the larger particles has led it to be one of the major forefronts in the development in this era.

Nanoparticles with its higher specific surface area offers far superior properties in several aspects compared to the commercially available textile products. Moreover, a notable advantage of nanoparticles lies in their inherent capacity for fundamental characteristic alterations corresponding to their morphology. The mechanical, thermal, physical, and bioactivity properties of nanoparticles exhibit variability contingent upon their size and shape [43–45]. Consequently, nanoparticles present

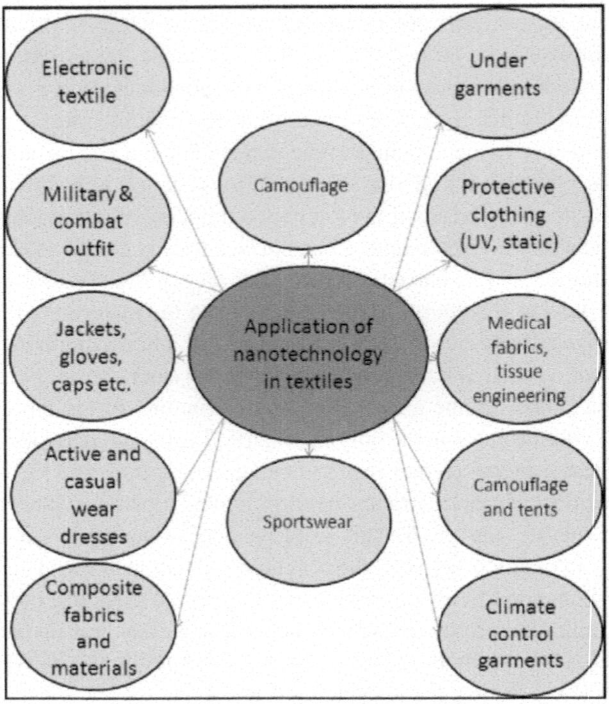

Fig. 2 Application of nanotechnology in textiles [45]

exceptional advantages in the development of engineered structures for applications in smart textiles. Hence, nanoparticles have been a major focal point of innovations in this era. Nanoparticles have seen a potential application in a plethora of smart textile applications as shown in Fig. 2.

Given that the characteristics of nanoparticles are contingent upon their morphology, and the morphology, in turn, is influenced by the manufacturing process and nanoparticle type. A profound comprehension of the operational mechanisms and manufacturing processes of nanoparticles is imperative for their effective utilization. Tailoring nanoparticles to specific tasks necessitates a meticulous consideration of their inherent strengths and limitations, contingent upon their diverse shapes produced by different synthesis methodologies. In addition to different synthesis techniques, the properties of nanoparticles are majorly governed by the type of nanoparticles synthesized. The classification of nano-materials are given in Fig. 3.

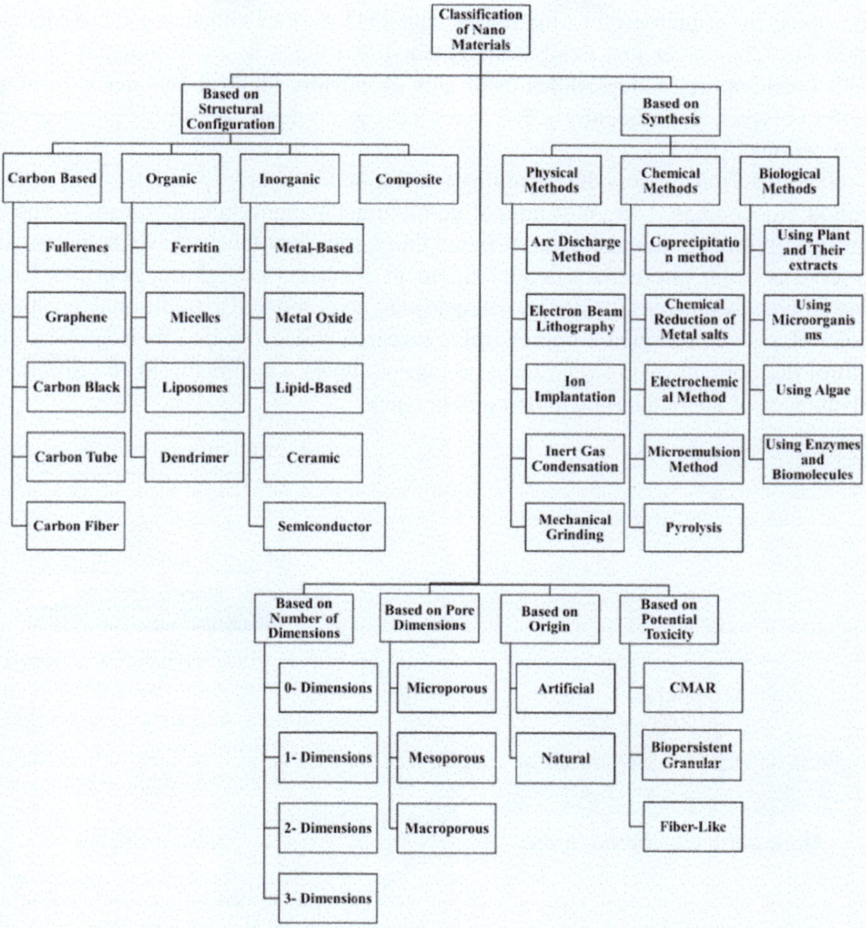

Fig. 3 Classification of nanomaterials

4 Why Nanoparticles in Flame-Retardant Finishing

Nanoparticles confer numerous advantages in the context of flame-retardant finishing applications. Since, nanoparticles with higher specific surface area correlate to higher surface energy [46]. The higher surface energy of nanoparticles amplifies the potential for energy absorption, thereby raising the energy absorption for surface ignition. Since the initial ignition occurs at the surface nearest to the heat source [47]; hence, a material with higher surface energy will require more energy to ignite.

Nanoparticles exhibit higher thermal conductivity compared to the bulk materials [48, 49]. Higher thermal conductivity of nanoparticles assists in efficient dissipation of heat throughout the material, contributing to the formation and sustenance of a protective char layer [50]. In addition to assisting the creation of char, nanoparticles contribute to an increased formation of char [51], further enhancing the development of a char layer that exhibits heightened resistance to environmental factors [52]. Furthermore, nanoparticles itself acts as barriers for heat and mass transfer to the polymer, consequently it has been used as coolants in various engineering applications [5, 8, 53].

Nanomaterials possess the capability to offer better comfort properties and lower weight [8] compared to conventional commercial flame-retardant finishes. Thus, through optimization, nanomaterial-based flame-retardant fabrics have the potential to serve as viable alternatives in the refinement of current firefighting gear [54]. The substantial advantages afforded by nanoparticles, as compared to traditional finishing materials, have prompted comprehensive research endeavors into their application within the domain of fire retardancy. Figure 4 shows a schematic of the different advantages of nanomaterials in flame retardancy.

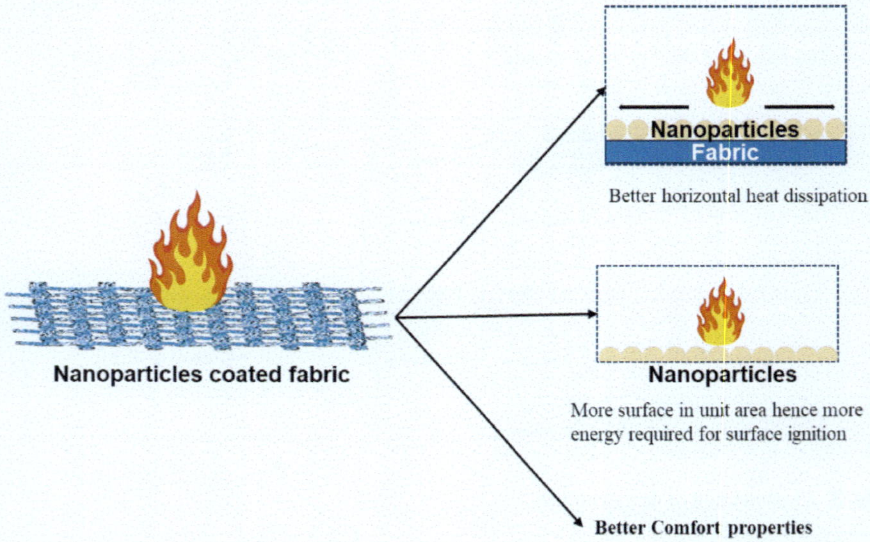

Fig. 4 Schematic advantages of nanoparticles in flame retardancy

5 Different Nanomaterial Finishes in Textiles

5.1 Nanoparticles

5.1.1 Oxide and Hydroxide Nanoparticles

Nanoparticles composed of hydroxides and oxides play a pivotal role in the gas phase flame retardation mechanism. During combustion, these nanoparticles hinder the formation of radicals necessary for combustion (Eqs. 1–6) [55]. Even some of the radicals released due to its combustion mixes with other compounds to release water vapor and water-based compounds. This hindrance in formation of desired radicles and emission of water vapor-based compounds leads to the dilution of combustion gases, thereby hindering the combustion process and impairing the propagation of flames [23, 56]. Oxide-based nanoparticle compounds of silica [55], titanium dioxide [57], zinc oxide [58], phosphorous [32], and aluminum [56] have been widely used as flame-retardant finishes.

A similar mechanism of gas phase flame retardancy is also followed by the halogens, rather halogens being highly electronegative, they attach with the electropositive ions generated during combustion, forming water molecules and stable chemical compounds [55]. This mechanism makes halogen better flame-retardant materials compared to oxides and hydroxides. However, the halogens have been linked to generation of poisonous gases and carcinogens, hence are not generally used as flame-retardant materials.

5.1.2 Natural-Based Nanomaterials

Naturally obtained polymeric materials and its derivatives have been used as an ecofriendly flame-retardant material. Lignin [59], tannin [60], chitosan [61], DOPO groups [12], sodium alginate [62], phytic acid [63], bio-based carboxylic acid [64], cyclodextrin [65], starch [66], and eggshell [67] have been among the natural sources for natural flame-retardant polymers (Fig. 4).

Within the polymers depicted in Fig. 5, a majority of naturally derived polymers feature substantial benzene-based polymeric groups or exhibit a cyclic polymer structure. Consequently, the degradation of these structures necessitates a considerable amount of energy. Moreover, natural polymeric materials are characterized (Fig. 5) by the presence of phosphorous, nitrogen, sulfur, or hydroxyl groups, serving as gas phase flame-retardant components as shown in Eqs. (9)–(15) [9, 61, 68].

$$NH/NH_2 \rightarrow^{\circ} N + n^{\circ}H \tag{9}$$

$$^{\circ}N + 3^{\circ}H/^{\circ}N \rightarrow NH_3/N_2 \tag{10}$$

$$\text{Substance} + \text{Flame} \rightarrow {}^{\circ}\text{PO} + {}^{\circ}\text{PO}_2 \quad (11)$$

$$^{\circ}\text{PO} + {}^{\circ}\text{O} \rightarrow {}^{\circ}\text{PO}_2 \quad (12)$$

$$^{\circ}\text{PO}_2 + {}^{\circ}\text{O} \rightarrow \text{PO}_3 \quad (13)$$

$$^{\circ}\text{PO}_2 + {}^{\circ}\text{H}/{}^{\circ}\text{OH} \rightarrow \text{HPO}_2/\text{HPO}_3 \quad (14)$$

$$\text{HPO}_2/\text{HPO}_3 + \text{H}_2\text{O} \rightarrow \text{H}_3\text{PO}_3/\text{H}_3\text{PO}_4 \quad (15)$$

In the aspect of naturally derived materials, only a limited polymers have been employed in nano-form. Additionally, in numerous investigations, derivatives of natural polymers have been utilized as nano-finishes. Nevertheless, in various instances, natural finishes have been implemented at the micro-level, and their application at the nanoscale remains an area yet to be explored in research.

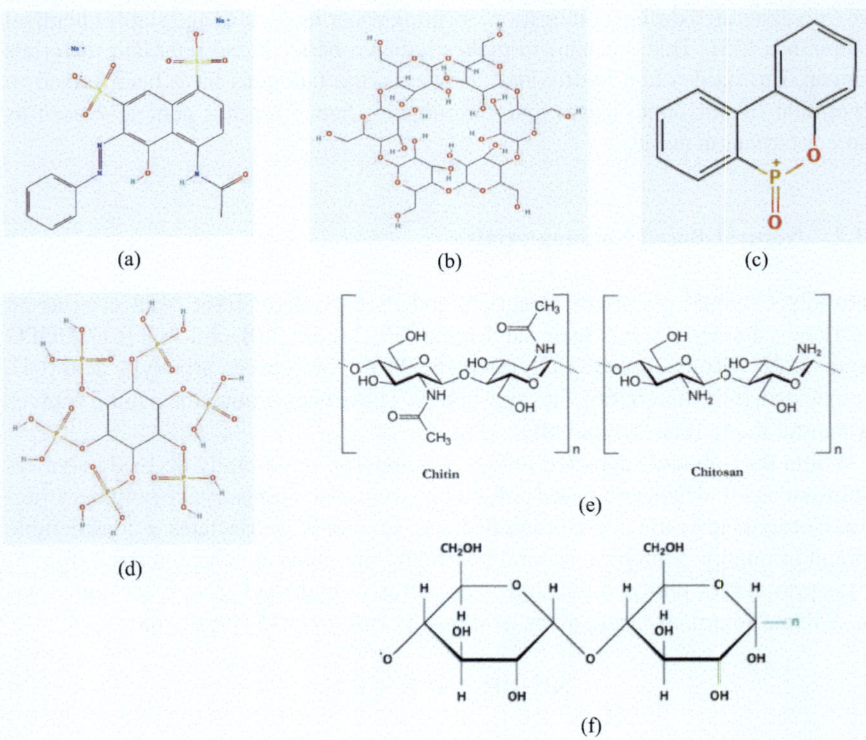

Fig. 5 Chemical structures of naturally derived polymers, **a** Lignin [69], **b** Cyclodextrin [70], **c** DOPO [71], **d** Phytic acid [72], **e** Chitin and chitosan [61], **f** Polymeric starch monomer [73]

5.2 Carbon Nanotubes (CNT)

Carbon nanotubes (CNTs) have crystalline tubular carbon structures, similar to rolled-up graphite sheets composed of a hexagonal network of carbon atoms [74, 75]. CNTs exhibit notable flexibility due to their exceptional Young's modulus and axial tensile strength [76], enabling them to be tightly knotted without breaking, a unique trait compared to conventional carbon fibers [77, 78]. Their exceptional thermal conductivity, mechanical strength, and electrical properties have positioned them as valuable additives in various polymer-based nanocomposites [5, 79]. Following Fig. 6 depicts the architecture of the CNTs.

CNT manufacturing techniques include high-temperature evaporation of solid carbon feedstock through electric arc-resistive heating (electric arc-discharge process), high-temperature laser evaporation, and medium-temperature catalytic decomposition of light hydrocarbons (from methane to benzene) via pyrolysis or chemical vapor deposition (CVD) processes. Processes such as plasma CVD and high-pressure CO conversion (HiPco) also yield single-wall CNTs (SWCNT) or multi-wall CNTs (MWCNT) [5, 41].

Various techniques, including dip coating, layer-by-layer assembly, chemical functionalization, electrospinning, sol–gel coating, and vapor deposition, are employed to incorporate carbon nanotubes into polymer substrates, enhancing their flame-retardant properties [5, 8]. In polymer/CNT systems, flame retardation primarily occurs through the formation of char layers, acting as heat barriers and thermal insulators that delay polymer degradation by re-emitting radiation to the gas phase. Additionally, CNTs enhance thermal conductivity, increase melt viscosity, and act as free radical scavengers in polymer nanocomposites. Research into polymer/CNT nanoarchitectures has unveiled a protective network layer, shielding polymers

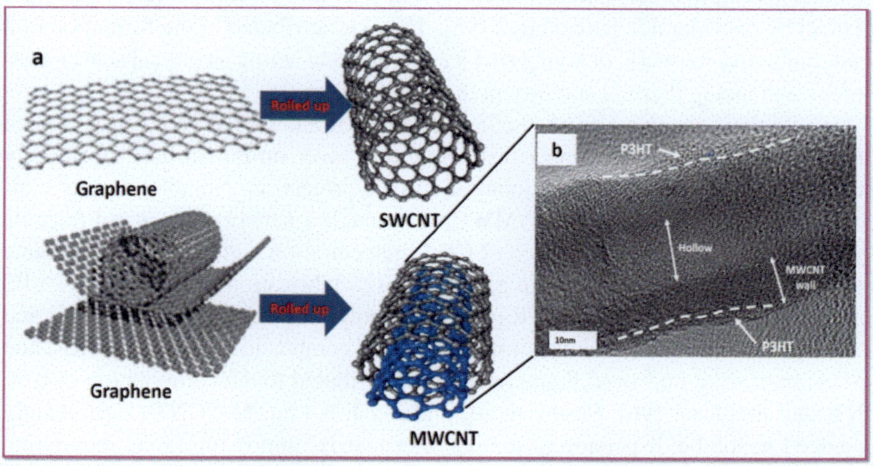

Fig. 6 CNT architecture [80]

from radiant heat flux during combustion [81, 82]. However, while conventional tests show complete polymer burning, leaving behind a CNT layer, this layer doesn't notably enhance char formation, unlike clays [83]. Cone calorimeter tests on PBT and its nanoarchitectures revealed reduced peak heat release rate (PHRR) values with functionalized diphenyl phosphoryl chloride (DPP(O_x)–A–CNT) inclusion, indicating up to a 29% decrease in flammability. These nanoarchitectures suppressed heat release and smoke production by promoting char formation [84]. Covalently grafted polysilicone FR (PA) on CNT surfaces reduced HRR in thermoplastic elastomers. CNT on PA inclusion significantly improved fire resistance compared to pristine materials [85, 86].

Surface-functionalized ammonium polyphosphate (APP) and carboxylic-functionalized MWCNTs in high-impact polystyrene (HIPS) reduced heat release and smoke production rates, creating dense char layers that enhanced fire resistance and thermal stability [85, 87]. PP@TiO_2@CNT nanoarchitectures exhibited enhanced fire resistance compared to pristine polypropylene. Higher load percentages of TiO_2@CNT led to reduced peak heat release rates, indicating improved fire resistance [88]. Polystyrene nanoarchitectures incorporating modified MWCNTs (PAC) showed significantly decreased HRR and smoke production rates, with compact char residues acting as barriers against combustible gases and thermal radiation, enhancing fire resistance [89, 90]. These findings underscore the potential of nanoarchitectures in significantly improving flame retardancy in polymers, offering promising pathways for enhanced fire safety [84, 91, 92].

Applications of CNT on textile-based polymers reported higher flame retardation with higher CNT content. Recent studies stabilized CNTs on cotton fabric, resulting in a polyvinylphosphonic/CNT nanocomposite that improved thermal stability and reduced flammability by leveraging CNTs' heat resistance, insulation effects, and mass transport barriers [93, 94]. Incorporating CNTs into PA6/glass fiber laminates significantly improved time to ignition (TTI) and the duration of peak heat release rate (pHRR) by considerable percentages [95]. This was attributed to the formation of a heat-conductive network, delaying surface ignition and acting as radical scavengers, thereby enhancing thermal stability in the nanocomposite.

Despite their positive effects, the impact on HRR and pHRR values was limited due to the absence of a compact insulating char layer on the surface, potentially requiring higher CNT concentrations for surface protection. Notably, reports indicated a reduction in pHRR of PP/MWCNT through a network-structured floccule layer, emphasizing the significance of CNT concentration and dispersion for flame retardation [78]. Advancements in nanocomposite fabrication introduced novel PP compositions by incorporating IFR (melamine polyphosphate and PER), CNT, and graphene. This composite demonstrated decreased combustion behavior, pHRR, and ASEA, indicating improved flame retardation attributed to the combination of IFR, CNT, and graphene [96]. Furthermore, grafting IFR onto MWCNTs significantly enhanced nanotube dispersion in the polymer matrix, improving flame retardation efficiency [5, 97].

5.3 Nanoclay

Nanoclay, primarily aluminum phyllosilicates, encompasses several metal oxides, including magnesium, calcium, and other metal oxides. Primarily, nanoclays can be divided into two types based on the arrangement of their alternating "SiO_2" and "AlO_6" sheets: 2:1 clay (like smectite and vermiculite) and 1:1 clay (such as kaolinite) [98, 99]. The 2:1 clay, exemplified by montmorillonite within the smectite family, possesses remarkable intercalation–exfoliation properties owing to its distinctive layered structure. Its high aspect ratio and unique interlayer action make it an influential reinforcement component for textile polymers. In contrast, 1:1 clay minerals are less favored in polymers due to their limited intercalation and exfoliation capabilities in individual layering [100, 101].

The crystalline structure of phyllosilicates entails the fusion of tetrahedral silica layers with metal atoms, generating octahedral metal oxide layers. These layers, separated by van der Waals gaps known as galleries or inter-strata, house cations that neutralize the negative charges resulting from isomorphous substitutions within the crystal. These metal cations, often alkaline or alkaline earth elements, are typically hydrated. Montmorillonite, a prominent member of the smectite group composition, predominantly consists of 1-nm-thick 2:1 silicate layers, where an aluminum octahedral sheet is sandwiched between two silica tetrahedral sheets, linked by shared oxygen atoms. Sodium and calcium ions typically reside within these layered galleries, contributing to the structure's characteristics. Figure 7 depicts such 2:1 layered silicate. The tight packing of layers in nanoclays, especially in montmorillonite, presents challenges for interlayer hydration or penetration by polymer chains due to increased charge density [41, 98, 102–104].

Nanoclay's inherent covalent bonds within its layers and its hydrophilic nature pose challenges for effective dispersion within polymer matrices. Modifying clay particles through surfactant intercalation increases interlayer spacing, imparting hydrophobic characteristics that facilitate fine dispersion within polymer matrices. Two primary modification techniques exist: physical and chemical. Physical modification involves chemical adsorption onto the nanoclay surface, offering marginal

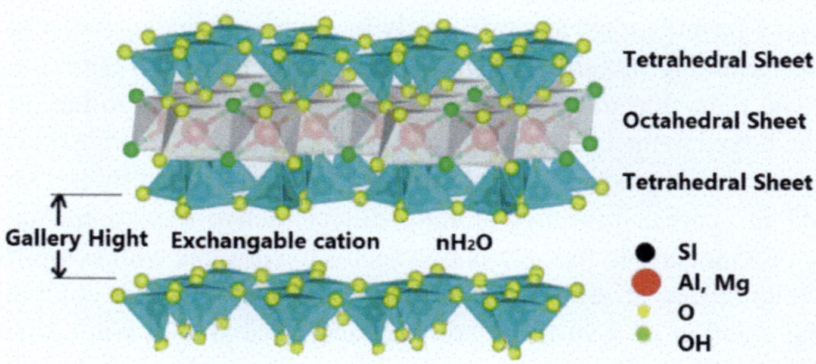

Fig. 7 Structure of 2:1 layered silicate [98]

enhancements to composite characteristics due to weak van der Waals forces. In contrast, chemical modification fosters chemical reactions between layered silicates and modifiers, often using cationic or anionic functional groups to enhance clay's dispersibility in polymers. This modification results in improved mechanical, thermal, optical, and barrier properties in polymer/nanoclay composites, attributed to the nano-thin clay layers, increased surface area, and higher aspect ratio [105–107]. Three common methods—solution blending, melt blending, and in situ polymerization—facilitate the incorporation of modified nanoclay into polymer matrices, each with distinct processing steps and advantages in achieving efficient nanocomposite formation [108–110].

Nanoclays, like montmorillonite (MMT), enhance polymer properties, particularly flame retardancy, due to their high aspect ratio, substantial surface area, and superior dispersion at the nanoscale. Surface modification with organic cations increases interlayer space, fostering compatibility with the polymer matrix. Flame retardation primarily occurs through intercalated or exfoliated structures. Exfoliated nanocomposites exhibit prolonged ignition times due to the lower decomposition temperature of organically modified MMT (OMMT), which forms stable char residues [20, 111].

Figure 8 illustrates the behavior of the polymer/MMT nanocomposite during combustion. As the temperature rises, the viscosity of the molten polymer/clay blend decreases, aiding the movement of nanoclays toward the surface. Heat initiates the breakdown of organomodifiers, creating protonic catalytic sites on the nanoclay surface, fostering stable char formation and polymer chain cross-linking. Gas bubbles generated from the decomposition of organomodifiers and polymer chains enhance the upward movement of MMT layers. This accumulation of nanoclays on the composite surface forms a protective barrier, limiting heat transfer, oxygen diffusion, and the release of combustible byproducts. Additionally, network formation within the composite enhances melt viscosity, constrains thermal polymer chain movement, and facilitates free radical entrapment during thermal degradation. These mechanisms collectively contribute to the flame retardation of the nanoclay-enhanced polymer [5, 102, 112, 113].

Despite their utility, technical and economic obstacles hinder the widespread use of polymer–clay nanocomposites. Challenges include addressing impact resistance, complex formulations, and effectively measuring clay dispersion in polymer matrices, necessitating investment in advanced equipment and research endeavors [98, 102]. The reduction in time to ignition (TTI) of polymers in the presence of MMT can be linked to the breakdown of MMT surfactants. This breakdown leads to earlier polymer decomposition, partly due to hydrolysis caused by H_2O molecules within MMT or generated through alumino-silicate dehydroxylation. Additionally, the efficacy of nanoclays in fire retardation seems to correlate with composite thickness. This relationship arises from a competitive process: thicker composites tend to foster the creation of a surface carbonaceous-silica shield, while thinner ones favor the volatilization of surrounding polymer, serving as fuel. Thicker composites encourage shield formation, inhibiting combustion, whereas in thinner ones, polymer volatilization dominates the process [113–115].

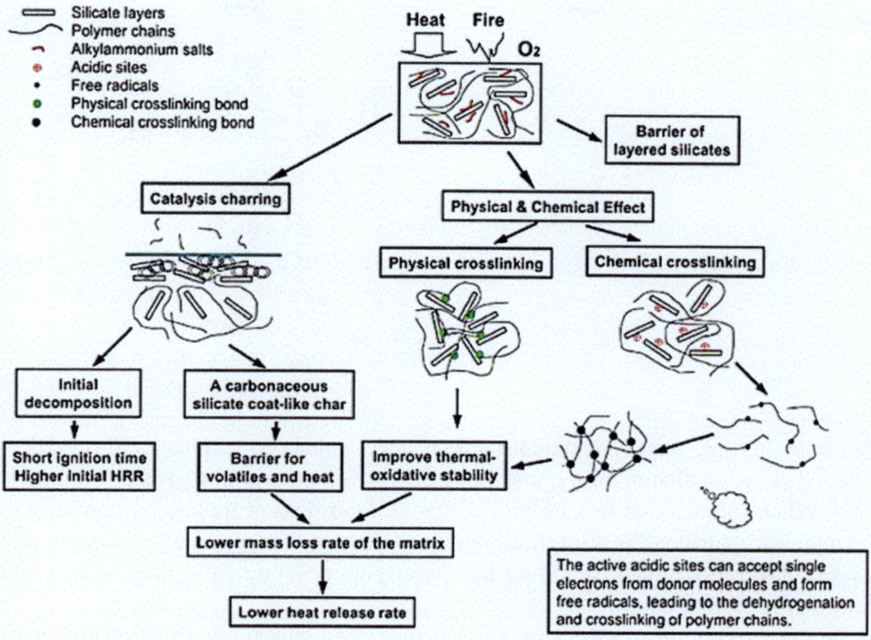

Fig. 8 Flame-retardant mechanism of nanoclay-treated polymer [113]

5.4 POSS

POSS, or polyhedral oligomeric silsesquioxane, has a cage-like structure (0.7–50 nm) of varying symmetry, primarily comprising silicon and oxygen (Si–O) molecules along with eight organic groups. When introduced into polymers, these molecules link to the organic components within the polymer matrix, creating elongated nanostructures. Their general formula, $(RSiO_{1,5})_n$, where n is an even number, and R represents an organic group, places them between silica (SiO_2) and silicones $(R_2SiO)_n$ [41, 116, 117]. Figure 9 charted POSS structure.

Unlike post-synthesis modifications in silica and clays (e.g., ion-exchange in clays), synthetic procedures for POSS enable the creation of chemically defined, crystalline hybrid nano-fillers through one-pot preparations. POSS can possess organic pendant groups tailored to interact favorably with the polymer matrix, copolymerize during polymer synthesis, or be grafted onto the polymer structure during processing [41]. POSS's chemical and affinity properties can be fine-tuned using a mix of differently functionalized precursors. Moreover, they can be molecularly modified to include various transition metals like Ti, Zr, V, Mo, Cr, Hf, Al, Sn, Sb, Y, and Nd within their polyhedral cage structures, creating potential catalytic centers with well-defined geometric characteristics [41, 119, 120].

In the realm of hybrid and nano-composite materials, POSS has proven effective in enhancing polymer properties. These enhancements encompass elevated

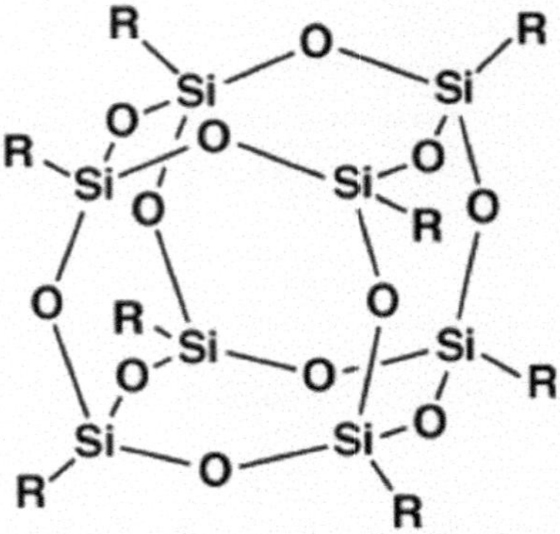

Fig. 9 POSS structure [118]

use temperatures, improved oxidation resistance, enhanced mechanical properties, reduced polymer flammability, and lowered viscosity during processing [41, 121, 122]. When subjected to fire, POSS undergoes controlled degradation, migrating to the material's surface. This migration process leads to the formation of a protective ceramic layer, effectively shielding the underlying polymer from intense heat and flames [123].

The incorporation of POSS into nanocomposites initiates a series of processes that enhance fire resistance. During combustion, POSS degradation occurs, releasing radicals that contribute to the formation of an intumescent char layer. This char layer expands into a multicellular structure, thanks to specific POSS variants like dodecaphenyl POSS (DP-POSS), creating foliated layers. These layers act as insulating barriers, significantly retarding the spread of flames and heat transfer [124, 125]. Moreover, the presence of metal atoms within POSS structures, as observed in PP/Al-POSS nanocomposites, influences the polymer degradation pathway during combustion. This interaction with metal moieties, such as aluminum, catalyzes secondary reactions that partially char the polymer instead of complete volatilization. The resulting carbonaceous char and graphene phase further contribute to improved fire retardation [5, 123].

The flame-retarding effects of POSS extend to various nanocomposites and textiles. Incorporating different types of POSS variants into polymer matrices (as filler material) like polypropylene (PP) or linking them to cotton fabrics consistently demonstrates enhanced flame resistance [5, 8, 126, 127]. This is evident through reduced peak heat release rates (pHRR), total heat evolved (THE), and improved thermal stability in textiles, surpassing the efficacy of conventional flame retardants like zinc phosphinate or boehmite [128]. Alongi et al. investigated cotton fabric treatment with POSS using dimethylol dihydroxy ethylene urea (DMDHEU) as a binder. Their findings showcased enhanced thermal stability, evident in prolonged time to ignition and reduced peak heat release rate (PHRR) of the treated fabric. The heightened carbonization at higher temperatures led to the formation of an insulating layer, impeding heat transfer and surpassing the effectiveness of bohemite nanoparticles [128]. Li et al. employed layer-by-layer deposition of octamethyl POSS and octaTMA POSS on cotton fabric, resulting in a substantial reduction in afterglow during vertical flame testing. This method significantly enhanced the fabric's flame retardancy, decreasing total and peak heat release rates compared to untreated fabric [129].

Carosio et al. utilized the layer-by-layer technique to treat polyester fabric with α-zirconium phosphate and n-propyl ammonium chloride POSS. This treatment notably improved time to ignition, remarkably reducing PHRR while mitigating smoke density and harmful gas emissions like carbon monoxide [130]. Carosio and Alongi targeted acrylic fabric treatment through layer-by-layer application of POSS and ammonium polyphosphate-like compounds. This strategy curbed melt dripping by combining char-forming properties of ammonium polyphosphate with the thermal insulation effect of POSS. Although this resulted in a slight reduction in Limiting Oxygen Index (LOI) values, it diminished HRR and pHRR values [131]. Chen et al. investigated cotton fabric treatment with branched poly(ethylenimine) (bPEI), ammonium polyphosphate (APP), and fluorinated-decyl polyhedral oligomeric silsesquioxane (F-POSS) via a solution dipping method. Their work demonstrated substantial char formation and an intumescent effect on the treated fabric, establishing a protective barrier against flames. The inclusion of POSS aimed to impart hydrophobic properties, enhancing flame resistance [132].

6 Application of Nanoparticles into Textile Substrate

6.1 *Surface Coating*

6.1.1 Pad-Dry-Cure

The pad-dry-cure method is widely used to enhance flame retardancy in textiles. This process involves creating a solution with flame-retardant nanoparticles, specifically designed for their fire-resistant properties. After carefully preparing this nanoparticle solution, it is applied to the fabric using a padding mechanism to ensure an even distribution (Fig. 10a). Following the padding phase, the fabric undergoes a drying process to remove the solvent, resulting in a fabric infused with flame-retardant nanoparticles. The final step is a crucial curing stage, where the fabric is exposed to elevated temperatures. This curing step is essential as it initiates cross-linking reactions between the padded compounds and the textile substrate, significantly enhancing the durability and effectiveness of the flame-retardant coating [133, 134].

The pad-dry-cure technique's adaptability and versatility make it advantageous for a variety of fabrics, including both natural and synthetic types. This flexibility allows the method to be applied to a wide range of polymers, highlighting its usefulness across diverse textile compositions. The continuous use of pad-dry-cure methods not only ensures consistent flame retardancy in textiles but also provides an advantage in terms of scalability. Due to its scalable nature, the pad-dry-cure method seamlessly

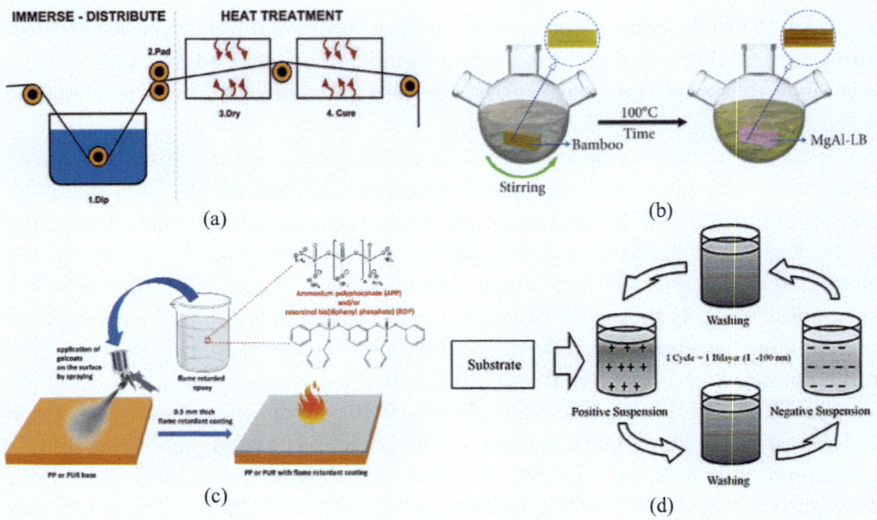

Fig. 10 Schematic process image of **a** Pad-dry-cure method [135], **b** In situ synthesis of Nano-MgAl-layered double hydroxide on bamboo [136], **c** Spray coating [137], and **d** layer-by-layer finishing [138]

integrates into existing textile manufacturing processes, offering a practical way to improve flame resistance in textiles across different industries. Table 1 contains various nanomaterials used for obtaining flame retardancy in textile substrate using pad-dry-cure.

6.1.2 In Situ Polymerization

In the context of flame retardancy, in situ polymerization is a method that involves the direct synthesis of nanoparticles onto the surface of fabric (Fig. 10b). The nanomaterial deposited onto the fabric significantly influences its flame-retardant properties. Specifically, in the mechanism, in situ nanomaterials play a crucial role in forming a char layer, acting as a physical barrier that prevents the degradation of the parent polymer chain [148].

The process of in situ polymerization entails immersing the fabric in a polymeric solution, typically a sol–gel solution. Subsequently, a cross-linker is applied to the solution, linking the polymers present in the solution with the substrate's structure [149]. This widely employed technique is particularly advantageous for producing layered double hydroxide nanocomposites. One notable benefit is its ability to provide a uniformly distributed coating onto the substrate, enhancing its overall flame-retardant performance [148]. Table 2 shows various flame-retardant nano-finished substrate produced by this method.

Table 1 Application of various nanomaterials through the pad-dry-cure method for flame-retardant purposes

Substrate	Nanomaterials	Particle size (diameter)	References
Cotton	Ethoxysilane-terminated phosphate polyurethane	90% Silica nanoparticles 240 ± 5 nm	[55]
	SiO_2	Average particle size 150 nm	[139]
	ZnO and SiO_2 nanoparticles	ZnO (<100 nm) and SiO_2 (<12 nm)	[140]
	TiO_2	100 nm	[141]
	Polyethyleneimine + silver nanoparticles	20 nm	[63]
	Ammonium polyphosphate modified nano-silica	100 nm	[142]
	Mixure of Tetrakis (hydroxymethyl) phosphonium hydroxide: urea: trimethylolmelamine in 2:4:1 ratio	–	[143]
	ZnO	100 nm	[144]
50/50 cotton/ polyester	ZnO and SiO_2 nanoparticles	ZnO (<100 nm) and SiO_2 (<12 nm)	[140]
35/65 cotton/ polyester	Bohemite nanoparticles	–	[145]
Jute	ZnO	Range 38–60 nm	[146]
Nylon	Chitosan/TiO_2/SiO_2	–	[147]

6.1.3 Spray Coating

Spray coating is a method employed for applying nanoparticle suspensions onto diverse surfaces. The process involves evaporating the suspension, thereby leaving the nanoparticles adhered to the surface (Fig. 10c). This technique, known for its simplicity and high efficiency, finds extensive use on an industrial scale. One crucial aspect of this coating method involves preparing sols that incorporate nanomaterials, considering factors such as viscosity, nozzle diameter, spray speed, and ambient temperature [162, 163]. Among these parameters, viscosity plays a particularly pivotal role in determining the final coating. Lower viscosity may lead to the dispersion of sols with nanoparticles, resulting in coagulation. Conversely, higher viscosity might cause higher droplet size and less cone area of spray [164]. Harrison and his colleagues noted an intriguing pattern in their observations: the cone spray angle undergoes an initial rise, followed by a subsequent decline, forming an inverted U-shaped curve in response to increasing viscosity [165]. The utility of spray coating on an industrial scale has been widespread owing to its user-friendly advantages

Table 2 Application of various nanomaterials through the in situ method for flame-retardant purposes

Substrate	In situ nanomaterials	Particle size (diameter)	References
Cotton	SiO_2	–	[150]
	TiO_2	4 nm	[151]
Polyester	Cu/Cu_2O nanoparticles	80.37 nm	[152]
Poly vinyl alcohol	Fe_3O_4	14 nm	[153]
Cellulose aerogel	Magnesium hydroxide	–	[154]
Nanocomposites			
Hydroxylated multi-walled carbon nanotubes	Modified barium phenolic resin		[155]
Polypropylene	Graphene oxide +4,4-diaminodiphenyl methane + phosphor amide oligomer		[156]
Epoxy resin	Fe/Zn-layered double hydroxide/ammonium polyphosphate hybrid		[157]
	Layered double hydroxide- α-zirconium phosphate		[158]
	Layered double hydroxide–carbon nanotubes		[159]
Silica aerogel	Layered double hydroxide		[160]
Ethylene–vinyl acetate	Layered double hydroxide–silicon dioxide hybrids		[161]

Table 3 Application of various nanomaterials through the spray coating method for flame-retardant purposes

Substrate	Nanomaterials	References
Polyester	Solution of carbon micro and nanoparticles	[167]
Cotton	Poly[1,4-diaminophenylene-tris(dimethyl hydroxymethyl)phosphine]	[168]
Polypropylene	Ammonium polyphosphate + resorcinol bis(diphenyl phosphate)	[137]

and ease of application [166]. The accessibility and practicality of this technique have contributed to its extensive adoption in various industrial contexts, as shown in Table 3.

6.1.4 Layer-By-Layer

The layer-by-layer (L-b-l) methodology involves the sequential deposition of alternating positive and negative layers onto the surface of the fabric. Initially, the fabric is immersed in deionized water to eliminate any surface charges, after which polymers with alternating charges are applied in a coating process (Fig. 10d). The application of two successive layers, characterized by positive and negative charges, constitutes a bilayer. The utilization of the layer-by-layer approach presents a distinct advantage owing to the precise control it affords over the number of bilayers. However, certain scholarly works also advocate for the maintenance of specific time intervals during the layering process. L-b-l technique can be implemented at room temperature, and in specific instances, it may necessitate mild environmental conditions. Furthermore, the coating produced by this l-b-l method is washable [169]. Layer-by-layer have been linked with the formation of a physical char layer, preventing further spread of fire to the inner polymer. It has been observed increasing number of bilayers of the l-b-l coating drastically increases the flame-retardant properties of the structure [169–171]. Hence, due to its various advantages, l-b-l has been widely adopted for producing flame-retardant coating. Table 4 provides a compilation of various nanomaterials employed in layer-by-layer coatings to attain flame retardancy for different textile substrates.

Each of this coating techniques has certain advantages and disadvantages which must be considered before its application (Table 5).

6.2 Modification of Non-Flame-Retardant Material to Obtain Flame-Retardant Material

6.2.1 Polymer Grafting

Polymer grafting is a method used to enhance the characteristics of a polymer by modifying its structure, physical attributes, and chemical properties. In this technique, a polymer, or a co-polymeric chain, or even another compound, is affixed to an existing parent polymeric chain. This attachment of a new material induces changes in the properties of the original polymeric chain (Fig. 11b). The objective is to achieve specific and desirable properties by altering the parent polymeric chain [186–188]. DOPO [12], phosphate [157], chitosan [178]-based grafted polymers have been used for obtaining flame retardancy as shown in Table 6.

Polymers derived from diverse sources may lack the requisite attributes for integration into the parent polymeric chain. In instances where desired characteristics are absent, chemical modifications are undertaken to yield a customized polymer. Reactive functional groups, including thio, hydroxyl, amino, and carboxylic acid groups inherent in the polymer structure, provide suitable sites for chemical modifications and grafting processes, enabling a tailored polymeric structure satisfying the required application.

6.2.2 Melt Blending (Nano-Composite)

Nanoparticle melt blending is a technique involving the introduction of nanoparticles into a material to augment specific desirable characteristics (Fig. 11a). Benefit of nanoparticle melt blending lies in its capacity to induce modifications in the

Table 4 Application of various nanomaterials through the layer-by-layer method for flame-retardant purposes

Substrate	Layer 1	Layer 2	Number of bilayers	References
Poly urethane	Poly (vinyl sulfonic acid sodium salt)	Chitosan	10	[172]
	Polyethylene imine-carbon nanofibrrs	Poly(acrylic acid)	4	[173]
	Phosphorylated cellulose nanofibrils	Chitosan	150	[168]
Ramie	Polyelectrolyte amino-functionalized multiwall carbon nanotube	Ammonium polyphosphate	5, 10, 20	[174]
Cotton	Phosphorylated chitin	Chitosan	–	[175]
	Alumina-coated silica	Silica	5	[176]
	Poly diallyl dimethyl ammonium chloride	Deoxyribo-nucleic acid	5, 10	[129]
	Chitosan/ ammonium phytate	Polydimethylsiloxane-zinc oxide	1	[134]
	Anionic octa(tetramethylammonium)-polyhedral oli-gomeric silsesquioxane	OctaAmmonium polyhedral oligomeric silsesquioxane	–	[177]
	OctaAmmonium polyhedral oligomeric silsesquioxane	Aminopropyl silsesquioxane oligomer	–	
	3-Aminopropyl triethoxysilane	Sodium phytate + chitosan	5, 10	[178]
	2,3-epoxy propyl trimethyl ammonium chloride	Al_2O_3	10, 16	[179]
	Branched polyethylenimine (BPEI) with urea and diammonium phosphate	Clay nanoparticles	–	[180]

(continued)

Table 4 (continued)

Substrate	Layer 1	Layer 2	Number of bilayers	References
35/65 cotton/polyester	γ-paperazinyl proplymethyl dimethoxy silane	Ammonium polyphosphate	15	[181]
Polyester	Positively charged silica	Negatively charged silica	–	[182]
	Ammonium polyphosphate	Chitosan	–	[183]
Nylon 6	Montmorillonite	Poly(allylamine)	40	[184]
Paper	Graphene nanoplatelets/multi-wall carbon nanotubes/lignin		25–30 μm in thickness	[185]

Table 5 Advantages and disadvantages of the different surface finish techniques [23]

Surface finish technique	Advantages	Disadvantages
Layer-by-Layer (l-b-l)	• Low temperature deposition • Simple and straightforward • Utilize an environmentally friendly solvent (e.g., water) • Control of number of BL • Multifunctionality (high)	• Surface treatment is sometimes required • Requires a proper selection of opposite charged polyelectrolytes • Limited pairs (usually organic) • Scalability (low to moderate)—not straightforward to develop continuous roll-to-roll plants
Pad-dry-cure	• Low temperature deposition • Suitable for deposition of sol–gel-based or organosilicon-based flame retardants • Amenable to scale up (continuous mode) • Multifunctionality (high)	• Chemical-based processes (impact of chemical precursors) • Inorganic materials layer cannot be deposited as thin film • Inorganic particles need to be prepared separately
In situ polymerization	• More generic approach than that of dip coating (able to deposit organic, inorganic, or hybrid coating) • Able to handle a wider range of substrates (types, shapes, etc.) • Ability to form fully inorganic flame-retardant coating	• Chemical-based processes (impact of chemical precursors) • Poor interaction between inorganic layer and substrates (low coating durability) • High temperature solvothermal/ hydrothermal synthesis (batch mode)
Brush/spray coating	• Low temperature deposition • Simple and straightforward • Excellent for deposition on rigid substrates • Allow for large area coating (amenable to scale up)	• Multifunctionality (low) • Narrow viscosity range to create fine droplets for spray coating • Amount of binder used is unnecessarily high (potentially undermining the flame retardancy effects)

crystalline structure, tensile properties, durability, and various other intrinsic characteristics of the material, owing to its integration into the material [201, 202]. For achieving flame retardancy, various types of nanoparticles melt blending have been used in the literature as shown in Table 7.

A major challenge in terms of incorporation of nanoparticles during melt spinning is the uniform dispersion of nanoparticles [218], even improper mixing may cause nucleation of nanoparticles [201]. The phenomenon of coagulation and uniform distribution of nanoparticles depends on particle size [219], viscosity of polymer melt [220], shape of nanoparticles, and temperature of polymer melt. In order for a nanoparticle to be completely immersed in a viscous liquid, its diameter must be smaller than a critical value, as described by Eq. (16) [219]. Additionally, the shape

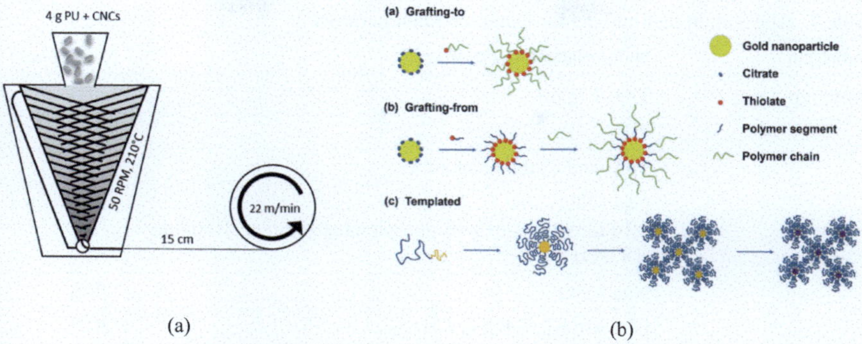

Fig. 11 Schematic process image of **a** Melt spinning set-up[1] [189], **b** Different grafting methods [190]

Table 6 Application grafting method for flame-retardant purposes

Substrate	Grafting material	References
Polyacrylonitrile (PAN)	Diethyl (acryloyloxyethyl) phosphate (DEAEP)	[191]
	Diethyl-2-(methacryloyloxyethyl)phosphate (DEMEP)	
	Diethyl(acryloyloxymethyl) phosphonate (DEAMP)	
	Dimethyl(acryloyloxymethyl)phosphonate (DMAMP)	
	Hydroxyethyl methacrylate	[192]
Polypropylene	CNT in situ and maleic anhydride-grafted	[193]
	POSS-DOPO-Graphene oxide	[194]
Cotton	Diethyl(acryloyloxyethyl)phosphate (DEAEP)	[195]
	Diethyl-2-(methacryloyloxyethyl)phosphate (DEMEP)	
	Diethyl(acryloyloxymethyl)phosphonate (DEAMP)	
	Dimethyl(acryloyloxymethyl)phosphonate (DMAMP)	
	Silica nano-particles and Zinc nanoparticles	[196]
Polystyrene	Graphene oxide	[197]
Acrylamide	Ethylene–vinyl acetate	[198]
Polyamide 66	DOPO-DAAM	[199]
Polylactic acid	10-(2,5-dihydroxyl phenyl)-DOPO-Graphene oxide	[200]

of these particles plays a role in determining the specific surface area, as highlighted by Datta and his coauthors [221]. This factor, in turn, influences the surface tension acting on the nanoparticle.

$$D_c \cong 2 \times 3^{\frac{3}{2}} \frac{l_T}{\theta_y^3} \tag{16}$$

[1] PU: Polyurethane, CNC: Cellulose Nano-crystals, RPM: Revolutions per minute.

Table 7 Application melt mixing for flame-retardant purposes

Substrate	Nano-material melt mixed	References
Polypropylene	TiO_2/Lignin	[203]
	TiO_2	[204]
	TiO_2 and Carbon nanotubes	[88]
	Nano-SiO_2 and 9,10-dihydro-9-oxa-10-phosphaphenanthrene-10-oxide (DOPO)	[205]
	Nano-SiO_2	[206]
	$[Zn_2Al(OH_6)][B_4O_5(OH)_4]_{0.5}$	[207]
	MnO and Mn_2O_3	[208]
70 Polypropylene/30 polyvinyl alcohol	Silica nanoparticles	[209]
Polyester	Polyhedral oligomeric silsesquioxanes (POSS)	[210]
	Layered hydrotalcite and brominated polystyrene	[211]
	Zinc phosphinate/nano-silica	[212]
Cellulose acetate butyrate	9,10-dihydro-9-oxa-10-phosphaphenanthrene-10-oxide (DOPO) and γ-cyclodextrins	[65]
Polyamide 12	Carbon nanotubes	[213]
Polyamide 66	Zn-DOPOx and melamine polyphosphate	[214]
Polyamide 6	Melamine cyanurate	[215]
	Ammonium sulphamate-dipentaerythritol	[216]
Polystyrene	Layered double hydroxide and graphene oxide	[217]

where D_c is critical diameter, l_T (tension length) $= \frac{\tau\text{(Line tension)}}{\gamma_s\text{(solid surface energy)}}$, and θ_y is the surface contact angle.

To comprehend the stability of nanoparticles within a solution, it is beneficial to employ the theoretical model proposed by Fuchs [222]. Equation (17) within this model addresses the coagulation frequency (J) for a nanoparticle dispersion system. This equation is pivotal in understanding the dynamics of nanoparticle interactions within the solution. The different factors mentioned must be optimized to obtain a fiber with uniform flame-retardant performance.

$$J = n_a 4\pi D / \int_{d_a}^{\infty} \frac{1}{r^2} \exp\left(\frac{U(r)}{k_B T}\right) dr \tag{17}$$

D is the aggregate diffusion coefficient, k_B is Boltzmann constant, T is temperature, n_a is temporal dynamics of aggregate concentration, and $U(r)$ is the potential field.

Both melt blending and polymer grafting are widely utilized to achieve polymers or fibers with improved flame-retardant properties. Although both methods result in

Table 8 Advantages and disadvantages of nanoparticle melt blending and grafting

Technique	Advantages	Disadvantages
Grafting	• Uniform flame-retardant properties • Since flame retardancy is applied in the polymeric layer, it very durable • Obtaining multifunctionality is easy • Very low chance of agglomeration	• More time-consuming and energy-consuming, hence it is difficult to have a high scalability • It changes base polymeric property, which may not be desirable • The polymeric chain grafted must be compatible. Furthermore, there is a possibility of side reaction and forming a different polymer than the desirable one
Nanoparticle melt blending	• Cost-effective process • Easy to scale process • No need to assess any chemical compatibility of the nanoparticles and the polymer	• Very difficult to obtain a homogeneous flame-retardant property • Possibility of agglomeration • Nanoparticles should have higher melting temperature than the polymer

the development of flame-retardant materials, they each possess distinct advantages and disadvantages, which are comprehensively delineated in Table 8.

7 Summary and Key Issues and Future Perspectives

The chapter in the book offers a comprehensive overview of distinct nano-species contributing to flame retardancy, accompanied by concise explanations of their respective mechanisms. Additionally, the chapter covers the various methodologies employed for achieving flame retardancy using nanomaterials, providing an analysis of the strengths and weaknesses inherent in each technique. It is noteworthy that the combustion behavior of flame-retardant materials treated with nanomaterials is contingent not only upon their species but also on the geometrical arrangement of these species. At higher temperatures, changes in morphology may occur due to the mobility of nano-fillers within the molten matrix, the diffusivity of polymer chain fragments, induced polarity in the polymer resulting from surface oxidation, decomposition of organic modifiers, and turbulence induced by volatiles diffusing toward the material surface. The amalgamation of these factors collectively influences the repeatability of performance in nano-finishes.

In addition to flame retardancy, use nanomaterials to achieve multifunctional flame-retardant textile have also been explored in several literatures. In a research Li and his co-researchers, they used silver nanoparticles with a cyclotriphosphazene/polydopamine/silver nanoparticles hybrid coatings for a flame-retardant, antimicrobial fabric [223]. Similar research has also been done to obtain multifunctional antibacterial, UV protection [224], wettable [140], toxic gas suppression [225]. The multifunctionality opens a new vista for more exploration of development of new novel materials.

In specific cases of fire protection, such as specialized gear like firefighting turnouts or attire for oil and gas field workers, nanomaterial integration significantly enhances flame resistance and overall safety [22, 226]. As discussed, nanoparticles and nanofibers, despite increasing polymer viscosity and posing processing challenges, notably augment textile properties. They improve mechanical strength, thermal insulation, and electrical characteristics while reducing flammability and enhancing anti-drip qualities [227]. Nanofibers, renowned for their high surface area per unit mass, afford breathability and exceptional thermal insulation. Coating methods effectively apply these nanofibers to clothing, ensuring robust protection. Additionally, smart clothing, equipped with temperature sensors and active injection systems, responds dynamically to fire exposure by releasing water, which evaporates to absorb heat, mitigating temperature rise and enhancing defense against flash fires [228]. Shape memory materials (SMM) and phase-change materials (PCM) embedded in clothing adapt to heat, expanding air gaps between layers for improved insulation [229]. This innovation reduces thickness while maintaining superior thermal protection. Recent studies spotlight nano-nonwoven fabrics in multi-layer clothing for outstanding heat resistance, air permeability, and water release rates [54, 230]. These fabrics rank among the lightest and thinnest, meeting stringent thermoprotective and thermophysiological standards [231, 232]. Furthermore, novel flame-retardant agents, utilizing nanoparticles such as TiO_2 and hydrogel-textile composites like acrylamide (AAM) and SiO_2, exhibit enhanced flame resistance and durability. These advancements revolutionize fireproof clothing production, ensuring exceptional safety for firefighters and personnel in high-risk environments like oil and gas fields [233, 234].

In addition to the pursuit of multi-functionality, significant research has been conducted in alignment with sustainability goals. Naturally derived nanomaterials have gained widespread adoption in flame-retardant applications. Moreover, the utilization of recycled materials, such as r-PET (recycled polyethylene terephthalate) [235–237], recycled PP (recycled polypropylene), recycled polyamide [235], and recycled cotton [238], for the development of nanomaterial-based flame-retardant materials has also been employed. A substantial body of research has been dedicated to exploring the potential of natural aerogels as innovative and lightweight flame-retardant materials. Both natural-based and synthetic aerogels have been synthesized across various literature sources. Despite in several literatures aerogels have been classified as non-nanomaterials [238], their utilization represents a promising frontier in the realm of flame retardancy. The implementation of aerogels introduces a novel perspective, offering a unique avenue for addressing fire safety challenges. While some literature explores the use of entirely natural-based nanomaterials, it is essential to acknowledge that for such technologies to attain mainstream acceptance, further advancements are still required. Nonetheless, a substantial body of work has been dedicated to advancing flame retardancy with a focus on sustainability considerations.

References

1. Hall S (2023) Fire loss in the United States. NFPA research. Accessed 28 Nov 2023. [Online]. Available: https://www.nfpa.org/education-and-research/research/nfpa-research/fire-statistical-reports/fire-loss-in-the-united-states
2. Mazumder N-U-S et al (2023) Firefighters exposure to per-and polyfluoroalkyl substances (PFAS) as an occupational hazard: a review. Front Mater 10. Accessed 23 Sept 2023. [Online]. Available: https://www.frontiersin.org/articles/10.3389/fmats.2023.1143411
3. Pálešová N et al (2023) Firefighters and the liver: exposure to PFAS and PAHs in relation to liver function and serum lipids (CELSPAC-FIREexpo study). Int J Hyg Environ Health 252:114215. https://doi.org/10.1016/j.ijheh.2023.114215
4. LeMasters GK et al (2006) Cancer risk among firefighters: a review and meta-analysis of 32 studies. J Occup Environ Med 48(11):1189–1202. https://doi.org/10.1097/01.jom.0000246229.68697.90
5. Norouzi M, Zare Y, Kiany P (2015) Nanoparticles as effective flame retardants for natural and synthetic textile polymers: application, mechanism, and optimization. Polym Rev 55(3):531–560. https://doi.org/10.1080/15583724.2014.980427
6. Paul R (2014) Functional finishes for textiles: improving comfort, performance and protection. Elsevier
7. Zhang J-Y et al (2022) Flame-retardant nanocoating towards high-efficiency suppression of smoke and toxic gases for polymer foam. Compos Part Appl Sci Manuf 159:107021. https://doi.org/10.1016/j.compositesa.2022.107021
8. Teles F, Martins G, Antunes F (2022) Fire retardancy in nanocomposites by using nanomaterial additives. J Anal Appl Pyrolysis 163:105466. https://doi.org/10.1016/j.jaap.2022.105466
9. Shen J, Liang J, Lin X, Lin H, Yu J, Wang S (2021) The flame-retardant mechanisms and preparation of polymer composites and their potential application in construction engineering. Polymers 14(1):82. https://doi.org/10.3390/polym14010082
10. Du A, Zhou B, Zhang Z, Shen J (2013) A special material or a new state of matter: a review and reconsideration of the aerogel. Materials 6(3):3. https://doi.org/10.3390/ma6030941
11. Vahidi G, Bajwa DS, Shojaeiarani J, Stark N, Darabi A (2021) Advancements in traditional and nanosized flame retardants for polymers—a review. J Appl Polym Sci 138(12):50050. https://doi.org/10.1002/app.50050
12. Ali W et al (2023) Flame-retardant finishing of cotton fabrics using DOPO functionalized alkoxy- and amido alkoxysilane. Cellulose 30(4):2627–2652. https://doi.org/10.1007/s10570-022-05033-3
13. Agarwal R, Jassal M, Agrawal AK (2021) Nano surface modification of poly(ethylene terephthalate) fabrics for enhanced comfort properties for activewear. J Ind Eng Chem 98:217–230. https://doi.org/10.1016/j.jiec.2021.03.050
14. Mandal S, Annaheim S, Capt A, Greve J, Camenzind M, Rossi RM (2019) A categorization tool for fabric systems used in firefighters clothing based on their thermal protective and thermo-physiological comfort performances. Text Res J 89(16):3244–3259. https://doi.org/10.1177/0040517518809055
15. Mandal S, Chowdhury IZ, Mazumder N-U-S, Agnew RJ, Boorady LM (2022) Characterization of sweat drying performance of single layered thermal protective fabrics used in high-risk sector workers clothing. Polymers 14(24):5393. https://doi.org/10.3390/polym14245393
16. Xiao F, Fontaine G, Bourbigot S (2021) Recent developments in fire retardancy of polybutylene succinate. Polym Degrad Stab 183:109466
17. Wang X, Kalali EN, Wan J-T, Wang D-Y (2017) Carbon-family materials for flame retardant polymeric materials. Top Vol Polym Chem 69:22–46. https://doi.org/10.1016/j.progpolymsci.2017.02.001
18. Yuen ACY et al (2020) Utilising genetic algorithm to optimise pyrolysis kinetics for fire modelling and characterisation of chitosan/graphene oxide polyurethane composites. Compos Part B Eng 182:107619. https://doi.org/10.1016/j.compositesb.2019.107619

19. Wilkie CA, Morgan AB (2009) Fire retardancy of polymeric materials. CRC press
20. Laoutid F, Bonnaud L, Alexandre M, Lopez-Cuesta J-M, Dubois P (2009) New prospects in flame retardant polymer materials: from fundamentals to nanocomposites. Mater Sci Eng R Rep 63(3):100–125
21. Lyon RE, Janssens ML (2005) Polymer flammability, United States. federal aviation administration. Office of aviation research
22. Song G, Mandal S, Rossi R (2016) Thermal protective clothing for firefighters. Woodhead Publishing
23. Wen OY, Tohir MZM, Yeaw TCS, Razak MA, Zainuddin HS, Hamid MRA (2023) Fire-resistant and flame-retardant surface finishing of polymers and textiles: a state-of-the-art review. Prog Org Coat 175:107330. https://doi.org/10.1016/j.porgcoat.2022.107330
24. Pearce E (2012) Flame-retardant polymeric materials. Springer Science & Business Media
25. Lu H, Song L, Hu Y (2011) A review on flame retardant technology in China. Part II: flame retardant polymeric nanocomposites and coatings. Polym Adv Technol 22(4):379–394
26. Papaspyrides C, Pavlidou S, Vouyiouka S (2009) Development of advanced textile materials: natural fibre composites, anti-microbial, and flame-retardant fabrics. Proc Inst Mech Eng Part J Mater Des Appl 223(2):91–102
27. Joseph P, Ebdon JR (2009) Phosphorus-based flame retardants. In: Wilkie CA, Morgan AB (eds) Fire retardancy polymeric materials, 2nd edn, pp 107–128
28. Feldman D (2013) Review polymer nanocomposites: flammability. J Macromol Sci Part A 50(12):1241–1249
29. Shan X, Zhang P, Song L, Hu Y, Lo S (2011) Compound of nickel phosphate with Ni $(OH)(PO_4)$ 2–layers and synergistic application with intumescent flame retardants in thermoplastic polyurethane elastomer. Ind Eng Chem Res 50(12):7201–7209
30. Alongi J, Han Z, Bourbigot S (2015) Intumescence: tradition versus novelty. A comprehensive review. Environ Relev Hybrid Polym Mater 51:28–73. https://doi.org/10.1016/j.progpolymsci.2015.04.010
31. Anderson CE, Ketchum DE, Mountain WP (1988) Thermal conductivity of intumescent chars. J Fire Sci 6(6):390–410. https://doi.org/10.1177/073490418800600602
32. Peng H-Q, Zhou Q, Wang D-Y, Chen L, Wang Y-Z (2008) A novel charring agent containing caged bicyclic phosphate and its application in intumescent flame retardant polypropylene systems. J Ind Eng Chem 14(5):589–595. https://doi.org/10.1016/j.jiec.2008.05.011
33. Stęplewski W, Kazimierczak J, Wawro D (2010) Novel method of preparing flame retardant cellulose-silicate fibres. Fibres Text East Eur 3(80):24–31
34. Yang F, Nelson GL (2011) Combination effect of nanoparticles with flame retardants on the flammability of nanocomposites. Polym Degrad Stab 96(3):270–276
35. Cinausero N, Azema N, Lopez-Cuesta J-M, Cochez M, Ferriol M (2011) Synergistic effect between hydrophobic oxide nanoparticles and ammonium polyphosphate on fire properties of poly (methyl methacrylate) and polystyrene. Polym Degrad Stab 96(8):1445–1454
36. Rault F et al (2009) Effect of manganese nanoparticles on the mechanical, thermal and fire properties of polypropylene multifilament yarn. Polym Degrad Stab 94(6):955–964
37. Wen OY, Tohir MZM, Yeaw TCS, Razak MA, Zainuddin HS, Hamid MRA (2023) Fire-resistant and flame-retardant surface finishing of polymers and textiles: a state-of-the-art review. Prog Org Coat 175:107330
38. Zhang Q, Li Z, Li X, Yu L, Zhang Z, Wu Z (2019) Zinc ferrite nanoparticle decorated boron nitride nanosheet: preparation, magnetic field arrangement, and flame retardancy. Chem Eng J 356:680–692. https://doi.org/10.1016/j.cej.2018.09.053
39. Ahmed L, Zhang B, Hatanaka LC, Mannan MS (2018) Application of polymer nanocomposites in the flame retardancy study. J Loss Prev Process Ind 55:381–391. https://doi.org/10.1016/j.jlp.2018.07.005
40. Mandal S, Song G, Zeng H, Agnew RJ, Chowdhury IZ, Boorady LM (2023) Polymer nanofibers. In: Encyclopedia of polymer science and technology, pp 1–39. https://doi.org/10.1002/0471440264.pst689

41. Duquesne S, Magniez C, Camino G (2007) Multifunctional barriers for flexible structure. Springer
42. Sikdar D, Katti DR, Katti KS, Bhowmik R (2006) Insight into molecular interactions between constituents in polymer clay nanocomposites. Polymer 47(14):5196–5205
43. Joshi M, Bhattacharyya A (2011) Nanotechnology—a new route to high-performance functional textiles. Text Prog 43(3):155–233. https://doi.org/10.1080/00405167.2011.570027
44. Khan I, Saeed K, Khan I (2019) Nanoparticles: properties, applications and toxicities. Arab J Chem 12(7):908–931. https://doi.org/10.1016/j.arabjc.2017.05.011
45. Patra JK, Gouda S (2013) Application of nano technology in textile engineering: an overview. J Eng Technol Res 5:104–111
46. Vollath D, Fischer FD, Holec D (2018) Surface energy of nanoparticles—influence of particle size and structure. Beilstein J Nanotechnol 9:2265–2276. https://doi.org/10.3762/bjnano.9.211
47. Andersen WH (1965) Theory of surface ignition energy of condensed explosives. Ind Eng Chem Process Des Dev 4(3):286–288. https://doi.org/10.1021/i260015a010
48. Coetzee D, Venkataraman M, Militky J, Petru M (2020) Influence of nanoparticles on thermal and electrical conductivity of composites. Polymers 12(4):742. https://doi.org/10.3390/polym12040742
49. Kalaoglu-Altan OI, Kayaoglu BK, Trabzon L (2022) Improving thermal conductivities of textile materials by nanohybrid approaches. iScience 25(3):103825. https://doi.org/10.1016/j.isci.2022.103825
50. Chen Z et al (2022) Effects of MWCNTs on char layer structure and physicochemical reaction in ethylene propylene diene monomer insulators. Polymers 14(15):3016. https://doi.org/10.3390/polym14153016
51. Nageswara Rao T, Naidu TM, Kim MS, Parvatamma B, Prashanthi Y, Heun Koo B (2019) Influence of zinc oxide nanoparticles and char forming agent polymer on flame retardancy of intumescent flame retardant coatings. Nanomaterials 10(1):42. https://doi.org/10.3390/nano10010042
52. Kačíková D et al (2021) The influence of nanoparticles on fire retardancy of pedunculate oak wood. Nanomaterials 11(12):3405. https://doi.org/10.3390/nano11123405
53. Hashim, Hamid A, Khan M (2020) Heat and mass transport phenomena of nanoparticles on time-dependent flow of Williamson fluid towards heated surface. Neural Comput Appl 32(8):3253–3263. https://doi.org/10.1007/s00521-019-04100-4
54. Mandal S, Mazumder N-U-S, Agnew RJ, Song G, Li R (2021) Characterization and modeling of thermal protective and thermo-physiological comfort performance of polymeric textile materials—a review. Materials 14(9):9. https://doi.org/10.3390/ma14092397
55. Liu X et al (2022) A superhydrophobic and flame-retardant cotton fabric fabricated by an eco-friendly assembling method. Text Res J 92(15–16):2873–2885. https://doi.org/10.1177/00405175211022628
56. Vo DK et al (2022) Effect of metal oxide nanoparticles and aluminum hydroxide on the physicochemical properties and flame-retardant behavior of rigid polyurethane foam. Constr Build Mater 356:129268. https://doi.org/10.1016/j.conbuildmat.2022.129268
57. El-Shafei A, ElShemy M, Abou-Okeil A (2015) Eco-friendly finishing agent for cotton fabrics to improve flame retardant and antibacterial properties. Carbohydr Polym 118:83–90. https://doi.org/10.1016/j.carbpol.2014.11.007
58. Bajwa DS, Rehovsky C, Shojaeiarani J, Stark N, Bajwa S, Dietenberger MA (2019) Functionalized cellulose nanocrystals: a potential fire retardant for polymer composites. Polymers 11(8):8. https://doi.org/10.3390/polym11081361
59. Chollet B, Lopez-Cuesta JM, Laoutid F, Ferry L (2019) Lignin nanoparticles as a promising way for enhancing lignin flame retardant effect in polylactide. Materials (Basel, Switzerland) 12(13):2132. https://doi.org/10.3390/ma12132132
60. Basak S, Raja ASM, Saxena S, Patil PG (2021) Tannin based polyphenolic bio-macromolecules: creating a new era towards sustainable flame retardancy of polymers. Polym Degrad Stab 189:109603. https://doi.org/10.1016/j.polymdegradstab.2021.109603

61. Malucelli G (2020) Flame-retardant systems based on chitosan and its derivatives: state of the art and perspectives. Molecules 25(18):4046. https://doi.org/10.3390/molecules25184046
62. Hou X, Xue Z, Xia Y (2018) Preparation of a novel agar/sodium alginate fire-retardancy film. Mater Lett 233:274–277. https://doi.org/10.1016/j.matlet.2018.09.026
63. Zhou Q, Chen J, Lu Z, Tian Q, Shao J (2022) In situ synthesis of silver nanoparticles on flame-retardant cotton textiles treated with biological phytic acid and antibacterial activity. Materials 15(7):2537. https://doi.org/10.3390/ma15072537
64. Huang Y, Ma T, Wang Q, Guo C (2019) Synthesis of biobased flame-retardant carboxylic acid curing agent and application in wood surface coating. ACS Sustain Chem Eng 7(17):14727–14738. https://doi.org/10.1021/acssuschemeng.9b02645
65. Park Y et al (2023) Organophosphate-cyclodextrin inclusion complex for flame retardancy in doped cellulose acetate butyrate melt-spun fibers. Ind Eng Chem Res 62(29):11595–11605. https://doi.org/10.1021/acs.iecr.3c00712
66. Wang D et al (2020) A bio-based flame-retardant starch based on phytic acid. ACS Sustain Chem Eng 8(27):10265–10274. https://doi.org/10.1021/acssuschemeng.0c03277
67. Xu Z, Chu Z, Yan L, Chen H, Jia H, Tang W (2019) Effect of chicken eggshell on the flame-retardant and smoke suppression properties of an epoxy-based traditional APP-PER-MEL system. Polym Compos 40(7):2712–2723. https://doi.org/10.1002/pc.25077
68. Douglas KM, Blitz MA, Mangan TP, Western CM, Plane JMC (2020) Kinetic study of the reactions $PO + O_2$ and $PO_2 + O_3$ and spectroscopy of the PO radical. J Phys Chem A 124(39):7911–7926. https://doi.org/10.1021/acs.jpca.0c06106
69. PubChem. Lignin. Accessed 27 Dec 2023. [Online]. Available: https://pubchem.ncbi.nlm.nih.gov/compound/175586
70. PubChem. Cyclohexaamylose. Accessed 27 Dec 2023. [Online]. Available: https://pubchem.ncbi.nlm.nih.gov/compound/320760
71. PubChem. 6H-Dibenzo[c,E][1,2]oxaphosphinine 6-oxide. Accessed 27 Dec 2023. [Online]. Available: https://pubchem.ncbi.nlm.nih.gov/compound/6328250
72. PubChem. Phytic acid. Accessed 27 Dec 2023. [Online]. Available: https://pubchem.ncbi.nlm.nih.gov/compound/890
73. Othman NK, Salleh E, Dasuki Z, Mat lazim A (2018) Dimethyl sulfoxide-treated starch of dioescorea hispida as a green corrosion inhibitor for low carbon steel in sodium chloride medium. https://doi.org/10.5772/intechopen.73552
74. Iijima S (1991) Helical microtubules of graphitic carbon. Nature 354(6348):56–58
75. Oberlin A, Endo M, Koyama T (1976) Filamentous growth of carbon through benzene decomposition. J Cryst Growth 32(3):335–349
76. Popov VN (2004) Carbon nanotubes: properties and application. Mater Sci Eng R Rep 43(3):61–102. https://doi.org/10.1016/j.mser.2003.10.001
77. Kashiwagi T, Grulke E, Hilding J, Harris R, Awad W, Douglas J (2002) Thermal degradation and flammability properties of poly (propylene)/carbon nanotube composites. Macromol Rapid Commun 23(13):761–765
78. Kashiwagi T et al (2004) Thermal and flammability properties of polypropylene/carbon nanotube nanocomposites. Polymer 45(12):4227–4239
79. Zare Y, Garmabi H (2014) Attempts to simulate the modulus of polymer/carbon nanotube nanocomposites and future trends. Polym Rev 54(3):377–400
80. Idumah CI (2023) Recent advancements in fire retardant mechanisms of carbon nanotubes, graphene, and fullerene polymeric nanoarchitectures. J Anal Appl Pyrolysis 174:106113. https://doi.org/10.1016/j.jaap.2023.106113
81. Geng H et al (2002) Fabrication and properties of composites of poly (ethylene oxide) and functionalized carbon nanotubes. Adv Mater 14(19):1387–1390
82. Ye L, Wu Q, Qu B (2009) Synergistic effects and mechanism of multiwalled carbon nanotubes with magnesium hydroxide in halogen-free flame retardant EVA/MH/MWNT nanocomposites. Polym Degrad Stab 94(5):751–756
83. Shaffer MS, Windle AH (1999) Fabrication and characterization of carbon nanotube/poly (vinyl alcohol) composites. Adv Mater 11(11):937–941

84. Zhu S-E et al (2018) Comparative studies on thermal, mechanical, and flame retardant properties of PBT nanocomposites via different oxidation state phosphorus-containing agents modified amino-CNTs. Nanomaterials 8(2):70
85. Idumah CI (2023) Recent advancements in fire retardant mechanisms of carbon nanotubes, graphene, and fullerene polymeric nanoarchitectures. J Anal Appl Pyrolysis 106113
86. Wang J (2021) Flame retardancy and dispersion of functionalized carbon nanotubes in thiol-ene nanocomposites. Polymers 13(19):3308
87. Ding L, Jia Z, Sun H, Pan Y, Zhao J (2019) Estimation of mechanical performance, thermal stability and flame retardancy of high-impact polystyrene/surface-modified APP/carboxylic-functionalized MWCNTs nanocomposites. Polymers 11(4):615
88. Cabello-Alvarado C et al (2019) Melt-mixed thermoplastic nanocomposite containing carbon nanotubes and titanium dioxide for flame retardancy applications. Polymers 11(7). https://doi.org/10.3390/polym11071204
89. Gao C et al (2023) Surface modification methods and mechanisms in carbon nanotubes dispersion. Carbon 212:118133. https://doi.org/10.1016/j.carbon.2023.118133
90. Huang R, Gao C, Shi Y, Fu L, Feng Y, Shui W (2022) Synergistic function between phosphorus-containing flame retardant and multi-walled carbon nanotubes towards fire safe polystyrene composites with enhanced electromagnetic interference shielding. Int J Mol Sci 23(21). https://doi.org/10.3390/ijms232113434
91. Chen F et al (2023) An adaptive framework to accelerate optimization of high flame retardant composites using machine learning. Compos Sci Technol 231:109818. https://doi.org/10.1016/j.compscitech.2022.109818
92. Nagaraj BK, Kalaivani A, Begum RS, Akila S, Sachdev HK, Kumar NS (2023) The emerging role of artificial intelligence in STEM higher education: a critical review. Int Res J Multidiscip Technovation 5(5):1–19. https://doi.org/10.54392/irjmt2351
93. Liu B-W, Zhao H-B, Wang Y-Z (2022) Advanced flame-retardant methods for polymeric materials. Adv Mater 34(46):2107905. https://doi.org/10.1002/adma.202107905
94. Parvinzadeh Gashti M, Almasian A (2013) UV radiation induced flame retardant cellulose fiber by using polyvinylphosphonic acid/carbon nanotube composite coating. Compos Part B Eng 45(1):282–289. https://doi.org/10.1016/j.compositesb.2012.07.052
95. Shen Z, Bateman S, Wu DY, McMahon P, Dell'Olio M, Gotama J (2009) The effects of carbon nanotubes on mechanical and thermal properties of woven glass fibre reinforced polyamide-6 nanocomposites. Compos Sci Technol 69(2):239–244. https://doi.org/10.1016/j.compscitech.2008.10.017
96. Huang G, Wang S, Song P, Wu C, Chen S, Wang X (2014) Combination effect of carbon nanotubes with graphene on intumescent flame-retardant polypropylene nanocomposites. Compos Part Appl Sci Manuf 59:18–25. https://doi.org/10.1016/j.compositesa.2013.12.010
97. Ma H-Y, Tong L-F, Xu Z-B, Fang Z-P (2008) Functionalizing carbon nanotubes by grafting on intumescent flame retardant: nanocomposite synthesis, morphology, rheology, and flammability. Adv Funct Mater 18(3):414–421. https://doi.org/10.1002/adfm.200700677
98. Abulyazied DE, Ene A (2021) An investigative study on the progress of nanoclay-reinforced polymers: preparation, properties, and applications: a review. Polymers 13(24):4401
99. Zanetti M, Lomakin S, Camino G (2000) Polymer layered silicate nanocomposites. Macromol Mater Eng 279(1):1–9
100. Murray HH (1991) Overview—clay mineral applications. Appl Clay Sci 5(5):379–395. https://doi.org/10.1016/0169-1317(91)90014-Z
101. Sabbagh F (2021) A comparative study on the clays incorporated with acrylamide-based hydrogels. Adv Appl NanoBio-Technol 2(4):15–23
102. Kiliaris P, Papaspyrides C (2010) Polymer/layered silicate (clay) nanocomposites: an overview of flame retardancy. Prog Polym Sci 35(7):902–958
103. Ramos Filho FG, Mélo TJA, Rabello MS, Silva SML (2005) Thermal stability of nanocomposites based on polypropylene and bentonite. Polym Degrad Stab 89(3):383–392. https://doi.org/10.1016/j.polymdegradstab.2004.12.011

104. Valapa RB, Loganathan S, Pugazhenthi G, Thomas S, Varghese T (2017) An overview of polymer–clay nanocomposites. Clay-Polym Nanocomposites 29–81
105. Fischer H (2003) Polymer nanocomposites: from fundamental research to specific applications. Curr Trends Nanosci—Mater Appl Proc Symp E-MRS Spring Meet 23(6):763–772. https://doi.org/10.1016/j.msec.2003.09.148
106. Kausar A (2020) A review of fundamental principles and applications of polymer nanocomposites filled with both nanoclay and nano-sized carbon allotropes—graphene and carbon nanotubes. J Plast Film Sheeting 36(2):209–228. https://doi.org/10.1177/8756087919884607
107. Wang R et al (2021) Montmorillonite-based two-dimensional nanocomposites: preparation and applications. Molecules 26(9):2521
108. Gao F (2004) Clay/polymer composites: the story. Mater Today 7(11):50–55. https://doi.org/10.1016/S1369-7021(04)00509-7
109. Najafi N, Heuzey MC, Carreau PJ (2012) Polylactide (PLA)-clay nanocomposites prepared by melt compounding in the presence of a chain extender. Compos Sci Technol 72(5):608–615. https://doi.org/10.1016/j.compscitech.2012.01.005
110. Rhim J-W, Hong S-I, Ha C-S (2009) Tensile, water vapor barrier and antimicrobial properties of PLA/nanoclay composite films. LWT—Food Sci Technol 42(2):612–617. https://doi.org/10.1016/j.lwt.2008.02.015
111. Gilman JW et al (2000) Flammability properties of polymer−layered-silicate nanocomposites. polypropylene and polystyrene nanocomposites. Chem Mater 12(7):1866–1873. https://doi.org/10.1021/cm0001760
112. Aslzadeh MM, Abdouss M, Sadeghi GM (2013) Preparation and characterization of new flame retardant polyurethane composite and nanocomposite. J Appl Polym Sci 127(3):1683–1690
113. Qin H, Zhang S, Zhao C, Hu G, Yang M (2005) Flame retardant mechanism of polymer/clay nanocomposites based on polypropylene. Polymer 46(19):8386–8395
114. Horrocks AR, Kandola BK (2007) Potential applications of nanocomposites for flame retardancy. Flame Retard Polym Nanocomposites 325–353
115. Lin J-S, Liu Y, Wang D-Y, Qin Q, Wang Y-Z (2011) Poly (vinyl alcohol)/ammonium polyphosphate systems improved simultaneously both fire retardancy and mechanical properties by montmorillonite. Ind Eng Chem Res 50(17):9998–10005
116. Harrison PG (1997) Silicate cages: precursors to new materials. J Organomet Chem 542(2):141–183
117. Provatas A, Matisons J (1997) Silsesquioxanes: synthesis and applications. Trends Polym Sci 10(5):327–332
118. Ayandele E, Sarkar B, Alexandridis P (2012) Polyhedral oligomeric silsesquioxane (POSS)-containing polymer nanocomposites. Nanomaterials 2(4):4. https://doi.org/10.3390/nano2040445
119. Feher FJ, Budzichowski TA (1995) Silasesquioxanes as ligands in inorganic and organometallic chemistry. Polyhedron 14(22):3239–3253
120. Murugavel R, Chandrasekhar V, Roesky H (1996) Acc Chem Res 29:183; Murugavel R, Voigt Walawalkar MG, Roesky HW (1996) Chem Rev 96:2205
121. Abbenhuis HC (2000) Advances in homogeneous and heterogeneous catalysis with metal-containing silsesquioxanes. Chem Eur J 6(1):25–32
122. Lu S-Y, Hamerton I (2002) Recent developments in the chemistry of halogen-free flame retardant polymers. Prog Polym Sci 27(8):1661–1712
123. Fina A, Tabuani D, Camino G (2010) Polypropylene–polysilsesquioxane blends. Eur Polym J 46(1):14–23
124. Didane N, Giraud S, Devaux E, Lemort G (2012) A comparative study of POSS as synergists with zinc phosphinates for PET fire retardancy. Polym Degrad Stab 97(3):383–391
125. Didane N, Giraud S, Devaux E, Lemort G, Capon G (2012) Thermal and fire resistance of fibrous materials made by PET containing flame retardant agents. Polym Degrad Stab 97(12):2545–2551
126. Baker ES, Gidden J, Anderson SE, Haddad TS, Bowers MT (2004) Isomeric structural characterization of polyhedral oligomeric silsesquioxanes (POSS) with styryl and epoxy phenyl capping agents. Nano Lett 4(5):779–785. https://doi.org/10.1021/nl049957g

127. Wu J, Mather PT (2009) POSS polymers: physical properties and biomaterials applications. Polym Rev 49(1):25–63. https://doi.org/10.1080/15583720802656237
128. Alongi J, Brancatelli G, Rosace G (2012) Thermal properties and combustion behavior of POSS- and bohemite-finished cotton fabrics. J Appl Polym Sci 123(1):426–436. https://doi.org/10.1002/app.34476
129. Li Y-C, Mannen S, Schulz J, Grunlan JC (2011) Growth and fire protection behavior of POSS-based multilayer thin films. J Mater Chem 21(9):3060–3069. https://doi.org/10.1039/C0JM03752D
130. Carosio F, Alongi J, Malucelli G (2011) α-Zirconium phosphate-based nanoarchitectures on polyester fabrics through layer-by-layer assembly. J Mater Chem 21(28):10370–10376. https://doi.org/10.1039/C1JM11287B
131. Carosio F, Alongi J (2016) Influence of layer by layer coatings containing octapropylammonium polyhedral oligomeric silsesquioxane and ammonium polyphosphate on the thermal stability and flammability of acrylic fabrics. J Anal Appl Pyrolysis 119:114–123. https://doi.org/10.1016/j.jaap.2016.03.010
132. Fang F et al (2015) Intumescent flame retardant coatings on cotton fabric of chitosan and ammonium polyphosphate via layer-by-layer assembly. Surf Coat Technol 262:9–14
133. Repon MdR, Sadia HT, Islam T, Rahman MM, Karim MdR (2022) Systematic pad dry cure technique for augmenting the flame resistance functionality of jute fabric. Clean Eng Technol 8:100492. https://doi.org/10.1016/j.clet.2022.100492
134. Vishwakarma A et al (2022) Construction of hydrophobic fire retardant coating on cotton fabric using a layer-by-layer spray coating method. Int J Biol Macromol 223:1653–1666. https://doi.org/10.1016/j.ijbiomac.2022.10.231
135. Yao X et al (2019) Flame-retardant and smoke suppression properties of nano MgAl-LDH coating on bamboo prepared by an in situ reaction. J Nanomater 2019:e9067510. https://doi.org/10.1155/2019/9067510
136. Carosio F, Laufer G, Alongi J, Camino G, Grunlan JC (2011) Layer-by-layer assembly of silica-based flame retardant thin film on PET fabric. Polym Degrad Stab 96(5):745–750. https://doi.org/10.1016/j.polymdegradstab.2011.02.019
137. Kovács Z, Pomázi Á, Toldy A (2023) Development of multifunctional flame-retardant gel coatings for automotive applications. Coatings 13(2):2. https://doi.org/10.3390/coatings13020345
138. Hackett AJ, Strover LT, Baek P, Malmström J, Travas-Sejdic J (2018) Polymer-grafted conjugated polymers as functional biointerfaces. In: Conjugated polymers for biological and biomedical applications. Wiley, pp 359–401. https://doi.org/10.1002/9783527342747.ch13
139. Alongi J, Tata J, Frache A (2011) Hydrotalcite and nanometric silica as finishing additives to enhance the thermal stability and flame retardancy of cotton. Cellulose 18(1):179–190. https://doi.org/10.1007/s10570-010-9473-z
140. Saleemi S et al (2020) Surface functionalization of cotton and PC fabrics using SiO_2 and ZnO nanoparticles for durable flame retardant properties. Coatings 10(2):2. https://doi.org/10.3390/coatings10020124
141. Lam YL, Kan CW, Yuen CWM (2011) Effect of titanium dioxide on the flame-retardant finishing of cotton fabric. J Appl Polym Sci 121(1):267–278. https://doi.org/10.1002/app.33618
142. Fanglong Z, Qun X, Qianqian F, Rangtong L, Kejing L (2016) Influence of nano-silica on flame resistance behavior of intumescent flame retardant cellulosic textiles: remarkable synergistic effect? Surf Coat Technol 294:90–94. https://doi.org/10.1016/j.surfcoat.2016.03.059
143. Beninate JV, Boylston EK, Drake GL, Reeves WA (1968) Conventional pad-dry-cure process for durable-flame and wrinkle resistance with tetrakis (hydroxymethyl) phosphonium hydroxide (THPOH). Text Res J 38(3):267–272. https://doi.org/10.1177/004051756803800307
144. Lam YL, Kan CW, Yuen CWM (2011) Effect of zinc oxide on flame retardant finishing of plasma pre-treated cotton fabric. Cellulose 18(1):151–165. https://doi.org/10.1007/s10570-010-9466-y

145. Guido E et al (2013) Thermal stability and flame retardancy of polyester fabrics sol–gel treated in the presence of boehmite nanoparticles. Polym Degrad Stab 98(9):1609–1616. https://doi.org/10.1016/j.polymdegradstab.2013.06.021
146. Samanta AK, Bhattacharyya R, Jose S, Basu G, Chowdhury R (2017) Fire retardant finish of jute fabric with nano zinc oxide. Cellulose 24(2):1143–1157. https://doi.org/10.1007/s10570-016-1171-z
147. Kundu CK, Song L, Hu Y (2020) Nanoparticles based coatings for multifunctional polyamide 66 textiles with improved flame retardancy and hydrophilicity. J Taiwan Inst Chem Eng 112:15–19. https://doi.org/10.1016/j.jtice.2020.07.013
148. Bifulco A, Imparato C, Aronne A, Malucelli G (2022) Flame retarded polymer systems based on the sol-gel approach: recent advances and future perspectives. J Sol-Gel Sci Technol. https://doi.org/10.1007/s10971-022-05918-6
149. Rane AV, Kanny K, Abitha VK, Thomas S (2018) Chapter 5—methods for synthesis of nanoparticles and fabrication of nanocomposites. In: Mohan Bhagyaraj S, Oluwafemi OS, Kalarikkal N, Thomas S (eds) Synthesis of inorganic nanomaterials. Micro and Nano Technologies. Woodhead Publishing, pp 121–139. https://doi.org/10.1016/B978-0-08-101975-7.00005-1
150. Talebi S, Montazer M (2020) Denim fabric with flame retardant, hydrophilic and self-cleaning properties conferring by in-situ synthesis of silica nanoparticles. Cellulose 27(11):6643–6661. https://doi.org/10.1007/s10570-020-03195-6
151. Noman MT et al (2018) In-situ development of highly photocatalytic multifunctional nanocomposites by ultrasonic acoustic method. Ultrason Sonochem 40:41–56. https://doi.org/10.1016/j.ultsonch.2017.06.026
152. Rezaie AB, Montazer M (2017) Polyester modification through synthesis of copper nanoparticles in presence of triethanolamine optimized with response surface methodology. Fibers Polym 18(3):434–444. https://doi.org/10.1007/s12221-017-6780-0
153. Ghanbari D, Salavati-Niasari M, Ghasemi-Kooch M (2016) In situ and ex situ synthesis of poly(vinyl alcohol)–Fe_3O_4 nanocomposite flame retardants. Particuology 26:87–94. https://doi.org/10.1016/j.partic.2015.12.003
154. Han Y, Zhang X, Wu X, Lu C (2015) Flame retardant, heat insulating cellulose aerogels from waste cotton fabrics by in situ formation of magnesium hydroxide nanoparticles in cellulose gel nanostructures. ACS Sustain Chem Eng 3(8):1853–1859. https://doi.org/10.1021/acssuschemeng.5b00438
155. Wang Y, Chai G, Zhu G, Sun Y (2020) Enhanced flame retardant efficiency of in situ polymerizing barium phenolic resin modified with carbon nanotubes. Mater Res Express 7(8):085602. https://doi.org/10.1088/2053-1591/abad06
156. Yu B et al (2014) Functionalized graphene oxide/phosphoramide oligomer hybrids flame retardant prepared via in situ polymerization for improving the fire safety of polypropylene. RSC Adv 4(60):31782. https://doi.org/10.1039/C3RA45945D
157. Shao Z-B, Cui J, Lin X-B, Li X-L, Jian R-K, Wang D-Y (2022) In situ coprecipitation formed Fe/Zn-layered double hydroxide/ammonium polyphosphate hybrid material for flame retardant epoxy resin via synergistic catalytic charring. Compos Part Appl Sci Manuf 155:106841. https://doi.org/10.1016/j.compositesa.2022.106841
158. Zhang H, Li K, Wang M, Zhang J (2021) The preparation of a composite flame retardant of layered double hydroxides and α-zirconium phosphate and its modification for epoxy resin. Mater Today Commun 28:102711. https://doi.org/10.1016/j.mtcomm.2021.102711
159. Yang Q, Jia Y, Zhou X, Zhang H (2022) Mechanically reinforced flame-retardant epoxy resins by layered double hydroxide in situ decorated carbon nanotubes. ACS Appl Polym Mater 4(9):6731–6741. https://doi.org/10.1021/acsapm.2c01110
160. Sun M et al (2022) In situ synthesis of layered double hydroxide/silica aerogel composite and its thermal safety characteristics. Gels 8(9):581. https://doi.org/10.3390/gels8090581
161. Zhou S, Qian Y, Chen X, Li L (2018) In situ synthesis of layered double hydroxides-silicon dioxide hybrids and its flame retardancy in EVA composites. J Therm Anal Calorim 134(2):1071–1082. https://doi.org/10.1007/s10973-018-7331-7

162. Han T, Zhao Z, Gillispie BA, Smith JR (2005) Effects of spray conditions on coating formation by the kinetic spray process. J Therm Spray Technol 14(3):373–383. https://doi.org/10.1361/105996305X59369
163. Xing L, Edward Glass J, Fernando RH (1999) Parameters influencing the spray behavior of waterborne coatings. J Coat Technol 71(3):37–50. https://doi.org/10.1007/BF02697898
164. Liu ZM et al (2020) Effect of viscosities on the spray characteristics of pressure swirl nozzle. J Appl Fluid Mech 13(3):861–870. https://doi.org/10.29252/jafm.13.03.30010
165. Harrison GM, Mun R, Cooper G, Boger DV (1999) A note on the effect of polymer rigidity and concentration on spray atomisation. J Non-Newton Fluid Mech 85(1):93–104. https://doi.org/10.1016/S0377-0257(98)00188-8
166. Del Secco B et al (2022) Particles emission from an industrial spray coating process using nano-materials. Nanomaterials 12(3):3. https://doi.org/10.3390/nano12030313
167. Faheem S, Baheti V, Behera P, Naeem S (2017) Development of flame retardant high loft polyester nonwovens. J Text Inst 108(8):1357–1364. https://doi.org/10.1080/00405000.2016.1251289
168. Carosio F, Ghanadpour M, Alongi J, Wågberg L (2018) Layer-by-layer-assembled chitosan/phosphorylated cellulose nanofibrils as a bio-based and flame protecting nano-exoskeleton on PU foams. Carbohydr Polym 202:479–487. https://doi.org/10.1016/j.carbpol.2018.09.005
169. Qiu X, Li Z, Li X, Zhang Z (2018) Flame retardant coatings prepared using layer by layer assembly: a review. Chem Eng J 334:108–122. https://doi.org/10.1016/j.cej.2017.09.194
170. Lee S, Seong D, Ju Y, Kwak HW, Kim WS, Lee D (2020) Revealing the flame retardancy mechanism of highly transparent cellulose nanopapers fabricated by vacuum filtration assisted layer-by-layer deposition. Carbohydr Polym 246:116628. https://doi.org/10.1016/j.carbpol.2020.116628
171. Laufer G, Kirkland C, Morgan AB, Grunlan JC (2013) Exceptionally flame retardant sulfur-based multilayer nanocoating for polyurethane prepared from aqueous polyelectrolyte solutions. ACS Macro Lett 2(5):361–365. https://doi.org/10.1021/mz400105e
172. Kim YS, Davis R, Cain AA, Grunlan JC (2011) Development of layer-by-layer assembled carbon nanofiber-filled coatings to reduce polyurethane foam flammability. Polymer 52(13):2847–2855. https://doi.org/10.1016/j.polymer.2011.04.023
173. Zhang T, Yan H, Peng M, Wang L, Ding H, Fang Z (2013) Construction of flame retardant nanocoating on ramie fabric via layer-by-layer assembly of carbon nanotube and ammonium polyphosphate. Nanoscale 5(7):3013. https://doi.org/10.1039/c3nr34020a
174. Pan H, Wang W, Pan Y, Song L, Hu Y, Liew KM (2015) Formation of self-extinguishing flame retardant biobased coating on cotton fabrics via layer-by-layer assembly of chitin derivatives. Carbohydr Polym 115:516–524. https://doi.org/10.1016/j.carbpol.2014.08.084
175. Alongi J, Carosio F, Frache A, Malucelli G (2013) Layer by layer coatings assembled through dipping, vertical or horizontal spray for cotton flame retardancy. Carbohydr Polym 92(1):114–119. https://doi.org/10.1016/j.carbpol.2012.08.086
176. Carosio F, Alongi J, Paravidino C, Frache A (2017) Improving the flame retardant efficiency of layer by layer coatings containing deoxyribonucleic acid by post-diffusion of hydrotalcite nanoparticles. Materials 10(7):7. https://doi.org/10.3390/ma10070709
177. Liu Y et al (2018) Effect of chitosan on the fire retardancy and thermal degradation properties of coated cotton fabrics with sodium phytate and APTES by LBL assembly. J Anal Appl Pyrolysis 135:289–298. https://doi.org/10.1016/j.jaap.2018.08.024
178. Uğur ŞS, Sarıışık M, Aktaş AH (2011) Nano-Al_2O_3 multilayer film deposition on cotton fabrics by layer-by-layer deposition method. Mater Res Bull 46(8):1202–1206. https://doi.org/10.1016/j.materresbull.2011.04.005
179. Chang S, Slopek RP, Condon B, Grunlan JC (2014) Surface coating for flame-retardant behavior of cotton fabric using a continuous layer-by-layer process. Ind Eng Chem Res 53(10):3805–3812. https://doi.org/10.1021/ie403992x
180. Wang B, Xu Y-J, Li P, Zhang F-Q, Liu Y, Zhu P (2020) Flame-retardant polyester/cotton blend with phosphorus/nitrogen/silicon-containing nano-coating by layer-by-layer assembly. Appl Surf Sci 509:145323. https://doi.org/10.1016/j.apsusc.2020.145323

181. Carosio F, Di Blasio A, Cuttica F, Alongi J, Frache A, Malucelli G (2013) Flame retardancy of polyester fabrics treated by spray-assisted layer-by-layer silica architectures. Ind Eng Chem Res 52(28):9544–9550. https://doi.org/10.1021/ie4011244
182. Jordanov I et al (2019) Flame retardant polyester fabric from nitrogen-rich low molecular weight additives within intumescent nanocoating. Polym Degrad Stab 170:108998. https://doi.org/10.1016/j.polymdegradstab.2019.108998
183. Apaydin K, Laachachi A, Fouquet T, Jimenez M, Bourbigot S, Ruch D (2014) Mechanistic investigation of a flame retardant coating made by layer-by-layer assembly. RSC Adv 4(82):43326–43334. https://doi.org/10.1039/C4RA08500K
184. Song K, Ganguly I, Eastin I, Dichiara AB (2017) Lignin-modified carbon nanotube/graphene hybrid coating as efficient flame retardant. Int J Mol Sci 18(11):11. https://doi.org/10.3390/ijms18112368
185. Atwah AA, Khan MA (2023) Influence of microscopic features on the self-cleaning ability of textile fabrics. Text Res J 93(1–2):450–467. https://doi.org/10.1177/00405175211069881
186. Maity N, Dawn A (2020) Conducting polymer grafting: recent and key developments. Polymers 12(3):3. https://doi.org/10.3390/polym12030709
187. Purohit P, Bhatt A, Mittal RK, Abdellattif MH, Farghaly TA (2023) Polymer grafting and its chemical reactions. Front Bioeng Biotechnol 10:1044927. https://doi.org/10.3389/fbioe.2022.1044927
188. Redondo A et al (2019) Melt-spun nanocomposite fibers reinforced with aligned tunicate nanocrystals. Polymers 11(12):12. https://doi.org/10.3390/polym11121912
189. Si KJ, Chen Y, Shi Q, Cheng W (2018) Nanoparticle superlattices: the roles of soft ligands. Adv Sci 5(1):1700179. https://doi.org/10.1002/advs.201700179
190. Tsafack MJ, Levalois-Grützmacher J (2006) Plasma-induced graft-polymerization of flame retardant monomers onto PAN fabrics. Surf Coat Technol 200(11):3503–3510. https://doi.org/10.1016/j.surfcoat.2004.11.030
191. Ren Y, Gu Y, Zeng Q, Zhang Y (2017) UV-induced surface grafting polymerization for preparing phosphorus-containing flame retardant polyacrylonitrile fabric. Eur Polym J 94:1–10. https://doi.org/10.1016/j.eurpolymj.2017.06.037
192. Song P, Xu L, Guo Z, Zhang Y, Fang Z (2008) Flame-retardant-wrapped carbon nanotubes for simultaneously improving the flame retardancy and mechanical properties of polypropylene. J Mater Chem 18(42):5083. https://doi.org/10.1039/b808309f
193. Yuan G, Yang B, Chen Y, Jia Y (2019) Synthesis of a novel multi-structure synergistic POSS-GO-DOPO ternary graft flame retardant and its application in polypropylene. Compos Part Appl Sci Manuf 117:345–356. https://doi.org/10.1016/j.compositesa.2018.12.006
194. Tsafack MJ, Levalois-Grützmacher J (2006) Flame retardancy of cotton textiles by plasma-induced graft-polymerization (PIGP). Surf Coat Technol 201(6):2599–2610. https://doi.org/10.1016/j.surfcoat.2006.05.002
195. Mohamed AL, El-Sheikh MA, Waly AI (2014) Enhancement of flame retardancy and water repellency properties of cotton fabrics using silanol based nano composites. Carbohydr Polym 102:727–737. https://doi.org/10.1016/j.carbpol.2013.10.097
196. Hu W, Yu B, Jiang S-D, Song L, Hu Y, Wang B (2015) Hyper-branched polymer grafting graphene oxide as an effective flame retardant and smoke suppressant for polystyrene. J Hazard Mater 300:58–66. https://doi.org/10.1016/j.jhazmat.2015.06.040
197. Shi L (2000) An approach to the flame retardation and smoke suppression of ethylene–vinyl acetate copolymer by plasma grafting of acrylamide. React Funct Polym 45(2):85–93. https://doi.org/10.1016/S1381-5148(00)00015-8
198. Kundu CK et al (2017) UV grafting of a DOPO-based phosphoramidate monomer onto polyamide 66 fabrics for flame retardant treatment. Ind Eng Chem Res 56(6):1376–1384. https://doi.org/10.1021/acs.iecr.6b04188
199. Shi X, Peng X, Zhu J, Lin G, Kuang T (2018) Synthesis of DOPO-HQ-functionalized graphene oxide as a novel and efficient flame retardant and its application on polylactic acid: thermal property, flame retardancy, and mechanical performance. J Colloid Interface Sci 524:267–278. https://doi.org/10.1016/j.jcis.2018.04.016

200. Heydari S, Tavanaie MA, Montazer M, Mahmoudi-Rad M (2024) A study on melt spinning of polyethylene terephthalate filament yarns containing various concentrations of synthesized CuO nanoparticles. J Appl Polym Sci 141(4):e54861. https://doi.org/10.1002/app.54861
201. Huang Y-P, Chen T-K, Tang J-W, Yeh C, Tien C-H (2007) Effect of PET melt spinning on TiO_2 nanoparticle aggregation and friction behavior of fiber surface. Ind Eng Chem Res 46(17):5548–5554. https://doi.org/10.1021/ie070248y
202. de Oliveira AD, Beatrice CAG, de Oliveira AD, Beatrice CAG (2018) Polymer nanocomposites with different types of nanofiller. In: Nanocomposites—recent evolutions, IntechOpen. https://doi.org/10.5772/intechopen.81329
203. Esthappan SK, Kuttappan SK, Joseph R (2012) Thermal and mechanical properties of polypropylene/titanium dioxide nanocomposite fibers. Mater Des 37:537–542. https://doi.org/10.1016/j.matdes.2012.01.038
204. Dong Q et al (2013) Synergistic effect of DOPO immobilized silica nanoparticles in the intumescent flame retarded polypropylene composites. Polym Adv Technol 24(8):732–739. https://doi.org/10.1002/pat.3137
205. Erdem N, Cireli AA, Erdogan UH (2009) Flame retardancy behaviors and structural properties of polypropylene/nano-SiO_2 composite textile filaments. J Appl Polym Sci 111(4):2085–2091. https://doi.org/10.1002/app.29052
206. Wang Q et al (2013) Synthesis of flame-retardant polypropylene/LDH-borate nanocomposites. Macromolecules 46(15):6145–6150. https://doi.org/10.1021/ma401133s
207. Rault F et al (2009) Effect of manganese nanoparticles on the mechanical, thermal and fire properties of polypropylene multifilament yarn. Polym Degrad Stab 94(6):955–964. https://doi.org/10.1016/j.polymdegradstab.2009.03.012
208. Yan X et al (2020) Porous fibers surface decorated with nanofillers: from melt-spun PP/PVA blend fibers with silica nanoparticles. J Appl Polym Sci 137(11):48470. https://doi.org/10.1002/app.48470
209. Didane N, Giraud S, Devaux E, Lemort G, Capon G (2012) Thermal and fire resistance of fibrous materials made by PET containing flame retardant agents. Polym Degrad Stab 97(12):2545–2551. https://doi.org/10.1016/j.polymdegradstab.2012.07.006
210. Wang M, Zhu M, Sun B (2006) A new nano-structured flame-retardant poly(ethylene terephthalate). J Macromol Sci Part A 43(11):1867–1875. https://doi.org/10.1080/10601320600941250
211. Salaün F, Lemort G, Butstraen C, Devaux E, Capon G (2017) Influence of silica nanoparticles combined with zinc phosphinate on flame retardant properties of PET. Polym Adv Technol 28(12):1919–1928. https://doi.org/10.1002/pat.4081
212. Perrot C, Piccione PM, Zakri C, Gaillard P, Poulin P (2009) Influence of the spinning conditions on the structure and properties of polyamide 12/carbon nanotube composite fibers. J Appl Polym Sci 114(6):3515–3523. https://doi.org/10.1002/app.30875
213. Klitsch J, Pfaendner R, Fasel C, Schönberger F (2022) Mode of action of Zn-DOPOx and melamine polyphosphate as flame retardants in glass fiber-reinforced polyamide 66. Polymers 14(18):18. https://doi.org/10.3390/polym14183709
214. Chen Y, Wang Q, Yan W, Tang H (2006) Preparation of flame retardant polyamide 6 composite with melamine cyanurate nanoparticles in situ formed in extrusion process. Polym Degrad Stab 91(11):2632–2643. https://doi.org/10.1016/j.polymdegradstab.2006.05.002
215. Dahiya JB, Kandola BK, Sitpalan A, Horrocks AR (2013) Effects of nanoparticles on the flame retardancy of the ammonium sulphamate-dipentaerythritol flame-retardant system in polyamide 6. Polym Adv Technol 24(4):398–406. https://doi.org/10.1002/pat.3095
216. Edenharter A, Feicht P, Diar-Bakerly B, Beyer G, Breu J (2016) Superior flame retardant by combining high aspect ratio layered double hydroxide and graphene oxide. Polymer 91:41–49. https://doi.org/10.1016/j.polymer.2016.03.020
217. Li Y et al (2018) Durable flame retardant and antibacterial finishing on cotton fabrics with cyclotriphosphazene/polydopamine/silver nanoparticles hybrid coatings. Appl Surf Sci 435:1337–1343. https://doi.org/10.1016/j.apsusc.2017.11.269

218. Liu S, Pandey A, Duvigneau J, Vancso J, Snoeijer JH (2018) Size-dependent submerging of nanoparticles in polymer melts: effect of line tension. Macromolecules 51(7):2411–2417. https://doi.org/10.1021/acs.macromol.7b02353
219. Thanh NTK, Maclean N, Mahiddine S (2014) Mechanisms of nucleation and growth of nanoparticles in solution. Chem Rev 114(15):7610–7630. https://doi.org/10.1021/cr400544s
220. Datta M, Nandi S, Nath D, Debnath S, Basu G (2022) Fiber cross sectional shape: a persuadable feature for assessment of acoustic noise control property of some natural fibers and its geometrical modeling. J Nat Fibers 19(17):15600–15613. https://doi.org/10.1080/15440478.2022.2131680
221. Boltachev GS, Ivanov MG (2020) Effect of nanoparticle concentration on coagulation rate of colloidal suspensions. Heliyon 6(2):e03295. https://doi.org/10.1016/j.heliyon.2020.e03295
222. Andrade-Guel M et al (2022) Green flame-retardant composites based on PP/TiO_2/Lignin obtained by melt-mixing extrusion. Polymers 14(7):7. https://doi.org/10.3390/polym14071300
223. Alongi J, Carosio F, Malucelli G (2014) Current emerging techniques to impart flame retardancy to fabrics: an overview. Polym Degrad Stab 106:138–149. https://doi.org/10.1016/j.polymdegradstab.2013.07.012
224. Attia NF, Mohamed A, Hussein A, El-Demerdash A-GM, Kandil SH (2023) Greener bio-based spherical nanoparticles for efficient multilayer textile fabrics nanocoating with outstanding fire retardancy, toxic gases suppression, reinforcement and antibacterial properties. Surf Interfaces 36:102595. https://doi.org/10.1016/j.surfin.2022.102595
225. Song G, Lu Y (2013) Handbook of fire resistant textiles. Sawston Woodhead, pp 520–541
226. Paul DR, Robeson LM (2008) Polymer nanotechnology: nanocomposites. Polymer 49(15):3187–3204
227. Chitrphiromsri P, Kuznetsov AV (2005) Modeling heat and moisture transport in firefighter protective clothing during flash fire exposure. Heat Mass Transf 41(3):206–215. https://doi.org/10.1007/s00231-004-0504-x
228. Chen ZCQ, Chen Y, Chen Y (2008) Textile bioengineering and informatics symposium proceedings
229. Mandal S, Song G (2012) Evaluating the physiological performance of thermal protective clothing to control heat stress, presented at the fiber society spring conference, St. Gallen, Switzerland
230. Mandal S, Chowdhury IZ, Mazumder N-U-S, Agnew RJ, Boorady LM (2022) Characterization of sweat drying performance of single layered thermal protective fabrics used in high-risk sector workers clothing. Polymers 14(24):5393
231. Mandal S, Song G (2012) Modeling for predicting the performances of thermal protective clothing, presented at the 5thEuropean conference on protective clothing and NOKOBETEF 10, Valencia, Spain
232. Illeperuma WR, Rothemund P, Suo Z, Vlassak JJ (2016) Fire-resistant hydrogel-fabric laminates: a simple concept that may save lives. ACS Appl Mater Interfaces 8(3):2071–2077
233. Nie Y, Mugaanire IT, Guo Y, Wang R, Hou K, Zhu M (2021) A hybrid hydrogel/textile composite as flame-resistant dress. Prog Nat Sci Mater Int 31(1):33–40
234. Danihelová A et al (2023) The effect of flame retardants on the fire technical characteristics of recycled textiles. J Ind Text 53:15280837231202526. https://doi.org/10.1177/15280837231202526
235. Nandi S, Mahish SS, Das SK, Datta M, Nath D (2023) A review of various recycling methods of PET waste: an avenue to circularity. Polym-Plast Technol Mater 62(13):1663–1683. https://doi.org/10.1080/25740881.2023.2222791
236. Ronkay F et al (2019) Development of flame-retarded nanocomposites from recycled PET bottles for the electronics industry. Polymers 11(2):2. https://doi.org/10.3390/polym11020233
237. Zhou X, Liu Y, Ren Y, Liu X (2022) A sustainable strategy for preparation of flame-retardant cotton fabric by phosphorylation of recycled cotton. Text Res J 92(19–20):3766–3781. https://doi.org/10.1177/00405175211068783

238. Keller JG et al (2021) Aerogels are not regulated as nanomaterials, but can be assessed by tiered testing and grouping strategies for nanomaterials. Nanoscale Adv 3(13):3881–3893. https://doi.org/10.1039/D1NA00044F

Chapter 11
Nano-finishing of Natural Fibres

Subhadeep Paul and Saikat Ghosh

1 Introduction

Nanofinishing of natural fibres refers to the enhancement of natural fibre properties through the chemical finishing of nanoparticles. Natural fibres obtained from plants (cotton, hemp, jute, flax, etc.) and animals (wool, silk, etc.) are known for their inherent fibre properties: moisture absorption, breathability, tactile feel and biodegradability [111]. Nanofinishing in natural fibres has the potential to improve various existing properties of the fibres like durability, comfort and functional [4]. Over the years, nanofinishing has created immense possibilities to provide new functional properties which cannot be obtained by conventional finishes.

High-performance finishes like UV-protection, flame-retardant, antibacterial, moisture management, water and oil-repellent, superabsorbent, etc., are possible via nanofinishing which has accounted for the various applications in technical textiles [43]. The preliminary work began through the initial research of Dr. Soane who started a nanotechnology-based company called nanotex in [96] (Nanotex). The products developed were specific to the needs of the textile industries. The nanofinishes developed were based on inspiration from natural substances like the "lotus effect", where the textile surface was designed to mimic the surface of a lotus leaf which is hydrophobic in nature. Over the years, biomimicry has been an inspiration to

S. Paul (✉) · S. Ghosh
Technical Textiles Research Centre, School of Arts and Humanities, University of Huddersfield, Huddersfield, UK
e-mail: subhadeep.paul@northumbria.ac.uk

S. Ghosh
e-mail: S.Ghosh3@hud.ac.uk

S. Paul
Living Construction Group, Hub for Biotechnology in the Built in Environment, Applied Sciences Department, Northumbria University, Newcastle-upon-Tyne, UK

S. Bairagi et al. (eds.), *Nanotechnology in Textile Finishing*, Materials Horizons: From Nature to Nanomaterials, https://doi.org/10.1007/978-981-97-2696-7_11

provide multiple technical finishes to textile materials such as superhydrophobicity, self-cleaning, energy conservation, insect repellent and so on [31].

Application of nanofinished textile materials has largely been in the field of technical textiles which are designed for specific functionalities required. Nanofinishing applied on natural fibre-based products has varied applications in industries, biomaterials (medical and healthcare), composites (automobiles, constructions, buildings, etc.), agriculture and horticulture, packaging materials and many more [149].

Functionalisation of textiles using nanoparticles has reached a revolutionary phase and this can be attributed to the significantly improved properties of the nanoparticles embedded on the surface and showing much better results compared to bulk chemicals or conventional finishes [128]. The advantage of nanoparticles is its surface to volume ratio, which is very high due to their small range of size (1–100 nm) thus provide an enhanced property of the material. There are various types of nanomaterials which are involved in imparting different functionalities to natural fibres and they can be classified as carbon-based nanomaterials (CNT, carbon black, etc.), inorganic and organic nanoparticles (Ag, Cu, TiO_2, ZnO, chitosan, alginate, etc.), composite nanomaterials, hybrid nanomaterials and polymeric nanomaterials [90]. The future trend in nanofinishing of technical textiles is to develop multi-functional textiles which are efficient, durable, cost effective and causes less impact on the environment. At the same time, nanomaterials that can impart more than one finish is beneficial in terms of reduced chemical consumption, requiring few processing steps and better outcomes.

Natural fibres like cotton, wool, silk, linen, hemp, etc., have been used for manufacturing textile materials for over 4000–5000 years. Natural fibres are mostly from plant and animal origin as shown in Fig. 1. They are known for their high strength, durability and absorbency, some are naturally antibacterial (hemp) and are comfortable against the skin [50]. Since the industrial revolution and the discovery of manmade fibres, the fibre market and its demand have shifted towards them. Synthetic fibres due to their low cost, faster production rate, high strength and durability, compared to natural fibres were preferred by consumers all over the world. Other properties like diameter, crimp, fibre length, crystallinity, etc., can be customised by varying several process parameters during preparation of the fibre. Eventually, the amount of research related to natural fibres was significantly reduced. Over the last few decades, a lot of focus has been towards imparting circularity and making sustainable solutions for existing and upcoming textile materials [62]. Manmade fibres contribute towards the depletion of fossil fuels, non-biodegradable waste generation, micro-plastic generation and emission of greenhouse gases during the manufacturing stage and life cycle of a product. This has shifted the global market towards the production of natural fibres which are biodegradable, easier to dispose of, do not generate microplastics and help in the completion of carbon cycles [11].

The advancement of nanotechnology has resulted in the introduction of various types of nanomaterial mainly inorganic and organic in nature. Among them, recent developments have made organic nanomaterials highly effective to be applied on textile materials to obtain finishes like antibacterial, flame-retardant, UV-protection,

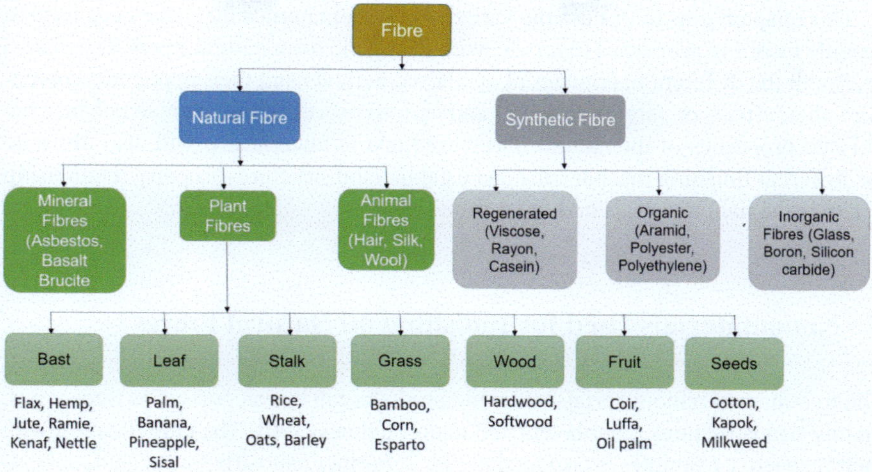

Fig. 1 Classification of natural fibres [71]

aromatherapy, thermoregulation and many more. The biocompatibility of organic nanoparticles with natural fibres is more compared to their synthetic counterparts.

Natural fibres are highly hygroscopic and possess good tactile and comfort properties which is why it is highly used in garment manufacturing [95]. Natural fibre is also highly breathable which aids in regulating the body temperature and moisture sorption properties. Thus, making textile materials made from them wearable. Currently significant amount of research is being oriented towards the development of nanofinished natural fibre-based textile materials. The challenge of nanofinishing of natural fibre products is to enhance these properties without compromising the tactile feel, aesthetics and breathability of the textile material.

In applications based on medical textiles and personal protective equipments, natural fibres are preferred due to their hypoallergenic properties and low possibilities of skin irritation. Nanofinishing has the capability to augment the performance of such medical products.

Researchers have also reported that the combination of natural and manmade fibres can enhance the capabilities of certain products and achieve a desired balance in properties. This section will be dealt in detail in the next chapter.

Challenges

There are multiple challenges involved with textile materials made from natural fibres and are finished using various types of nanomaterials. Among them some major challenges are summarised which are being perceived for nanofinishing in natural fibres; (i) compatibility of nanomaterials with the fibre structure and composition, (ii) maintaining biodegradability of natural fibres after incorporation of nanomaterials in the substrate, (iii) reduction in tensile strength of the textile materials after treatment, (iv) durability of nanomaterials, so that they don't leach out to the environment during washing.

This chapter aims to discuss the varieties of nanomaterials that have been used to provide finishing properties to textile materials made from natural fibres. It will also deal with the different nanomaterial synthesis methods and their application procedure, their effects on fabric properties and various process parameters to enhance the finished properties of the natural fibre-based textile materials. It will also illustrate the different finishing mechanisms and establish the structure–property relationship between the nanomaterials and the finished textiles.

2 Nanomaterials Used for Finishing on Natural Fibres

The treatment of nanomaterials on textile products can be categorised as three types, mainly nanostructures, nanolayers and nanoroughness [90]. The classification with their respective examples is provided in Fig. 2. Further nanostructures can be subcategorised to nanoparticles and nanocomposites. Nanoparticles can be of various shapes and dimensions, mainly zero dimension (nanosized particles), one dimension (filament or rods), two dimension (films) two multidimensional nanoparticles [105]. These nanoparticles are of two types mainly inorganic and organic. Inorganic nanoparticles that have been used for finishing are mostly metal and metal oxides [18, 88]. Organic nanoparticles majorly used are chitosan, cellulose, carbon nanotubes, graphene and many more. Researchers have abundantly used various types and shapes of nanoparticles for obtaining various technical finishes and end-use properties for textile materials [135, 137].

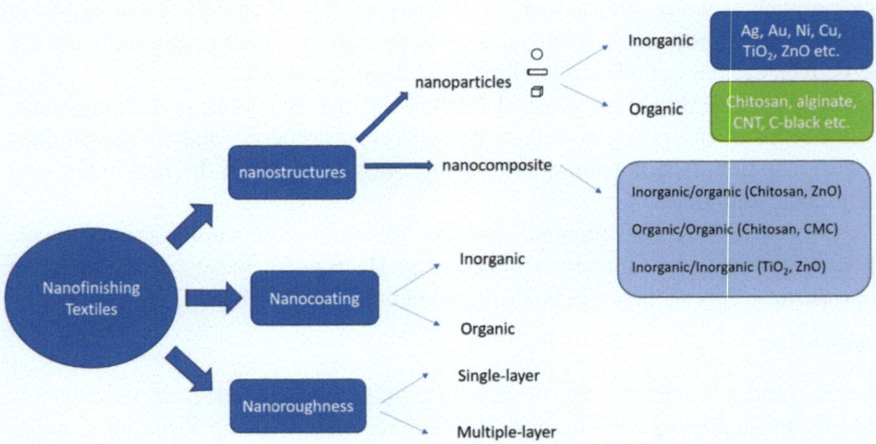

Fig. 2 Schematic representation of types of nanomaterials involved in nanofinishing of textiles

2.1 Types of Nanoparticles

Nanoparticles of dimensions between 1 and 100 nm are embedded in the fibre structure of textile products by different application procedures. The nanoparticles are synthesised using certain reaction conditions and process parameters. Depending on the reaction temperature, time, auxiliaries, precursor concentration, type of nucleation, pH, etc., the nanoparticles synthesised can alter in shape and size [64]. Both size and shape affect the properties of the nanoparticles as well as the textile structure where they are embedded [39, 94].

2.1.1 Metal Nanoparticles

Today the utilisation of metal nanoparticles is drawing a lot of attention to impart various functionalities and enhance the fabric properties [137]. The metal nanoparticles like Ag, Au, Pd, Pt, Cu, Ni are known for imparting finishes like antimicrobial, water and stain repellence and thermal properties [28], El-Din Zakaria [33, 85, 104]. Additionally, metal nanoparticles can also exhibit unique optical properties which can further change the colouration of the textile materials [46]. For example, the colour of Ag nanoparticles can vary from metallic yellow to dark pink depending upon the size of the nanoparticles. This happens due to the surface plasmons which exhibit a differential light scattering due to the change in shape and size [103]. Metal nanoparticles like silver and copper can be used to impart conductivity within a textile structure. They can be used to develop conductive and electronic textiles [49].

Silver (Ag)

Silver nanoparticles due to their small particle size and high surface area are susceptible to close interaction with bacterial cells. The silver nanoparticles get attached to the cell wall via electrostatic bonding and damage the lipids, proteins and DNA of the microbial cells. This results in the death of the bacterial cell which is an established mechanism for the antibacterial finishing of textile materials [134].

The finish provided by silver has been attempted to make it less toxic and biodegradable. Thus silver has been used alongside poly (*N*-vinyl pyrrolidone) PVP, SiO_2, chitosan and carboxymethyl cellulose (CMC) as nanocomposite to augment the antibacterial activity or provide other functionality to the textile material [20, 93, 98].

Gold (Au)

Gold nanoparticles due to their high surface plasmon resonance (SPR) can absorb and scatter light very strongly depending upon the size of the nanoparticles. Recent studies have reported that adjusting the surface functionalities of gold nanoparticles can make them efficient against Gram positive and negative bacteria [78]. The application of gold nanoparticles on textiles is also known for its high electrical conductivity, catalytic activity and optical properties. Gold nanoparticles are heavily used in the field of medical science specifically for drug deliveries in diseases like cancer, cardiovascular disease and diabetes mellitus [69].

Copper (Cu)

Copper nanoparticles are also known for their high thermal properties, catalytic activity, optical and electronic properties and antimicrobial properties. Copper nanoparticles have also drawn attention for their utilisation as conductive inks and further application in natural fibre-based textile materials [24].

2.1.2 Metal Oxides

Metal oxides in the field of inorganic chemistry have contributed heavily towards the finishing of natural fibre-based textiles through their application. The advent of nanometal oxides in textile finishing was due to its UV-blocking ability [142]. With further progress of research, metal oxides became renowned for its multi-functional properties, which means that they can provide more than one functionality in a single bath treatment. Nano-TiO_2 due to their high chemical stability, non-toxicity and high temperature resistant are very successfully used in the finishing areas [148]. They are heavily used as semi-conductors which can generate pair of electrons and holes under UV radiation resulting in a flow of electron towards the positively charged hole [139]. This creates the photo-excitation state of the metal oxides that can be applied for degradation of aromatic chemical structures, also known as photocatalytic effect. All metal oxides show very high antibacterial properties due to their highly active metal state, which can deposit across the bacterial cell wall to target its internal lipids, DNA and cell membrane [30].

ZnO is an *n*-type semiconductor known for its UV-blocking ability thus it has a very wide application area in sunscreens, wastewater treatments, piezoelectricity, solar cells, sensors and so on. The UV absorbency range of ZnO nanoparticles is typically in the range of 350–365 nm but this range is subject to vary depending on the particle size of the ZnO nanostructures [122]. The mechanism of ZnO nanoparticles being involved as a flame-retardant finishing agent has also been established. Metal oxide due to its high thermal stability or high bond energy does not allow the propagation of flame through its structure [147].

Metal oxides like silica have been used to provide water-repellent and hydrophobic finishes to natural fibre-based fabrics. SiO_2 can enhance the surface roughness of a textile substrate which can result in an increased barrier for water droplets to penetrate the structure of SiO_2 [19].

Researchers in recent times have established that metal oxides like Zinc stannate ($ZnSnO_3$) exhibit a greater range of UV absorbency properties compared to ZnO and TiO_2 [132]. Zinc stannate nanoparticles possess multi-functional finishing ability mainly UV-absorbent, antibacterial, photocatalytic activity as well as flame retardancy [29, 32]. This could be considered an efficient and sustainable way of using zinc stannate nanoparticles in the multi-functional finishing of textiles.

Several metal salts and their nanostructures have been used for providing flame-retardant, antibacterial and hydrophobic finishes to conventional as well as technical textiles. Aluminium hydroxide is readily used in the form of nanocomposites to impart flame-retardant finishes in cotton fabric. They are also added as fillers along with conventional fire retardant chemicals to aid the flame-retardant property of the textile substrate. Aluminium hydroxide is known to decompose at 180 °C and release high amount of water vapour molecules to decrease the intensity of the fire and itself get converted to Al_2O_3 which further does not allow the fire to propagate [129]. Other salts like silver nitrate, nickel nitrate, copper chloride and its respective nanostructures have also been used to obtain very high antibacterial properties on natural fibre-based textile materials.

In recent times, there has been a surge in the investigation of Mxene in material science and its application on textiles. Mxene is a type of two-dimensional inorganic compound (mostly metal carbides and nitrides) which has a lamellar stacking structure and contributes towards excellent electrical conductive properties [131]. In situ synthesis of titanium carbide was done on silk fabric in the form of 2D nanosheets to obtain excellent photo-thermal properties, under natural light irradiation the fabric can reach a temperature of 60 °C within 120 s [143]. Further, the same fabric provided outstanding UV-protection and antibacterial performance against *E. coli* bacteria. Mxene nanostructures are very innovative material which has found its application in multiple disciplines and its implication in textile finishing and electronic textiles could revolutionise this specific field.

2.1.3 Organic Nanoparticles

Organic nanoparticles are known to be biocompatible with natural fibres. Since toxicity of inorganic metals and their respective salts is a significant concern for the aquatic life, more focus is shifting towards organic-based nanoparticles which are biodegradable as well as compatible with natural fibres. Nanocellulose in recent times has been used in diversified application area starting from medical fields to textiles and this is attributed to its low cost, biodegradability and ease of manufacturing [124]. Among the subcategories of nanocellulose, cellulose nanocrystals and cellulose nanofibrils have been applied on textile substrates with the help of coating. Nanocellulose contains a high amount of OH groups and their surface groups can be modified and used against all cellulosic and protein-based fibres. The incorporation of nanocellulose on textile substrate has shown to increase the mechanical behaviour of the fibres. At the same time, nanocellulose has been used alongside other additives like chitosan, silica, zinc oxide to achieve properties like anti-UV, antibacterial, resistance to degradation on cotton fabric [68, 76, 144].

Chitosan is obtained from prawn or crab shells and is a derivative of chitin (exoskeleton of crabs) which is produced by deacetylation (enzymatic or chemical). Chitosan is biodegradable and susceptible to chemical or enzymatic catalysis. Thus, synthesis of chitosan nanoparticles and its application on natural fibre-based textile materials has its benefits. Chitosan nanoparticles are known for imparting properties antimicrobial properties, and they are used extensively in medical grade cotton fibres and fabrics which are used for wound dressing, surgical gowns, threads, bandages, nappies and tampons. Chitosan nanoparticles are also known to impart finishes like UV-protection, flame retardancy and moisture management. The amine functionality in chitosan allows it to interact with microbial cell membranes and disrupt the structure which leads to the antibacterial effect [110].

Cyclodextrins are a type of cyclic oligomers which are obtained from the degradation of starch. They are shaped in the form of hollow cavities and are made up of various glucose structures joined together. These materials can be used to encapsulate different nanoparticles as well as other finishing agents which can provide complex finishes like aroma, mosquito repellent and moisture management. Due to the availability of chlorine in the cyclodextrin structures, they can react easily with cellulose through nucleophilic substitution and thus obtain excellent durability [123].

The application of organic nanofibers generated through electrospinning can enhance many functional properties. Nanofiber applications include creating breathable and lightweight textiles for sportswear or medical textiles with improved barrier properties. Polylactic acid (PLA) and cellulose acetate-based nanofibers can be used to enhance filtration properties of various medical products or even apply moisture management finish in sports textiles [133].

Carbon-based materials like carbon nanotubes (CNT), carbon black, graphene and graphene oxides have been extensively used in textile finishing for their excellent mechanical, thermal and electrical properties. The incorporation of CNT into natural fibre-based textile materials like cotton, wool, jute, flax, etc., has resulted in a significant improvement in the tensile strength of the natural fibres [6]. Apart from that, CNT is also known to impart properties like flame-resistant, UV-resistant and superhydrophobicity to natural fibres. Grafting of cotton fabric with CNT alongside poly-butyl acrylate and creating a network armour on the fabric showed a great improvement in the tensile strength of the material [79]. This CNT-coated textiles could be a possible replacement for Aramid and Kevlar fabrics which are used for military grade equipment. Modification of natural fibres using CNTs can also find usage in medical textiles in wound dressings and implants. CNTs are also used to improve the dyeing and printing ability of natural fibre-based materials.

2.1.4 Nanocomposites

Nanocomposites were introduced in textile finishing applications keeping in mind the multi-functional properties that more than one type of nanoparticles could provide. Nanocomposites are mainly of three types organic-organic, organic–inorganic and inorganic-inorganic frameworks. Nanocomposites like organic silicone and fluorine could be applied to obtain superhydrophobic fabrics and enhance the hydrophobicity of a material. Polymers like polyvinyl acetate (PVA) alongside TiO_2 or ZnO nanoparticles can be used to provide high antibacterial as well as photocatalytic properties to cotton fabrics [89]. Thus, nanocomposites can enhance the properties of a single finish (synergistic effect) at the same time impart multi-functional finishes to natural fibre-based textile materials.

Polymer nanocomposites mostly consist of a confined matrix of polymer, nanoparticles and an arrangement which is of nanodimension. They are made by homogenous mixing of monomers with additives which are mostly inorganic or organic nanoparticles and applied on textile substrates by various application methods. It is possible to alter the type of nanostructures and impart multi-functional properties like UV-resistant, flame-retardant, antibacterial, moisture management, antistatic properties and many more on textile substrates [41].

Metal organic frameworks (MOF) are crystalline materials which consist of metal ions embedded in a framework with organic ligands (Fig. 3). These hybrid materials can be of different dimensions. They also possess a high surface area and porous structure along with diversified functionalities. This versatility of the material and its incorporation into textile substrates lead to their application in fuel cells and wastewater treatments. The metals present in the system can impart antimicrobial properties at the same time due to their highly crystalline structure as well as organic links they can be used as UV-blockers in textile finishing [121]. The structure of the MOF can be altered to obtain functionalities like high electrochemical activity and stability (requirements for smart textiles).

Since the advent of nanotechnology, various types of nanoparticles have been applied to natural fibre-based textile materials, among them the relevant ones are listed in Table 1. This shows the immense potential of the various types of nanomaterials that can be used in textile finishing. At the same time, the toxicology of these nanomaterials is an important consideration. Deposition of nanoparticles in the ecosystem has affected various life forms and the current trend is progressing more towards designing nanoparticles in such a way that they are biocompatible as well as biodegradable on their own.

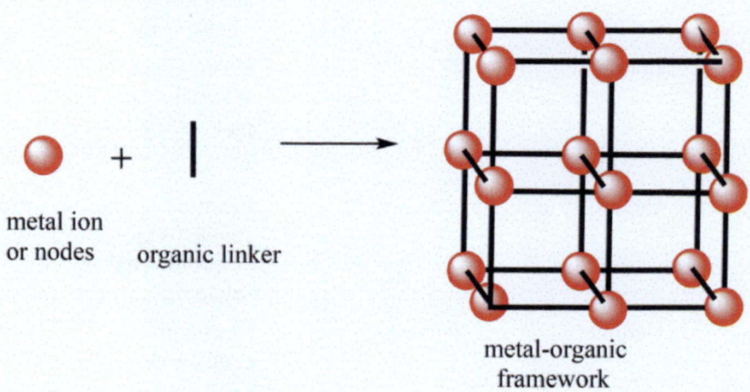

Fig. 3 Schematic representing metal organic framework (CC by 4.0 Deed http://creativecommons.org/licenses/by/4.0/) [121]

Table 1 Types of nanoparticles, their size and application properties used in natural fibre-based textile substrate

Type of nanostructures	Textile substrate	Size (nm)	Properties	References
Metal				
Ag	Jute, cotton, flax	47–80	Antimicrobial, electrical conductivity	[37, 47]
Cu	Cotton	60 and 100	Antimicrobial, catalytic activity	[119]
Au	Cotton	8–21	UV-resistant, catalytic, electrical conductivity	[127]
Metal oxides				
ZnO	Cotton, jute	40, 140 (nanorods)	UV-resistant, antibacterial, flame-retardant and hydrophobic	[13, 142]
TiO_2	Cotton	12 nm	UV-resistant, antibacterial, flame-retardant, self-cleaning	[74, 126]
CuO	Wool	36	Antibacterial, UV-protection and colouration	[109]
Al_2O_3	Wool, cotton	100–200, < 50	Flame-retardant, hydrophobic	[70, 140]
SiO_2	Cotton	150	Flame retardancy	[8]
Fe_3O_4	Cotton	19	UV-resistant, electrical conductivity	[120]
$ZnSnO_3$	Cotton	65	UV-resistant, antimicrobial, photocatalytic	[101]
Organic				
Carbon nanotubes	Cotton		Electrical conductivity, electromagnetic shielding	[80]
Graphene	Cotton	Nanosheet	Electrical conductivity	[59]
Nanocellulose	Hemp		Hydrophilicity, tensile strength	[27]

(continued)

Table 1 (continued)

Type of nanostructures	Textile substrate	Size (nm)	Properties	References
Chitosan	Wool, cotton	75, 100	Antibacterial, hydrophilicity	[26, 112]
Nanocomposites	Cotton, jute	38, 65	Antibacterial, UV-resistant, hydrophobic	[10, 136]

2.2 *Synthesis Techniques of Nanoparticles for Finishing of Natural Fibre-Based Textile Materials*

There are two methods of nanoparticle synthesis, top-down and bottom-up approaches. The schematic representation of these two phenomena is provided in Fig. 4. Further they are subdivided into several methods and these methods are chosen depending on the type, shape, size and material yield of the nanoparticles [105]. The techniques which are majorly used for the synthesis of nanoparticles for finishing natural fibre-based textile materials are discussed below.

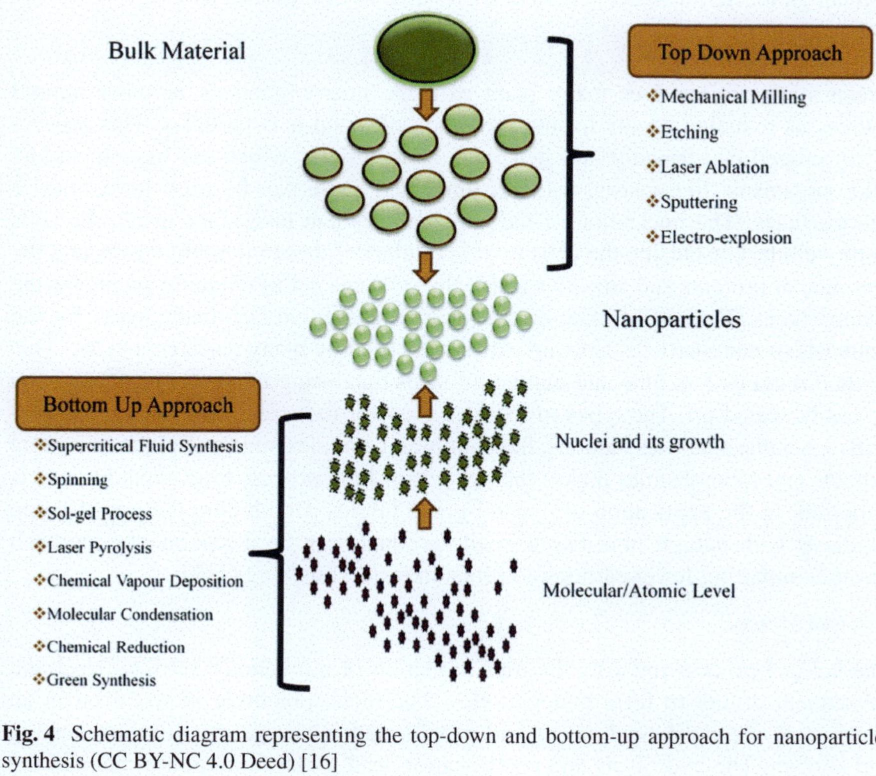

Fig. 4 Schematic diagram representing the top-down and bottom-up approach for nanoparticle synthesis (CC BY-NC 4.0 Deed) [16]

In other words, nanoparticle synthesis can be divided into two parts, (i) Ex situ where the nanoparticles are synthesised separately and then they are incorporated on the textile substrate; (ii) In situ where the nanoparticles grown/synthesised inside the polymer or fibre matrix, this is a one-step process. This includes the integration of nanoparticles within the fibre structure thus avoiding any other additional treatment on the textile materials. The textile substrate acts as a matrix to the nanoparticles which could further stabilise the nucleation of the nanoparticles, control the size of the nanoparticles and agglomeration. In situ synthesis of nanoparticles on textile substrate can be done through various other methods like sol–gel, hydrothermal or solvothermal, polymerisation, chemical reduction, etc.

Chemical Reduction

This is a chemical reduction process which takes place in a reaction involving metal salts. The metal salts are reacted with a reducing agent to form metal nanoparticles. The properties of the metal nanoparticles are dependent on their size, shape and size distribution. Using reducing agents and stabilisers provides precision and control of these parameters. This is a bottom-up approach and is easily scalable for bulk production. The size and distribution of the nanoparticles can be altered by adjusting the type of reducing agent and concentration of the reducing agent. The size, shape and distribution of metal nanoparticles like Ag, Cu, Au, etc., are reflected in their optical, antibacterial, water-repellent and conductive properties [63].

Green Synthesis

Green synthesis involves using plant extracts, microorganisms or other natural sources as reducing agents for the formation of metal nanoparticles. This process is considered eco-friendly and generates nanoparticles which are biocompatible. Microorganisms like yeasts, bacteria fungi and algae can be used to synthesise nanoparticles. The nucleation of the nanoparticles can take place inside the cells (intra-cellular) or outside the cells (extra-cellular) of these microorganisms and the presence of proteins and enzymes inside the cells can act as reducing agents for the nanoparticles. The use of plant extract as a reducing and stabilising agent for the synthesis of nanoparticles is being extensively used by many researchers [25]. This method requires less time and number of steps than using microorganisms and can be easily scaled up. The types of extracts used for reducing the metals are amino acids, phenols, alkaloids, tannins, flavonoids and complex sugars. The use of natural extracts and biomolecules makes the textile finishing process biocompatible and is important in the application of apparel-grade fabrics. Combining the use of green synthesis with natural fibre-based textile products can be a sustainable approach towards imparting fewer carbon footprints in textile finishing [117].

Sol–Gel Method

The sol–gel process involves the transformation of a precursor sol into a gel and subsequent drying to form nanoparticles. The metal precursor is dissolved in an aqueous or alcoholic medium and is transformed into a gel by constant heating and stirring. The hydrolysis and condensation lead to the formation of gel and the solvent is evaporated from the gel system to obtain the nanoparticles using freeze drying techniques [97].

Co-precipitation

Co-precipitation involves the simultaneous precipitation of two or more salts, resulting in the formation of nanoparticles. This method is commonly associated with the synthesis of metal and metal oxide nanoparticles. This involves the mixing of two metal salts in a solution and adjusting the appropriate pH which leads to precipitation of metal oxide nanoparticles, which is further obtained by centrifuge and purification of the material. This method generates a high yield and uniform size of the metal oxide nanoparticles.

Hydrothermal or Solvothermal

This method involves synthesis of nanoparticles in a hydrothermal reactor under subsequent pressure and high temperature. The textile materials are dipped in a solution containing precursors (e.g. metal oxides) and reducing and stabilising agents. Then they are transferred in an autoclave chamber at a designated pressure and temperature. Textile materials are subjected to degradation at higher temperatures, so the temperature range is kept within 80–200 °C depending on the type of fibre. The reaction time determines the formation of the nucleation and nanoparticles on the textile substrate. The yield of this method is comparatively lesser than other ex situ methods, but hydrothermal synthesis can be used to manufacture complex and 3D nanostructures which can provide elevated properties and specific functionalities [53].

2.3 Nanoparticle Application Techniques on Natural Fibre-Based Textile Materials

Many conventional and advanced techniques are used to apply nanoparticles on natural fibre-based textile materials. The most effective and common one among them is in the form of coating and lamination. Coating is the formation of a single or more than one chemical layer (natural or synthetic) either on one side or both sides of the textile substrate. Conventionally after the coating, the textile substrate is dried or cured depending on the type of materials. The thickness of the coating, concentration, chemical nature and number of layers depend on the end-use application of the product.

Conventional coating methods like knife, dip, roller, coating, calendar coating, etc., had certain limitations in terms of application. These methods were also associated with strength loss, poor abrasion properties, uneven adhesion and finally low durability. With the advancement of nanocoating, minimising of the coating-to-weight ratio was done and implementation of nanoparticles of varied dimensions via different new techniques were introduced. Techniques like vapour deposition, sputter coating, dip coating, sol–gel coating, layer-by-layer coating electrochemical deposition and plasma-assisted coating are now been extensively used for the application of nanoparticles in the form of coating [91].

2.3.1 Electroless Plating

Electroless deposition involves the surface coating of fibres with a uniform metallic layer via chemical reduction of metallic ion solution in an aqueous medium. The metallic deposition on the electroplated fibre makes them conductive and provides high shielding against electromagnetic interference (EMI). Electroless deposition of copper on Cu- and Ag-coated cotton fabric showed very high EMI shielding effect compared to Cu-coated cotton fabric samples. This method also showed good durability in terms of EMI shielding properties and antimicrobial behaviour after washing the fabric [5].

The electroless plating mechanism begins with surface activation using a sensitising agent like palladium or tin. After the catalytic seed layer is deposited, it behaves as a catalyst for the metal deposition. A plating bath is prepared which consists of metal ions in a reduced form, complexing agents, stabilisers and other chemicals. The surface-coated substrate is immersed in the bath allowing the catalytic activity to take place and the metal ions subsequently reduce to metal atoms and are deposited. The reactions involved in electroless plating can be altered by various process parameters like concentration of the metal salt, concentration of reducing agent, concentration of complexing agent and temperature and pH [40].

The concentration of the metal salt can be changed to alter the deposition rate of the metal. The concentration of the complexing agent can affect the stability of the solution bath. Variation in temperature of the bath is associated with the rate of deposition which can be enhanced or reduce stability at elevated temperatures. The pH of the bath solution affects the stability of the metals, effectiveness of the catalytic activity and reduction potential of the metal ions. These alternations can be done to achieve different properties at the surface level of the finished textile materials. Ensuring good adhesion between the metal coating and the natural fibres can be a challenge. The surface chemistry of natural fibres and metals is different thus several pre-treatments of natural fibre-based textile materials are done to enhance the surface coating ability of the metals through chemical treatments, plasma technology, surface etching, etc.

2.3.2 Vapour Deposition

Vapour deposition involves the deposition of material in vapourised form via physical (PVD) and chemical (CVD) means through condensation and chemical reaction. PVD pertains to an atom-by-atom transfer from a solid phase to a vapour phase and further deposition of a uniform layer on the surface of the textile substrate. PVD has a thickness range of a few nanometres to micrometres, whereas CVD can go till the range of millimetres. The techniques involved in PVD are vacuum evaporation, implantation and sputtering. Various nanostructures like ZnO, silver, copper, TiO_2, metal-doped oxides in the form of films have been sputter coated on cotton fabrics to obtain EMI shielding, antibacterial, conductivity, photocatalytic and multiple other properties [52]. Yarn obtained from Merino wool was used alongside silver nitrate solution to generate silver nanoparticles with the help of a reducing agent, and this composite yarn was further sputter coated with platinum and then carbon to provide a conductive layer [61]. The advantage of sputter coating is that nanostructures with high bond energy could be easily coated on a textile substrate at a relatively low temperature range.

CVD is associated with high temperature chemical reactions to form a coating from a vapour, where the by-products from the reaction are volatile in nature. CVD in recent times has been modified with plasma-enhanced CVD and photo-CVD to provide chemical reaction formation at lower temperature. Cotton fabric was reacted with tricholoromethyl silane and followed by pyridine at room temperature to form a nano-level silicone coating via this method to impart super hydrophobic and self-cleaning finishes [77]. The thickness of the coated layer can be altered by parameters like gas flow rates, temperature and deposition time in CVD. The uniformity of the coating depends on the gas flow rate of the precursor gases. The textile substrate temperature can be altered to improve the adherence of the coating and affect the reaction kinetics of the gases. And finally, deposition time directly links to the thickness of the coating [82].

Obtaining a uniform thickness across the natural fibre-based textile substrates through sputter coating (both PVD and CVD) is a challenge. A complex natural fibre like wool shows high variation in a single fibre diameter, scale structure as well as surface morphology [102]. Thus, it is always a challenge to impart a uniform sputter coating layer on natural fibres which contain variation in their natural shape and structures. Scaling up of PVD and CVD is also a major challenge due to the technology involved and is not feasible for industry scaled manufacturing.

2.3.3 Layer-by-Layer Method

Layer-by-layer method is used with oppositely charged polyelectrolytes coated in layers and they react within themselves by electrostatic forces. Each layer of coating is followed by a rinsing step to avoid any contamination with the next layer and remove any unattached adsorbent molecules. Multi-layer structures can be coated on the textile substrate using this method alternatively charged adsorbents. Process parameters like concentration of the electrolytes, temperature, adsorption time, rinsing can be altered to impart layers with various nanoparticles and layers of nanodimension. The advantage of this method is that any charged nanoparticles can be utilised to make the nanolayer coating on textile substrates. This technique has been efficiently used to manufacture natural fibre-based textiles with multi-functional finishes due to the flexibility of the type of nanoparticles that can be used during application [91].

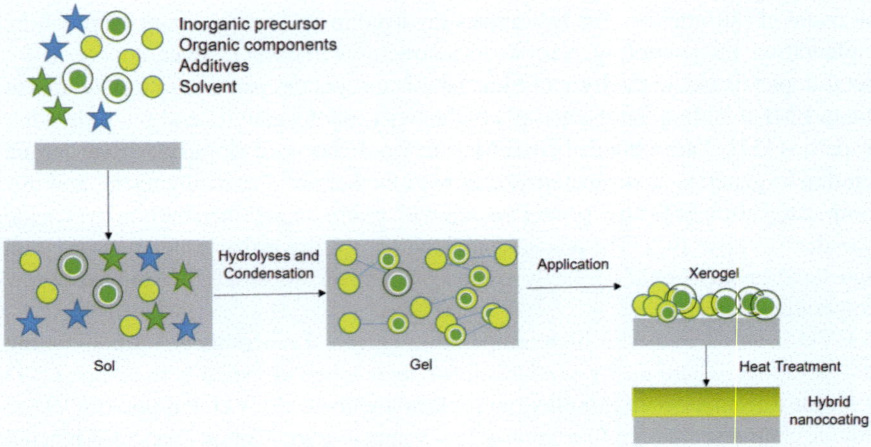

Fig. 5 Schematic diagram of sol–gel nanocoating [87]

Natural fibres with porous structures can be challenging to have a uniform penetration and coating on the fibre structure. Also, multiple layers of coating on natural fibre-based textiles could affect the flexibility, strength and handle of the fabric.

2.3.4 Sol–Gel Method

Sol–gel method for imparting nano-level coating on the textile substrate has been widely used in textile finishing. This method involves a conversion of solution to a gel form followed by conversion to its solid form by chemical reaction and drying of the gel (Fig. 5). This method can be applied to textiles using a pad-dry-cure technique and dip coating. Sol–gel coatings are typically associated with good uniformity, allowing for precise control over thickness and composition [22]. This technique can be used to prepare nanocomposite hybrid coatings on the surface of natural fibre-based textiles, which gives the flexibility to provide multi-functional finishes as well selective finishes like insect repellent (can be incorporated as additives within the coating). The coated layer is mostly transparent in nature which also provides a very high reflectance property to the material. The specific adhesion and the compatibility of the sol–gel nanocoating with the natural fibres is one of the major challenges.

2.3.5 Plasma Technology

Conventional finishes in textiles require very high usage of water, which further needs to go through wastewater treatment before they are re-used or released to different water bodies. Plasma technology has attracted more attention in textile finishing as it requires no water, no solvent and less treatment time compared to conventional finishing methods. Plasma treatment on textile substrates is known to functionalise the surface properties of the textile substrate rather than the bulk properties which keeps the intrinsic properties of the material almost intact. The change in functionalisation at the surface level happens due to the activation of the gas phase and the type of interactions with the substrate [118]. In plasma technology, different types of gases (oxygen, argon, hydrogen, methylene, acetylene, etc.) are used at high energy and electromagnetic waves are used to produce an ionised state of the gases, which further modifies the surface chemistry of the textile substrate. Depending on the type of gas, the chemical structure of the textile substrate changes, e.g. oxygen gas can make the hydrophobic surface of wool fabric hydrophilic and a fluorocarbon gas can make the surface of a hydrophilic cotton fabric hydrophobic. Treatment time, power density of the plasma and pressure can be altered to adjust the thickness of the layer of coating [54]. Another means of imparting nanoparticles on natural fibre-based textiles is to achieve a plasma etching on the surface of the textile which increases the adhesion of the fabric and then apply a pad-dry-cure technique with the respective nanoparticle solution. This method has shown very effective results in producing textile materials coated by AgNP, TiO_2, ZnO nanoparticles resulting in finishes like antibacterial, UV-resistant, water-repellent and many more such finishes. The durability of metal and metal oxide nanoparticles on natural fibres has always been a challenge, whereas this method has shown highly effective finishes even after several washing cycles of the textile substrate.

Non-thermal plasma treatment can be associated with radical polymerisation which further forms thin films on textile substrates. Natural fibre-based textiles like cotton or wool fabrics were padded with a monomer solution. The plasma technology is used to activate the polymerisation within the coated fabric which results in a highly crosslinked thin film on the surface of the fabrics. This can lead to graft polymerisation of the coating along with nanoparticles on the wool fabrics. This method improves the adhesion of the coating of the polymer film on the fabric and consumes no additional chemicals for the process to take place [118]. Plasma treatment on natural fibre-based textiles has resulted in creating rough surfaces which affects the feel or handle of the fabric. Variability in natural fibre properties, e.g. moisture contents, surface morphology and impurities in fibre can affect the uniformity of the plasma treatment.

There are multiple application methods of nanofinishing on natural fibre-based textile products as mentioned in Table 2, and the most relevant and recent technologies are discussed in this section.

Table 2 Different application form and techniques of nanofinishing on natural fibre-based textile materials

Type of nanoparticles	Application form and techniques on textile substrate	Type of textile substrate	Properties	References
TiO_2	Dip coating	Flax	Mechanical	[38]
SiO_2	Spray coating	Cotton	Superhydrophobic	[115]
Al_2O_3	Vapour deposition	Cotton	Super hydrophobic/ oleo phobic	[9]
CNT, Chitosan	Layer-by-layer coating	Cotton	Flame-retardant conductive	[84]
Chitosan, TiO_2, ZnO	pad-dry-cure	Cotton	Antimicrobial UV-protection	[107]
Nanoclay	Pad batch	Wool	Flame-retardant	[57]
Silver	In situ synthesis	Cotton	Antibacterial	[15]
Paraffin wax	Nanoencapsulation	Cotton	Thermoregulation	[60]
Silicone and Pine oil	Nanoemulsion	Cotton	Mosquito repellent, antibacterial	[106]
$AgNO_3$	Plasma	Cotton	Antibacterial	[66]

3 Nanofinishing of Cellulosic-Based Textiles

3.1 Cellulose Chemical Structure

Cellulose is the most abundant biopolymer in nature. This polymer is known for its biocompatibility and biodegradability. Cellulose is represented by a chemical formula of $(C_6H_{10}O_5)_n$. The basic repeating unit in cellulose is the β-*D*-glucopyranose which are linked by β-1,4-gucosidic linkage (Fig. 6). The degree of polymerisation in cellulose ranges from 300 to 15,000 units approximately, in wood they are reported as 10,000; whereas in cotton fibres, they are 15,000. The primary (–CH_2OH) hydroxyl group is attached to C-6, whereas the secondary is attached to the C-3 and C-4. Cellulose possesses a high crystallinity and does not dissolve in organic and inorganic solvents. Cellulose structure is prone to microorganism attacks, degradation under UV rays, dimensional stability, degradation at high temperatures [48].

These drawbacks of the cellulosic fibres (cotton) could be mitigated through various finishes that are applied to them. These are mainly targeted towards modification of the cellulose structure by reacting the finished chemicals with its highly reactive primary hydroxyl groups.

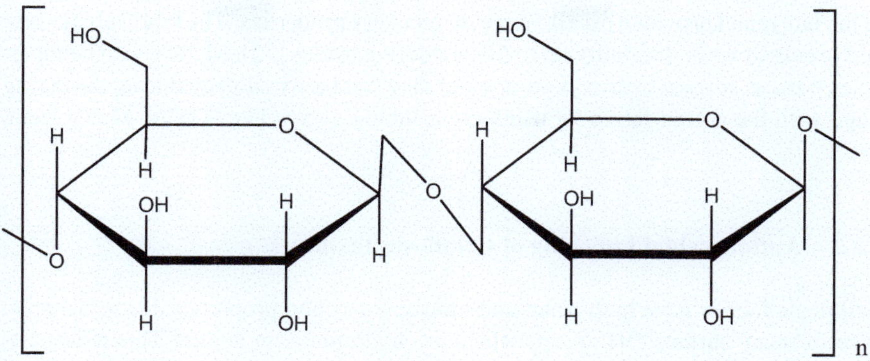

Fig. 6 Chemical structure of cellulose

3.2 Finishing of Cellulosic Textiles Using Nanoparticles

Before the beginning of the finishing treatment the cellulosic textile materials are thoroughly cleaned (scouring), bleached, mercerised and then finally dried. They further go through the colouration process where they are dyed with various dyes (reactive, vat, direct, etc.) and finally ready for finishing treatment. To overcome the drawbacks of cellulosic textiles and impart functional properties on them, they undergo softening, water and oil-repellent, UV-protection, easy-care, flame-retardant and self-cleaning finishes. All these finishes can be applied by different techniques, imparted by various types of nanoparticles and evaluated in terms of the functional behaviour of the finished textiles.

3.2.1 Crosslinking of Cellulosic Textiles

The easy-care finish was introduced in finishing of cotton fabrics to improve its crease recovery property. This property provides a wrinkle free fabric and allows it to recover from folding and deformations. This is attained by crosslinking of cellulosic molecules with various crosslinking agents. Several researchers have worked with various crosslinking agents like DMU, DMDHEU derivatives, PCA over the years to obtain formaldehyde-free crosslinking agents like BTCA which shows excellent durable press rating of cotton fabrics. The introduction of nanofinishing in textiles showed the researchers various ways to crosslink nanoparticles with the help of crosslinking agents on cotton fabrics [67]. This resulted in crease recovery of the cotton fabrics as well as improvement in the durability of nanofinishes on cotton fabric. The treatment of cotton fabric with ZnO, CaO and MgO nanoparticles along with cross-linkers like maleic acid and citric acid showed very high crease recovery angles at very low concentrations of the metal oxide nanoparticles. Crease recovery finishes can show a decrease in tear strength of the cotton fabric. Silica nanoparticles were attached to cotton fabric by plasma treatment which increased the hydrophilicity of the fabric and provided excellent crease recovery properties. The fabric also exhibited excellent wash durability up to 50 laundering cycles [75]. Metal oxide nanoparticles possess surface hydroxyl groups and they can be crosslinked during the curing stage with the cotton fabric by using crosslinking agents like BTCA, Maleic acid, etc.

3.2.2 Antimicrobial Finishing of Cellulosic Textiles

Cellulosic fabrics have high water and moisture sorption properties due to the presence of many surface OH groups which are hydrophilic in nature. This makes the fabric prone to microbial attacks and further degradation in properties and can result in generating body odour for the users. Over the years, researchers have worked with various antimicrobial agents and explored both organic and inorganic finishing agents to provide antimicrobial finishes to cotton fabrics. Nanoparticles like Ag, Cu, ZnO, TiO_2, chitosan, etc., showed relatively higher antimicrobial properties compared to their bulk counterparts [42]. The antimicrobial mechanism usually works in three counterparts, mainly (i) formation of highly reactive oxidative species (ii) release of ions and (iii) reactivity of the nanoparticles with the bacterial cellular membranes (Fig. 7). Metal nanoparticles depending on their grain size are known to reach the cell wall of the bacterial membrane. If the particle size is relatively small, the nanoparticles can diffuse through the cell wall of the bacterial cell, while the larger nanoparticles remain attached to the outside of the cell wall. The generation of metal ions leads to a disruption of the intra-cellular structures which ends up destroying the bacterial cell components [30].

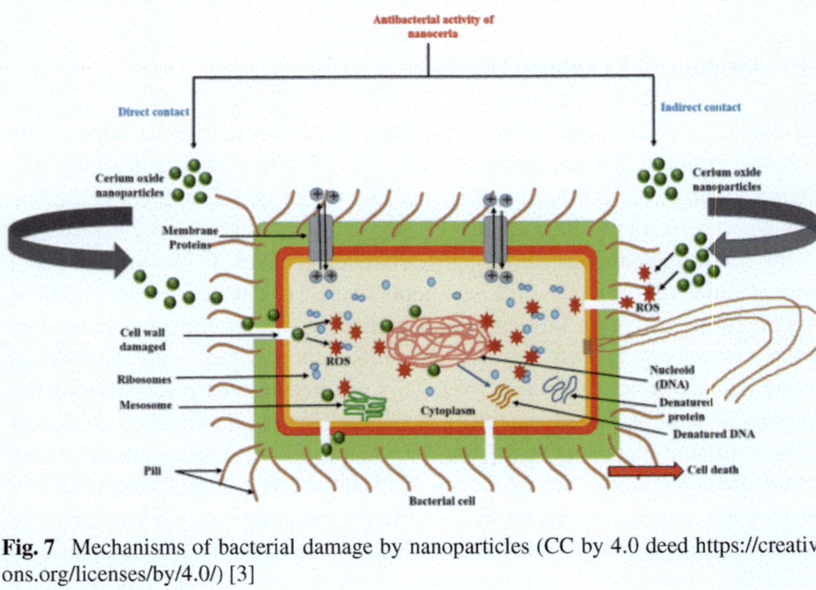

Fig. 7 Mechanisms of bacterial damage by nanoparticles (CC by 4.0 deed https://creativecommons.org/licenses/by/4.0/) [3]

Another mechanism proposed by Kumar et al. indicated that TiO_2 NPs induce considerable oxidative stress causing damage of DNA and *E. coli* cell death. This can generate reactive oxidative species like OH·, H_2O_2 and O_2^- which are responsible for the oxidation of polyunsaturated phospholipids, DNA, mRNA, ribosomes. Further oxidation of lipids imparts damage to the cell membrane morphology [72].

The final mechanism shows the tendency of the nanoparticles to bond with the thiol groups present in the enzymes. This does not allow the enzyme to multiply thus preventing further growth of the bacterial cell. This mechanism has been reported by researchers in finishing of cotton fabric.

Silver nanoparticles on cellulosic textiles are used extensively in medical applications for antibacterial protection against both Gram positive and Gram negative bacteria. Rehan et al. used a UV-irradiation, green synthesis method to apply silver nanoparticles on cotton fabric. The AgNP nanoparticles had a size range of 50–100 nm and resulted in 99% protection against *E. coli* bacteria [108]. In situ synthesis of silver nanoparticles on cotton fabric was performed by amination of the fabric and further dipping in silver nitrate solutions, followed by pad-dry-cure (150 °C). The fabric colour turned yellow indicating formation of silver nanoparticles, which was further confirmed through XRD and UV–visible spectroscopy. The treated fabric (scoured and bleached) showed very high bacterial reduction percentage (97–99%) for *B. subtilis*, *S. aureus*, *E. coli*, *P. aeruinosa* bacteria. They also showed good antibacterial properties after washing (20 cycles) [34]. The cellulose structure accounted for the growth of silver nanoparticles and scouring and bleaching of the fabric lead to an increase in surface adhesion of the nanoparticles.

Another in situ synthesis of AgNPs were performed by dipping cotton fabric titanium isopropoxide and urea nitrate and then subsequently padded, dried at 80 °C for 5 min and cured at 130 °C for 3 min. The TiO_2 nanoparticles are formed by hydrolysis of the titanium isopropoxide followed by peptization of Titanium hydroxide via release of nitric acid. The coated fabric showed very high antibacterial properties and UPF greater than 27 and did not show a great decrease in the properties after 20 wash cycles. The tensile strength and elongation at break of the cotton fabric showed a steady decrease with an increase in concentration of the urea nitrate solution. This may be caused by acid degradation of cellulose structure during drying conditions due to the presence of nitric acid [35]. Thus in situ synthesis of various inorganic-based nanoparticles like Ag, Cu, TiO_2, ZnO, $AgNO_3$ on cotton fabric has shown excellent reduction of bacterial growth and they have similar interaction with cotton fabric.

Organic nanoparticles like chitosan have also been used extensively for antibacterial textile finishing. Ali et al. in their research about incorporation of chitosan nanoparticles via LBL method on cotton fabric showed that the chitosan nanoparticle-coated fabric demonstrates higher antibacterial properties (41%) compared to its bulk counterparts. The average particle size of chitosan nanoparticle involved in this experiment was 115 nm which showed a bacterial reduction of $\approx 70\%$ against *Staphylococcus aureus* [7]. Over the years, researchers have shown how the incorporation of silver, ZnO, TiO_2 and other nanoparticles alongside chitosan nanoparticles provide better antibacterial efficacy to cellulosic fabrics [2].

Jute is a lignocellulosic-based fibre which shows an inherent antibacterial property due to the presence of phenolic groups in the lignin structure. Jute fabric was functionalised with CaO nanoparticles to attain 99% bacterial reduction against *Staphylococcus aureus* and *Escherichia coli* [12]. Further in situ synthesis of silver nanoparticles and subsequent coating on the surface of jute fabric showed excellent antibacterial properties against both Gram negative and Gram positive bacteria. The jute fabric retained the finishes after 15 wash cycles.

Mostly the effectivity of the antibacterial properties is dependent on factors like size, shape, distribution surface charge, release rate of the ions. Lower size of nanoparticles is associated with a higher surface area which provides much higher efficiency towards the antimicrobial properties. Also, metal nanoparticles of < 100 nm can get released from the textile substrate and enter the bacterial cell membrane to interact with the intra-cellular elements. Incorporation of nanoparticles on textile substrates is often associated with agglomeration, and this happens if the growth of the nanoparticles is not homogeneous. This can also affect the effectivity of the antibacterial property [30].

The other factors which are related to treatment conditions are concentration of treatment of the nanoparticles, synergistic effect and durability of the finishes. Higher treatment concentration is associated with higher antibacterial efficacies. Finishes where nanocomposites, combination of different types of nanoparticles are used, work as a synergistic effect and enhance the reactivity of the antibacterial mechanism.

3.3 Water and Oil-Repellent Finishes of Cellulosic Textiles

Water and oil-repellent finishes are important for various technical textile products. The factors affecting the wettability of a textile substrate are the type of fibre, geometry and roughness, pore size within the textile structure. These finishing agents are mainly fluorocarbon-based and non-fluorocarbon-based in nature. Nanoparticles applied on the surface of textiles create an extremely rough surface which decreases the wettability of the textile substrates. This is also known as the lotus leaf effect derived from biomimicry [73].

Silica nanoparticles alongside other fluorocarbon-based chemicals were applied via sol–gel process followed by dip-dry-cure on cotton fabric to obtain a water-repellent finish. The silica nanoparticles were crosslinked (Si–O–Si) with the fluoropolymer and they were further crosslinked with the cellulosic fabric via condensation. The fabric showed a high contact angle of 154° for water but not enough to call it superhydrophobic [73]. Cotton fabric was coated by LBL method with poly diallyldi-methylammonium chloride and nanocomposite of ZnO and SiO_2 in a colloid medium. Further post-treatment of the fabric was done with stearic acid. The number of layers and ZnO/SiO_2 weight ratio was varied to obtain a hydrophobic finish which gave a contact angle in the range of 115°–136° [1]. A combination of ZnO and SiO_2 nanoparticles has shown to impart water and oil-repellent finishes on cotton fabric [141]. The roughness created on the surface of the textile substrates through various nanofinishes can enhance the contact angle of water or oil drops on the fabric. Silicone-based compounds are extensively used in nano form because of its flexibility in chemical forumulation, where the surface OH groups can be utilised to develop a crosslinking with cellulosic fabric. Water-repellent finishes were obtained on jute fabric by pad-dry-cure application of silica nanoparticles. The silica nanoparticles (2% conc.) alongside NUVA (commercial water-repellent) showed an enhancement of the contact angle (138°) compared to only NUVA-treated jute fabric. This method also produced high wash durability due to crosslinking of the silica nanoparticles during curing method with the OH groups of cellulosic molecules [10].

In recent times, plasma technology has been used to impart nanoroughness on cellulosic fabrics to create hydrophobic fabrics. Gases based on fluorine chemistry (hexafluroethane) can be used during plasma treatment to provide a hydrophobic finish to the fabric. The application of hydrophobic or superhydrophobic finishes to cellulosic textiles should not affect the breathability of the material which is an inherent property of the material. Nanofinishes due to their high effectivity at low add-on% can keep the bulk properties of the fabric intact.

3.4 UV-Resistant Finishing of Cellulosic Textiles

The UV rays entering the earth are divided into three categories mainly UV-A (320–400 nm), UV-B (280–320) and UV-C (100–290 nm). UV-C rays are usually blocked by the ozone layer, while the others reach the earth. Long exposure to UV-A and UV-B rays could cause serious damage to skin and can further be diagnosed as skin cancer. UV-blockers or various types have both organic and inorganic which have been used in textile finishing. Inorganic blockers are much more effective than organic blockers due to their high stability and non-toxicity. ZnO and TiO_2 are used extensively in the field of cosmetics as UV-blockers. Organic blockers also release the absorbed energy in the form of heat which is a concern for sensitive skin and they do not show photo stability for prolonged periods [116]. Nanofinishing of cellulosic fabrics (garments) using TiO_2, ZnO, CuO, CeO_2, Al_2O_3, $ZnSnO_3$ shows very promising UV-protection against UV rays, due to their high absorption capability (UV-range). The capability of the nanoparticles to absorb in various UV-range varies with the shape, size, distribution, band gap and crystal structure of the nanoparticles.

TiO_2 nanoparticle has three crystal structures mainly anatase, rutile and brukide. Anatase is better in terms of antibacterial activity and the most photoactive form of TiO_2, having a higher photocatalytic activity than rutile which makes it usable as a UV-resistant materials in sunscreen [130].

A comparative study of ZnO and TiO_2 nanoparticles application on cotton fabric with particle size ranging from 20 to 40 nm was performed. The nanoparticles were applied by pad-dry-cure method and tested for multi-functional finishing properties. The UPF values of the fabric increased with the increase in concentration of the treatment Table 3. Also TiO_2 showed much higher UPF value compared to ZnO

Table 3 UPF values of TiO_2 and ZnO treated cotton fabric [65]

Sample	ZnO dispersion (g/l)	TiO2 dispersion (g/l)	Mean UPF	UV-A blocking (%)	UV-B blocking (%)
Pristine cotton	–	–	5.7	75.4	83.9
2	–	–	10.0	77.6	93.5
5	30	–	35.7	91.5	97.8
6	20	–	20.9	88.4	96.0
7	10	–	15.7	86.1	94.8
8	–	30	147.7	94.0	99.8
9	–	20	72.9	91.2	99.4
10	–	10	44.8	89.0	98.7

nanoparticle this can be attributed to the wide absorbance range of TiO_2 compared to ZnO nanoparticles [65].

UV-protection of a fabric depends on several parameters such as type of fibre, colour (dye), presence of UV absorbers and additives, porosity of fabric, thickness, aerial density, washing and the wearing conditions.

Metal oxide nanoparticles can be implemented in re-generated cellulosic fibre during melt spinning techniques to produce UV-resistance finished fibres. This incorporation of the nanoparticles in the fibre form provides very high wash durability and better properties as they are embedded inside the fibre structure. Economically viable UV-resistant finishes are still a challenge as the application process involves the use of inorganic nanoparticles, which does require additional synthesis methodologies. Application of green synthesis and plant extract nanoparticles are also looked into as future propositions for such finishes on natural fibre-based textiles.

3.5 *Self-cleaning Finishing of Cellulosic Textiles*

There are two approaches to obtain self-cleaning textiles, first, the surface of the textile materials can be modified and provided with multiple levels of roughness to produce a superhydrophobic fabric. This is also known as the "lotus effect" which is the mimic of the lotus leaf structure on the surface of the textiles. The lotus leaf contains multiple levels of roughness (surface) and a wax coating on the surface of the leaf which allows dirt particles to roll off through the mechanism shown in Fig. 8. The essential properties to attain this finish are high contact angles, different levels of surface roughness and low tilt angle [146].

A similar hierarchical structure was generated by Joshi et al. using nanosilica via LBL coating on cotton fabric, the surface roughness was confirmed through AFM studies and the finished fabric (0.5% weight nanosilica) showed contact angle greater than 150° [58]. Deposition of single-walled carbon nanotubes on cotton

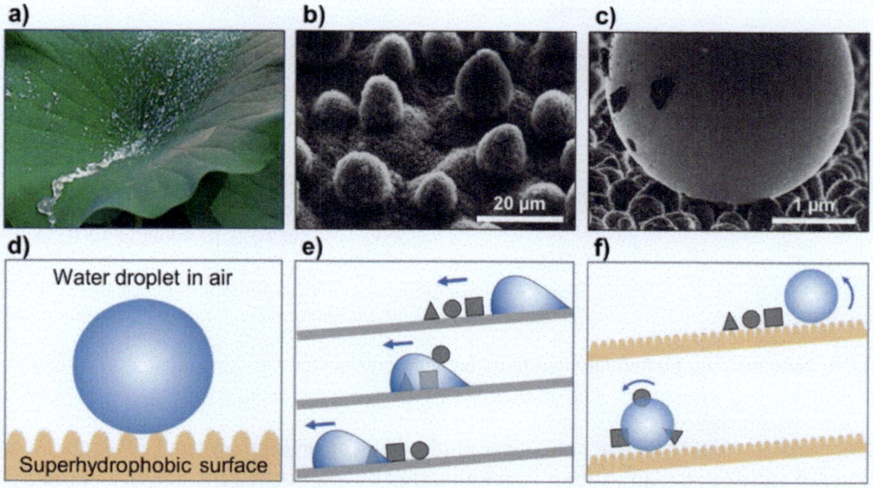

Fig. 8 **a** Water droplets on lotus leaf, **b** SEM image of surface of lotus leaf, **c** SEM image of accumulation of dirt particle **d** schematic diagram of superhydrophobic surface, **e**, **f** rolling off mechanism of water droplets on a smooth and structured surface (Copyright 2020, Elsevier) [146]

fabric was done through CVD technique which was further treated with perfluorooctyltriethoxysilane to establish a superhydrophobic surface with a contact angle of 163°. This fabric also showed very high oil/water separation efficiency [150]. TiO_2 nanoparticles along with hexadecyltrimethoxysilane were applied on jute fabric to obtain a contact angle of 159° and at the same time showed high absorbency for oil droplet [125]. The surface roughness can be enhanced by various plasma treatment methods which can be obtained using fluorocarbon chemistry-based gases and provide multi-level roughness to cellulosic fabrics. This in turn increases the surface area and helps in lowering of the surface energy which increases the self-cleaning efficiency.

The secondary approach to obtain self-cleaning textile materials is through the involvement of metal oxide nanoparticles (semi-conductors) which provide high photocatalytic effect. When metal oxides are irradiated by UV rays (appropriate wavelength) the electrons present in their valence band jumps to the conduction band (higher energy state) which further generates holes. This results in the production of free radical oxygen and hydroxyl radicals which can degrade stains via oxidation and reduction mechanism which leads to a decomposition of the complex unsaturated organic structures in them (Fig. 9).

Metal oxides like ZnO and TiO_2 are widely used due to their wide band gap of 3.22 and 3.37 eV, respectively [45]. The anatase crystal structure of TiO_2 is known to provide the highest photocatalytic activity compared to rutile. The particle size of TiO_2 also affects the photocatalytic activity as with the change in particle size, there is a change in band gap energy which directly affects the UV-absorption range of the

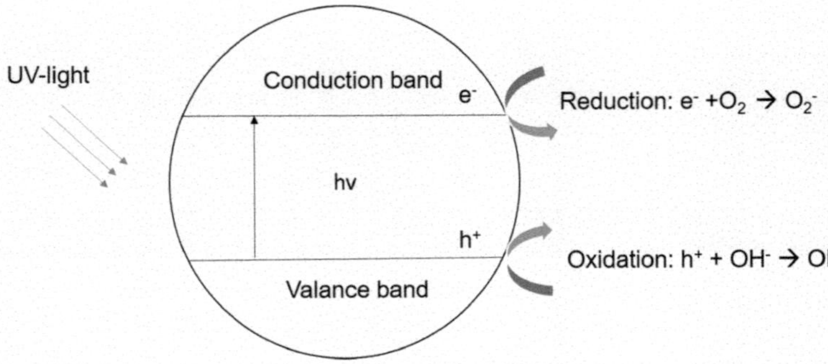

Fig. 9 Schematic for photocatalytic activity for semi-conductors

nanoparticles. ZnO nanoparticles were applied on cotton fabric by pad-dry-cure technique of varied concentration. It was found that the photocatalytic activity of methylene blue degradation increased with the increase in ZnO nanoparticle concentration of the solution [151]. The methylene blue stains disappeared after 24 h exposure to sunlight. The wash durability of such coatings can be improved by modification of the cotton fabric by cationisation or via crosslinking between the metal oxide nanoparticles and cellulosic fabric with the help of silicon-based binders. The evaluation of the photocatalytic activity can be investigated with different types of stains like dyes (methylene blue, methyl orange, Congo red), stearic acid, coffee, red wine, etc. Other factors that can affect the photocatalytic activity of the nanoparticles are degree of crystallinity, doping of metal oxide structure and surface area of the particles. Expanding the absorption range of the metal oxides using different doping could enhance the self-cleaning effect under day light. Integration of metal oxides on natural fibre-based textiles has been found to affect the tensile strength and flexibility of the material. These challenges can be overcome by various new application techniques (plasma, sol–gel coating, crosslinking) of nanofinishing on natural fibres.

3.6 Flame-Retardant Finished Cellulosic Textiles

Flame-retardant textiles have developed in terms of the chemistry of finishing agents and their preferred mechanism over the span of many years. Flame-retardant finished textile is a form of protective textile which should have the capability to protect the wearer. Statistical studies in the US have reported that most of the materials from which house-hold fire commence are textile materials (44%) [51]. These finishes are also applied in house-hold materials, for the protection of home. The flame-retardant chemistry at the twenty-first century is targeted towards the development of eco-friendly and green approaches thus creating less hazards and environmental toxicity. Halogen-based flame-retardants are very effective but can produce halogen gas during the burning phase thus more toxic to the environment. Phosphorous-based flame-retardants promote char formation and act as a barrier to gases, protecting the material surface from heat and air. In recent times, flame-retardant finishes on cellulosic fabrics have been achieved by bio-macromolecules generated natural extracts. Imparting nanotechnology in flame-retardant finishing has improved the physical and chemical properties and thermal stability of the finishes [17].

The thermal degradation behaviour of cellulosic textiles can be classified into 3 stages: (i) Initial cracks: The glucose present in the cellulose structure is associated with dehydration reaction, which increases carboxyl and carbonyl content in the molecular structure. This happens at temperatures lower than 300 °C and results in a small amount of weight loss. (ii) Major cracks: This happens when the temperature ranges between 300 and 450 °C, when there is a rapid decrease in mass of the fibre. The cellulose structure dehydrates to generate *L*-glucose, water and CO_2. This is an important stage of pyrolysis, as fire retardants are designed so that the generation of L-glucose is prevented. This enhances the pyrolysis temperature of the cellulosic material. (iii) Residual cracks: This stage occurs at temperature higher than 450 °C where dehydration and condensation of the structure lead to decarboxylate formation and generation of CO_2 and water. Eventually, a black carbon layer is formed on the structure. Thus, the flame-retardant chemistry can help in decreasing the heating rate of the fibre, which delays the pyrolysis stage. Also, generation of combustible gases can be prevented by producing less tar and more solid residue (char formation) [138].

The most used nano-fire retardants are ZnO, CNT, graphite and nanoclay on textile fabrics. Polypyrrole-ZnO and polypyrrole-ZnO-CNT were applied on cotton fabric, in situ synthesis and used as an intumescent flame-retardant alongside BTCA and sodium hypophosphite (catalyst). ZnO-CNT-coated fabric showed a lower char length than ZnO-coated fabric and a very high char yield (98%). The crosslinking of cotton fabric with BTCA and the composite polymeric structure enhanced the wash durability of the cotton fabric [145]. Cotton fabric was coated via sol–gel method with ZnO and SiO_2 nanoparticles, individual and in combination. The combination of both the nanoparticles showed higher efficacy of flame-retardant properties compared to the individual one suggesting a synergistic affect [113].

Promising LOI value of 30 was achieved by treating cotton fabric with $ZnCO_3$ nanoneedles, which provided very high thermal stability, the finished cotton fabric showed self-extinguishing properties [119]. Metal hydroxides like aluminium and magnesium have very high thermal stability and can absorb large amounts of heat before they are decomposed. These metal hydroxide nanomaterials can work with other fire retardants to provide a decrease in flammability of cellulosic materials. ZnO nanoparticles (0.01%) were synthesised using co-precipitation method followed by annealing, and then they were coated on jute fabric by pad-dry methods with potassium methyl silicone as a binder. The coated jute fabric showed excellent LOI value of (35) with a char length of 1 cm [114]. The challenges of such nanomaterials providing flame-retardant finishes can be associated with their toxicity level, compatibility with other fire retardants and wash durability in cellulosic materials.

3.7 Superabsorbent Finished Cellulosic Textiles

Superabsorbent finishing plays a major role in hygiene products, wipes, surgical dressings, wound dressing, diapers, sportswear, etc. These textile materials are involved with treatments that enhance the fabric's ability to absorb and retain large amounts of water or other aqueous solutions. Primarily such finishes are associated with cellulosic-based polymers like cotton, viscose, polyacrylates which has the capability to absorb large amounts of water (intrinsic property) in their fibre structure. Hydrophilic polymers, such as acrylic acid-based polymers or sodium polyacrylates, have been applied on cotton fabric to enhance the absorbency of the material. Superabsorbent finishes can be targeted towards absorbing of organic liquids and gases as well. The basic superabsorbents are hydrogels or polymer composites. Hydrogels are three-dimensional networks of hydrophilic polymers that can absorb and swell up (100 times its own volume) to retain large amounts of water. These polymeric materials are biocompatible as well as biodegradable in nature. The superabsorbent mechanism can be explained by hydration, capillary action, diffusion of water molecules and formation of hydrogen bonds. Initially, the water molecules migrate through the porous and fibrous structure via capillary action and they wet the fabric. Further the water molecules interact with the hydrophilic functional groups of the superabsorbent and create hydrogen bonds. The osmotic pressure difference leads to the swelling up of the polymer chains and the gel structure. This whole process can be driven by various stimuli, e.g. changes in temperature, state of matter, etc. [100]. Modern technologies like electrospinning of cellulose-based nanofibers, which can absorb high amounts of water due to their high surface area, can be used to coat cellulose fabric with nanofiber layers to enhance the absorbent behaviour. Incorporation of such finishes with naturally biodegradable polymers is considered for eco-friendly and sustainable finishes in textile industries.

4 Nanofinishing of Wool Textiles

Wool is a keratin-based fibre known for its multi-functional properties like thermal insulation, water absorption, elongation and high ignition temperature. These unique properties of wool fibres are associated with the complex macroscopic and microscopic structure of the fibre. Wool is also chemically resistant due to its hydrophobic surface cuticle. At the same, time wool fibre is susceptible to moth attacks and fungal growth under appropriate conditions. Accumulation of Gram positive and negative bacteria on skin can cause odour and in certain cases skin ulcers. Further, the mechanical and functional properties of wool fibres can be affected by such bacterial growth. Nanoparticles which are known for its antimicrobial properties are Ag, Cu, ZnO, TiO_2, Fe_3O_4, chitosan, plant extracts (cinnamon, neem, thyme). Kaolinite and Alumina nanoparticles were used to study the anti-moth treatment on wool fabrics [56, 92].

The antibacterial activity of green synthesised Ag nanoparticles was investigated on wool fabric. The AgNP was synthesised with sodium alginate as the reducing agent and the wool fabric was coated with AgNP resulted in high antimicrobial and UV-resistant values [81]. Wool fabric was treated by pad-dry-cure method with chitosan, chitosan NPs and chitosan-AgNPs. The chitosan AgNP treated fabric showed excellent bacterial reduction of 96.38% for *S. aureus* and 97.66% for *E. coli* which was higher than all other treated samples [112].

Pectin and gelatin-based biopolymers were used for encapsulation of metal NPs (ZnO, TiO_2 an Ag) and lemon oil. This was applied on wool fabric by pad-dry and pacing method. The lemon oil treated wool fabrics could maintain the fragrance up to 2 months. The presence of nanoparticles on the gelatin encapsulated samples provided high UV-protection and antimicrobial properties [86].

Wool fabric was treated by low pressure plasma in presence of hexafluoroethane (C_2F_6) gas. The effect of treatment time power and gas flow rate was evaluated on the hydrophobic behaviour of the wool fabric. Wool fabric treated 150W and 12 sccm (standard cubic centimetre per minute) for 20 min showed superhydrophobic behaviour with a contact angle of 150°. The fabric also showed good abrasion resistance [99].

Wool fabric due to high ignition temperature is readily used for flame-retardant finishes. Wool fabric treated with phytic acid, BTCA and ZnO/TiO_2 nanoparticles using the pad-dry-cure process. The treated fabric showed excellent flame-retardant properties LOI (> 30) and retained them after 30 washes. TGA analysis confirmed the involvement of the intumescent behaviour of the coating and formation of char which limited the smoke generation. The thermal degradation behaviour of untreated wool fibre showed two stages of weight loss. Below 100 °C wool fibre is subjected to dehydration of water molecules from their internal structure. When the temperature range is between 220 and 425 °C, a second stage of weight loss takes place. This is due to the breaking of the peptide bonds, H-bonds within the helical keratin structure of wool, and finally the S–S (cystine linkage). The fibre disintegrates to form aromatic structure which is a semi-solid and liquid stage with further release of methane, carbon monoxide and carbon dioxide. The phytic acid present in the coated wool fabric changes the decomposition behaviour by forming a protective char due to the phosphate group present in their structure. Further, the char is thermally stable due to presence of TiO_2 NPs which has high thermal stability and does not allow further decomposition of the wool fabric [23].

Wool fabric also showed LOI values of 31 after treatment with 2.5% nano-kalonite (clay). The char length of the wool fabrics (two different application method exhaust and pad-batch) was 2.5 and 3.3 cm, respectively. Finally, the wool fabric showed an increase in tensile strength by $\approx$ 8% [56]. Thus, this provides an effective way for manufacturing of flame-retardant wool fabric. The protein chemistry involved in the keratinous structure of the wool fibre has shown to high durability of finishes with metal and metal oxide nanoparticles. The presence of acidic and basic groups in wool fibre provides the flexibility to form electrostatic bonds with the fibre. Another challenging aspect is the hydrophobic epicuticle layer of the wool fibre, which can inhibit the deposition of nanoparticles on the fibre structure. Plasma treatment of wool fibre can cause nanoroughness on the epicuticle layer which can further enhance the adhesion of nanoparticles on wool fabrics.

Nanoparticle finishes have also been applied on other animal-based fibres like silk, cashmere, angora, alpaca to obtain various functional properties. Keeping in mind the aspects of Net Zero natural fibre finishing will have a huge role to play to impart circularity in textile industries.

5 Multi-functional Finishing of Natural Fibre-Based Textiles

Multi-functional textile finishing is the current need of the hour considering the aspects of sustainability, lower carbon impact and net zero policies all around the globe. Textile industries contribute towards the generation of huge amount of effluents and chemical finishes are a big part of it. Multi-functional finishing can be obtained either by combination of multiple finishing agent or by utilising one finishing agent which can value add several finishing properties to the textile substrate. Since the advent of nanotechnology various strategies have been looked into for application of multi-functional finishing using one or more than type of nanomaterials. The added advantage is that it requires lesser amount of water consumption compared to conventional finishes at the same time provides efficient properties at lower add-on (lesser consumption of chemicals). Generally, the properties associated with multi-functional finishes are antimicrobial, UV-resistant, flame retarding, water-repellent, self-cleaning, crease recovery and antistatic. ZnO nanoparticles along with BTCA cross-linker and an acrylic binder were applied on cotton fabric using the pad-dry-cure technique. The effect of ZnO nanoparticle concentration was significant on the UPF, which increased (to 45) with the increase in concentration (15–45 g/l). The fabric also showed very good antimicrobial efficacy against *S. aureus* and *E. coli* bacteria and the zone of inhibition increased with the concentration of Zn nanoparticle concentration. The treated fabric also retained its finishing properties up to 10 wash cycles [14]. ZnO and TiO_2 nanoparticles synthesised via sol–gel method were applied on cotton fabric using pad-dry-cure technique. The finished fabric showed excellent self-cleaning effect alongside UV-resistant and antibacterial properties. The combination of ZnO and TiO_2 nanoparticles treated fabric showed much higher UPF (> 50) compared to only ZnO and TiO_2 finished fabric. The TiO_2 coating followed by ZnO coating resulted in very high deposition of metal oxides on the cotton fabric and is the possible reason for showing highest UPF and photocatalytic activity [21]. Another combination of silica and titania nanoparticles was applied via sol–gel coating method on bamboo fabric. The fabric showed UV-resistant, hydrophobic and soil release properties and wash durability up to 5 washes [55].

Paul et al. used co-precipitation method to synthesise zinc stannate nanoparticles and applied to cotton fabric via dip-pad-dry technique. The zinc stannate nanoparticle finished fabric showed UV-resistant, self-cleaning, antibacterial as well as flame-retardant properties. Zinc stannate showed wider range of UV-absorption (290–400 nm, both UV-A and B) compared to commercial ZnO and TiO_2 nanoparticles. The finished cotton fabric showed high self-cleaning effect by degrading coffee stain after 20 h of UV-irradiation (300–350 nm) and a significant decrease in K/S value of methylene blue dyed fabric [101].

Linen fabric when coated with chitosan and further in situ synthesis of ZnO nanoparticles was obtained by flower extract of *Bombax ceiba*. The finished fabric showed antibacterial, anti-oxidant and UV-resistant properties [44]. Multi-functional properties were obtained on printed wool fabric using selenium nanoparticles. The results showed that the printed wool fabric showed UPF (> 50), high antimicrobial properties alongside colouration of the fabric. They also exhibited washing fastness properties up to 10 washing cycles [36]. Cotton fabric was treated with chitosan, AgNP and lignin nanoparticles to produce superior antimicrobial and UV-resistant properties. Application of lignin in a chitosan/lignin/AgNP nanocomposite by LBL can act as UV protective agent due to its highly unsaturated structure. Lignin also functionalised as reducing and stabilising agent for AgNP nucleation and growth. The treated fabric showed 100% reduction in bacterial growth and 50 + UPF [83]. Achieving multi-functional finishing through synergistic effect of different nanoparticles has been achieved on various natural fibre-based textiles, at the same time it involves an incredible effort towards trying out various combinations of nanoparticles which can affect or deteriorate the intrinsic fibre properties of these natural fibres. The bioaccumulation and toxicity levels of such nanoparticles are also a concern which leads towards improving the wash durability of such finishes and considering the environmental impact from cradle to grave.

6 Conclusion

Nanofinishing of natural fibres and their textile products has developed numerous possibilities to enhance the natural fibre properties and apply multiple functionalities to conventional textiles. Despite such wide-ranging possibilities commercialisation of nanofinishing in natural fibre-based textiles is an area of progress. Many nanofinishes application technique lack potential to be and are not possible to scale up to an industrial level. The new applications techniques like plasma, supercritical CO_2 finishing is still on the development stage to counter the problem of major usage of water in textile finishing.

Nanofinishing in natural fibres has the potential to enhance their tensile strength, dimension stability, moisture sorption capabilities and other inherent properties. Still, biocompatibility of the nanoparticles with natural fibre is an area of concern and various alternative approaches are coming forward to mitigate this problem. This has resulted in an increase in green synthesis techniques of nanoparticles, incorporation of organic nanoparticles in textiles (bio-macromolecules) and implementation of multi-functional finishing agents. The biodegradability and toxicity of nanoparticles are considered an important aspect while converting the textile supply chains to net zero.

Globally all textile sectors are moving towards circularity which has already seen an increase in demand of natural fibre-based products. This provides a platform for these various technologies to be up scaled and become extremely beneficial in terms of conventional finishing techniques. Another important aspect is to maintain the core properties of the textile substrate and impart washing fastness on the finished fabric. The low add-on% of finishes do provide a huge benefit towards nanofinishing, and at the same time, many researchers have shown that the inherent property of the fabric is not affected by this phenomenon. Finally, enhancement of social sustainability will provide the consumers assurance to buy natural fibre-based nanofinished products which can provide multiple-functionality.

References

1. Abd El-Hady MM, Sharaf S, Farouk A (2020) Highly hydrophobic and UV protective properties of cotton fabric using layer by layer self-assembly technique. Cellulose 27:1099–1110. https://doi.org/10.1007/s10570-019-02815-0
2. AbdElhady MM (2012) Preparation and characterization of chitosan/zinc oxide nanoparticles for imparting antimicrobial and UV protection to cotton fabric. Int J Carbohydr Chem 2012:840591. https://doi.org/10.1155/2012/840591
3. Abebe B, Zereffa EA, Tadesse A, Murthy HCA (2020) A review on enhancing the antibacterial activity of ZnO: mechanisms and microscopic investigation. Nanoscale Res Lett 15:190. https://doi.org/10.1186/s11671-020-03418-6
4. Adak B, Mukhopadhyay S (2023) Smart and functional textiles, 1st edn. De Gruyter, Berlin
5. Ali A, Baheti V, Vik M, Militky J (2020) Copper electroless plating of cotton fabrics after surface activation with deposition of silver and copper nanoparticles. J Phys Chem Solids 137:109181. https://doi.org/10.1016/j.jpcs.2019.109181
6. Ali SW, Chowdhury A, Banerjee S, Bairagi S (2022) Bio resources mediated technological advancements in chemical finishing of textiles. In: Shanmugasundaram OLBT-A of B for STP (ed) The textile institute book series. Woodhead Publishing, pp 187–220 (Chapter 9)
7. Ali SW, Joshi M, Rajendran S (2011) Novel, self-assembled antimicrobial textile coating containing chitosan nanoparticles. AATCC Rev 11:49–55
8. Alongi J, Tata J, Frache A (2011) Hydrotalcite and nanometric silica as finishing additives to enhance the thermal stability and flame retardancy of cotton. Cellulose 18:179–190. https://doi.org/10.1007/s10570-010-9473-z
9. Aminayi P, Abidi N (2013) Imparting super hydro/oleophobic properties to cotton fabric by means of molecular and nanoparticles vapor deposition methods. Appl Surf Sci 287:223–231. https://doi.org/10.1016/j.apsusc.2013.09.132
10. Ammayappan L, Chakraborty S, Pan NC (2021) Silica nanocomposite based hydrophobic functionality on jute textiles. J Text Inst 112:470–481. https://doi.org/10.1080/00405000.2020.1764779
11. Anandjiwala RD, Blouw S (2007) Composites from bast fibres-prospects and potential in the changing market environment. J Nat Fibers 4:91–109. https://doi.org/10.1300/J395v04n02_07
12. Araújo JC, Ferreira DP, Teixeira P, Fangueiro R (2021) In-situ synthesis of CaO and SiO_2 nanoparticles onto jute fabrics: exploring the multifunctionality. Cellulose 28:1123–1138. https://doi.org/10.1007/s10570-020-03564-1

13. Arfaoui MA, Dolez PI, Dubé M, David É (2017) Development and characterization of a hydrophobic treatment for jute fibres based on zinc oxide nanoparticles and a fatty acid. Appl Surf Sci 397:19–29. https://doi.org/10.1016/j.apsusc.2016.11.085
14. Aslam S, Hussain T, Ashraf M et al (2018) Multifunctional finishing of cotton fabric. Autex Res J 19. https://doi.org/10.1515/aut-2018-0048
15. Bacciarelli-Ulacha A, Rybicki E, Matyjas-Zgondek E et al (2014) A new method of finishing of cotton fabric by in situ synthesis of silver nanoparticles. Ind Eng Chem Res 53:4147–4155. https://doi.org/10.1021/ie4011113
16. Baig N, Kammakakam I, Falath W (2021) Nanomaterials: a review of synthesis methods, properties, recent progress and challenges. Mater Adv 2:1821–1871. https://doi.org/10.1039/D0MA00807A
17. Basak S, Ali SW (2016) Sustainable fire retardancy of textiles using bio-macromolecules. Polym Degrad Stab 133:47–64. https://doi.org/10.1016/j.polymdegradstab.2016.07.019
18. Bhandari V, Jose S, Badanayak P et al (2022) Antimicrobial finishing of metals, metal oxides, and metal composites on textiles: a systematic review. Ind Eng Chem Res 61:86–101. https://doi.org/10.1021/acs.iecr.1c04203
19. Borah MP, Jose S, Kalita BB et al (2020) Water repellent finishing on eri silk fabric using nano silica. J Text Inst 111:701–708. https://doi.org/10.1080/00405000.2019.1659470
20. Bozaci E, Akar E, Ozdogan E et al (2015) Application of carboxymethylcellulose hydrogel based silver nanocomposites on cotton fabrics for antibacterial property. Carbohydr Polym 134:128–135. https://doi.org/10.1016/j.carbpol.2015.07.036
21. Butola BS, Garg A, Garg A, Chauhan I (2018) Development of multi-functional properties on cotton fabric by in situ application of TiO_2 and ZnO Nanoparticles. J Inst Eng Ser E 99:93–100. https://doi.org/10.1007/s40034-018-0118-3
22. Caruso RA, Antonietti M (2001) Sol−gel nanocoating: an approach to the preparation of structured materials. Chem Mater 13:3272–3282. https://doi.org/10.1021/cm001257z
23. Cheng X-W, Guan J-P, Yang X-H, Tang R-C (2018) Durable flame retardant wool fabric treated by phytic acid and TiO_2 using an exhaustion-assisted pad-dry-cure process. Thermochim Acta 665:28–36. https://doi.org/10.1016/j.tca.2018.05.011
24. Crisan MC, Teodora M, Lucian M (2022) Copper nanoparticles: synthesis and characterization, physiology, toxicity and antimicrobial applications. Appl Sci 12
25. Čuk N, Šala M, Gorjanc M (2021) Development of antibacterial and UV protective cotton fabrics using plant food waste and alien invasive plant extracts as reducing agents for the in-situ synthesis of silver nanoparticles. Cellulose 28:3215–3233. https://doi.org/10.1007/s10570-021-03715-y
26. da Silva DJ, Ferreira RR, Ferreira GS et al (2023) Multifunctional cotton fabrics with novel antibacterial coatings based on chitosan nanocapsules and polyacrylate. J Coat Technol Res 20:1541–1555. https://doi.org/10.1007/s11998-023-00761-y
27. Dai D, Fan M (2013) Green modification of natural fibres with nanocellulose. RSC Adv 3:4659–4665. https://doi.org/10.1039/C3RA22196B
28. Dastjerdi R, Montazer M, Shahsavan S (2010) A novel technique for producing durable multifunctional textiles using nanocomposite coating. Colloids Surf B Biointerfaces 81:32–41. https://doi.org/10.1016/j.colsurfb.2010.06.023
29. Davodiroknabadi A (2023) Modification of polyester fabrics with Zn_2SnO_4 nanorods for superior self-cleaning, UV-protection and antibacterial performance. J Text Inst 114:1091–1098. https://doi.org/10.1080/00405000.2022.2105119
30. Dizaj SM, Lotfipour F, Barzegar-Jalali M et al (2014) Antimicrobial activity of the metals and metal oxide nanoparticles. Mater Sci Eng C 44:278–284. https://doi.org/10.1016/j.msec.2014.08.031
31. Eadie L, Ghosh TK (2011) Biomimicry in textiles: past, present and potential. An overview. J R Soc Interface 8:761–775
32. Ebrahimi K, Valipour P, Mirjalili M, Tayebi H-A (2023) In situ synthesis of reduced graphene oxide and zinc stannate nanoparticles on cotton fabric. Starch Stärke 75:2300063. https://doi.org/10.1002/star.202300063

33. El-Din Zakaria El-Sayed H, Mowafi S, Abou El-Kheir A, El-Khatib EM (2018) A comprehensive critique on wool grease extraction, properties and applications. Egypt J Chem 61:1151–1159. https://doi.org/10.21608/ejchem.2018.4214.1372
34. El-Naggar ME, Shaarawy S, Hebeish AA (2018) Bactericidal finishing of loomstate, scoured and bleached cotton fibres via sustainable in-situ synthesis of silver nanoparticles. Int J Biol Macromol 106:1192–1202. https://doi.org/10.1016/j.ijbiomac.2017.08.127
35. El-Naggar ME, Shaheen TI, Zaghloul S et al (2016) Antibacterial activities and UV protection of the in situ synthesized titanium oxide nanoparticles on cotton fabrics. Ind Eng Chem Res 55:2661–2668. https://doi.org/10.1021/acs.iecr.5b04315
36. Elmaaty TA, Raouf S, Sayed-Ahmed K, Plutino MR (2022) Multifunctional dyeing of wool fabrics using selenium nanoparticles. Polymers (Basel) 14
37. Ferreira DP, Costa SM, Felgueiras H, Fangueiro R (2019) Smart and sustainable materials for military applications based on natural fibres and silver nanoparticles. In: Advanced materials for defense. Trans Tech Publications Ltd., pp 66–74
38. Foruzanmehr M, Vuillaume PY, Robert M, Elkoun S (2015) The effect of grafting a nano-TiO_2 thin film on physical and mechanical properties of cellulosic natural fibers. Mater Des 85:671–678. https://doi.org/10.1016/j.matdes.2015.06.105
39. Fu H-B, Yao J-N (2001) Size effects on the optical properties of organic nanoparticles. J Am Chem Soc 123:1434–1439. https://doi.org/10.1021/ja0026298
40. Gan X, Wu Y, Liu L et al (2007) Electroless copper plating on PET fabrics using hypophosphite as reducing agent. Surf Coat Technol 201:7018–7023. https://doi.org/10.1016/j.surfcoat.2007.01.006
41. Gowri S, Almeida L, Amorim T et al (2010) Polymer nanocomposites for multifunctional finishing of textiles—a review. Text Res J 80:1290–1306. https://doi.org/10.1177/0040517509357652
42. Granados A, Pleixats R, Vallribera A (2021) Recent advances on antimicrobial and anti-inflammatory cotton fabrics containing nanostructures. Molecules 26. https://doi.org/10.3390/molecules26103008
43. Gulrajani ML (2013) The use of nanotechnology in the finishing of technical textiles. In: Gulrajani MLBT-A in the D and F of TT (ed) Woodhead publishing series in textiles. Woodhead Publishing, pp 280–308 (Chapter 12)
44. Gupta M, Sheikh J, Annu SA (2021) An eco-friendly route to develop cellulose-based multi-functional finished linen fabric using ZnO NPs and CS network. J Ind Eng Chem 97:383–389. https://doi.org/10.1016/j.jiec.2021.02.023
45. Haghighatzadeh A, Hosseini M, Mazinani B, Shokouhimehr M (2019) Improved photocatalytic activity of ZnO–TiO_2 nanocomposite catalysts by modulating TiO_2 thickness. Mater Res Express 6:115060. https://doi.org/10.1088/2053-1591/ab49c4
46. Hasan KMF, Horváth PG, Kóczán Z et al (2021) Colorful and facile in situ nanosilver coating on sisal/cotton interwoven fabrics mediated from European larch heartwood. Sci Rep 11:22397. https://doi.org/10.1038/s41598-021-01914-y
47. Hassan MM (2019) Enhanced colour, hydrophobicity, UV radiation absorption and anti-static properties of wool fabric multi-functionalised with silver nanoparticles. Colloids Surf A Physicochem Eng Asp 581:123819. https://doi.org/10.1016/j.colsurfa.2019.123819
48. Heinze T (2016) Cellulose: structure and properties BT—cellulose chemistry and properties: fibers, nanocelluloses and advanced materials. In: Rojas OJ (ed) Springer International Publishing, Cham, pp 1–52
49. Hong HR, Kim J, Park CH (2018) Facile fabrication of multifunctional fabrics: use of copper and silver nanoparticles for antibacterial, superhydrophobic, conductive fabrics. RSC Adv 8:41782–41794. https://doi.org/10.1039/C8RA08310J
50. Ibrahim M, Mondal H (2021) Fundamentals of natural fibres and textiles. Woodhead Publishing
51. Illinois.gov (2023) State fire marshal encourages everyone to think fire safety when decorating for fall and Halloween. https://www.illinois.gov/news/press-release.22230.html. Accessed 22 Dec 2023

52. Irfan M, Perero S, Miola M et al (2017) Antimicrobial functionalization of cotton fabric with silver nanoclusters/silica composite coating via RF co-sputtering technique. Cellulose 24:2331–2345. https://doi.org/10.1007/s10570-017-1232-y
53. Javed A, Wiener J, Tamulevičienė A et al (2021) One step in-situ synthesis of zinc oxide nanoparticles for multifunctional cotton fabrics. Materials (Basel) 14
54. Jelil RA (2015) A review of low-temperature plasma treatment of textile materials. J Mater Sci 50:5913–5943. https://doi.org/10.1007/s10853-015-9152-4
55. Jhatial AK, Khatri A, Ali S, Babar AA (2019) Sol–gel finishing of bamboo fabric with nanoparticles for water repellency, soil release and UV resistant characteristics. Cellulose 26:6365–6378. https://doi.org/10.1007/s10570-019-02537-3
56. Jose S, Nachimuthu S, Das S, Kumar A (2018) Moth proofing of wool fabric using nano kaolinite. J Text Inst 109:225–231. https://doi.org/10.1080/00405000.2017.1336857
57. Jose S, Shanmugam N, Das S et al (2019) Coating of lightweight wool fabric with nano clay for fire retardancy. J Text Inst 110:764–770. https://doi.org/10.1080/00405000.2018.1516529
58. Joshi M, Bhattacharyya A, Agarwal N, Parmar S (2012) Nanostructured coatings for super hydrophobic textiles. Bull Mater Sci 35:933–938. https://doi.org/10.1007/s12034-012-0391-6
59. Karimi L, Yazdanshenas ME, Khajavi R et al (2014) Using graphene/TiO_2 nanocomposite as a new route for preparation of electroconductive, self-cleaning, antibacterial and antifungal cotton fabric without toxicity. Cellulose 21:3813–3827. https://doi.org/10.1007/s10570-014-0385-1
60. Karthikeyan M, Ramachandran T, Shanmugasundaram OL (2014) Synthesis, characterization, and development of thermally enhanced cotton fabric using nanoencapsulated phase change materials containing paraffin wax. J Text Inst 105:1279–1286. https://doi.org/10.1080/00405000.2014.886368
61. Kelly FM, Johnston JH (2011) Colored and functional silver nanoparticle−wool fiber composites. ACS Appl Mater Interfaces 3:1083–1092. https://doi.org/10.1021/am101224v
62. Keßler L, Matlin SA, Kümmerer K (2021) The contribution of material circularity to sustainability—recycling and reuse of textiles. Curr Opin Green Sustain Chem 32:100535. https://doi.org/10.1016/j.cogsc.2021.100535
63. Khan A, Rashid A, Younas R, Chong R (2016) A chemical reduction approach to the synthesis of copper nanoparticles. Int Nano Lett 6:21–26. https://doi.org/10.1007/s40089-015-0163-6
64. Khan I, Saeed K, Khan I (2019) Nanoparticles: properties, applications and toxicities. Arab J Chem 12:908–931. https://doi.org/10.1016/j.arabjc.2017.05.011
65. Khan MZ, Baheti V, Ashraf M et al (2018) Development of UV protective, superhydrophobic and antibacterial textiles using ZnO and TiO_2 nanoparticles. Fibers Polym 19:1647–1654. https://doi.org/10.1007/s12221-018-7935-3
66. Khatabi H, Bidoki SM, Haji A (2022) A green approach for in-situ synthesis of silver nanoparticles on cotton fabric by low pressure cold plasma. Mater Chem Phys 290:126548. https://doi.org/10.1016/j.matchemphys.2022.126548
67. Khoddami A, Shokohi SS, Morshed M, Abedi D (2011) Simultaneous application of silver nanoparticles with different crease resistant finishes. Fibers Polym 12:635–641. https://doi.org/10.1007/s12221-011-0635-x
68. Kolman K, Nechyporchuk O, Persson M et al (2018) Combined nanocellulose/nanosilica approach for multiscale consolidation of painting canvases. ACS Appl Nano Mater 1:2036–2040. https://doi.org/10.1021/acsanm.8b00262
69. Kong F-Y, Zhang J-W, Li R-F et al (2017) Unique roles of gold nanoparticles in drug delivery, targeting and imaging applications. Molecules 22. https://doi.org/10.3390/molecules22091445
70. Korkmaz N, Aksoy SA (2016) Enhancing the performance properties of ester-cross-linked cotton fabrics using Al_2O_3-NPs. Text Res J 86:636–648. https://doi.org/10.1177/0040517515592806
71. Kumar A, Biswal M, Mohanty S, Nayak SK (2021) Recent developments of lignocellulosic natural fiber reinforced hybrid thermosetting composites for high-end structural applications: a review. J Polym Res 28:459. https://doi.org/10.1007/s10965-021-02788-4

72. Kumar A, Pandey AK, Singh SS et al (2011) Engineered ZnO and TiO_2 nanoparticles induce oxidative stress and DNA damage leading to reduced viability of *Escherichia coli*. Free Radic Biol Med 51:1872–1881. https://doi.org/10.1016/j.freeradbiomed.2011.08.025
73. Kwon J, Jung H, Jung H, Lee J (2020) Micro/nanostructured coating for cotton textiles that repel oil, water, and chemical warfare agents. Polymers (Basel) 12. https://doi.org/10.3390/polym12081826
74. Lam YL, Kan CW, Yuen CWM (2011) Effect of titanium dioxide on the flame-retardant finishing of cotton fabric. J Appl Polym Sci 121:267–278. https://doi.org/10.1002/app.33618
75. Lao L, Fu L, Qi G et al (2017) Superhydrophilic wrinkle-free cotton fabrics via plasma and nanofluid treatment. ACS Appl Mater Interfaces 9:38109–38116. https://doi.org/10.1021/acsami.7b09545
76. Li M, Farooq A, Jiang S et al (2021) Functionalization of cotton fabric with ZnO nanoparticles and cellulose nanofibrils for ultraviolet protection. Text Res J 91:2303–2314. https://doi.org/10.1177/00405175211001807
77. Li S, Xie H, Zhang S, Wang X (2007) Facile transformation of hydrophilic cellulose into superhydrophobic cellulose. Chem Commun 4857–4859. https://doi.org/10.1039/B712056G
78. Li X, Robinson SM, Gupta A et al (2014) Functional gold nanoparticles as potent antimicrobial agents against multi-drug-resistant bacteria. ACS Nano 8:10682–10686. https://doi.org/10.1021/nn5042625
79. Liu Y, Wang X, Qi K, Xin JH (2008) Functionalization of cotton with carbon nanotubes. J Mater Chem 18:3454–3460. https://doi.org/10.1039/B801849A
80. Ma J, Zhao Q, Zhou Y et al (2021) Hydrophobic wrapped carbon nanotubes coated cotton fabric for electrical heating and electromagnetic interference shielding. Polym Test 100:107240. https://doi.org/10.1016/j.polymertesting.2021.107240
81. Mahmud S, Pervez MN, Habib MA et al (2018) UV protection and antibacterial treatment of wool using green silver nanoparticles. Asian J Chem 30:116–122. https://doi.org/10.14233/ajchem.2018.20924
82. Makhlouf ASH (2011) Current and advanced coating technologies for industrial applications. In: Makhlouf ASH, Tiginyanu IBT-N and U-TF (eds) Woodhead publishing series in metals and surface engineering. Woodhead Publishing, pp 3–23 (Chapter 1)
83. Marković D, Petkovska J, Mladenovic N et al (2023) Antimicrobial and UV protective chitosan/lignin multilayer nanocoating with immobilized silver nanoparticles. J Appl Polym Sci 140:e53823. https://doi.org/10.1002/app.53823
84. Mateos AJ, Cain AA, Grunlan JC (2014) Large-scale continuous immersion system for layer-by-layer deposition of flame retardant and conductive nanocoatings on fabric. Ind Eng Chem Res 53:6409–6416. https://doi.org/10.1021/ie500122u
85. Mehravani B, Ribeiro AI, Zille A (2021) Gold nanoparticles synthesis and antimicrobial effect on fibrous materials. Nanomater (Basel, Switz) 11. https://doi.org/10.3390/nano11051067
86. Mohamed AL, El-Zawahry M, Hassabo AG, Abd El-Aziz E (2023) Encapsulated lemon oil and metal nanoparticles in biopolymer for multifunctional finishing of cotton and wool fabrics. Ind Crops Prod 204:117373. https://doi.org/10.1016/j.indcrop.2023.117373
87. Mohamed H, EL-Halwagy AA (2021) Plasma-based nanotechnology for textile coating. J Text Color Polym Sci 18:11–31. https://doi.org/10.21608/jtcps.2021.67022.1049
88. Mondal MIH, Ahmed F, Islam MM et al (2022) Metal and metal oxides nanoparticles in healthcare and medical textiles. In: Mondal MIH (ed) Medical textiles from natural resources. Woodhead Publishing, pp 341–371 (Chapter 11)
89. Montaser AS, Mahmoud FA (2019) Preparation of chitosan-grafted-polyvinyl acetate metal nanocomposite for producing multifunctional textile cotton fabrics. Int J Biol Macromol 124:659–666. https://doi.org/10.1016/j.ijbiomac.2018.11.254
90. Montazer M, Harifi T (2018) Nanofinishing: fundamental principles. In: Montazer M, Harifi T (eds) Nanofinishing of textile materials. Woodhead Publishing, pp 19–34 (Chapter 2)
91. Montazer M, Harifi T (2018b) Nanocoating and lamination. In: Montazer M, Harifi TBT-N of TM (eds) The textile institute book series. Woodhead Publishing, pp 95–107 (Chapter 7)

92. Montazer M, Harifi T (2018c) Nanofinishes for protective textiles. In: Montazer M, Harifi TBT-N of TM (eds) The textile institute book series. Woodhead Publishing, pp 265–294 (Chapter 18)
93. Montazer M, Shamei A, Alimohammadi F (2012) Synthesizing and stabilizing silver nanoparticles on polyamide fabric using silver-ammonia/PVP/UVC. Prog Org Coat 75:379–385. https://doi.org/10.1016/j.porgcoat.2012.07.011
94. Montes-Hernandez G, Di Girolamo M, Sarret G et al (2021) In situ formation of silver nanoparticles (Ag-NPs) onto textile fibers. ACS Omega 6:1316–1327. https://doi.org/10.1021/acsomega.0c04814
95. Morton WE, Hearle JWS (2008) An introduction to fibre structure. In: Morton WE, Hearle JWSBT-PP of TF (Fourth E) (eds) Woodhead publishing series in textiles, 1st edn. Woodhead Publishing, pp 1–81 (Chapter 1)
96. Nanotex (1998) Nanotex. https://nanotex.com/. Accessed 22 Dec 2023
97. Pal S, Mondal S, Pal P et al (2020) Applications of nanotechnology for antibacterial finishing textiles: a review. Sens Lett 18:437–448
98. Panwar K, Jassal M, Agrawal A (2014) In situ synthesis of Ag–SiO_2 Janus particles with epoxy functionality for textile applications. Particuology 19. https://doi.org/10.1016/j.partic.2014.06.007
99. Parveen S, Rana S, Goswami P (2021) Developing super-hydrophobic and abrasion-resistant wool fabrics using low-pressure hexafluoroethane plasma treatment. Mater (Basel, Switz) 14. https://doi.org/10.3390/ma14123228
100. Paul R (2015) Functional finishes for textiles: an overview. In: Paul RBT-FF for T (ed) Woodhead publishing series in textiles. Woodhead Publishing, pp 1–14 (Chapter 1)
101. Paul S, Basak S, Ali W (2019) Zinc stannate nanostructure: is it a new class of material for multifunctional cotton textiles? ACS Omega 4:21827–21838. https://doi.org/10.1021/acsomega.9b02719
102. Paul S, Hewitt A, Rana S, Goswami P (2023) Development of novel parameters for characterising scale morphology of wool fibre and its correlation with dye diffusion coefficient of acid dye. Sci Rep 13:18444. https://doi.org/10.1038/s41598-023-45689-w
103. Penninkhof JJ, Graf C, van Dillen T et al (2005) Angle-dependent extinction of anisotropic Silica/Au core/shell colloids made via ion irradiation. Adv Mater 17:1484–1488. https://doi.org/10.1002/adma.200401742
104. Pérez-Alvarez M, Cadenas-Pliego G, Pérez-Camacho O et al (2021) Green synthesis of copper nanoparticles using cotton. Polymers (Basel) 13. https://doi.org/10.3390/polym13121906
105. Periyasamy AP, Militky J, Sachinandham A, Duraisamy G (2020) Nanotechnology in textile finishing: recent developments. In: Kharissova OV, Martínez LMT, Kharisov BI (eds) Handbook of nanomaterials and nanocomposites for energy and environmental applications. Springer International Publishing, Cham, pp 1–31
106. Pukale DD, Bansode AS, Pinjari DV et al (2017) Development of nanoemulsion of silicone oil and pine oil using binary surfactant system for textile finishing. J Surfactants Deterg 20:1061–1073. https://doi.org/10.1007/s11743-017-1970-8
107. Refaee AA, El-Naggar ME, Mostafa TB et al (2022) Nano-bio finishing of cotton fabric with quaternized chitosan Schiff base-TiO_2-ZnO nanocomposites for antimicrobial and UV protection applications. Eur Polym J 166:111040. https://doi.org/10.1016/j.eurpolymj.2022.111040
108. Rehan M, Barhoum A, Van Assche G et al (2017) Towards multifunctional cellulosic fabric: UV photo-reduction and in-situ synthesis of silver nanoparticles into cellulose fabrics. Int J Biol Macromol 98:877–886. https://doi.org/10.1016/j.ijbiomac.2017.02.058
109. Rezaie AB, Montazer M, Rad MM (2017) A cleaner route for nanocolouration of wool fabric via green assembling of cupric oxide nanoparticles along with antibacterial and UV protection properties. J Clean Prod 166:221–231. https://doi.org/10.1016/j.jclepro.2017.08.046
110. Rouhani Shirvan A, Shakeri M, Bashari A (2019) Recent advances in application of Chitosan and its derivatives in functional finishing of textiles. In: Shahid-ul-Islam, Butola BSBT-TI and P of GC for TT (eds) The textile institute book series. Woodhead Publishing, pp 107–133 (Chapter 5)

111. Rowell RM (2008) Natural fibres: types and properties. In: Pickering KLBT-P and P of N-FC (ed) Woodhead publishing series in composites science and engineering. Woodhead Publishing, pp 3–66 (Chapter 1)
112. Şahan G, Demir A (2016) A green application of nano sized chitosan in textile finishing. Text Appar 26:414–420
113. Saleemi S, Naveed T, Riaz T et al (2020) Surface functionalization of cotton and PC fabrics using SiO_2 and ZnO nanoparticles for durable flame retardant properties. Coatings 10
114. Samanta A, Bhattacharyya R, Jose S et al (2017) Fire retardant finish of jute fabric with nano zinc oxide. Cellulose 24. https://doi.org/10.1007/s10570-016-1171-z
115. Sasaki K, Tenjimbayashi M, Manabe K, Shiratori S (2016) Asymmetric superhydrophobic/superhydrophilic cotton fabrics designed by spraying polymer and nanoparticles. ACS Appl Mater Interfaces 8:651–659. https://doi.org/10.1021/acsami.5b09782
116. Schneider SL, Lim HW (2019) A review of inorganic UV filters zinc oxide and titanium dioxide. Photodermatol Photoimmunol Photomed 35:442–446. https://doi.org/10.1111/phpp.12439
117. Shahid-ul-Islam BBS, Kumar A (2020) Green chemistry based in-situ synthesis of silver nanoparticles for multifunctional finishing of chitosan polysaccharide modified cellulosic textile substrate. Int J Biol Macromol 152:1135–1145. https://doi.org/10.1016/j.ijbiomac.2019.10.202
118. Shahidi S, Ghoranneviss M, Moazzenchi B (2014) New advances in plasma technology for textile. J Fusion Energy 33:97–102. https://doi.org/10.1007/s10894-013-9657-2
119. Sharma V, Basak S, Ali SW (2022) Synthesis of copper nanoparticles on cellulosic fabrics and evaluation of their multifunctional performances. Cellulose 29:7973–7988. https://doi.org/10.1007/s10570-022-04709-0
120. Sharouf H, Saffour Z (2023) Synthesizing and using iron oxide nanoparticles as nanocomposite in cotton fabrics nanofinishing. Baghdad Sci J. https://doi.org/10.21123/bsj.2023.7743
121. Singh C, Mukhopadhyay S, Hod I (2021) Metal–organic framework derived nanomaterials for electrocatalysis: recent developments for CO_2 and N_2 reduction. Nano Converg 8:1. https://doi.org/10.1186/s40580-020-00251-6
122. Singh D, Pandey D, Yadav R, Singh D (2012) A study of nanosized zinc oxide and its nanofluid. Pramana 78:759–766. https://doi.org/10.1007/s12043-012-0275-8
123. Singh N, Yadav M, Khanna S, Sahu O (2017) Sustainable fragrance cum antimicrobial finishing on cotton: indigenous essential oil. Sustain Chem Pharm 5:22–29. https://doi.org/10.1016/j.scp.2017.01.003
124. Spagnuolo L, D'Orsi R, Operamolla A (2022) Nanocellulose for paper and textile coating: the importance of surface chemistry. ChemPlusChem 87:e202200204. https://doi.org/10.1002/cplu.202200204
125. Srishti KA (2023) Sustainable approach to oil recovery from oil spills through superhydrophobic jute fabric. Mar Pollut Bull 197:115701. https://doi.org/10.1016/j.marpolbul.2023.115701
126. Sundaresan K, Sivakumar A, Vigneswaran C, Ramachandran T (2012) Influence of nano titanium dioxide finish, prepared by sol-gel technique, on the ultraviolet protection, antimicrobial, and self-cleaning characteristics of cotton fabrics. J Ind Text 41:259–277. https://doi.org/10.1177/1528083711414962
127. Tang B, Lin X, Zou F et al (2017) In situ synthesis of gold nanoparticles on cotton fabric for multifunctional applications. Cellulose 24:4547–4560. https://doi.org/10.1007/s10570-017-1413-8
128. Torkamani R, Aslibeiki B (2023) Bulk ZnO, nanoparticles, nanorods and thin film: a comparative study of structural, optical and photocatalytic properties. J Cryst Growth 618:127317. https://doi.org/10.1016/j.jcrysgro.2023.127317
129. Uğur ŞS, Sarıışık M, Aktaş AH (2011) Nano-Al_2O_3 multilayer film deposition on cotton fabrics by layer-by-layer deposition method. Mater Res Bull 46:1202–1206. https://doi.org/10.1016/j.materresbull.2011.04.005

130. Ünlü H, Horing NJM, Dabrowski J (2015) Low dimensional and nanostructured materials and devices, 1st edn. Springer, Cham
131. Uzun S, Han M, Strobel CJ et al (2021) Highly conductive and scalable $Ti_3C_2T_x$-coated fabrics for efficient electromagnetic interference shielding. Carbon N Y 174:382–389. https://doi.org/10.1016/j.carbon.2020.12.021
132. Valipour P (2023) Preparation of multifunctional polyester fabrics using zinc stannate nanoparticles. Indian J Fibre Text Res 48. https://doi.org/10.56042/ijftr.v48i2.56296
133. Varesano A, Rombaldoni F, Tonetti C et al (2014) Chemical treatments for improving adhesion between electrospun nanofibers and fabrics. J Appl Polym Sci 131. https://doi.org/10.1002/app.39766
134. Verkhovskii R, Kozlova A, Atkin V et al (2019) Physical properties and cytotoxicity of silver nanoparticles under different polymeric stabilizers. Heliyon 5:e01305. https://doi.org/10.1016/j.heliyon.2019.e01305
135. Vigneshwaran N, Arputharaj A (2020) Functional finishing of cotton textiles using nanomaterials. In: Shahid M, Adivarekar R (eds) Advances in functional finishing of textiles. Springer Singapore, Singapore, pp 43–56
136. Vigneshwaran N, Kumar S, Kathe AA et al (2006) Functional finishing of cotton fabrics using zinc oxide–soluble starch nanocomposites. Nanotechnology 17:5087. https://doi.org/10.1088/0957-4484/17/20/008
137. Vigneshwaran N, Varadarajan PV, Balasubramanya RH (2010) Application of metallic nanoparticles in textiles. In: Nanotechnologies for the life sciences. Wiley
138. Wang Q, Song H, Pan S et al (2020) Initial pyrolysis mechanism and product formation of cellulose: an experimental and density functional theory (DFT) study. Sci Rep 10:3626. https://doi.org/10.1038/s41598-020-60095-2
139. Wang Y, Sun C, Zhao X et al (2016) The application of nano-TiO(2) photo semiconductors in agriculture. Nanoscale Res Lett 11:529. https://doi.org/10.1186/s11671-016-1721-1
140. Xiao X, Cao G, Chen F et al (2015) Durable superhydrophobic wool fabrics coating with nanoscale Al_2O_3 layer by atomic layer deposition. Appl Surf Sci 349:876–879. https://doi.org/10.1016/j.apsusc.2015.05.061
141. Xu B, Cai Z, Wang W, Ge F (2010) Preparation of superhydrophobic cotton fabrics based on SiO_2 nanoparticles and ZnO nanorod arrays with subsequent hydrophobic modification. Surf Coat Technol 204:1556–1561. https://doi.org/10.1016/j.surfcoat.2009.09.086
142. Yadav A, Prasad V, Kathe AA et al (2006) Functional finishing in cotton fabrics using zinc oxide nanoparticles. Bull Mater Sci 29:641–645. https://doi.org/10.1007/s12034-006-0017-y
143. Yan B, Zhou M, Liao X et al (2021) Developing a multifunctional silk fabric with dual-driven heating and rapid photothermal antibacterial abilities using high-yield MXene dispersions. ACS Appl Mater Interfaces 13:43414–43425. https://doi.org/10.1021/acsami.1c12915
144. Yang X, Wang Z, Zhang Y, Liu W (2020) A biocompatible and sustainable anti-ultraviolet functionalization of cotton fabric with nanocellulose and Chitosan nanocomposites. Fibers Polym 21:2521–2529. https://doi.org/10.1007/s12221-020-1339-x
145. Yazhini KB, Prabu HG (2015) Study on flame-retardant and UV-protection properties of cotton fabric functionalized with ppy–ZnO–CNT nanocomposite. Rsc Adv 5:49062–49069
146. Yu C, Sasic S, Liu K et al (2020) Nature-Inspired self–cleaning surfaces: mechanisms, modelling, and manufacturing. Chem Eng Res Des 155:48–65. https://doi.org/10.1016/j.cherd.2019.11.038
147. Zhang D, Zhang G, Chen L et al (2013) Multifunctional finishing of cotton fabric based on in situ fabrication of polymer-hybrid nanoparticles. J Appl Polym Sci 130:3778–3784. https://doi.org/10.1002/app.39634
148. Zhang Q, Li C (2020) High temperature stable anatase phase titanium dioxide films synthesized by mist chemical vapor deposition. Nanomater (Basel, Switz) 10. https://doi.org/10.3390/nano10050911
149. Zhang Y, Shahid-ul-Islam RLJ, Li Q (2022) Recent advances in the surface modification strategies to improve functional finishing of cotton with natural colourants—a review. J Clean Prod 335:130313. https://doi.org/10.1016/j.jclepro.2021.130313

150. Zhang Z, Dong H, Liao Y (2023) Facile fabrication of SWCNT film-based superhydrophobic cotton fabrics for oil/water separation and self-cleaning. J Environ Chem Eng 11:109570. https://doi.org/10.1016/j.jece.2023.109570
151. Zhu C, Shi J, Xu S et al (2017) Design and characterization of self-cleaning cotton fabrics exploiting zinc oxide nanoparticle-triggered photocatalytic degradation. Cellulose 24:2657–2667. https://doi.org/10.1007/s10570-017-1289-7

Chapter 12
Nanofinishing of Synthetic Fibres

Manisha Yadav, Nagender Singh, Shelly Khanna, and Amit Madhu

1 Introduction

In the domain of modern textile engineering, the pursuit of enhancing the performance, functionality, and aesthetics of synthetic fibres has given rise to cutting-edge technologies. One of these standout approaches is nanofinishing, a revolutionary technique that has reshaped the landscape of modifying textile surfaces. This innovative method involves applying nanomaterials, typically within the range of 1–100 nm in size, to synthetic fibres, resulting in a multitude of desirable characteristics [1]. These attributes may include enhanced durability, water repellency, antimicrobial features, UV resistance, and a variety of other specialized functionalities [2]. Nanofinishing holds the potential to transform synthetic fibres into versatile materials that can meet the dynamic demands of contemporary industries, spanning from fashion and sportswear to medical textiles and aerospace applications. The focus of this chapter is to unveil the profound impact of nanofinishing on synthetic fibres, elucidating its influence on their properties and diverse applications.

2 Nanomaterials in Textile Field

Nanomaterials are materials with structures or properties characterized by nanoscale dimensions, typically in the range of 1–100 nm. At this scale, materials can exhibit unique and often enhanced properties compared to their bulk counterparts. These

M. Yadav (✉) · S. Khanna
DLC State University of Performing and Visual Arts, Rohtak, Haryana, India
e-mail: Manishayadav3089@gmail.com

N. Singh · A. Madhu
The Technological Institute of Textile and Sciences, Bhiwani, Haryana, India

S. Bairagi et al. (eds.), *Nanotechnology in Textile Finishing*, Materials Horizons: From Nature to Nanomaterials, https://doi.org/10.1007/978-981-97-2696-7_12

materials can be composed of a variety of substances like metals, semiconductors, polymers, ceramics, and composites. They can take on various forms, such as nanoparticles, nanotubes, nanowires, nanosheets, and more. Nanomaterials offer a wide range of functional properties when integrated into textiles. Some of the key functional properties they provide are: antimicrobial properties, water and stain resistance, UV protection, thermal regulation, electrical conductivity, mechanical strength and durability, flame retardancy, barrier properties, gas and vapour adsorption, self-cleaning and easy-care, sensory and smart properties, drug delivery, and therapeutic effects. These functional properties provided by nanomaterials expand the potential applications of textiles across various industries, from health care and sports to aerospace and electronics. Furthermore, nanotechnology presents substantial economic potential by reducing production costs compared to traditional methods. This spans various domains, including economics, energy conservation, environmental sustainability, precise substance release, advancements in packaged food technologies, microscale material separation and storage for future utilization, and controlled deployment. Additionally, nanotechnology provides the capacity to create adaptable textile systems while maintaining the intrinsic qualities of textiles, such as washability, softness, and elasticity [3].

Nanofibres

The latest technique for manufacturing nanofibers is the electrospinning method, as illustrated in Fig. 1. This cost-effective method generates synthetic nanofibers from a range of materials, including polymers, inorganic substances, and inorganic/organic hybrids. These nanofibers are commonly referred to as electrospun nanofibers (ENFs) [4]. They encompass a broad spectrum of cross-sectional diameter, spanning from ten nanometre to hundred nanometre, featuring substantial surface area, and notable porosity. Nevertheless, a significant drawback of electrospinning technology has been the limited ability to precisely control pore size, a crucial factor given its impact on fibre diameter [5, 6]. Both natural and synthetic polymers have been used depending on the final application. Some electrospun natural nanofibers such as collagen, chitin, chitosan, hyaluronic acid, silk, and cellulose are already reported [7]. For the electrospinning of synthetic fibres, various compounds have been used such as poly vinyl alcohol (PVA), poly vinyl acetate (PVA), poly ethylene oxide (PEO), and others [8]. Typically, these polymers are dissolved in a suitable solvent to create the electrospinning solution. Commonly employed solvents include water, ethanol, N,N-dimethylformamide (DMF), isopropanol, and various others. Key parameters influencing the production of electrospun nanofibers include the solution's viscosity, flow rate, applied voltage level, and the distance between the syringe and the collector. The optimum selection of above parameters significantly affects the nanofibre fabrication [9]. Electrospun nanofibres find applications in a diverse array of emerging fields, including but not limited to water treatment, antibacterial uses, medical uses such as wound dressing and drug delivery, tissue engineering, microwave absorption, flame retardancy, and other applications [4, 5].

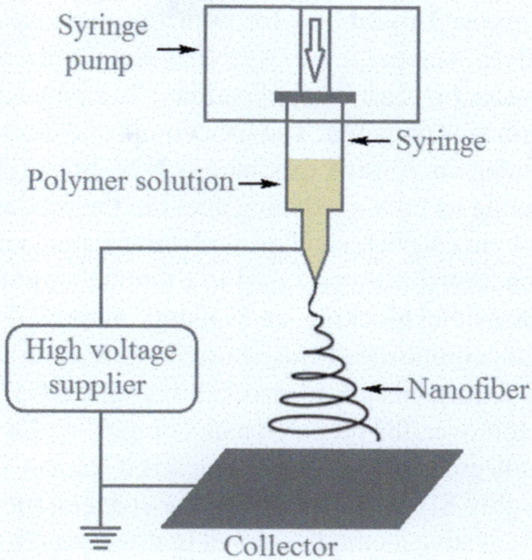

Fig. 1 Schematic of electrospinning method [6]

Nano-coating

In the process of nano-coating procedure, a thin layer, typically under 100 nm in thickness, is applied onto the substrate to enhance specific characteristics or add advanced functionalities. The conventional coatings come with certain drawbacks such as reduced durability, limited resistance to abrasion, diminished strength, decreased flexibility, and inadequate adhesion between the substrate and the coating layer [10]. Therefore, functionalization methodologies should be used to overcome these shortcomings. Nano-coating of thin layers of nanoparticles can be done on synthetic fibres through various techniques including cross-linking, thin film deposition, and sol–gel techniques [11]. It is reported that the nanomaterial coating application on fabric does not affect the hand feel or breathability of fabric [12].

Nanocomposites

Nanocomposites refer to composite materials wherein at least one of the phases exhibits dimensions within the nanometre range. The polymer-based nanocomposites contain nanomaterials which are dispersed in polymer matrices. Incorporating nanoparticles represents a significant advancement in enhancing properties like mechanical strength, toughness, as well as electrical and thermal conductivity [13]. A nanocomposite is created by integrating metals, metal oxides, or nanoclay into polymeric fibres. This application in synthetic fibres serves to imbue them with specific desired attributes, including antimicrobial, water and soil release, UV protection, water resistance, flame retardancy, and antistatic properties. Nanocomposites are also known to improve mechanical, optical, electrical as well as biological activity of textiles [14]. Figure 2 e.g. a study was conducted where a nanocomposite was created by dispersing nanokaoline, a type of nanoclay, in a sodium polyacrylate/kaoline mixture capable of forming soluble polymer. This nanocomposite was then utilized on viscose fabric, resulting in enhanced tear strength, tensile strength, and improved dyeability for both reactive and direct dyes [15]. In another study, the silver nanoparticles (Ag NPs) were embedded in the polymer to form nanocomposite for various applications. The polyacrylonitrile (PAN)/silver nanocomposite film was prepared. The nanocomposite films were prepared by reducing Ag^+ to Ag^0 under microwave irradiation. N,N-dimethylformamide (DMF) served a dual role, acting as both a reducing agent in the production of silver nanoparticles from silver nitrate ($AgNO_3$) and as a solvent for polyacrylonitrile (PAN). The developed multifunctional film was used to examine antimicrobial activity, electrical conductivity, ultraviolet blocking, and catalytic activity. The results showed that the PAN/Ag NPs nanocomposites films showed significant antimicrobial activities against different microorganisms. It also exhibits good UV properties and electrical conductivity. Moreover, the prepared nanocomposite film was used for identification of biological molecules using surface enhanced Raman scattering and results showed that it has highly SERS sensitivity at low concentration as well as high catalytic activity. The innovative method explored in this research has the potential to be applied in intelligent medical applications, such as promoting scar-free healing of burn wounds and developing protective substrates with biological properties. [16]. It was found that carbon nanotubes are good filler for nanocomposites. The polymer/carbon nanotubes nanocomposites are used for textile finishing. They produce multifunctional textile which can which can be used in various applications as shown in Fig. 2.

Nanoparticles

The International Organization for Standardization characterizes nanoparticles as nano-entities whose outer dimensions are within the nanoscale, with minimal disparity between the lengths of the longest and shortest axes of the nano-entity [17]. Nanoparticles present a superior option compared to traditional micron-sized particles utilized in surface treatments for dirt removal. They offer benefits such as resistance to staining, flame resistance, wrinkle resistance, moisture control, antimicrobial properties, antistatic properties, UV protection, and improved dyeing capabilities [18]. Inorganic nanoparticles like TiO_2, ZnO, SiO_2, Cu_2O, Ag, Au, and nanoclay find application in textile functionalization due to their larger surface area, enduring thermal and chemical stability, consistent processability, and non-toxic nature in comparison with organic materials. Nanoparticles find diverse applications in composite materials, including fields like construction and tissue engineering [19]. Padding, spraying, and sol–gel are some of the techniques which are used for the application of nanoparticles. Among all techniques, padding is the most used method for applying nanoparticles onto textile materials using a padding solution at suitable speed and pressure. Following this step, the fabric undergoes drying and curing [20].

Sol–gel coating methods have recently become more popular in nano-coating due to their capability to operate at low processing temperatures, achieve high chemical uniformity, and control particle size and shape [21].

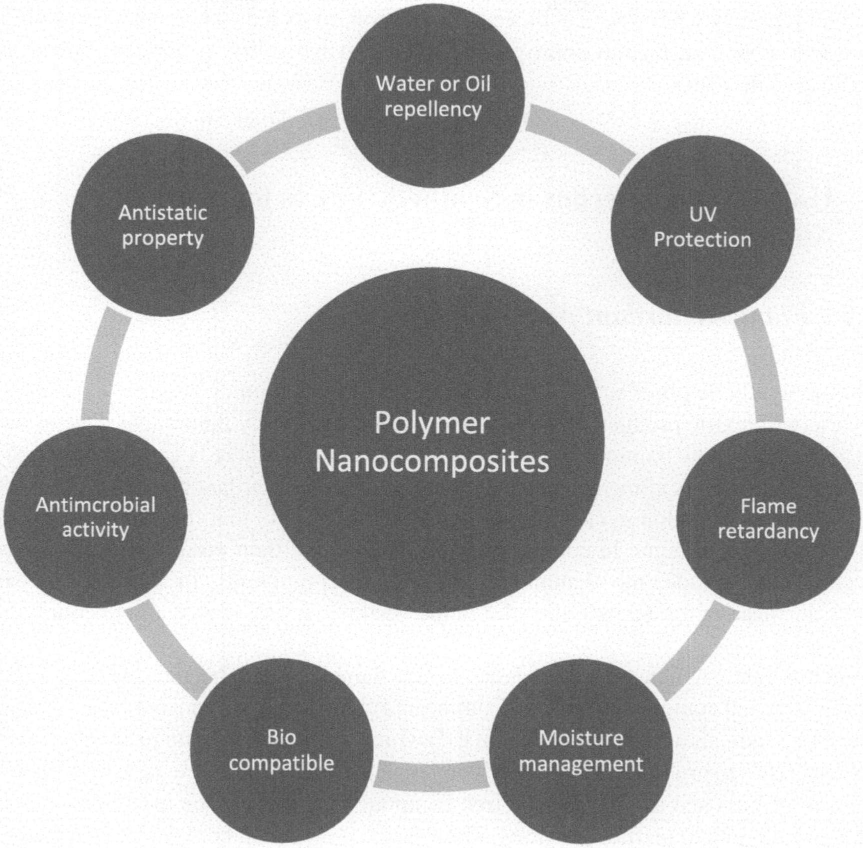

Fig. 2 Application of nanocomposites in textile finishing

2.1 Traditional Synthetic Fibres

The ongoing progress in polymer chemistry has facilitated the production of synthetic fibres composed of polymers or tiny molecules, with petroleum serving as the crucial primary material. Among popular synthetic fibres like polyester, polyolefin and acrylic, polyamide (specifically PA6, also called as nylon), polyester (particularly PET, called polyethylene terephthalate), polypropylene is predominantly utilized in industrial and residential uses. However, the main disadvantages of these fabrics are poor wettability and hydrophobicity. This can have an impact on the ease of processing the polymer and the ability to dye the fibres [22]. With the advancement of nanotechnology, various forms of nanoparticles, primarily inorganic or metal oxide in nature, have been effectively integrated into artificial fibres. This integration aims to create functionalized fibres endowed with necessary properties. Properties like self-cleaning, flame retardancy, antistatic, antibacterial, UV protection, and water and oil repellency have been achieved. In addition, trials have been taken to modify synthetic fibres' surface to improve properties like dyeability, absorbency, softness and wettability [2].

3 Use of Nanomaterials in Synthetic Fibres for Various Applications

3.1 Flame-Retardant Application

Various synthetic fibres such as nylon, polyester, and polyamide are used to produce a variety of textile products. However, these fibres are highly flammable. Due to the requirement of low flammable products in textile industry, there is various investigation and research is going on the development of flame-retardant products. Figure 3 shows the combustion cycle of synthetic fibres. It depicts that the synthetic fibres firstly melt when come in contact with fire source and then production of species occurs that degrade into flammable and volatile compounds. In the presence of oxygen, these compounds ignite the flame, making it easy for synthetic fibres to catch fire.

In this context, many compounds are used as flame retardant but they are composed of halogenated compounds such as bromine and chlorine. In the conventional system, particular radicals are released ($Cl^\bullet$ and $Br^\bullet$) which then combine with reactive species ($H^\bullet$ and $OH^\bullet$) to produce less reactive halogen atoms [23]. Therefore, new regulations have been imposed to use halogen free compounds which can be used as effective flame retardants in textile industries. Phosphorus is one such compound which uses chemicals free from halogen. These compounds create species like $PO^\bullet$, $PO_2^\bullet$, and $OHPO^\bullet$ which further react with $H^\bullet$ and $OH^\bullet$ and stop the fibres from oxidation. This leads to below shown combustion process.

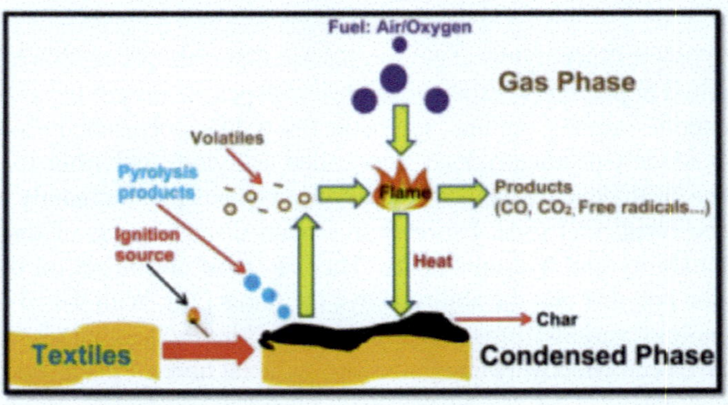

Fig. 3 Schematic of combustion process of a synthetic fibre [23]

$$PO^{\bullet} + H^{\bullet} \rightarrow HPO$$
$$PO^{\bullet} + OH^{\bullet} \rightarrow HPO_2$$
$$HPO_2^{\bullet} + H^{\bullet} \rightarrow H_2O + PO$$

The use of nanoparticles as fire-retardant materials is an interesting approach and gaining lots of interest. The benefit of employing nanoparticles is that its low amount can be effective in providing flame retardancy. Further, to obtain better results in terms of flame retardancy, factors like chemical nature, concentration and dispersion on textile fabrics, morphology (shape and size) need to be considered [24]. The various nanoparticles used for fire-retardant application are silicon, zinc oxide (ZnO), and titanium oxide (TiO_2).

The treatment with silicon nanoparticles onto synthetic fibres protect the fibres from contact with oxygen and leads to reduction in heat transfer. Carosio and colleagues applied a layer-by-layer (LbL) assembly technique to coat polyester (PET) fabrics. This process involved utilizing negatively charged colloidal SiO_2 nanoparticles, characterized by two distinct sizes, i.e.10 nm and 30 nm along with positively charged alumina-coated silicon dioxide. The findings showed that the smallest nanoparticles increase the ignition time and reduce pHRR using cone calorimetry test [25]. It was also reported that thin film of silica was applied on polyester textile (PET) in multi-layers and the finished textile showed improved flame-retardant properties. Moreover, the plasma technique was employed to activate polyester fabric's surface, enhancing the adsorption of clay nanoparticles. This activation contributes to an augmentation in the thermal stability and flame-retardant properties of the materials. It was verified that the flame-retardant impact of SiO_2 on selected textile fibres enhances their thermal stability, and thus requires a higher amount of energy to start combustion process [26, 27]. The integration of SiO_2 nanoparticles with polypropylene (PP) resulted in enhanced flame-retardant and mechanical properties. The tensile strength of the PP composite showed improvement compared to the original PP [27]. The study investigates the impact of immersing fabrics with various nanoparticles, including clay nanoparticles, TiO_2, or silica, either individually or in combination. Employing silica through the pad-dry-cure technique leads to flame retardancy, accompanied by a reduction in the amounts of CO_2 and CO. Furthermore, combining hydrotalcite ($Mg_6A_{12}(OH)_{16}(CO_3) \cdot 4H_2O$) with SiO_2 nanoparticles enhances the flame-retardant properties [28]. The combination of nanoparticles through the pad-dry and cure method was noted to significantly improve flame retardancy compared to the use of a single type of nanoparticles. In the study carried out by Abou El-Kheir et al., polyester fabrics treated with varying concentrations of SiO_2 dioxide and nanoparticles (bentonite) using the padding-drying-curing method. The outcomes revealed a substantial improvement in the thermal stability of polyester following nanoparticle treatment. Additionally, several characteristics, such as antibacterial property, moisture regain, and UV protection, were significantly enhanced [29, 30].

The polyester fabric was given flame retardancy application through the application of a pad-dry-cure method incorporating zinc oxide nanoparticles. Results show that increasing the dosage of ZnO (from 0.25 to 0.5%) decreased the flammability of the fabric [4]. El-Hady et al., introduced an innovative method for flame retardancy, employing zinc oxide nanoparticles for the application on blends of cotton and polyester. The technique pad-dry-cure was used for incorporating ZnO nanoparticles into textile. Two distinct polycarboxylic acids, namely succinic acid and 1,2,3,4-butane tetracarboxylic acid (BTCA), have been utilized in conjunction with sodium hypophosphite as a catalyst. The study investigated that as the curing temperature increases from 160 to 180 °C, the new ester bonds were formed which cross links cellulose chains and polycarboxylic acids. Moreover, char yield was also measured to study the effect of flame retardancy. The findings suggested that both succinic acid (SA) and BTCA serve as effective agents in decreasing the flame retardancy of treated fabrics when used in conjunction with sodium hypophosphite (SHP) [31]. Apaydin et al. made a flame-retardant coating from poly(allylamine), sodium polyphosphates, and TiO_2 nanoparticles which was applied polyamide and polyester fabric using layer-by-layer assembly. Nano-coatings were applied to both polyamide and polyester fabrics, revealing notable distinctions in surface characteristics, such as wettability and surface energy. This observation was confirmed through SEM images, EDX analysis, and contact angle measurements. The resulting thin films uniformly coated all polyamide fabrics, while polyester sample exhibited partial coverage with some aggregates. The crucial distinction in morphology serves as the focal point for comprehending the diverse responses to flammability tests. It is the underlying reason why the coating effectively diminishes only the peak heat release rate value for polyamide fabrics [32]. In another study, polyamide 66 was treated with novel nanoparticles which is titanium dioxide (TiO_2), silicon dioxide (SiO_2), along with naturally occurring phytic acid (PA), and chitosan (CS). The coating was done with the aim to improve the flame retardancy and hydrophilic properties of polyamide 66 fabric using pad-dry-cure method. The use of PA-TiO_2 increased the limiting oxygen index value to 24.5% for the treated fabrics, compared to the 19.5% observed in the untreated fabric, and also increased the char yield percentage. On the contrary, incorporating chitosan (CS) into the PA-TiO_2/SiO_2 formulations resulted in a notable decrease in the peak heat release rate (pHRR), with the PA66-PA-CS-T fabric sample achieving a maximum reduction of 25%. However, there was a substantial enhancement in the hydrophilicity of the treated fabrics, particularly with the PA-SiO_2 formulation, as evidenced by a decrease in the water contact angle from 112° for pure PA66 to 29° for the treated fabrics. Furthermore, the inclusion of nanoparticles (TiO_2/SiO_2) in conjunction with PA elevated the tensile strength of the treated textiles [33]. Conzalves et al. used multi-walled carbon nanotubes (MWCNTs) for the application of flame retardancy on cotton and polyester fabrics. The nanotubes were functionalized with oxygen-containing surface groups. The results showed that the polyester fabrics treated with carbon nano-tubes halved the burning rate and double the time of burning when compared with the control specimen. This may be due to higher MWCNTs loading in polyester relative to cotton fabric and more compact coating which facilitate stronger interactions between MWCNTs-polyester. Thus, polyester fabric is more advantageous to achieve flame retardancy [34]. In another study, globular nanoparticles such as titania (anatase crystalline form), silica, and octapropylammonium POSS have also been used on polyester and cotton to achieve flame-retardant properties. It was noted that these nanoparticles reduce CO and CO_2 yields [35].

3.2 Antimicrobial Application

The increased utilization of NPs in the medical domain has created avenues for in-depth exploration of their antibacterial mechanisms. Research revealed that nanoparticles possessing characteristics such as extensive surface area, charge, capacity to deliver a multitude of antibiotics or other substances, as well as specific size and shape, render them an ideal antimicrobial arsenal [36, 37]. Typically, silver nanoparticles of small sizes (5, 9, 10, 12, and 13.5 nm), gold nanoparticles (8.4 nm), zinc oxide nanoparticles (12 nm), and titanium oxide nanoparticles (12 and 17 nm) exhibit potent antimicrobial properties [38]. As these particles exert their effects only upon contacting bacterial cell walls, diverse methods of enhancing the interaction between nanoparticles and bacteria, such as electrostatic attraction, Van der Waals forces, receptor–ligand interaction, and hydrophobic interactions have been investigated. Upon contact, nanoparticles have the ability to traverse microbe membranes, disrupt metabolic pathways, and bring about alterations in membrane structure and functionality. Once within cells, nanoparticles engage with microbial cellular machinery, inhibiting enzymes, deactivating proteins, inducing oxidative stress and electrolyte imbalance, and altering gene expression levels [39]. Here are some key mechanisms by which nanoparticles exert their antimicrobial effects on bacterial cells: generation of reactive oxygen species (ROS), interaction with intracellular components, release of metal ions [39], disruption of cell membrane [40], inhibition of enzymatic activity, DNA damage [41], and induction of apoptosis-like processes [42]. It is crucial to emphasize that the antimicrobial mechanisms of nanoparticles can differ based on factors like nanoparticle size, shape, surface charge, and the particular type of bacteria under consideration. Furthermore, the concentration of nanoparticles and the duration of exposure are pivotal factors influencing their efficacy against bacterial cells.

Antimicrobial Activity by Silver Nanoparticles

Silver nanoparticles typically consist of 20–15,000 silver atoms, with diameters generally falling below 100 nm. Their significant surface-to-volume ratio imparts noteworthy antimicrobial efficacy, even when present in low concentrations. Moreover, they are economical and exhibited minimal cytotoxicity and immunological reactivity [43]. Because of its exceptional antibacterial properties, it finds application in diverse sectors such as healthcare, food preservation, wastewater treatment, and the production of textile materials. The exact mechanism pf silver nanoparticles' antimicrobial activity has not been clarified. Figure 4 shows the various antibacterial actions.

- Because of electrostatic attraction and a strong affinity for sulphur proteins, silver ions have the ability to attach to both the cell wall and cytoplasmic membrane. This binding can increase the permeability of cytoplasmic membrane, ultimately causing disruption to the bacterial envelope [44].
- After the absorption of free silver ions into cells, respiratory enzymes may be deactivated, resulting in the production of reactive oxygen species and the interruption of adenosine triphosphate (ATP) synthesis [45].
- Reactive oxygen species play a significant role in initiating the disruption of cell membranes and the modification of deoxyribonucleic acid (DNA). Since sulphur and phosphorus components are crucial for DNA, the interaction between silver ions and the sulphur-phosphorus elements of DNA can inhibit DNA replication, hinder cell reproduction, or lead to the termination of microorganisms. Additionally, silver ions have the ability to hinder synthesis of protein by denaturing cytoplasmic ribosomes [46].
- Apart from the release of silver ions, silver nanoparticles have the ability to directly eliminate bacteria. They can gather in the indentations formed on the cell wall once anchored to the cell surface. The accumulated silver nanoparticles contribute to the denaturation of the cell membrane, leading to the rupture of organelles and even cell lysis [47].

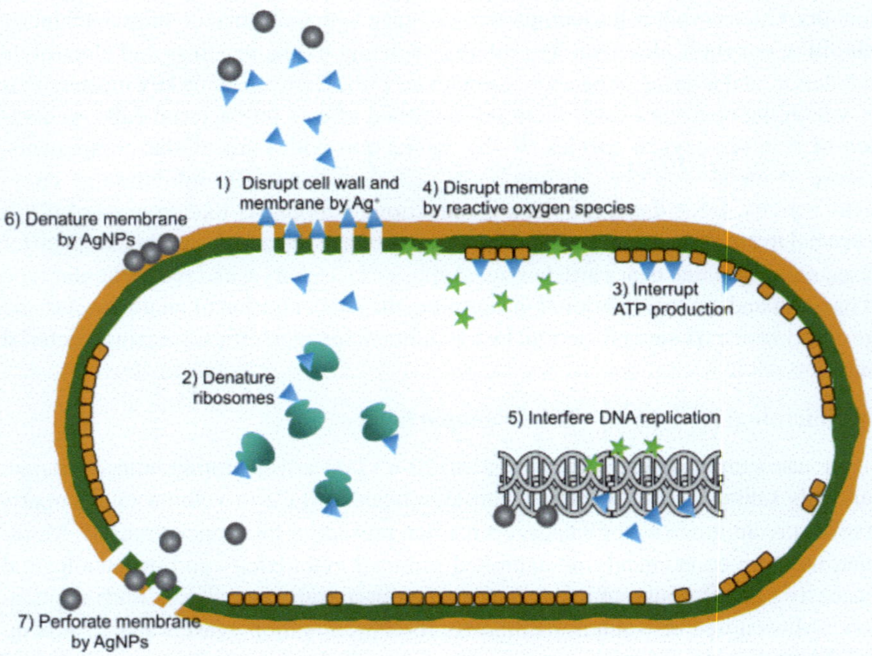

Fig. 4 Antibacterial action of silver nanoparticles [40]

Various research have been carried out on the effect of Ag nanoparticles on antimicrobial properties of synthetic fibres. For example, silver-tricalcium phosphate (Ag/TCP) nanoparticles was used to provide antimicrobial properties to polyamide 6 fibres. When bacteria are present, the TCP carrier particles undergo biodegradation, initiating the release of silver. In this study, Ag/TCP nanoparticles were generated through flame spray synthesis and later transformed into fibres using extrusion and melt spinning with the assistance of polyamide 6 for subsequent antibacterial assessments. The fibres were exposed to clinically significant strains, namely *Escherichia coli* or *Streptococcus sanguinis*, respectively. The reactive fibres showed significant reduce plate count within 24 h. The number of colonies formed was reduced by 99.9% with *E. coli* and 99.6% with *S. sanguinis* as compared to pure polyamide 6 fibres [48]. In another study, surface functionalization of polyester fibre with nanostructured silver film by magnetron sputtering to achieve superior property against UV radiation, hydrophobicity, and antibacterial performance. The results showed that 99.8% of the Staph is destroyed on the silver coated polyester fabric. The antibacterial reduction rate for E. coli was more than 99.7% [49].

Plasma jet deposition system was used to activate nonwoven polyethylene terephthalate (PET) fabric surface. This method helps the binding of silver nanoparticles (AgNPs) onto fabric surface. The treated fabrics showed bacterial reduction against *Pseudomonas aeruginosa*, *Staphylococcus aureus*, and *Candida albicans* [50]. The antibacterial polyester fabric was developed via green impregnation of silver nanoparticles. This was achieved firstly, by coating the polyester fabric with a dopamine aqueous solution. Subsequently, silver nanoparticles were produced in situ on the surface of dopamine-modified polyester textiles at room temperature, employing an aqueous solution of silver nitrate. The resulting fabrics consisting of silver nanoparticles, exhibited strong, and long-lasting antibacterial effectiveness [51, 52]. Bagriche et al. investigated the effect of DC-magnetron sputtering (DC) and pulsed DC-magnetron sputtering (DCP) on deposition of uniform nanoparticulate Ag-thin films on polyester textiles. Results revealed that silver sputtered with DCP penetrated the polyester fibre more extensively compared to silver sputtered using DC. The Ag-coverage on the polyester fibre using DCP seems to have a higher effect on the *E. coli* inactivation kinetics [53]. In another study, the silver nanoparticles were incorporated using in situ polymerization method in polyamide 6. The study has prepared nanocomposites, where $AgNO_3$ is used as filler and was directly reduced to silver nanoparticles which results in uniform dispersed nanoparticles in the PA6 matrix. The antimicrobial resistance properties of nano-Ag/PA6 nanocomposites were improved when compared to those of the native PA6 [54]. Gawish et al., investigated melt spinning technique for polypropylene fibres containing silver and zinc nanoparticles. The antimicrobial property for both PP/Ag and PP/Zn composites was examined. The results revealed that percentage bacterial reduction of *E. coli* and *S. aureus* increased with an increase in nanometal content. Modifying the silver content in PP fibres, ranging from 0.028 to 0.081%, resulted in an elevation of the *E. coli* reduction percentage from 88.4 to 90.7%, reaching 99.99% at 0.721% Ag. Additionally, enhancing the silver content in PP fibres from 0.028 to 0.721% led to an improvement in the percentage reduction of *S. aureus*, increasing it from 80.9% to

89.4%. Enhanced antibacterial effectiveness was noted in the PP/1% Zn fibre, where a bacterial reduction of 99.99% for *E. coli* and 91.4% for *S. aureus* occurred with 0.601% Zn. Moreover, at a higher zinc content of 1.316%, a reduction of 99.99% for *E. coli* and 92.6% for *S. aureus* was observed [55]. In another study, needle-punch nonwoven polypropylene (PP) was treated with powder activated carbon and Ag nanoparticle through pad-dry and spray techniques. The produced nonwoven filters were used as filters for removing harmful air pollutants. The results showed that the PP on woven layer coated with silver nanoparticles could deactivate the microorganism like *S. aureus* and *E. coli* [56].

Antimicrobial Activity by ZnO Nanoparticles

The proposed mechanisms for antimicrobial activity of ZnO nanoparticles include: (a) The disruption of cellular integrity resulting from the direct interaction between ZnO-NPs and cell membranes, (b) the release of antimicrobial ions, specially Zn^{2+} ions, and (c) ROS formation. Nevertheless, as the chemical characteristics of dissolved zinc are influenced by the components of the media, it is probable that the toxicity mechanism of ZnO-NPs varies depending on the specific media used [57, 58]. The photocatalytic property of ZnO is high, and found to be more biocompatible than TiO_2. When subjected to UV radiation in a water-based solution, ZnO-NPs produce reactive oxygen species (ROS), hydrogen peroxide (H_2O_2), and superoxide ions (O_2^-), offering potential for directing their impact towards microorganisms [59]. For example, Nirmala et al. synthesized ZnO-NPs through DC thermal plasma and observed in a photocatalytic investigation that these nanoparticles effectively decomposed methylene blue. Additionally, the ZnO-NPs demonstrated inhibitory effects on the growth of bacteria, including *Bacillus subtilis*, *E. coli*, *Salmonella typhi*, and *S. aureus* [60]. It is reported that zinc oxide nanoparticles were applied onto the synthetic fabric's surface, combined with different surfactants to uphold and condense the coating. This approach aimed to extend the lifespan of the zinc oxide nanoparticles and minimize their release [61]. Hydroentangled nonwoven fabrics polypropylene fabrics were coated with transition metal oxides. Following the application of RF plasma etching on nonwoven fabrics, nanoscale layers of ZnO and CuO were deposited using the pulsed laser deposition method. Antimicrobial studies were conducted to investigate the mechanism underlying the antibacterial properties of the coated fabrics. The findings indicated noteworthy antibacterial effectiveness in ZnO and CuO-coated hydroentangled nonwoven polypropylene (PP), exhibiting superior activity against Gram-positive S. aureus compared to Gram-negative *E. coli*. Additionally, these coatings are inherently non-toxic. PP nonwoven fabrics coated with ZnO and CuO present potential for utilization as wound dressings, allowing for the incorporation of appropriate antibiotic medications [62]. Zinc oxide was employed to create an antibacterial layer through surface modification of polyamide 6 (PA), polyethylene terephthalate (PET), and polypropylene (PP) textiles. Zinc oxide microrods were generated on zinc oxide nanoparticles (NPs) as nucleation centres through the utilization of chemical bath deposition (CBD). It was noted that greater amounts of zinc oxide deposition occurred on materials that were rougher and exhibited higher hydrophilicity. The ZnO modified textiles show significant bactericidal activity against *E. coli* and *S. aureus*. The PA material exhibited the highest concentration of ZnO microrods, followed by PET and PP. Consequently, PA-ZnO demonstrated the most significant bactericidal effect, attributed to its higher ZnO content [63]. In another study, polyester/cotton woven fabric was functionalized with ZnO nanoparticles to add antibacterial properties to them for workwear and sportswear applications. It was seen that the samples treated with smaller particles have bigger zone of inhibitions as compared to bigger particle size suspensions [64]. Farouk et al. prepared inorganic–organic hybrid polymers based on ZnO nanoparticles and 3-glycidyloxypropyltrimethoxysilane (GPTMS). The hybrid material prepared was applied to cotton/polyester blend to investigated antibacterial performance investigated against Gram-negative bacterium *Escherichia coli* and Gram-positive *Micrococcus luteus*. The findings suggest that the treated fabric samples showed significant antibacterial activity and was inhibited by *E. coli* and *M. luteus* because of low ZnO content and minute particle size [65].

Antimicrobial Activity by TiO_2 Nanoparticles

Titanium dioxide (TiO_2) has the capability to impart various attributes to artificial fibers, such as self-cleaning properties, antimicrobial effects, and protection against ultraviolet (UV) radiation. TiO_2 nanoparticles were synthesized through the sol–gel method using two distinct reaction mediums. These nanoparticles were then applied to Nylon 6,6 fabric using the layer-by-layer (LBL) technique to confer antimicrobial activity. Polymer thin film can be deposited on the textile fabric by using LBL technique to produce polyelectrolyte multi-layers (PEM). The findings revealed that TiO_2 nanoparticles synthesized by two different methods showed similar antimicrobial property. It was also observed that increasing the concentration of nanoparticles in the polyelectrolyte solution allows achieving the desired outcome with fewer layers, leading to time and chemical savings. Additionally, as the number of PEM increases from the 1st to the 25th, there is a continuous enhancement in antimicrobial activity. This improvement occurs because with each additional layer, there is a corresponding increase in the concentration of TiO_2 on the fabric [66]. In another study, the polyester fabric was modified with natural polysaccharide alginate (ALG) and colloidal TiO_2 nanoparticles to investigate its multi-functionality against UV protection, antibacterial, and photocatalytic activity. The modified polyester fabric (PES) was tested against Gram-negative bacterium *E. coli*. The finding revealed that the antibacterial performance of the PES + TiO_2 fabric was unsatisfactory. The addition of alginates to the PES fabric (PES + ALG) minimally contributed to its antibacterial effectiveness ($R = 9.6\%$). Nevertheless, incorporating alginates into PES fabrics and then loading them with TiO_2 NPs (PES + ALG + TiO_2) resulted in the highest reduction of bacteria. The exceptional antibacterial efficacy of this nanocomposite textile material remained intact ($R = 99.8\%$) even after undergoing five washing cycles, demonstrating its robust laundering durability [67]. Montazer et al., treated the polyester/wool blend fabric with two different types of enzymes to increase the surface activity for the higher absorption of TiO_2. The initial step involved subjecting the fabric to proteases and lipases to enzymatically breakdown the wool and polyester surfaces, respectively. The cross-linking agent, butane tetracarboxylic acid (BTCA), additionally contributed to improving the adsorption and stabilization of nanoparticles on the fabric surface. The findings suggested that the sample treated with nano-TiO_2 exhibited outstanding antibacterial effectiveness against *E. coli*. The reduction in bacteria reached 100% with 0.75% nano-TiO_2 and 99% with 0.25% nano-TiO_2. Importantly, nano-TiO_2 demonstrated the ability to eliminate bacteria rather than merely inhibiting their growth [68]. In another study, TiO_2 nanoparticles were applied to polyester (PES) fabric to provide desirable level of antibacterial activity, UV protection, and self-cleaning property. The binding efficiency of TiO_2 nanoparticles was enhanced by activating the surface of polyester fabric by low-pressure RF air plasma or corona. The results showed that applying corona or air RF plasma to PES fabrics before introducing TiO_2 NPs had a beneficial impact on the antibacterial effectiveness. Both the corona and plasma treated fabrics reached maximum bacterial reduction, which was preserved after five washing cycles, i.e. 99.9% [69].

Antimicrobial Activity by Other Nanoparticles

The other nanoparticles like silicon dioxide (SiO_2), copper (Cu), gold (Au), magnesium oxide (MgO), and gallium oxide can also be used for providing antibacterial activity. For example, the SiO_2 nanoparticles were imparted onto polyester fabric to investigate the antibacterial activity against Gram-positive (*S. aureus*), Gram-negative (*E. coli*), and pathogenic fungi (*Candida albicans* and *Aspergillus flavus*). The fabric underwent chemical activation through alkali hydrolysis to enhance the binding affinity of SiO_2 NPs by creating carboxylic and hydroxyl groups on the fabric surface. The findings suggest that alkali treated sample treated with SiO_2 nanoparticles have higher reduction % (ranges from 91 to 97%) than that pre-treated with NaOH only (ranges from 38 to 45%) for all the tested bacterial and fungi. This could be attributed to the strong interaction between the cationic surfaces of the nanoparticles and the cells of microorganisms [29]. Gold nanoparticles are used in medical sector like antimicrobial agent, drug delivery, cosmetic, and medical materials. In one of the study, gold nanoparticles (AuNPs) were synthesized, where chloroauric acid was used as precursor, $NaBH_4$ as reducing chemical, and sodium citrate act as a capping agent. The prepared nanoparticles were coated on natural and synthetic fabric (nylon and polyester) using a drop-wise method. The outcome demonstrates the effectiveness of gold nanoparticles against the Gram-positive bacteria *S. aureus* [70]. Mehravani et al. discussed the most used methods for AuNPs preparations with their properties for antimicrobial applications [71]. Copper nanoparticles exhibit antibacterial properties by attaching to the bacterial envelope and releasing ions within the cell wall. The released copper and copper oxide nanoparticles subsequently form bonds with DNA, intercalate within nucleic acid strands, and ultimately initiate the oxidation of lipids. This process results in the degradation of the bacterial wall and membrane [72]. One of the studies showed that gallium oxide nanoparticles having 100 nm size have antibacterial properties against *E. coli* and *S. aureus* and therefore can be used for inhibition of HIV [73]. Similarly, MgO nanoparticles exhibit a significant antibacterial activity against pathogenic bacteria [74].

3.3 Nanotechnology for UV Protection Applications

The concept of UV blocking finishes and coatings on textiles hold a significant role in preventing the users from the harmful effects of the natural as well as artificial UV radiations. The energy associated with the UV rays is significantly higher than the visible and infra-red spectrum that holds the capability of causing potential hazards to the human health up to the level of cells and tissues as skin burns, hyper-pigmentation, pre-ageing of skin, disturbances with the skin collagen, skin cancer and also in the extremity, DNA damage. In addition, the prolonged exposure and absorption of UV radiations in the UVA (320–400 nm), UVB (280–320 nm), and UVC (100–280 nm) range results in molecular degradation of the chemical bonds followed by severe drop in physical and mechanical properties as poor colour fastness, photodegradation of fibres/filaments of fabrics [75, 76]. Thus, it becomes imperative to provide UV protective finishes to the specialty textiles as cold weather and high-altitude mountain trekking sportswear, exterior clothing, outerwear, covering textiles, buildtech technical textiles, and many more. The main objectives of the UV protective finishing are to provide a shield against the detrimental effects of the invisible high frequency EMRs for humans, vehicles as well as the exterior of buildings. Thus, it becomes imperative for the textile substrates to be treated with an extra layer of protection from UV rays in the form of external coatings, laminates, microencapsulating agents, and nanoparticles.

The main classes of UV protective films are UV absorbers and UV blockers. The main differentiating factors between the two classes are their chemical nature, i.e. organic or inorganic nature. The organic UV absorbers mainly work on the principle of absorbing the high intensity EMRs and converting the UV rays into harmless energy without any photodegradation. These are additionally termed as anti UV agents or photostabilising agents, that helps to refrain the textile materials from the chain reactions of UV degradation. On the other hand, inorganic UV blockers work on the textile materials by either refraction or scattering of the UV rays due to the higher refractive indices of the textiles [76]. The application of anti UV agents can work successfully equally on human derma and textiles as shown in Fig. 5. As most of the current market segmentation is based on either the use of 100% synthetic materials or their blended counterparts due to enhance properties of the end products according to the diversity needed [77]. Thus, the impact of different categories of finishing agents at variable molecular levels as macro-coatings, micro-coatings, nanomaterials on the treated materials needs to be studied scientifically and in a systematic manner to explore the potential of the innovative nanofinishing.

The nanomaterials are widely available in a variety of forms as nano-tubes, nanofibres, nano-clays, nanocomposites, nanospheres, and nano-powders. Nano-structured materials that are used on the synthetic substrates in the context of physical, mechanical, comfort, and hand properties need to be explored. The value addition of the textiles with NPs meant for UV barrier can offer multitude of property enhancement of the treated surfaces as shown in Fig. 6.

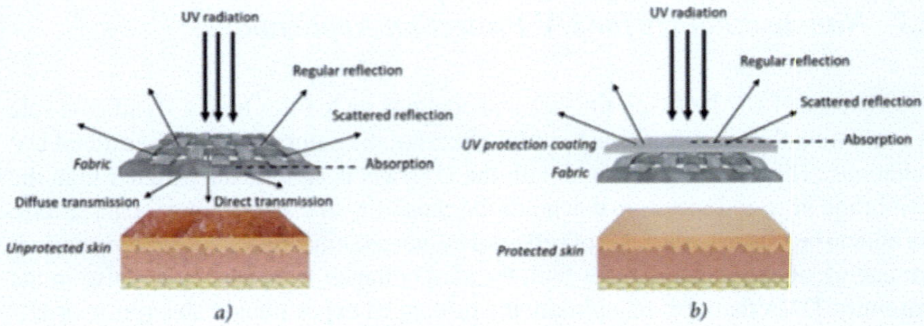

Fig. 5 Representation of the effect of UV radiations on the textiles: **a** unprotected with direct exposure of skin, **b** protected with UV blocking agents [75]

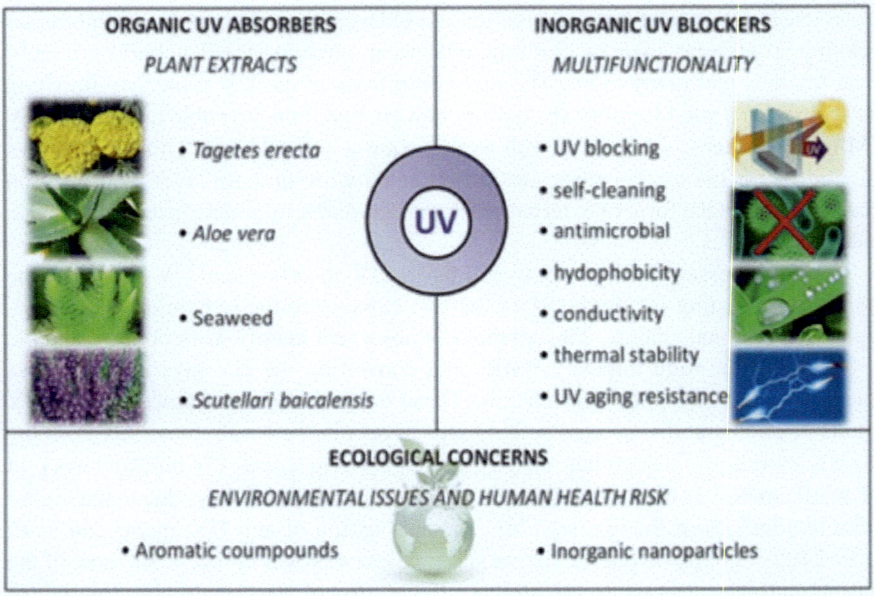

Fig. 6 Pathway of UV protection to textiles offering multitude of value added functionalities to textile surfaces [75]

Effect of UV Radiations on Textiles

Ultraviolet Electromagnetic Radiation (UV EMR) stands out as a primary factor contributing to the deterioration of textile materials. This degradation occurs due to excitations in specific segments of the polymer molecule, leading to a gradual loss of integrity. The extent of damage depends on the nature of the fibres [78]. The penetration of UV radiation in nylon-based textiles results in photodegradation, causing a reduction in elasticity and tensile strength. In the absence of UV filters, the decline in tensile strength is notably more pronounced in nylon (experiencing a 100% loss), followed by wool, cotton, and polyester, with approximate losses of 23%, 34%, and 44%, respectively. There are plentiful polymers that are susceptible to degradation due to UVRs as polyolefins (POs), polyketones (PKOs), PVC, polycarboxylic acids, PAN, and POS. Among other heterochain polymers are polyesters, polyamides, aramids, polyethers, polyimides, polyurethanes, and polysulphides, that are highly susceptible to UV degradation. Even high strength and performance fibres such as PBO, Spectra, and Kevlar have proven to be sensitive to UV damage when exposed for prolonged time spans.

Nanofinishing for UV Protection

The nanofinishing is carried out on the synthetics and its blends either by sol–gel or plasma polymerization to impart multitude of properties to a sole textile as combination of flame resistant, UV protection and breathability to quote as commercial market viability. A wide gamut of inorganic materials as metal oxides in the form of ZnO, TiO_2, Ag, Au, SiO_2, and nanoclay have been preferred due to diverse advantages offered by them as increased surface area, ease in processability, chemically stable towards the UV rays and non-toxicity as compared to the conventional finishing agents. Numerous methods as spraying, washing, dyeing, printing, and padding can be employed for applying the nanofinishes to textiles. ZnO and TiO_2 were chosen for their photocatalytic properties, exhibiting enhanced durability, heightened absorption, and superior UV blocking capabilities [2].

Several attempts have been done on the synthetic fibres with nanofinishing to either impart or enhance the UV blocking property. Nylon fabric modified with Au nanoparticles, nylon fabrics treated with Si NPs, ZnO-NPs treated polyester fabrics [4, 79–83], acrylic fibres treated in situ with Ag NPs, polyester/cotton blends treated with the combination of Ag/Cu and TiO_2 NPs have shown excellent improvement in the properties of the functionalized substrates [2]. But the systematic approach of the application of nanofinishing on textiles has not been reported as yet. Novel research had shown the effect of TiO_2 NPs on a blend of polyester/wool to enhance the self-cleaning characteristic in the presence of UV-A light with citric acid as the fixing agent on oxygen plasma pre-treated blend. It was observed that the plasma treatment has a great outcome on the loading of the NPs on the fibre surfaces to enhance the self-cleaning properties of the blends. Also, the plasma functionalized textiles depicted better water absorbency, enhanced tensile strength and better handle in comparison to untreated ones [20, 84].

Nano-coatings for UV Protection

Nano-coating is the process of application of a super-fine invisible layers or coating of nanoparticles with varying methods as thin-film deposition, cross-linking, and sol–gel technique [11, 85]. Various natural and synthetic materials that find extensive use in our daily routines are susceptible to UV degradation. Nano-coatings designed to provide UV protection play a crucial role in preventing sun damage in a range of products and structures, particularly those crafted from PVC, rubber, and wood. The application of UV protection nano-coatings to surfaces regularly exposed to sunlight creates an imperceptible shield. This layer serves to avert issues such as fading, wear, and cracking on the exposed surfaces. These can be widely applied to textiles and leathers [85]. These coatings reveal exceptional adhesion to various substrates to provide durable and enduring protection against EMRs. The ultrafine nature of the coatings safeguards the native facade and functionality of the coat treated surfaces without affecting the tactile characteristics, lucidity, and other properties [86]. The diverse types of nano-coatings include metal-based coatings, metal oxide coatings, organic UV absorbers, and multi-layer coatings comprised of ZnO, TiO_2, Indium tin oxide (ITO), fraphene, and dual layered sulphur oxide. In a study, knit polyester fabrics were coated with graphene oxide, and reduced graphene oxide/silver nanoparticles (rGO/AgNPs) with the dip-coating approach. The graphene oxide treated fabrics have demonstrated high resilience properties and mouldability. On the other hand, the polyester fabric coated with nanocomposite of rGO/AgNPs with little concentration of AgNPs achieved a UV protection ability superior to that of the polyester fabric coated with graphene oxide alone [87].

Nanofibres for UV Protection

Nanofibers have nanometric diameter, which can be synthesized from a multiplicity of polymers. In result of that, the physical distinctiveness and their application potential varies depending on the used polymer. The crucial features entailed with nanofibers include high degree of permeability, exceptional filtration, thin layering, large and extended surface area to react and extremely fine fibre diameter. For the fabrication of nanofibers, different techniques as electrospinning, melt blowing, flash spinning, self-assembly, force spinning, phase separation, and bicomponent materials are the commercially employed methods [2]. A variety of synthetic polymers are employed for the production of nanofibers as PVA, PCL, PU, poly(lactic-co-glycolicacid) (PLGA), poly(3-hydroxybutyrate-co-3-hydroxyvalerate) (PHBV), and poly(ethylene-co-vinyl acetate) (PEVA) with polymerization and sonochemical synthesis, electro hydrodynamic writing, solution blow spinning, centrifugal jet spinning, CO_2 laser supersonic drawing synthesis, and plasma-induced synthesis [88].

Nanocomposites for UV Protection

Nanocomposites are intricate blends created by integrating materials such as metal oxides or nano-clays into a polymeric fibre state. These are desired due to the preferential requirement of the treated substrate for the functionalization with desired properties from the treated textiles. Composites can be designed and developed with a predilection of the required characteristics as an outcome from the textiles. Its application in synthetic fibres is to provide some essential properties like antimicrobial, odour control/aroma impregnation, water and soil release, UV protection, stain repellent, flame retardancy, biodegradability, and antistatic properties [89]. These materials are better described as nanofilled polymer composites. Controlled structures of the nanofilled composites can introduce as well as enhance the physical properties that are lacking in the unfilled matrices [90]. Clays have been used as the conventional filler material in the fibre matrices for the custom designed composites. In the recent times, clays have been replaced by the nano-clays as the efficient filler for the property enhancement of some of the synthetic fibres as Nylon 6, polyester, polypropylene, etc. Nanocomposites have also been researched to improve the biological and other properties of textile materials according to the functionality required [89]. In a novel research work, acrylic fabric was treated with a sodium polyacrylate nanocomposite as patented by A. Abou El-Kheir and his co-authors [2].

UV Protection in Smart/Functional/Protective Textiles

A smart textile is the category of textiles that can sense alterations in the surrounding environment and act in response by the modification of its performance variables to execute a set of functions [91]. These textiles are the novel combination of conventional as well as modern textile materials with innovative smart surface coatings [81]. In the present day, textile manufacturers have wide scope to play with the gamut of materials to incorporate the smart and functional nanomaterials. The application of the nanonmaterials is employed in a manner that the physical properties of the textiles are not hampered yet the functional properties are enhanced up to the required level such as weight, thickness, tensile properties, resilience, surface texture are retained, and in addition to it, special characteristics are inculcated as desired by the end consumer [4, 89]. The incorporation of nanofinishes into the textiles can offer a multitude of innovative features to suit the applications of health, defence, fashion, sports, and advanced protection. The improved functionalities of textiles include the UV protection as the foremost characteristic desired for a range of consumer markets like kidswear, sportswear, workwear and leisurewear [92]. One class of functional textiles includes the pProtective textiles that entail the purpose of protection of professionals against the hostile physical environment, treacherous working conditions and danger from the opponents in the personal and professional arenas [93].

3.4 *Nanofinishing for Self-cleaning and Superhydrophobicity*

Self-cleaning refers to the ability of the textile to initiate and accomplish the cleaning phenomenon without the necessity of any laundering agents or any other external stimuli. These are the need of the time for diverse areas of applications as defence, high-altitude clothing, hospitality industry, kid swear, health care and hygiene products, home-textiles, anti-microbial protective textiles, medical textiles, sportswear, innerwear and outdoor construction materials [94]. This has aided the consumers to save their time, material, energy, money and repeated experiences to clean, maintain and disinfect the textiles against dirt, dust, contaminants and microbes. The different mechanisms of self-cleaning that are used for the functionalization of textiles include "the lotus effect", "the photocatalytic coatings", and "easy to clean finishing" [95].

Above all, the nature has itself given the insight to create, design and develop the concepts of self-cleaning as wings of butterfly, scales of the marine animals, and also the lotus leaves to enumerate a few. The combined effect of microscopic cum nanostructures and a highly hydrophobic surface of the lotus leaf surface provide the characteristic of highly superhydrophobicity and self-cleanability to the leaf. The self-cleaning is initiated by the minimization of the contact area between water and dirt particles, and also, the establishment of high contact angles to drop off water from the surfaces at slightest of inclinations. In addition, the self-cleaning can be achieved either by hydrophobicity or hydrophilicity of surfaces. The lotus effect/hydrophobicicity of self-cleaning is based on the superhydrophobicity mechanism, in which the surface is designed with microscopic ridges and crests along with the nanomaterial layer to accentuate the effect of self-cleaning. A synthetic poly (dimethysiloxane) polymer was synthesized on the practical implication of the lotus leaf as an artificial lotus leaf surface [96].

The second approach for the self-cleaning is photocatalytic coatings/hydrophilicity effect that are based on the phenomenon of acceleration of photoreactions in the presence of suitable catalysts and sunlight (primarily UV rays) with the decomposition of dirt particles into simpler compounds [97, 98].

Advent of Nanofinishing in the Production of Self-cleaning Textile Surfaces

The advent of nanofinishing in the development of self-cleaned textiles has reached to the inconceivable pinnacle. The main techniques involved in the production of the self-cleaned textiles include use of microwaves, photocatalytic agents, colloids of metal oxides, carbon nanotubes, and silver nanocomponents for the production of fuctionalized textiles.

Use of Microwaves for Adhering Nanoparticles to Synthetic Surfaces

The nanoparticles and the active agents for warding off dirt, contaminants, water, and other fluids are adhered to the synthetic surfaces by the means of nanofinishing to enhance the synergy of textiles with the effect materials, Fig. 7. This facilitates the textile to be devoid of contaminants of both hydrophilic and hydrophobic nature by beading cum running off the defile components from the treated surfaces. The functionalized textiles find the main market for the defence personnel by maintaining the hygiene of uniforms, undergarments and ancillary equipment for a long duration without interfering with their field performance, even without washing for days [99].

Use of Photocatalytic Agents for Infusion of Nanoparticles into Synthetic Surfaces

The photocatalytic effect is based on the principle of hydrophilic self-cleaning that activates the photo-activated purification function of the treated surfaces by inducing the chemical alterations of the textile surfaces, Fig. 8. This is accomplished by breaking down of the contaminants, microbes and pollutants by photoreduction method into simpler compounds as CO_2 and H_2O (Fig. 5). The photocatalysis initiated self-cleaning characteristics are complemented by additional features of de-odourization, antimicrobial, mosquito repellence, and UV blocking properties. The most common photocatalysts being employed are TiO_2 and ZnO for imparting functional features to the textile substrates.

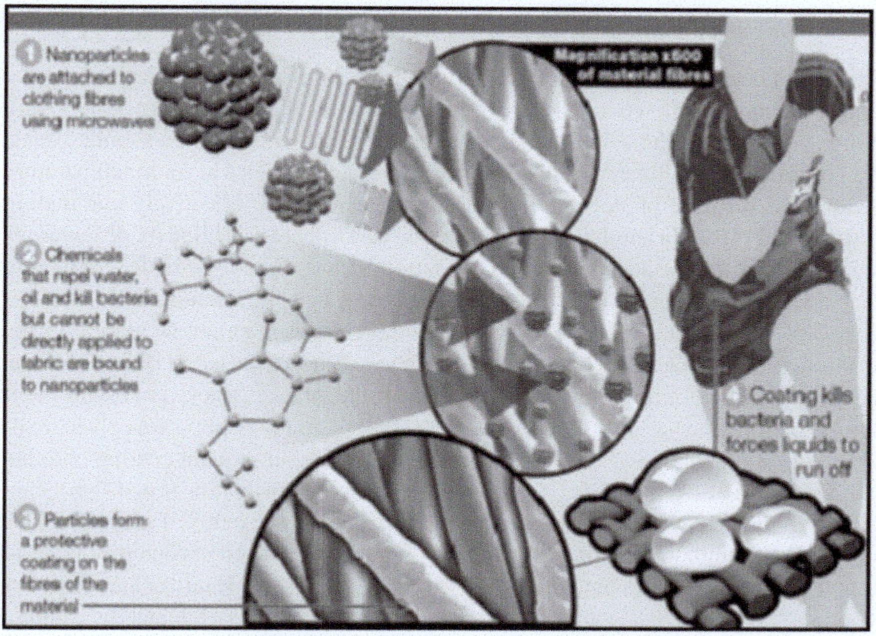

Fig. 7 Mechanism of nanofunctionalization with microwave technology [96]

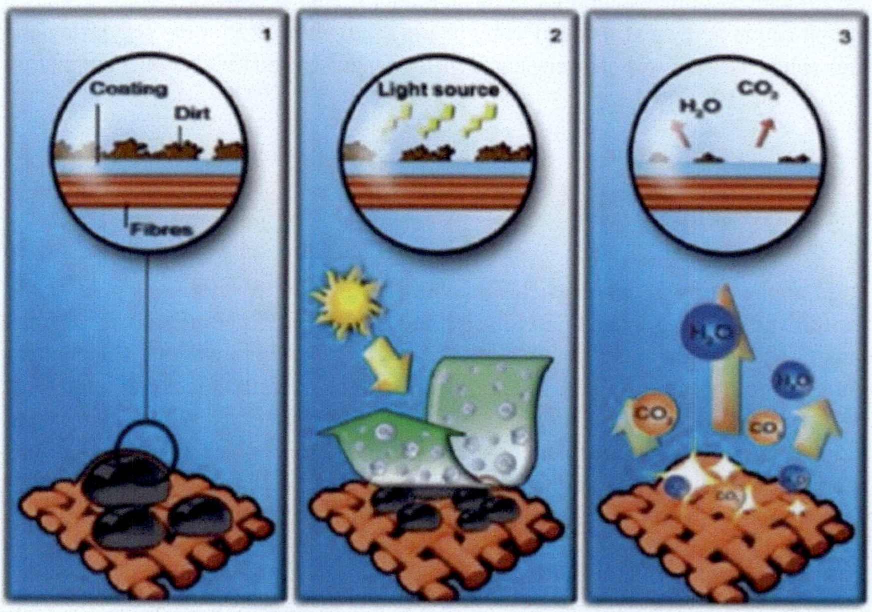

Fig. 8 Conceptualization of photocatalysis with nanofinishing on textile substrates [100]

Studies on Use of Nanoparticles in Synthetic Fibres for Self-cleaning Properties

Rani and co-authors utilize binders, different levels of TiO_2 nanoparticles in the anatase form were affixed to polyester through a pad-dry-cure method to develop thin films with photocatalytic self-cleaning properties. Following eight washing cycles, the polyester fabrics treated with sonicated TiO_2 nanoparticles at room temperature demonstrated robust photocatalytic self-cleaning attributes, effectively eliminating coffee stains [101]. In another study, zinc oxide nanospheres and zinc oxide nanorods are derived from zinc acetate through a reaction with sodium hydroxide, employing varying bath treatment conditions such as stirring and ultrasound. Following the treatment of polyester fabric with caustic soda, there was an augmentation in the adsorption of nanoparticles onto the treated fabric. When exposed to UV irradiation, treated polyester fabrics showed improved self-cleaning efficacy, improving their ability to eliminate various types of stains [102]. Polyamide 6 fabric was given treatment with ZnO which was dispersed in methanol, with subsequent coating onto the fabric surface. The polyamide 6 fabric with ZnO deposition demonstrated significant potential for utilization in the textile industry, particularly in terms of acquiring self-cleaning and antibacterial properties [103]. Emami et al. used corona discharge to apply TiO_2 nanoparticles onto nylon 6 fabric, conferring self-cleaning characteristics to the sample. The self-cleaning effectiveness of the fabric was evaluated by staining the nylon 6 fabric treated with corona/TiO_2 using methylene blue dye. The treated fabrics exhibited highly effective self-cleaning properties [104]. Polypropylene was blended with varying concentrations of TiO_2 nanoparticles, and fibres were produced through single-screw extrusion. The inclusion of 20 wt% TiO_2 in the polypropylene fibres enhanced their self-cleaning capabilities when exposed to 20 W of UV radiation for up to 5 h [105].

4 Future Scope

Future applications of nanoparticles in synthetic fibres have enormous potential to transform a number of sectors. This discipline is positioned to uncover advanced functionality and applications through continued research and innovation. The creation of multifunctional fibres by the integration of various nanoparticles is an interesting new direction that promises materials with increased conductivity, self-cleaning capabilities, and customized mechanical strengths. One exciting development that may happen soon is the idea of smart fabrics, which might sense changes in their surroundings and adjust to suit the demands of users. The investigation of biocompatible nanoparticles may also prove advantageous for biomedical applications, since it could facilitate the production of synthetic fibres appropriate for uses such as tissue engineering and drug delivery. Sustainable synthesis techniques and ecologically friendly nanoparticles are becoming more and more important as the area develops.

5 Conclusions

In summary, a paradigm change in material science is presented by the incorporation of nanoparticles into synthetic fibres for a variety of uses, such as flame retardancy, antibacterial qualities, UV protection, and self-cleaning. The safety and functionality of synthetic fibres have been significantly improved by the inclusion of nanoparticles. When it comes to flame retardancy, nanoparticles help create materials that are more fire resistant, which is important for situations where user safety is top priority. The antibacterial characteristics of nanoparticles solve hygienic issues and are used in consumer goods and medical textiles. Furthermore, the addition of nanoparticles for UV protection improves the textile's defence against damaging UV rays, meeting requirements for both functionality and aesthetics. Furthermore, the self-cleaning properties of fibres reinforced with nanoparticles mark a significant advancement in low-maintenance and sustainable textiles. Future prospects for nanoparticle-infused synthetic fibres are promising, as research into these materials continues to overcome difficulties and optimize manufacturing methods, bringing in a new era of high-performance and multifunctional materials across a range of industries.

Author's Contribution 1. Dr. Manisha Yadav: Conceptualization, Original draft, abstract formation, Introduction, conclusions, future scope, Overall editing and formatting

2. Dr. Nagender Singh: Wrote flame retardant and antimicrobial applications, Over all formatting.

3. Dr. Shelly Khanna: Wrote UV protection and self-cleaning applications.

4. Dr. Amit Madhu: Work on overall editing, formatting and removal of plagiarism.

References

1. Khan Y, Sadia H, Ali Shah SZ, Khan MN, Shah AA, Ullah N, Ullah MF, Bibi H, Bafakeeh OT, Khedher NB, Eldin SM, Fadhl BM, Khan MI (2022) Classification, synthetic, and characterization approaches to nanoparticles, and their applications in various fields of nanotechnology: a review. Catalysts 12(11):1386
2. El-Kheir A, El-Gabry LK (2022) Potential applications of nanotechnology in functionalization of synthetic fibres (a review). Egypt J Chem 65(9):67–85
3. Sawhney APS, Condon B, Singh KV, Pang SS, Li G, Hui D (2008) Modern applications of nanotechnology in textiles. Text Res J 78(8):731–739
4. Rivero PJ, Urrutia A, Goicoechea J, Arregui FJ (2015) Nanomaterials for functional textiles and fibers. Nanoscale Res Lett 10:1–22
5. Lim CT (2017) Nanofiber technology: current status and emerging developments. Prog Polym Sci 70:1–17
6. Aldalbahi A, El-Naggar ME, El-Newehy MH, Rahaman M, Hatshan MR, Khattab TA (2021) Effects of technical textiles and synthetic nanofibers on environmental pollution. Polymers 2021(13):155
7. Venugopal J, Ramakrishna S (2005) Applications of polymer nanofibers in biomedicine and biotechnology. Appl Biochem Biotechnol 125:147–157
8. Huang ZM, Zhang YZ, Kotaki M, Ramakrishna S (2003) A review on polymer nanofibers by electrospinning and their applications in nanocomposites. Compos Sci Technol 63(15):2223–2253

9. Li D, Xia Y (2004) Electrospinning of nanofibers: reinventing the wheel? Adv Mater 16(14):1151–1170
10. Nguyen-Tri P, Nguyen TA, Carriere P, Ngo Xuan C (2018) Nanocomposite coatings: preparation, characterization, properties, and applications. Int J Corrosion 2018
11. Gashti MP, Pakdel E, Alimohammadi F (2016) Nanotechnology-based coating techniques for smart textiles. In: Active coatings for smart textiles. Woodhead Publishing, pp 243–268
12. Temesgen AG, Turşucular ÖF, Eren R, Ulcay Y (2018) Novel applications of nanotechnology in modification of textile fabrics properties and apparel. Int J Adv Multidiscip Res 5(12):49–58
13. Fu S, Sun Z, Huang P, Li Y, Hu N (2019) Some basic aspects of polymer nanocomposites: a critical review. Nano Mater Sci 1(1):2–30
14. Gowri S, Almeida L, Amorim T, Carneiro N, Pedro Souto A, Fátima Esteves M (2010) Polymer nanocomposites for multifunctional finishing of textiles—a review. Text Res J 80(13):1290–1306
15. Abou El-Kheir AA, Ezzat M, Bassiouny F, El-Gabry LK (2018) Development of some functional properties on viscose fabrics using nano kaolin. Cellulose 25:4805–4818
16. Rehan M, Nada AA, Khattab TA, Abdelwahed NA, Abou El-Kheir AA (2020) Development of multifunctional polyacrylonitrile/silver nanocomposite films: antimicrobial activity, catalytic activity, electrical conductivity, UV protection and SERS-active sensor. J Market Res 9(4):9380–9394
17. Joudeh N, Linke D (2022) Nanoparticle classification, physicochemical properties, characterization, and applications: a comprehensive review for biologists. J Nanobiotechnol 20(1):262
18. Hassan BS, Islam GMN, Haque ANMA (2019) Applications of nanotechnology in textiles: a review. Adv Res Text Eng 4(2):1038
19. Syafiuddin A (2019) Toward a comprehensive understanding of textiles functionalized with silver nanoparticles. J Chin Chem Soc 66(8):793–814
20. Damuluri R, Kiran K, Chakravarthy DP (2017) Review studies on application of nanotechnology in textiles. Int J Textile Fashion Technol 7(6):1–4
21. Sakka S (2005) The outline of applications of the sol-gel method. In: Klein L, Aparicio M, Jitianu A (eds) Handbook of sol-gel science and technology
22. Perepelkin KE (2005) Principles and methods of modification of fibres and fibre materials. A review. Fibre Chem 37(2):123–140
23. Kim Y, Lee S, Yoon H (2021) Fire-safe polymer composites: flame-retardant effect of nanofillers. Polymers 13(4):540
24. Norouzi M, Zare Y, Kiany P (2015) Nanoparticles as effective flame retardants for natural and synthetic textile polymers: application, mechanism, and optimization. Polym Rev 55(3):531–560
25. Carosio F, Laufer G, Alongi J, Camino G, Grunlan JC (2011) Layer-by-layer assembly of silica-based flame retardant thin film on PET fabric. Polym Degrad Stab 96(5):745–750
26. Carosio F, Di Blasio A, Cuttica F, Alongi J, Frache A, Malucelli G (2013) Flame retardancy of polyester fabrics treated by spray-assisted layer-by-layer silica architectures. Ind Eng Chem Res 52(28):9544–9550
27. Doğan M, Bayramlı E (2013) Effect of boron phosphate on the mechanical, thermal and fire retardant properties of polypropylene and polyamide-6 fibers. Fibers Polym 14:1595–1601
28. Tata J, Alongi J, Frache A (2012) Optimization of the procedure to burn textile fabrics by cone calorimeter: part II. Results on nanoparticle-finished polyester. Fire Mater 36(7):527–536
29. El-Gabry LK, El-Ghany A, Nahed A, Aboras SE, El-Kheir A, Adel A (2021) Biocidal activity of polyester fabrics modified with SiO_2 NPs. Egypt J Chem 64(3):1411–1419
30. El-Kheir A, El-Ghany A, Fahmy MM, Aboras S, El-Gabry L (2020) Functional finishing of polyester fabric using bentonite nano-particles. Egypt J Chem 63(1):85–99
31. Abd El-Hady MM, Farouk A, Sharaf SJCP (2013) Flame retardancy and UV protection of cotton based fabrics using nano ZnO and polycarboxylic acids. Carbohyd Polym 92(1):400–406

32. Apaydin K, Laachachi A, Ball V, Jimenez M, Bourbigot S, Ruch D (2015) Layer-by-layer deposition of a TiO_2-filled intumescent coating and its effect on the flame retardancy of polyamide and polyester fabrics. Colloids Surf, A 469:1–10
33. Kundu CK, Song L, Hu Y (2020) Nanoparticles based coatings for multifunctional polyamide 66 textiles with improved flame retardancy and hydrophilicity. J Taiwan Inst Chem Eng 112:15–19
34. Gonçalves AG, Jarrais B, Pereira C, Morgado J, Freire C, Pereira MFR (2012) Functionalization of textiles with multi-walled carbon nanotubes by a novel dyeing-like process. J Mater Sci 47:5263–5275
35. Alongi J, Tata J, Carosio F, Rosace G, Frache A, Camino G (2014) A comparative analysis of nanoparticle adsorption as fire-protection approach for fabrics. Polymers 7(1):47–68
36. Sharmin S, Rahaman MM, Sarkar C, Atolani O, Islam MT, Adeyemi OS (2021) Nanoparticles as antimicrobial and antiviral agents: a literature-based perspective study. Heliyon 7(3)
37. Bhosale SV, Ekambe PS, Bhoraskar SV, Mathe VL (2018) Effect of surface properties of $NiFe_2O_4$ nanoparticles synthesized by dc thermal plasma route on antimicrobial activity. Appl Surf Sci 441:724–733
38. Slavin YN, Asnis J, Hńfeli UO, Bach H (2017) Metal nanoparticles: understanding the mechanisms behind antibacterial activity. J Nanobiotechnol 15:1–20
39. Shaikh S, Nazam N, Rizvi SMD, Ahmad K, Baig MH, Lee EJ, Choi I (2019) Mechanistic insights into the antimicrobial actions of metallic nanoparticles and their implications for multidrug resistance. Int J Mol Sci 20(10):2468
40. Yin IX, Zhang J, Zhao IS, Mei ML, Li Q, Chu CH (2020) The antibacterial mechanism of silver nanoparticles and its application in dentistry. Int J Nanomed, 2555–2562
41. Weir E, Lawlor A, Whelan A, Regan F (2008) The use of nanoparticles in anti-microbial materials and their characterization. Analyst 133(7):835–845
42. Jiang J, Pi J, Cai J (2018) The advancing of zinc oxide nanoparticles for biomedical applications. Bioinorganic Chem Appl 2018
43. Oves M, Aslam M, Rauf MA, Qayyum S, Qari HA, Khan MS, Alam MZ, Tabrez S, Pugazhendhi A, Ismail IM (2018) Antimicrobial and anticancer activities of silver nanoparticles synthesized from the root hair extract of *Phoenix dactylifera*. Mater Sci Eng C 89:429–443
44. Khorrami S, Zarrabi A, Khaleghi M, Danaei M, Mozafari MR (2018) Selective cytotoxicity of green synthesized silver nanoparticles against the MCF-7 tumor cell line and their enhanced antioxidant and antimicrobial properties. Int J Nanomed, 8013–8024
45. Ramkumar VS, Pugazhendhi A, Gopalakrishnan K, Sivagurunathan P, Saratale GD, Dung TNB, Kannapiran E (2017) Biofabrication and characterization of silver nanoparticles using aqueous extract of seaweed *Enteromorpha compressa* and its biomedical properties. Biotechnol Rep 14:1–7
46. Durán N, Nakazato G, Seabra AB (2016) Antimicrobial activity of biogenic silver nanoparticles, and silver chloride nanoparticles: an overview and comments. Appl Microbiol Biotechnol 100:6555–6570
47. Liao C, Li Y, Tjong SC (2019) Bactericidal and cytotoxic properties of silver nanoparticles. Int J Mol Sci 20(2):449
48. Gerber LC, Mohn D, Fortunato G, Astasov-Frauenhoffer M, Imfeld T, Waltimo T, Zehnder M, Stark WJ (2011) Incorporation of reactive silver-tricalcium phosphate nanoparticles into polyamide 6 allows preparation of self-disinfecting fibers. Polym Eng Sci 51(1):71–77
49. Jiang SX, Qin WF, Guo RH, Zhang L (2010) Surface functionalization of nanostructured silver-coated polyester fabric by magnetron sputtering. Surf Coat Technol 204(21–22):3662–3667
50. Deng X, Yu Nikiforov A, Coenye T, Cools P, Aziz G, Morent R, De Geyter N, Leys C (2015) Antimicrobial nano-silver non-woven polyethylene terephthalate fabric via an atmospheric pressure plasma deposition process. Sci Rep 5(1):10138
51. Raza ZA, Rehman A, Anwar F, Usman A (2016) Development and antibacterial performance of silver nanoparticles incorporated polydopamine–polyester-knitted fabric. Bull Mater Sci 39:391–396

52. Xu H, Shi X, Lv Y, Mao Z (2013) The preparation and antibacterial activity of polyester fabric loaded with silver nanoparticles. Text Res J 83(3):321–326
53. Baghriche O, Ruales C, Sanjines R, Pulgarin C, Zertal A, Stolitchnov I, Kiwi J (2012) Ag-surfaces sputtered by DC and pulsed DC-magnetron sputtering effective in bacterial inactivation: testing and characterization. Surf Coat Technol 206(8–9):2410–2416
54. Xu Q, Feng X, Liu J, Gong P (2017) Preparation of Ag nanoparticles-reinforced polyamide 6 nanocomposites by in situ polymerization and investigation of its properties. Pol J Chem Technol 19(4):36–40
55. Gawish SM, Avci H, Ramadan AM, Mosleh S, Monticello R, Breidt F, Kotek R (2012) Properties of antibacterial polypropylene/nanometal composite fibers. J Biomater Sci Polym Ed 23(1–4):43–61
56. Khajavi R, Bahadoran MMS, Bahador A, Khosravi A (2013) Removal of microbes and air pollutants passing through nonwoven polypropylene filters by activated carbon and nanosilver colloidal layers. J Ind Text 42(3):219–230
57. Li M, Zhu L, Lin D (2011) Toxicity of ZnO nanoparticles to *Escherichia coli*: mechanism and the influence of medium components. Environ Sci Technol 45(5):1977–1983
58. Lipovsky A, Nitzan Y, Gedanken A, Lubart R (2011) Antifungal activity of ZnO nanoparticles—the role of ROS mediated cell injury. Nanotechnology 22(10):105101
59. Sivakumar P, Lee M, Kim YS, Shim MS (2018) Photo-triggered antibacterial and anticancer activities of zinc oxide nanoparticles. J Mater Chem B 6(30):4852–4871
60. Nirmala M, Nair MG, Rekha K, Anukaliani A, Samdarshi S, Nair RG (2010) Photocatalytic activity of ZnO nanopowders synthesized by DC thermal plasma. Afr J Basic Appl Sci 2(5–6):161–166
61. El-Nahhal IM, Elmanama AA, El Ashgar NM, Amara N, Selmane M, Chehimi MM (2017) Stabilization of nano-structured ZnO particles onto the surface of cotton fibers using different surfactants and their antimicrobial activity. Ultrason Sonochem 38:478–487
62. Ramamurthy P, Chellamani KP, Dhurai B, ThankaRajan SP, Subramanian B, Santhini E (2017) Antimicrobial characteristics of pulsed laser deposited metal oxides on polypropylene hydroentangled nonwovens for medical textiles. Fibres Text Eastern Eur
63. Fiedot-Toboła M, Ciesielska M, Maliszewska I, Rac-Rumijowska O, Suchorska-Woźniak P, Teterycz H, Bryjak M (2018) Deposition of zinc oxide on different polymer textiles and their antibacterial properties. Materials 11(5):707
64. Ashraf M, Siyal MI, Nazir A, Rehman A (2017) Single-step antimicrobial and moisture management finishing of PC fabric using ZNO nanoparticles. Autex Res J 17(3):259–262
65. Farouk A, Moussa S, Ulbricht M, Schollmeyer E, Textor T (2014) ZnO-modified hybrid polymers as an antibacterial finish for textiles. Text Res J 84(1):40–51
66. Kale RD, Meena CR (2012) Synthesis of titanium dioxide nanoparticles and application on nylon fabric using layer by layer technique for antimicrobial property. Adv Appl Sci Res 3(5):3073–3080
67. Mihailović D, Šaponjić Z, Radoičić M, Radetić T, Jovančić P, Nedeljković J, Radetić M (2010) Functionalization of polyester fabrics with alginates and TiO_2 nanoparticles. Carbohyd Polym 79(3):526–532
68. Montazer M, Seifollahzadeh S (2011) Enhanced self-cleaning, antibacterial and UV protection properties of nano TiO_2 treated textile through enzymatic pretreatment. Photochem Photobiol 87(4):877–883
69. Mihailović D, Šaponjić Z, Molina R, Račić M, Esquena J, Jovančić P, Nedeljković J, Radetić M (2011) Multifunctional properties of polyester fabrics modified by corona discharge/air RF plasma and colloidal TiO_2 nanoparticles. Polym Compos 32(3):390–397
70. Chan KL, Fawcett D, Poinern GEJ (2016) Gold nanoparticle treated textile-based materials for potential use as wearable sensors. Int J Sci 2(05):82–89
71. Mehravani B, Ribeiro AI, Zille A (2021) Gold nanoparticles synthesis and antimicrobial effect on fibrous materials. Nanomaterials 11(5):1067
72. Sánchez-López E, Gomes D, Esteruelas G, Bonilla L, Lopez-Machado AL, Galindo R, Cano A, Espina M, Ettcheto M, Camins A, Silva AM, Durazzo A, Santini A, Garcia ML, Souto

EB (2020) Metal-based nanoparticles as antimicrobial agents: an overview. Nanomaterials 10(2):292
73. Narayanasamy P, Switzer BL, Britigan BE (2015) Prolonged-acting, multi-targeting gallium nanoparticles potently inhibit growth of both HIV and mycobacteria in co-infected human macrophages. Sci Rep 5(1):8824
74. Ibănescu M, Muşat V, Textor T, Badilita V, Mahltig B (2014) Photocatalytic and antimicrobial Ag/ZnO nanocomposites for functionalization of textile fabrics. J Alloy Compd 610:244–249
75. Mavrić Z, Tomšič B, Simončič B (2018) Recent advances in the ultraviolet protection finishing of textiles. Tekstilec 61(3)
76. Singha K, Maity S, Pandit P (2020) UV protection via nanomaterials. In: Frontiers of textile materials: polymers, nanomaterials, enzymes, and advanced modification techniques, pp 153–166
77. Saleem H, Zaidi SJ (2020) Sustainable use of nanomaterials in textiles and their environmental impact. Materials 13(22):5134
78. Malik SK, Arora T (2003) UV radiations: problems and remedies. Man Made Textiles in India 46(5):164–169
79. Verbič A, Gorjanc M, Simončič B (2019) Zinc oxide for functional textile coatings: recent advances. Coatings 9(9):550
80. Abo El-Ola SM, Elshakankery MH, Kotb RM (2022) Integration of nanocomposite finishing on polyester fabric for enhanced UV protection, performance, and comfort properties. J Eng Fibers Fabr 17:15589250221119448
81. Allam OG (2011) Improving functional characteristics of wool and some synthetic fibres
82. Sun L, Rippon JA, Cookson PG, Wang X, King K, Koulaeva O, Beltrame R (2008) Nano zinc oxide for UV protection of textiles. Int J Technol Transf Commer 7(2–3):224–235
83. Sun L, Rippon JA, Cookson PG, Koulaeva O, Wang X (2009) Effects of undoped and manganese-doped zinc oxide nanoparticles on the colour fading of dyed polyester fabrics. Chem Eng J 147(2–3):391–398
84. Haji A, Mousavi Shoushtari A, Mazaheri F, Tabatabaeyan SE (2016) RSM optimized self-cleaning nano-finishing on polyester/wool fabric pretreated with oxygen plasma. J Text Inst 107(8):985–994
85. https://www.nasiol.com/nanoblog/working-principles-of-nano-coatings/
86. https://nanografi.com/blog/nano-coatings-the-ultimate-barrier-for-uv-and-nir-radiation-nanografi/
87. Attia NF, Osama R, Elashery SE, Kalam A, Al-Sehemi AG, Algarni H (2022) Recent advances of sustainable textile fabric coatings for UV protection properties. Coatings 12(10):1597
88. Sista D (2021) New perspective of nano fibers: synthesis and applications. In: Nanofibers-synthesis, properties and applications. Intechopen
89. Dhineshbabu NR, Bose S (2018) Smart textiles coated with eco-friendly UV-blocking nanoparticles derived from natural resources. ACS Omega 3(7):7454–7465
90. Mandot AA, Chaudhari SB, Patel BH (2012) Nanocomposite: manufacturing and applications in textiles. Melliand Int 18(3 (D 45169 F)):188–189
91. Jagadeshvaran PL, Panwar K, Ramakrishnan I, Bose S (2021) Smart textiles coated with functional particles derived from sustainable sources that can block both UV and EM. Prog Org Coat 159:106404
92. Shah MA, Pirzada BM, Price G, Shibiru AL, Qurashi A (2022) Applications of nanotechnology in smart textile industry
93. Bashari A, Shakeri M, Shirvan AR (2019) UV-protective textiles. In: The impact and prospects of green chemistry for textile technology. Woodhead Publishing, pp 327–365
94. Afroz S, Azady MAR, Akter Y, Al Ragib A, Hasan Z, Rahaman MS, Islam JM (2021) Self-cleaning textiles: structure, fabrication and applications. In: Fundamentals of natural fibres and textiles. Woodhead Publishing, pp 557–597
95. Kiron MI (2021) Self-cleaning textiles: new concept of textile finishing. Textile Learner
96. https://www.technicaltextile.net/articles/self-cleaning-textile-an-overview-2646
97. https://textilevaluechain.in/in-depth-analysis/self-cleaning-of-textiles/

98. Saad SR, Mahmed N, Abdullah MMAB, Sandu AV (2016) Self-cleaning technology in fabric: a review. IOP Conf Ser Mater Sci Eng 133(1):012028
99. Chin CDW, Treadwell LJ, Wiley JB (2021) Microwave synthetic routes for shape-controlled catalyst nanoparticles and nanocomposites. Molecules 26(12):3647
100. Tung WS, Daoud WA (2011) Self-cleaning fibers via nanotechnology: a virtual reality. J Mater Chem 21(22):7858–7869
101. Rani S, Sharma R, Rani N (2020) Application of TiO_2 Nano-particles to incorporate self cleaning in polyester fabric with the help of binder and to study changes in physical properties. Int J Mod Trends Sci Technol 6:142–146
102. Mohammadi M, Karimi L, Mirjalili M (2016) Simultaneous synthesis of nano ZnO and surface modification of polyester fabric. Fibers Polym 17:1371–1377
103. Ummartyotin S, Pechyen C (2016) Role of ZnO on nylon 6 surface and the photocatalytic efficiency of methylene blue for wastewater treatment. Colloid Polym Sci 294:1217–1224
104. Emami F, Shekarriz S, Shariatinia Z, Moridi Mahdieh Z (2018) Self-cleaning properties of nylon 6 fabrics treated with corona and TiO_2 nanoparticles under both ultraviolet and daylight irradiations. Fibers Polym 19:1014–1023
105. Panutumrong P, Metanawin S, Metanawin T (2015) The effect of nano-titanium dioxide on the self-cleaning properties of TiO_2-PP composite fibers. Int J Adv Cult Technol 3(2):25–33

Chapter 13
Nanofinishing of Nonwoven Fabrics

Chintan R. Madhu

1 Introduction

Modernization has resulted in the highly developed use of textiles, which date back to earlier civilizations. The purpose of it has shifted from covering the naked body, protecting against adverse weather, and appealing to the beholder through aesthetics to becoming functional and even multifunctional. Multifunctionality can be integrated into textiles in various ways. When functionalities are introduced, it is possible to classify the addition of multifunctionality at the textile specimen stage. It was established that polymerization can incorporate multifunctionality [88]. Various radicals can be attached to polymers to offer various functionalities, which means that the bifunctional polymer will be at least bifunctional. At the fiber spinning level, it is possible to introduce multifunctionality by using two or more different polymers while spinning. During yarn spinning, it is possible to introduce multifunctionality by blending two or more different fibers that offer different functionalities into one yarn. At the weaving stage, it is possible to introduce yarns with different functions into the same fabric. Any nonwoven manufacturing technique can be utilized to produce multifunctional nonwovens. Research details the process for producing needle-punched nonwovens with several layers of varying characteristics by combining molten polymers with distinct functions, spinning or melt-blowing them together, and stitching the layers together [88, 93]. The introduction of multifunctionality in textiles is mostly during later stages, such as finishing and coating. Most manufactured finishes are added after dyeing or printing, even if they enhance usefulness to materials during the dyeing stage. Coating textiles is a relatively young area that provides several opportunities to include multifunctionality. Various methods can make the fabric multifunctional, but nanocoating is receiving a lot of praise.

C. R. Madhu (✉)
Department of Textile Processing Technology, R C Technical Institute, Ahmedabad, India
e-mail: chintan.madhu@gujgov.edu.in

S. Bairagi et al. (eds.), *Nanotechnology in Textile Finishing*, Materials Horizons: From Nature to Nanomaterials, https://doi.org/10.1007/978-981-97-2696-7_13

Nanotechnology is the study of matter at the nanoscale or very close to the molecular level, including its properties, properties changes, controllability, and efficient use. A nanometer is a linear measure that is one millionth of a meter, or, more accurately, one billionth of a meter. The use of nanotechnology and development has been employed to enhance the qualities of fibers and textiles, including woven, knitted, and nonwoven fabrics. Rapid growth in application sectors has been achieved due to the commercial viability of the technology. At the molecular level, it is possible to enhance properties such as physical, mechanical, chemical, biological, light spectrum, electronic energy, and similar ones. Some materials with its nano size when applied on cotton fiber or fabric, the softness, wrinkle resistance, abrasion resistance, antimicrobial properties, and dimensional stability have been improved. The rapidly growing nanotechnology is expected to have a major impact on textiles, with cotton textiles and nonwovens being the main beneficiaries.

The development of nanostructures has led to the creation of a novel range of materials with a wide class of performance in concern with functionality. Nanotechnology applications in textiles have been the focus of research scholars for the past two decades. Consumer products in both the private and public sectors can benefit from the application of research results. Nanotechnology has a vast potential for textile and apparel applications, but it is still unexploited to its fullest extent. Novel textile products majorly attract researchers like life-saving caring sartorial for ground zero emergency response people [12, 66], smart clothing [12, 100, 118], and medical textiles for preventing infection and improving healing [48, 90]. There is a growing amount of work being done to exploit this potential to create innovative nano-based smart and environmentally adaptive textiles. This chapter provides a summary of advances in nanotechnology as they are mentioned in the research literature pertaining to the textile materials covered in this report. This chapter proposed particulate and fibrous structures to improve the nanoscale functionalities of textiles. This chapter also offers a foundation that facilitates the connection between the potential of nanotechnology in textile research and the contemporary consumer market.

2 Nonwoven Textiles

Nonwovens, a type of fabric made from fibers, are innovative, high-tech, and engineered. In both consumer and industrial products, they are utilized either in combination with other materials or alone. The individual uses of nonwovens are taken into account, and the available alternatives vary from nonwovens that are thin and light to nonwovens that are robust and durable. This is true regardless of whether the nonwovens are used in consumer or industrial applications. By choosing the right raw materials, employing appropriate forming and bonding techniques, and utilizing finishing treatments like printing, embossing, laminating, and so on, it is possible to combine their unique qualities. It allows for the delivery of high-performance products. Checking out nonwovens in daily life is a good place to start if one is wondering

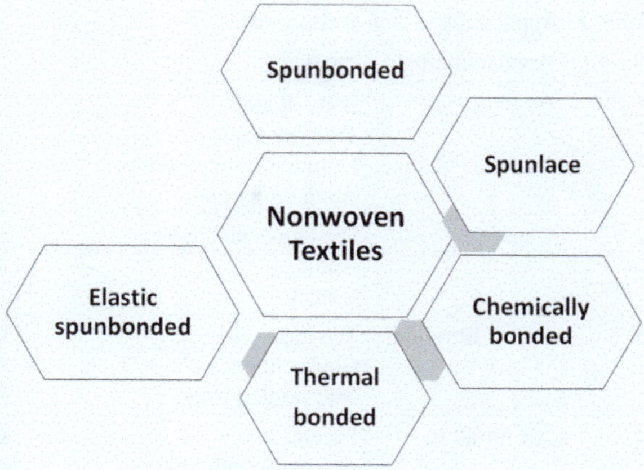

Fig. 1 Manufacturing techniques for nonwoven textiles

where to find nonwovens or want to see specific examples. Nonwoven textiles are prepared by different techniques suggested in Fig. 1.

The modern nonwoven industry is divided into several segments, such as those dealing with roll goods production, raw material suppliers, end-use goods converters and fabricators, machinery supporting these processes, auxiliary material suppliers, and winding, slitting, and packaging equipment manufacturers. Due to the wide range of vertical and horizontal integration within the sector, this list fails to provide the clear image that would be expected. The local market and subtle economic differences add to the overall global picture's complexity.

Nonwoven fabric markets reported 40.5 billion USD in 2020 and targeted to reach 53.5 billion till 2025, which is projected to growth with 5.73% from 2020 to 2025. Increasing in the demand of nonwoven products such as hygiene, medical, construction, and filtration application increase the nonwoven market. Rise in birth rate, increase in population, and development of modern health care also played a role for increasing the demand of nonwoven products (https://www.marketsandmarkets.com/Market-Reports/non-woven-fabrics-market-101727296.html).

The application of nonwoven textiles in various fields has been suggested in Table 1.

3 Nanotechnology for Nonwoven

Textiles have played a significant role in human history and are present in all facets of contemporary life. Technical textiles and traditional textiles, which are used in fashion apparel, are the two main categories into which textiles are divided based on their intended uses. Nonwoven technical textiles find applications in a wide range of industries, including automotive, agricultural, civil engineering, health care, sportswear, packaging, medicine, and hygiene. It is crucial to realize that the bulk phase's composition and that of manufactured or natural polymer surfaces often

Table 1 Application of nonwoven textiles

S. No.	Application field	Product	References
1	Hygiene	Baby diapers Training pants Feminine care products Adult under pads	Liu et al. [73], Ma et al. [76], Venkataraman et al. [134]
2	Wipes	Pre-moistened baby Adult Others	Hadley et al. [38], Kwon et al. [65], Li et al. [70])
3	Liquid filtration	Tea-bags Coffee filter Micro-rated bags	Gharaei and Russell [34], Ke et al. [60], Venkataraman et al. [134]
4	Air filtration	Facemasks Airconditioning filters Carburetor air filters Vacuum cleaner bags Passenger car cabin filter	Gharaei and Russell [34], Ke et al. [60], Tronci and Russell [129]
5	Medical	Surgical packs and gowns Sterilization wraps Caps Masks Shoe covers Sponges and dressings Patient and staff apparel	Gharaei and Russell [34], Tronci and Russell [129], Venkataraman et al. [134]
6	Disposable	Fabric softener substrate Sterile and nonsterile packaging Napkins and table top materials Packaging materials Cosmetic pads	Gharaei and Russell [34], Hadley et al. [38], Tronci and Russell [129]
7	Home	Carpet Upholstery Wall covering materials Bedding materials	Gharaei and Russell [34], Nag and Kumar [94]
8	Building	Roofing materials Insulation Wall protection	Gharaei and Russell [34], Karimi et al. [58], Nag and Kumar [94], Ventura et al. [135]
9	Geotextiles	Geomembranes Geogrids Fabricated drains	Prambauer et al. [108]
10	Automotive	Trunk liners Kick panels Package trays Headliners Seat construction Engine hood liners	Boominathan et al. [15], Gharaei and Russell [34], Mewoli et al. [89], Tronci and Russell [129]

(continued)

Table 1 (continued)

S. No.	Application field	Product	References
11	Agriculture	Crop covers Seed blankets Greenhouse shading Root bags Biodegradable plant pots	Kopitar et al. [63]), Manickam and Kandhavadivu [81], Periyasamy et al. [105]
12	Others	Interlinings Shoe, leather goods Paper making felts Coated bale wrap Electronics Abrasive pads	Gharaei and Russell [34], Tronci and Russell [129]

diverge. A number of chemical interactions, including those involving the low surface energy of molecular groups, microbes, the oxidation or sulphonation of surface molecules, and the adhesion of environmental pollutants like dirt or soil, may produce this change in surface properties. Physical defects, such as scratches or abrasions, can lead to the degradation of textiles in certain applications. To improve the efficiency of nonwoven textiles in different application field, nanofinishes can be applied on such textiles [115, 121].

3.1 Nanofinishing for Nonwoven Healthcare and Hygiene Products

The medical and associated healthcare and hygiene industries make up a significant and growing share of the textile industry. Growth has been attributed to continuous innovation and improvement in both medical procedures and textile technology. A significant and expanding portion of the fabric industry is the clinical and associated healthcare and hygiene sectors. Textile has traditionally been associated with health care. The ward and operating theater both require personal hygiene care and protection. The programs are diverse, with the most popular being the easy cleaning wipe and the superior barrier fabric used for running rooms. Scientific applications require medical textile systems to be designed and executed. Textiles have been engineered to suit the needs of medical, surgical, health care, and hygiene materials. Strength, flexibility, moisture, and air permeability are necessary for these applications [37].

A nonwoven medical product having different important and unique parts. These are responsible for rapid water absorption, clothing against microbes, disposability, washability, and user-friendliness.

The use of nonwoven fabrics made from polypropylene (PP) and polyester (PES) was paired with woven fabrics made from polyamide (PA) and polyamide (PA). The combination was created through the use of antimicrobial silver nanoparticles and plasma surface activation. The antibacterial properties of nano-coated PP nonwoven fabric against Staphylococcus aureus were shown to be higher after surface activation by low-temperature plasma [114].

Even though there is no direct layer-by-layer nanocoating of textiles, it has been shown that coating fabrics could be an effective method to prevent viruses. The outer layer of surgical gowns was made from a threefold layer of polypropylene, followed by polytetrafluoroethylene (PTFE) in the middle, and a nonwoven polyester layer in the innermost. The fabric's viral adhesion and penetration could not be stopped using PTFE.

The masks can be divided in various sectors like cloth face masks, surgical face masks, and respirators [92]. Comfort, biocompatibility, breathability, fluid resistance, flammability, and filtration capacity must be considered in the account for evaluating the mask [20, 57]. Those properties are taken into account when manufacturing the recommended masks. The simplest mask is basic cloth face masks which are generally homemade face masks had an efficiency as per the raw materials used for making them [92, 124, 127]. Surgical masks consist of three nonwoven textile layers, each of which serves a distinct purpose and is disposed of after use. The external layer's hydrophobic properties allow it to repel fluids, including mucosal fluid. The intermediate layer serves as a filter, keeping pathogens and other undesirable substances from penetrating the system. The mucosal fluid and moisture from exhaled air are absorbed by the absorbent material used in the internal layer. Respirators can be constructed with four layers of filters. The respirator's inner and outer layers, both made of nonwoven polypropylene (PP), function as hydrophobic layers, preventing the absorption of moisture. Modacrylic support is employed as an intermediate layer to provide thickness and shape, while nonwoven PP is employed for filtering [98]. From reported work, it can be clearly seen that simple face mask is reusable, comfortable, and possessing required breathability also, but it has low filtration efficiency. The comfort, breathability, and filter efficiency of surgical masks surpass that of homemade masks. On the other hand, micro and nanoparticles do not have very good filtering capabilities. Furthermore, surgical masks are not recyclable. Reusable respirators are better at filtering out particles of all sizes, but they are not very breathable and cannot be reused. Respirators operate more effectively overall [71, 124, 127].

Some of the personal protective equipment (PPE), especially face masks, have had issues. More effective, reusable, and able to filter out microbes, face masks are also great for protecting against airborne infections. Cotton and synthetic masks having larger pore sizes are not very effectively function for filtering nanosized contaminants, pathogens, and tiny virus [20, 57]. This non-reusable PPE's negative impact on the environment is also a concern. During the COVID-19 epidemic, healthcare professionals throughout the globe used around 44 million disposable PPEs daily, according to current studies. Polypropylene is widely used for PPEs as it is economical and having good performance, but these are single used masks, and therefore, at the end they are used for landfilling which affect environment [57, 124].

For improving the performance of medical textiles, researches had carried out many research works, as filter layers modification by adding nanofibers or by applying nanoparticles on the surface. The works have been strongly supported to coating of antimicrobial agents to protect the textiles from viruses, fungi, and bacteria [57, 98].

There are physical and chemical ways to encourage surface changes in fabrics. Many researchers have shown interest in nonwoven surface modification techniques such as physical vapor deposition (PVD), chemical grafting, dip coating, spray coating, and sol–gel over solution-based treatments [52, 117]. A noble gas (typically Ar) is used to bombard the coating material (target) through spitting. It travels through the vapor phase until it reaches the substrate, where it condenses to form a coating. The addition of a reactive gas allows it to react with the growing film, resulting in a compound coating [117]. The implementation of this technology has allowed for the modification of various material surfaces, with particular emphasis on textiles. Diamond like carbon (DLC) is utilized for adding hydrophobic characteristic to the textiles by modifying the surface without toxicity [104]. Coating the fabric with a non-toxic proportion of silver nanoparticles (AgNPs) enhances its antibacterial capabilities and makes it very efficient against microbial colonization [17, 61]. Table 2 suggests some nanofinishing materials for nonwoven health care and hygiene textile.

Pharmacotherapeutic applications can be used to combat the pathogen through the use of antiviral polysaccharides. A coating approach based on layer-by-layer nanocoating is being considered to cover surfaces that could harbor pathogenic coronaviruses that are exposed to humans. The spread of viruses is passively prevented in wet markets by coating masks, clothing, and work surfaces with these antiviral polysaccharides. A measure called "eradicate-in-place" is proposed as a solution to the virus. The use of antiviral polysaccharides as a green chemistry pathway to eradicating viruses is due to their primarily biological origin and the ability to be modified through minimal synthetic approaches [99].

3.2 *Nanofinishing for Nonwoven Filtration Products*

During fluid flow, a nonwoven filter medium with a random fiber structure is used to separate various phases (Fig. 2). Filtration is often understood to be the act of eliminating particulate matter from the flowing fluid by entrapping it in the filter media. The filtering process will include adsorption, absorption, extraction coalescence, electrostatic effects, and antibacterial activity [47].

Air, gas, and liquid filtration are just a few of the many applications for nonwoven filters. Approximately 65–70% of the nonwoven filtration medium is used for air and gas filtration, with the remaining 30–35% going toward liquid filtration. The HVAC systems of homes, offices, and businesses may benefit from nanofilters for air and water filtration; the chemical, pharmaceutical, microelectronic, food, and medical sectors all make use of liquid filtration. Power stations, cement, and metallurgical industries use air and water filtration [82].

Nanofiltration is a relatively new method for purifying water by removing molecular and ionic chemical species as well as big visible particles. Many people consider

Table 2 Nanofinishing materials for nonwoven health care and hygiene product

S. No.	Nonwoven textile	Nano material	Detail	References
1	Polypropylene	TiO_2 NPs	Environmental protection	Li et al. [69]
2	Cellulose pulp	Ag NPs	Antibacterial	Kavitha Sankar et al. [59]
3	Cotton Viscose Polyester	Chitosan	Antimicrobial	Vellingiri et al. [132]
4	Polylactic	Polyurethane/ triclosan	Medical cover	Zhang et al. [145]
5	Lyocell	ZnO	Surgical face mask	Pragadheeswari et al. [107]
6	Polyester	Ag NPs	Atmospheric pressure plasma deposition Antimicrobial	Deng et al. [25]
7	Polylactic acid/ low melt polylactic acid	Polyurethane	Spray coating Medical cover	Zhang et al. [146]
8	Polyacrylonitrile Carbon	Bioactive ceramic CaO, SiO_2, P_2O_5	Bioactivity	Fraczek-Szczypta et al. [29]
9	Polyester	SiO_2 Ag NPs	Medical packing materials	Liu et al. [74]
10	Nonwoven fabric filter with polypropylene	Silica-resin coating	Plastic-free prevention of environment	Tsutsumi-Arai et al. [130]
11	Needle-punched fabric	Polyester staple Ag NPs	Antibacterial	Gadkari et al. [31]
12	Polypropylene	Ag NPs Polyethyleneimine	Antiviral activity	Baselga et al. [11]
13	Polypropylene	$CaSiO_3$/CuO	Antibacterial protective textile	Hamoda et al. [40]
14	Polypropylene/ polyethylene	AgCu NPs	Antibacterial activity	Zhou et al. [148]
15	Polyester	Carbon black NPs Vinyl acrylic dispersion	Antibacterial EMI shielding	Chemeh et al. [18]
16	Polypropylene	Ag NPs Graphene oxide	Antibacterial	Zhou et al. [149]

the most important component in filtering procedures to be the selection of the right filter medium or membrane material. A wide variety of materials are at your disposal, including rayons, wool, linen, glass fiber, fiber metal, porous carbon, and many more. Smart textiles that can detect their environment, take action, communicate with one another, and even clean themselves have been developed with the

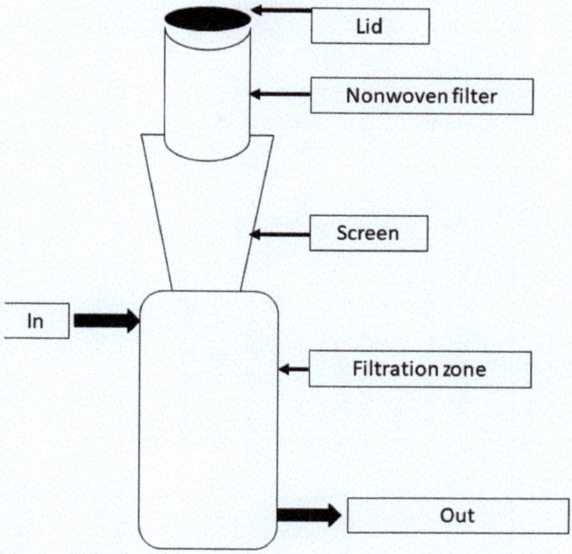

Fig. 2 Filtration process using nonwoven textiles

use of nanotechnology. This was made feasible by innovations such as novel materials, fibers, conducting nanocoatings, carbon nanotubes, and so on. Many industries, including medicine, athletics, the armed forces, and the fashion industry, make use of these supplementary features. A revolutionary shift in the practicality of clothing and everyday textiles may be possible as smart nanotextiles develop [142].

Polyethyleneimine (PEI) and terephthalate (TA) were finished with Ag and Cu nanoparticles by layer-by-layer technique for microfiltration of drinking water. At the passing of 2500 L/m^2h for 30 min, researchers received 4.5–5 LRV antiviral capacity of the finished nonwoven PEI/TA [22, 120]. For virus removal from water by filtration, spray-dried Al_2O_3 granules modified with CuO nanoparticles were applied on nonwoven materials, and antiviral capacity of 0.2–3.2 LRV (1 h) has been received. The study suggests that Cu(I) oxide was the active element for virus removal when teste with MS2 bacteriophages [22, 85]. Nanofibrillated cellulose (NFC) with the diameter of 20–60 nm and Cu nanoparticles with the spherical diameter of 15 nm resulted in antiviral capacity more than 5 LRV (20 min), and the virus removing capacity was tested with MS2 bacteriophages [22, 126].

For textile wastewater decolorization needle-punched nonwoven polyester was finished with polyethyleneimine (PEI) utilizing glow discharge plasma, and the resulted nonwoven filter resulted in removal of color from wastewater. Needle-punched PET coated with PEI increased dye retention capacity of 5.3–5.76 mg/g as a filter media as compared with commercially available product [95]. Examples of nonwoven nanofilters are shown in Table 3.

Table 3 Nanofinished nonwoven filtration materials

S. No.	Nonwoven textiles	Nano material	Detail	References
1	Amyloid nanofibrils	–	H1N1 SARS-CoV-2 Filtration membrane	Palika et al. [102]
2	PPE	TiO_2	Reusable protective masks	Horváth et al. [45]
3	Polyester	CNTs	Air purification Air filtration	Issman et al. [50]
4	Poly(vinyl alcohol-coethylene) (EVOH)/ poly[5,5-dimethyl-3-(30-triethoxysilylpropyl)-hydratoin] (PSPH)/Cl	EVOH grafted-PSPH nanofibers (EPNMs)	Water filtration Antiviral capacity 5 LRV (2 min)	Ma et al. [77]
5	Glycidyltrimethylammonium chloride/nanofibrillated cellulose	NFC/GTAC aerogels	Water purification Antiviral capacity 1.2–3 LRV (93.6– 99.9%, pH 7.0)	Watts et al. [137]
6	Carbon	Polydimethylsiloxane (PDMS) zeolite imidazole Framework-8 (ZIF-8)	Oil–water separation	Pakdel et al. [101]
7	Poly-ether ether-ketone	TiO_2/SiO_2 Nanocomposites	VOC removal Air purification	Bai et al. [9]
8	Polyester composite sieve membranes (PET@Cs/Gx)	Chitosan and graphene Oxide	Antifouling Protein rejection Oil–water emulsion separation	Ahmad et al. [3]
9	Polypropylene	Ag Multiwall CNTs	Air filtration	Haider et al. [39]
10	PTFE foam-coated glass fabric	PTFE nanofibers	Air filtration	Ahn et al. [4]

(continued)

Table 3 (continued)

S. No.	Nonwoven textiles	Nano material	Detail	References
11	Polysulfone	Ag NPs	Ultrafiltration Reverse osmosis	Fiaschini et al. [28]
12	Polyamide 6	Polyvinylidene fluoride nanofiber	Oil–water separation	Mahmud et al. [80]

3.3 Nanofinishing for Nonwoven Household Products

Cleaning, filtering, and adding an aesthetic touch to the house are just a few examples of household applications where nonwovens are used (Fig. 3). When used in the kitchen, living room, bedroom, or bathroom, high-performance nonwovens may provide solutions that are practical, sanitary, safe, attractive, and intelligent for contemporary living. Nonwovens have the ability to be extremely durable, or single-use, limited-life fabrics, depending on what needs to be done.

Nonwovens have unique and irreplaceable properties that no other material can match, including excellent flexibility, resilience, liquid repellency, strength, softness, flame retardancy, cushiness, washability, bacterial barriers, filtration, and sterility.

Nonwoven manufacturers are capable of combining these qualities in multiple ways to create products that have the unique and specific characteristics a client requests or needs, usually at a very affordable price.

Nanofinishing of nonwoven or nanoparticles embedding or impregnating with nonwoven fiber fabrics improves the aesthetic values as well application values. Specifically, self-cleaning mats, table cloth, and table cover are the major areas where nanotechnology can perform well by incorporating with textiles. Also, curtains, bed linings, bedsheets, bed covers, etc., must be antimicrobial, biocompatible, and breathable.

New approaches to creating environmentally benign and long-lasting triboelectric nanogenerators (TENGs) have emerged as a result of technological advancements in the processing of natural biomaterials. Many people are interested in TENGs made of bacterial cellulose (BC), even though they do not contain the crucial element for temporary wearable electronics. The article explains how to make breathable, biodegradable, and biocompatible triboelectric nanogenerators (TENG) using a novel and easy method for in situ chemical modification of basic carbon (BC). Nanocoatings of SiO_2, polydopamine, or polypyrrole applied on BC nanofibrils allow for the

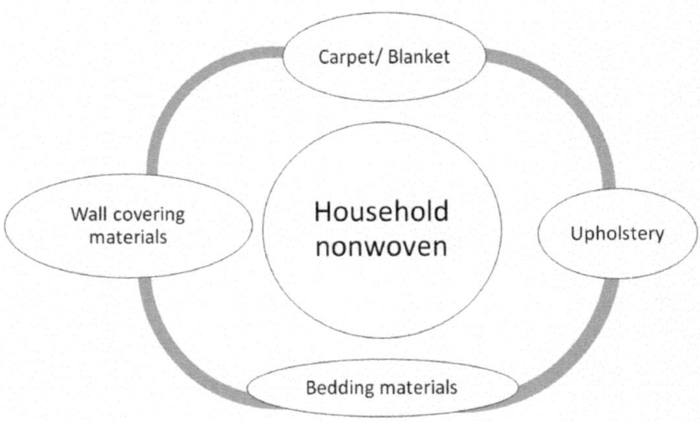

Fig. 3 Nonwoven applications in household nonwoven

manipulation of the BC layer's surface potential. In order to reposition BC in the triboelectric series, this alteration may be used. This enables the fabrication of a variety of various kinds of TENG devices that are based on BC. It has been found that all BC films coated with nanoparticles are biocompatible, biodegradable, and breathable, which are the primary characteristics needed for temporary electronics [27].

Using a sol–gel nanocoating technique, ultra-long SiO_2 nanotubes (SNTs) can be synthesized to assemble nonwoven mats without success. Although a sol–gel nanocoating technique can be used to synthesize ultra-long SiO_2 nanotubes (SNTs), it has not been successful in assembling nonwoven mats. Measurement of the mineralization process of tetraethoxysilane (TEOS) on the surface of bacterial cellulose (BC) is done using SEM and TEM. At their surface, BC's hydroxyl groups serve as nucleation sites for silica mineralization and catalyze silica polymerization directly. SNTs have diameters of 100 and 15 nm on average, respectively. The blue shift in photoluminescence of SNTs was evident compared to SiO_2 nanoparticles [143].

Viscose nonwoven textile was treated with copper nanoparticles for improving various characteristics. Nonwoven fabric treated with Cu has modest anti-HSV-1 action. The antiviral action is affected by the structure of the textile material in addition to its copper content. The researchers found that even when exposed to the identical viruses, various modified textiles exhibited varying degrees of protection. When it comes to the impact of Cu sputtering, the nonwoven's spatial 3-D structure and its alignment with the sputtering target are crucial factors. The formation of a uniform copper layer can be achieved in flat textile materials with high density structures, but if the structure is too tight, changes to them may increase their airflow resistance. This type of coated viscose can be utilized in surgical mask, bed linen, and bed covering [21].

3.4 Nanofinishing for Nonwoven Medical Products

Nonwoven textiles have enormous advantages for the medical sector. The medical field makes a great deal of a wide variety of wearable items, including surgical drapes, pads, dressings, filtration materials, inserted textiles, surgical gowns, masks, and many more. The majority of nonwoven items that are used externally are single-use, disposable items that have the benefit of not needing to be cleaned or sterilized in order to be reused. Nevertheless, some can be used again to carry out the necessary task for a brief amount of time [111, 112].

The use of nonwovens for medical textiles is applicable to wound care, surgical, and ostomy dressing. The present synthesis and preparation have enabled nonwovens to create more absorbent, bacterial barriers, soft and stretchable, and liquid repellents. Sheet or web structures made of long fibers that are bonded together through mechanical, thermal, or chemical means. Medical and surgical nonwovens have always relied on wound care as their primary use. Nonwovens are used in many different products, including incontinence products, absorbent pads, and staff and patient clothing. Baby wipes, bed linens, blankets, burn dressings, gowns, disposable pants, medication delivery devices, face masks, nasal strips, pillows, shoe covers, sponges, sutures, tissue scaffolds, towels, and wraps are just a few examples of the many medical nonwoven materials available [26, 68].

Common medical textiles are modified or enhanced with the help of nanotechnology by researchers. The use of nanofiber-based or nano-coated medical textiles are more attractive in the pandemic situation due to their environmentally friendly, biodegradable, and reusable materials. The primary objective of using a mask is to prevent pollutants from entering the respiratory system. The protection from more contaminants particles is achieved by using nanofibers or nanocoating in the medical nonwoven textiles [2, 10].

A copolymer containing an antimicrobial tertiary amine moiety and a crosslinker of ethylene glycol dimethylacrylate (EGDMA) makes up the nanocoating. The nanocoating is applied to the nanofiber filter using a chemical vapor deposition procedure so that the necessary functions may be imparted in a conformal way without deforming its very porous structure. The produced non-cytotoxic non-EGDMA or pDE film was very effective in eradicating Staphylococcus aureus and *Escherichia coli* O157:H7 (*E. coli* O157:H7), with biocidal effectiveness values of > 99.985% and > 99.995%, respectively. The filtering efficiency of the pDE-coated filter is 91%, and its low air resistance is 50 Pa. The surface-modified filter can reclaim dead microorganisms with a simple water rinse and is highly reusable up to six washes [122].

A nanolayer is formed by combining hydrophobic electronegative molecules from 1H, 1H, 2H, and 2H-perfluorooctyltriethoxysilane (PFOTES), which contain no electronegative molecules at all, with Si–Al sol–gel and crystal violet (CV) photosensitizer. The molecules of photosensitizer are spatially dispersed and firmly immobilized in the transparent Si–Al matrix, which prevents them from self-quenching. The anisotropic rearrangement of PFOTES in the Si–Al matrix during nanolayer production allowed the Si–Al/PFOTES-CV (SAPC)-coated filter to achieve water resistance and triboelectric charging. The photosensitizer was able to maintain its photoexcited state and promote redox reactions thanks to the SAPC nanolayer. Additionally, the SAPC filter may be reused as its filtering and photo biocidal capabilities were preserved even after five detergent washes [53].

Nonwoven masks made of electrospun microfibers, nanofibers, and melt-blown polypropylene microfibers enhanced with biocide-loaded polymer fibers incorporating nanoparticles of silver, copper, titanium oxide (TiO_2), and zinc oxide (ZnO) can be used to shield individuals from airborne contaminants [86]. Table 4 discusses the various applications of medical textiles involving nonwoven textiles and nanofinishes.

Table 4 Nanofinished nonwoven medical textiles

S. No.	Nonwoven textiles	Nano material	Detail	References
1	Cotton	Ag NPs	Wipes	Hamouda et al. [41]
2	Cotton	Ag NPs	Hydrophobic Antimicrobial	Islam et al. [49]
3	Poly(methyl methacrylate) (PMMA) nanofibers	ZnO nanorods and Ag nanoparticles	Antibacterial Antiviral Self-cleaning	Karagoz et al. [56]
4	Polypropylene	Single-walled carbon nanotube (SWCNT)	Self-sterilize	Soni et al. [123]
5	Cotton	ZnO	Antibacterial	Petkova et al. [106]
6	Cotton	Sodium alginate Ag NPs (SA@AgNPs)	Prevention against infectious disease	Velusamy et al. [133]
7	PPE	Ag NPs	SARS-CoV-2 protection	Abulikemu et al. [1]
8	PPE	Graphene	SARS-CoV-2 protection	De Luca et al. [23]
9	Aramid	TiO_2 $Sn(Sb)O_2$ TiO_2@ATO	Thermal protection	Liu et al. [75]
10	Polyester/ polyurethane matrix	CNTs Fe_3O_4	Absorbency	Gu et al. [36]
11	Polypropylene	Polytetrafluoroethylene (PTFE) nanoparticles	Durability Wash stability	Galante et al. [33]
12	Polyviscose	AgNPs Reduced graphene oxide (rGO)	Antimicrobial	Noor et al. [97]
13	Polyester	Chitosan-cinnamaldehyde cross-linked nanoparticles	Wet purification	Gadkari et al. [32]
14	Cotton Viscose Polyester	Hibiscus curry leaves Coconut oil	Antimicrobial	Korgaonkar and Sayed [64]
15	Wool	Sericin	Biomedical Cosmetics	Lee et al. [67]

3.5 Nanofinishing for Nonwoven Geotextiles

The yearly use of geotextiles exceeds 1.4 billion square meters and is continuing to rise. Polyolefin, polyester, or polyamide polymers are the mainstays of geotextiles, which are not biodegradable. Geotextiles can face complex environmental conditions, like complex acid–base conditions, when they are used in geotechnical engineering, resulting in higher performance requirements. The development of geotextiles should focus on high performance and multifunctionality. Many academics are researching the possibility of switching from nonbiodegradable polymer geotextiles to natural geotextiles as a result of the green movement. In 50% of all applications, synthetic geotextiles are being replaced by natural geotextiles at present. The use of optical fiber sensors is increasing the pace of geotextile development toward multifunctional intelligence. A number of new high-performance geotextiles, including as wicking and basalt fiber needling varieties, have emerged as a result of developments in composite fiber technology [138].

Geotextiles, which are mostly thermoplastic polymers, are used to reinforce or protect various landscapes by processing them into flexible material sheets. Nonbiodegradable polymers, such as polyesters and polyolefins, make up the majority of the materials used in the industry and can contribute to environmental problems like microplastic buildup and soil contamination. These drawbacks necessitate the time-consuming material collection for some applications. For these and other application-related reasons, it is therefore intriguing to develop more environmentally and biodegradable-friendly geotextiles. There are two commonly available material classes: natural fibers and biodegradable polymers. While there are a lot of geotextiles made of natural fibers currently available on the market, there is still research being done on the potential applications of industrial biopolymers. Plant fiber geotextile fabrics represent a promising approach that has demonstrated success in numerous applications to date. The use of natural fibers has some limitations when it comes to water absorption and stability [108].

The electrical surface resistivity of the coated nonwoven polyester is higher than that of the polypyrrole-coated woven polyester, and the nonwoven polyester is coated with crystalline graphene oxide to create high-performance nonwoven textiles. This coated nonwoven polyester has the range of application in geotextiles [62]. Nanocomposite layer of silver nanoparticles and polyaniline (PANi) has been coated on nonwoven polyester fabric for producing antimicrobial active geotextiles. Furthermore, the coated nonwoven polyester suggested excellent electrical conductivity and UV ray blocking effect [5].

The 15–20% expansive soil in India seriously damages the existing structures, such as retaining walls, pavements, dams, and multistory buildings. Materials used for the study were a polypropylene fiber net, coir geotextile mat (both treated and untreated), lime, and silica fume. Using a single or double layer of coir geotextile mat at varying depths is one method to evaluate the effect of reinforcement on the expansive soil subgrade. The goal of the treatment was to improve the coir geotextile's ability to reduce the swelling pressure of the expansive soil subgrade. According to the findings of the experiments, the upward swelling pressure is reduced by 81.89% in double-layer systems and by 52.19% in single-layer systems when the coir geotextile mat is treated with limestone. The expansive soil subgrade's shear strength and bearing capacity were discovered to be enhanced by using lime-treated coir geotextile. Implementing a single or double layer of coir geotextile treated with lime helps stabilize the expansive soil subgrade. Paved constructions may benefit from thinner subbase and base course layers thanks to reinforced expansive soil subgrade, which improves the structure's overall quality [128]. With the aid of oxygen plasma, the surface of nonwoven polyester was transformed into a nanostructure, and after that, the polyester was covered in nanofilm. Water retention in geotextiles is impacted positively by the nonwoven polyester treated with plasma because of its enhanced water-repellent properties [119].

3.6 *Nanofinishing for Nonwoven Automotive Textiles*

Automotive upholstery, interior coverings and panels, air/liquid filtration, tyre application, and structural body components are all heavily reliant on textiles, which the industry regards with considerable importance. Because of its material-intensive nature, the automobile sector uses a wide variety of materials, including metals, textiles, polymers, and fillers. The commercialization of any automobile component constructed of these materials depends on meeting specified standards, as well as overall performance and cost. The goal of using nanotechnology into automobile fabrics is to level the playing field between price fluctuations and better technical performance [55]. Needle punch, spunbond, and spunlace nonwoven textiles are commonly applied for decorating interior portion of automobiles (Fig. 4).

Polyester nonwoven fiber was needle punched for modifying its surface with increment in nanolayer for affecting its hydrophobicity. Changing adhesion and electrophysical properties is possible through plasma physicochemical treatment, but it also costs time by forming new active centers that are not present in the native fiber. Different gases (Ar, O_2, and air) and electrical conditions were utilized for study. By measuring the cation exchange capacity (CEC) and monitoring its impact on the increase of mass nanofibers, surface changes of PET fibers were evaluated. Plasma

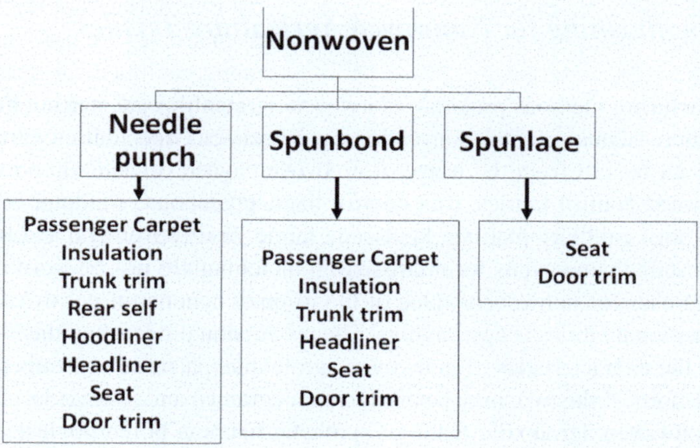

Fig. 4 Nonwovens as automotive interior

treatment was observed to have a significant impact on the surface wettability of PET fibers and also improves the capacity of air filter used in automotive [96].

The main cause of dirt and dust accumulation on vehicle compartment surfaces is the generation of static electricity. This energy accumulation in insulators may result in fire. Due to the Coulomb blockade effect, charges are drained from surfaces when long-chain aliphatic amines, amides, and polyols include silver nanoparticles. The use of carbon nanotubes as fillers in automobile parts has antistatic properties introduced [84].

Particulate and gas pollutants need to be filtered to maintain air quality. The use of nanotechnology ensures that foam does not pollute and dirt particles do not stick to it. Transporting air via nanofilters with nanofibers added reduces air resistance and eliminates pressure loss. Activated carbon and fiberglass are widely used in the air filtration sector. Since nanofibers are much smaller than regular fibers, they find widespread use in these kinds of technologies. Nanofibers' enhanced filtering performance, low resistance, and high surface-to-volume ratio make them useful in many different contexts. The reduction of pollution in the air we breathe can prevent lung diseases and asthma. Nanofiber filters are crucial in removing VOCs and reducing air pollution [42, 125].

The tires' condition is crucial to the vehicle's safety. The tire's grip is determined by its wear. Good grip and low rolling resistance are the requirements for the tire. Nanostructured soot can be used as a filler material in the tire to achieve high fuel efficiency and durability. Tires use soot particles because they have a coarser surface. Nanoparticles and natural rubber increase the surface area, and so inner friction is reduced and provides smooth rolling of tires. The addition of 10% nano-silica to natural rubber results in an increase in wear resistance and grip. The tensile strength and hardness are increased when natural rubber is combined with 3% of multiwall carbon nanotube [150], and other applications of nonwoven are suggested in Table 5.

3.7 Nanofinishing for Nonwoven Agriculture Textiles

As manufacturing methods progress, nonwoven agrotextiles are starting to provide more and more advantages over conventional agrotextiles. Greenhouses, gardens, and crops may all benefit from the usage of nonwoven agrotextiles. Crop covers, seed blankets, weed control fabrics, root control bags, greenhouse shading, biodegradable plant pots, capillary matting, landscape fabric, lawn coverings, and bio-based and compostable nonwovens for multi-season mulching are just a few of the many agricultural uses for nonwovens. One of the primary benefits of nonwoven fabrics over conventional fabrics is cost savings. The production process and the fibers used determine the characteristics of nonwoven agrotextiles. Natural or synthetic fibers, or combinations of the two, may be used to create nonwoven agrotextiles. The most popular nonwoven agrotextile made of synthetic fibers is polypropylene, whereas jute is the most widely used natural fiber. The advantages and benefits of using jute nonwoven agrotextile have been demonstrated by numerous studies. Polypropylene

Table 5 Nanofinished nonwoven automotive textiles

S. No.	Nonwoven textiles	Nano material	Detail	References
1	Cotton	Si–P–N complex	Rail seat cover Hydrophobicity	Li et al. [72, 73]
2	Cotton Polyester	Nanoparticles of polyanions, rennet, and casein Nanoparticles made of polycations and chitosan	Layer-by-layer Flame retardancy Antibacterial Toxic gases suppression	Attia et al. [8]
3	Cotton Polyaniline	Aniline/HCl Ammonium persulfate/HCl	Surface conductivity	Medagedara and Pemachandra [87]
4	Cellulose-based filter	Nylon-66 nanofiber	Engine air filtration	Johnson et al. [54]
5	Nylon-66	PVA and SiO_2 NPs	Air beg coating	Ali et al. [7])
6	Cellulose nonwoven composite with carbon fiber	CNTs (MWCNT)	Piezoresistive sensors Touch sensor	Jang et al. [51]
7	Wool	Chitosan nanofiber	Air filter	Mahajan and Tyagi [79]
8	Cotton	CNT Carbon black NPs	Fire warning sensor	Xia et al. [140]
9	Polyurethane	Nano-silica CNT	Thermal stability Flame retardants	Biswal et al. [14]
10	Flexible polyurethane foam	Molybdenum disulfide nanosheet Sodium alginate	Surface adsorption Physical barrier	Zhi et al. [147]

and jute are blended to create nonwoven fabrics that are more robust, flexible, and tough [24, 83].

Multifunctional textiles with a metallic look were created by coating a polypropylene nonwoven fabric substrate with a single layer of copper or copper oxide. Fabrics coated with dark green and dark red oxide film stood out from the copper film-coated fabric, which was orange-red in color, and the grey uncoated nonwoven fabric. A drop in Cu concentration and subsequent conversion to Cu_2O occurred as a result of the increase in oxygen flow; a further transformation into CuO occurred considerably later. The color became lighter as the brightness increased, whereas the K/S ratio decreased as the CuO content increased. The coated nonwoven fabric demonstrated an increase in hydrophobicity [46].

After being heat-stabilized, two commercially available nonwoven fabrics—one made of pure cotton (PCF) and the other of polyester/viscous (PVF)—were impregnated with AgNPs using an easy adsorption (dipping) technique. The antibacterial effects of PCF were less effective against Staphylococcus aureus and Escherichia coli. At all doses measured here (6.25–400 ppm), AgNP nanoemulsions and textiles were cytotoxic to L929-fibroblasts in vitro cell culture investigations. Although the concentration of 6.25 ppm of AgNPs loaded into the fabrics via nanoemulsion was less than that of cytotoxicity, the fabrics maintained their bacterial resistance. Tried cloth is used to make plant bags [91].

Research was conducted for modifying viscose rayon with nanoparticles to make the biodegradable crop cover. Modified cellulose nonwoven fabric may be made by staining it with NB-BT helion brown, padding it with a potassium nitrate (KNO_3) solution (fertilizer), and then coating it with a polylactide (PLA) solution. The mechanical testing, structural analysis, and physicochemical characteristics of the nonwoven fabric were all characterized. A heat-retardant, water-absorbing, and anti-weed mulching material that avoids soil infestation and releases fertilizers slowly was used in tunnel tomato growing using modified cellulose nonwovens [30].

4 Nano Application Method for Finishing Nonwoven

Nonwoven fibers and textiles are transformed into nano-enabled conventional textiles by applying nano-enabled coatings and finishes. Nanofinishing of nonwoven is performed to achieve required characteristic for particular end use. In order to maximize the efficacy of the coating and finishing treatments that follow, fibers are often treated with wet chemicals or plasma. Although there are many other finishing and coating techniques, the most popular ones for nanofinishing and nanocoating are layer-by-layer, sol–gel, and padding (Fig. 5).

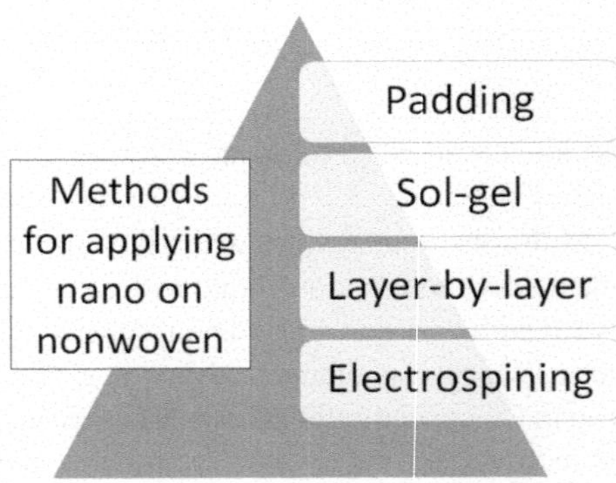

Fig. 5 Methods of application nano on nonwoven

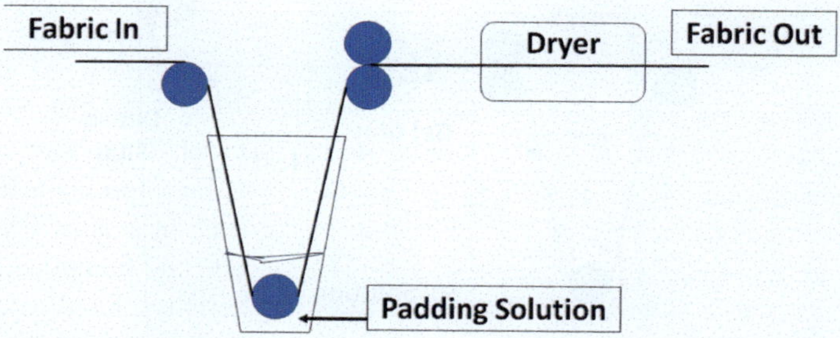

Fig. 6 Padding process

4.1 Padding Operation

The process of pressing a fabric or fiber while it is being treated with a chemical is called "pad-dry-cure." After the fabric has been dried and cured, the active components, such as dyes and finishing chemicals, are added to it from a liquid state using pressure rollers. Figure 6 is a schematic diagram depicting a typical padding method. The amount of coating and finishing material kept in the fabric is controlled by the spacing between the rollers. Inline drying and face-specific coating are accomplished with the use of heated rollers and several rollers in succession. Adding interesting nanomaterials to the finishing solution makes achieving nanofinishes easy [44].

4.2 Sol–Gel

Using the sol–gel method and a thermal process, nonwoven textiles are coated with a colloidal solution or suspension of metal particles, organometallic materials, and nano metal oxides. The treated fabric is then allowed to dry. Textile surfaces are coated using the sol–gel process to provide properties like flame retardancy, abrasion resistance, and antimicrobial activity. Several different sol–gel coating methods for textiles and fibers are now available for use in large-scale industrial processes [110, 116]. Figure 7 shows a process combination of sol–gel and continuous dip coating in action.

4.3 Layer-by-Layer (LbL)

The process of layer-by-layer (LbL) thin-film deposition (Fig. 8) involves attaching oppositely charged polyelectrolytes onto a charged planar substrate in a sequential fashion, resulting in multilayer coatings of nanometer-thick layers. Functionalized thin-film coating is a popular method for adding new physicochemical properties to the surface of textiles and fibers. This method is easy to use, affordable, and gives

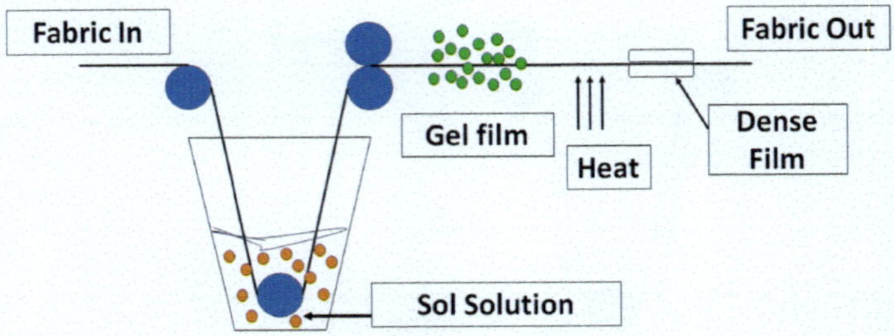

Fig. 7 Sol–gel process

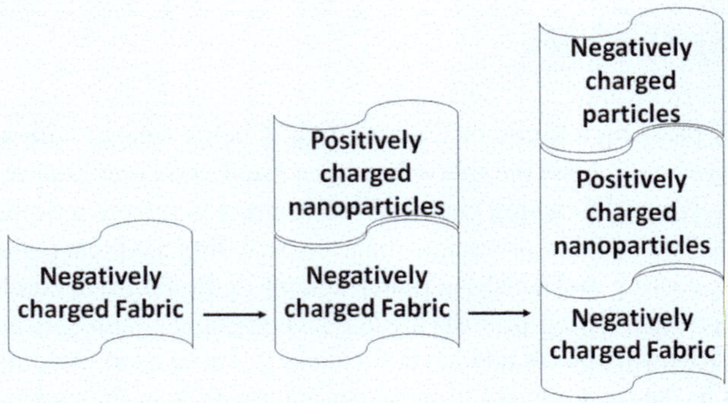

Fig. 8 Layer-by-layer finishing process

you great control over the film's thickness despite its slow speed. LbL has the ability to apply continuous polymer thin films, polymer nanocomposites, and nanoparticles. The immersive technique and spray LbL are both available for automated industrial scale roll-to-roll processing in textile applications [78, 103].

4.4 Electrospinning

The majority of polymer nanofibers and nanocomposite fibers are made using the electrospinning process. The method involves using an electric field to draw polymer fibers with sizes in the nanoscale range from a polymer melt or blend. As shown in Fig. 9, a typical electrospinning setup involves collecting the nanofibers that are produced on a rotating drum. Nanofibers' shape and quality may be altered by adjusting processing factors such as solution concentration, viscosity, spinning atmosphere, acceleration voltage, and tip-to-collector distance. Nanofibers work very well on surfaces and have great mechanical characteristics because of their high surface-to-volume ratio. They can therefore be used in filtration, wound care, packaging, and protective apparel. To enhance the characteristics of electrospun nanofibers and make them compatible with a wide range of polymer–solvent systems, nanomaterial

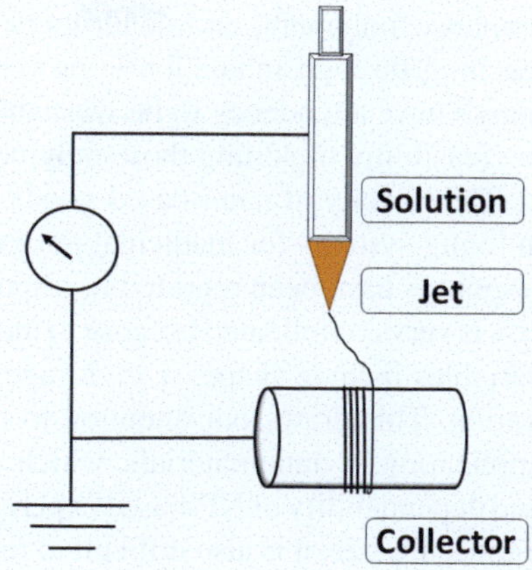

Fig. 9 Sketch diagram for electrospinning

fillers may be added to the polymer solution. The creation of highly functionalized nanocomposite fibers relies on the proper dispersion of nanoparticles in the polymer solution [139, 141].

5 Future Aspects

Using traditional masks to combat the COVID-19 pandemic is not a recommended course of action. Nanofabricated masks are here to stay, according to a plethora of research. In order to reduce the penetration rate in harsh environments, the primary objective is to develop more efficient filtration masks. Standard surgical masks offer more protection than N95 respirators do. Surgical masks and N95 masks may not provide protection against non-polluting particles (NPs), such as viruses, according to a number of studies. The WHO and the Centers for Disease Control and Prevention encouraged people to wear face masks during the COVID-19 pandemic. Conventional masks have a lot of disadvantages. The physical characteristics of nanomaterials can be changed to improve their functionality. By wearing a nanofabricated mask, the user can be shielded from incoming ultrafine viruses as small as 50 nm. Face masks already contain a variety of nanoadditives and nanomaterials like graphene. These nanomaterial-based face masks are antimicrobial, lightweight, biodegradable, comfortable, affordable, and have a long shelf life. They are also good for the environment [16, 109, 113].

The masks are still undergoing improvements in their properties and functions, but there is still a lot of room for improvement. The masks' self-cleaning properties may be reduced by the high hydrophilicity of cotton or cellulose-based fabrics. As a result, it is necessary to modify those biodegradable fabrics according to our specifications. The mask may be treated with three layers; outer and inner can be formulated with cellulose-based nonwoven grafting with polyester chain, and middle layer decorated with antiviral agents or polymers like aminoglycosides, polysaccharides, peptides, branched polymers, and dendrimers [13, 19, 43, 136]. The ingredient's potential leaching through inhalation is prevented by covalent grafting. Furthermore, these masks have a tendency to be washable and reusable easily. It is necessary to assess the feasibility of adding those polymers to the masks further.

The toxicity of nano-based goods is a major concern when it comes to their use in living systems for medicinal purposes. Allergies and other potentially hazardous responses have been reported in response to certain nanomaterials. The idea of toxicity is very complicated because to the large number of factors that influence it. These variables include shape, size, dosage, surface area, route, and duration of administration. This draws our attention to the need for standardizing or customizing the application of nanomaterials, which is another area. Furthermore, the repeatability and dependability of NP-based experiments can still be strengthened. Because of their minuscule size, it is also difficult to regulate their behavior in delicate settings. Some of the other challenges include their expensive price, the existence of contaminants, the way they affect the environment, and so on.

When compared to more conventional cementation preparations, nanotechnology improves the compressive flexural characteristics and decreases the porosity of cement, leading to less water absorption. The reason for this is the high surface-to-volume ratio of nanosized particles. This method, known as "green concrete," can increase the strength, durability, and affordability of concrete by lowering the cement content. Nanorods, nanofibers, and carbon nanotubes are quickly taking the place of steel structures. Improving the mechanical characteristics of nanostructures and nanoclay formations paves the way for nanoengineering to emerge as a distinct subfield of civil engineering. For building fractures to mend more quickly, repair mortars and self-healing concrete use nanoparticles in lieu of traditional cement combinations. In addition, titanium dioxide, zinc, and nanostructures are being used to create photocatalytic items that have built-in technologies that can resist water and germs, clean themselves, and guard against diseases. The use of quantum dot technologies for solar energy generation is becoming more prevalent [6, 35, 144].

Nanotechnology's advantages can be utilized to enhance the efficiency and durability of automobiles on the road. Nanomaterials can be used to enhance the strength, lightweight, flame retardant, and UV resistance of aerospace materials in a significant way. Despite the many positive things that have been reported, such as improvements, safety, and durability linked to nanotechnology, there is still much to be learned. Still, nanotechnology application for nonwoven in automotive industries is not fully developed, but it can be implemented in many ways by adding ecofriendly process and precautioned use. Once the technology is fully developed, it is expected to be affordable, user-friendly, and environmentally friendly.

Environmental cleanup agents, nanobiosensors, nanofertilizers, and nanopesticides are just a few of the many agricultural applications of nanotechnology. The agricultural and environmental sciences still face a significant challenge in understanding the fate and environmental impacts of nanomaterials. Developing efficient, versatile, stable, cost-effective, and ecologically friendly nanomaterials requires collaborating with institutions to study their many uses. This would round up our understanding of

NMs' function, trajectory, behavior, and ecotoxicity evaluation. Although the reaction of various plant species may vary, it is possible that applying NMs to crops may enhance their development and productivity. It is vital to carry out in-depth research into screening and optimizing nanomaterials for various plant species before using them commercially [131]. It is possible to control the stability and properties of nanomaterials to alter their behavior and efficiency. To increase their efficiency, it would be very helpful to develop new and improved synthesis techniques that allow for precise control over the composition of the final product. In order to develop integrated remediation strategies for bioremediation, it is imperative to investigate the role of natural resources (NRs). In addition, there is a lack of information on their use in the field as most research on nano-assisted agriculture has relied on controlled tests. For the large-scale use of nano-based techniques, more field-level understanding would be very advantageous.

Overall, nanofinishing or nanocoating on nonwoven textile materials has great impact in the field of health care, hygiene, medical, filtration, geotextiles, agrotextiles, and automotive textiles. The scholars have ample opportunities in the area of agrotextils, automotive textiles, and geotextiles. As the safety and personal hygiene concern removing the toxicity of nanoparticles and formulating a novel approach is the most focused zone on recent time.

References

1. Abulikemu M, Tabrizi BEA, Ghobadloo SM, Mofarah HM, Jabbour GE (2022) Silver nanoparticle-decorated personal protective equipment for inhibiting human coronavirus infectivity. ACS Appl Nano Mater 5(1):309–317. https://doi.org/10.1021/acsanm.1c03033
2. Adanur S, Jayswal A (2022) Filtration mechanisms and manufacturing methods of face masks: an overview. J Ind Text 51(3_suppl):3683S–3717S. https://doi.org/10.1177/152808372098 0169
3. Ahmad S, Siddiqi WA, Ahmad S (2022) Facile hydrophilic chitosan and graphene oxide modified sustainable non-woven fabric composite sieve membranes (NWF@Cs/Gx): antifouling, protein rejection, and oil-water emulsion separation studies. Chem Eng Res Des 181:220–238. https://doi.org/10.1016/j.cherd.2022.03.012
4. Ahn S, Shim E, Kim Y, Bae Y-S, Eom H (2022) Air filtration performance enhancement of PTFE foam–coated filters at high temperatures via secondary strongly adhering PTFE nanofiber coatings. Process Saf Environ Prot 162:914–922. https://doi.org/10.1016/j.psep.2022.04.067
5. Alatawi RAS, Bukhari AAH, Al-Sayed HMA, Alenazi DAK, Alnawmasi JSh, Abomuti MA, Faridi U (2023) Production of biologically active non-woven textiles from recycled polyethylene terephthalate. Luminescence 38(3):350–359. https://doi.org/10.1002/bio.4462
6. Ali AA (2020) Nanotechnology in civil engineering construction. Int J Struct Civil Eng Res, 87–90. https://doi.org/10.18178/ijscer.9.1.87-90
7. Ali S, Abdalla I, Gibril ME, Ahmed HA, Jiang X (2022) Enhancing the properties of nylon 66 fabric coated with a combination of PVA and SiO_2 nanoparticles composite for vehicle airbag application. J Ind Text 51(2_suppl):3330S–3346S. https://doi.org/10.1177/152808372 1988961
8. Attia NF, Mohamed A, Hussein A, El-Demerdash A-GM, Kandil SH (2023) Greener bio-based spherical nanoparticles for efficient multilayer textile fabrics nanocoating with outstanding

fire retardancy, toxic gases suppression, reinforcement and antibacterial properties. Surf Interf 36:102595. https://doi.org/10.1016/j.surfin.2022.102595
9. Bai W, Pakdel E, Wang Q, Tang B, Wang J, Chen Z, Zhang Y, Hurren C, Wang X (2022) Synergetic adsorption-photocatalysis process of titania-silica photocatalysts and their immobilization on PEEK nonwoven filter for VOC removal. J Environ Chem Eng 10(6):108920. https://doi.org/10.1016/j.jece.2022.108920
10. Barhoum A, Pal K, Rahier H, Uludag H, Kim IS, Bechelany M (2019) Nanofibers as new-generation materials: from spinning and nano-spinning fabrication techniques to emerging applications. Appl Mater Today 17:1–35. https://doi.org/10.1016/j.apmt.2019.06.015
11. Baselga M, Uranga-Murillo I, de Miguel D, Arias M, Sebastián V, Pardo J, Arruebo M (2022) Silver Nanoparticles–polyethyleneimine-based coatings with antiviral activity against SARS-CoV-2: a new method to functionalize filtration media. Materials 15(14):4742. https://doi.org/10.3390/ma15144742
12. Bhuiyan MAR, Wang L, Shaid A, Shanks RA, Ding J (2019) Advances and applications of chemical protective clothing system. J Ind Text 49(1):97–138. https://doi.org/10.1177/1528083718779426
13. Bianculli RH, Mase JD, Schulz MD (2020) Antiviral polymers: past approaches and future possibilities. Macromolecules 53(21):9158–9186. https://doi.org/10.1021/acs.macromol.0c01273
14. Biswal T, Sethy TR, Sahoo PK (2023) Role of flame-retardants on the physico-mechanical properties of polyurethanes, pp 53–70. https://doi.org/10.1021/bk-2023-1453.ch004
15. Boominathan S, Bhuvaneshwari M, Sangeetha K, Pachiyappan KM, Devaki E (2022) Influence of fiber blending on thermal and acoustic properties nonwoven material. J Nat Fibers 19(15):11193–11203. https://doi.org/10.1080/15440478.2021.2021123
16. Cano-Vicent A, Tuñón-Molina A, Martí M, Muramoto Y, Noda T, Takayama K, Serrano-Aroca Á (2021) Antiviral face mask functionalized with solidified hand soap: low-cost infection prevention clothing against enveloped viruses such as SARS-CoV-2. ACS Omega 6(36):23495–23503. https://doi.org/10.1021/acsomega.1c03511
17. Carvalho I, Curado M, Palacio C, Carvalho S, Cavaleiro A (2019) Ag release from sputtered Ag/a:C nanocomposite films after immersion in pure water and NaCl solution. Thin Solid Films 671:85–94. https://doi.org/10.1016/j.tsf.2018.12.010
18. Chemeh ZK, Montazer M, Tavanaie M, Mahmoudirad M (2023) Effective method for multi-functionalization of non-woven polyester fabric using carbon black nanoparticles/vinyl acrylic dispersion by padding technique. Fibers Polym 24(12):4253–4268. https://doi.org/10.1007/s12221-023-00399-y
19. Chen L, Huang G (2018) The antiviral activity of polysaccharides and their derivatives. Int J Biol Macromol 115:77–82. https://doi.org/10.1016/j.ijbiomac.2018.04.056
20. Chua MH, Cheng W, Goh SS, Kong J, Li B, Lim JYC, Mao L, Wang S, Xue K, Yang L, Ye E, Zhang K, Cheong WCD, Tan BH, Li Z, Tan BH, Loh XJ (2020) Face masks in the new COVID-19 normal: materials, testing, and perspectives. Research 2020. https://doi.org/10.34133/2020/7286735
21. Cieślak M, Kowalczyk D, Baranowska-Korczyc A, Kamińska I, Krzyżowska M, Janicka M, Kubacki J (2023) Viscose nonwoven fabric with copper and its multifunctional properties. Cellulose 30(15):9843–9859. https://doi.org/10.1007/s10570-023-05454-8
22. Dahanayake MH, Athukorala SS, Jayasundera ACA (2022) Recent breakthroughs in nanostructured antiviral coating and filtration materials: a brief review. RSC Adv 12(26):16369–16385. https://doi.org/10.1039/D2RA01567F
23. De Luca, P., B.Nagy, J., & Macario, A. (2023). Nanomaterials Used in the preparation of personal protective equipment (PPE) in the fight against SARS-CoV-2. Inorganics 11(7):294. https://doi.org/10.3390/inorganics11070294
24. Debnath S (2014) Jute-based sustainable agrotextiles, their properties and case studies, pp 327–355. https://doi.org/10.1007/978-981-287-065-0_10
25. Deng X, Yu Nikiforov A, Coenye T, Cools P, Aziz G, Morent R, De Geyter N, Leys C (2015) Antimicrobial nano-silver non-woven polyethylene terephthalate fabric via an atmospheric pressure plasma deposition process. Sci Rep 5(1):10138. https://doi.org/10.1038/srep10138

26. El-Ghazali S, Khatri M, Kobayashi S, Kim IS (2022) An overview of medical textile materials. In: Medical textiles from natural resources. Elsevier, pp 3–42. https://doi.org/10.1016/B978-0-323-90479-7.00026-9
27. Fatma B, Andrabi SM, Gupta S, Verma V, Kumar A, Pitsalidis C, Garg A (2023) Biocompatible, breathable and degradable microbial cellulose based triboelectric nanogenerator for wearable transient electronics. Nano Energy 114:108628. https://doi.org/10.1016/j.nanoen.2023.108628
28. Fiaschini N, Giuliani C, Vitali R, Tammaro L, Valerini D, Rinaldi A (2022) Design and manufacturing of antibacterial electrospun polysulfone membranes functionalized by Ag nanocoating via magnetron sputtering. Nanomaterials 12(22):3962. https://doi.org/10.3390/nano12223962
29. Fraczek-Szczypta A, Rabiej S, Szparaga G, Pabjanczyk-Wlazlo E, Krol P, Brzezinska M, Blazewicz S, Bogun M (2015) The structure and properties of the carbon non-wovens modified with bioactive nanoceramics for medical applications. Mater Sci Eng, C 51:336–345. https://doi.org/10.1016/j.msec.2015.03.021
30. Gabryś T, Fryczkowska B, Grzybowska-Pietras J, Biniaś D (2021) Modification and properties of cellulose nonwoven fabric—multifunctional mulching material for agricultural applications. Materials 14(15):4335. https://doi.org/10.3390/ma14154335
31. Gadkari RR, Ali W, Das A, Alagirusamy R (2022) Configuration of a unique antibacterial needle-punched nonwoven fabric from silver impregnated polyester nanocomposite fibres. J Ind Text 51(1_suppl):1511S–1527S. https://doi.org/10.1177/1528083720924727
32. Gadkari RR, Gupta A, Teke U, Awadhiya A, Shahadat M, Ali W, Das A, Alagirusamy R (2021) A sustainable way for surface functionalisation of PET nonwoven with novel chitosan-cinnamaldehyde cross-linked nanoparticles. J Ind Eng Chem 99:214–223. https://doi.org/10.1016/j.jiec.2021.04.031
33. Galante AJ, Haghanifar S, Romanowski EG, Shanks RMQ, Leu PW (2020) Superhemophobic and antivirofouling coating for mechanically durable and wash-stable medical textiles. ACS Appl Mater Interfaces 12(19):22120–22128. https://doi.org/10.1021/acsami.9b23058
34. Gharaei R, Russell SJ (2022) Overview of nonwoven product applications. In: Handbook of nonwovens. Elsevier, pp 13–47. https://doi.org/10.1016/B978-0-12-818912-2.00012-4
35. Ghernaout D (2018) Nanotechnology phenomena in the light of the solar energy. J Energy Environ Chem Eng 3(1):1. https://doi.org/10.11648/j.jeece.20180301.11
36. Gu W, Zhan R, Li R, Liu J, Zhang J (2021) Preparation and characterization of PU/PET matrix gradient composites with microwave-absorbing function. Coatings 11(8):982. https://doi.org/10.3390/coatings11080982
37. Guru R, Kumar A, Kumar R (2023) Healthcare and hygiene products application in medical textile. In: Next-generation textiles. IntechOpen. https://doi.org/10.5772/intechopen.103662
38. Hadley T, Hickey K, Lix K, Sharma S, Berretta T, Navessin T (2022) Flushed but not forgotten: the rising costs and opportunities of disposable wet wipes. BioResources 18(1). https://doi.org/10.15376/biores.18.1.Hadley
39. Haider AJ, Sukkar KA, Abdalsalam AH, Ali AF, Jaber SH, Ridha TTA (2021) Enhancement of the air quality and heat transfer rate of an air-conditioning system using a hybrid polypropylene nanofilter. Process Saf Environ Prot 149:56–66. https://doi.org/10.1016/j.psep.2020.10.041
40. Hamoda DM, Kanawy Ibrahim S, Aboalmaged HH, Rashed U, Almetwally A, El-Masry HM (2022) Plasma technique application for coating non-woven fabric (SMS) by ($CaSiO_3$/CuO) nano particles for biomedical sector. Egypt J Chem. https://doi.org/10.21608/ejchem.2022.128815.5705
41. Hamouda T, Ibrahim HM, Kafafy HH, Mashaly HM, Mohamed NH, Aly NM (2021) Preparation of cellulose-based wipes treated with antimicrobial and antiviral silver nanoparticles as novel effective high-performance coronavirus fighter. Int J Biol Macromol 181:990–1002. https://doi.org/10.1016/j.ijbiomac.2021.04.071
42. He H, Yan Y, Qiu Z, Tan X (2017) A novel antistatic polyurethane hybrid based on nanoscale ionic material. Prog Org Coat 113:110–116. https://doi.org/10.1016/j.porgcoat.2017.09.004

43. He X, Fang J, Guo Q, Wang M, Li Y, Meng Y, Huang L (2020) Advances in antiviral polysaccharides derived from edible and medicinal plants and mushrooms. Carbohyd Polym 229:115548. https://doi.org/10.1016/j.carbpol.2019.115548
44. Hoque E, Acharya S, Shamshina J, Abidi N (2023) Review of foam applications on cotton textiles. Text Res J 93(1–2):486–501. https://doi.org/10.1177/00405175221107400
45. Horváth E, Rossi L, Mercier C, Lehmann C, Sienkiewicz A, Forró L (2020) Photocatalytic nanowires-based air filter: towards reusable protective masks. Adv Funct Mater 30(40). https://doi.org/10.1002/adfm.202004615
46. Huang M-L, Lu S-G, Zhou J-J, Luo B-S, Li Y-H (2022) Metallic coloration with Cu/CuO coating on polypropylene nonwoven fabric via a physical vapor deposition method and its multifunctional properties. J Text Inst 113(7):1345–1354. https://doi.org/10.1080/00405000.2021.1928995
47. Hutten IM (2007) Introduction to nonwoven filter media. In: Handbook of nonwoven filter media. Elsevier, pp 1–28. https://doi.org/10.1016/B978-185617441-1/50016-0
48. Ignatova M, Manolova N, Rashkov I (2013) Electrospun antibacterial chitosan-based fibers. Macromol Biosci 13(7):860–872. https://doi.org/10.1002/mabi.201300058
49. Islam MdT, Mamun MdAAl, Hasan MdT, Shahariar H (2021). Scalable coating process of AgNPs-silicone on cotton fabric for developing hydrophobic and antimicrobial properties. J Coat Technol Res 18(3):887–898. https://doi.org/10.1007/s11998-020-00451-z
50. Issman L, Graves B, Terrones J, Hosmillo M, Qiao R, Glerum M, Yeshurun S, Pick M, Goodfellow I, Elliott J, Boies A (2021) Filtration of viral aerosols via a hybrid carbon nanotube active filter. Carbon 183:232–242. https://doi.org/10.1016/j.carbon.2021.07.004
51. Jang SJ, Kim M, Lim JY, Park YK, Ko JH (2022) Development of mode-switchable touch sensor using MWCNT composite conductive nonwoven fabric. Polymers 14(8):1545. https://doi.org/10.3390/polym14081545
52. Jelil RA (2015) A review of low-temperature plasma treatment of textile materials. J Mater Sci 50(18):5913–5943. https://doi.org/10.1007/s10853-015-9152-4
53. Jeong SB, Lee DU, Lee BJ, Heo KJ, Kim DW, Hwang GB, MacRobert AJ, Shin JH, Ko HS, Park SK, Oh YS, Kim SJ, Lee DY, Lee S-B, Park I, Kim SB, Han B, Jung JH, Choi DY (2022) Photobiocidal-triboelectric nanolayer coating of photosensitizer/silica-alumina for reusable and visible-light-driven antibacterial/antiviral air filters. Chem Eng J 440:135830. https://doi.org/10.1016/j.cej.2022.135830
54. Johnson J, Muzwar M, Ramakrishnan S, Chetty R, Karaiyan AP (2023) Electrospun nylon-66 nanofiber coated filter media for engine air filtration applications. J Appl Polym Sci 140(44). https://doi.org/10.1002/app.54618
55. Joshi M, Adak B (2019) 6. Nanotechnology-based textiles: a solution for the emerging automotive sector. In: Rubber nanocomposites and nanotextiles. De Gruyter, pp 179–230. https://doi.org/10.1515/9783110643879-006
56. Karagoz S, Kiremitler NB, Sarp G, Pekdemir S, Salem S, Goksu AG, Onses MS, Sozdutmaz I, Sahmetlioglu E, Ozkara ES, Ceylan A, Yilmaz E (2021) Antibacterial, antiviral, and self-cleaning mats with sensing capabilities based on electrospun nanofibers decorated with ZnO nanorods and Ag nanoparticles for protective clothing applications. ACS Appl Mater Interfaces 13(4):5678–5690. https://doi.org/10.1021/acsami.0c15606
57. Karim N, Afroj S, Lloyd K, Oaten LC, Andreeva DV, Carr C, Farmery AD, Kim I-D, Novoselov KS (2020) Sustainable personal protective clothing for healthcare applications: a review. ACS Nano 14(10):12313–12340. https://doi.org/10.1021/acsnano.0c05537
58. Karimi F, Soltani P, Zarrebini M, Hassanpour A (2022) Acoustic and thermal performance of polypropylene nonwoven fabrics for insulation in buildings. J Build Eng 50:104125. https://doi.org/10.1016/j.jobe.2022.104125
59. Kavitha Sankar PC, Ramakrishnan R, Rosemary MJ (2016) Biological evaluation of nanosilver incorporated cellulose pulp for hygiene products. Mater Sci Eng, C 61:631–637. https://doi.org/10.1016/j.msec.2015.12.072
60. Ke Z, Yu L, Wang G, Sun R, Zhu M, Dong H, Xu Y, Ren M, Fu S, Zhi C (2023) Three-dimensional modeling of spun-bonded nonwoven meso-structures. Polymers 15(3):600. https://doi.org/10.3390/polym15030600

61. Kitahara N, Sato T, Isogawa H, Ohgoe Y, Masuko S, Shizuku F, Hirakuri KK (2010) Antibacterial property of DLC film coated on textile material. Diam Relat Mater 19(7–9):690–694. https://doi.org/10.1016/j.diamond.2010.03.013
62. Kongahge D, Foroughi J, Gambhir S, Spinks GM, Wallace GG (2016) Fabrication of a graphene coated nonwoven textile for industrial applications. RSC Adv 6(77):73203–73209. https://doi.org/10.1039/C6RA15190F
63. Kopitar D, Marasovic P, Jugov N, Schwarz I (2022) Biodegradable nonwoven agrotextile and films—a review. Polymers 14(11):2272. https://doi.org/10.3390/polym14112272
64. Korgaonkar S, Sayed U (2021) Nano herbal oils application on nonwovens. Int J Pharmaceut Res 13(02). https://doi.org/10.31838/ijpr/2021.13.02.245
65. Kwon S, Zambrano MC, Venditti RA, Frazier R, Zambrano F, Gonzalez RW, Pawlak JJ (2022) Microfiber shedding from nonwoven materials including wipes and meltblown nonwovens in air and water environments. Environ Sci Pollut Res 29(40):60584–60599. https://doi.org/10.1007/s11356-022-20053-z
66. Laufer G, Kirkland C, Cain AA, Grunlan JC (2012) Clay-chitosan nanobrick walls: completely renewable gas barrier and flame-retardant nanocoatings. ACS Appl Mater Interfaces 4(3):1643–1649. https://doi.org/10.1021/am2017915
67. Lee HG, Jang MJ, Um IC (2023) Fabrication, structural characteristics, and properties of sericin-coated wool nonwoven fabrics. Int J Mol Sci 24(19):14750. https://doi.org/10.3390/ijms241914750
68. Lequeux A, Maze B, Laroche G, Heim F (2022) Non-woven textiles for medical implants: mechanical performances improvement. Biomedical Engineering/Biomedizinische Technik 67(4):317–330. https://doi.org/10.1515/bmt-2022-0017
69. Li C, Ren L, Liu X, Zhang C, Chen D, Xu W, Qin Y (2021) Superhydrophilic and underwater superoleophobic poly(propylene) nonwoven coated with TiO_2 by atomic layer deposition. Adv Mater Interfaces 8(1). https://doi.org/10.1002/admi.202001485
70. Li F, Ni Y, Cong J, Shen C, Ji P, Wang H, Yin L, Xu C (2022) Wiping conditions and fabric properties influenced the microfiber shedding from non-woven products. Environ Sci Process Impacts 24(10):1855–1866. https://doi.org/10.1039/D2EM00292B
71. Li G, Liu H, Li T, Wang J (2012) Surface modification and functionalization of silk fibroin fibers/fabric toward high performance applications. Mater Sci Eng, C 32(4):627–636. https://doi.org/10.1016/j.msec.2011.12.013
72. Li M, Prabhakar MN, Song J (2022) Synthesis of novel Si–P–N complex coating for the fabrication of flame-retardant and hydrophobic cotton textile potentially suitable for rail seat cover. Progr Organ Coat 172:107144. https://doi.org/10.1016/j.porgcoat.2022.107144
73. Liu M, Ma C, Zhou D, Chen S, Zou L, Wang H, Wu J (2022) Hydrophobic, breathable cellulose nonwoven fabrics for disposable hygiene applications. Carbohyd Polym 288:119367. https://doi.org/10.1016/j.carbpol.2022.119367
74. Liu R, Qu M, Qiu X, Wang H, Fan M, Zhang A, Chen Q, Bin Y (2022) Poly (ethylene terephthalate) nonwoven fabrics-based membranes modified by electrospinning of thermoplastic polyurethane, nano SiO_2 and Ag particles as medical packing materials. Packag Technol Sci 35(7):557–567. https://doi.org/10.1002/pts.2651
75. Liu W, Sun Q, Xue C, Long X, Zhang W (2022) Evaluation on thermal protection performance of TiO_2@ATO coated aramid nonwoven. Coatings 12(5):657. https://doi.org/10.3390/coatings12050657
76. Ma J, Zhang N, Cheng Y, Kou X, Niu Y, Jin X, Ke Q, Zhao Y (2022) Green fabrication of multifunctional three-dimensional superabsorbent nonwovens with thermo-bonding fibers. Adv Fiber Mater 4(2):293–304. https://doi.org/10.1007/s42765-021-00108-5
77. Ma Z, Yan Z, Yin X, Si Y, Yu J, Ding B (2021) Antibacterial and antiviral nanofibrous membranes with renewable oxidative function for high-efficiency and super-throughput water disinfection. Compos Commun 27:100875. https://doi.org/10.1016/j.coco.2021.100875
78. Magovac E, Vončina B, Jordanov I, Grunlan JC, Bischof S (2022) Layer-by-layer deposition: a promising environmentally benign flame-retardant treatment for cotton, polyester, polyamide and blended textiles. Materials 15(2):432. https://doi.org/10.3390/ma15020432

79. Mahajan C, Tyagi S (2019) Preparation and evaluation of wool keratin based chitosan and water filtration. Int J Innov Res Adv Eng 6(2):37–44
80. Mahmud MB, Rahman SS, Fashandi M, Lee PC, Park CB (2023) Hydrophilic-to-hydrophobic conversion of polyamide 6-based mat via polyvinylidene fluoride nanofiber integration with enhanced oil-water separation. Mater Today Sustain 24:100559. https://doi.org/10.1016/j.mtsust.2023.100559
81. Manickam P, Kandhavadivu P (2022) Development of banana nonwoven fabric for eco-friendly packaging applications of rural agriculture sector. J Nat Fibers 19(8):3158–3170. https://doi.org/10.1080/15440478.2020.1840479
82. Mao N (2017) Nonwoven fabric filters. In: Fibrous filter media. Elsevier, pp 133–171. https://doi.org/10.1016/B978-0-08-100573-6.00005-8
83. Marasović P, Kopitar D (2019) Overview and perspective of nonwoven agrotextile. Text Leather Rev 2(1):32–45. https://doi.org/10.31881/TLR.2019.23
84. Markarian J (2008) New developments in antistatic and conductive additives. Plastics Addit Compound 10(5):22–25. https://doi.org/10.1016/S1464-391X(08)70172-7
85. Mazurkow JM, Yüzbasi NS, Domagala KW, Pfeiffer S, Kata D, Graule T (2020) Nano-sized copper (oxide) on alumina granules for water filtration: effect of copper oxidation state on virus removal performance. Environ Sci Technol 54(2):1214–1222. https://doi.org/10.1021/acs.est.9b05211
86. Meda US, Soundarya VG, Madhu H, Bhat N (2023) Nano-engineered textiles: development and applications. Mater Sci Eng, B 296:116636. https://doi.org/10.1016/j.mseb.2023.116636
87. Medagedara H, Pemachandra K (2020) Investigating the feasibility of a conductive textile structure to indicate pre-tensioning of seat-belts. From Innov Impact (FITI) 2020:1–5. https://doi.org/10.1109/FITI52050.2020.9424874
88. Memon H, Kumari N (2016) Study of multifunctional nanocoated cold plasma treated polyester cotton blended curtains. Surf Rev Lett 23(05):1650036. https://doi.org/10.1142/S0218625X16500360
89. Mewoli A, Segovia C, Soppie AG, Ebanda FB, Njom AE, Ateba A, Brosse N (2023) Investigation of the performance of needle-punched nonwoven fabrics using *Triumfetta cordifolia* and thermoplastic fibers, compared to other commercial bast fibers used in preformed biosourced composites. Heliyon 9(7):e17888. https://doi.org/10.1016/j.heliyon.2023.e17888
90. Mirzaei H, Darroudi M (2017) Zinc oxide nanoparticles: biological synthesis and biomedical applications. Ceram Int 43(1):907–914. https://doi.org/10.1016/j.ceramint.2016.10.051
91. Moghtader F, Salouti M, Türk M, Pişkin E (2014) Nanoemulsions and nonwoven fabrics carrying AgNPs: antibacterial but may be cytotoxic. Artificial Cells Nanomed Biotechnol 42(6):392–399. https://doi.org/10.3109/21691401.2013.834908
92. Morais D, Guedes R, Lopes M (2016) Antimicrobial approaches for textiles: from research to market. Materials 9(6):498. https://doi.org/10.3390/ma9060498
93. Musschoot J, Dendooven J, Deduytsche D, Haemers J, Buyle G, Detavernier C (2012) Conformality of thermal and plasma enhanced atomic layer deposition on a non-woven fibrous substrate. Surf Coat Technol 206(22):4511–4517. https://doi.org/10.1016/j.surfcoat.2012.02.038
94. Nag MK, Kumar P (2023) Fabrication and characterization of woven and comingled nonwoven sheet polypropylene hybrid composite by recycling and alkali-treated jute waste fibers. Proc Inst Mech Eng C J Mech Eng Sci 237(16):3630–3649. https://doi.org/10.1177/09544062221149388
95. Neznakomova MP, Gospodinova DN, Dineff PD (2020) Glow-discharge plasma enhanced coating and functionalization of needle-punched nonwoven PET filter media for textile wastewater decolorization. IOP Conf Ser Mater Sci Eng 878(1):012017. https://doi.org/10.1088/1757-899X/878/1/012017
96. Neznakomova MP, Klotz M-L, Gospodinova DN (2019) Non-woven composites intensification properties for air filters by plasma pre-treatment. IOP Conf Ser Mater Sci Eng 659(1):012044. https://doi.org/10.1088/1757-899X/659/1/012044

97. Noor N, Mutalik S, Younas MW, Chan CY, Thakur S, Wang F, Yao MZ, Mou Q, Leung PH (2019) Durable antimicrobial behaviour from silver-graphene coated medical textile composites. Polymers 11(12):2000. https://doi.org/10.3390/polym11122000
98. O'Dowd K, Nair KM, Forouzandeh P, Mathew S, Grant J, Moran R, Bartlett J, Bird J, Pillai SC (2020) Face masks and respirators in the fight against the COVID-19 pandemic: a review of current materials, advances and future perspectives. Materials 13(15):3363. https://doi.org/10.3390/ma13153363
99. Otto DP, de Villiers MM (2020) Layer-by-layer nanocoating of antiviral polysaccharides on surfaces to prevent coronavirus infections. Molecules 25(15):3415. https://doi.org/10.3390/molecules25153415
100. Pakdel E, Naebe M, Sun L, Wang X (2019) Advanced functional fibrous materials for enhanced thermoregulating performance. ACS Appl Mater Interfaces 11(14):13039–13057. https://doi.org/10.1021/acsami.8b19067
101. Pakdel E, Wang J, Varley R, Wang X (2021) Recycled carbon fiber nonwoven functionalized with fluorine-free superhydrophobic PDMS/ZIF-8 coating for efficient oil-water separation. J Environ Chem Eng 9(6):106329. https://doi.org/10.1016/j.jece.2021.106329
102. Palika A, Armanious A, Rahimi A, Medaglia C, Gasbarri M, Handschin S, Rossi A, Pohl MO, Busnadiego I, Gübeli C, Anjanappa RB, Bolisetty S, Peydayesh M, Stertz S, Hale BG, Tapparel C, Stellacci F, Mezzenga R (2021) An antiviral trap made of protein nanofibrils and iron oxyhydroxide nanoparticles. Nat Nanotechnol 16(8):918–925. https://doi.org/10.1038/s41565-021-00920-5
103. Pan Y, Fu L, Du J, Zhang D, Lü T, Zhang Y, Zhao H (2022) Layer-by-layer self-assembly coating for multi-functionalized fabrics: a scientometric analysis in CiteSpace (2005–2021). Molecules 27(19):6767. https://doi.org/10.3390/molecules27196767
104. Peralta-Videa JR, Zhao L, Lopez-Moreno ML, de la Rosa G, Hong J, Gardea-Torresdey JL (2011) Nanomaterials and the environment: a review for the biennium 2008–2010. J Hazard Mater 186(1):1–15. https://doi.org/10.1016/j.jhazmat.2010.11.020
105. Periyasamy AP, Karunakaran G, Rwahwire S, Kesari K (2023) Nonwoven fabrics developed from agriculture and industrial waste for acoustic and thermal applications. Cellulose 30(11):7329–7346. https://doi.org/10.1007/s10570-023-05285-7
106. Petkova P, Francesko A, Perelshtein I, Gedanken A, Tzanov T (2016) Simultaneous sonochemical-enzymatic coating of medical textiles with antibacterial ZnO nanoparticles. Ultrason Sonochem 29:244–250. https://doi.org/10.1016/j.ultsonch.2015.09.021
107. Pragadheeswari R, Arunkumar KV, Rajakumar R, Sangeetha K (2014) Nano membrane fibrous on lyocell non-woven fabric for bacterial filitration efficiency in surgical face mask. J Nanosci Nanotechnol 2(5):598–599
108. Prambauer M, Wendeler C, Weitzenböck J, Burgstaller C (2019) Biodegradable geotextiles—an overview of existing and potential materials. Geotext Geomembr 47(1):48–59. https://doi.org/10.1016/j.geotexmem.2018.09.006
109. Pullangott G, Kannan U, Gayathri S, Kiran DV, Maliyekkal SM (2021) A comprehensive review on antimicrobial face masks: an emerging weapon in fighting pandemics. RSC Adv 11(12):6544–6576. https://doi.org/10.1039/D0RA10009A
110. Raduly FM, Răditoiu V, Răditoiu A, Frone AN, Nicolae CA, Răut I, Constantin M, Grapin M (2023) Multifunctional finishing of cotton fabric with curcumin derivatives coatings obtained by sol-gel method. Gels 9(5):369. https://doi.org/10.3390/gels9050369
111. Rajendran S, Anand SC (2002) Developments in medical textiles. Text Prog 32(4):1–42. https://doi.org/10.1080/00405160208688956
112. Rohani Shirvan A, Nouri A (2020) Medical textiles. In: Advances in functional and protective textiles. Elsevier, pp 291–333. https://doi.org/10.1016/B978-0-12-820257-9.00013-8
113. Rowan NJ, Moral RA (2021) Disposable face masks and reusable face coverings as non-pharmaceutical interventions (NPIs) to prevent transmission of SARS-CoV-2 variants that cause coronavirus disease (COVID-19): role of new sustainable NPI design innovations and predictive mathematical modelling. Sci Total Environ 772:145530. https://doi.org/10.1016/j.scitotenv.2021.145530

114. Ščasníková K, Dubec A (2023) The influence of low-temperature plasma on permanence of antimicrobial nano-finish. Fibres Text 30(4):41–49. https://doi.org/10.15240/tul/008/2023-4-005
115. Schwarz I, Kovačević S (2017) Textile application: from need to imagination. In: Textiles for advanced applications. InTech. https://doi.org/10.5772/intechopen.68376
116. Sfameni S, Hadhri M, Rando G, Drommi D, Rosace G, Trovato V, Plutino MR (2023) Inorganic finishing for textile fabrics: recent advances in wear-resistant, UV protection and antimicrobial treatments. Inorganics 11(1):19. https://doi.org/10.3390/inorganics11010019
117. Shahidi S, Moazzenchi B, Ghoranneviss M (2015) A review-application of physical vapor deposition (PVD) and related methods in the textile industry. Eur Phys J Appl Phys 71(3):31302. https://doi.org/10.1051/epjap/2015140439
118. Shastri JP, Rupani MG, Jain RL (2012) Antimicrobial activity of nanosilver-coated socks fabrics against foot pathogens. J Text Inst 103(11):1234–1243. https://doi.org/10.1080/00405000.2012.675680
119. Shin B, Lee K-R, Moon M-W, Kim H-Y (2012) Extreme water repellency of nanostructured low-surface-energy non-woven fabrics. Soft Matter 8(6):1817–1823. https://doi.org/10.1039/C1SM06867A
120. Sinclair TR, Patil A, Raza BG, Reurink D, van den Hengel SK, Rutjes SA, de Roda Husman AM, Roesink HDW, de Vos WM (2019) Cationically modified membranes using covalent layer-by-layer assembly for antiviral applications in drinking water. J Membr Sci 570–571:494–503. https://doi.org/10.1016/j.memsci.2018.10.081
121. Singh M, Vajpayee M, Ledwani L (2020) Eco-friendly surface modification and nanofinishing of textile polymers to enhance functionalization, pp 529–559. https://doi.org/10.1007/978-3-030-33774-2_23
122. Song Y, Lee Y, Lee Y, Hwang W-T, Lee J, Park S, Park N, Song H, Kim H, Lee KG, Kim I-D, Kim Y, Im SG (2023) Anti-viral, anti-bacterial, but non-cytotoxic nanocoating for reusable face mask with efficient filtration, breathability, and robustness in humid environment. Chem Eng J 470:144224. https://doi.org/10.1016/j.cej.2023.144224
123. Soni R, Joshi SR, Karmacharya M, Min H, Kim S-K, Kumar S, Kim G-H, Cho Y-K, Lee CY (2021) Superhydrophobic and self-sterilizing surgical masks spray-coated with carbon nanotubes. ACS Appl Nano Mater 4(8):8491–8499. https://doi.org/10.1021/acsanm.1c01082
124. Sousa-Pinto B, Fonte AP, Lopes AA, Oliveira B, Fonseca JA, Costa-Pereira A, Correia O (2020) Face masks for community use: an awareness call to the differences in materials. Respirology 25(8):894–895. https://doi.org/10.1111/resp.13891
125. Sundarrajan S, Tan KL, Lim SH, Ramakrishna S (2014) Electrospun nanofibers for air filtration applications. Procedia Eng 75:159–163. https://doi.org/10.1016/j.proeng.2013.11.034
126. Szekeres GP, Németh Z, Schrantz K, Németh K, Schabikowski M, Traber J, Pronk W, Hernádi K, Graule T (2018) Copper-coated cellulose-based water filters for virus retention. ACS Omega 3(1):446–454. https://doi.org/10.1021/acsomega.7b01496
127. Tcharkhtchi A, Abbasnezhad N, Zarbini Seydani M, Zirak N, Farzaneh S, Shirinbayan M (2021) An overview of filtration efficiency through the masks: mechanisms of the aerosols penetration. Bioactive Mater 6(1):106–122. https://doi.org/10.1016/j.bioactmat.2020.08.002
128. Tiwari N, Satyam N (2020) An experimental study on the behavior of lime and silica fume treated coir geotextile reinforced expansive soil subgrade. Eng Sci Technol Int J 23(5):1214–1222. https://doi.org/10.1016/j.jestch.2019.12.006
129. Tronci G, Russell SJ (2022) Raw materials and polymer science for nonwovens. In: Handbook of nonwovens. Elsevier, pp 49–88. https://doi.org/10.1016/B978-0-12-818912-2.00005-7
130. Tsutsumi-Arai C, Iwamiya Y, Hoshino R, Terada-Ito C, Sejima S, Akutsu-Suyama K, Shibayama M, Hiroi Z, Tokuyama-Toda R, Iwamiya R, Ijichi K, Chiba T, Satomura K (2022) Surface functionalization of non-woven fabrics using a novel silica-resin coating technology: antiviral treatment of non-woven fabric filters in surgical masks. Int J Environ Res Public Health 19(6):3639. https://doi.org/10.3390/ijerph19063639
131. Usman M, Farooq M, Wakeel A, Nawaz A, Cheema SA, ur Rehman H, Ashraf I, Sanaullah M (2020) Nanotechnology in agriculture: current status, challenges and future opportunities. Sci Total Environ 721:137778. https://doi.org/10.1016/j.scitotenv.2020.137778

132. Vellingiri K, Ramachandran T, Senthilkumar M (2013) Eco-friendly application of nano chitosan in antimicrobial coatings in the textile industry. Nanosci Nanotechnol Lett 5(5):519–529. https://doi.org/10.1166/nnl.2013.1575
133. Velusamy P, Kiruba K, Rajnish KN, Madhavan T, Anbu P (2021) Recent advances in the development of antimicrobial nanotextiles for prevention of infectious diseases transmission in healthcare workers. In: Green chemistry for sustainable textiles. Elsevier, pp 17–26. https://doi.org/10.1016/B978-0-323-85204-3.00019-1
134. Venkataraman D, Shabani E, Park JH (2023) Advancement of nonwoven fabrics in personal protective equipment. Materials 16(11):3964. https://doi.org/10.3390/ma16113964
135. Ventura H, Álvarez MD, Gonzalez-Lopez L, Claramunt J, Ardanuy M (2022) Cement composite plates reinforced with nonwoven fabrics from technical textile waste fibres: mechanical and environmental assessment. J Clean Prod 372:133652. https://doi.org/10.1016/j.jclepro.2022.133652
136. Vilas Boas LCP, Campos ML, Berlanda RLA, de Carvalho Neves N, Franco OL (2019) Antiviral peptides as promising therapeutic drugs. Cell Mol Life Sci 76(18):3525–3542. https://doi.org/10.1007/s00018-019-03138-w
137. Watts S, Maniura-Weber K, Siqueira G, Salentinig S (2021) Virus pH-dependent interactions with cationically modified cellulose and their application in water filtration. Small 17(30). https://doi.org/10.1002/smll.202100307
138. Wu H, Yao C, Li C, Miao M, Zhong Y, Lu Y, Liu T (2020) Review of application and innovation of geotextiles in geotechnical engineering. Materials 13(7):1774. https://doi.org/10.3390/ma13071774
139. Wu S, Dong T, Li Y, Sun M, Qi Y, Liu J, Kuss MA, Chen S, Duan B (2022) State-of-the-art review of advanced electrospun nanofiber yarn-based textiles for biomedical applications. Appl Mater Today 27:101473. https://doi.org/10.1016/j.apmt.2022.101473
140. Xia L, Lv Y, Miao Z, Luo L, Luo W, Xu Y, Yuan C, Zeng B, Dai L (2022) A flame retardant fabric nanocoating based on nanocarbon black particles@polymer composite and its fire-alarm application. Chem Eng J 433:133501. https://doi.org/10.1016/j.cej.2021.133501
141. Xu X, Si Y, Zhao Y, Ke Q, Hu J (2023) Electrospun textile strategies in tendon to bone junction reconstruction. Adv Fiber Mater 5(3):764–790. https://doi.org/10.1007/s42765-022-00233-9
142. Yusuf M, Madhu A (2022) Smart nanotextiles for filtration. In: Smart nanotextiles. Wiley, pp 341–358. https://doi.org/10.1002/9781119654872.ch10
143. Zeng W, Wan T, Gou PP (2013) Ultralong silica nanotubes assembling macroscopic non-woven mat from bacterial cellulose template. Mater Res Innovations 17(sup1):205–209. https://doi.org/10.1179/1432891713Z.000000000216
144. Zhang P, Wang L, Wei H, Wang J (2021) A critical review on effect of nanomaterials on workability and mechanical properties of high-performance concrete. Adv Civil Eng 2021:1–24. https://doi.org/10.1155/2021/8827124
145. Zhang Y, Li T-T, Shiu B-C, Sun F, Ren H-T, Zhang X, Lou C-W, Lin J-H (2021) Eco-friendly versatile protective polyurethane/triclosan coated polylactic acid nonwovens for medical covers application. J Clean Prod 282:124455. https://doi.org/10.1016/j.jclepro.2020.124455
146. Zhang Y, Li T-T, Shiu B-C, Sun F, Ren H-T, Zhang X-F, Lou C-W, Lin J-H (2020) Processing and characterizations of Short fluoroalkyl chain /polyurethane-polylactic acid/low melt polylactic acid Janus nonwoven Medical covers using spray coating. Prog Org Coat 147:105736. https://doi.org/10.1016/j.porgcoat.2020.105736
147. Zhi M, Chen H, Jia X, Zhu X, Feng S, Liu Q (2019) Suppression effect of MoS_2 nanosheets for CO and CO_2 production during combustion of flexible polyurethane foams. In: 2019 9th international conference on fire science and fire protection engineering (ICFSFPE), pp 1–4. https://doi.org/10.1109/ICFSFPE48751.2019.9055827
148. Zhou F, Peng J, Tao Y, Yang L, Yang D, Sacher E (2023) The enhanced durability of AgCu nanoparticle coatings for antibacterial nonwoven air conditioner filters. Molecules 28(14):5446. https://doi.org/10.3390/molecules28145446
149. Zhou J, Yan Y, Zhang Y, Zhang Y, Zhang D, Lin H, Chen Y, Zhao B, Xiong J (2023) Amino-terminated hyperbranched polymer functionalized graphene oxide with in situ trapped silver

nanoparticles for high-performance antibacterial nonwoven fabric. J Appl Polym Sci 140(7). https://doi.org/10.1002/app.53466
150. Zhou X, Zhu Y, Liang J, Yu S (2010) New fabrication and mechanical properties of styrene-butadiene rubber/carbon nanotubes nanocomposite. J Mater Sci Technol 26(12):1127–1132. https://doi.org/10.1016/S1005-0302(11)60012-1

Chapter 14
Nano-finishing in Apparel Textiles

Nagender Singh, Amit Madhu, and Manisha Yadav

1 Introduction

The fusion of science and fashion has resulted in a revolution in apparel technology in modern textiles. The application of nano-finishing to apparel textiles is a cutting-edge field in which the minuscule realm of nanotechnology is utilized to create materials that provide us with clothing and improve our quality of life [1–3]. This chapter takes the reader on a tour of the complex and revolutionary field of nano-finishing in clothing textiles, where minuscule particles can significantly impact the functioning and performance of fabrics.

The limitations of conventional fabrics have been surpassed by the advancement of nanotechnology's application to textiles. These days, clothing materials play a more active role in our everyday lives by offering answers to various complex problems rather than just being passive coverings that cover our bodies. The building blocks of this textile revolution are nanoparticles, which have enormous potential despite their tiny size and many benefits that go well beyond aesthetics [4, 5].

The advent of nano-finishing processes has created clothing resistant to stains, repels water and oil, blocks damaging UV rays, and gives customized thermal comfort. Furthermore, by opening new functional realms, these advancements have transformed clothing from static accessories into dynamic pieces that adjust to the wearer's needs [6, 7].

As we commence this investigation, it is imperative to consider the diverse aspects of nano-finishing in apparel and clothing materials. Beyond the countless ways to improve the quality of clothing, concerns about environmental sustainability and

N. Singh (✉) · A. Madhu
The Technological Institute of Textile and Sciences, Bhiwani, Haryana, India
e-mail: jangra.nagender@gmail.com

M. Yadav
DLC State University of Performing and Visual Arts, Rohtak, Haryana, India

S. Bairagi et al. (eds.), *Nanotechnology in Textile Finishing*, Materials Horizons: From Nature to Nanomaterials, https://doi.org/10.1007/978-981-97-2696-7_14

safety also need to be considered [8]. This chapter thoroughly analyzes the developments, difficulties, and consequences of nano-finishing within this framework in the textile and fashion industries. It is a tale of invention, change, and the marriage of fashion and science that is changing the fabric of our clothes. Consumer expectations in the dynamic and quickly changing fashion world go beyond aesthetics to encompass sustainability and functionality [9]. In the textile business, nano-finishing has become a game-changer, offering creative ways to improve the appearance, comfort, and functionality of apparel materials [10]. This chapter explores nano-finishing's uses, benefits, difficulties, and environmental consequences in apparel textiles.

2 The Fundamentals of Nano-finishing

A novel process known as "nano-finishing" uses nanotechnology, which studies material manipulation at the nanoscale, usually between 1 and 100 nm. Textiles with remarkable qualities can be engineered by utilizing the unique qualities of materials at this size, increasing their functionality, toughness, and sustainability [11–13].

The fundamental principle of nano-finishing is the integration of nanomaterials, including nanoparticles, nanofibers, and nano coatings, into textile fibers or fabrics. These nanomaterials provide textiles with their unique functions. They can be made of metals, polymers, or ceramics [14].

2.1 Nano-finishing Techniques

Nano-finishing employs various techniques to infuse nanomaterials into textiles. Some of the most common methods include:

- **Dip Coating**: Textiles are immersed in a solution containing nanomaterials, allowing the nanomaterials to adhere to the textile's surface, as shown in Fig. 1. The thickness of the deposited films is a crucial factor in the dip-coating process, as it forms the foundation for a range of physical and chemical characteristics and applications. It is well recognized that several factors, such as temperature, composition, coating solution concentration, withdrawal speed, and dip-coating duration, significantly impact the thickness of deposited films. The impact of processing factors on the thickness of deposited coatings has been the subject of several published papers [14–18]. Various dip-coating methods include simple solution dipping, sol–gel-based approaches, vacuum assistance, spin assistance, photo assistance, and multi-layered methods. [19].
- **Spray Coating**: Nanomaterials are sprayed onto the textile, creating a thin, uniform coating. This method is particularly suitable for large-scale production. Analyzing the various process parameters, such as spray rate, atomization air pressure, inlet and exhaust air temperature, inlet and exhaust airflow, nozzle size, and

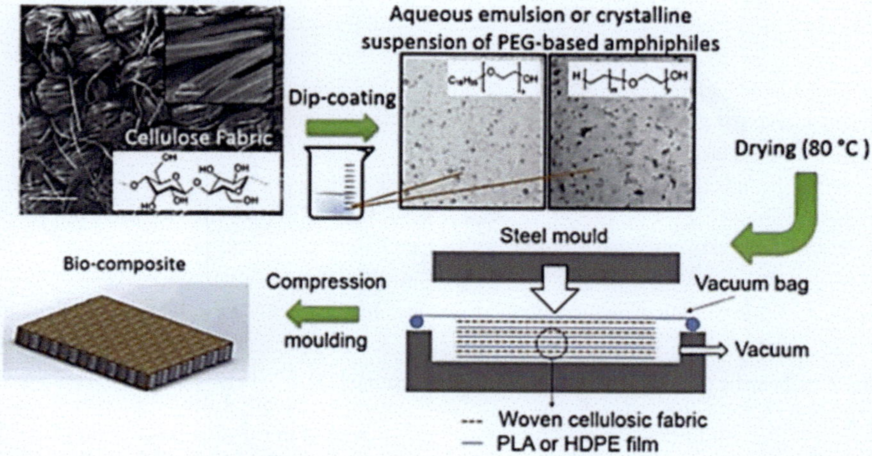

Fig. 1 Process of fabricating a fabric-reinforced bio-composite through compression molding and dip-coating cellulose fabric [20]

nozzle-to-bed distance, is necessary for spray-coating method monitoring [21]. A crucial element that influences coating quality that is influenced by spray pressure and rate is spray droplet size. When the spray rate is increased, the droplet size is bigger, but as the spray pressure rises, it gets smaller. In addition to being adjusted to guarantee the ideal spray droplet size, the spray rate must be calibrated to avoid sticking. The equipment and technique employed during the application process have a significant impact on the coating quality as well [22]. Various spray coating can be employed for the application of nanoparticles on textiles; however, ultrasonic spray coating offers significant improvements in coating efficiency. This new approach coats huge areas without stabilizers or fillers by spraying the loaded suspension onto a substrate without using an ultrasonic nozzle. It is paired with an ultrasonic syringe to maintain an evenly distributed solution throughout the procedure and prevent the flakes or particles from settling. As the suspension is poured into the nozzle, it is sonicated and placed into the syringe. Capillary waves are produced when the nozzle's ultrasonic vibration oscillates the suspension inside. These waves grow in amplitude until they reach the nozzle's atomizing surface (tip), at which point they separate from the surface to create the spray, as shown in Fig. 2 [23, 24].

- **Electrospinning**: This technique involves electrostatic forces to draw nanofibers from a solution or melt, depositing them onto a textile substrate. Electrospinning is employed for adding strength and specific functionalities to textiles. Three parts comprise a standard electrospinning setup: a collecting screen, a capillary tube with a needle, and a high-voltage supply [25, 26]. A typical setup for electrospinning is shown in Fig. 3a, b. Taylor states that an applied voltage of 6 kV is

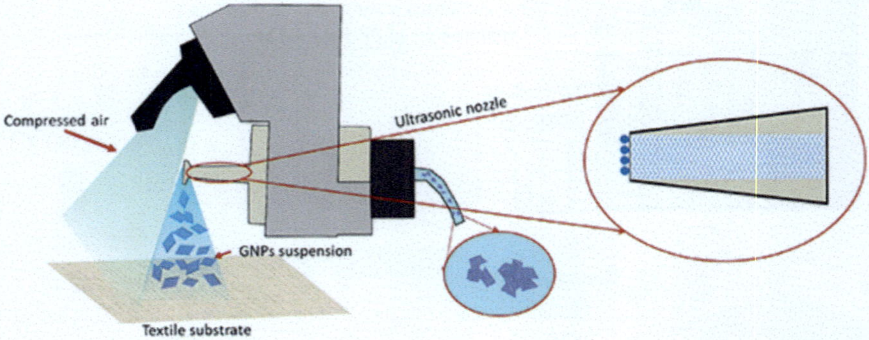

Fig. 2 Schematic depiction of the ultrasonic spray coating [23]

necessary for the electrospinning process to begin [27]. Nevertheless, the electrospinning process can occur at a lower applied voltage if a grounded target is inserted closer to the spinneret [28].

- **In-situ Synthesis**: Nanoparticles can be synthesized directly within the textile matrix, embedding them deeply into the fabric for lasting effects. Various research studies have discussed the in-situ synthesis of nanoparticles on textile substrates. Nina Čuk et al. have synthesized the AgNP synthesis directly (in situ) on cotton fabrics to develop antibacterial and UV-protective fabrics. The raw cotton fabrics were first wrung out on padder rollers (foulard) (Mathis, Switzerland) with a wet pick-up of 100% and then immersed for 60 min in the reducing agent bath (plant extract) at 60 °C. The precursor solution of 0.01 M silver nitrate ($\geq$ 99.8% $AgNO_3$, Sigma-Aldrich, UK) was used for the wringing process. Twenty to one was the mass ratio of the fabric to the mass of the reducing agent bath. Once the samples had been functionalized, they were air-dried at room temperature and washed in cold deionized water. In Fig. 4, the procedure is schematically represented [31].

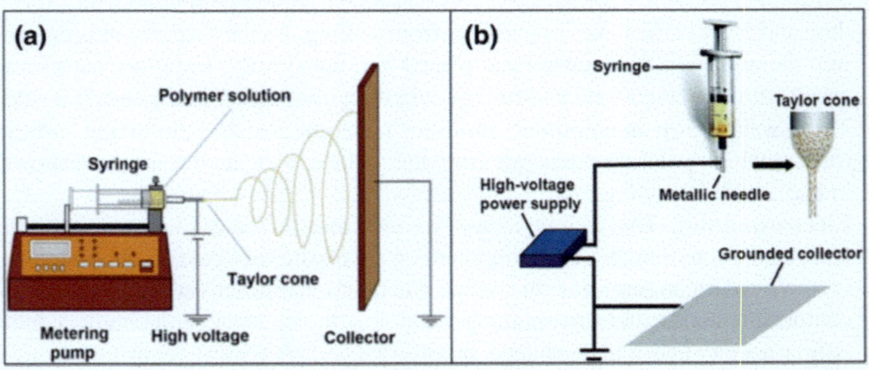

Fig. 3 Diagrammatic representation of the electrospinning apparatus configuration, with a horizontal setup [29] and a vertical setup [30]

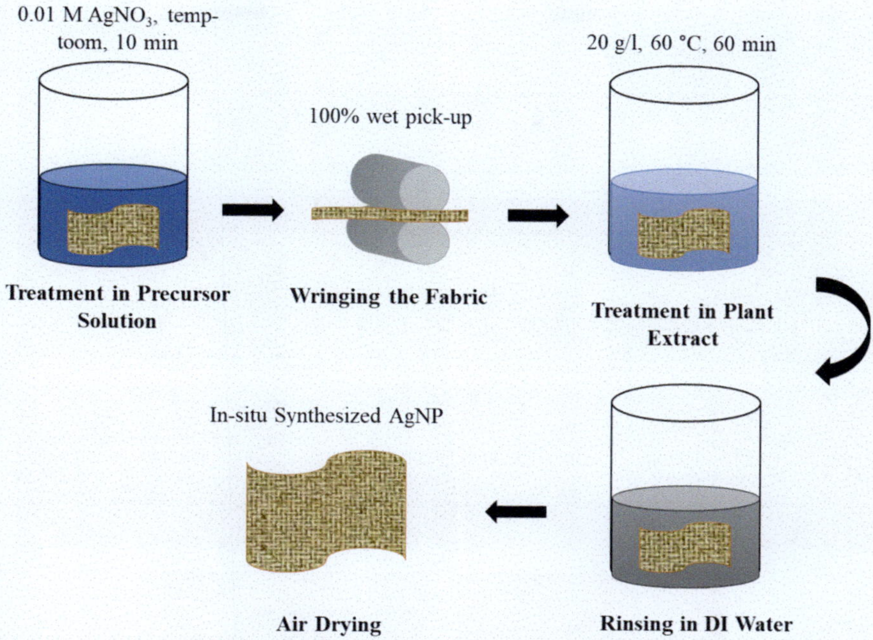

Fig. 4 Diagrammatic representation of silver nanoparticles synthesized in-situ on cotton fabric [31]

3 Applications of Nano-finishing in Apparel Textiles

Nano-finishing in apparel textiles refers to the application of nanotechnology to enhance fabrics' functional and aesthetic properties. This cutting-edge technology has revolutionized the textile industry by offering many applications. One prominent application is the development of water-repellent and stain-resistant fabrics, ensuring durability and ease of maintenance for clothing items. Nano-finishing also enables the creation of self-cleaning textiles, as nanoscale particles inhibit the adhesion of dirt and contaminants. Additionally, antimicrobial nano-finishes have been instrumental in producing textiles with enhanced resistance to bacteria, fungi, and odors, thereby promoting hygiene and comfort [32–34]. Figure 5 shows the applications of nano-finishing in apparel textiles.

3.1 Antimicrobial Properties

The growth of bacteria and fungi is inhibited when antimicrobial nanoparticles are included in textile materials. Undergarments, sportswear, and medical textiles will benefit from this. Incorporating nanoparticles to provide clothing textiles with

Fig. 5 Applications of nano-finishing in apparel textiles

antibacterial qualities entails introducing elements at the nanoscale into the fabric to prevent the growth of microbes [35, 36]. The antibacterial properties of silver nanoparticles (AgNPs) are well established. Adding silver nanoparticles to textiles is a widespread technique. This can be accomplished by several methods, including ion exchange, in-situ synthesis, and using textiles coated with silver nanoparticles.

Hebeish et al. investigated the antibacterial properties of cotton fabric treated with AgNPs against gram-positive and gram-negative bacteria. The result shows that the green synthesized AgNPs can significantly increase the antibacterial activity of cotton fabric [37]. Various studies have been conducted to investigate the antibacterial properties of silver nanoparticles on textiles [38–41]. Applying zinc oxide nanoparticles (ZnONPs) to textiles via dip coating, sol–gel coating, or direct inclusion during fiber formation can all result in antibacterial activity. Shaheen et al. synthesized the ZnONPs in situ onto cotton fabrics, and the resultant offers durable antibacterial and UV protective properties [42]. The intrinsic antibacterial qualities of copper nanoparticles allow for their application on textiles through methods such as coating or electrospinning [43]. When exposed to ultraviolet radiation, titanium dioxide nanoparticles (TiO_2NPs) release reactive oxygen species with antibacterial properties. The sol–gel coating is one technique that can be used to coat textiles with these nanoparticles [44]. Through techniques like encapsulation or blending during fiber manufacture, textiles can incorporate nanoparticles derived from polymers having intrinsic antibacterial characteristics, such as chitosan or polyhexamethylene biguanide [45].

3.2 *Stain and Water Resistance and Self-cleaning*

Textiles with high hydrophobicity can be enhanced by nanofinishing, which stops liquids from being absorbed. This is especially useful for outdoor and sportswear clothing. By adding nanomaterials that form a protective barrier on the fabric surface, clothing fabrics can be made water and stain-resistant using nanoparticles [46]. Fluoropolymer-based nanoparticles can offer exceptional stain and water resistance, including polyvinylidene fluoride (PVDF) or polytetrafluoroethylene (PTFE). Applications for these nanoparticles include spray coating, dip coating, and cushioning [47].

Textiles can have a hydrophobic surface treated with silica nanoparticles. Sol–gel procedures can be used to apply silica-based treatments, where the nanoparticles thin out and coat the fabric's surface [48]. Coatings can be developed to make fabrics extremely hydrophobic and self-cleaning by emulating the micro/nanostructure of lotus leaves. Techniques like dip coating and electrospinning can be used to accomplish this. Textiles can be made more stained and water-resistant by using polyurethane nanoparticles. They can be applied during textile processing using coating or finishing techniques [49]. Nanoparticles can be used as a vehicle for hydrophobic substances, such as silicone-based compounds or waxes.

After that, textiles can be treated with these nano-encapsulated compounds using standard finishing techniques [50]. Graphene nanoparticles have unique qualities that can help them resist stains and water. During the fiber-making process, these nanoparticles can be administered via direct integration, spray coating, or dip coating [51]. Textiles can be treated with hydrophobic nanoemulsions to provide a stain- and water-resistant barrier. Processes for finishing textiles frequently employ this technique [52]. The regulated release of water and stain-resistant chemicals into the textile surface is made possible by the ability of cyclodextrin nanoparticles to encapsulate hydrophobic compounds [53]. The layer-by-layer deposition process entails applying alternating layers of nanoparticles with opposing charges to the fabric's surface. The protective layer that is produced by this stacking improves stain and water resistance [54]. On the textile surface, hydrophobic molecules or nanoparticles can form self-assembled monolayers (SAMs) that provide a coating resistant to stains and water. This approach is frequently combined with other methods of deposition [55].

3.3 UV Protection

A textile's capacity to block damaging UV radiation can be improved by nanocoatings, which makes it essential for swimwear and outdoor apparel. Adding elements that can absorb, reflect, or block ultraviolet (UV) radiation is how nanoparticles are used to provide apparel textiles with UV protection [56, 57]. The unique ability of titanium dioxide nanoparticles (TiO_2 NPs) to absorb and scatter ultraviolet light is well recognized [58]. It is possible to insert these nanoparticles into textiles by dip coating, sol–gel coating, or direct inclusion during the fiber-making process [59, 60]. According to M Rashid et al., the photocatalytic activity mechanism of TiO_2 is based on its electronic structure and can be broken down into six steps (Fig. 6): surface photooxidation and photoreduction reactions, charge separation and transfer to the particle surface, photoexcitation, and recombination of bulk and surface charge. Under UV light irradiation, TiO_2 absorbs photons with energies equivalent to or greater than the band gap energy (Eg). This excites the electrons in the valence band (VB) to the vacant conduction band (CB), creating holes in the VB [61].

The UV-blocking abilities of zinc oxide nanoparticles (ZnONPs) are quite good. They can be applied during textile manufacturing using techniques like direct blending, dip coating, and sol–gel coating [62]. ZnO exhibits excellent chemical stability when exposed to UV radiation and has a high efficacy in blocking UV-A and UV-B radiation, contributing to its UV protective qualities [63]. According to the suggested mechanism of UV protection in ZnO, UV radiation is absorbed, refracted, and/or scattered by the material's high refractive index, which stops UV rays from diffusing or directly penetrating textile material and reaching the skin, as shown in Fig. 7. Additionally, studies on several textile substrates have been presented [64, 65].

Cerium oxide nanoparticles (CeO_2NPs) have also been explored in various research studies for their potential to present UV protection in apparel textiles [66]. Rohit Tripathi and co-authors have modified the linen fabric using chitosan and in-situ

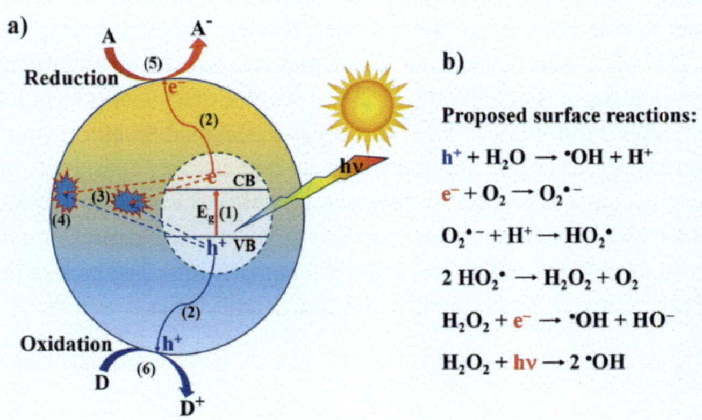

Fig. 6 Mechanism of TiO_2 photocatalytic activity [61]

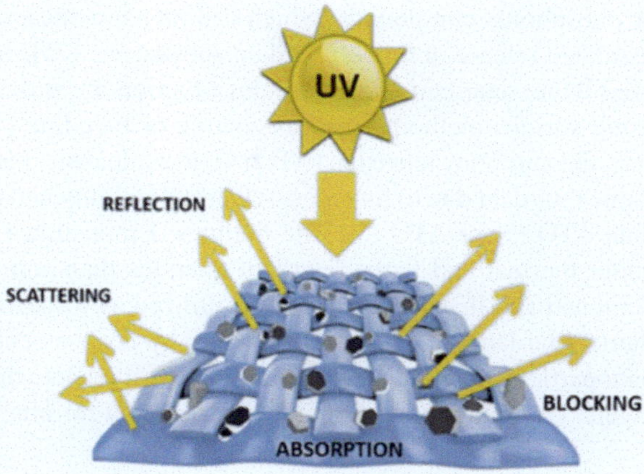

Fig. 7 Illustration of ZnO nanoparticle protection against UV radiation on textile fibers

synthesized CeO_2 nanoparticles. The results reveal excellent UV protection of linen fabric and functional properties retained even after repeated usage and washing [67]. Moreover, nanoparticles developed from UV-absorbing polymers can be another possibility to impart UV protection in textiles [68].

3.4 Flame Retardant

The process of adding elements that can prevent or suppress the ignition and spread of flames to clothing textiles using nanoparticles is known as flame retardancy. The incorporation of nanoclays, such as montmorillonite, into textiles can improve their ability to withstand flames. The barrier that the clay nanoparticles create can lessen the fabric's flammability. They can be used in techniques like melt mixing and solution casting [69]. Phosphorus-containing nanoparticles, such as phosphorus-containing polymers or metal phosphinates, are effective flame retardants [70]. Apart from that, silica nanoparticles can be employed as flame retardants by developing a shield on the fabric's surface. F. Salaün et al. have investigated a novel flame-retardant (FR) method for PET textiles developed by blending zinc phosphinate (Exolit OP950), nano-silica particles, and PET. The mixture was then melt-spun. The wettability, dispersibility, and thermal characteristics of the two types of nano-silica fillers were examined to ascertain the impact of hydrophilicity level on the PET matrix. From cone calorimeter investigations, it was found that the addition of nanoparticles nearly never improved the PET fire reaction [71]. Carbon nanotubes (CNTs) have also shown promise in improving flame retardancy [72]. Cotton fabric, for example, has carbon nanotubes implanted in it, making it more thermally stable and flame-retardant than raw textiles [73]. According to Parvinzadeh Gashti et al., the direct use and stabilization of CNT led to the improvement of cotton fabrics' thermal qualities as well as the achievement of highly effective flame-retardant finishing [74]. Textiles can be coated with nanocoating using flame-retardant additives like metal oxides or phosphorus compounds, which can be nano-encapsulated. This method offers a regulated release of flame-retardant substances. [75]. Antimony trioxide is a widely used flame retardant. Nanoparticles of antimony trioxide can be applied to textiles using various methods, such as coating or blending [76]. Hyelim Kim et al. have selected antimony tetroxide (ATO (4)) to evaluate its potential usage as an alternative flame retardant due to its superior thermal stability and resilience to acids and alkalis than ATO (3) or ATO (5). The findings demonstrated that antimony tetroxide had better thermal and flame-retardant qualities than antimony trioxide, and they also demonstrated that ATO (4) may be utilized as a substitute for ATO (3) in terms of flame retardancy [77]. Using the layer-by-layer assembly process, flame-retardant nanoparticles and polyelectrolytes are deposited onto the fabric surface in alternating layers. Layering improves flame resistance by forming a protective layer [78].

3.5 Thermoregulation

Textile clothing that adjusts to the wearer's body temperature can be made with nanomaterials. The ability to heat or cool can be provided by phase-change materials and nanoparticles, which makes clothes more comfortable. Using nanoparticles to give apparel textiles, thermoregulation qualities entail adding substances that can improve the fabric's capacity to control temperature, resulting in comfort under various environmental circumstances [79, 80]. Textile materials can incorporate PCM nanoparticles, which assist in controlling body temperature. PCMs are the basis for cooling clothing because they can absorb and store thermal energy as latent heat within a specific temperature range.

Thermal energy transfer occurs when a substance undergoes a phase transition [79]. In several research, scientists have looked at the various applications of nanoPCMs, such as their use in solar water systems, air conditioning, building materials, heat exchangers, heat sinks, heat pipes, textiles, and helmets. The thermal conductivity properties of carbon nanotubes (CNTs) are exceptional. By adding CNTs to textiles, the materials' capacity for heat transfer is improved, resulting in effective cooling. Techniques like electrospinning and mixing can be used to accomplish this [81, 82]. There is an intense heat conductivity in the graphene nanoparticles. When incorporated into textiles, they can improve the fabric's ability to regulate temperature and promote effective heat transfer. During the fiber manufacturing process, methods like direct integration or dip coating might be applied [7, 83]. Developing fabrics that respond to temperature changes can be developed using nanoparticles composed of thermoresponsive polymers. Variations in temperature cause these polymers to change structurally, which affects the fabric's ability to transport heat [84]. Hydrogel nanoparticles can also help with cooling and moisture management by absorbing and releasing water [85]. Inorganic nanoparticles with high thermal conductivity, like aluminum oxide, can be integrated into textiles to enhance heat transfer [86].

3.6 Color Enhancement

The brightness and durability of fabric colors can be improved by nanoparticles, preserving the aesthetic appeal of garments. In the textile technology world, applying nanoparticles for color enhancement in textile dyes is a cutting-edge and novel strategy. Increasing color brightness and intensity in textile dyes is an important nanoparticle use. Researchers and textile engineers can obtain a more uniform and vivid coloration of fabrics by adding nanoparticles to the dye mix. This is especially helpful for creating vibrant and durable colors, which is an important feature in the textile business where durability and aesthetics are vital. One intriguing approach to obtaining better color characteristics, increased efficiency, and novel textile features is the integration of nanoparticles into textile dyes. Continued research and development in this area will likely result in the introduction of cutting-edge, environmentally friendly textile dyeing techniques with enhanced efficiency [87–89]. Metal nanoparticles, such as gold or silver, can have unique optical qualities that improve color. For instance, reddish coloration can be achieved using gold nanoparticles [90], and a yellowish look can be achieved with silver nanoparticles [91]. During the fiber-making process, these nanoparticles can be administered using coating or direct blending methods. Quantum dots exhibit the size-dependent color characteristics of semiconductor nanoparticles. They can be designed to release colors when lit up [92, 93]. Nano-pigments are microscopic particles that can improve the vibrancy and intensity of color. These diversely constituted nanoparticles can be used on textiles for printing or dying procedures [94]. Various studies have suggested using TiO_2 nanoparticles for color enhancement of dyed fabric. Through their ability to scatter and reflect light, titanium dioxide nanoparticles (TiO_2NPs) can affect how color is perceived [95, 96].

3.7 Energy Storage

The application of materials that can store and release energy—often through nanotechnology—enables fabrics used in clothing to have energy-storage qualities. Textiles can be more conducive for energy storage by incorporating carbon nanotubes (CNTs), which have outstanding electrical conductivity. Supercapacitors, which store and release electrical energy, can be created using fabrics incorporating carbon nanotubes. Techniques like electrospinning and blending can be used to apply CNTs [97, 98]. Graphene nanoplatelets exhibit excellent electrical conductivity and a large surface area, making them a viable material for energy storage. Textiles can have their capacitance increased by adding graphene nanoplatelets [99]. Textiles can also be given energy-storage capabilities by applying nanoparticles derived from conductive polymers like polyaniline or polypyrrole [100]. Electrodes for textile-based energy-storage devices can be made with metal oxide nanoparticles, such as titanium or manganese oxide [101]. Carbon-based nanomaterials and metal oxides are two examples of materials that can be used to create nanostructured supercapacitors for textile engineering. These nanostructured components can be developed via layer-by-layer deposition or electrospinning [102]. Textiles can also be made flexible and wearable by incorporating nanoparticles, such as lithium cobalt oxide or other battery components. One can employ methods such as direct blending or layer-by-layer construction [103]. Energy-harvesting textiles can be developed with nanoparticles, such as triboelectric or piezoelectric nanomaterials. These substances change frictional or motion-related mechanical energy into electrical energy [104].

3.8 Sensing and Monitoring

A novel nexus between wearable technology and nanotechnology is represented by the sensing and monitoring of clothing with nanoparticles. Through the incorporation of nanoparticles into the fabric, these smart textiles can function as sophisticated sensors for an array of uses. Wearables that can monitor the wearer's health, the environment, or even certain compounds in the surroundings can be made possible by using nanoparticles with sensing capabilities, such as those that respond to physiological factors or environmental changes. As an illustration, nanoparticles can be engineered to alter color in response to changes in humidity, temperature, or the presence of specific compounds. This breakthrough creates opportunities for safety, environmental sensing, and real-time health monitoring applications. The use of nanoparticles in sensing and monitoring clothing highlights the development of textiles into responsive, intelligent interfaces for a networked and data-driven society and the possibility for proactive, individualized health care [105–108].

3.9 Miscellaneous Applications

- **Infrared (IR) camouflage**: The traditional use of camouflage clothing in military environments has given way to new uses in various industries, such as outdoor recreation and wildlife photography. Nanotechnology has enhanced textile camouflage, which has become a potent tool. As a means of hiding from thermal imaging equipment, infrared (IR) camouflage is essential to military strategy [109]. A new era of improvements in infrared camouflage fabrics has been brought about by the integration of nanoparticles into textile engineering. Nanoparticles with special optical characteristics, such as carbon nanotubes, graphene, and nanocomposites, are used to absorb and dissipate infrared radiation. Samolov et al. have looked at the possibilities of enhancing the camouflage qualities of regular cotton fabric by using fullerene-like tungsten disulfide nanoparticles, PVB/IF-WS2, and poly(vinyl butyral), or PVB, in the camouflage print. The results obtained indicate that the PVB/IF-WS2 impregnation system enhanced the materials' camouflage qualities, meaning that IF-WS2 improved the fabric's spectrophotometric characteristics [110].

Individuals using such textiles are less noticeable to thermal imaging sensors because these materials efficiently control the heat signatures released by the human body. For textile surfaces to have the best IR absorption properties, nanoparticles make precise surface engineering possible. This precision makes effective hiding in various climates and terrains possible, enabling adaptation to changing environmental conditions. To make the clothing blend in with the surroundings, quantum dots, for example, help match color. Military technology has advanced significantly with the use of nanoparticles in the creation of infrared camouflage materials. The precision engineering of nanomaterials allows for superior IR absorption, adaptability, and multifunctionality in textiles, offering enhanced concealment and protection for military personnel [108, 111]. Siadat and Mokhtari coated nano ZrO_2 and magnesium oxide (MgO) on cotton/nylon fabrics printed with patterns of jungles and deserts [112]. The concentrations of metal oxide nanoparticles in the padding solution were 0.5, 1, 1.5, and 2% (w/v (%)), together with 6% (w/v (%)) of citric acid crosslinking agent. According to this study, 1.5% ZrO_2 and MgO nanoparticles successfully lowered reflectance levels over NIR and SWIR areas to mimic a jungle environment. The addition of citric acid cross-linker further reduced reflectance. The polyethylene/nanoparticle composite was knitted into a fabric after being extruded as a multi-filament yarn. The fabrics exhibited approximately 80% infrared transparency (Figs. 8 and 9), a passive cooling effect of 1.6–1.8 °C, and satisfactory tensile strength and washability [113]. The future of military and defense technologies is bright, as research on nanoparticle-enhanced infrared camouflage materials shows promise for scalable and sustainable applications.

- **Antioxidant**: An innovative approach to practical and health-conscious textiles is represented by antioxidant clothing enhanced by the transforming potential of nanoparticles [114, 115]. These cutting-edge clothes actively fight oxidative stress by using naturally occurring antioxidant nanoparticles like silver and zinc oxide. As potent free radical scavengers, the nanoparticles shield the fabric against environmental deterioration and may benefit the wearer's health [116]. This innovative technology creates a new dimension to personal well-being by continuously neutralizing damaging free radicals while extending the garment's lifespan by protecting it from external stressors. Antioxidant clothing is becoming increasingly popular, especially in sportswear and outdoor apparel. It represents a promising future where functional textiles provide comfort, durability, and holistic well-being for people. They exemplify how textile engineering and health-conscious design combine [117, 118].
- **Self-healing**: A new era of durable and long-lasting apparel textiles is being ushered in by the incorporation of nanoparticles. In this innovative technique, nanoparticles like polymers or nanocapsules are selectively inserted within the fabric to facilitate self-healing. These nanoparticles are activated when the textile

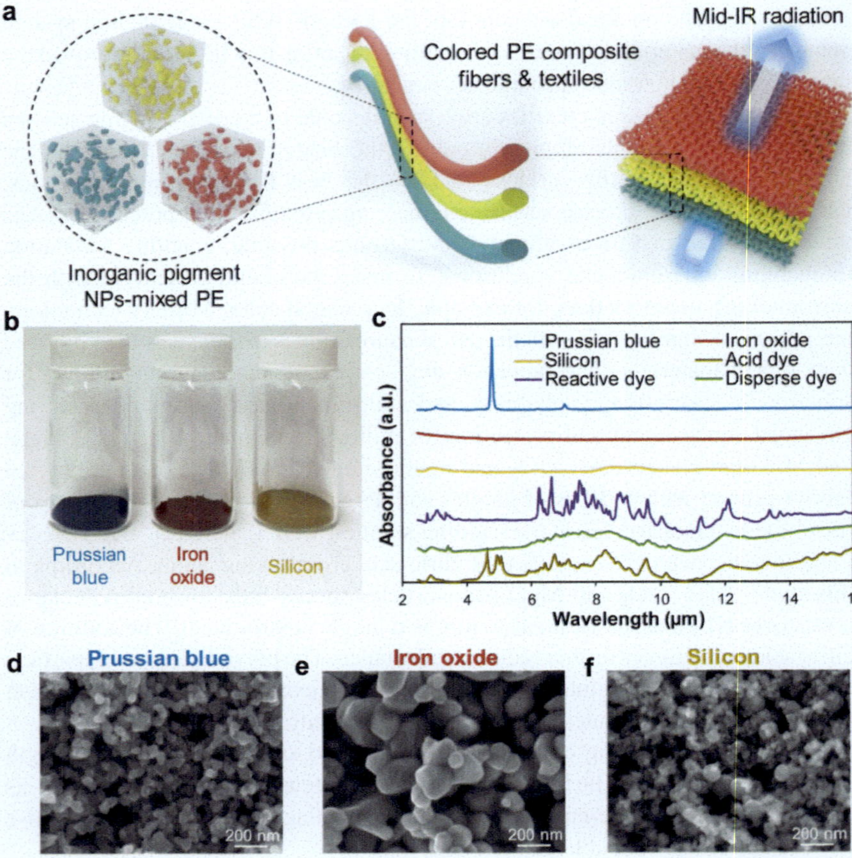

Fig. 8 Schematic representation of coloration design for infrared-transparent polyethylene textiles [112]

sustains damage, from abrasions, cuts, or other types of wear and tear. This initiates a self-repair mechanism. In response to mechanical stress or environmental cues, the nanoparticles release healing chemicals that repair the fabric's damaged areas. This ground-breaking invention lowers the frequency of replacements while extending the clothing's lifespan and durability. Self-healing fabrics for apparel result from an amazing collaboration between textile engineering and nanotechnology. This technology opens the door to clothing that can self-heal and adapt, giving customers a long-term, sustainable solution to the problems associated with daily wear [6, 119, 120]. By incorporating paraffin wax and PDMS into polyester textiles, Zhao et al. created self-healing superhydrophobic textiles for water–oil separation [121]. Chen et al. used dip coating of ODA-loaded nickel hydroxide ($Ni(OH)_2$) micro-nanoparticles to create cotton textiles that are superhydrophobic and capable of self-cleaning and self-healing [122]. A self-healing silicon rubber

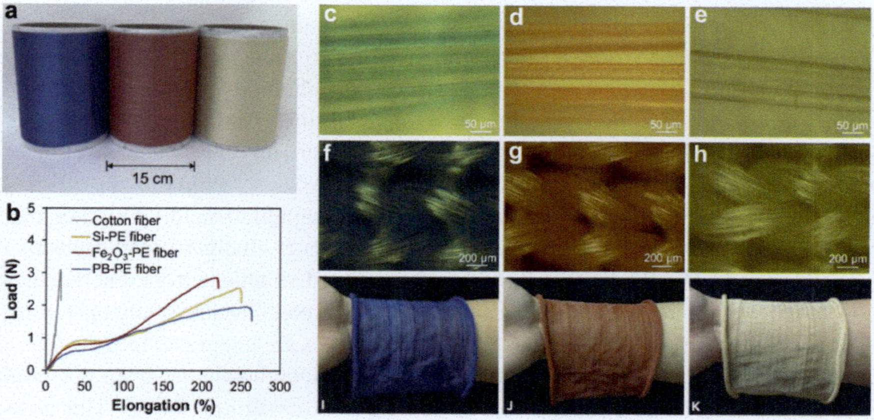

Fig. 9 Illustration of the development of colored polyethylene fibers and textiles [112]

copolymer (SR-Co), developed by attaching pyridine groups to polysiloxane and then reversibly coordinating with cobalt ions, was added to PET fabrics by Liu et al. [123]. Iron oxide nanoparticles (IONPs) treated with oleic acid (OA) and ODA nanocapsules modified with PDA were used by Liu et al. to functionalize a cotton fabric. Superhydrophobicity was imparted through the hierarchical structuring of the PDA@ODA–Fe_3O_4 (polydopamine-coated octadecyl amine-modified iron oxide) nanocapsules, in combination with the natural microscale roughness of the fabric [124]. To give mechano-responsive superhydrophobicity to wearable products, the same team developed a microemulsion of PDA nanocapsules loaded with hydrophobic chemicals ODA and octadecanethiol (ODT) as a coating for polyester and cotton fabrics [125, 126].

Chen et al. coated cotton, glass, aluminum sheet, paper, and wood with FAS-loaded pH-responsive capsules combined with SiO_2 and TiO_2 nanoparticles. While the TiO_2 nanoparticles enabled photocatalytic destruction of organic pollutants, they also conferred self-cleaning characteristics. The FAS-releasing particles demonstrated self-healing behavior [127]. By sequentially coating polyester fabrics with fluoroalkyl-modified silica nanoparticles (FS-NPs) and fluorinated-decyl polyhedral oligomeric silsesquioxane (FD-POSS), Wang et al. presented an alternate technique for creating self-healing fabrics. In addition to plasma treatment, the surface roughness that the NPs added greatly improved the fabric's liquid-repellent properties and allowed it to self-heal from abrasion damage [128]. The method for growing zirconium phosphate (ZrP) nanoplates in situ on polyester-cotton textiles was demonstrated by Zeng et al. [129]. The incorporation of ODA into the hierarchical ZrP structure gave the material hydrophobicity that could mend itself. To provide self-healing superamphiphobicity against low surface tension liquids (< 34 mN/m), Guo et al. [130] used SiO_2 nanoparticles with FOTS. Song et al. produced self-healing pressure sensors that repaired mechanical damage when exposed to UV light in the sun by dip-coating PU and silver nanowires (AgNWs) onto wool yarn [131]. Wu et al. exposed an F-POSS/silver nanoparticles (AgNPs)/PEI-coated cotton fabric to moderate relative humidity (RH ~ 55%) for 30 min to recover the superhydrophobicity of the material. The prolonged migration time required for F-POSS to resurface from the interior of the fibers led to longer recovery times (up to 10 h after 16 cycles) when the number of damage-repair cycles increased [119].

- **Electromagnetic shielding**: Textiles for electromagnetic shielding clothes that have been strengthened with nanoparticles are a game-changer in the protective clothing industry. Because of their small size and distinct electromagnetic characteristics, nanoparticles are effective agents for protecting fabrics against unintended radiation exposure. When these nanoparticles are integrated into the fabric matrix, they function as powerful barriers that protect the wearer from potential injury by absorbing, reflecting, and dispersing electromagnetic waves. This technique is very important in places like hospitals, labs, or even regular metropolitan settings where there are a lot of electronic gadgets and electromagnetic fields [132–134]. In addition to providing increased protection, nanoparticle-infused textiles retain the comfort and flexibility of regular apparel, allowing the wearer to continue being mobile and comfortable. The application of nanoparticles in electromagnetic shielding textiles for clothing heralds a hopeful new era in which wearable technology and safety are seamlessly integrated, providing a valuable and adaptable solution for people in various locations and sectors [135, 136].
- **Anti-wrinkle**: The textile industry has undergone a paradigm change with the integration of nanotechnology, resulting in a dramatic metamorphosis of wrinkle-resistant products. With its ability to regulate materials at a scale never before possible, nanotechnology opens up new possibilities for improving the functionality and performance of textiles. Researchers and producers have turned to the accuracy and adaptability of nanoparticles in their quest to create textiles that resist wrinkles. Nanoencapsulation is one of the main strategies for using nanotechnology to create wrinkle resistance. This method entails packing active anti-wrinkle substances into nanoparticles. As carriers, these nanoparticles offer a defense shield surrounding the agents, anything from state-of-the-art polymers to conventional fabric treatments. One of the benefits of nanoencapsulation is that it can manage how these compounds are released onto the fabric surface, so the effect is gradual and well-maintained [137–139]. This helps to prolong the anti-wrinkle effects and saves on the number of reapplications required by making the treatment more effective overall. Surface modification is another area where nanotechnology is essential to create wrinkle-resistant fabrics. Textiles can be coated with a protective nanocoating by putting a thin layer of nanoparticles on their surface. By acting as a barrier, this coating stops the pressures that cause wrinkling and creasing in fibers. Because of the accuracy with which nanotechnology can create this protective layer, the breathability and organic feel of the cloth are guaranteed to remain intact. The product is a fabric that is still aesthetically pleasing but has a practical advantage over wrinkling. A creative way to achieve wrinkle resistance is to add nanoparticles into the fibers' structure during manufacturing. Improved mechanical qualities are seen in nanocomposite fibers,

which are created during spinning by integrating nanoparticles with the underlying materials. This increases the fibers' resistance to the physical strain that causes wrinkles and adds to the fabric's overall durability and strength. Beyond only being wrinkle-resistant, nanocomposite fibers can be used to create fabrics that perform better in various ways. There are several ways to apply nanoparticles to fabrics, from spraying and padding to electrospinning. Each technique is selected based on the nanoparticles' unique characteristics and the fabric's intended result. For example, the process of electrospinning yields nanofibers, which, when added to textiles, improve their mechanical strength and anti-wrinkle characteristics even further. The flexibility of its application methods demonstrates the adaptation of nanotechnology to many textile manufacturing processes. As the textile sector adopts nanotechnology to create materials resistant to wrinkles, thorough testing and assessment become essential. Manufacturers must guarantee that their anti-wrinkle finishes are efficacious and long-lasting, holding their characteristics even after frequent washings and use [33, 139–143].

4 Advantages of Nano-finishing for Apparel Textiles

The apparel industry has revolutionized because of nano-finishing, a cutting-edge textile engineering technology that offers numerous benefits beyond conventional finishing techniques. The improved functionality that nanofinishing gives textiles is one of the main advantages. Despite dimensions at the nanoscale, nanoparticles can be evenly dispersed across the fabric's surface to provide a barrier against UV rays, stains, and water. As a result, textiles have improved stain resistance, UV protection, and water repellency, all of which contribute to the lifetime and durability of clothing [50].

Additionally, the environmental issues raised by traditional textile treatments are addressed by nano-finishing. Conventional finishing techniques frequently utilize chemicals, which can potentially be detrimental to the environment and human health. On the other hand, the environmental impact is generally more negligible when using nano-finishing procedures because they usually require fewer chemicals in lower quantities. Furthermore, nano-finishing accuracy enables focused application, reducing waste and guaranteeing the best use of available resources. Given the growing emphasis on sustainability in the textile sector, nano-finishing's environmentally benign characteristics make it a forward-thinking solution [144, 145].

One other important factor that nanofinishing improves is the breathability of textiles. Improved breathability is achieved in textiles by controlling the permeability of air and moisture vapor by incorporating nanoparticle-sized particles. This is especially helpful for sportswear and activewear, where moisture management and comfort are top priorities. Effectively wicking away moisture, fabrics with nano-finishes keep the wearer dry and comfortable while engaging in physical activity. Because nano-finished textiles are more breathable, they are a better option for customers who value comfort and practicality in their everyday clothing [146, 147].

Moreover, textiles could become surfaces that clean themselves with the help of nano-finishing. In the presence of light, organic materials such as dirt and stains can be broken down more quickly by nanoparticles like titanium dioxide. Its ability to clean itself lessens the need for frequent washing and helps preserve water supplies. Certain nanoparticles' antibacterial qualities also prevent the growth of germs that cause odors on textiles, enhancing hygiene and prolonging the freshness of clothing between washes. Enhancing textile aesthetics is also greatly aided by nano-finishing. Vibrant and durable graphics are produced by the technology's exact control over hue and print application. Nano-pigments can be designed to stick to fibers more securely, preventing fading and preserving the textile's aesthetic appeal over time. This helps to address the changing needs of fashion-conscious consumers by producing high-quality, aesthetically beautiful clothing [32, 148, 149].

5 Challenges and Considerations

Even though nano-finishing has many advantages, specific essential issues must be considered.

- *Environmental Impact*: Nanomaterials can be energy-intensive to produce and dispose of, and their release into the environment may have unforeseen effects. It is imperative to adopt sustainable procedures and manage nanomaterials responsibly.
- *Regulatory Compliance*: Manufacturers of nano-finishing technologies must make sure they abide by safety and environmental requirements, as they may be subject to restrictions.
- *Cost Implications*: Adding nanomaterials to textiles may raise production costs, affecting how reasonably priced nano-finished clothing is.
- *Consumer Awareness*: For the market to accept nano-finished textiles, customers must be aware of their advantages and associated restrictions.

6 The Future of Nano-finishing in Apparel Textiles

As the textile industry progressively adopts high-performance and sustainable solutions, the future of nano-finishing in clothing textiles is promising. It is projected that nano-finishing technologies will be crucial in influencing future textiles, solving current issues, and satisfying changing customer demands. Growing environmental awareness will probably lead to more developments in eco-friendly formulas for nano-finishing, which will lessen the environmental impact of textile treatments. Furthermore, it is anticipated that introducing smart textiles into the market, made possible by nanotechnology, will completely change how functional clothing may be. This includes the development of textiles that respond to their environment, improving user experience through features like adaptive heat regulation and self-healing capabilities. The customization of nano-finished textiles for uses, such as space exploration suits or medical textiles, is expected to increase in popularity as research and development in nanotechnology advance. Future developments in nano-finishing for apparel textiles have the potential to improve textile performance and sustainability while also promoting creativity and innovation in the fast-paced fields of textile engineering and fashion.

7 Conclusion

The apparel textile business has undergone a revolution thanks to nano-finishing, which has opened new and exciting possibilities for producing clothes with remarkable performance and attractive appearance. Its benefits include increased sustainability and comfort, and its applications range from improving stain resistance to UV protection. The future of fashion will be significantly influenced by the responsible incorporation of nano-finishing technologies as the textile industry develops. In the end, nano-finishing makes it possible to create stylish, long-lasting, practical, and environmentally beneficial clothing. Furthermore, nano-finishing garment textiles offer a variety of transformative benefits. With its improved breathability, self-cleaning qualities, and increased utility and sustainability, nano-finishing has established itself as a leader in textile innovation. Nano-finishing technologies are set to redefine performance, comfort, and environmental responsibility requirements in manufacturing apparel textiles as the textile industry continues to change.

References

1. Morshed AMA, Ikbal MH, Hasan SMK (2010) Impact of nanotechnology in the arena of textile apparel finishing. Bangladesh Textile Today 3(3):53–55
2. Paramsothy M (2021) Nanotechnology in clothing and fabrics. Nanomaterials 12(1):67
3. Tamulevičius T (2022) Nanotechnologies in textiles. Materials 15(4):1466
4. Rahman M, Islam KS, Dip TM, Chowdhury MFM, Debnath SR, Hasan S, Houshyar S (2023) A review on nanomaterial-based additive manufacturing: dynamics in properties, prospects, and challenges. Progr Addit Manuf, 1–28
5. Saleem H, Zaidi SJ (2020) Sustainable use of nanomaterials in textiles and their environmental impact. Materials 13(22):5134
6. Shaheen TI (2022) Nanotechnology for modern textiles: highlights on smart applications. J Text Inst 113(10):2274–2284
7. Shah MA, Pirzada BM, Price G, Shibiru AL, Qurashi A (2022) Applications of nanotechnology in smart textile industry: a critical review. J Adv Res 38:55–75
8. Periyasamy AP, Militky J, Sachinandham A, Duraisamy G (2020) Nanotechnology in textile finishing: recent developments. In: Handbook of nanomaterials and nanocomposites for energy and environmental applications, pp 1–31
9. Fletcher K (2012) Durability, fashion, sustainability: the processes and practices of use. Fash Pract 4(2):221–238

10. Syduzzaman M, Hassan A, Anik HR, Akter M, Islam MR (2023) Nanotechnology for high-performance textiles: a promising frontier for innovation. ChemNanoMat 9(9):e202300205
11. Sawhney APS, Condon B, Singh KV, Pang SS, Li G, Hui D (2008) Modern applications of nanotechnology in textiles. Text Res J 78(8):731–739
12. Saleh TA, Gupta VK (2016) Nanomaterial and polymer membranes: synthesis, characterization, and applications. Elsevier
13. Dolez PI (ed) (2015) Nanoengineering: global approaches to health and safety issues. Elsevier
14. Rivero PJ, Urrutia A, Goicoechea J, Arregui FJ (2015) Nanomaterials for functional textiles and fibers. Nanoscale Res Lett 10:1–22
15. Tang X, Yan X (2017) Dip-coating for fibrous materials: mechanism, methods and applications. J Sol-Gel Sci Technol 81:378–404
16. Lii DF, Huang JL, Tsui LJ, Lee SM (2002) Formation of BN films on carbon fibers by dip-coating. Surf Coat Technol 150(2–3):269–276
17. Wu YC, Parola S, Marty O, Mugnier J (2004) Elaboration, structural characterization and optical properties of the yttrium alkoxide derived Y_2O_3 planar optical waveguides. Opt Mater 27(1):21–27
18. Hynek J, Kalousek V, Žouželka R, Bezdička P, Dzik P, Rathouský J, Demel J, Lang K (2014) High photocatalytic activity of transparent films composed of ZnO nanosheets. Langmuir 30(1):380–386
19. Ye Z, Li S, Zhao S, Deng L, Zhang J, Dong A (2021) Textile coatings configured by double-nanoparticles to optimally couple superhydrophobic and antibacterial properties. Chem Eng J 420:127680
20. Church JS, Voda AS, Sutti A, George J, Fox BL, Magniez K (2015) A simple and effective method to ameliorate the interfacial properties of cellulosic fibre based bio-composites using poly (ethylene glycol) based amphiphiles. Eur Polymer J 64:70–78
21. Agrawal AM, Pandey P (2015) Scale up of pan coating process using quality by design principles. J Pharm Sci 104(11):3589–3611
22. Trabucco S, Ortelli S, Del Secco B, Zanoni I, Belosi F, Ravegnani F, Nicosia A, Blosi M, Costa AL (2021) Monitoring and optimisation of ag nanoparticle spray-coating on textiles. Nanomaterials 11(12):3165
23. Sadanandan KS, Bacon A, Shin DW, Alkhalifa SF, Russo S, Craciun MF, Neves AI (2020) Graphene coated fabrics by ultrasonic spray coating for wearable electronics and smart textiles. J Phys Mater 4(1):014004
24. Shin DW, Barnes MD, Walsh K, Dimov D, Tian P, Neves AI, Wright CD, Yu SM, Yoo J-B, Russo S, Craciun MF (2018) A new facile route to flexible and semi-transparent electrodes based on water exfoliated graphene and their single-electrode triboelectric nanogenerator. Adv Mater 30(39):1802953
25. Islam MS, Ang BC, Andriyana A, Afifi AM (2019) A review on fabrication of nanofibers via electrospinning and their applications. SN Appl Sci 1:1–16
26. Jaeger R, Schönherr H, Vancso GJ (1996) Chain packing in electro-spun poly (ethylene oxide) visualized by atomic force microscopy. Macromolecules 29(23):7634–7636
27. Taylor GI (1964) Disintegration of water drops in an electric field. Proc R Soc Lond Ser A Math Phys Sci 280(1382):383–397
28. Lai X, Wu D, Sun D (2010) Electro spinning under sub critical voltage. In: 2010 IEEE 5th international conference on nano/micro engineered and molecular systems. IEEE, pp 1170–1173
29. Riazi K, Kübel J, Abbasi M, Bachtin K, Indris S, Ehrenberg H, Kádár R, Wilhelm M (2016) Polystyrene comb architectures as model systems for the optimized solution electrospinning of branched polymers. Polymer 104:240–250
30. Zhang CL, Yu SH (2014) Nanoparticles meet electrospinning: recent advances and future prospects. Chem Soc Rev 43(13):4423–4448
31. Čuk N, Šala M, Gorjanc M (2021) Development of antibacterial and UV protective cotton fabrics using plant food waste and alien invasive plant extracts as reducing agents for the in-situ synthesis of silver nanoparticles. Cellulose 28:3215–3233

32. Gokarneshan N, Velumani K (2018) Significant trends in nano finishes for improvement of functional properties of fabrics. Scrivener Publishing LLC, Beverly, MA, USA, pp 387–434
33. Temesgen AG, Turşucular ÖF, Eren R, Ulcay Y (2018) Novel applications of nanotechnology in modification of textile fabrics properties and apparel. Int J Adv Multidiscip Res 5(12):49–58
34. Aguda ON, Lateef A (2022) Recent advances in functionalization of nanotextiles: a strategy to combat harmful microorganisms and emerging pathogens in the 21st century. Heliyon
35. Meda US, Soundarya VG, Madhu H, Bhat N (2023) Nano-engineered textiles: development and applications. Mater Sci Eng, B 296:116636
36. Nawab R, Iqbal A, Niazi F, Iqbal G, Khurshid A, Saleem A, Munis MFH (2023) Review featuring the use of inorganic nano-structured material for anti-microbial properties in textile. Polym Bull 80(7):7221–7245
37. Hebeish A, El-Naggar ME, Fouda MM, Ramadan MA, Al-Deyab SS, El-Rafie MH (2011) Highly effective antibacterial textiles containing green synthesized silver nanoparticles. Carbohyd Polym 86(2):936–940
38. Zahran MK, Ahmed HB, El-Rafie MH (2014) Surface modification of cotton fabrics for antibacterial application by coating with AgNPs–alginate composite. Carbohyd Polym 108:145–152
39. Ahmed MJ, Murtaza G, Mehmood A, Bhatti TM (2015) Green synthesis of silver nanoparticles using leaves extract of *Skimmia laureola*: characterization and antibacterial activity. Mater Lett 153:10–13
40. Hassabo AG, Reda EM, Ghazal H, Othman HA (2023) Synthesis of AgNPs and ZnONPs using tea leaves extract and their utilization to improve dyeability, printability and functionality of cotton and wool fabrics. Inorg Chem Commun 150:110525
41. Rodrigues AG, de Oliveira R, Gonçalves PJ, Ottoni CA, de Cássia Ruiz R, Morgano MA, de Araújo WL, de Melo IS, De Souza AO (2019) Functional textiles impregnated with biogenic silver nanoparticles from *Bionectria ochroleuca* and its antimicrobial activity. Biomed Microdevice 21:1–10
42. Shaheen TI, El-Naggar ME, Abdelgawad AM, Hebeish A (2016) Durable antibacterial and UV protections of in situ synthesized zinc oxide nanoparticles onto cotton fabrics. Int J Biol Macromol 83:426–432
43. Phan DN, Dorjjugder N, Khan MQ, Saito Y, Taguchi G, Lee H, Mukai Y, Kim IS (2019) Synthesis and attachment of silver and copper nanoparticles on cellulose nanofibers and comparative antibacterial study. Cellulose 26:6629–6640
44. Ziental D, Czarczynska-Goslinska B, Mlynarczyk DT, Glowacka-Sobotta A, Stanisz B, Goslinski T, Sobotta L (2020) Titanium dioxide nanoparticles: prospects and applications in medicine. Nanomaterials 10(2):387
45. Tanasa F, Teaca CA, Nechifor M, Ignat M, Duceac IA, Ignat L (2023) Highly specialized textiles with antimicrobial functionality—advances and challenges. Textiles 3(2):219–245
46. Harifi T, Montazer M (2017) Application of nanotechnology in sports clothing and flooring for enhanced sport activities, performance, efficiency and comfort: a review. J Ind Text 46(5):1147–1169
47. Mhaske ST, Mohanty JD, Chugh KW (2023) Fluoropolymers: brief history, fundamental chemistry, processing, structure, properties, and applications. In: Advanced fluoropolymer nanocomposites. Woodhead Publishing, pp 1–27
48. Ciriminna R, Fidalgo A, Pandarus V, Beland F, Ilharco LM, Pagliaro M (2013) The sol–gel route to advanced silica-based materials and recent applications. Chem Rev 113(8):6592–6620
49. Abo El-Ola SM, El-Bendary MA, Mohamed NH, Kotb RM (2022) Substantial functional finishing and transfer printing of polyester fabric using zinc oxide/polyurethane nanocomposite. Fibers and Polymers 23(10):2798–2808
50. Montazer M, Harifi T (2018) Nanofinishing of textile materials. Woodhead Publishing
51. Ghazal H, Khaleed N, Abdelaziz E (2023) Significance advantages, and disadvantages of nanotechnology in textile finishing. Egypt J Chem 66(12):467–482
52. Silva AC, Silvestre AJ, Freire CS, Vilela C (2021) Modification of textiles for functional applications. In: Fundamentals of natural fibres and textiles. Woodhead Publishing, pp 303–365

53. Teli MD (2018) Finishing of carpets for value addition. In: Advances in carpet manufacture. Woodhead Publishing, pp 175–211
54. Mohamed AM, Abdullah AM, Younan NA (2015) Corrosion behavior of superhydrophobic surfaces: a review. Arab J Chem 8(6):749–765
55. Kumar A, Nanda D (2019) Methods and fabrication techniques of superhydrophobic surfaces. In: Superhydrophobic polymer coatings. Elsevier, pp 43–75
56. Dutta MM, Goswami M (2021) Coating materials: nano-materials. In: Advanced surface coating techniques for modern industrial applications. IGI Global, pp 1–30
57. Kibria G, Repon MR, Hossain MF, Islam T, Jalil MA, Aljabri MD, Rahman MM (2022) UV-blocking cotton fabric design for comfortable summer wears: factors, durability and nanomaterials. Cellulose 29(14):7555–7585
58. Smijs TG, Pavel S (2011) Titanium dioxide and zinc oxide nanoparticles in sunscreens: focus on their safety and effectiveness. Nanotechnol Sci Appl, 95–112
59. Riaz S, Ashraf M, Hussain T, Hussain MT, Younus A (2019) Fabrication of robust multifaceted textiles by application of functionalized TiO_2 nanoparticles. Colloids Surf, A 581:123799
60. Kafafy H, Helmy H, Zaher A (2021) Treatment of cotton and wool fabrics with different nanoparticles for multifunctional properties. Egypt J Chem 64(9):5255–5267
61. Rashid MM, Simončič B, Tomšič B (2021) Recent advances in TiO_2-functionalized textile surfaces. Surf Interfaces 22:100890
62. Verbič A, Gorjanc M, Simončič B (2019) Zinc oxide for functional textile coatings: recent advances. Coatings 9(9):550
63. Becheri A, Dürr M, Lo Nostro P, Baglioni P (2008) Synthesis and characterization of zinc oxide nanoparticles: application to textiles as UV-absorbers. J Nanopart Res 10:679–689
64. Wang M, Zhang M, Pang L, Yang C, Zhang Y, Hu J, Wu G (2019) Fabrication of highly durable polysiloxane-zinc oxide (ZnO) coated polyethylene terephthalate (PET) fabric with improved ultraviolet resistance, hydrophobicity, and thermal resistance. J Colloid Interface Sci 537:91–100
65. Huang J, Yang Y, Yang L, Bu Y, Xia T, Gu S, Yang H, Ye D, Xu W (2019) Fabrication of multifunctional silk fabrics via one step in-situ synthesis of ZnO. Mater Lett 237:149–151
66. Nadeem M, Khan R, Afridi K, Nadhman A, Ullah S, Faisal S, Ul Mabood Z, Hano C, Abbasi BH (2020) Green synthesis of cerium oxide nanoparticles (CeO_2 NPs) and their antimicrobial applications: a review. Int J Nanomed, 5951–5961
67. Tripathi R, Narayan A, Bramhecha I, Sheikh J (2019) Development of multifunctional linen fabric using chitosan film as a template for immobilization of in-situ generated CeO_2 nanoparticles. Int J Biol Macromol 121:1154–1159
68. Mishra A, Butola BS (2019) UV Protective clothing. In: Advanced functional textiles and polymers: fabrication, processing and applications, pp 33–64
69. Rault F, Giraud S, Salaün F (2015) Flame retardant/resistant based nanocomposites in textile. In: Flame retardants: polymer blends, composites and nanocomposites, pp 131–165
70. Mihajlović I (2015) Recent development of phosphorus flame retardants in thermoplastic blends and nanocomposites. Flame retardants: polymer blends, composites and nanocomposites. Springer International Publishing, Cham, pp 79–114
71. Salaün F, Lemort G, Butstraen C, Devaux E, Capon G (2017) Influence of silica nanoparticles combined with zinc phosphinate on flame retardant properties of PET. Polym Adv Technol 28(12):1919–1928
72. Shahidi S, Moazzenchi B (2018) Carbon nanotube and its applications in textile industry—a review. J Text Inst 109(12):1653–1666
73. Alimohammadi F, Parvinzadeh M, Shamei A (2011) Carbon nanotube embedded textiles. US20110171413 A1, 14
74. Gashti MP, Almasian A (2013) UV radiation induced flame retardant cellulose fiber by using polyvinylphosphonic acid/carbon nanotube composite coating. Compos B Eng 45(1):282–289
75. Pakdel E, Fang J, Fang J, Sun L, Wang X, Wang X (2018) Nanocoatings for smart textiles. In: Smart textiles: wearable nanotechnology, pp 247–300

76. Guo M, Wu D, Wang X, Huang Y, Guo Z, Jiang Z (2023) Step-by-step assembly of antimony tin oxide/polymer coatings on cotton fabrics for enhanced electrical conductivity and fire resistance properties. J Text Inst, 1–11
77. Kim H, Kim JS, Jeong W (2021) Study on the flame retardancy and hazard evaluation of poly (acrylonitrile-co-vinylidene chloride) fibers by the addition of antimony-based flame retardants. Polymers 14(1):42
78. Magovac E, Vončina B, Jordanov I, Grunlan JC, Bischof S (2022) Layer-by-layer deposition: a promising environmentally benign flame-retardant treatment for cotton, polyester, polyamide and blended textiles. Materials 15(2):432
79. Rahimi E, Babapoor A, Moradi G, Kalantari S, Esmaeelpour MM (2024) Personal cooling garments and phase change materials: a review. Renew Sustain Energy Rev 190:114063
80. Pakdel E, Naebe M, Sun L, Wang X (2019) Advanced functional fibrous materials for enhanced thermoregulating performance. ACS Appl Mater Interfaces 11(14):13039–13057
81. Ji T, Feng Y, Qin M, Feng W (2016) Thermal conducting properties of aligned carbon nanotubes and their polymer composites. Compos A Appl Sci Manuf 91:351–369
82. Dong ZJ, Sun B, Zhu H, Yuan GM, Li BL, Guo JG, Li XK, Chong Y, Zhang J (2021) A review of aligned carbon nanotube arrays and carbon/carbon composites: fabrication, thermal conduction properties and applications in thermal management. New Carbon Mater 36(5):873–892
83. Zhu Q, Ong PJ, Goh SHA, Yeo RJ, Wang S, Liu Z, Loh XJ (2023) Recent advances in graphene-based phase change composites for thermal energy storage and management. Nano Mater Sci
84. Ward MA, Georgiou TK (2011) Thermoresponsive polymers for biomedical applications. Polymers 3(3):1215–1242
85. Ullah HMK, Lejeune J, Cayla A, Monceaux M, Campagne C, Devaux É (2022) A review of noteworthy/major innovations in wearable clothing for thermal and moisture management from material to fabric structure. Text Res J 92(17–18):3351–3386
86. Farooq AS, Zhang P (2021) Fundamentals, materials and strategies for personal thermal management by next-generation textiles. Compos A Appl Sci Manuf 142:106249
87. Selvam S, Sundrarajan M (2012) Functionalization of cotton fabric with PVP/ZnO nanoparticles for improved reactive dyeability and antibacterial activity. Carbohyd Polym 87(2):1419–1424
88. Du H, Yue M, Huang X, Duan G, Yang Z, Huang W, Shen W, Yin X (2022) Preparation, application and enhancement dyeing properties of ZnO nanoparticles in silk fabrics dyed with natural dyes. Nanomaterials 12(22):3953
89. Hasan KMF, Pervez MN, Talukder ME, Sultana MZ, Mahmud S, Meraz MM, Bansal V, Genyang C (2019) A novel coloration of polyester fabric through green silver nanoparticles (G-AgNPs@ PET). Nanomaterials 9(4):569
90. Tang B, Lin X, Zou F, Fan Y, Li D, Zhou J, Chen W, Wang X (2017). In situ synthesis of gold nanoparticles on cotton fabric for multifunctional applications. Cellulose 24:4547–4560
91. Pulit-Prociak J, Chwastowski J, Kucharski A, Banach M (2016) Functionalization of textiles with silver and zinc oxide nanoparticles. Appl Surf Sci 385:543–553
92. Kalaiarasan S, Uthirakumar P, Shin DY, Lee IH (2021) The degradation profile of high molecular weight textile reactive dyes: a daylight induced photocatalytic activity of ZnO/carbon quantum dot photocatalyst. Environ Nanotechnol Monit Manage 15:100423
93. Murugadoss G, Prakash J, Rajesh Kumar M, Alothman AA, Habila MA, Peera SG (2022) Controlled synthesis of europium-doped SnS quantum dots for ultra-fast degradation of selective industrial dyes. Catalysts 12(10):1128
94. Diaa M, Othman H, Hassabo AG (2022) Printing wool fabrics with natural dyes curcuma and alkanet (a critique). J Text Coloration Polym Sci 19(1):11–16
95. Piri N, Shams-Nateri A, Mokhtari J (2019) The effect of TiO_2 nanopigment on the optical properties of polyester fabric in UV–VIS–NIR regions. Color Res Appl 44(2):257–263
96. Rathi VH, Jeice AR (2023) Green fabrication of titanium dioxide nanoparticles and their applications in photocatalytic dye degradation and microbial activities. Chem Phys Impact 6:100197

97. Maheswaran R, Shanmugavel BP (2022) A critical review of the role of carbon nanotubes in the progress of next-generation electronic applications. J Electron Mater 51(6):2786–2800
98. Lu Z, Raad R, Safaei F, Xi J, Liu Z, Foroughi J (2019) Carbon nanotube based fiber supercapacitor as wearable energy storage. Front Mater 6:138
99. Lima RMAP, De Oliveira MCA, de Oliveira HP (2019) Wearable supercapacitors based on graphene nanoplatelets/carbon nanotubes/polypyrrole composites on cotton yarns electrodes. SN Appl Sci 1:1–11
100. Heme HN, Alif MSN, Rahat SSM, Shuchi SB (2021) Recent progress in polyaniline composites for high capacity energy storage: a review. J Energy Storage 42:103018
101. Samuel E, Joshi B, Kim Y, Park C, Aldalbahi A, El-Newehy M, Lee H-S, Yoon SS (2021) Cotton fabric decorated with manganese oxide nanorods as a supercapacitive flexible electrode for wearable electronics. Appl Surf Sci 568:150968
102. Dhanabalan SC, Dhanabalan B, Chen X, Ponraj JS, Zhang H (2019) Hybrid carbon nanostructured fibers: stepping stone for intelligent textile-based electronics. Nanoscale 11(7):3046–3101
103. Marriam I, Tebyetekerwa M, Xu Z, Chathuranga H, Chen S, Chen H, Zheng J-C, Du A, Yan C (2021) Techniques enabling inorganic materials into wearable fiber/yarn and flexible lithium-ion batteries. Energy Storage Mater 43:62–84
104. Bairagi S, Shahadat M, Mulvihill DM, Ali W (2023) Mechanical energy harvesting and self-powered electronic applications of textile-based piezoelectric nanogenerators: a systematic review. Nano Energy, 108414
105. Hossain MR, Ahmed MR, Alam MS (2023) Smart textiles. Text Prog 55(2):47–108
106. Stoppa M, Chiolerio A (2014) Wearable electronics and smart textiles: a critical review. Sensors 14(7):11957–11992
107. Isapour G, Lattuada M (2018) Bioinspired stimuli-responsive color-changing systems. Adv Mater 30(19):1707069
108. Yetisen AK, Qu H, Manbachi A, Butt H, Dokmeci MR, Hinestroza JP, Skorobogatiy M, Khademhosseini A, Yun SH (2016) Nanotechnology in textiles. ACS Nano 10(3):3042–3068
109. Su Y, Yu B, Zhao X (2023) Research status and development of infrared camouflage textile materials. Text Res J, 00405175231170323
110. Samolov AD, Simić DM, Fidanovski BZ, Obradović VM, Tomić LD, Knežević DM (2021) Improvement of VIS and IR camouflage properties by impregnating cotton fabric with PVB/IF-WS2. Defence Technol 17(6):2050–2056
111. Pereira C, Pereira AM, Freire C, Pinto TV, Costa RS, Teixeira JS (2020) Nanoengineered textiles: from advanced functional nanomaterials to groundbreaking high-performance clothing. In: Handbook of functionalized nanomaterials for industrial applications. Elsevier, pp 611–714
112. Cai L, Peng Y, Xu J, Zhou C, Zhou C, Wu P, Lin D, Fan C, Cui Y (2019) Temperature regulation in colored infrared-transparent polyethylene textiles. Joule 3(6):1478–1486
113. Degenstein LM, Sameoto D, Hogan JD, Asad A, Dolez PI (2021) Smart textiles for visible and IR camouflage application: state-of-the-art and microfabrication path forward. Micromachines 12(7):773
114. Shabbir M, Mohammad F (2018) Multifunctional AgNPs@ wool: colored, UV-protective and antioxidant functional textiles. Appl Nanosci 8:545–555
115. Ghaheh FS, Khoddami A, Alihosseini F, Jing S, Ribeiro A, Cavaco-Paulo A, Silva C (2017) Antioxidant cosmetotextiles: cotton coating with nanoparticles containing vitamin E. Process Biochem 59:46–51
116. Butola BS, Verma D (2019) Facile synthesis of chitosan-silver nanoparticles onto linen for antibacterial activity and free-radical scavenging textiles. Int J Biol Macromol 133:1134–1141
117. Zhou Y, Tang RC (2018) Facile and eco-friendly fabrication of AgNPs coated silk for antibacterial and antioxidant textiles using honeysuckle extract. J Photochem Photobiol, B 178:463–471
118. Sadeghi-Kiakhani M, Hashemi E, Gharanjig K (2019) Inorganic nanoparticles and natural dyes for production of antimicrobial and antioxidant wool fiber. 3 Biotech 9(12):456

119. Wu M, Ma B, Pan T, Chen S, Sun J (2016) Silver-nanoparticle-colored cotton fabrics with tunable colors and durable antibacterial and self-healing superhydrophobic properties. Adv Func Mater 26(4):569–576
120. Zhou H, Wang H, Niu H, Gestos A, Lin T (2013) Robust, self-healing superamphiphobic fabrics prepared by two-step coating of fluoro-containing polymer, fluoroalkyl silane, and modified silica nanoparticles. Adv Func Mater 23(13):1664–1670
121. Zhao Y, Liu E, Fan J, Chen B, Hu X, He Y, He C (2019) Superhydrophobic PDMS/wax coated polyester textiles with self-healing ability via inlaying method. Prog Org Coat 132:100–107
122. Chen L, Liu W, Wang P, Zhao Z, Wu F, Wang T, Zhang T (2021) Facile fabrication of robust and healable superhydrophobic cotton fabric with flower-like $Ni(OH)_2$@ODA micro-nanoparticles. Cellulose 28:581–592
123. Liu L, Shan Y, Pu M, Zhao X, Huang Y (2020) Preparation of superhydrophobic fabrics via chemical self-healing strategy and their high oil/water separation performance and enhanced durability. Macromol Chem Phys 221(2):1900356
124. Liu Y, Pei X, Liu Z, Yu B, Yan P, Zhou F (2015) Accelerating the healing of superhydrophobicity through photothermogenesis. J Mater Chem A 3(33):17074–17079
125. Liu Y, Liu Y, Hu H, Liu Z, Pei X, Yu B, Yan P, Zhou F (2015) Mechanically induced self-healing superhydrophobicity. J Phys Chem C 119(13):7109–7114
126. Ramesh S, Khan S, Park Y, Ford E, Menegatti S, Genzer J (2022) Self-healing and repair of fabrics: a comprehensive review of the application toolkit. Mater Today 54:90–109
127. Chen K, Gu K, Qiang S, Wang C (2017) Environmental stimuli-responsive self-repairing waterbased superhydrophobic coatings. RSC Adv 7(1):543–550
128. Wang H, Zhou H, Gestos A, Fang J, Lin T (2013) Robust, superamphiphobic fabric with multiple self-healing ability against both physical and chemical damages. ACS Appl Mater Interfaces 5(20):10221–10226
129. Zeng M, Wang P, Luo J, Peng B, Ding B, Zhang L, Wang L, Huang D, Echols I, Abo Deeb E, Bordovsky E, Choi C-H, Ybanez C, Meras P, Situ E, Mannan MS, Cheng Z (2018) Hierarchical, self-healing and superhydrophobic zirconium phosphate hybrid membrane based on the interfacial crystal growth of lyotropic two-dimensional nanoplatelets. ACS Appl Mater Interfaces 10(26):22793–22800
130. Li D, Guo Z (2017) Versatile superamphiphobic cotton fabrics fabricated by coating with SiO_2/FOTS. Appl Surf Sci 426:271–278
131. Song YX, Xu WM, Rong MZ, Zhang MQ (2020) A sunlight self-healable fibrous flexible pressure sensor based on electrically conductive composite wool yarns. Express Polym Lett 14(11)
132. Tunakova V, Tunak M, Tesinova P, Seidlova M, Prochazka J (2020) Fashion clothing with electromagnetic radiation protection: aesthetic properties and performance. Text Res J 90(21–22):2504–2521
133. Liu YS, Chen GH (2013) Design and shielding effectiveness of electromagnetic shielding textiles. Adv Mater Res 796:653–656
134. Wang X, Hang G, Liu Z, He S (2022) Study on finishing and electromagnetic properties of electromagnetic shielding fabric based on multilayer Ti_3C_2T x medium. J Text Inst 113(12):2704–2713
135. Pušić T, Šaravanja B, Malarić K (2021) Electromagnetic shielding properties of knitted fabric made from polyamide threads coated with silver. Materials 14(5):1281
136. Asfand N, Daukantienė V (2023) Electrical properties and electromagnetic shielding effectiveness of cotton/antistatic polyester knitted fabrics treated with antibacterial finish. Text Res J, 00405175231158820
137. Lao L, Fu L, Qi G, Giannelis EP, Fan J (2017) Superhydrophilic wrinkle-free cotton fabrics via plasma and nanofluid treatment. ACS Appl Mater Interfaces 9(43):38109–38116
138. Gulati S, Kumar S, Kumar S, Wadhawan V, Batra K (2022) Wrinkle-resistant fabrics: nanotechnology in modern textiles. In: Handbook of consumer nanoproducts. Springer Nature Singapore, Singapore, pp 911–928
139. Roy Choudhury AK (2017) Easy-care finishing. In: Principles of textile finishing, pp 245–284

140. Tarek D, Mohmoud A, Essam Z, Sayed R, Maher A, Maamoun D, Abdel Salam S, Mohamed H, Hassabo A, Khattab T (2023) Development of wrinkle free and self-cleaning finishing of cotton and polyester fabrics. J Text Coloration Polym Sci 20(2):175–180
141. Ji B, Sun G (2020) Advances in durable press finishing of fabrics with formaldehyde-free agents. In: Advances in functional and protective textiles. Woodhead Publishing, pp 81–116
142. Jadoun S, Verma A, Arif R (2020) Modification of textiles via nanomaterials and their applications. In: Frontiers of textile materials: polymers, nanomaterials, enzymes, and advanced modification techniques, pp 135–152
143. Abdellatif FHH, Abdellatif MM, Ahmed HM (2022) Advanced nanotechnologies for finishing of cellulosic textiles. In: Fundamentals of nano-textile science. Apple Academic Press, pp 73–99
144. Radetić M, Marković D (2019) Nano-finishing of cellulose textile materials with copper and copper oxide nanoparticles. Cellulose 26:8971–8991
145. Radetić M, Marković D (2022) A review on the role of plasma technology in the nano-finishing of textile materials with metal and metal oxide nanoparticles. Plasma Processes Polym 19(4):2100197
146. Wang R (2007) Multi-functional protective nano-finishing of textile materials
147. Rujido-Santos I, Herbello-Hermelo P, Barciela-Alonso MC, Bermejo-Barrera P, Moreda-Piñeiro A (2022) Metal content in textile and (nano) textile products. Int J Environ Res Public Health 19(2):944
148. Haule LV, Nambela L (2022) Sustainable application of nanomaterial for finishing of textile material. Green Nanomater Ind Appl, 177–206
149. Dolez PI (2019) Application of nanomaterials in textile coatings and finishes. In: Nanomaterials-based coatings. Elsevier, pp 139–169

Chapter 15
Nanotechnology in Smart Textiles

Danmei Sun, Madiha Ahmad, Muhammad Owais Raza Siddiqui, and Kashif Iqbal

1 Introduction to Nanotechnology in Textiles

'Nano' is derived from the Greek word 'nanos', meaning 'very small'. It denotes one billionth of a unit. In nanotechnology, nano refers to the manipulation and control of a material to its nanoscale, at dimensions less than 100 nm. Under this scale, materials generally emerge enhanced properties. This makes nano a crucial field for various advanced applications such as medicine, electronics, and materials science, and it continues to be an area of active research and development. Nanotechnology involves the manipulation of materials at nanoscale for the creation of new technologies and materials offering novel properties. At nanoscale, the properties of the material would perform very differently from its larger scales, leading to the unique and enhanced characteristics. Applications of nanotechnology in textiles include but are not limited to the areas of health care, defense and protection, sports, leisure, transportation, construction, etc., shown in Fig. 1. Table 1 summarizes various ways in which nanomaterials were integrated within textiles.

D. Sun (✉) · M. Ahmad
School of Textiles and Design, Heriot-Watt University, Galashiels, UK
e-mail: d.sun@hw.ac.uk

M. O. R. Siddiqui
Department of Textile Engineering, NED University of Engineering and Technology, Karachi, Pakistan

K. Iqbal
Department of Textile Engineering, National Textile University, Faisalabad, Pakistan

S. Bairagi et al. (eds.), *Nanotechnology in Textile Finishing*, Materials Horizons: From Nature to Nanomaterials, https://doi.org/10.1007/978-981-97-2696-7_15

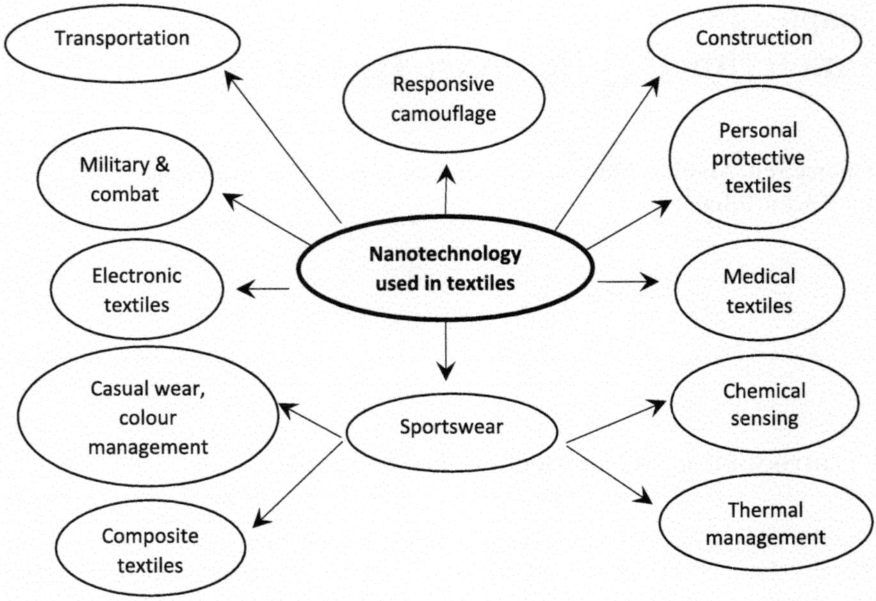

Fig. 1 Product domains of nanotechnology in textiles

2 Nanomaterials for Smart Textiles

2.1 Nanoparticles

The integration of nanoparticles into textiles can enhance their properties in numerous ways. Zinc oxide nanoparticles, for instance, can be added to textiles to provide UV protection. Quantum dots, on the other hand, enable textiles to emit light, enabling applications in illuminated clothing and safety gear. Phase change chemicals with stored latent heat can be used for protective textiles in extreme conditions.

Fouda et al. synthesized nanoparticles of ZnO-NPs encapsulated by protein and applied them onto cotton fabrics for antibacterial and UV protective textiles. Figure 2 shows the image from scanning transmission electron microscopy (STEM) that was used to determine the morphology and the approximate size of the biosynthesized ZnO-NPs, and the size distribution of the nanoparticles in colloidal solution. They reported that the particles polydispersed mix with 95.4% particles are in an average size of 175.85 nm [17]. Iqbal and Sun [18] synthesized nano-scaled composite with the active core of phase change material Glauber's salt covered by poly (methyl methacrylate) (PMMA), shown in Fig. 3. Furthermore, Sun et al. reported a type of non-Newtonian materials composed by nanoparticles of silica n-SiO_2 with 7 nm and viscous liquid of polyethylene glycol (PEG) [19].

Table 1 summarizes various methods used in incorporating nanomaterials with textiles [1]

Nanoparticles	Textile materials	Incorporation process	Other materials used	Refs.
ZnO	Cotton	Dip coating	Zinc acetate and 2-methoxy-ethanol	[2]
Graphene Ag	Cotton bandage	Dip coating	H_2SO_4 (98%, $NaNO_3$, $KMnO_4$, H_2O_2 (30%), HCl	[4]
Ag	Cellulosic fabric	Pad-dry-cure	$NaBH_4$, $NaH_2PO_2{\cdot}H_2O$, $C_6H_5Na_3O_7{\cdot}2H_2O$	[5]
Al_2O_3	Linen	Dip coating	Modified glyoxylic resin	[6]
O_4SiTi	Cotton	Dip coating	C6-Fluorocarbon resin	[7]
Ag	Cotton	Dip coating	Nonionic surfactant	[8]
Cu	Cotton	Dip coating	Maleic acid and sodium hypophosphite	[9]
Cu	Denim	Pad-dry-cure	Copper (II) sulfate and glucose	[10]
ZnO	Cotton	Spraying or dipping	Zinc nitrate and sodium hydroxide	[11]
Ag	Viscose rayon	Dip coating	Chitosan	[12]
Ag	Silk	Dip coating	1,2,3,4 butane tetra carboxylic acid	[13]
Ag-doped TiO_2	Nonwoven Co/PET or PP and woven cotton fabrics	Electrospray processing	Glacial acetic acid and sodium hydroxide	[14]
Mn-doped TiO_2	Cotton	Spray coating	Silicon binder	[15]
TiO2	Cotton	Spray coating sol–gel and hydrothermal	Silicon binder tetrabutyl titanate and ethanol	[16]

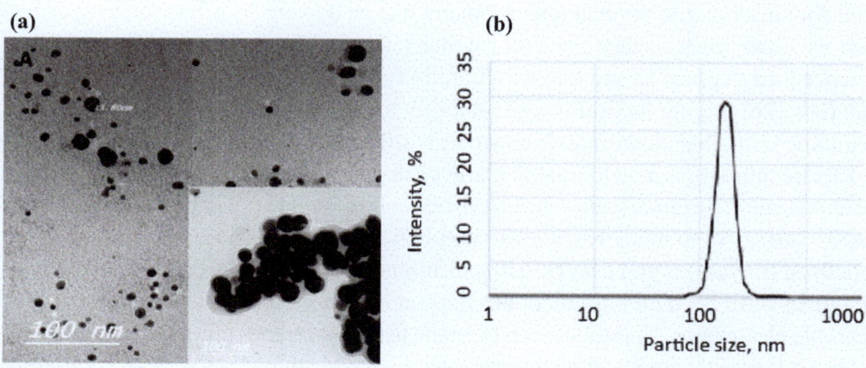

Fig. 2 Particle size of $Zn(CH_3CO_2)_2$: **a** STEM image and **b** size and size distribution [17]

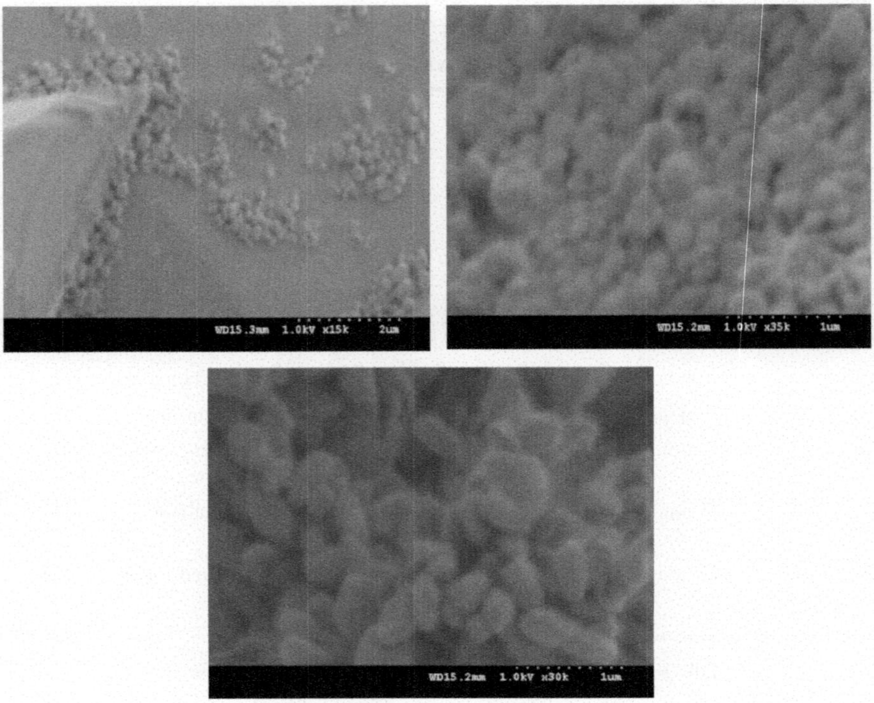

Fig. 3 Nano-scaled phase change composite Glauber's salt encapsulated by PMMA [18]

2.2 Nanofibers and Composites

Research on nanofibers may be traced back to the nineteenth century, and much research work have been reported since then [20–24]. Nanofibers have been developed for smart textile applications by many due to its various advanced properties, such as super high surface area to volume ratio for air and water filtration [25], enhanced mechanical strength to weight ratio for reinforcing ultra-strong composite materials [26], highly flexibility and pliability to meet the requirements of wearable technology, flexible electronics, and smart textiles [27].

Electrospinning is a technique that allows the production fibers with diameters in the nanoscale. The remarkable characteristics and versatile applications of nanofibers made it attractive to many researchers and industries alike. Nanofibers can be functionalized for various end uses by using different nanoparticles. For example, carbon nanotubes have been incorporated into the conductive nanofiber-making process for wearable electronics. Nanofibers can be made using various techniques, each offering distinct advantages for specific applications. Electrospinning is the most commonly used technique, involving the application of an electric field to a polymer solution or melt. This process forces the material to form ultrafine fibers that can have diameters ranging from a few nanometers to a few micrometers. Electro-spun nanofibers possess an extraordinarily high surface area, making them ideal for applications requiring filtration, drug delivery, and tissue engineering. Researchers have utilized electrospinning to create nanofibrous scaffolds for regenerative medicine, such as skin grafts and bone implants [28].

Bing et al. devised composite nanofibers incorporating MXene and hydrogel, showcasing photo-stimulus responsiveness and possessing optical monitoring for drug release. Figure 4a outlines the preparation process for the MXene@Hydrogel composite nanofiber membrane, along with in-light drug release, real-time monitoring of drug release, and SEM images of the composite nanofibers. The surface morphology and structure were scrutinized using optical microscopy, scanning electron microscopy (SEM), and transmission electron microscope (TEM). The composite nanofiber membrane demonstrated heat production within the temperature range from room temperature to 50 °C at the phase transition point. The researchers assessed the recycling capability of the composite nanofiber membrane through repeated immersion in tissue fluid. Subjecting the membrane to alternate exposure and non-exposure to light multiple times confirmed its robust thermal stability. As depicted in Fig. 4b, even under illumination, the composite nanofiber membrane could achieve a temperature exceeding the phase transition temperature of MXene@Hydrogel. As illustrated in Fig. 5, the heat production of the fiber membrane is adjustable within the range of room temperature to 50 °C at the phase transition point. Consequently, the fiber membrane can be exposed to light to generate heat, thereby controlling the phase transition of MXene@Hydrogel.

With the challenges of declining traditional energy and increasing generating energy from environmentally friendly new energy sources [30], a type of ultra-robust separator made of polyimide nanofiber with a shutdown function has been explored to be used for advanced lithium-ion batteries [31]. The morphology of both the nanofiber and separators was analyzed, as depicted in Fig. 6. Specifically, Fig. 6a illustrates the pristine PI nanofiber membrane characterized by incompact and randomly accumulated nanofibers, while Fig. 6b showcases the PIE nanofiber membrane featuring a well-defined and compact microstructure with evident bonding and crosslinking morphologies. Additionally, Fig. 6c and e provide magnified views of the pristine PI and PIE separators, respectively, while Fig. 6d presents the cross-section of the PIE nanofiber, corresponding to the schematic representation of the organic sheath of the PI nanofibers. The use of nanofiber membranes can ensure the wettability of the PIE separators due to their polarity and high porosity structure that the commercial separators lack. A contact angle of 17.7 °C was found for the PIE separator in comparison with 43.9 °C for the commercial Celgard separator. As shown in Fig. 7a, the temperature change in endothermic reaction for the nanofiber-made PEI separator is much higher than that of the traditional Celgard separator. This means the nanofiber material is much more stable under high temperatures. It was also found that the PIE nanofiber separator possessed excellent thermal stability at a very high temperature of 470 °C. Conversely, the commercial Celgard separator initiates decomposition at 250 °C and undergoes complete weight loss at 360 °C, presented in Fig. 6b. Both the separators were put in an oven for half an hour at different temperatures of 25, 100, and 200 °C. The nanofiber PEI separator showed no apparent shrinkage and

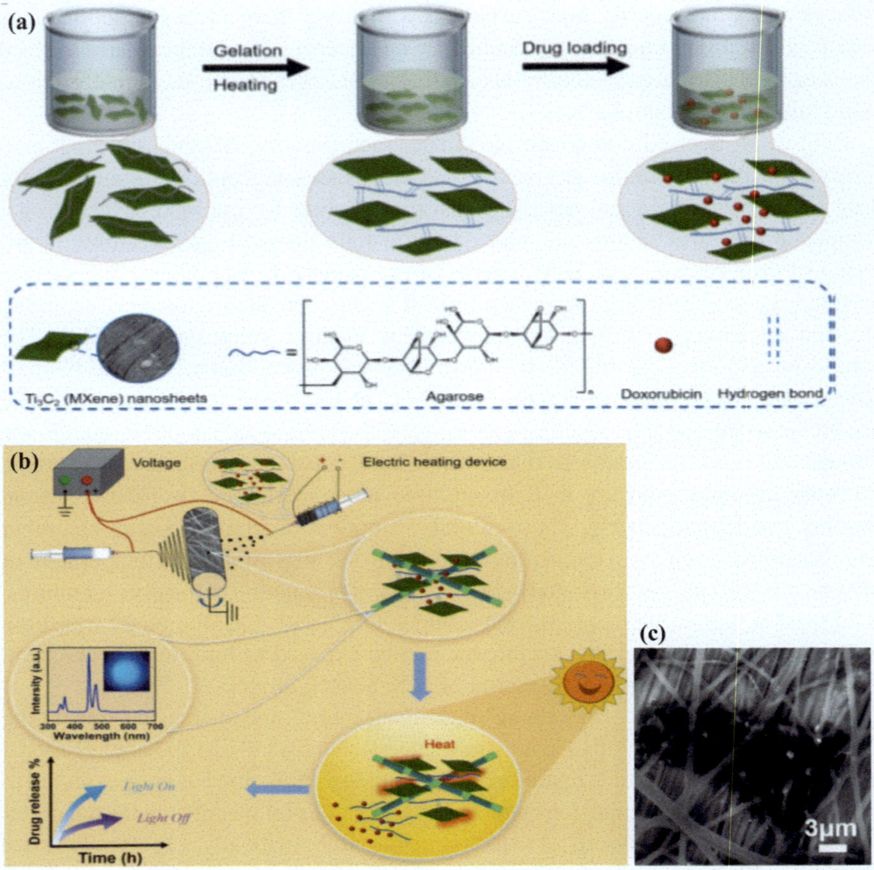

Fig. 4 **a** Schematic steps of composite nanofiber membrane MXene@Hydrogel preparation, under-the-light drug release, **b** real-time release monitoring, and **c** SEM image of the developed composite nanofibers [29]

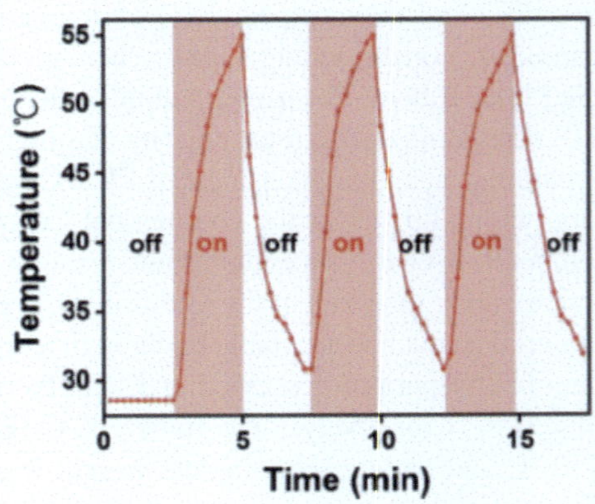

Fig. 5 Cyclic heating profile of the composite nanofiber membrane involves repetitive cycles of light activation and deactivation [29]

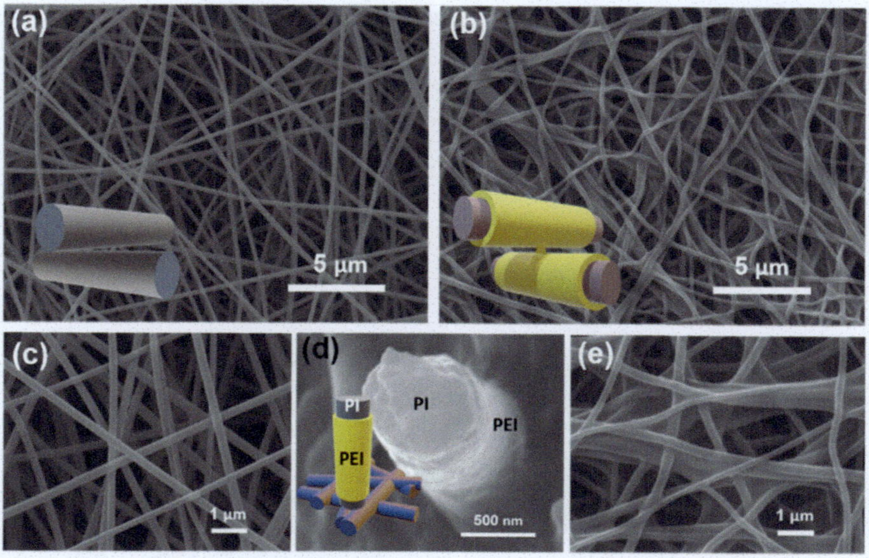

Fig. 6 **a**, **b** shows that images from SEM, **c**, **e** are magnified pristine PI and PIE separator images from SEM, and **d** is the SEM image of the fiber cross-section [31]

deformation for all temperatures, while the Celgard separator started shrinking at 150 °C and melted entirely at 200 °C, as shown in Fig. 7c. Furthermore, the fire retardancy behavior of the two types of separators was tested, and the results shown in Fig. 7d show the nanofiber PIE separator can significantly improve the security of LIBSs due to its self-extinguishing performance, while the commercial Celgard separator was highly flammable. Furthermore the PEI-layer blocked the micropore of the separator at the temperature of 290 °C, leading to a 'shutdown' function. This implies that lithium-ion batteries (LIBs) with nanofiber PIE separators can operate safely across a broad temperature range.

2.3 *High-Performance Smart Nanostructures*

Chen et al. [32] have successfully created flexible, high-strength smart fibers with multiple functions utilizing biomass resources. In a nutshell, a 1% solution of holocellulose nanofibrils (HCNF) was prepared using peracetic acid (PAA). A delignification technique was used by adding Birch wood chips into the PAA solution containing CH_3COOH and H_2O_2 [33]. The HCNF solution was obtained by stirring for a few minutes, after the resultant holocellulose fibers were added to Ndimethylformamide (DMF), containing urea and NH_2SO_3H. Secondly the cellulose solution was made using bamboo pulp immersed in LiOH/urea/H_2O solution at 13 °C, before it was added to HCNF solution (20%) to make spinning solution for hydrogel fiber

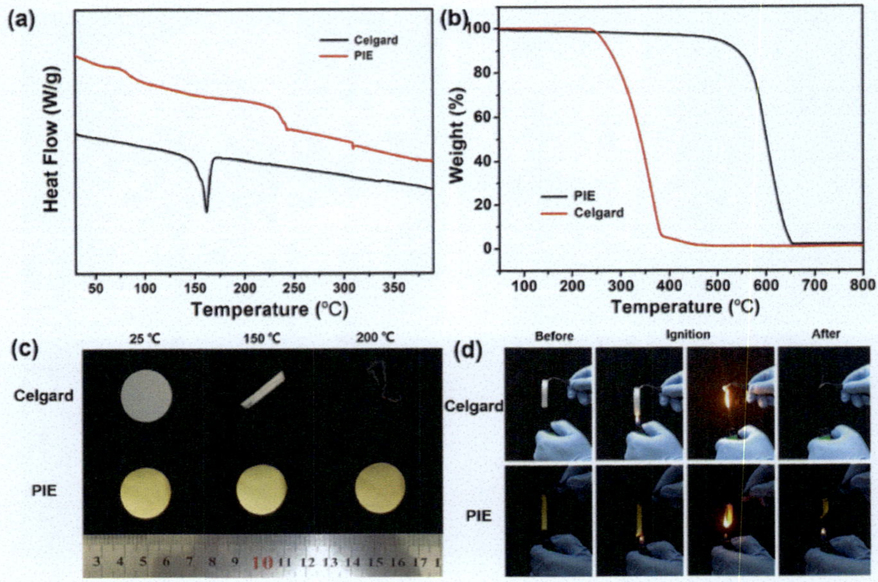

Fig. 7 **a** Differential scanning calorimetry (DSC) analysis of both Celgard and PIE separators. **b** Thermogravimetric analysis (TGA) of the Celgard and PIE separators. **c** Thermal shrinkage behavior illustrated through digital photos. **d** Ignition tests conducted on both Celgard and PIE separators [31]

extrusion. Holocellulose nanofibril/CNT (conductive carbon nanotubes)/cellulose hydrogel fibers were also achieved by ultrasonically dispersing CNT into the spinning solution, using the same wet-spinning technique. The hydrogel fibers obtained were introduced into a solvent, dehydrated, and immersed in liquid CO_2. Subsequently, holocellulose nanofibrils/cellulose aerogel fibers (HCAFs) and holocellulose nanofibrils/carbon nanotubes/cellulose aerogel fibers (HCCAFs) were successfully acquired via supercritical fluid drying. In Fig. 8, images of HCAFs and HCCAFs are presented along with their flexibility after folding (Fig. 8a–e). The figures also depict surface and cross-sectional morphologies (f–h), the porous network structure, and pore size in the fibers (i, j). Image (k) illustrates the uniform dispersion of carbon nanotubes (CNTs) on cell walls. Additionally, certain physical and mechanical properties of the fibers are showcased in images (l–p). It was evidenced that the developed fibers presented excellent thermal insulation over a wide range of temperatures between − 187 and 190 °C. In addition, the HCCAFs fibers could provide self-cleaning capabilities and electrical conductivity attributed by the CNTs added in the solution for fiber extruding. These versatile fibers, displaying multiple functionalities, are proposed for application in self-powered wearable electronics. This assertion is substantiated by the effective incorporation of HCCAFs textile into thermoelectric devices, showcasing stable and reversible solar-thermo-electrical power generation.

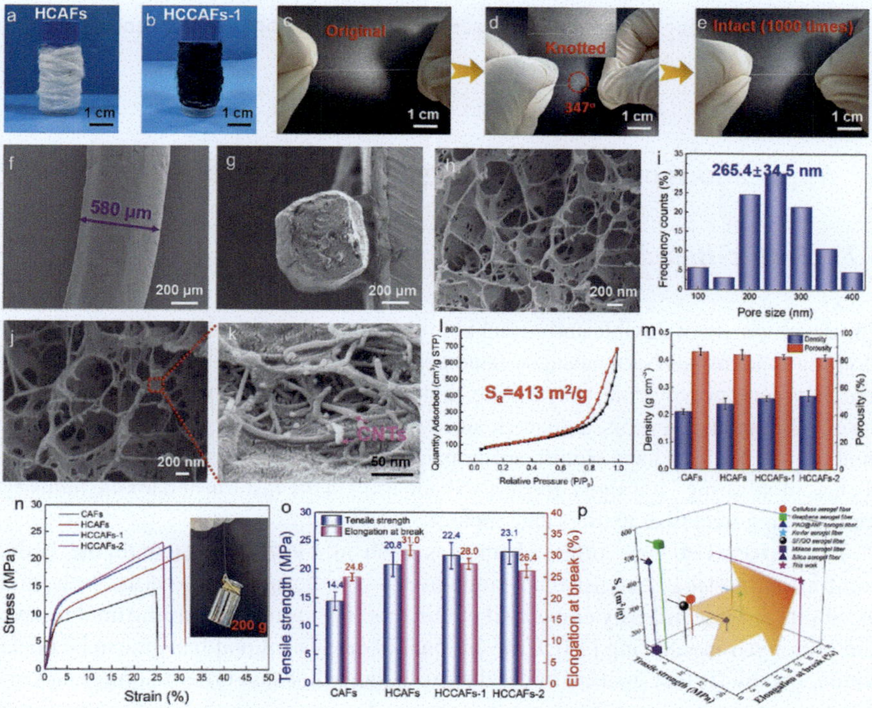

Fig. 8 **a, b** shows depictions of white holocellulose nanofibrils/cellulose aerogel fibers (HCAFs) and black holocellulose nanofibrils/CNTs (10%)/cellulose aerogel fibers (HCCAFs-1) following wet spinning and supercritical fluid drying. **c, d** are illustrations of the original and knotted HCAF. **e** is an image of an intact HCAF after undergoing 1000 folding and unfolding cycles. **f, g** are surface and cross-sectional scanning electron microscopy (SEM) images of the aerogel fiber. **h, i** are cross-sectional SEM image and pore size analysis of the porous HCAFs. **j, k** are cross-sectional SEM image of HCCAFs-1 with varying magnifications. **l** shows the results of nitrogen adsorption/desorption isotherms for HCAFs. **m** is density and porosity comparisons for CAFs, HCAFs, HCCAFs-1, and holocellulose nanofibrils/CNTs (20%)/cellulose aerogel fibers (HCCAFs-2). **n** is the stress–strain curves for CAFs, HCAFs, HCCAFs-1, and HCCAFs-2; the insert depicts a bundle of HCAFs supporting a tensile load of 200 g. **o** is tensile strength and elongation at breakage of CAFs, HCAFs, HCCAFs-1, and HCCAFs-2. **p** is comparative analysis of tensile strength, elongation at breakage, and surface area (Sa) between fibers produced in this study and those reported previously for aerogel fibers [32]

Feng et al. [34] developed novel membranes made of AOPAN/RC blend nanofibers through electrospinning process. Figure 3 shows the processes of making the membranes including solution preparation, nanofiber spinning, and surface modification.

Nanotechnology has ushered in a new era of scientific exploration and innovation, with nanofibers emerging as a particularly exciting development in this field. Electrospinning, a nanofiber production technique, allows for the creation of incredibly thin fibers with diameters on the nanoscale. This section delves into the world of nanofibers, exploring their unique characteristics, fabrication methods, and a plethora of real-world applications.

3 Applications of Nanotechnology in Smart Textiles

3.1 *Textile-Based Electronics for Smart Sensing*

The development of flexible and stretchable electronics being integrated in textile and clothing systems has been paved by nanotechnology. Nanomaterials such as conductive polymers (polypyrrole, polyaniline, and polythiophene, etc.) and graphene made the development of sensors, actuators, and even energy storage devices possible. Such innovative creations have significant impact in the healthcare area for applications such as monitoring vital signs. It has also been used in the fashion and textile industry for creating interactive textiles and apparel.

Textile-based sensors offer advantages such as close contact with the body, comfort during wear, extensive coverage over a significant portion of the body, and the ability to be seamlessly embedded within clothing systems, enabling unobtrusive and prolonged monitoring [35]. Conventional fibers, yarns, and fabrics can be made conductive by surface coating of conductive nanoparticles in order to create electrically conductive textiles. This approach can produce textiles with good conductivity without considerably changing the coated substrate properties such as flexibility, mechanical properties, and handle. However, great attention needs to be paid to the strength of coating, washability, etc.

3.2 *Biomedical and Health Care*

Smart textiles have been innovatively applied in health care, encompassing real-time monitoring of physiological properties and wearer health requirements through diverse sensors. The integration of nanotechnology further enhances the overall performance of these textiles. It is crucial to address fundamental requirements such as comfort and perspiration management to ensure the optimal functionality of textile items, while integrating sensors and nanomaterials in a fabric and clothing system [36]. Nano conductive pigment has been used for developing conductive coils through screen printing, to be integrated in a clothing system for charging electronics worn by patients [37]. Figure 9 shows the developed conductive coils.

It has been reported that the unique structure and properties of the ceria nanoparticles made them become the promising materials as catalysis for biomedical applications [38]. Ali et al. [39] investigated nano copper particles to be applied on chitosan coated cotton fabric surface for toxic dye reduction. Figure 10(1) outlines the experimental procedures conducted. The Cu/Chi-CC catalyst prepared was evaluated for

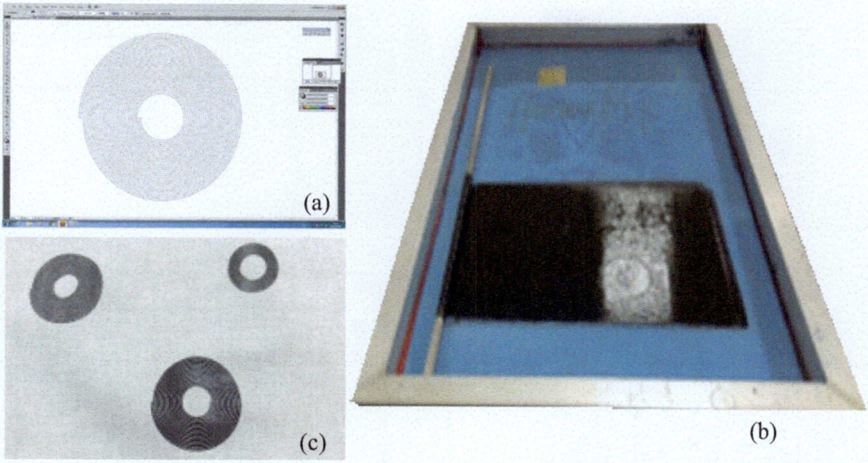

Fig. 9 **a** A pre-design coil pattern, coil pattern printed on a fabric beneath the printing screen on a semi-auto printing machine, and **c** conductive coils printed on a woven fabric

the degradation of Congo red (CR) dye using $NaBO_4$. The results indicated that the dye degradation was completed in less than 12 min, with a faster degradation observed when a higher amount of Cu nanoparticles was present in Cu/Chi-CC. Furthermore, the Cu catalyst remained active on the Cu/Chi-CC fabric even after three uses, demonstrating reliable recyclability and stability during the process of degradation process, as depicted in Fig. 10(2). In this investigation, chitosan served as a supporting material to enhance the affinity of the copper nanoparticles and prevent their agglomeration. Additionally, it played a role in augmenting the recyclability of the copper nanoparticle catalyst employed for the reductive elimination of Congo red dye from wastewater, with sodium borohydride serving as the reducing agent.

In the context of wound rehabilitation, the substantial porosity of nanofibers offers additional structural space for hosting transplanted cells, fostering cell migration and proliferation, and enhancing oxygen exchange and waste removal. The minute pore size of nanofibrous scaffolds plays a role in preventing dehydration and wound infection throughout the entire healing process. An antimicrobial peptides integrated nanofiber for wound dressings has been developed by Afshar et al. [40]. The incubation of epidermidis and nanofibers and the antibacterial process is illustrated in Fig. 11. The antibacterial effect was of the developed smart wound dressing material was studied with different concentrations of peptides against epidermidis using AlamarBlue cell viability assay. It was found that there is significant bacterial reduction in the nanofiber dressing with the increase of concentrations.

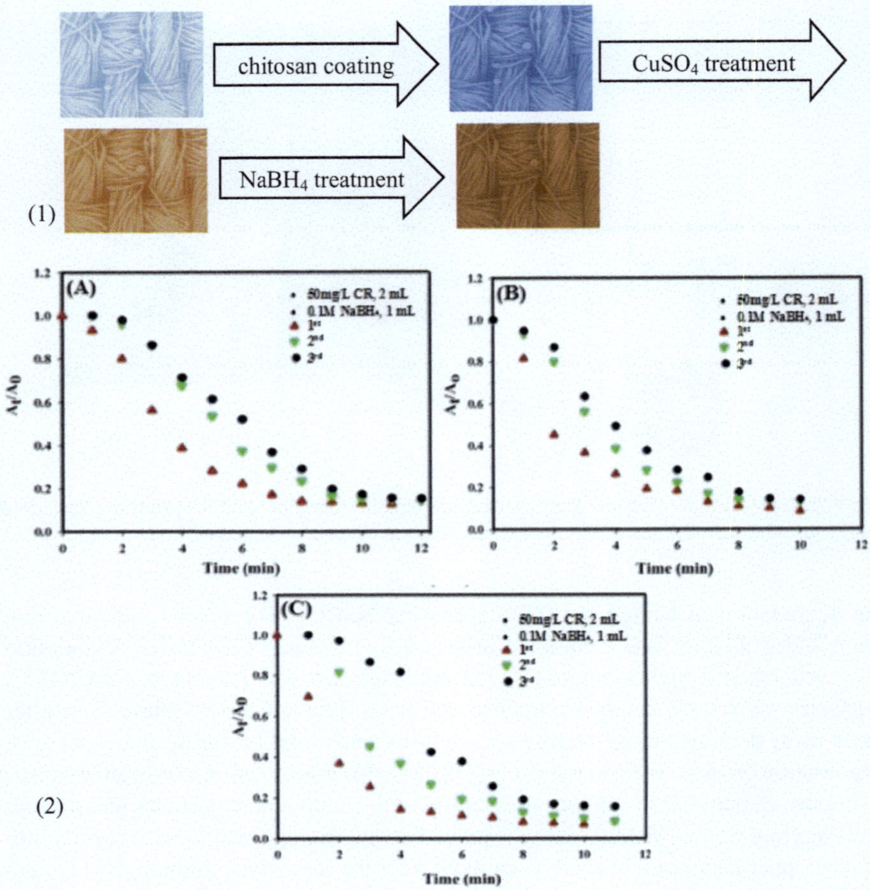

Fig. 10 (1) Steps of developing Cu nanoparticles impregnated Cu/Chi-CC catalyst, and (2) Congo red degradation affected by the concentration of $CuSO_4$ with 3×2 cm^2 Cu/Chi-CC catalyst, 0.05 (**A**), 0.1 (**B**), and 0.2 M (**C**) [39]

3.3 Environmental Sensing and Protection

Environmental factors such as temperature and toxic chemicals can be detected and monitored by incorporating nano-sensors into smart textiles. Lee et al. [40] engineered a graphene-based electronic sheet (GES) on polyester, and the fabrication process is depicted in Fig. 12a. Initially, the polyester sheet underwent treatment with a dopamine solution (2 mg/l) using a dip coating technique, creating a dopamine-coated sheet. This coating facilitated the interaction of graphene oxide (GO) flakes through mechanisms like π–π and electrostatic interaction, as well as hydrogen bonding. The dopamine-coated sheet was then immersed in a 0.3 g/l GO solution for a specific duration to uniformly coat the sheet with GO, resulting in a GO-dopamine sheet. Subsequent to the conventional chemical reduction of GO [41, 42], a reduced

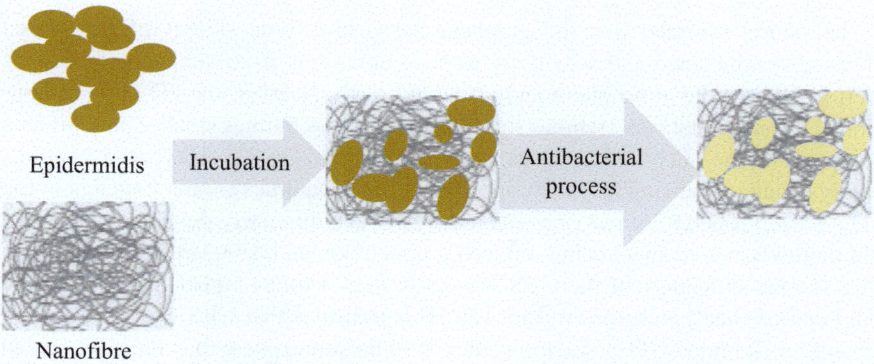

Fig. 11 Schematic diagram illustrating the incubation of epidermidis and nanofiber and antibacterial process

GO (rGO) flakes sheet was generated. The final product, an rGO-dopamine sheet or GES, was obtained after drying the sheet overnight using a heating device.

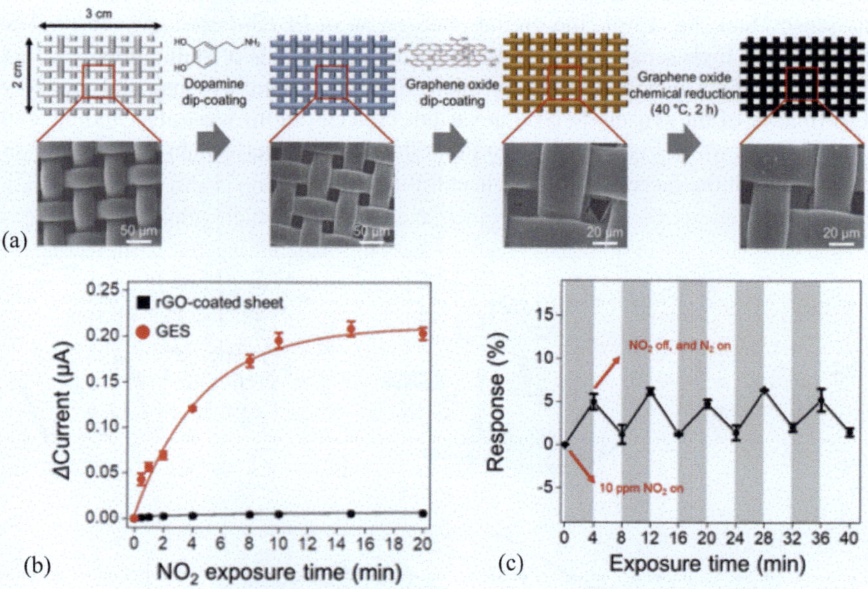

Fig. 12 **a** Schematic illustration of the graphene-based e-textile sheet (GES) fabrication process. The electrical assessment of both the sheet coated with reduced graphene oxide (rGO) and the GES in reaction to NO_2: **b** Fluctuations in electric current over an approximately 20 min exposure to 100 ppm NO_2, **c** Evaluation of repeatability when exposing GES to 10 ppm NO_2 under the room temperature. The changes in electrical response portray repeatability over time in a continuous five-cycle sequence. A triplicate measurements was made for each data point represtented [40]

The study revealed that the graphene-based electronic skin (GES) exhibited elevated conductance and sensitivity to NO_2 emissions from vehicle exhausts, as well as NH_3 in the atmosphere and artificial breath. Notably, the 2D GES demonstrated a weight that was 19 times lighter and thickness 5 times smaller per unit area compared to the 1D e-textile yarns. Additionally, the 2D GES showcased a significantly enhanced sensing response, approximately three times greater than that of the 1D e-yarn. Property comparison was made between the newly developed GES and old methods coated sheet using reduced graphene oxide (rGO), shown in Fig. 11a. The sensing efficiency of the GES was more than 4 times higher than that of the rGO-coated sheet, presented in Fig. 12b. This indicates that GES would be able to absorb much higher NO_2 toxic molecules from the atmosphere than the rGO-coated sheet. When the GES was repeatedly exposed to NO2, it exhibited stable response shown in Fig. 12c.

In their work, Nair et al. [43] created a gas sensor through the fabrication of a CNFs@Ni-Pt nanohybrid on a flexible polyester substrate, employing electrospinning and a chemical reduction method. The nanocatalysts, comprising bimetallic Ni and Pt on CNF, provided heightened active sites for gas pollutant sorption, leading to improved sensing performance. Khalifa and Anandhan [44] also developed a gas sensor using nanocomposite films. Figure 13a demonstrates the setup of the gas sensor, and Fig. 13b is the digital photograph of PBNF-based gas sensor. The response and high sensitivity performance of the sensor kept stable after 30 days in 100 operation cycles, proving excellent durability. The response and recovery performance of the sensors to NO_2 at various concentrations were also studied, and results are shown in Fig. 13c. It shows that the response performance of the sensor increased with the increase of gas concentration.

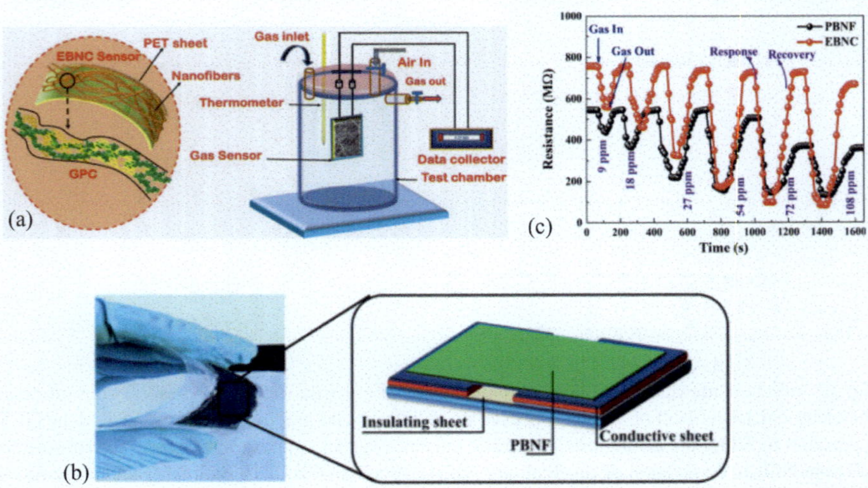

Fig. 13 **a** Gas sensing setup used for the performance evaluation of the gas sensors, **b** digital photograph and schematic of gas sensor, and **c** response transients to gas NO2 at different concentrations [44]

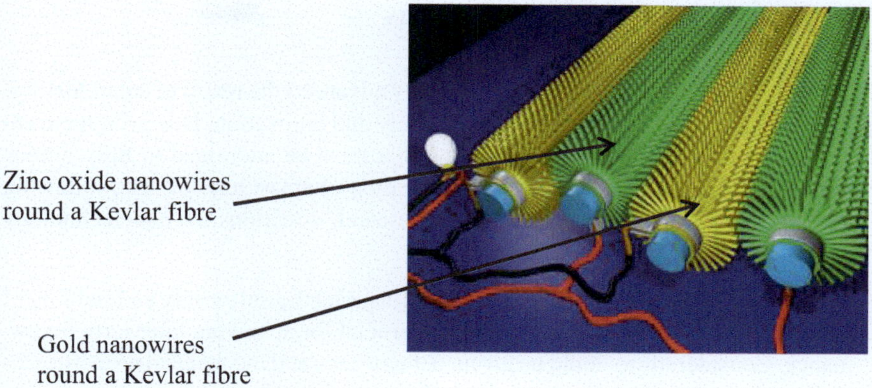

Fig. 14 Schematic illustration of microfiber-nanowire hybrid nanogenerator for generating electricity [45]

3.4 Sports and Fitness

Nano sensors can be integrated into sportswear that brings benefit to athletes by track recording of body movement, muscle fatigue, and hydration levels. Figure 14 illustrates the work principle of smart sportswear which include textile sensors based-on piezo-resistive materials for motion capture in the training of athletes. Other applications exist such as smart textile products that actively provide thermal regulation allowing the wears stay in a temperature stable microenvironment. These are used in sporting activities that involve high levels of exertion, where it is necessary to remove body heat effectively, followed by a period of relaxation, when the body cools down and there is a requirement to put heat back into the body, to maintain a constant temperature. High-performance textiles, which have low surface friction in air or water, are required by swimmers and athletes. Other sports required protective equipment, such as sailing, mountaineering, and racing car driving. In addition, health monitoring of athletes, for measuring levels of fitness and performance, is becoming an established technology. These textiles enhance performance, reduce the risk of injury, and aid in post-workout recovery.

3.5 Defense

An early work at Georgia Institute Technology investigated in generating electricity in boots of soldiers, hikers, and other users using piezoelectric materials. The fundamental idea involved utilizing piezoelectric nanowires as bottlebrush spikes to generate electricity through their movement. Although the generated electricity proved insufficient to power devices like MP3 players or mobile phones, it holds potential for charging sensors employed in monitoring physiological properties.

4 Challenges and Future Directions

Undoubtedly, nanotechnology has significantly advanced the realm of smart textiles, opening up diverse possibilities for functionality and innovation. However, there are challenges and areas for future exploration that must be addressed to fully unlock the potential of nanotechnology in the application of smart textiles. The following aspects highlight the challenges and future research directions for nanotechnology in smart textiles:

- Uniformity and Stability of Nanoparticles: Ensuring the uniformity and stability of nanoparticles incorporated into textiles is crucial for achieving long-term performance. Consideration should be given to the surface and mechanical properties, as well as properties related to physiological comfort when integrating nanomaterials into textile structures.
- Cost-Effectiveness of Nanomaterials: The current cost of producing nanomaterials may limit the widespread potential use of nanotechnology in smart textiles. Developing more cost-effective techniques and production processes for large-scale production is key to making nanotechnology accessible for smart textiles.
- Washability and Durability: Smart textiles used in harsh conditions may undergo regular wear and washing. Ensuring that integrated nanomaterials in textiles are washable, durable against repeated wearing and washing, and maintain stable functionality is essential.
- Health and Environmental Impact: Further study is needed to assess the health and environmental impacts of certain nanomaterials. This is critical for ensuring consumer safety and regulatory compliance in the textile industry.
- Efficient Power Systems: Electricity is crucial for powering sensors and actuators in smart textiles. Developing lightweight and efficient power systems, such as using energy harvesting technologies, remains a challenge.
- Multifunctionality and Advanced Nanomaterials: With the increasing demand for multifunctional smart textiles, there is a need for the continued development of advanced nanomaterials with improved properties such as conductivity and durability.
- Computational Approaches and Nanoscale Structure Design: Integrating computational approaches and nanoscale structure design is necessary for the accurate and efficient production of nanomaterials with unique properties to meet specific functionality requirements of smart textiles.
- Communication Devices and 3D Printing: Nanomaterials can be explored further for the development of nanoscale devices facilitating communication within different components of smart textiles and with external devices. Additionally, existing technologies like 3D printing and nanofiber-making require further exploration for smart textile applications.

In conclusion, the potential for nanotechnology in smart textiles is substantial. Collaboration across disciplines and continuous research efforts are essential to fully realize the capabilities of this field. Researchers from diverse subject areas must work together to explore new materials and fabrication techniques, pushing the boundaries to revolutionize products across various industries and enhance our daily lives.

References

1. Syduzzaman M, Hassan A, Anik HR, Akter M, Islam MR (2023) Nanotechnology for high-performance textiles. In: Chemistry of nanomaterials for energy, biology and more, pp 1–35
2. Roy TS, Shamim UD, Rahman MK, Ahmed F, Gafur MA (2020) The development of ZnO nanoparticle coated cotton fabrics for antifungal and antibacterial applications. Mater Sci Appl 11(9):601–610
3. Karim N, Afroj S, Malandraki A, Butterworth S, Beach C, Rigout M, Novselov KS, Casson AJ, Yeates SG (2017) All inkjet-printed graphene-based conductive patterns for wearable e-textile applications. J Mater Chem C 4(44):11640–11648
4. Yin B, Wen Y, Hong T, Xie Z, Yuan G, Ji Q, Jia H (2017) Highly stretchable, ultrasensitive, and wearable strain sensors based on facilely prepared reduced graphene oxide woven fabrics in an ethanol flame. ACS Appl Mater Interfaces 9:32054–32064
5. Habib S, Kishwar F, Raza ZA (2024) Citrate-mediated impregnation of silver nanoparticles for durable antibacterial cellulosic fabric. Pigment Resin Technol 53(2):240–248
6. Yılmaz F, Bahtiyari Mİ (2021) Antibacterial finishing of linen fabrics by combination with nano-aluminum oxide and crosslinking agent. Fibers Polym 22:1830–1836
7. Yılmaz F, Bahtiyari Mİ (2022) An approach for multifunctional finishing of cotton with the help of nano titanium silicon oxide. J Nat Fibers 19:3467–3474
8. Shirvan AR, Kordjazi S, Bashari A (2021) Environmentally friendly finishing of cotton fabric via star-like silver micro/nano particles synthesized with neem/salep. J Nat Fibers 18(10):1472–1480
9. Nourbakhsh S, Habibi S, Rahimzadeh M (2017) Copper nano-particles for antibacterial properties of wrinkle resistant cotton fabric. Mater Today: Process 4:7032–7037
10. Zarbaf D, Montazer M, Maryan AS (2017) In-situ synthesis of nano-copper on denim garment along with nano-clay for antibacterial and decoloration purposes. Cellulose 24(9):4083–4095
11. Prasad V, Arputharaj A, Bharimalla AK, Patil PG, Vigneshwaran N (2016) Durable multi-functional finishing of cotton fabrics by in situ synthesis of nano-ZnO. Appl Surf Sci 390:936–940
12. Raza ZA, Bilal U, Noreen U, Munim SA, Riaz S, Abdullah MU, Abid S (2019) Chitosan mediated formation and impregnation of silver nanoparticles on viscose fabric in single bath for antibacterial performance. Fibers Polym 20:1360–1367
13. Nadiger V, Shukla SR (2016) Antibacterial properties of silk fabric treated with silver nanoparticles. J Text Inst 107:1543–1553
14. Aksit A, Camlibel N, Zeren E, Kutlu B (2017) Development of antibacterial fabrics by treatment with Ag-doped TiO_2 nanoparticles. J Text Inst 108:2046–2056
15. Zahid M, Papadopoulou EL, Suarato G, Binas VD, Kiriakidis G, Gounaki I, Moira O, Venieri D, Bayer IS, Athanassiou A (2018) Fabrication of visible light-induced antibacterial and self-cleaning cotton fabrics using manganese doped TiO_2 nanoparticles. ACS Appl Bio Mater 1(4):1154–1164
16. Zhang G, Wang D, Yan J, Xiao Y, Gu W, Zang C (2019) Study on the photocatalytic and antibacterial properties of TiO_2 nanoparticles-coated cotton fabrics. Materials 12(2010):1–10
17. Fouda A, Hassan SE-D, Salem SS, Shaheen TI (2018) In-vitro cytotoxicity, antibacterial, and UV protection properties of the biosynthesized zinc oxide nanoparticles for medical textile applications. Microb Pathog 125:252–261
18. Iqbal K, Sun D (2018) Synthesis of nanoencapsulated Glauber's slat using PMMA shell and its application on cotton for thermoregulating effect. Cellulose 25:2103–2113

19. Sun D, Zhu F, Stylios G (2019) Investigation of composite fabric impregnated with non-Newtonian fluid for protective textiles. J Compos Mater 54(8):1013–1021
20. Tucker N, Stanger JJ, Staiger MP, Razzaq H, Hofman K (2012) The history of the science and technology of electrospinning from 1600 to 1995. J Eng Fibers Fabrics 7:63–74
21. Zhang X, Malekahmadi O, Sajadi SM, Li Z, Abu-Hamdeh NH, Rawa MJH, Al-Ebrahim MA, Karimipour A, Viet H (2023) Thermomechanical properties of coated PLA-3Dprinted orthopedic plate with PCL/Akermanite nano-fibers: experimental procedure and AI optimization. J Market Res 27:1307–1317
22. Zhang Y, Sun T, Zhang D, Sun S, Liu J, Li B, Shi Z (2023) Preparation of controllable micro-nano fiber bundle asymmetric porous PLA-PPC ethyl acetate driving film by pre-freezing treatment. Chem Eng J 469(1–12):143959
23. Wang C, Zhang P, Guo J, Wang J, Zhang T (2023) Durability and microstructure of cementitious composites under the complex environment: Synergistic effects of nano-SiO_2 and polyvinyl alcohol fiber. Constr Build Mater 400(1–14):132621
24. Qais DS, Islam MN, Othman MHD, Mahmud E, Quayum ME, Islam MA, Ismail M, Habib A (2023) Nano-zinc oxide fibers: synthesis, characterization, adsorption of acid blue 92 dye, isotherms, thermodynamics and kinetics. Emerg Contam 9(1–13):100224
25. Lee KY, Mooney DJ (2001) Hydrogels for tissue engineering. Chem Rev 101(7):1869–1879
26. Ramalingam M, Ramakrishna, S (2017) An introduction to nanofibers and nanocomposites. In: Nanofiber compoistes for biomedical applications, Elsevier Ltd, pp 3–29
27. Jung YH, Chang TH, Zhang H, Yao C, Zheng Q, Yang VW, Ma Z (2015) High-performance green flexible electronics based on biodegradable cellulose nanofibril paper. Nat Commun 6(1):7170
28. Li D, Xia Y (2004) Electrospinning of nanofibers: reinventing the wheel? Adv Mater 16(14):1151–1170
29. Yan BY, Cao ZK, Hui C, Sun TC, Xu L, Ramakrishna S, Yang M, Long YZ, Zhang J (2023) MXene@Hydrogel composite nanofibers with the photo-stimulus response and optical monitoring functions for on-demand drug release. J Colloid Interface Sci 648:963–971
30. Chen NJ, Wang HH, Kim SP, Kim HM, Lee WH, Hu C, Bae JY, Sim ES, Chung YC, Jang JH, Yoo SJ, Zhuang YB, Lee YM (2021) Poly(fluorenyl arlpiperidinium) membranes and ionomers for anion exchange membrane fuel cells. Nat Commun 12:2367–2379
31. Sun G, Jiang S, Feng X, Shi X, Zhang X, Li T, Chen N, Hou L, Qi S, Wu D (2022) Ultra-robust polyimide nanofiber separators with shutdown function for advanced lithium-ion batteries. J Membr Sci 645:1–8
32. Chen Y, Zhang C, Tao S, Chai H, Xu D, Li X (2023) High-performance smart cellulose nanohybrid aerogel fibers as a platform toward multifunctional textiles. Chem Eng J 466:1–11
33. Chen Y, Li Y, Zhang C, Qi H, Hubbe M (2022) Holocellulosic fibers and nanofibrils using peracetic acid pulping and. Carbohyd Polym 295:1–10
34. Feng Q, Wu D, Zhao Y, Wei A, Wei Q, Fong H (2018) Electrospun AOPAN/RC blend nanofiber membrane for efficient removal of heavy metal ions from water. J Hazard Mater 344:819–828
35. Honarvar MG, Latifi M (2017) Overview of wearable electronics and smart textiles. J Text Inst 108(4):631–652
36. Barman J, Tirkey A, Batra S, Paul A, Panda K, Deka R, Babu PJ (2023) The role of nanotechnology based wearable electronic textiles in biomedical and healthcare applications. Mater Today Commun 32(1–19):104055
37. Sun D, Chen M, Podilchak S, Georgiadis A, Abdullahi S, Joshi R, Yasin S, Rooney J, Roneey J (2019) Investigating flexible textile-based coils for wireless charging wearable electronics. J Ind Text 50(3):333–345
38. Tripathi R, Narayan A, Bramhecha I, Sheikh J (2019) Development of multifunctional linen fabric using chitosan film as a template for immobilization of in-situ generated CeO_2 nanoparticles. Int J Biol Macromol 121:1154–1159
39. Ali N, Kamal T, Ul-Islam M, Khan A, Shah SJ, Zada A (2018) Chitosan-coated cotton cloth supported copper nanoparticles for toxic dye reduction. In J Biol Nacromol 111:832–838

40. Lee SW, Jung HG, Jang JW, Park D, Lee D, Kim I, Kim YCDY, Hwang KS, Lee G, Yoon DS (2021) Graphene-based electronic textile sheet for highly sensitive detection of. Sens Actuat: B Chem 345(1–10):130361
41. Yun Y, Hong W, Kim W-J, Jun Y, Kim B (2013) A novel method for applying reduced graphene oxide directly to electronic textiles from yarns to fabrics. Adv Mater 25:5701–5705
42. Lee S, Jung H, Kim I, Lee D, Kim W, Kim S (2020) Highly conductive and flexible dopamine–Graphene hybrid electronic textile yarn for sensitive and selective NO_2 detection. ACS Appl Mater Interface 12:46629–46638
43. Nair KG, Vishnuraj R, Pullithadathil B (2021) Highly sensitive, flexible H2 gas sensors based on less platinum bimetallic Ni–Pt nanocatalyst-functionalized carbon nanofibers. ACS Appl Electron Mater 3(4):1621–1633
44. Khalifa M, Anandhan S (2021) Highly sensitive and wearable NO2 gas sensor based on PVDF nanofabric containing embedded polyaniline/g-C3N4 nanosheet composites. Nanotechnology 32(48):485504
45. Wang ZI, Wang XD (2009) New nanogenerator could charge iPods and cell phones with a wave of the hand, New Atlas, 31 03 2009. [Online]. https://newatlas.com/nanogenerator-ipod-charger/11364/. Accessed 31 Dec 2023
46. SENSATEX, "SENSATEX," SENSATEX, [Online]. https://www.sensatex.com/smartshirt_as.html. Accessed 31 Dec 2023
47. Shathi MA, Chen M, Khoso NA, Rahman MT, Bhattacharjee B (2020) Graphene coated textile based highly flexible and washable sports bra for human health monitoring. Mater Des 193:108792

Chapter 16
Biocompatibility and Toxicity of Nanoparticles in Textile Finishing

Mandira Mondal, Sourav Banerjee, Swagata Banerjee, S. N. Naik, and S. Wazed Ali

1 Introduction

Textile is one of the basic needs of human beings. Formerly, textiles were only used for apparel and home furnishing purposes whereas currently, it is used for functional purposes in the textile sector as well as other sectors. Textile finishing is required to enhance aesthetic, comfort and functional properties of a textile substrate. Textile finishings are evolving with the advancement of technologies and nano-finishing of textiles is in prime attention to prepare highly functional textiles [65]. The nano-finishing technique is to form nano-dimensional materials using nanotechnology and then imparting it to textile substrate by different finishing methods. The functional properties get magnified in the case of nano-finished textiles compared to bulk-finished textile using a particular add-on of a similar finishing agent. The range of textile applications is getting wider by adding such distinctive qualities to textiles through nanomaterials [2]. Nano-finishing leads to less add-on on textile substrates compared to bulk finishing on textile substrates using the same finishing agents. This results in more durability due to the increment of surface area to volume ratio of nanostructured material and nano-finishing also maintains the inherent physical properties of textile substrates [55]. Despite several advantages of nano-finishing of textiles, there are downside of using nanomaterials for nano-finishing of textiles such

M. Mondal · S. Banerjee · S. W. Ali (✉)
Department of Textile and Fibre Engineering, Indian Institute of Technology Delhi, Hauz Khas, New Delhi 110016, India
e-mail: wazed@iitd.ac.in

S. Banerjee · S. N. Naik · S. W. Ali
School of Interdisciplinary Research (SIRe), Indian Institute of Technology Delhi, Hauz Khas, New Delhi 110016, India

S. N. Naik
Centre for Rural Development and Technology, Indian Institute of Technology Delhi, Hauz Khas, New Delhi 110016, India

S. Bairagi et al. (eds.), *Nanotechnology in Textile Finishing*, Materials Horizons: From Nature to Nanomaterials, https://doi.org/10.1007/978-981-97-2696-7_16

as toxicity, biocompatibility, and hike in cost. There are different finishing methods by which nano-finishing agents can be incorporated into textile substrates. Layer-by layer (L-b-L), physical vapour deposition, chemical vapour deposition method, sol-gel method, plasma method, etc., are the various nano-finishing techniques [4]. The functional properties which are introduced to textile substrates by nano-finishing are flame retardancy [38], crease recovery [75], UV protection [52], antioxidant [41], anti-microbial [72], [48], superhydrophobicity [44], self-cleaning [58], etc. There are vast types of nano-finishing agents that are explored as textile finishing agents such as nano-silica [23], ZnO nanoparticle with TiO_2 [13], nano-Titanium Silicon Oxide [7] ZnO @SiO_2 nanoparticles [46] silver nanoparticles [7], nano-chitosan [69], nano-lignin [29], coconut fibre nano-lignin [30], caffeine loaded nanoparticle [41], selenium nanoparticles [20], and nano-chitosan-polyurethane [45].

Toxicity and biocompatibility are two major concerns that arise from the usage of nanomaterials in nano-finishing methods of textiles. These two factors are still challenging even after so much advancement in the textile sector. The topic of toxicity is associated with two segments, one is toxicity related to the environment and the other is toxicity related to human beings [55]. Toxicity is inducing oxidative stress into the environment or living cells triggered by a foreign material or increment of a specific material [19]. Biocompatibility is the response of a specific host after interaction with a foreign material. It also describes the biological mechanism by which the living cells are reacting to the external materials or nanomaterials [68]. The nano-finished textile has a huge contribution towards toxicity and biocompatibility issues. The factors of nano-finished textile which deals with the biocompatibility and toxicity issues are preparation techniques of the nanostructures, application methods of nanomaterials to the textile substrates, durability of nano-finished textile, storage, and transportation conditions of the textile substrates (which also associates with durability) and finally disposing and recycling techniques of the nano-finished textile substrates. The overall factors of toxicity and biocompatible issues are explained in Fig. 1 [55].

This book chapter is a detailed discussion about the impact of nanostructured materials used in textile finishing on environment and living beings in terms of toxicity and biocompatibility.

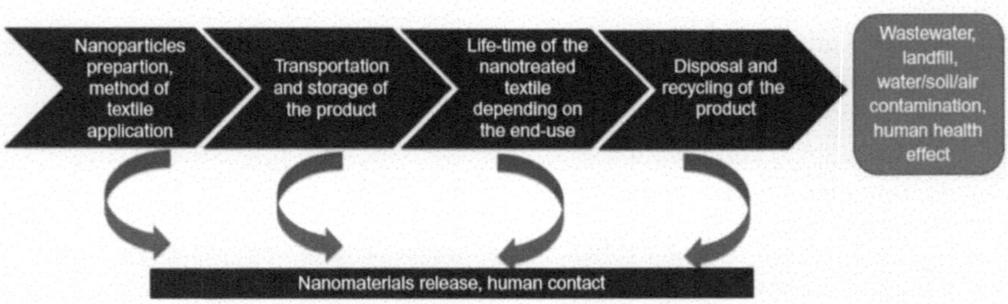

Fig. 1 Overall factors of toxicity and biocompatible issues from nano-finished textiles associated with the lifecycle of nano-finished textiles [55]

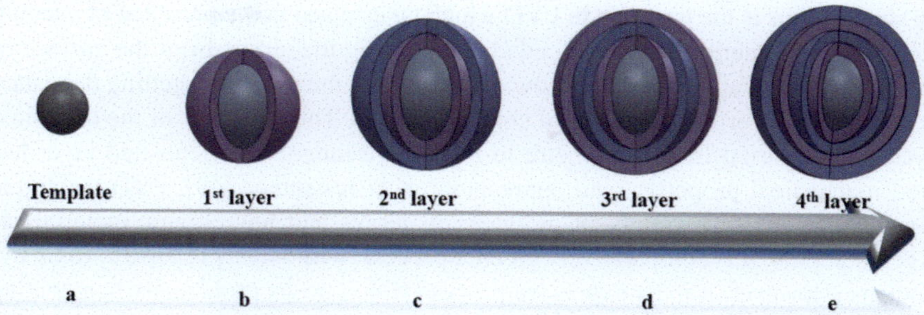

Fig. 2 Multilayer nanoparticle formation using layer-by-layer deposition technique [57]

2 Application Techniques of Nano-dimensional Materials to Textile Substrates

2.1 Layer-By-Layer Assembly

The layer-by-layer assembly is a technique of depositing films with controllable thickness, on textile substrates. The main mechanism behind the layer-by-layer technique is the alternate exposure to oppositely charged solutions leading to adsorption of oppositely charged ions. The films can be composed of nanostructures like nanoparticles or finishing agents. Textiles are suitable for layer-by-layer deposition method due to its ability to hold and support the ultrathin film growing on its surface [50]. This is a facile, economical, and controllable method for deposition of ultrathin films on textile substrates [16]. Apart from thin film formation on textile substrates, nanoparticles with multiple layers can also be developed. To do this, a nanoparticle template is immersed in a polyelectrolyte solution. Excess solution is removed by centrifugation, followed the first step. Next, the oppositely charged colloid solution is deposited on the nanoparticle forming the second layer. In this manner, the steps can be repeated depending on the number of layers required to be deposited on the nanoparticle to form a multilayered nanoparticle. The process for the same has been illustrated in Fig. 2 below. One of the primary advantages of this method is its ability to monitor the growth rate of the film in nanometre range. Further, it increases the colloidal stability of the nanocarriers.

2.2 Vapour Deposition Techniques

The vapour deposition techniques are widely explored for the development of thin films on textile substrates. The two most commonly used processes include the physical vapour deposition (PVD) and chemical vapour deposition (CVD). PVD involves the removal of atoms from the target material with the help of a high-energy ion source. This is the first step in PVD which takes place in the presence of vacuum [37]. The high-energy ion source aids in vaporizing the atoms from the surface of the target. These vaporized atoms now move towards the substrate, getting deposited on its surface to form a thin layer of coating (Fig. 3). The thickness of the deposited layer can be varied from nanometre to millimetre range. The deposited layer has good mechanical properties and appreciable wear resistance [49]. PVD technique allows the usage of a variety of inorganic materials and few organic materials as target materials. It is comparatively greener process, without any release of toxic by-products. However, it is an expensive process which poses problems when coating complex shaped substrates.

Chemical vapour deposition technique is a commonly used method for the deposition of thin metallic, alloy or intermetallic compounds. Unlike PVD, CVD involves the deposition of target material from its gaseous form. The volatile precursors react with the heated substrate to form a solid deposition of the target material on the substrate [40]. The quality of the coatings developed through this technique is

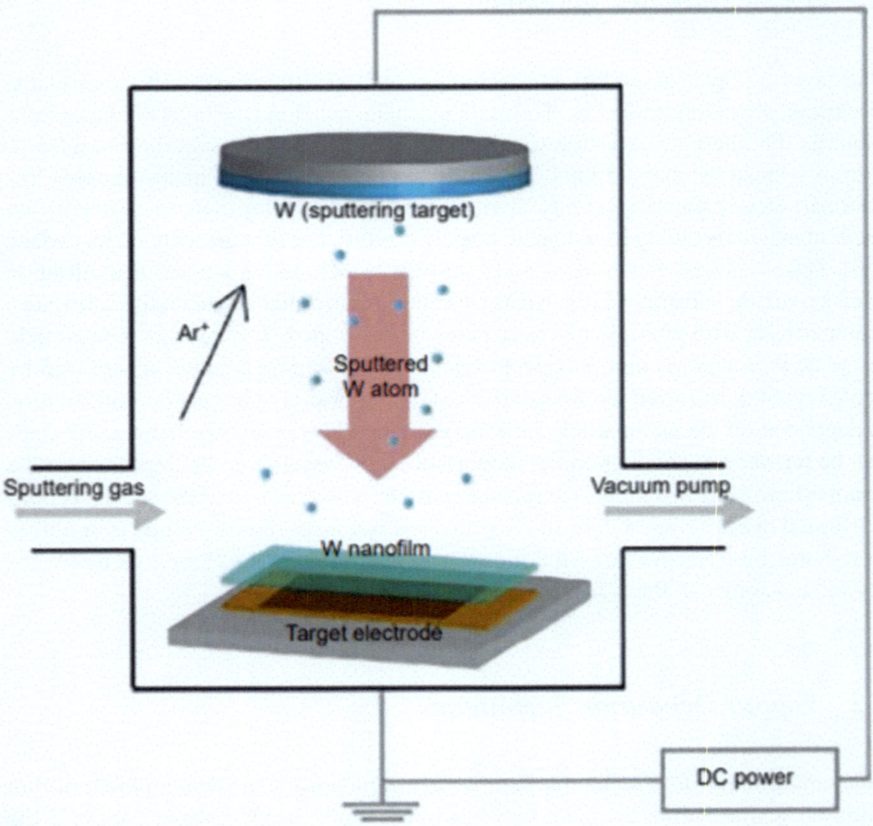

Fig. 3 Process of physical vapour deposition technique [49]

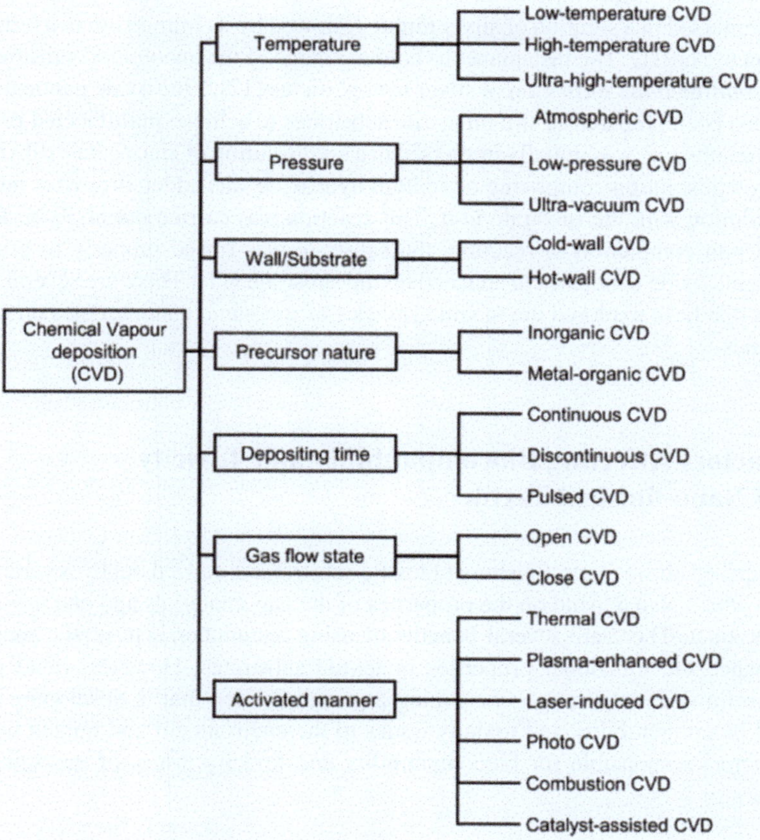

Fig. 4 Types of CVD technique [54]

commendable. Further, the deposition rate can be controlled by controlling the rate of chemical reaction. The different types of CVDs [62] have been compiled in the diagram below (Fig. 4).

2.3 *In Situ Nanostructure Formation*

Often the preparation of nanostructures involves a series of chemical reactions. The preparation followed by incorporation of such structures in/onto textiles creates a lengthy process. The adherence of the nanostructures with the substrate often becomes a concern with the conventional techniques of nanomaterial incorporation. Hence, the strategy of in situ nanomaterial synthesis has been adopted that overcomes the problems of adherence between the nanostructure and the substrate. Silver nanoparticles were prepared in situ on textile substrates. The textile substrates were immersed in a solution of silver nitrate followed by its immersion in a reducing solution of $NaBH_4$. The instantaneous colour change of the uncoloured substrates to black confirmed the formation of silver nanoparticles [12]. Zinc oxide nanoparticles have also been prepared in situ on textile substrates to achieve multifaceted properties. The substrate was initially immersed in a seed solution of zinc acetate dihydrate. The growth solution comprising of sodium hydroxide was added dropwise into the seed solution with the substrate in it. This reaction was carried out at 50 °C for an hour. Upon completion of reaction, the substrate was rinsed properly to produce in situ zinc oxide nanoparticle attached textile substrate [11]. There are several other studies that have explored the in situ synthesis of inorganic nanoparticles on textile substrates.

3 Factors Affecting Biocompatibility and Toxicity of Nano-finished Textile

As described above, nano-finishing of textile substrates is carried out by several techniques which also depend on the properties of the nanomaterials and end use of the final product. There are several benefits of using nanomaterial in textile finishing to enhance the functional properties of textile substrates. However, rapid usage of nano-finished substrates is becoming a prime concern that is developing unintended biocompatibility and toxicity issues to the environment and human bodies. The factors responsible for biocompatibility and toxicity issues of nano-finished textiles are:

3.1 Properties of Nanomaterials

The properties of nanomaterials that are involved in toxicity and biocompatibility issues are size, morphology, aspect ratio, agglomeration, type of material, dissolution, crystallinity, and coating of nanomaterials [19], [55]. The properties of nanomaterials are shown in Fig. 5.

Some factors about how these properties affect toxicity and biocompatibility issues are described below:

3.1.1 Size

Nanoparticles which are used for textile finishing are mostly below or almost 100 nm. The surface area is also related to the size of nanostructured material. The decrease in size of nanomaterial results in increase in surface area of nanomaterial [19]. The

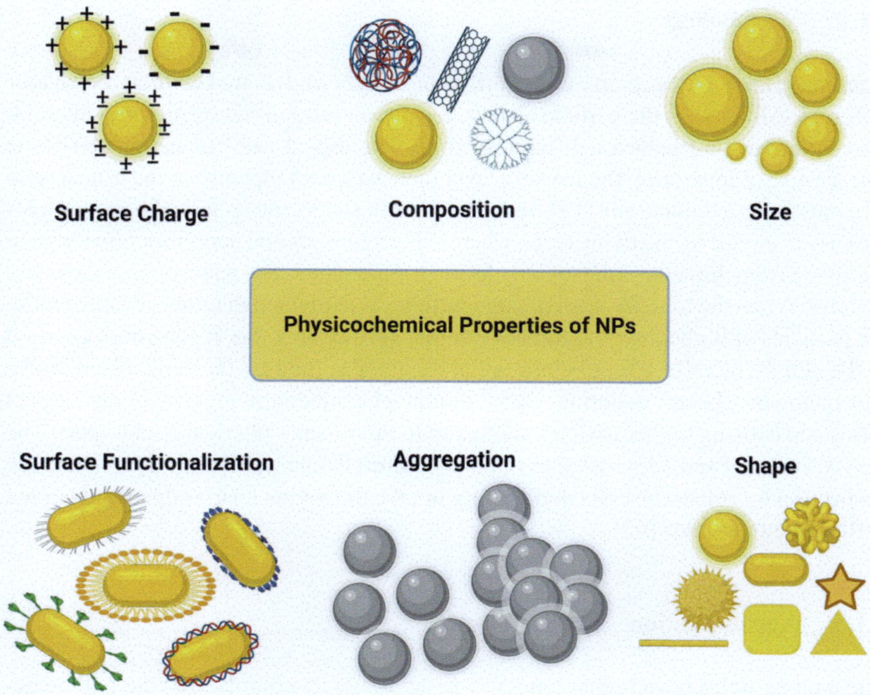

Fig. 5 Properties of nanostructured materials responsible for toxicity and biocompatibility [8]

mobility of nanomaterials through cells increases with decrease in the size of nanomaterials used in textile finishing. The nanomaterials enter the living cells and react with the macromolecules or biomolecules and then the level of toxicity gets increased. The toxicity level decides the biocompatible nature due to interaction between nanomaterial and living cells. Generally, the nanomaterial having size of 100 nm is permeable to cells as the size of nanomaterials is same as size of DNA, protein globules, and thickness of cell membrane. The invasion of nanoparticles (below 20 nm) into cells is even more toxic as it can penetrate and accumulate in the nucleus of living cells [63]. The surface area is higher in nanoparticles so there will be more number of nanoparticles per unit mass compared to bulk particles of the same material. The greater number of nanoparticles in a unit mass makes the body cells react differently compared to the bulk bigger particles as the higher surface area and higher surface reactivity of the nanoparticles leads to more toxicity to cells. The larger number of reactive oxygen species from higher surface reactive nanoparticles is the root cause of cytotoxicity and genotoxicity [19].

3.1.2 Morphology

Nanomaterials can be prepared in different shapes and it makes their behaviour different. Along with the particle shapes, the aspect ratio of nanostructured material also functions and behaves differently [63]. The aspect ratio of nanomaterials is surface-to-volume ratio. The toxicity level increases with increase in the aspect ratio of nanostructured materials [19]. In terms of only shape, the spherical nanoparticles are easy to engulf by the living cells, whereas the nanorods and nanofibres are tough to eliminate from the cells so the toxicity level is tremendous. The nanospheres show less toxicity. When the toxicity level of nanoparticles is explained in terms of aspect ratio, the particles of higher aspect ratio have higher surface area. So, it gets more exposed to the cell surfaces of living beings and is easily absorbed by the cells. Nano-worm and nanowire shaped materials show similar phenomena as having higher aspect ratios and causing higher toxicity compared to their nano-spherical counterpart. The curved surfaces and edgy surfaces of nanostructured materials also play a role in the absorption by cells. The cells don't easily uptake the higher sharp-edged and curved surface nanoparticles [63].

3.1.3 Agglomeration

The nanoparticles have higher tendency to agglomerate compared to the bulk materials as the surface area is high which leads to higher free surface energy. There is a high chance of nanoparticles in agglomerated form rather than in dispersed form when nanoparticles are added to a solution. This is the reason due to which additional agent is added to provide shield to the nanoparticle to avoid agglomerations. The agglomeration is also one of the reasons of toxicity and biocompatibility issues due to nanoparticles [19]. It can be concluded from a study that oxide nanoparticles and nanotubes of carbon show more toxicity in agglomerated form compared to dispersed form [66].

3.1.4 Surface Charge

There is always a possibility of having charge polarity and charge density present in nanoparticles which can cause cytotoxicity. The metallic nanoparticles such as silica, zinc oxide, and gold of positive charge show more toxic effects than their counterpart. Even similar types of nanoparticles act differently to different cells of living bodies. The porosity of nanoparticles is also a factor for the toxicity and biocompatibility to living beings. When the cell membrane and nanoparticles interact, some sort of ionic interaction goes through between them. The mechanism of ionic interaction between them is a very vital factor. The behaviour of cell membranes having negative charge to positively charged nanoparticles is different depending on the density of charge and other factors. The biomolecules which have many benefits to living bodies show toxic nature when changes in surface charge are observed [63]. In a study, it is reported that nano-curcumin is coated with polyvinylpyrrolidone, polyvinyl alcohol, and dextran to achieve positive, negative, and neutral surface charges, respectively. The higher cytotoxic level is observed in case of positively charged nano-curcumin through apoptosis and reactive oxygen species (ROS) generation [39].

3.2 Properties of Nano-finished Textiles

The fabrication procedures of nanomaterials, different application techniques to impart nanomaterials to the textile substrates, washing durability of nano-finished textiles, rubbing fastness of the nano-finished textile, and other durability properties are the factors responsible for biocompatibility and toxicity of nano-finished textiles [55]. In case of application of nanoparticles to textile substrate, several factors act in biocompatibility and toxicity such as techniques followed for application of nanoparticles to textile substrate and solvent used for dispersion and incorporation of nanoparticles to textile substrates. The technique required for imparting nanoparticles to textile substrate and the interaction mechanism between nanoparticle and textile substrate defines the durability of the nano-finished textile. Generally, the chemical bond formation between textile substrates and finishing agents is responsible for durability. However, in many cases especially for nano-finishing, adhesion of nanoparticles to textile substrates is weak and this causes leaching out from textile substrates. The tendency of leaching out of nanoparticles from the textile substrates during any events such as washing, perspiration, rubbing with skin, or other textile substrates is also important factor of toxicity and biocompatibility issues [10].

4 Toxicity and Biocompatibility Issues of Nano-finished Textile

Nanostructured materials which are used in nano-finished textiles are either metallic, organic, inorganic, or polymeric nanoparticles. The types and properties of nanoparticles determine the biocompatibility and toxicity issues of nano-finished textiles. The methods to impart nanoparticles to textile substrates also play a vital role in biocompatibility and toxicity of nano-finished textiles.

The nanomaterials which are used in nano-finished textiles are intake by human being or living beings through several routes. The several routes through which nanomaterials are uptake by human beings are listed below [22]:

- Cellular uptake (from skin)
- Oral uptake
- Inhale uptake

In case of textile industries, the workers or textile technocrates are exposed to nanoparticles for a longer period. So, the route through which the body mostly uptake the prepared nanoparticles are skin as well as respiratory tract. The accumulation of nanoparticles in different human organs takes place after the entry of nanoparticles to human body [51].

Textile substrates are mostly exposed to the skin and the nanoparticles from nano-finished textile are mostly absorbed by skin [55]. Skin consists of several layers and the main routes through which nanoparticles enter through skin are stratum corneum and skin appendages. In the outermost layers of skin, the nanoparticles get accumulate in the roots of hair follicle and the nanoparticles enter to the deep layers of skin when the hair follicle opens up (shown in Fig. 6). The nanoparticles penetrate the skin along the intercellular spaces in between the lipid layers. The penetration rate is higher for the metallic nanoparticles compared to other types of nanoparticles. The physical state of skin also plays a very important role in penetration of nanoparticles through the skin. The penetration of nanoparticles through damaged skin is easier compared to healthy skin. The skin is also considered as a super absorptive organ because of its connection with lymph vessels, dendrites, tissues, blood vessels, and different types of sensory nerve endings [55].

As per previous studies, it can be concluded that leaching out of nanoparticles from nano-finished textile is very common and it can occur during several events such as laundering, perspiration, and rubbing (dry or wet) [10], [42]. These types of events can lead to absorption of nanoparticles when nano-treated textile is directly in contact with human bodies. The nanoparticles even leach out during washing and get disposed of in waterbodies and lands. The abrasion of nano-finished textile can even lead to discharge of nano-finishing agents into air but in very lesser quantities [55]. The uttermost important concern is the data acquisition of the emission of nanoparticles and environmental concentrations of nanoparticles of the nanotextiles. The textile finishing agents used in nano-forms are majorly disposed of in waste

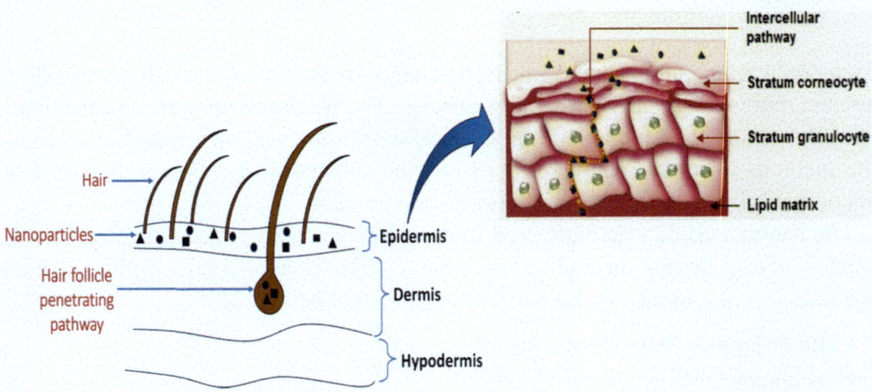

Fig. 6 Easiest route followed by nanoparticles to penetrate through skin [Adapted from [63]]

incineration, wastewater, and sewage sludge. These all are major causes of environmental degradation [42]. Many studies have reported that the release of nanoparticles from textiles causes environmental hazards. The antibacterial textiles and conductive textiles are widely prepared using silver nanoparticles. The silver nanoparticles are released from textile substrate during perspiration and laundering. The release of silver nanoparticles in artificial sweat varies from 0 to 322 mg/kg of fabric weight which depends on other factors such as sweat formulation, fabric quality, and silver concentration [34]. Similarly, titanium dioxide nanoparticles are highly used in textile substrates for UV absorption. However, the wash-off of TiO_2 nanoparticles from textile substrates is reported in many cases. It is reported that TiO_2 nanoparticles are released from textile substrates used for UV protection and anti-microbial properties is 0.01–0.06 wt% and 3.4 wt% of total Ti, respectively, after 1st washing cycle [67]. Most of the washed nanoparticles get off into water and then accumulate in lands and water bodies. In various cases, the durability of nanoparticles in nano-finished textiles is good and nanoparticles are not completely leached out during its lifetime and that is finally dumped on land or water bodies [42]. Figure 7 shows the emission and way of uptake of nanoparticles from nano-finished textiles by environment and human bodies.

The nanoparticles after entering into human bodies get accumulate in different body parts and the transportation of nanoparticles into body parts is shown in Fig. 8. The nanoparticles penetrate the cells and damage the living cells (shown in Fig. 9). The accumulated nanomaterials in different body parts cause serious health issues explained in Table 1.

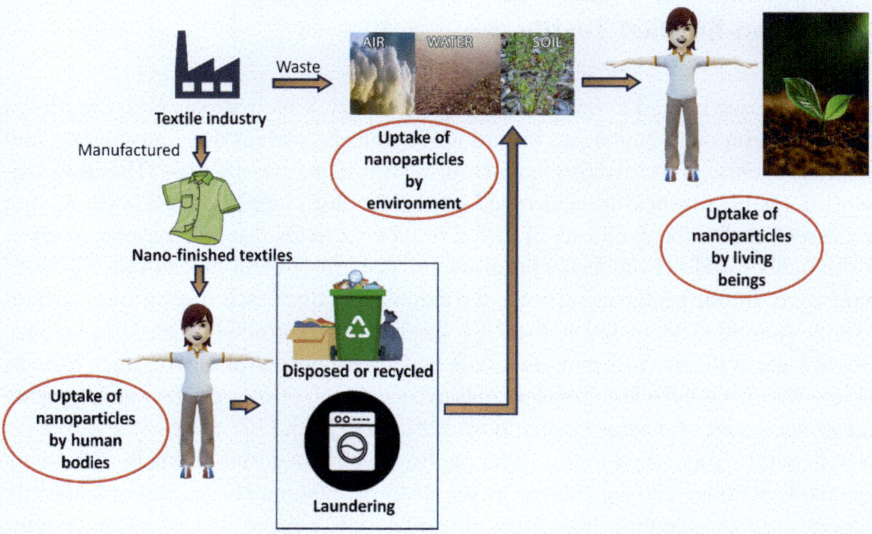

Fig. 7 Pathways of uptake of nanoparticles from nano-finished textile by human bodies and environment [Redraw from [51]]

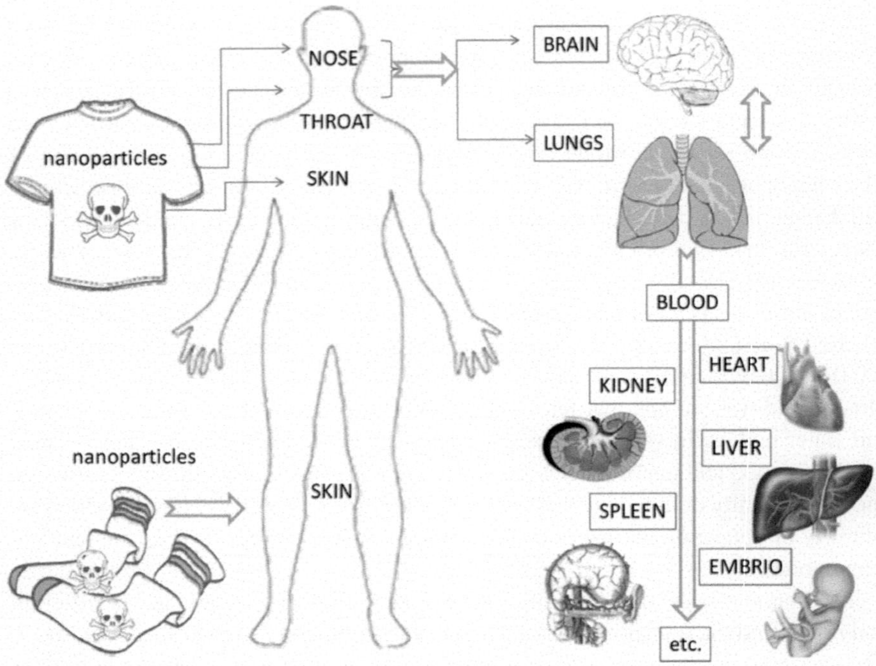

Fig. 8 Pathway followed by nanoparticles throughout the human body [51]

5 Assessment of Toxicity and Biocompatibility of Nano-finished Textile

One major issue related to textile finishing is toxicity, which affects both the person and the environment. Numerous hazardous chemicals, such as dyes, auxiliaries, and finishing agents, primarily originate from textile industries [3], [9]. The advancement of textile finishes has leveraged nanotechnology applications, and ongoing processes involve the synthesis of different nanoparticles through various green or in situ methods. However, as we progress, the need for greater precautions becomes imperative. In our previous sections, we extensively discussed the various types of toxicity stemming from finished textile products. It is crucial to strike a balance between the utilization of nanomaterials and their concentration for safer use. To achieve this, toxicity assessment emerges as a vital tool, ensuring safety and securing a healthier society for future generations [25], [26], [32], [70], [56].

In general, 'Biocompatibility' refers to how well a material interacts with living organisms without causing harm. In the case of nanomaterials, biocompatibility focuses on understanding how these tiny materials behave in biological systems like cells, tissues, or living organisms. We can claim a nanomaterial is biocompatible if they shouldn't cause adverse effects like toxicity, inflammation, or cell damage when they encounter with living tissues or cells. Biocompatible nanomaterials should

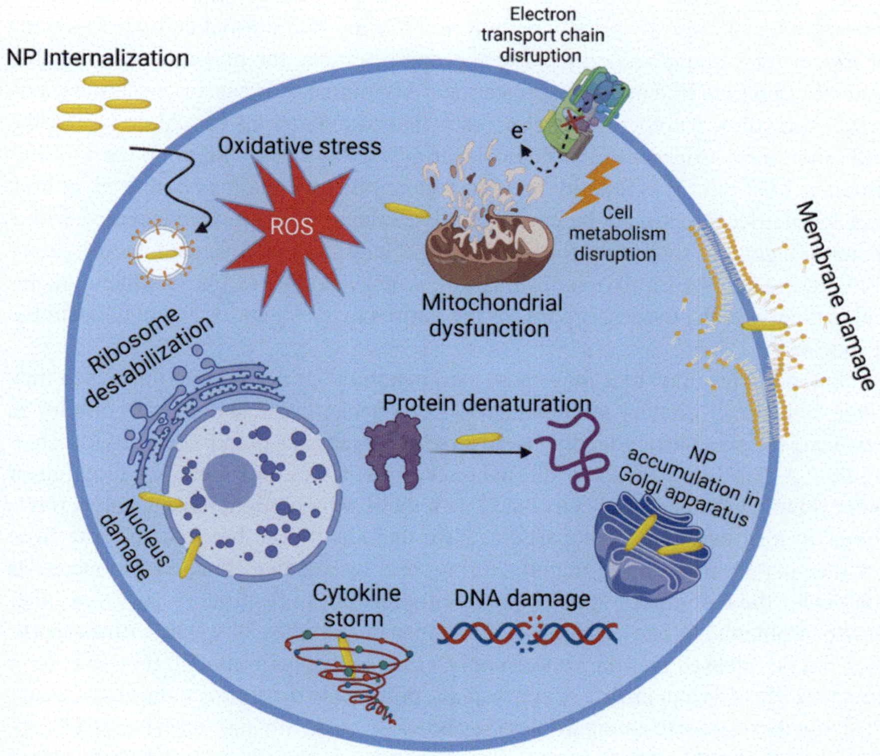

Fig. 9 Different modes of action followed after entering of nanoparticles into living cells [8]

Table 1 Diseases associated with exposure of nanoparticles to different parts of the body [14]

Body part	Diseases
Brain	Neurological diseases: Parkinson's, Alzheimer diseases
Lungs	Asthma, bronchitis, emphysema, cancer
Circulatory system	High blood pressure, thrombus, vasoconstriction, artheriosclerosis
Heart	Arrythmia, other heart diseases
Gastro-intestinal system	Crohn's disease, colon cancer
Orthopaedic implant wear debris	Auto-immune diseases, dermatitis, urticaria, vascultitis
Lymphatic system	Podoconiosis, Kaposi's sarcoma
Skin	Auto-immune diseases, dermatitis

interact with biological systems without interfering with normal cellular functions or triggering immune responses. These properties make the nanomaterials capable for emerging field like medical applications, environmental uses, or consumer products like textiles. Another major concern is that nanomaterials should remain stable and retain their properties in biological settings without degrading into harmful by-products [31] [36, 71]. Toxicity and biocompatibility evaluations both look at how a material affect health, but biocompatibility evaluations are more concerned with a material's general suitability for living things, with the goal of guaranteeing safety for certain uses. As for toxicity evaluation, it measures the extent to which a material can harm cells, tissues, organs, or the entire environment, or create undesirable responses.

Nanoparticles in textiles may cause skin irritation or allergic reactions when they come into direct contact with the skin. Certain nanoparticles have the ability to permeate the skin barrier, leading to adverse skin reactions in individuals with sensitivities. A baby may swallow the nanomaterials if they lick items of clothing or other infant gear that have been coated with them, which might have negative physiological repercussions. Nanoparticles have the potential to become airborne from textiles, particularly during manufacturing, use, or washing. Inhaled nanoparticles may enter the respiratory system, potentially leading to respiratory issues or other health problems. When clothing containing nanoparticles is washed, these particles can be released into the water supply. Once in the environment, they may have adverse effects on aquatic life or ecosystems, potentially disrupting natural processes. Prolonged exposure to nanoparticles may be associated with long-term health effects, as nanoparticles themselves may undergo changes or chemical reactions over time, especially when exposed to different environmental conditions. These altered forms could possess different toxicological properties compared to their original state. Some nanoparticles have the potential to accumulate in the body over time, and their long-term exposure on human health can be dangerous, affecting various organs and the nervous system (as shown Fig. 10) [6, 43].

In this segment, we aim to provide a detailed layout of how to assess the toxicity of different materials, particularly those can be used in textile nano-finishes (Fig. 11).

5.1 In Vitro Method

In vitro methods for assessing toxicity are widely regarded as the most dependable, cost-efficient, versatile, and ethically superior compared to animal studies. This approach is primarily employed to evaluate the toxicity patterns of modified textiles. These techniques function by replicating cellular elements and projecting potential outcomes within an organism's body. Reviews play a pivotal role in establishing guidelines for safe dosage levels and the handling of xenobiotics. Various cell lines are exposed to nanomaterials in a controlled environment, and subsequent assessments of the exposed component's proliferation and metabolism are conducted through

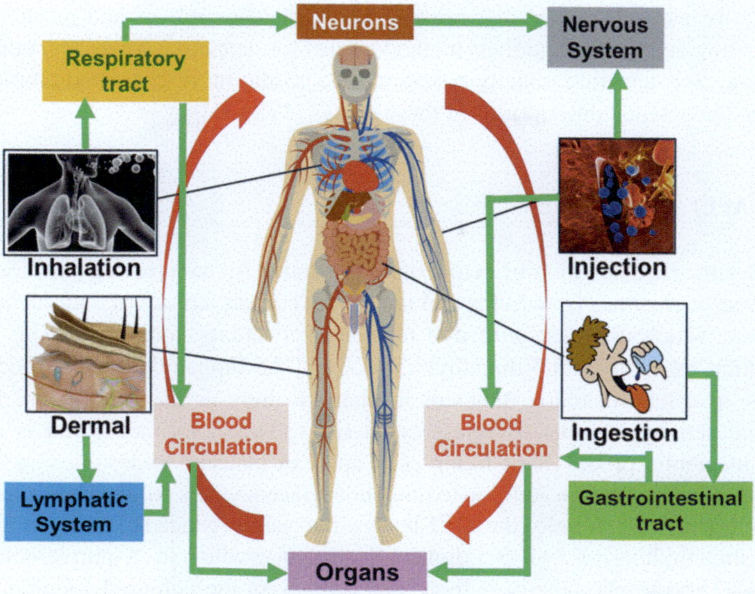

Fig. 10 Various routes of exposure to nanomaterials exist in humans (Self-drawn)

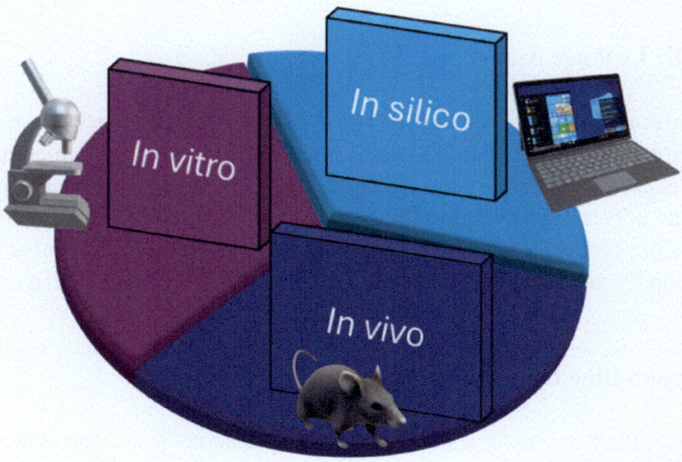

Fig. 11 Nanomaterial toxicity assessment methods (Self-drawn)

diverse assays following an incubation period. However, accurately predicting physiological effects and outcomes of xenobiotics remains a significant challenge. Nevertheless, initial evaluations lean towards in vitro procedures due to their simplicity and ready availability, despite some minor obstacles. Instant and detailed identification via in vitro methods has been demonstrated as the predominant approach in cytotoxicity research, employing various assays that differ only in their mechanisms for detecting cell deaths and their methodologies for detection. There exist diverse approaches for detecting toxicity based on the specific in vitro method employed. Here are the detailed descriptions of these assays [27], [35].

5.1.1 MTT Assay

The colorimetric evaluation of cytotoxicity generated by nanoparticles is based on the release of enzymes from living and non-living cells as a result of their metabolic processes, which combine with dye molecules to create complexes. The MTT test, which uses the 3-[4,5-dimethylthiazol-2-yl]-2,5-diphenyltetrazolium) cation (MTT+) as a helpful redox indicator in pharmacology, is used to analyse mitochondrial activity in order to assess cytotoxicity. This colorimetric test is based on the metabolic processes of living cells and can measure endosomes and non-mitochondrial enzymes in addition to mitochondrial enzymes. After passing through the membrane of active cells, the MTT tetrazolium salt is reduced to create formazan (1-[4,5-dimethylthiazol-2-yl]-3,5-diphenylformazan), which is a purple-coloured substance. Spectrophotometry is then used to analyse the coloured solution. This test allows for the quantitative measurement of live cells since the colour's intensity matches the quantity of living cells [47], [60].

5.1.2 LDH Leakage Assay

One internal enzyme that is present in all live cells is lactate dehydrogenase (LDH). As a sign of cytotoxicity, LDH leaks into the extracellular space when cellular membrane disruption caused by nanoparticle. The liberated LDH outside the cell converts β-nicotinamide adenine dinucleotide (NADH) into NAD+ and helps convert pyruvate to lactate. Since NADH absorbs light at 340 nm, it is possible to determine the LDH concentration by measuring how much lactate and NADH initially weighed [59].

5.1.3 Trypan Blue Dye + Assay

An azo dye called trypan is used to stain non-viable cells in cytotoxic analyses. Trypan cannot be absorbed by viable cells; hence, their cytoplasm is unaffected. On the other hand, trypan-treated non-viable cells have blue cytoplasm, which makes it possible to identify these cells using colorimetry [53], [59].

5.1.4 Apoptosis Assay

Several biological signals control apoptosis, a type-I cell death process that is a kind of controlled cell death. Cell death occurs when these signals abruptly stop.

A series of processes leading to cell death are set in motion by the activation of external proteases called caspases, which marks the beginning of apoptosis. DNA fragmentation and chromatin and nucleus condensation are features of this process.

Apostain, Lamina-B, and TUNEL are a few of the assays that are available to detect apoptosis. The TUNEL method includes combining assay-derived deoxyuridine triphosphate nucleotides with a fluorescent dye, which binds to broken DNA. Fluorescence microscopy or immunohistochemistry staining can be used for quantitative analysis. This approach, which only looks at the process's conclusion, is efficient and affordable, but it is unable to distinguish between necrosis and apoptosis. However, the apostain method does not rely on DNA fragments to identify caspase-3 early in the cytoplasm. This technique helps with early detection by causing brown colouration because of activating apoptosis and releasing protease, whereas healthy cells appear blue under light microscopy. A popular approach for analysing apoptosis, the Apostain technique is well-known for its sensitivity and specificity [24].

5.1.5 2′, 7′-Dichlorofluorescein Diacetate (DCFH-DA) Assay

Internalization of reactive oxidative species (ROS) results in oxidative stress in living cells. A non-polar dye known as 2,7-dichlorofluorescin diacetate (DCFH-DA) can pass through damaged cell membranes and be hydrolyzed by an intracellular esterase to produce non-fluorescent dichlorofluorescin. Then, in the presence of ROS, it oxidizes to fluorescent dichlorofluorescein. Therefore, ROS levels may be quantified by measuring fluorescence intensity [24].

5.1.6 Comet Assay

This extensively used and reasonably priced in vitro analytical method is known as single-cell gel electrophoresis (SCGE) or the comet test because of its striking visual similarity to a comet (as shown in Fig. 12). It is quite reliable. The analysis of DNA damage brought on by nanoparticle toxicity is done using this test. Gel electrophoresis is used in this procedure to separate negatively charged DNA fragments. Toxic cells are embedded in agarose gel and then lysed (salts and detergents) to release the cytoplasm and other components of the cell, except the nucleoids. A structure resembling a comet is created by further electrophoresis at high pH levels, with the intact DNA represented in the head and the fragmented part in the tail. As a result, the tail's fluorescent labelling and intensity show the damaged DNA segment, making it possible to estimate toxicity [5], [64], [15], [33].

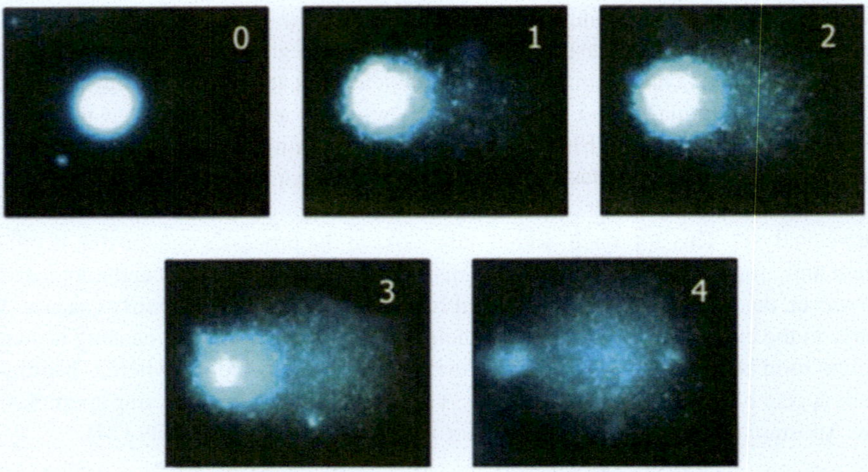

Fig. 12 Comet assay of 5 different classes (0 to 4) by visual scoring based on 0 to 100% tail DNA [15]

5.2 *In Vivo Method*

These approaches uphold their optimal and fundamental benchmarks for toxicity assessment. However, studies reliant on living animals, although effective, raise ethical concerns. In vivo investigations necessitate the introduction of nanomaterials into test animals' bodies, monitoring progression through various techniques. This real-time analysis produces results more aligned with human bodily functions, reducing time and cost impacts. Factors like hormonal changes and cell-matrix interactions significantly impact in vivo assessments but cannot be replicated in in vitro studies. Long-term chronic effects, for instance, are challenging to simulate in in vitro environments, leading to certain impacts being overlooked during analysis. Due to the complex nature of doing in vivo investigations, more care must be taken. Since nanoparticles are so small and have unique characteristics in biological systems, determining the in vivo dosage depends on the actual exposure of the materials to the body, which presents technological difficulties. Important factors to consider in in vivo investigations include choosing a non-reactive carrier to deliver the nanoparticle dosage and making sure it is properly dispersed inside it. Because of their extensive surface area, nanoparticles are highly prone to agglomeration, which results in inadequate dispersion and improper biological distribution, yielding undesired outcomes. Once these nanomaterials are within the body, they can engage with protein counterparts, triggering the formation of a protein corona. This leads to changes in properties, their interactions, and how they are distributed biologically. Additionally, the protein structure undergoes conformational alterations, affecting biological functions and signalling paths. Therefore, it is important to consider the several ways that nanomaterials could interfere with other substrates before assessing their toxicity within a biological system [17].

5.3 *In Silico Method*

In silico toxicology is the study of the toxicity of chemicals and nanomaterials using computational methods. Scientists use specific techniques to model, estimate, and comprehend the potential toxicity of these substances. For nanomaterials to be used effectively, scientists must have a thorough understanding of their chemical and physical characteristics as well as possible environmental interactions. Developing models that illustrate the relationship between the structure, characteristics, and effects of these nanomaterials and their products on living organisms is the primary objective of this computer-based research approach. According to the European Chemicals Agency (ECHA), if these models are accurate, they can aid in the first step of evaluating how dangerous nanoparticles may be. They can also assist in the next phase by classifying these components into several categories according to their hazard level.

There is a major benefit to using computers for research: fewer trials, including those involving animals, may be conducted. It saves money and time to do this kind of examination. Computer techniques such as this are quick and can analyse several nanomaterials simultaneously. However, creating these assessment models is not simple. They must be resilient, robust, and capable of making accurate predictions in a variety of circumstances. They require a tonne of accurate and comprehensive experiment data to achieve this. The issue is that there is a lot of conflicting information on nanoparticles' risk factors, exposure, and characterization. There may be gaps in the data or some missing information. Furthermore, there might be wide variations in the quality of the data, including the number of repetitions of an experiment, the clarity of the results, and whether or not the standards were the same for each test. As a result, building these models may be challenging.

Scientists have developed multiple models to forecast the harmful effects of nanomaterials, although there are some restrictions. These models utilize machine learning tools along with different mathematical and statistical approaches to comprehend the impact of nanomaterials' structure, toxicokinetic, and molecular characteristics on living organisms. They employ different algorithms used in machine learning, including principal component analysis, artificial neural networks (ANN), decision trees, regression analysis, and support vector machines (SVM) algorithms. Furthermore, researchers use a technique called 'molecular dynamics' to simulate and anticipate the behaviour of nanomaterials in living organisms [18], [61].

5.3.1 Molecular Docking

Computer simulations used in molecular docking research predict how nanomaterials and biomolecules would interact based on their three-dimensional structural details.

This technique has also been used to evaluate the potential toxicity of a variety of nanoparticles, including Ag, Fe_3O_4, ZnO, TiO_2, and CuO, when they interact with biological molecules [28].

5.3.2 Quantitative Structure–Activity Relationship (QSAR)

A QSAR model's main goal is to create a suitable function that links the chemical structure to biological activity. This work is promising in terms of distilling large amounts of physiochemical and biological data to predict possible adverse effects or aid in the production of ideal nanomaterials. For QSAR models to function, a mathematical link must be established between the biological activity or toxicity, of nanomaterials which are mainly size, shape, aggregation state, surface charge, solubility, etc., and their molecular descriptors, which include different physicochemical properties like cytotoxicity, genotoxicity, cell uptake, oxidative stress, etc. A molecular descriptor can be derived through experimentation or computation. It is crucial to validate calculated descriptors by confirming their correlation with experimental values. Otherwise, these descriptors might significantly affect the model's accuracy. Hence, it is advisable to craft and customize descriptors based on the specific data obtained from physicochemical characterization, rather than relying on theoretical structures. When assembling data sets for building QSAR models, it is essential to include a diverse range of nanomaterials in sufficient quantities. The primary constraint of QSAR models lies in their reliance on molecular descriptors that typically pertain to the original state of nanomaterials. However, when these nanoparticles encounter with biological environments, their characteristics undergo changes in response to that environment. Likewise, since nanomaterials can elicit diverse biological impacts under varying biochemical conditions such as cell lines or species, replicating the practical scenario becomes challenging in QSAR modelling [28].

5.3.3 Grouping/Read Across

The use of grouping or read-across, which is mostly used to estimate the toxicity of substances using computational techniques, may be expanded to estimate accumulation, deterioration, physicochemical features, or prospective applications of a particular material. Its objective is to forecast the effects of an unknown material (called the target material) by utilizing data from compounds that have been classified together because of apparent similarities (known as the source material). These similarities often revolve around structural and physicochemical traits. Essentially, this method operates under the assumption that materials sharing similarity tend to exhibit comparable toxicological behaviour. Consequently, available toxicological data from a source material can be utilized to extrapolate the toxicological properties of a target substance lacking substantial experimental data [28].

Recently, a machine learning model, termed the ML-PEMST (Machine Learning Properties-Environmental Conditions-Multi Species-Toxicity) model, has been utilized in an in silico setting to forecast the toxicity levels of various metallic nanomaterials across multiple aquatic species [74]. They have chosen 3 different machine learning algorithms—random forest (RF), support vector machine (SVM), and artificial neural network (ANN). Multiple machine learning (MML) approaches have been employed, as it yields more accurate prediction which proves advantageous by surpassing the limitations posed by a single algorithm. Another study has highlighted the utilization of nano-SAR (Nanostructure-Activity Relationship) models, specifically designed to delve into potential mechanistic insights concerning the toxicity pathways of both metal and metal oxide nanoparticles (NPs) [21]. The outcomes have indicated that the energy variance between the nanoparticles conduction band and the biological pairs' redox potential, along with the released metal cation's hydration enthalpy, stood out as the most effective descriptors. These findings shed light on potential primary pathways governing the toxicity of metal and metal oxide NPs. The hydration enthalpy served as an indirect indicator of metal ion release and charge potential, directly depicting water's affinity for the released metal cation. This affinity significantly influences the transport of cations through cell membranes. A study in ecological risk assessment has carried out on seven engineered metallic nanoparticles, examining their impact on Escherichia coli at various mixing ratio [73]. A Quantitative Structure-Activity Relationship (QSAR) model was utilized to forecast the cytotoxicity of combinations of metal oxide nanoparticles. The investigation revealed that a model incorporating two machine learning methods, namely support vector machine and neural network, alongside nano-descriptors such as metal electronegativity and metal oxide energy, played pivotal roles in predicting the toxicity levels of these mixtures.

A universal standard for evaluating the toxicity of nanomaterials in textile products is currently lacking. Instead, a combination of various methods is needed to describe the toxicological characteristics of the material accurately. It is improbable that a single-cell-based test or computational model will completely replace existing animal-based clinical trials. Therefore, the most suitable methodologies should be used together in an integrated assessment and testing approach. Both experimental and computational methods complement each other, and their combination is expected to advance the field of nanotoxicology. The biological impact of nanoparticle toxicity presents a significant challenge, requiring high-quality equipment, advanced culture mediums, and dependable in vitro and in vivo assessments [24]. Currently, most studies assess toxicity using synthesized nanoparticles, but it is crucial to consider the products or degraded materials that enter the environment after or during their use. Assessing the toxicity, particularly in the case of carbon-based nanomaterials on biological substrates, remains a significant challenge due to the transformed or degraded materials in the environment [1].

6 Conclusions

Nanomaterials used as a finishing agent are a perfect solution for developing high-tech functional textiles. There are different finishing techniques that are used to impart nanomaterials to textile substrates. Recently, finishing using nanomaterials is usually done to impart functional properties such as ultraviolet resistance, water repellent, antibacterial, electroconductive, antioxidant, flame retardant, and wrinkle resistance. Nano-finishing shows additional advantages rather than bulk finishing on textile substrates such as the requirement of less add-on, enhancement of functional properties, and maintaining the inherent properties of textile material after nano-finishing. Despite effective advantages, there are several concerning issues to which attention is required. The cons are that there are leaching out tendencies of nanomaterials from nano-finished textiles which leads to biocompatibility and toxicity issues for human beings as well as the environment. The nanoparticles are added to wastes or wastewater and dumped in landfills or water bodies during the manufacturing of nano-finished textiles. These nanoparticles finally end up into plants, animals, and human beings. Additionally, some events such as washing, abrasion, rubbing, and perspiration can lead to leaching out of nanoparticles during the usage of nano-finished textiles and absorbed by human beings and environment. The overdose of nanoparticles inside the body causes cell damage and several diseases including life-threatening diseases. The engineered nanoparticles in environments cause ecological imbalances and the ecosystem is getting disturbed.

There must be some immediate efforts and strict regulations to tackle these problems. After analysing serious problems of nanoparticles, the nanoparticles must be developed or modified in such a way that it should be very eco-friendly. It is observed that in vivo toxicity studies are less performed compared to in vitro studies as in vitro toxicity studies are not lengthier, easier, and less costly whereas in vivo studies are equally relevant. So, analysing of nanoparticles is very much necessary. Even, analysing and studying the effect of nanoparticles present in water and land must be carried out accurately.

There must be proper guidelines or SOP to handle the nano-finishing treatments to prepare nano-finished textiles. The awareness regarding harmful effects of nanoparticles is necessary in the textile sector and nanoparticles leaching out must be monitored which is present in textiles as well as in effluents. The details information regarding nanomaterials used in nano-finished textiles must be available in the ecolabel, while some industries are already applying it. In case of textile industry, green chemistry must be preferred to develop nanomaterials and application techniques of nanomaterials to textile substrates. The application techniques where strong chemical bonds among nanoparticles and textile substrate must be focused. Consequently, less effluent can be achieved by in situ application techniques. There should be proper guidelines and instructions to dispose the nano-finished textile. Globally, 3 R's waste management (reduce, reuse, and recycle) is preferred, so recycling of textile products is in high demand. However, recently various challenges are been faced by industries regarding recycling, so it will be easier if guidelines to recycle nano-finished textiles are mentioned beforehand.

References

1. Adiseshaiah PP, Hall JB, McNeil SE (2010) Nanomaterial standards for efficacy and toxicity assessment. Wiley Interdiscip Rev: Nanomed Nanobiotechnol 2(1):99–112. https://doi.org/10.1002/WNAN.66
2. Ahmed A et al (2020) Recent advances in 2D MXene integrated smart-textile interfaces for multifunctional applications. Chem Mater 32:10296–10320. https://doi.org/10.1021/acs.chemmater.0c03392
3. Ali SW, Banerjee S, Bairagi S (2023) Sustainable textile finishing using natural materials. In: Sustainable textile chemical processing, pp 153–182. https://doi.org/10.1201/9781032629933-7
4. Ali SW, Banerjee S, Mondal M (2023) Nanotechnology in textile finishing: an approach towards imparting manifold functionalities. In Adak B, Mukhopadhyay S (eds) Smart and functional textiles. Berlin: De Gruyter, pp 63–96. https://doi.org/10.1515/9783110759747-002
5. Al-Khazali ZK, Alghanmi HA (2023) Environmental toxicity of nano iron oxides (Fe_2O_3 NPs) on Algal Growth Klisinema persicum and cellular DNA damage using comet assay. Egypt J Aquat Biol Fish 27(1):431–453
6. Almeida L, Ramos D (2017) Health and safety concerns of textiles with nanomaterials. IOP conference series: materials science and engineering 254(10). https://doi.org/10.1088/1757-899X/254/10/102002
7. Atta AM, Abomelka HM (2021) Multifunctional finishing of cotton fibers using silver nanoparticles via microwave-assisted reduction of silver alkylcarbamate. Mater Chem Phys 260:124137. https://doi.org/10.1080/15440478.2020.1848704
8. Awashra M, Młynarz P (2023) The toxicity of nanoparticles and their interaction with cells: an in vitro metabolomic perspective. Nanoscale Adv R Soc Chem 5:2674–2723. https://doi.org/10.1039/d2na00534d
9. Banerjee S, et al (2023) Recent advances in nanobiotechnology for medical textiles. In: Advances in healthcare and protective textiles, pp 151–171. https://doi.org/10.1016/B978-0-323-91188-7.00012-1
10. Bhandari V et al (2022) Antimicrobial finishing of metals, metal oxides, and metal composites on textiles: a systematic review. Ind Eng Chem Res 61:86–101. https://doi.org/10.1021/acs.iecr.1c04203
11. Branquinho R et al (2018) Fibres for antibacterial finishing using a rapid and low cost in situ synthesis †. New J Chem 42:1052–1060. https://doi.org/10.1039/c7nj03418k
12. Bureau S et al (2021) In situ formation of silver nanoparticles (Ag-NPs) onto textile fibers. ACS Omega 6:1316–1327. https://doi.org/10.1021/acsomega.0c04814
13. Butola BS et al (2018) Development of multi-functional properties on cotton fabric by in situ application of TiO_2 and ZnO nanoparticles. J Inst Eng India Ser E 99(1):93–100. https://doi.org/10.1007/s40034-018-0118-3
14. Buzea C, Pacheco II, Robbie K (2007) Nanomaterials and nanoparticles: sources and toxicity. Biointerphases 2(4):MR17–MR71. https://doi.org/10.1116/1.2815690
15. Cardoso R et al (2022) In vivo mammalian alkaline comet assay: method adapted for genotoxicity assessment of nanomaterials. Front Toxicol 4:903896. https://doi.org/10.3389/FTOX.2022.903896/BIBTEX
16. Chen J et al (2018) Enhanced physical and biological properties of silk fibroin nanofibers by layer-by-layer deposition of chitosan and rectorite. J Colloid Interface Sci 523:208–216. https://doi.org/10.1016/j.jcis.2018.03.093

17. Clichici S, Assessment AF-N-T, R. 2015, undefined (no date) In vivo assessment of nanomaterials toxicity, books.google.com [Preprint]
18. Concu R, et al (2017) Probing the toxicity of nanoparticles: a unified in silico machine learning model based on perturbation theory. Taylor & Francis 11(7): 891–906. https://doi.org/10.1080/17435390.2017.1379567
19. Egbuna C et al (2021) Toxicity of nanoparticles in biomedical application: nanotoxicology. J Toxicol. https://doi.org/10.1155/2021/9954443
20. Elmaaty TA et al (2022) Multifunctional dyeing of wool fabrics using selenium nanoparticles. Polymers 14(1):191. https://doi.org/10.3390/polym14010191
21. Gakis G, et al. (2023) Metal and metal oxide nanoparticle toxicity: moving towards a more holistic structure–activity approach. pubs.rsc.org 10: 761. https://doi.org/10.1039/d2en00897a
22. Ganguly P, Breen A, Pillai SC (2018) Toxicity of nanomaterials: exposure, pathways, assessment, and recent advances. In: ACS biomaterials science and engineering, pp 2237–2275. https://doi.org/10.1021/acsbiomaterials.8b00068
23. Gao LZ et al (2020) Multifunctional silk fabric via surface modification of nano-SiO_2. Text Res J 90(13–14):1616–1627. https://doi.org/10.1177/0040517519897112
24. Gupta A, et al (2019) Challenges for assessing toxicity of nanomaterials, biochemical toxicology—heavy metals and nanomaterials [Preprint]. https://doi.org/10.5772/INTECHOPEN.89601
25. Islam T, et al (2023) Impact of textile dyes on health and ecosystem: a review of structure, causes, and potential solutions, Springer 1(4): 3. https://doi.org/10.1007/s11356-022-24398-3
26. Jain VK, et al (no date) Nano-finishing techniques: a review. https://doi.org/10.1177/0954406211426948
27. Jones C, reviews DG-A drug delivery and 2009, undefined (no date) In vitro assessments of nanomaterial toxicity, Elsevier [Preprint]
28. Journal AI, Fröhlich E, Fr€ E (2022) Experimental and computational nanotoxicology—complementary approaches for nanomaterial hazard assessment. Nanomaterials 12(8):1346. https://doi.org/10.3390/NANO12081346
29. Juikar SJ, Nadanathangam V (2020) Microbial production of nanolignin from cotton stalks and its application onto cotton and linen fabrics for multifunctional properties. Waste Biomass Valoriz 11(11):6073–6083. https://doi.org/10.1007/s12649-019-00867-8
30. Juikar SJ, Vigneshwaran N (2017) Microbial production of coconut fiber nanolignin for application onto cotton and linen fabrics to impart multifunctional properties. Surfaces Interfaces 9:147–153. https://doi.org/10.1016/j.surfin.2017.09.006
31. Kenry, Lim CT (2017) Biocompatibility and nanotoxicity of layered two-dimensional nanomaterials. ChemNanoMat 3(1):5–16. https://doi.org/10.1002/CNMA.201600290
32. Khan W, et al (no date) A critical review of hazardous waste generation from textile industries and associated ecological impacts, Elsevier [Preprint]
33. Kizilkaya D, et al (2023) investigation of iron oxide nanoparticles and microparticles using the in vitro bacterial reverse mutation and in vivo Allium chromosome aberration and comet assays. Springer 25(9): 173. https://doi.org/10.1007/s11051-023-05819-x
34. Kulthong K et al (2010) Determination of silver nanoparticle release from antibacterial fabrics into artificial sweat. Part Fibre Toxicol 7(1):1–9. https://doi.org/10.1186/1743-8977-7-8/TABLES/3
35. Kumar V, et al (2017) In vitro and in vivo toxicity assessment of nanoparticles. Springer 7(4): 243–256. https://doi.org/10.1007/s40089-017-0221-3
36. Li X et al (2012) Biocompatibility and toxicity of nanoparticles and nanotubes. J Nanomater. https://doi.org/10.1155/2012/548389
37. Li Z, Aik Khor K (2019) Preparation and properties of coatings and thin films on metal implants. Encycl Biomed Eng. https://doi.org/10.1016/B978-0-12-801238-3.11025-6
38. Ling C, Guo L (2021) An eco-friendly and durable multifunctional cotton fabric incorporating ZnO and a branched polymer. Cellulose 28(9):5843–5854. https://doi.org/10.1007/s10570-021-03868-w

39. Loo CY et al (2022) Toxicity of curcumin nanoparticles towards alveolar macrophage: effects of surface charges. Food Chem Toxicol. https://doi.org/10.1016/j.fct.2022.112976
40. Makhlouf ASH (2011) Current and advanced coating technologies for industrial applications. In: Nanocoatings and ultra-thin films. Woodhead Publishing Limited. https://doi.org/10.1533/9780857094902.1.3
41. Massella D, et al (2018) Preparation of bio-functional textiles by surface functionalization of cellulose fabrics with caffeine loaded nanoparticles. IOP conference series: materials science and engineering 460(1). https://doi.org/10.1088/1757-899X/460/1/012044
42. Mazari SA, et al (2020) Environmental impact of using nanomaterials in textiles. In: Nanosensors and nanodevices for smart multifunctional textiles. Elsevier Applied Science, pp 321–342. https://doi.org/10.1016/B978-0-12-820777-2.00018-2
43. Mishra R, Militky J, Arumugam V (2018) Nanorisks and nanohazards, nanotechnology in textiles: theory and application. Elsevier Ltd. https://doi.org/10.1016/B978-0-08-102609-0.00008-0
44. Mohamed AL, Hassabo AG (2022) Core–shell titanium@silica nanoparticles impregnating in poly (itaconic acid)/poly (N-isopropylacrylamide) microgel for multifunctional cellulosic fabrics. J Polym Res. https://doi.org/10.1007/s10965-022-02921-x
45. Muzaffar S et al (2021) Enhanced mechanical, UV protection and antimicrobial properties of cotton fabric employing nanochitosan and polyurethane based finishing. J Mater Res Technol 11:946–956. https://doi.org/10.1016/j.jmrt.2021.01.018
46. Nozari B, Montazer M, Mahmoudi Rad M (2021) Stable ZnO/SiO_2 nano coating on polyester for anti-bacterial, self-cleaning and flame retardant applications. Mater Chem Phys 267:124674. https://doi.org/10.1016/j.matchemphys.2021.124674
47. Rabiei H et al (2023) Antimicrobial activity and cytotoxicity of cotton-polyester fabric coated with a metal–organic framework and metal oxide nanoparticle. Appl Nanosci (Switzerland) 13(8):5765–5776. https://doi.org/10.1007/S13204-023-02823-4
48. Radetić M, Marković D (2019) Nano-finishing of cellulose textile materials with copper and copper oxide nanoparticles. Cellulose. Springer Netherlands, pp 8971–8991. https://doi.org/10.1007/s10570-019-02714-4
49. Rafique MS, et al (2020) Synthesis methods of nanostructures, nanotechnology and photocatalysis for environmental applications. Elsevier Inc. https://doi.org/10.1016/b978-0-12-821192-2.00003-6
50. Rawtani D, Agrawal YK (2014) Emerging strategies and applications of layer-by-layer self-assembly. Nanobiomedicine 1:1–15. https://doi.org/10.5772/60009
51. Rezić I (2011) Determination of engineered nanoparticles on textiles and in textile wastewaters. TrAC—Trends in Analytical Chemistry, pp 1159–1167. https://doi.org/10.1016/j.trac.2011.02.017
52. Riaz S et al (2022) Cationization of TiO_2 nanoparticles to develop highly durable multifunctional cotton fabric. Mater Chem Phys 19:1–11. https://doi.org/10.1080/15440478.2020.1870625
53. Roszak J, et al (no date) A strategy for in vitro safety testing of nanotitania-modified textile products. Elsevier [Preprint]
54. Saeed M et al (2020) Chemical vapour deposition of graphene—synthesis, characterisation, and applications: a review. Molecules. https://doi.org/10.3390/molecules25173856
55. Saleem H, Zaidi SJ (2020) Sustainable use of nanomaterials in textiles and their environmental impact. Materials 13:1–28. https://doi.org/10.3390/ma13225134
56. Saleem H, Zaidi SJ (2020) Sustainable use of nanomaterials in textiles and their environmental impact. Materials 13(22):1–28. https://doi.org/10.3390/ma13225134
57. Shende P, Patil A, Prabhakar B (2020) Layer-by-layer technique for enhancing physicochemical properties of actives. J Drug Del Sci Technol 56(January):101519. https://doi.org/10.1016/j.jddst.2020.101519
58. Singh A, Chakraborty JN (2022) 'Multifunctional finishing of cotton using zinc oxide and silicon dioxide nanoparticles along with DMDHEU. Res J Text Apparel. https://doi.org/10.1108/RJTA-11-2021-0130

59. Singh G, et al (2017) Cytotoxicity study of textile fabrics impregnated with CuO nanoparticles in mammalian cells. Journals.sagepub.com 36(6): 478–484. https://doi.org/10.1177/1091581817736712
60. Su T, et al (no date) Green in-situ synthesis of silver coated textiles for wide hygiene and healthcare applications, Elsevier [Preprint]
61. Subramanian N, omega AP-A, 2021, undefined (2021) NanoTox: development of a parsimonious in silico model for toxicity assessment of metal-oxide nanoparticles using physicochemical features, ACS Publications 6(17): 11729–11739. https://doi.org/10.1021/acsomega.1c01076
62. Synthesis G et al (2020) Chemical vapour deposition of graphene synthesis, characterisation, ADN application: a review. Molecules 25(3856):2–62
63. Thomas NG, et al (2023) Toxicity evaluation and biocompatibility of nanostructured biomaterials. www.intechopen.com
64. Varet J, et al (no date) For testing the in vitro genotoxicity of nanomaterials: comparison of 2D and 3D HepaRG liver cell models and classical and high throughput comet assay, Elsevier [Preprint]
65. Wang Y, et al (2021) Structured fabrics with tunable mechanical properties, Nature 596(7871): 238–243. https://doi.org/10.1038/s41586-021-03698-7
66. Wick P et al (2007) The degree and kind of agglomeration affect carbon nanotube cytotoxicity. Toxicol Lett 168(2):121–131. https://doi.org/10.1016/j.toxlet.2006.08.019
67. Windler L et al (2012) Release of titanium dioxide from textiles during washing. Environ Sci Technol 46(15):8181–8188. https://doi.org/10.1021/es301633b
68. Witika BA et al (2020) Biocompatibility of biomaterials for nanoencapsulation: current approaches. Nanomaterials. https://doi.org/10.3390/nano10091649
69. Yang HC et al (2010) Preparation and application of nanochitosan to finishing treatment with anti-microbial and anti-shrinking properties. Carbohydr Polym 79(1):176–179. https://doi.org/10.1016/j.carbpol.2009.07.045
70. Yetisen AK et al (2016) Nanotechnology in textiles. ACS Nano 10(3):3042–3068. https://doi.org/10.1021/acsnano.5b08176
71. Yoshioka Y, Higashisaka K, Tsutsumi Y (2016) Biocompatibility of nanomaterials. Methods Pharmacol Toxicol 39:185–199. https://doi.org/10.1007/978-1-4939-3121-7_9
72. Zayed M et al (2022) Psidium guajava leave extract for improving ultraviolet protection and antibacterial properties of cellulosic fabrics. Biointerface Res Appl Chem 12(3):3811–3835. https://doi.org/10.33263/BRIAC123.38113835
73. Zhang F et al (2023) Machine learning-driven QSAR models for predicting the mixture toxicity of nanoparticles. Environ Int 177:108025. https://doi.org/10.1016/J.ENVINT.2023.108025
74. Zhou Y et al (2022) Using machine learning to predict adverse effects of metallic nanomaterials to various aquatic organisms. Environ Sci Technol 57:17786–17795. https://doi.org/10.1021/ACS.EST.2C07039/SUPPL_FILE/ES2C07039_SI_001.PDF
75. Zhu J et al (2021) Preparation of Ag NPs and its multifunctional finishing for cotton fabric. Polymers 13(1):1338. https://doi.org/10.3390/polym13081338

Chapter 17
Challenges and Opportunities in Nanotechnology for Textile Finishing

Amit Madhu, Nagender Singh, and Manisha Yadav

1 Introduction

In recent years, the textile industry has witnessed a remarkable transformation due to nanotechnology. Nanotechnology, the science and application of materials at the nanoscale, has opened up a world of possibilities in textile finishing. From enhancing fabric properties to introducing novel functionalities, the integration of nanomaterials into textiles has brought about numerous benefits [1–3]. In textiles, nanotechnology can work with three fundamental approaches:

1. In upgradation of the already existing functionality and performance characteristics of the textile materials, like enhanced strength and abrasion resistance, insulating behaviour, hydrophilic or hydrophobic characteristics, and flame resistance [4, 5].
2. Introduction of new innovative functionalities in textiles like antibacterial, self-cleaning, UV protection, electrical conductivity, controlled release for active agents, etc. [5, 6].
3. Development of smart and intelligent textiles, that are responsive and show/modify specific properties or have sensing and actuating capabilities [1, 6, 7].

The combination of textile science with nanotechnology has created new opportunities for the engineering of fabric surfaces at the atomic and molecular scale, leading to notable enhancements in utility, performance, and appearance. There exist two primary reasons why materials may exhibit distinct behaviour at the nanoscale. Firstly, nanoparticles have a comparatively larger surface area compared to the same mass of material generated in a bigger form. This can change a material's strength or

A. Madhu (✉) · N. Singh
The Technological Institute of Textile and Sciences, Bhiwani, Haryana, India
e-mail: amit.madhu5@gmail.com

M. Yadav
DLC State University of Performing and Visual Arts, Rohtak, Haryana, India

S. Bairagi et al. (eds.), *Nanotechnology in Textile Finishing*, Materials Horizons: From Nature to Nanomaterials, https://doi.org/10.1007/978-981-97-2696-7_17

electrical properties as well as increase its chemical reactivity (sometimes materials that are inert in bigger form become reactive when generated in their nanoscale form). Second, quantum effects supersede the laws of conventional physics below 50 nm, resulting in optical, electrical, and magnetic behaviours that diverge from larger-scale properties of the same material. These phenomena can endow materials with highly advantageous physical characteristics, such as remarkable electrical resistance or conduction, or a substantial capacity for heat storage or transfer and can even modify biological properties, e.g. silver becoming a bactericide on a nanoscale [7–9].

Emulsification at the nanoscale represents an advanced technique for applying finishes to textiles, resulting in an unparalleled enhancement of textile performance attributes such as resistance to stains, improved hydrophilicity, anti-static properties, wrinkle resistance, and shrink-proof characteristics. The utilisation of nanoparticles, specifically metal oxides and ceramics at the nano-level, yields advantages due to their increased surface area, leading to higher efficiency compared to larger particles. Additionally, these nanoparticles offer transparency without compromising the colour and brightness of the textile substrates [10–13]. Fabrics treated with nanoparticles like TiO_2 and MgO have replaced traditional active carbon materials in applications related to chemical and biological protection, owing to their photocatalytic properties capable of decomposing harmful chemicals and biological agents [12, 14–16]. The potential uses of different nanoparticles in textile finishing are virtually limitless, as shown in Table 1.

Finishing with nanoparticles can also convert fabrics into sensor-based materials. If nanocrystalline piezoceramic particles are incorporated into fabrics, the finished fabric can convert exerted mechanical forces into electrical signals enabling the monitoring of bodily functions such as heart rhythm and pulse if they are worn next to skin [17–19].

Self-assembled nanolayer (SAN) coating is a cutting-edge technology in textile finishing that involves the formation of ultra-thin nanoscale layers on textile surfaces. These self-assembled layers are composed of functional molecules, which arrange themselves spontaneously to create a uniform and precisely engineered coating [20, 21]. SAN coatings offer remarkable advantages, as they enable enhanced water repellency, stain resistance, and improved durability while preserving the natural feel and breathability of the fabric. This innovative approach to textile finishing is gaining prominence for its ability to tailor textile properties with great precision, opening up a world of possibilities for the development of advanced, high-performance textiles across various industries [21, 22].

An extensive array of products, such as fire-resistant clothing, medicinal and protective equipment with antibacterial and antiviral qualities, high-strength army jackets, uniforms, and boots, and nano-sensors in clothing and blankets that provide cooling or heating effects, have been made possible by nanotechnology in textiles [20, 23]. This technology is beneficial to a nation's development both now and in the future. It has been wisely employed these days to create masks and other goods that can capture harmful viruses like COVID-19 [18–20, 24].

Table 1 Potential uses of different nanoparticles in textile finishing

Nanoparticle	Textile finishing
Ag (Silver)	Antibacterial Super-hydrophobicity
TiO_2 (Titanium dioxide)	UV protection Self-cleaning Antibacterial Super-hydrophobicity Antireflective
SiO_2 (Silicone dioxide/Silica)	Super-hydrophobicity
ZnO (Zinc oxide)	UV protection Antibacterial/Antimicrobial Additive for dyes
CNT (Carbon nanotubes)	Self-cleaning Fire-proof Electrical conductivity/Antistatic Chemical resistance Mechanical reinforcement
Clay	Flame retardant Anticorrosive Heat and chemical resistance
Al_2O_3 (Aluminium oxide)	Super-hydrophobicity Flame retardancy
Fe (Iron)	Conductive magnetic property Remote heating
$Ag + TiO_2$	Durable antibacterial Self-cleaning UV Protective Stain photo-degradability Air-permeable
$TiO_2 + ZnO$	Photocatalytic application
$ZnO + SiO_2$	Durable antibacterial UV Protection Flame retardant
$TiO_2 + MgO$	Chemical and biological protection Self-sterilising

However, these unprecedented opportunities are associated with an array of challenges: ranging from regulatory and safety concerns to the scalability of production processes and the need for sustainable practices. This chapter explores the intricate tapestry of challenges and opportunities that define the dynamic field of nanotechnology in textile finishing. We will explore the complexities of this fascinating journey as we read on, aiming to comprehend the breakthroughs that are changing the textile sector and the challenges that need to be addressed to fully realise the potential of nanotechnology in creating tomorrow's textiles.

2 Challenges Associated with Nano-finishing of Textiles

Nanotechnology has greatly enhanced textile finishing to improve the functional and aesthetic properties of fabrics. However, the challenges are also associated with the application of nano-finishing of textiles (Fig. 1).

2.1 Safety and Environmental Concerns

One of the foremost challenges in nanotechnology-based textile finishing is the potential impact on human health and the environment. Since the majority of textile materials come into direct and extended contact with human skin, the impact of any carcinogenic or hazardous compounds present in textiles must be thoroughly investigated. Nanoparticles, often used in textile finishings, have potential risks for human health, safety, the environment, and sustainability [25, 26]. The assessment of risks associated with the nanoparticles used in finishing is very challenging. The spectrum of nanomaterials, production processes, size, shape, crystallinity, porosity, agglomeration, and aggregation of nanoparticles, among other characteristics, are responsible for the aforementioned. Ensuring that these materials are safe for consumers and workers is crucial. Additionally, the disposal of nanomaterial-treated textiles raises questions about their environmental impact and biodegradability [27, 28]. Let's discuss these in detail.

Nanomaterials used in textile finishing have the potential to become airborne during manufacturing processes. Workers in textile factories may be exposed to these airborne particles, and inhalation of nanomaterials can pose health risks. The small size of nanoparticles allows them to penetrate deep into the respiratory system,

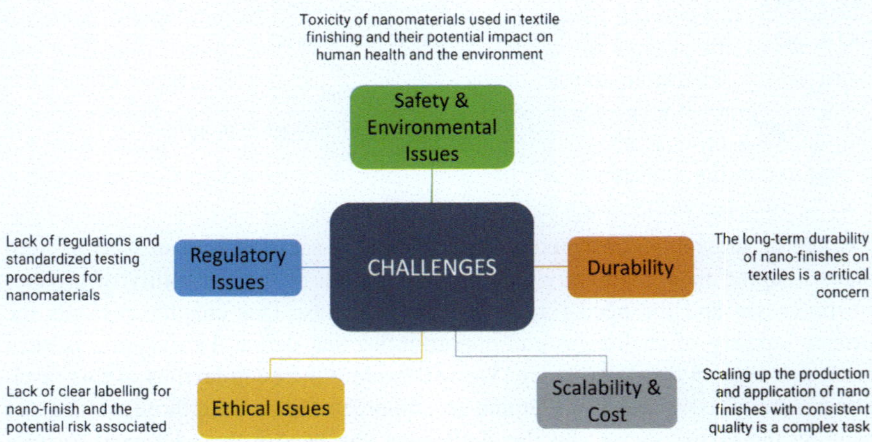

Fig. 1 Challenges associated with nanotechnology in textile finishing

potentially causing lung and systemic health issues [25, 27, 29]. This exposure can be especially problematic if adequate safety measures are not in place. There is still limited information regarding the long-term health effects of exposure to nanomaterials, and their toxicity can vary depending on the specific nanomaterial and its properties. The potential for nanomaterials to induce inflammation, oxidative stress, and other adverse health effects is a significant concern.

Assessing the toxicity of nanomaterials used in textiles is challenging, as these materials may interact with human biology differently compared to their bulk counterparts. Some nanomaterials, when incorporated into textiles, may cause allergic reactions or skin sensitisation in consumers. This is particularly concerning when nano-coatings or finishes come into direct contact with the skin, as is the case with apparel and undergarments [30, 31]. Nanoparticles used in textile finishing may be released into the environment during washing or disposal of textiles. These nanoparticles can potentially accumulate in water bodies and ecosystems, raising concerns about their impact on aquatic life and the environment. The persistence and bioaccumulation of certain nanomaterials can have far-reaching consequences [26, 30]. Figure 2 makes it simple to comprehend the point of nanoparticle emission or exposure to humans and the environment.

The impact of nanoparticles on the environment can be easily understood with the comprehensive knowledge of their identification, their source of emission in the environment, and their toxicity [32]. The ability of nanoparticles to penetrate and infiltrate soil, water, and air is what essentially determines their natural potential toxicity, and the amount of nanoparticles released into the environment directly affects this ability. So, once they reach the environment, the impact of nanoparticles

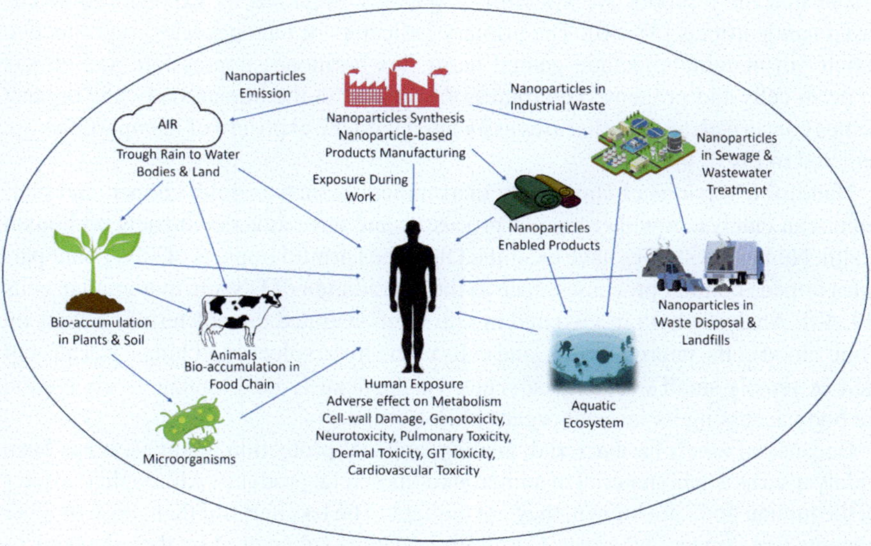

Fig. 2 Risk of nanoparticle exposure and emission associated with human and environment

is directly related to their ability to accumulate in organisms, harming them [33]. Therefore, the risks associated with the use of nanomaterials are determined by all those processes that control their release or emission into the environment.

Compared to common metal materials (such as Li and Na), metal-based nanoparticles have distinct physical and chemical characteristics that result in distinct biological activities and toxicities. Because metal-based nanoparticles are similar in size to protein molecules, they can readily cross cell membranes and interact with subcellular organelle structures or protein molecules, which can lead to genotoxicity, immunotoxicity, and neurotoxicity [33, 34]. The biosafety of these novel materials and technologies has garnered significant attention due to the widespread application of nanotechnology and nanoparticles. Therefore, to understand the possible hazards, it is necessary to outline the common toxicity features of nanoparticles at the cellular and tissue levels.

Textile nanoparticles present a risk of inhalation during production, usage, and disposal. Airborne nanoparticles have the potential to cause adverse health effects when breathed into the respiratory system. Additionally, direct skin contact with nanotextiles may cause unknown diseases in humans due to the possibility that nanoparticles may absorb through the skin and then travel through the bloodstream to other tissues and vital organs. These particles cause the production of respiratory tract inflammation, which is detrimental to tissues, and raises the risk of cardiopulmonary disorders [35, 36]. One of the organs targeted by nanoparticles is the central nervous system, and the most susceptible region of this system is the hippocampus [35–37]. The nervous system is stimulated by nanoparticles to release cytokines and reactive oxygen species, these substances have the potential to harm the blood–brain barrier and cause nervous system dysfunction. Numerous animal studies have demonstrated that the reproductive system is negatively impacted by exposure to metal-based nanoparticles [38–40]. The primary indicators of nanoparticles' reproductive toxicity in mammals include genital harm, sex hormone suppression, and effects on germ cells and progeny. The routes of exposure in the human body, the organs/tissues concerned, and the diseases associated with the exposure of nanoparticles are depicted in Fig. 3 [41].

Nanoform versions of chemically inert minerals, such as gold, copper, and platinum, can catalyse a wide range of processes and have adverse impacts on human health. Numerous studies have examined how the harmful impacts of silver nanoparticles impede cellular processes, such as the degradation of DNA in mammalian cells [39, 40]. Analyses both in vivo and in vitro demonstrated the deleterious effects on white blood cells, neuronal cells, and cells in the liver, spleen, and lungs. Because of their extremely small size, they can enter via the nostrils and circulate freely around the body accessing its internal organs [42].

Because of zinc's antibacterial, healing, and UV protection qualities, it has been widely used as a nanomaterial in numerous commercial products. Although its effects on the human body are known, they are thought to be less harmful than those of other reagents like copper and silver. Currently, concerns are centred on the effects of its overall toxicity brought on by excessive intake [43, 44], which results in significant

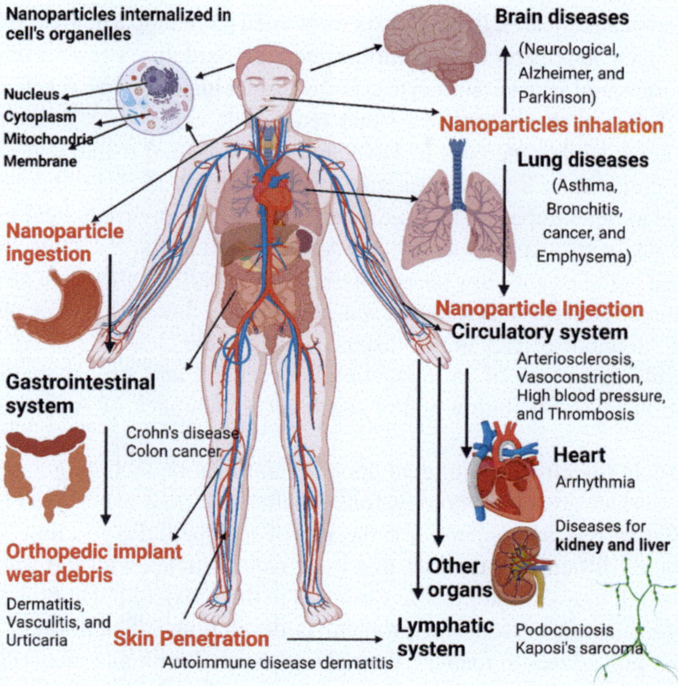

Fig. 3 Representation of nanoparticle exposure routes in the human body, the organs/tissues concerned and the diseases associated with exposure (reprinted Reference 42)

enzyme excretion and pancreatitis and pancreatic necrosis. Renal tubular epithelial necrosis and glomerular damage are brought on by the increased production of proteins (such as creatinine, amylase, and urine protein) [45]. Long-term overconsumption of zinc can cause oxidative immune system damage that can include fever, chills, lethargy, exhaustion, inflammation, excitation, and cell death. It can also cause infections in the urinary tract and there is an increased risk of prostate cancer [46, 47].

Nanomaterials based on silica are gaining considerable interest for use in textile applications [47]. Silica nanoparticle toxicity varies according on the cell. According to a paper, cancer epithelial cells exhibit strong resistance to silica nanoparticles, and their toxicity mostly affects macrophages, or white blood cells, based on the surface charge [47, 48]. The toxicity of titanium dioxide nanoparticles has been the subject of several investigations [49, 50]. To improve surface reactivity and adsorption, a smaller size is preferred. In the workplace, workers may inhale aerosolised nanoparticles or they may diffuse through their skin and enter their blood vessels and veins. Toxic effects of breathing nano-titanium dioxide include physical illnesses such as spleen damage, immunological dysfunction, oxidative stressors, apoptosis, abnormalities in the metabolic process, and cell growth [51].

The synergistic effect of hypertension and inhalation of nanoparticles can lead to irreversible haemodynamic damage associated with cardiac structural damage and may ultimately cause heart failure after six weeks of TiO_2 nanoparticle exposure [49]. Conversely, TiO_2 nanoparticles can further increase oxidative stress in mouse skin and human immortal skin keratinocyte cells by producing excessive reactive oxygen species [50]. Researchers discovered that renal cells exposed to ZnO nanoparticles saw a considerable increase in DNA and chromosomal damage along with a decrease in glutathione S-transferase and catalase activity [51]. Furthermore, their biodynamics may be impacted by co-exposure to various environmental nanoparticles. When rats were exposed to Au and Ag nanoparticles at the same time, their lungs showed both fast and slow phases in the Ag nanoparticle clearance process, but the Au nanoparticle clearance process was consistently linked to a slow phase. The presence of Au nanoparticles in this instance changed the clearance of Ag nanoparticles, possibly as a result of an interaction between Ag and Au nanoparticles that affected the in vivo dissolution and/or mechanical clearance of Ag nanoparticles [52, 53].

Nanoparticle toxicity is contingent upon several factors, including composition, size, shape, surface area, morphology, functionality, and dose exposure (Fig. 4) [37, 41] Quantitative risk assessments for the use of nanoparticles in clinical settings are made more difficult by the fact that even nanoparticles with the same chemical composition may exhibit notable variances in their toxicological properties [40, 41]. Therefore, a major technical problem is developing efficient and consistent development procedures to manage the risks associated with nanomaterials. It can be costly, time-consuming, and laborious to research the characteristics of specific nanoparticles, their exposure pathway, their exposure duration, the appropriate dose, and the appropriate model system. Here's where computational methodologies and high throughput screening techniques come to the rescue. These can expedite the process of establishing a link between nanoparticles and toxicology assays by quickly screening and prioritising them.

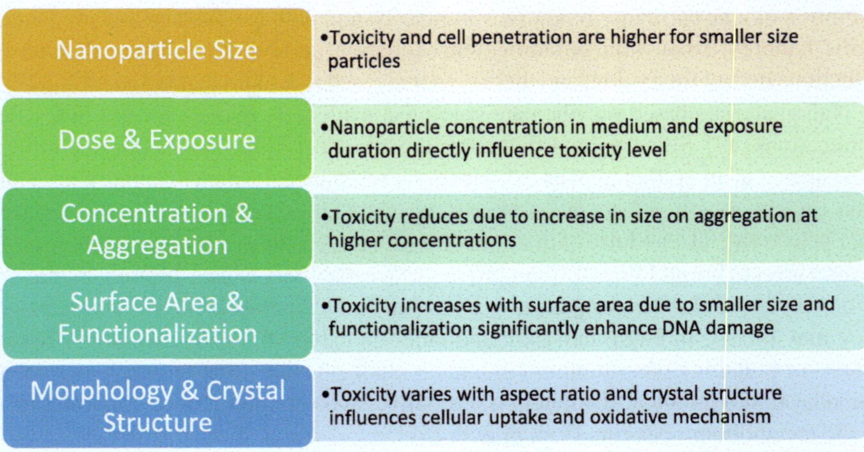

Fig. 4 Factors influencing nanoparticle toxicity

We can anticipate the cytotoxicity of various metal nanoparticles with the aid of a quantitative nanostructure–activity relationship [46]. A thorough characterisation of the relevant nanomaterial is another crucial factor that must be taken into account for the field to move forward. This will enable researchers to use the data to create toxicology assays, appropriately interpret the results, and make sure the data is repeatable and comparable to other researchers. Understanding material qualities, interactions, and toxicity mechanisms has been sparked by research, but understanding physical and chemical properties, interactions, and responses will remain a significant problem in the years to come.

The risk of nanoparticle release in the environment is associated with its entire life cycle: starting from the nanomaterial's synthesis, how they are incorporated into the products, how those products are utilised and ending with the recycling and disposal procedures [52, 54]. Although nanomaterial release could happen in any stage of the life cycle, a major concern occurs when these are employed in the products (e.g. fabric or clothes padding or nanoparticle-containing aerosol employment, etc.) [55]. The emission and exposure risk of nanoparticles during their life cycle is depicted in Fig. 5.

Thus, an extensive understanding of the product's life cycle is necessary for analysing the risks associated with the nanoparticle released from the nanotextiles and nano-finishing techniques. This considers the basic characteristics of the nanomaterials, the processes involved in fabricating them, and the various application methods used to impart nanocomposites and nanoparticles on textile substrates or insert nanoparticles into fibre formation processing, the durability of nano-finishing on the surface, storage, and transportation of the treated textiles, the final product's

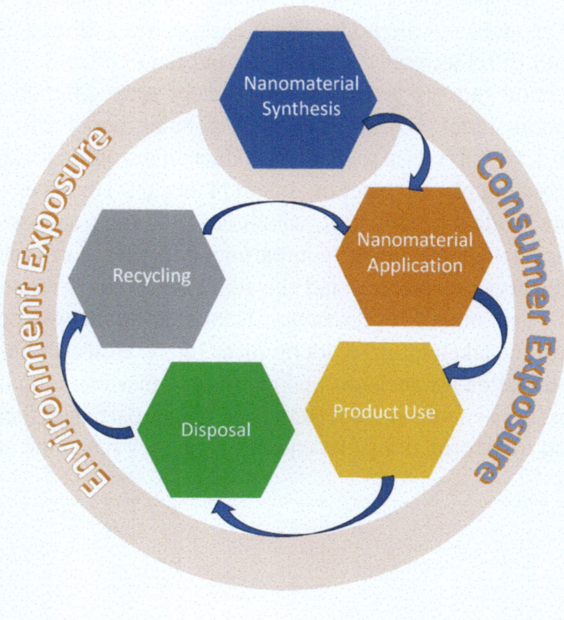

Fig. 5 Emission and exposure risks associated with the nanoparticles during their life cycle

use, the environments to which it will be subjected, and product recycling or disposal. It is now nearly universally accepted that the product life cycle is the proper perspective for thinking about emitted nanomaterials released from nano-coated fabrics of textile products [56]. The product life cycle of nano-coated textiles determines in what life cycle phases and in what environmental compartments or in what technological facilities nanoparticles are released [31, 57].

Subsequent research revealed that, due to the extraordinarily high degree of uncertainty surrounding the characterisation of nanotechnology, comparative approaches to life cycle assessment (LCA) that stress analysis within a specific decision context may be more helpful than attempts to determine life cycle features explicitly. An LCA usually includes some midpoints about human health or ecotoxicology. The characteristics of toxicological midpoints are determined by the relationship between source terms and health midpoints concerning selected benchmarks. Important elements in characterisation include the distribution and dispersion of the chemical in the environment and the relationship between a biological dosage and a health response [56, 58].

Life cycle assessment (LCA) has proven to be an all-encompassing method of system and product evaluation; but, at present, there is not enough data to apply LCA to the environmental impacts of nanoparticle development. The safety issues associated with nanotechnology sometimes lead to the neglect of normalisation and weighting phases, making impact assessment in life cycle assessment (LCA) extremely difficult. The approach for risk assessment of released nanoparticles should also be considered with life cycle concepts and vice versa to reduce the possibility of having just a limited understanding of the risks inherent in released nanoparticles [31, 54, 59].

Nanoparticle emissions into the environment can also be evaluated using a variety of models and methods. These approaches and models come in a variety of forms, with different conceptualisation and data collection strategies, ranging from more general models to models utilised for a specific situation involving an identified nanomaterial [60]. The two most popular models are material flow analysis (MFA) [59], which is used to predict product emissions to technical compartments (landfills, recycling plants, water treatment plants, etc.) to the environment, and environmental fate models (EFM) [61], which describe the transport and distribution through the various environmental compartments (water, earth, and air). Even with the recent advances in research and the removal of study-related limitations, the prediction of these models is restricted. Hence, any outcome derived from these techniques must be taken into consideration as likely. However, these models could offer a solid foundation for decision-making to cut emissions soon if more research and studies in this area enable an understanding of nanoparticles (their life cycle and the processes and transformations that endure when they reach the environment) [62].

2.2 Regulatory Issues

The textile industry lacks comprehensive regulations and standardised testing procedures for nanomaterials used in finishing processes. This regulatory gap may result in inconsistent quality control and safety procedures throughout manufacturers and geographical areas. The absence of definitive rules makes it challenging to regulate and ensure the safe use of nanotechnology in textiles [25, 28, 53]. A particular concern is the durability of nano-finishes and coatings on fabrics. The nanostructures may disintegrate over time and with successive washings, releasing nanoparticles onto the skin or into the surrounding environment. This raises questions about the long-term safety and stability of nanomaterials in textile products [59, 61].

After nanoproducts are synthesised and characterised, comprehensive research into the toxicity of specific nanomaterials and their exposure risks is essential to assess potential hazards. Governments, research institutions, and manufacturers should collaborate on studies to better understand the safety implications. The manufacture, distribution, storage, and transportation of nanoproducts are not governed by any such international organisations. Regulatory agencies should establish clear guidelines and standards for the use of nanomaterials in textiles, encompassing manufacturing, labelling, and disposal. These guidelines should take into account the unique properties of nanomaterials [61, 63].

However, certain nations continue to use the regulatory agencies that already supervise the food, medicine, cosmetic, and pesticide industries to enforce their laws and policies regarding the marketing of nanoparticle products. Food and Drug Administration (FDA) in the USA and Department of Toxic Substances Control (DTSC) within the California Environmental Protection Agency look after the regulations of nano-products and nanomaterials [61]. A new Classification, Labelling, and Packaging (CLP) Regulation for nanoproducts was put into effect by the European Union [63]. In Australia, seven distinct organisations oversee the use of nanotechnology or nanomaterials in commercial products. They set rules for workplace safety and control the handling, storage, and packing of nanoproducts. Additionally, they consider how using these nanomaterials may affect the ecosystem [26, 53, 60]. Health Canada has developed a working definition of nanomaterials and urges manufacturers to communicate with the applicable regulatory body early in the development phase to identify and assess the product's hazards and qualities. Under the direction of the Taiwan Nanotechnology Industry Development Association (TANIDA), the Nano-mark system in Taiwan handles product certification and regulations.

Risk assessment is an essential preliminary step to control levels, limit emissions, and prevent toxic substances from nano-coated textile products to public health and the environment [24]. Evaluations of the safety of nanoparticle products must be conducted and incorporated into the current safety and quality policy throughout the supply chain management and life cycle assessment. An intricate risk assurance management system, such as ONR 49,000 or ISO/TC 229, is the best way to guarantee the security and calibre of textiles coated with nanotechnology. The

ISO/TC 229 works to establish and govern rules about measurement and characterisation, nomenclature and terminology, material requirements, and environmental and health safety standards for nanotechnologies [64, 65]. FERMA (UK) and AS/NZS 4360:2004 (Australia/New Zealand) also outline the essential categories of preventative and restricted actions performed regarding the use and marketing of nanoproducts that pose a major danger to the health and safety of consumers. The development of nano-risk assessment, such as CENARIOS®, was closely aligned with industrial requirements, encompassing communication, knowledge management, risk monitoring, and risk evaluation [53, 56]. Assuming a solid understanding of the nature and toxicity of nano-textile goods will lower the possibility of a product being recalled by customers or removed from the market.

The majority of developed nations have joined these international organisations such as the OECD WPMN (Organisation for Economic Co-operation and Development the Working Party on Manufactured Nanomaterials), REACH Annex (Registration, Evaluation, Authorization and Restriction of Chemicals Annex), ISO/TC 229 (International Organisation for Standardisation/Technical Committee 229), and others in response to the risk associated with nanomaterials [41, 64]. NIOSH, the National Institute of Occupational Safety and Health. This organisation is primarily in charge of researching how workers' exposure to nanotechnology impacts their health and safety at work and offering recommendations for lowering that exposure. NIOSH determined an airborne respirable 8 h time-weighted average (TWA) concentration of 0.9 μg per cubic metre ($\mu g/m^3$) as the recommended exposure limit (REL) for silver nanoparticles (<100 nm primary particle size). Furthermore, NIOSH still advises an 8 h TWA for total silver (metal dust, fume, and soluble compounds, like Ag) with a REL of 10 $\mu g/m^3$ [64, 65]. To safeguard employees, NIOSH also suggests using engineering controls, safe work practices, training and education, and well-established medical surveillance methods in addition to workplace exposure evaluations. These international organisations control the development of nanomaterial standardisation and test recommendations. Additionally, they guarantee the security of nanoproduct usage, storage, transit, and manufacture [57, 63].

India still lacks a well-developed regulatory framework to address potential risks associated with nanotechnology. There are standards for the safe use of nanoparticles in industry and research, but none are specific to nanotechnology. The Environment Protection Act, of 1986, Hazardous Waste (Management, Handling, and Transboundary Movement) Rules, of 2008, Chemical Accidents (Emergency Planning, Preparedness, and Response) Rules, of 1996, and the Manufacture, Storage, and Import of Hazardous Chemical Rules, 1989 can be the foundation for controlling waste related to nanotechnology and classifying nanotechnology as a hazardous material. However, they may not be equipped to handle risks associated with nanomaterials [64, 66]. Various applications of nanotechnology may also be subject to various regulations, such as the Drug and Cosmetics Act, of 1940, the Insecticide Act, of 1968, and the Food Safety and Standards (Contaminants, Toxins, and Residues) Regulations, of 2011. However, these laws currently do not address the potential risks posed by nanoparticles. Occupational health and safety regulations are also crucial for ensuring the safe handling and disposal of nanoparticles in industrial and research settings. Thus, India needs a comprehensive regulatory framework that addresses potential risks specific to nanotechnology [57, 59].

A committee was formed by the Indian Government and the Department of Science and Technology to supervise nanotechnology and create draft laws. To assist policymakers in charting a route for the regulation of nanoproducts, these recommendations devised a three-tiered governance architecture [63, 65]. In summary, while creating a strong legislative framework for nanotechnology, India must address the issues of investment, research quality, and private sector engagement to fully realise the promise of this burgeoning subject. However, India can take the lead in nanotechnology research and development if different stakeholders commit to the project and concentrate on creating useful applications.

2.3 *Scalability and Cost*

Scalability challenges in the case of nanotechnology applications in textile finishes illustrate the difficulty in upscaling nanotechnological processes for commercial production of the textile industry. Although nanotechnology offers promising advantages to improve fabric properties and performance through functionalities, the transfer from laboratory-scale research and development to large-scale industrial production presents several significant challenges. One of the main challenges in scaling up nanotechnology for textile finishing is the production of nanomaterials. Many nanomaterials are produced using batch methods in laboratories, which are not suitable for the continuous and high-volume production required by the textile industry. Scaling up the production of nanomaterials while maintaining quality control and consistency is a complex task [30, 61, 65, 67].

Consistency and repeatability are important in industrial textile manufacturing. Expanding from small-scale research to large-scale manufacturing can lead to changes in the properties and performance of nanomaterials and coatings. Achieving precise and uniform control of nanomaterial applications in textiles is challenging [56, 61]. Moreover, scaling up nanotechnological processes often requires large investments in specialised equipment capable of handling large amounts of material. To achieve these goals, the textile industry must invest in new machinery and technology, which can be costly and time-consuming [64, 68].

Large-scale production of nanomaterials can lead to increased environmental and safety concerns. Waste management, hazardous chemical management, and addressing potential health and safety hazards associated with high production volumes should be importantly considered. Scaling up production can be energy-intensive and resource-intensive. The textile industry needs to consider the environmental impact of increased energy consumption and raw materials used in scaling up nanotechnological processes [66]. As productivity increases, quality and testing become more complex. Large-sized nanotextiles require rigorous testing and quality control programmes to ensure quality and consistent performance. This includes addressing durability, wash resistance, and functional integrity issues [57, 59]. The economics of scaling up nanotechnological processes should be thoroughly examined. Manufacturers must balance the increased costs associated with high-volume production against the potential gains in textile properties and functionalities [62, 64].

The challenge of scalability is a major obstacle in the application of nanotechnology for textile finishing. Meeting this challenge requires a combination of innovation, investment, and cooperation to fully realise the benefits of nanotechnology in the textile industry, and meet the demand for large-scale production role [63]. Continued R&D efforts to optimise and streamline nanomaterial synthesis and production methods. This includes developing more efficient, continuous, and cost-effective manufacturing processes, and before full implementation, conducting pilot demonstration projects can help identify and address potential issues related to manufacturing processes. These tasks can serve as test cases for refining production processes [69].

To develop best practices for scaling up nanotechnology in textile production, industry stakeholders, academic institutions, and government agencies can work together to share expertise. Investments in infrastructure, such as facilities built to handle nanotechnology processes and specialised production machinery, should be made by the textile industry. Manufacturers should continually refine and improve their scaling processes based on feedback and experience to enhance efficiency and reduce costs. Sustainable production methods, such as recycling waste and employing renewable energy sources, are crucial for addressing resource and environmental issues. Regulations and standards about nanotechnology in textiles can offer direction for large-scale production that is reliable and secure [28, 55, 67].

2.4 Ethical Issues

Consumers may not be aware of the presence of nanotechnology in the textiles they purchase. There is often a lack of clear labelling to inform consumers about the use of nanomaterials in textile products. Ensuring consumers are informed about the presence of nanotechnology and any potential associated risks is essential for transparency and safety [31, 70]. Educating consumers about the unique care requirements and potential allergenic reactions associated with nanotechnology in textiles can help address compatibility challenges related to user acceptance. Collaboration among material scientists, textile manufacturers, and regulators can facilitate the development of best practices and guidelines for ensuring compatibility and safety [29, 62, 71].

Manufacturers should also be transparent about the use of nanotechnology in their textile products, and consumers should be educated about potential health and environmental impacts. Textile manufacturing facilities should implement adequate worker safety measures, including the use of personal protective equipment and ventilation systems to minimise exposure to airborne nanoparticles. Also, efforts should be made to develop environmentally friendly methods for the disposal of textiles containing nanomaterials, ensuring that nanoparticles are not released into the environment [27, 53, 65].

Governments, industry stakeholders, and researchers should collaborate to share knowledge, best practices, and safety data related to nanotechnology in textiles. Therefore, a well-informed and collaborative approach is required to strike a balance between innovation and safety in the application of nanotechnology in textile finishing. As discussed in an earlier section the FERMA (UK), AS/NZS 4360:2004, OECD WPMN, REACH Annex, ISO/TC 229, the different organisations are working for the standardisation of characterisation, testing procedures, transition to handling, and environmental and health risks associated with nanotechnology-based products. These standards can help textile industries to develop special care labelling for nano-finished products.

2.5 Durability or Long-Lasting Performance

The long-term durability of nano-coatings on textiles is a critical concern in the application of nanotechnology to enhance textile properties and functionalities. Textiles are subjected to several severe circumstances, such as machine washing, ultraviolet light, and mechanical stress. Ensuring that nanotechnology-based textile finishes remain effective and durable under these conditions is a critical challenge. There is continuous research being conducted to improve the durability of these finishes to satisfy consumer demands for long-lasting performance. These coatings, which can provide various benefits such as water repellency, stain resistance, and UV protection, are expected to maintain their performance over extended periods. However, maintaining the long-term durability of nano-coatings on textiles is fraught with challenges [54, 56].

Textiles with nano-coating are prone to abrasion and wear, particularly in high-contact areas such as outdoor gear, clothing, and upholstery. The constant friction and mechanical stress can cause the nano-coatings to degrade, leading to a loss of functionality. Textiles are exposed to a wide range of environmental factors, such as moisture, variations in temperature, and UV radiation. These elements have the potential to degrade nano-coatings over time, rendering them ineffective. Specifically, UV radiation has the potential to cause photodegradation of the nanomaterials [60, 63].

The stability and durability of nano-coatings can be impacted by the various chemicals that textiles come into contact with, including detergents and cleaning agents. The nanomaterials may degrade or lose some of their effectiveness as a result of interactions with certain chemicals. Frequent washing is a major challenge for the durability of nano-finishing on textiles. The majority of nano-finishes are not made to endure the rigorous cleaning cycles. During laundry, the coatings may be weakened or completely removed due to abrasion, chemical exposure, and mechanical stress. Some nano-coatings may require special care, such as gentle washing, to maintain their functionality [62].

Furthermore, colour changes or altered aesthetics may arise from incompatibilities between nanomaterials and textile dyes or pigments. Maintaining the desired colour and appearance of textile products is challenging when applying nano-finishes. Maintaining the flexibility and drape of textiles is essential for user comfort. Numerous nanocoating can stiffen or deform textiles, which may reduce their wearability and comfort [26, 31, 57].

The fibre compositions and structures of various textiles vary. It is difficult to guarantee that nano-coatings work well with a variety of textiles, including those made of both natural and synthetic fibres. The durability of coatings can be impacted by how differently they adhere to different textile substrates. It can be difficult to establish uniform quality standards for textiles coated with nanotechnology. Variability in the application process, material quality, and durability can lead to differences in product quality among manufacturers. It is crucial to make sure that textiles and nanocoating adhere to safety and environmental impact regulations. Non-compliance can lead to restrictions and challenges in product development and marketing [60, 63].

Textiles often require multiple functions, such as water repellency, UV protection, and antimicrobial properties. Ensuring compatibility between different types of nano-coatings and finishes to achieve these multifunctional requirements is a challenge. Incorporating nanotechnology into existing textile manufacturing processes can be challenging. Compatibility issues may arise when adapting nanomaterial application methods to established textile production lines [55, 61, 69].

The compatibility of nano-finishes in textiles refers to the ability to seamlessly integrate nanomaterials and nano-coatings into textile structures, ensuring they maintain their functionality, durability, and appearance. The goal of this challenge is to maximise the adherence of nanoparticles to textile surfaces, which can differ based on the kind, structure, and finish of the fibre. For long-lasting performance, the textile and nanomaterials must form a strong and long-lasting bond [69, 71].

Developing more durable nano-coatings that can withstand wear, environmental exposure, and washing is crucial. Materials scientists are constantly researching and innovating to improve the life of nano-coatings. The development of robust test systems that mimic real-world conditions, including degradation, washing, and environmental exposure, can help to measure the durability of nano-coatings [34, 72]. Implementing multi-layer coatings can enhance durability. By adding protective layers or encapsulating nanomaterials, the coatings can better withstand wear and tear. Collaboration between materials scientists, textile engineers, and chemists can lead to the development of durable and flexible nano-coating that is compatible with different fabrics. Educating consumers on the care and maintenance of nanocoated fabrics is also important. Proper care can help prolong the life and durability of nanocoating [28, 60, 71]. In summary, solving the long-term challenges of nano-coatings on textiles requires a combined effort of materials science, test design, and manufacturing research and innovation among researchers, manufacturers, and regulatory agencies is essential to ensure sustainable and effective development of nano-coatings.

3 Opportunities with the Application of Nano-finishing in Textiles

Nanotechnology is poised to reshape the textile industry by providing an exciting opportunity to significantly enhance the functionality of textiles [73]. The ability to engineer materials and structures at the nanoscale is fundamental to this transformation and opens up a world of possibilities for textiles (Fig. 6).

3.1 Improved Functionality and Enhanced Performance

The ability of nanotechnology to increase the strength and durability of fabrics is one of its most amazing features. Manufacturers can produce textiles with increased resistance to abrasion by adding nanofibers and nanoparticles. This not only prolongs the lifespan of textiles but also makes them suitable for demanding applications, such as those in aerospace and construction, where durability is paramount [73, 74].

Another area where nanotechnology shines in enhancing textile functionality is in its capacity to create water-repellent and stain-resistant fabrics. Applied to textiles, nanoparticles create a shield that protects against liquids and stains at the nanoscale. This development has many uses, such as rain-proof outdoor apparel that keeps users dry during downpours and stain-resistant upholstery. Furthermore, nanotechnology has astounding potential for controlling temperature. Textiles that contain phase change materials (PCMs) and nanoparticles can regulate heat well, absorbing and releasing it as necessary to keep users comfortable. With the use of this technology, clothing can be made to adjust to changing environmental conditions, eliminating the need for bulky or heavy clothing [18, 22, 75, 76].

Nanotechnology's impact on textiles goes even further by addressing health and safety concerns. Textiles embedded with nanoparticles can offer effective protection against harmful UV rays, reducing the risk of skin damage and skin cancer [72–74]. Additionally, antimicrobial nanomaterials like silver nanoparticles can be integrated

Fig. 6 Opportunities for the application of nanotechnology in textile finishing

into textiles, making them resistant to bacterial and fungal growth [16]. This development is especially valuable in healthcare, sportswear, and undergarments, as it helps maintain hygiene and comfort. Furthermore, nanotechnology empowers the creation of "smart textiles" by incorporating sensors and nanoelectronics [63–65]. These textiles are capable of monitoring vital signs, environmental conditions, and even biometric data. In healthcare, they can play a pivotal role in continuous patient monitoring, allowing for early disease detection and improving patient outcomes [76, 77].

3.2 Customisation

Nanotechnology holds the promise of revolutionising the textile industry by offering the potential for customised textiles tailored to individual preferences. At the heart of this innovation is the remarkable ability of nanomaterials to be precisely engineered at the nanoscale, allowing for the manipulation of textile properties with unprecedented precision [32, 78]. Whether it's enhancing the strength and durability of fabrics, altering their colours and appearances, or tailoring their comfort and fit, nanotechnology empowers consumers to have a say in the textiles they wear. This level of customisation offers an exciting opportunity to cater to a wide range of individual needs and preferences [31, 78–81].

One of the most significant advantages of nanotechnology-enabled customisation in textiles is the ability to create clothing that not only fits perfectly but also prioritises comfort. Nanomaterials can be used to enhance breathability, moisture-wicking properties, and stretchability, resulting in clothing that conforms to an individual's unique body shape and specific requirements. Smart textiles embedded with nano-sensors can further elevate this personalisation, adjusting clothing in real time to ensure an optimal fit and comfort level [82–84]. Beyond just fit and comfort, nanotechnology also allows for personalisation in terms of aesthetics. Fabrics can be adorned with customisable colours, patterns, and appearances through the incorporation of nanoparticle-based technologies [25, 84]. This means that consumers can choose textiles that change colour in response to temperature changes, lighting conditions, or even their mood, promoting individuality and self-expression in fashion [85–89].

Customised textiles are not limited to just aesthetics and comfort; they extend to health monitoring and protection. In healthcare, nanotechnology can lead to textiles that are tailored to an individual's skin type and sensitivities [86, 90]. For patients with chronic illnesses, customised textiles with embedded sensors can monitor and transmit data specific to their health conditions, improving their quality of life. Additionally, in hazardous environments, individuals can benefit from protective textiles customised to the specific risks they face, ensuring their safety while working in challenging conditions [90–92].

The potential for customised textiles through nanotechnology is a testament to the ever-evolving synergy between science and fashion. It represents a transformative shift in the textile industry, placing the power of choice and personalisation firmly in the hands of consumers. As this technology continues to develop, it has the potential to redefine the way we think about clothing, making it not only a functional necessity but also a form of personal expression and empowerment [93, 94].

3.3 *Market Opportunities*

The application of nanotechnology in textile finishing presents a promising frontier for industry growth. This convergence of nanoscience and textiles creates fresh market opportunities across multiple fronts. Research and development efforts are flourishing, leading to the formulation of cutting-edge nanomaterials and processes that enhance textile properties [85, 95, 96]. These innovations drive product differentiation and innovation, opening doors to specialised functional textiles and the emerging field of smart textiles. As a result, new avenues for job creation are being forged, from scientists and engineers specialising in nanotechnology to skilled manufacturing workers. This symbiotic relationship between nanotechnology and textiles not only fosters progress but also augments the overall dynamism of both industries, contributing to their enduring evolution [97–99].

3.4 *Sustainability*

Nanotechnology plays a significant role in contributing to the sustainability of the textile industry, addressing some of the environmental challenges associated with textile manufacturing. One of the primary ways in which nanotechnology contributes to sustainability is by reducing resource consumption. Traditional textile dyeing and finishing processes are notorious for their high water and energy usage. Nanotechnology has enabled the development of more efficient dyeing and finishing methods that require significantly fewer resources. For instance, the use of nanoscale particles and eco-friendly dyes allows for reduced water consumption and shorter processing times, thereby lessening the environmental impact [92, 100, 101].

Furthermore, nanotechnology offers the potential to reduce the use of harmful chemicals in textile production. Traditional textile treatments often involve toxic chemicals that can harm both the environment and workers. By incorporating nanomaterials, such as nanoparticles and nano-coatings, textiles can achieve desired properties (e.g. water repellency or stain resistance) without the need for toxic chemicals. This shift towards more environmentally friendly processes not only safeguards the health of workers but also minimises the release of pollutants into the environment [98, 102]. Nanotechnology also facilitates the development of recyclable and biodegradable textiles. Nanomaterials can be used to create textiles with enhanced biodegradability, meaning they break down more easily when disposed of, reducing the burden on landfills [103, 104]. Additionally, nanotechnology enables the design of textiles that can be more easily recycled, fostering a more circular economy in the textile industry. By reimagining the composition of textiles and incorporating nanoscale components, manufacturers can produce fabrics that can be disassembled, recycled, and reused [93, 104–107].

Lastly, nanotechnology offers a pathway to reduce waste in textile production. With traditional processes, significant amounts of textiles are often discarded as defective or excess material. Nanotechnology can enhance the precision of manufacturing, minimising defects and waste [108–110]. Moreover, the ability to engineer textiles at the nanoscale allows for the creation of highly efficient filtration materials for air and water purification, providing additional environmental benefits. Nanotechnology is a vital tool in the quest for sustainability in the textile industry. By reducing resource consumption, eliminating the need for harmful chemicals, and promoting biodegradability and recycling, nanotechnology is helping to create a more environmentally responsible and eco-friendly textile manufacturing sector [111, 112]. This transition towards sustainable practices not only benefits the environment but also aligns with the growing demand for eco-conscious and ethical products in the textile market.

4 Conclusion

The application of nanoparticles in textile finishes offers a promising strategy to revolutionise the industry, delivering improved functionality, productivity, customisation, and sustainability; however, the journey of integrating nanotechnology in textile finishes is not without obstacles. The challenges of toxicity, health, and environmental hazards associated with the use of nanoparticles emphasise the need for careful consideration and responsible practices. It is imperative that we establish stringent rules and standards systems to protect consumers and the environment, and that ethical concerns associated with nano-finished textiles must be carefully considered. A balance between technological progress and ethical responsibility is needed to ensure that nanotechnological progress meets societal norms and expectations. In conclusion, although the challenges and concerns are inherent in any emerging technology, the opportunities offered by nanotechnology in textiles are too significant to ignore. Advanced research, innovation, and the responsible use of nanotechnology in textiles hold the promise of ushering in a new era of safe, efficient, and environmentally sustainable clothing, and ultimately benefit individuals and the planet. The careful navigation of these challenges and the embrace of these opportunities can embark the textile industry on a transformative journey towards a more advanced, ethical, and sustainable future.

References

1. Shabbir M, Mohammad F (2017) Insights into the functional finishing of textile materials using nanotechnology. In: Textiles and clothing sustainability: nanotextiles and sustainability, pp 97–115
2. Periyasamy AP, Militky J, Sachinandham A, Duraisamy G (2020) Nanotechnology in textile finishing: recent developments. In: Handbook of nanomaterials and nanocomposites for energy and environmental applications, pp 1–31
3. Mahmud R, Nabi F (2017) Application of nanotechnology in the field of textile. IOSR J Polym Text Eng 4(1):1
4. Zille A, Almeida L, Amorim T, Carneiro N, Esteves MF, Silva CJ, Souto AP (2014) Application of nanotechnology in antimicrobial finishing of biomedical textiles. Mater Res Express 1(3):032003
5. Bashari A, Shakeri M, Shirvan AR, Najafabadi SAN (2018) Functional finishing of textiles via nanomaterials. In: Nanomaterials in the wet processing of textiles, pp 1–70
6. Vhanbatte SB, Landage SM, Wasif AI, Dansena B, Karche NV (2017) Nanotechnology for antimicrobial finishing of textiles. Int J Adv Res Eng Appl Sci 6(6):14–23
7. Pal S, Mondal S, Pal P, Das A, Mondal D, Chaudhuri A, Maity J (2020) Applications of nanotechnology for antibacterial finishing textiles: a review. Sens Lett 18(6):437–448
8. Nascimento JHO, Felipe BHS, Cabral RLB, Ahmad A, da Silva AB, Neto NFA, Teófilo, ALC (2021) New advances in nanotechnology in textile engineering: functional finishing with quantum dots and other nanoparticles.In: Nanomaterials and Nanotechnology: Biomedical, Environmental, and Industrial Applications, pp 239–281
9. Ijaz M (2013) Role of nanotechnology in the field of textile finishes. Res J Soc Sci Manag 3(3):201–207
10. Patra K (2013) Application of nanotechnology in textile engineering: an overview. J Eng Technol Res 5(5):104–111
11. Tarafder N (2018) Applications of nanotechnology for textile products: a review. Nanoscale Rep 1(3):15–22
12. Samanta KK, Basak S, Chowdhury P (2014) Nano-technology for advanced textile application. Man-Made Textiles in India, 42(2)
13. Shaarawy S (2019) A review of the development of innovative capabilities in the textile finishing of natural fibres. Egypt J Chem, 62(Special Issue (Part 2) Innovation in Chemistry), 857–579
14. Temesgen AG, Turşucular ÖF, Eren R, Ulcay Y (2018) Novel applications of nanotechnology in modification of textile fabrics properties and apparel. Int J Adv Multidiscip Res 5(12):49–58
15. Faraz A, Faizan M, Hayat S (2022) Functional value-added finishing of textile substrates using nanotechnology: a review. In: Emerging technologies for textile coloration, pp 101–109
16. Verma A, Arif R, Jadoun S (2021) Impact of nanotechnology on sustainable textile material and its application. In: Green chemistry for sustainable textiles, Woodhead Publishing, pp 165–172
17. Shaheen TI (2022) Nanotechnology for modern textiles: highlights on smart applications. J Text Inst 113(10):2274–2284
18. Abdul-Reda Hussein U, Mahmoud ZH, Abdalaziz KM, Alid ML, Yasin Y, Ali FK, Kianfar E (2023) Antimicrobial finishing of textiles using nanomaterials. Braz J Biol 84:e264947
19. Gokarneshan N, Gopalakrishnan PP, Jeyanthi B (2012) Influence of nano finishes on the functional properties of textile materials. Int J Basic Appl Chem Sci ISSN, pp 2277–2073
20. Hossain M, Biswas AK, Parvez S, Bain S (2013) Advanced functionalization of textiles by nano finishing: a review. In: Proceedings of the International Conference on Mechanical, Industrial and Materials Engineering, Rajshahi, Bangladesh, pp 1–3
21. Uddin F (2008) Some aspects of the advancement of textile nanofinishing. In: 86th textile-institute world conference. Hong Kong: polytechnic university, institute of textiles and clothing press, pp U291–U307

22. Black S (2009) The role of nanotechnology in sustainable textiles. In: Sustainable textiles, Woodhead Publishing, pp 302–328
23. Kaounides L, Yu H, Harper T (2007) Nanotechnology innovation and applications in textiles industry: current markets and future growth trends. Mater Technol 22(4):209–237
24. Nazeer AA, Dhandapani S, Vijaykumar SD (2018) Advanced technologies for coloration and finishing using nanotechnology. In: Handbook of renewable materials for coloration and finishing, p 473
25. Chen J, Lu YG, Sun C (2011) Safety and health assessment of manufactured nanoparticles in nano-coated textile products. Adv Mater Res 175:722–728
26. Chakrabarty S, Jasuja K (2022) Applications of nanomaterials in the textile industry. In: Nanoscale engineering of biomaterials: properties and applications,Springer Nature Singapore, Singapore, pp 567–587
27. Afzali A, Maghsoodlou SH (2016) Modern application of nanotechnology in textile. In: Nanostructured polymer blends and composites in textiles, pp 41–85
28. Bhatia SC, Devraj S (2017) Pollution control in the textile industry. WPI publishing, pp 203–218
29. Shabbir M, Kaushik M (2020) Engineered nanomaterials: scope in today's textile industry. In: Handbook of nanomaterials for manufacturing applications, Elsevier, pp 249–263
30. Liliana I, Sabina G, Steven M (2023) Nanomaterials in textiles: performance, health, and environmental aspects. In: Handbook of nanofibers and nanocomposites, Jenny Stanford Publishing, pp 169–191
31. Singh D, Sharma P, Pant S, Dave V (2022) Health safety and environment aspect of nanotextiles. In: Fundamentals of nano-textile science, Apple Academic Press, pp 317–342
32. Perera S, Wijesekara D, Thiripuranathar G, Menaa F (2022) The use of nanoparticles to enhance performance in the textile industry-a concise review. Curr Nanosci 18(3):319–335
33. Singh S, Nalwa HS (2007) Nanotechnology and health safety–toxicity and risk assessments of nanostructured materials on human health. J Nanosci Nanotechnol 7(9):3048–3070
34. Cattaneo AG, Gornati R, Sabbioni E, Chiriva-Internati M, Cobos E, Jenkins MR, Bernardini G (2010) Nanotechnology and human health: risks and benefits. J Appl Toxicol 30(8):730–744
35. Mu X, Li W, Ze X, Li L, Wang G, Hong F, Ze Y (2020) Molecular mechanism of nanoparticulate TiO_2 induction of axonal development inhibition in rat primary cultured hippocampal neurons. Environ Toxicol 35(8):895–905
36. Mehrbeheshti N, Esmaili Z, Ahmadi M, Moosavi M (2022) A dose response effect of oral aluminum nanoparticle on novel object recognition memory, hippocampal caspase-3 and MAPKs signaling in mice. Behav Brain Res 417:113615
37. Song B, Zhang Y, Liu J, Feng X, Zhou T, Shao L (2016) Unraveling the neurotoxicity of titanium dioxide nanoparticles: focusing on molecular mechanisms. Beilstein J Nanotechnol 7(1):645–654
38. Chaudhary K, Masram DT (2020) Biological activities of nanoparticles and mechanism of action. Model organisms to study biological activities and toxicity of nanoparticles, pp 19–34
39. Nasrollahzadeh M, Sajadi SM (2019) Risks of nanotechnology to human life. In: Interface science and technology, Elsevier, 28: 323–336
40. Shatkin JA (2017) Nanotechnology: health and environmental risks. CRC Press
41. Barhoum A, García-Betancourt ML, Jeevanandam J, Hussien EA, Mekkawy SA, Mostafa M, Bechelany M (2022) Review on natural, incidental, bioinspired, and engineered nanomaterials: history, definitions, classifications, synthesis, properties, market, toxicities, risks, and regulations. Nanomaterials 12(2):177
42. Williams D (2005) The risks of nanotechnology. Med Device Technol 16(9):6–9
43. Xiong P, Huang X, Ye N, Lu Q, Zhang G, Peng S, Liu Y (2022) Cytotoxicity of metal-based nanoparticles: from mechanisms and methods of evaluation to pathological manifestations. Adv Sci 9(16):2106049
44. Naseer B, Srivastava G, Qadri O, Faridi S, Islam R, Younis K (2018) Importance and health hazards of nanoparticles used in the food industry. Nanotechnol Rev 7(6):623–641. https://doi.org/10.1515/ntrev-2018-0076

45. Kuempel ED, Roberts JR, Roth G, Zumwalde RD, Nathan D, Hubbs AF, Holdsworth G (2021) Health effects of occupational exposure to silver nanomaterials. https://doi.org/10.26616/NIOSHPUB2021112
46. Riaz U, Iqbal S, Shahzad L, Samreen T, Abbasi WM (2021) Health and safety hazards of nanomaterials. In: Nanomaterials: synthesis, characterization, hazards and safety, Elsevier, pp 223–240
47. Ramanathan A (2019) Toxicity of nanoparticles_challenges and opportunities. Appl Microsc 49(1):2
48. Saif S, Tahir A, Chen Y (2016) Green synthesis of iron nanoparticles and their environmental applications and implications. Nanomaterials 6(11):209
49. Rossi S, Savi M, Mazzola M, Pinelli S, Alinovi R, Gennaccaro L, Miragoli M (2019) Subchronic exposure to titanium dioxide nanoparticles modifies cardiac structure and performance in spontaneously hypertensive rats. Part Fibre Toxicol 16(1):1–18
50. Sun X, Chang Y, Cheng Y, Feng Y, Zhang H (2018) Band alignment-driven oxidative injury to the skin by anatase/rutile mixed-phase titanium dioxide nanoparticles under sunlight exposure. Toxicol Sci 164(1):300–312
51. Kononenko V, Repar N, Marušič N, Drašler B, Romih T, Hočevar S, Drobne D (2017) Comparative in vitro genotoxicity study of ZnO nanoparticles, ZnO macroparticles and $ZnCl_2$ to MDCK kidney cells: size matters. Toxicol Vitro 40:256–263
52. Kim JK, Kim HP, Park JD, Ahn K, Kim WY, Gulumian M, Yu IJ (2021) Lung retention and particokinetics of silver and gold nanoparticles in rats following subacute inhalation co-exposure. Part Fibre Toxicol 18(1):1–12
53. Mottier A, Mouchet F, Pinelli E, Gauthier L, Flahaut E (2017) Environmental impact of engineered carbon nanoparticles: from releases to effects on the aquatic biota. Curr Opin Biotechnol 46:1–6
54. Aileni RM, Surdu L, Oksuz L (2015) Life cycle and environment impact for textile materials functionalized by using plasma technology. In: 2015 IEEE international conference on plasma sciences (ICOPS), IEEE, pp 1–10
55. Jatoi AS, Khan FSA, Mazari SA, Mubarak NM, Abro R, Ahmed J, Sabzoi N (2021) Current applications of smart nanotextiles and future trends. In: Nanosensors and nanodevices for smart multifunctional textiles, pp 343–365
56. Bisht D, Yadav S, Rose NM, Grover A, Rukhaya S (2021) Sustainable textile finishes: a boon for reducing toxic disposal to the environment. 10(11): 861–867
57. Ali T, Najam-ul-Haq M, Mohyuddin A, Musharraf SG, Hussain D (2023) Next-generation functional nanotextiles—Prospects and challenges. In: Sustainable materials and technologies, p e00640
58. Pillai MM, Senthilkumar R, Selvakumar R, Bhattacharyya A (2018) Characterization methods of nanotechnology-based smart textiles. Smart Text: Wearable Nanotechnol 1:347–377
59. Fuentes KM, Gómez M, Rebolledo H, Figueroa JM, Zamora P, Naranjo-Briceño L (2022) Nanomaterials in the future biotextile industry: a new cosmovision to obtain smart biotextiles. Front Nanotechnol 4:1056498
60. Saleem H, Zaidi SJ (2020) Sustainable use of nanomaterials in textiles and their environmental impact. Materials 13(22):5134
61. Martínez G, Merinero M, Pérez-Aranda M, Pérez-Soriano EM, Ortiz T, Villamor E, Alcudia A (2020) Environmental impact of nanoparticles' application as an emerging technology: a review. Materials 14(1):166
62. Stensberg MC, Wei Q, McLamore ES, Porterfield DM, Wei A, Sepúlveda MS (2011) Toxicological studies on silver nanoparticles: challenges and opportunities in assessment, monitoring and imaging. Nanomedicine 6(5):879–898
63. Barker TF, Fatehi L, Lesnick MT, Mealey TJ, Raimond RR (2011) Nanotechnology and the poor: opportunities and risks for developing countries. Springer, Netherlands, pp 277–290
64. Hoyt VW, Mason E (2008) Nanotechnology: emerging health issues. J Chem Health Saf 15(2):10–15

65. Resnik DB (2019) How should engineered nanomaterials be regulated for public and environmental health?. AMA J Ethics, 21(4): 363–369
66. Iavicoli I, Leso V, Beezhold DH, Shvedova AA (2017) Nanotechnology in agriculture: opportunities, toxicological implications, and occupational risks. Toxicol Appl Pharmacol 329:96–111
67. Rickerby DG (2013) Nanotechnology for more sustainable manufacturing: opportunities and risks. In: Sustainable nanotechnology and the environment: advances and achievements, American Chemical Society, pp 91–105
68. Shah MA, Pirzada BM, Price G, Shibiru AL, Qurashi A (2022) Applications of nanotechnology in smart textile industry: a critical review. J Adv Res 38:55–75
69. Iavicoli I, Leso V, Ricciardi W, Hodson LL, Hoover MD (2014) Opportunities and challenges of nanotechnology in the green economy. Environ Health 13(1):1–11
70. Islam SU, Butola BS (eds) (2018). The impact and prospects of green chemistry for textile technology. Woodhead Publishing, pp 190–212
71. Karn B, Kuiken T, Otto M (2009) Nanotechnology and in situ remediation: a review of the benefits and potential risks. Environ Health Perspect 117(12):1813–1831
72. Borm PJ, Berube D (2008) A tale of opportunities, uncertainties, and risks. Nano Today 3(1–2):56–59
73. Bergeson LL (2008) The new business of nanotechnology: exploring commercial opportunities and risks. Environ Claims J 20(2):144–159
74. Marchant GE, Sylvester DJ, Abbott KW (2008) Risk management principles for nanotechnology. NanoEthics 2:43–60
75. Petersen A, Anderson A, Wilkinson C, Allan S (2007) Nanotechnologies, risk and society. 9(2): 117–124. https://doi.org/10.1080/13698570701306765
76. Romig AD Jr, Baker AB, Johannes J, Zipperian T, Eijkel K, Kirchhoff B, Walsh S (2007) An introduction to nanotechnology policy: opportunities and constraints for emerging and established economies. Technol Forecast Soc Chang 74(9):1634–1642
77. Barman J, Tirkey A, Batra S, Paul AA, Panda K, Deka R, Babu PJ (2022) The role of nanotechnology-based wearable electronic textiles in biomedical and healthcare applications. Mater Today Commun, p 104055
78. Landage SM, Wasif AI (2012) Nanosilver–an effective antimicrobial agent for finishing textiles. Int J Eng Sci Emerg Technol 4(1):66–78
79. Ali SW, Banerjee S, Mondal M (2023) Nanotechnology in textile finishing: an approach towards imparting manifold functionalities. Smart Func Text. https://doi.org/10.1515/9783110759747-002
80. Sawhney APS, Condon B, Singh KV, Pang SS, Li G, Hui D (2008) Modern applications of nanotechnology in textiles. Text Res J 78(8):731–739
81. Yadav A, Prasad V, Kathe AA, Raj S, Yadav D, Sundaramoorthy C, Vigneshwaran N (2006) Functional finishing in cotton fabrics using zinc oxide nanoparticles. Bull Mater Sci 29:641–645
82. Montazer M, Harifi T (2018) Nanofinishing of textile materials. Woodhead Publishing
83. Shabbir M, Sheikh JN (2020) Introduction to textiles and finishing materials. Frontiers of Textile Materials: Polymers, Nanomaterials, Enzymes, and Advanced Modification Techniques, pp 1–11
84. Hassabo AG, Elmorsy H, Gamal N, Sediek A, Saad F, Hegazy BM, Othman H (2023) Applications of nanotechnology in the creation of smart sportswear for enhanced sports performance: efficiency and comfort. J Text, Colorat Polym Sci 20(1):11–28
85. Romagnoli MJ, Gonzalez JS, Martinez MA, Alvarez VA (2020) Micro-and nanotechnology applied on eco-friendly smart textiles. In: Handbook of nanomaterials and nanocomposites for energy and environmental applications, pp 1–19
86. Sbai SJ, Boukhriss A, Majid S, Gmouh S (2020) The recent advances in nanotechnologies for textile functionalization. In: Advances in functional and protective textiles, pp 531–568
87. Jadoun S, Verma A, Arif R (2020) Modification of textiles via nanomaterials and their applications. In: Frontiers of textile materials: polymers, nanomaterials, enzymes, and advanced modification techniques, pp 135–152

88. Krishna A, Chaudhary SN (2014) Nanotechnology in textiles. Man-Made Text India 42(3):10–17
89. Omietimi HB, Afolalu SA, Kayode JF, Monye SI, Lawal SL, Emetere ME (2023) An overview of nanotechnology and its application. In: E3S web of conferences, EDP sciences, 391: 01079
90. Gashti MP, Pakdel E, Alimohammadi F (2016) Nanotechnology-based coating techniques for smart textiles. In: Active coatings for smart textiles, Woodhead Publishing, pp 243–268
91. Anand G, Pramanik AM (2022) An introduction to sustainable finishing techniques of textiles. Int J Home Sci 8(2):217–220
92. Ngô C, Van de Voorde M, Ngô C, Van de Voorde MH (2014) Nanotechnology for the textile industry. In: Nanotechnology in a nutshell: from simple to complex systems, pp 321–329
93. Abd E-Aty M, Ghazal H (2023) Nanotechnology and application of nano titanium dioxide, nano zinc oxide, and nano copper oxide on textile for high performance. Egypt J Chem, 66(3): 309–322
94. Zhang F, Wu X, Chen Y, Lin H (2009) Application of silver nanoparticles to cotton fabric as an antibacterial textile finish. Fibers Polym 10:496–501
95. Ali SW, Chowdhury A, Bairagi S, Banerjee S (2022) Green nanomaterials for multifunctional textile finishes. In: green functionalized nanomaterials for environmental applications, Elsevier, pp 343–364
96. Damuluri R, Kiran K, Chakravarthy DP (2017) Review studies on the application of nanotechnology in textiles. Int J Text Fashion Technol 7(6):1–4
97. Parmar MS (2022) Application of nanotechnology-based UV finishes on textiles and their evaluation. In: Fundamentals of nano-textile science, Apple Academic Press, pp 127–144
98. Tyagi G, K, Madhu A (2023) Nanotechnology in textiles: unleashing innovation and value addition. Fashion Technol. 9(1). https://doi.org/10.31031/TTEFT.2023.09.000702
99. Hassan MSB (2020) Nano technology applications in textiles to improve puncture resistant of fabrics. Int Des J 10(1):147–154
100. Chinta SK, Landage SM, Swapnal J (2013) Water repellency of textiles through nanotechnology. Int J Adv Res IT Eng 2(1):36–57
101. Paul R (ed) (2014) Functional finishes for textiles: improving comfort, performance and protection. Elsevier
102. Gharehaghaji AA (2019) Nanotechnology in sport clothing. In: Materials in sports equipment, Woodhead Publishing, pp 521–568
103. Sarvalkar PD, Barawkar SD, Karvekar OS, Patil PD, Prasad SR, Sharma KK, Vhatkar RS (2022) A review on multifunctional nanotechnological aspects in modern textile. J Text Inst 114(3):470–487
104. Andra S, Balu SK, Jeevanandam J, Muthalagu M (2021) Emerging nanomaterials for antibacterial textile fabrication. Naunyn Schmiedebergs Arch Pharmacol 394:1355–1382
105. Mishra R, Militky J, Baheti V, Huang J, Kale B, Venkataraman M, Wang Y (2014) The production, characterization and applications of nanoparticles in the textile industry. Text Prog 46(2):133–226
106. Chowdhury KP, Susan MABH, Ahmed S (2023) Nanomaterials for multifunctional textiles. Emerging Applications of Nanomaterials 141:169–217
107. Kumar PS, Joshiba GJ (2020) Sustainability in dyeing and finishing. In: Sustainability in the textile and apparel industries: production process sustainability, pp 165–177
108. Mishra R, Militky J (2018) Future outlook in the context of nanoscale textiles as a technology for the twenty-first century. In: Nanotechnology in textiles: theory and application, pp 387–398
109. Ghajarieh A, Habibi S, Talebian A (2021) Biomedical applications of nanofibers. Russ J Appl Chem 94:847–872
110. Syduzzaman M, Hassan A, Anik HR, Akter M, Islam MR (2023) Nanotechnology for high-performance textiles: a promising frontier for innovation. Chemnanomat 9(9):e202300205
111. Subashini JM, Ramakrishnan DG (2021) Sustainable finishes for healthcare apparels. Res J Text Leather 2(1):66–70
112. Syduzzaman MD, Patwary SU, Farhana K, Ahmed S (2015) Smart textiles and nanotechnology: a general overview. J Text Sci Eng 5(1):1–7

Chapter 18
Nanofinishing and Circular Economy

Monabbir Rafsan Fahim, Mahamu-Ul Islam, Dipa Saha, Humayra Akhter Himu, Shamima Akter Smriti, and Ummul Khair Fatema

1 Introduction

1.1 Nanotechnology, Nano-Finishing, and Adverse Impacts

Nanotechnology is the new approach of science which goal is to build and flourish devices, machines, different kinds of product and processes using nanomaterials as well as nanoparticles [1, 2]. The element having a nano range size, 1–100 nm, is called nanoparticles show several types of physical and chemical feature which are generally absent in other natural or simulated materials [1]. Nanoparticles can obtain from nature such as soils, waters, or volcanic dust. As like as natural nanoparticles, NPs are also manufactured using modern technology called engineered nanoparticles broadly have very effective practice in agriculture, electronics, medicine, pharmacy, cosmetology, textile industry, waste water management or sewage treatment, and so on [1]. Natural nanoparticles are varied from engineered nanoparticles as it is designed for any particular use with a certain property [1]. Nanomaterials have an excellent construction that make them resistant to chemical attacks, mechanically stiffer, self-cleaning, and self-healing to the thermal and mechanical damages [2]. Nanoparticles are very harmful, and to build nanomaterials, a super clean room and authentic origin of materials are very essential. Otherwise, contamination occurs and

M. R. Fahim
Department of Textile Engineering Management, Bangladesh University of Textiles, Tejgaon Industrial Area, Dhaka 1208, Bangladesh

M.-U. Islam · D. Saha · H. A. Himu (✉) · U. K. Fatema
Department of Environmental Science and Engineering, Bangladesh University of Textiles, Tejgaon Industrial Area, Dhaka 1208, Bangladesh
e-mail: humayra@ese.butex.edu.bd

S. A. Smriti
Department of Fabric Engineering, Bangladesh University of Textiles, Tejgaon Industrial Area, Dhaka 1208, Bangladesh

S. Bairagi et al. (eds.), *Nanotechnology in Textile Finishing*, Materials Horizons: From Nature to Nanomaterials, https://doi.org/10.1007/978-981-97-2696-7_18

causes the generation of hazardous crucial problems [2]. The International Standardization Organization (ISO) prescribes a nanomaterial as a "material with any external dimension in the nanoscale or having internal structure or surface structure in the nanoscale", with nanoscale being the "length range approximately from 1 to 100 nm" [3]. The advanced nanoscale materials or particles which is called smart nanoparticles have an ability to provide a certain characteristic suitable to the environmental condition such as pH, temperature, different chemicals and solvents, light, air, electric and magnetic fields, and mechanical forces. These smart nanoparticles are very diversified and produce a very perplexing composition with a very tiny structure which imposes diametrical problems on the way of detecting its variable and alternating properties, and feasible poisonous action or consequences after the exemption into the environment and during their whole life cycle [3]. The growing application of nanotechnology in commercial and industrial sector without any specific provision and management causes environmental and human health problems along with affecting the workers in indoor and outdoor workplaces [4]. Nanoparticles are used for both, making new advanced materials or goods and recycling or removing pollutants from the environment [5]. Carbon based nanoparticles such as fullerenes, nanotubes, metal oxides- oxide of iron, titanium and zinc, naturally and unintentionally produced inorganic nanoparticles, soot (the tiny particles which is also called ultra-fine particles), polymeric nanoparticles have several detrimental impacts on the biological, chemical and physiological functions of the environment and also on human body, microorganisms, plants, soil, marine environment, living organisms etc. [4]. From various sources, NPs are released to the environment and revolve in the ecosystem seamlessly [1]. Though it has legionary amenities, it exposes a dangerous effect on the safety either of the environment or the other living and non-living things. The evaluation or estimation of the hazardous nature, ability to posing toxicity, purpose of use, etc., are directly or indirectly depend on their durability in the environment or in the organisms' body, mobility, and their reactivity [4]. NPs are highly stable, and it creates aggregate due to the lack of feasible stabilization [5]. That is why it is mandatory to stabilize the NPs using three principal stabilization path shown in Fig. 1 so that aggregation or congregation cannot take place [5].

After deliverance, NPs generally mixed with the terrestrial and aquatic biodiversity, where they create their own habitat into the living organisms as they take nanoparticles into the body by swallowing or absorbing [1]. Inside the organism's body, NPs act as an external substance which generates complications in the physiological cycle of the fetus, growth of bacteria, algae, invertebrates, and vertebrates by exhibiting their own physiochemical properties [1]. Due to the tiny size, large surface area, and the capability of generating reactive oxygen species, they can easily enter into the body cells of organism, gather in a particular place of cells (mitochondria), and provide certain toxic reaction because of their unusual composition (physical and chemical properties) [4]. Sometimes the cellular absorption of nanoparticles takes place by endocytosis and phagocytosis (Fig. 2) [4].

Nanoparticles penetrate into the cell and sabot the pores which interrupt the chemical exchange processes between the cell and the circumambient medium, and as a result cell becomes damage and inhibits the growth [6]. ZnO NPs destroy the cell

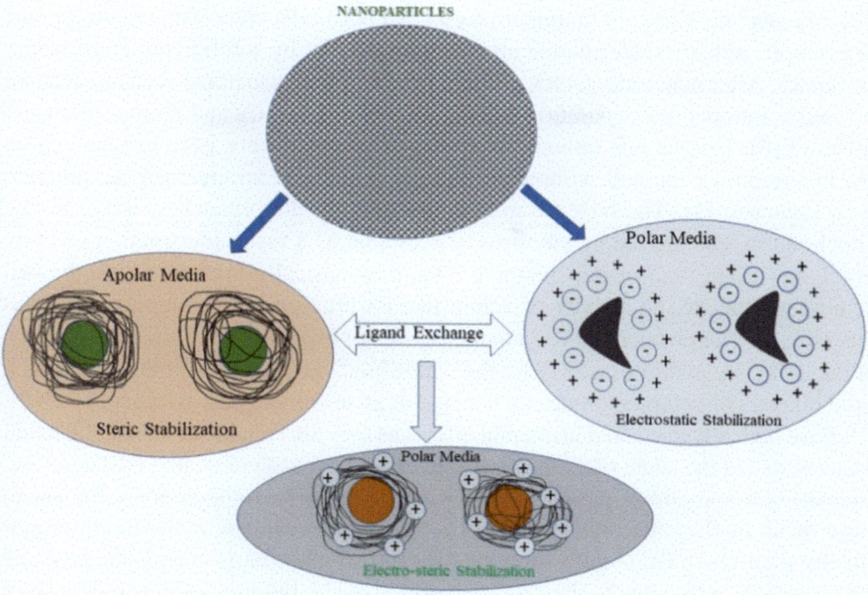

Fig. 1 Mechanism of stabilization of nanoparticles in media with different polarity

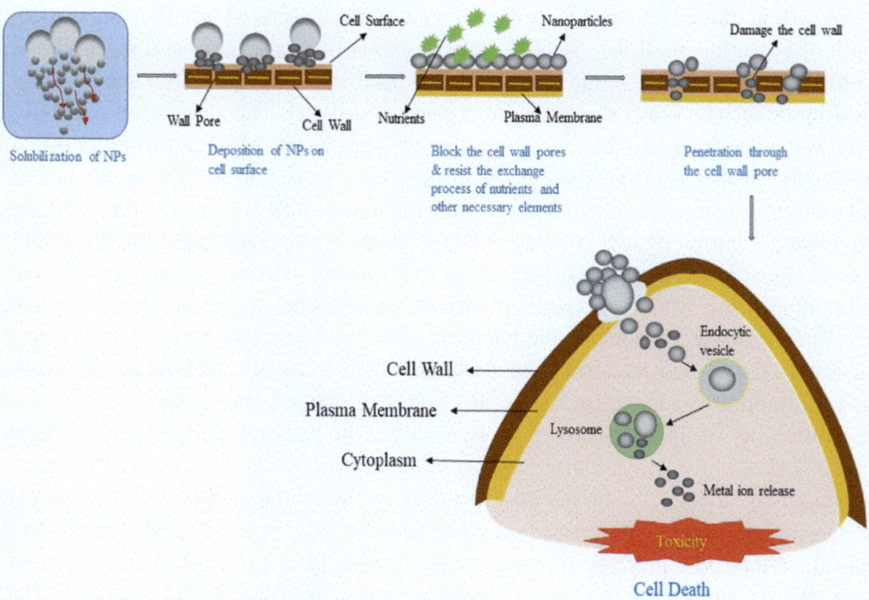

Fig. 2 Venomous accession of NPs, causing the cell damage

wall of yeast and algae. In mammalian cell, MOx of NPs enter various alterations, for example, redox transformation and complexation or by solubilizing in the acidic lysosomes. After penetrating, they produce toxicity by Trojan horse-type mechanism and causes intracellular separation due to the production of Cu and Zn ion from CuO and ZnO [6]. Due to this unnoticeable size, it becomes very easy to impenetrate into the velum of the cell without facing any protection arrangement or function of an organism [1]. The NPs are moved from cell to the organelles, for example, mitochondria, change the metabolism system, and as a result, destruction of body cells happens. In case of impenetrable NPs, they basically intervene with the cell membrane, impede the activity of velum like ion transportation, signal transportation, etc. [1]. The physiochemical characteristics of NPs are undoubtedly poisonous for the living organisms as well as for the whole biodiversity directly or indirectly [1]. Lipid bilayers of cell membrane can damage or attack by the positive charge of NPs. NPs have both adverse and detrimental effects as they are capable of vanish the lethal diseases, and at the same time, they are able to be the reason of death [1]. Therefore, nanoparticles sometimes cause many serious health issues which are sufficient to cause death as they not only interrupt the natural mechanism of the living organisms by their chemical composition but also they impart many chemicals, physical and biological action due to their undetectable size [1]. Inflammation and fibrosis in multicellular creature; cytotoxicity in unicellular organism; respiratory and cardiovascular diseases, lung diseases, premature death of workers occur respectively due to the carbon black NP, small particles of soot and NP polluted air. Thereby causing death of bacteria, small fish, other aquatic organisms for the toxic action of NP [4]. Nanoparticles of metal oxide like TiO_2 are used in sun screen, cosmetics, paint, building materials, seafood, aquaculture, and so forth [7]. The stability or durability of TiO_2 in water disturbs the marine ecosystem seamlessly. Momentous reduction in superoxide dismutase (SOD), CAT, POD function, increment in LOP levels in cell, and oxidative aspiration of fish occurs in the presence of TiO_2 just only 100–200 mg/L in water. The toxicity level of TiO_2 NPs on Daphnia magna increased with the integration of exposure time. TiO_2 NPs show long-lasting effect on marine ecosystem, and it inhibits the growth of aquatic plants and changes the chemical cycle of aquatic life [7]. The silver NP attacks the bacterial cell and destroy it. Silver NPs are very poisonous for *E. coli* bacteria than silver ions [4]. Inorganic NPs of metal oxides such as titanium oxide, sulfur oxide, and zinc oxide attack the bacterial cell, and in the availability of light it becomes more toxic for the bacterial cell and causes death (Fig. 3) [4].

Oxidative stress (OS) arises due to the disequilibrium of the level of reactive species and antioxidant defenses (enzymatic or non-enzymatic) [6]. Metals released from the MOx NPs produce reactive species and deplete the antioxidant defenses. Basically, the generation of reactive oxygen species (ROS), hydroxyl radicles ($OH^{\cdot}$), and hydrogen peroxide (H_2O_2) happens by the interaction between NPs and biological processes. OS is connected with the loss of biological molecules, for example, cellular lipid through lipid peroxidation (LPO), carbohydrates, proteins, and DNA [6].

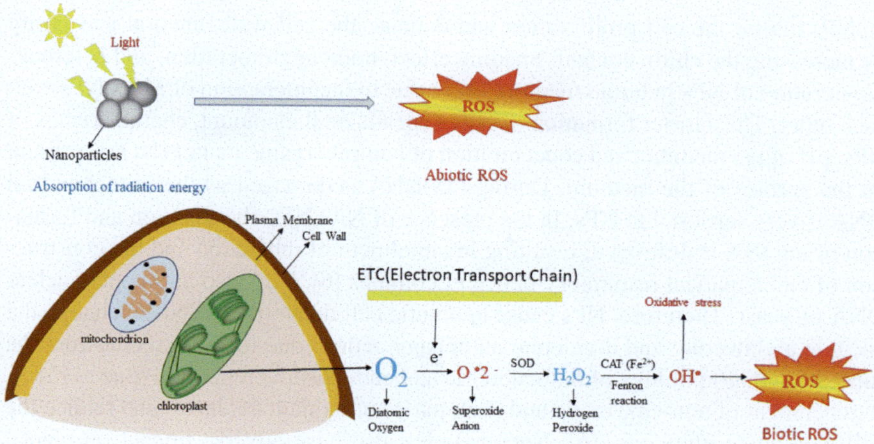

Fig. 3 Generation of reactive oxygen species (ROS): biotic and abiotic ROS; SOD (superoxide dismutase)

NPs use in various fields of science and technology, the deliverance of NPs in the indoor and outdoor environment as well as workplaces creates air pollution and various toxic chemical reactions with chemical substances present in our atmosphere [4]. The uncontrolled percolation of NPs in the environment and in the human body accelerates the destructive impact of NPs. Their exceptional chemical and physical features change the biochemical system of human body [4]. NOMs take up the NPs and help to migrate the NPs from the algae to the higher trophic level, in such a way NPs are involved in food cycle and impart obscure environmental and health danger [6]. Due to the layer formation on the surface, sunlight cannot reach to the underwater photosynthetic organisms and plants. As a result, growth and chlorophyll content of those organisms and plants are impeded (Fig. 4).

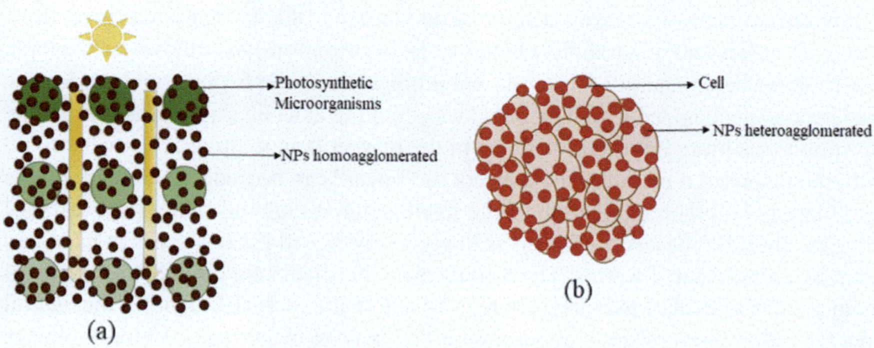

Fig. 4 **a** Shading effect (on photosynthetic microorganisms): homoagglomeration of NPs. **b** Physical repression: heteroagglomeration of NPs

NPs inhibit the cell proliferation and damage the cell wall function and purity by increasing the chitin content. Shading effect, homoagglomeration, and heteroagglomeration of NPs in liquid medium occur due to the interaction of MOx NPs with each other. This cluster formation of NPs depends on the amount, characteristics of NPs, pH of the medium, and concentration of natural organic matter (NOM) present on the surface of the medium. During OS DNA is damaged as the mitochondrial DNA is very sensorial to NPs. In the presence of NiO NPs, the function and formation of mtDNA is destroyed, resulting in annulment of inhalation and the construction of emblematical respiratory-deficient colonies [6]. NiO also harms the nuclear DNA of yeast. Therefore, NPs cause apoptotic cell death (due to NiO), change the vacuole architecture and accelerate autophagy action (due to Zno), reconstruct the function and morphology of mitochondria and endoplasmic reticulum (due to ZnO), retrenchment of non-enzymatic and enzymatic antioxidant defenses, and reduce the cell membrane unity, etc. [6]. Nanoparticles also have adverse impact on plants. Uncoated aluminum NPs slow down the growth of root [4]. On account of their shape, size, and particular properties, they react with the root of plants and penetrate into the cell, intercellular space as well as the woody tissue. Very interesting thing is that the toxicity of NPs either rises or reduces due to the inorganic or toxic materials. If the amalgamation of NPs and contaminants is toxic, then it is very pernicious, and if not, then no poisonous activities will be arisen [4]. In this advanced era, nanofinishing is largely used in many fields, and textile is one of them. According to textile, finishing is the last process or the last scope to give the determined quality or properties to the textile product.

1.2 Circular Economy Model

1.2.1 Linear Economy

The world has been on so many developments since the 1800's when newer inventions made life easier and remarkable. On the verge of inventions and innovations, people searched for new materials, systems, and production methods that could modernize society in a mechanized way. The world was thinking in a one way journey where the invention was more important to place in the market than to think about what to do with the disposed materials. The concept of "linear" can be understood on the basis that there is no other application of the manufactured material rather than disposal after the shelf life has expired. This is likely to mining all the resources in hand and creating unimaginable wastes. The resources on this planet are limited, and plenty of them are in the level of red alert. There is no automatic replenishment for the natural sources rather humans think about a modified way of using them. Before going to the solution, it is necessary to have a deeper understanding of linear economy itself.

The value chain of the economy relies on the concept of "extract, manufacture, consume, and dispose" where the life cycle of the products ends in the environmental sources, and this is known as linear economy (LE). After the third industrial revolution, the linear economy prevailed in society largely with the excessive use of plastics or synthetic materials that threatened the environment and its livings [8]. The businesses that are running in this economy use unreplenishable sources, non-biodegradable components, lack of collaboration in production methods, lower cost consumption, and more [9]. The negative impression of LE gets into the mind firstly due to the wastes generated which are represented as harmful for the environment, not thinking about reusing them as sources. Second most important aspect worked against it is the use of scarce resources, though the world always expected cheap products in higher quality [10]. Though maximum industries are built on this concept, the global textile and fashion industry is the pioneer in exploring the issue, the landfills, and rivers all are full of waste materials and disposed garments pieces [9, 10]. The fast fashion trends, season-based clothes, cultural paradigm, and huge profit margins all have revoked the traditional way of people seeing clothes as a material of safety to a material of exhibiting class differences in society. The extensive buying habits in the clothing industry compel manufacturers to produce more in the linear fashion. Customers all over the world are demanding for more quality at a lower price, making it difficult for the manufacturers and retailers to think on the sustainability portion of this. As a result, the textile industry becomes the second largest polluter in the industrial sector worldwide, comprising almost 10% of the numbers [11].

1.2.2 Adverse Impact of Linear Economy

Nikolay et al. described an environmental model in their work where they take into account seven different key processes to address the negative impacts of LE onto nature [12]. The processes are termed as greenhouse gas from both the industries and agricultural sector, CO_2 emissions, rise in global temperature, interruption in fresh water supply, deforestation, and natural disasters [12]. The textile industry is not far behind in contributing to the adverse impact the industries are leading the world. The trend of fast fashion evolved concepts of low cost manufacturing and rapid changes in the habits, which generated excessive use of chemicals, fuels, and consumption of resources [11]. The contamination is somehow larger in the manufacturing countries spreading into the air, water, and landfills. Again with the production of CO_2, water consumption is a great concern for this industry as the processes from dyeing to finishing require a lot of them. The conventional dyeing process consumes a lot of water and releases the hard water in the rivers or water bodies. The effluents and wastes contain heavy metals that ultimately get into the food chain and become the causes of destruction of water species [13]. At the end of the life cycle, the textile products are mostly disposed of in the landfills, and they are mostly manufactured from synthetic materials. The extensive use of the synthetic materials leads to micro-plastic pollution especially in the ocean bodies. Almost 35% of the microplastics pollution arises from regularly used clothes by washing rigorously, and it was estimated to be of 2–13 million tons of release per year globally (Fig. 5) [14].

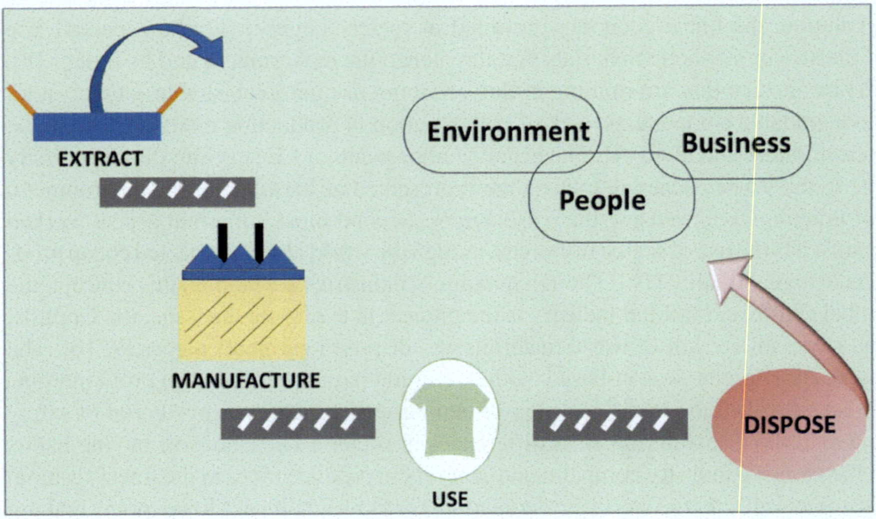

Fig. 5 Representation of the linear economy in a summarized way

1.2.3 How Nanoparticles Can Join in Environmental Degradation

The application of nanoparticles (NPs) is widely spread in the medical, electronics, cosmetics, biosensor, environmental remediation, photo-imaging, and many more industries [15, 16]. With the use of nanomaterials, there is always a risk of exposing to a health hazard due to the surface area and nano level size of the particles that might easily penetrate in medium or react with human skin. The NPs might get into the environment from the production process, landfills, sewage system, wastewater treatment plants, etc. [15].

Water is an important aspect in everyone's life, but the availability of freshwater is not unlimited, raising a concern in generating new sources of water along with restricted use of the available ones. The water pollution is considered when any particle in excess amount that might be harmful for humans and water bodies is found in water. Groundwater is mostly contaminated by natural nanoparticles, and abundant particles in the environment like NPs arise from volcanic ash, with a low concentration of them. This can be an alarming issue in future as the contamination in water is increasing rapidly. The rapid development of silver NPs, gold NPs, titanium dioxide, zinc oxide NPs, etc., has enabled engineered nanomaterials to spread onto the water bodies from the sources mentioned before [17]. Ultra-fine particles getting into the environment may raise to inflammatory, oxidative stress, etc., and nanoparticles may well work in this sense for their sizes [18]. The air pollution in the urban cities is caused from nitrogen oxides, particulate matters, and transition metals which may directly affect the central nervous system of humans along with neuroinflammation (Fig. 6) [19].

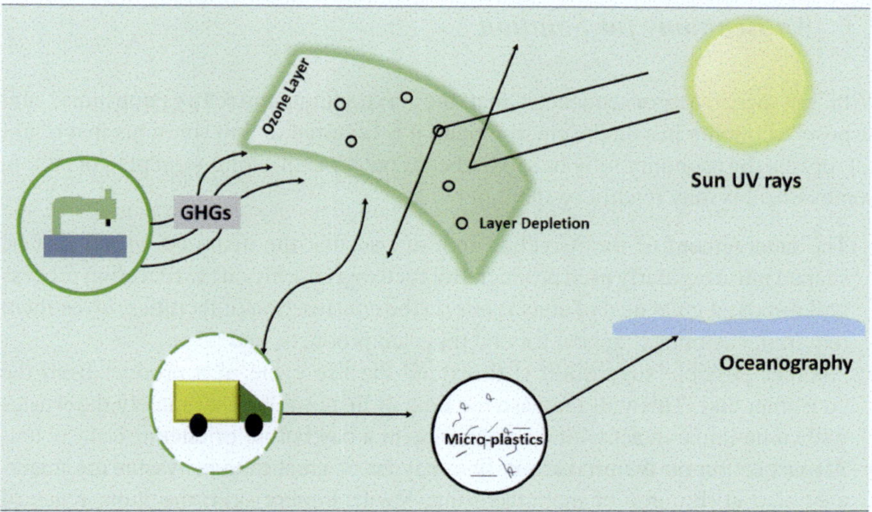

Fig. 6 Widespread allocation of nanoparticles and pollution into the environment

The widespread consumer use of nanotextiles, a traditional textile product incorporated with nanoparticles, has increased the chances of NPs release into the environment [20]. The release can be triggered from the life span stages like rubbing or mechanical pressure, washing, finishing treatments, sweat, irradiation, and so on [21, 22]. Lorenz et al. focused on the study of silver nanoparticles released in the washing stage from eight different consumer nanotextile products. They stated that the release of silver in case of washing depends on the amount, form in which the NPs are coated, and the medium which is used for washing like tap water, distilled water, washing solution, etc. Their work concluded that 85% of the silver was released in the ion form as Ag^+, and there was some sort of agglomeration of the NPs found in the washing solution [20]. Another study by McGillicuddy et al. revealed similar conclusions that the release of Ag^+ and particulate Ag depends on the incorporation method of Ag to the textiles [22]. The binding strength and bond of silver to the fabrics can be of great importance as stronger binding strength will help in less release of Ag^+. This strength mainly relies on the fabric types, NPs types, fabric-NPs interaction, pretreatment of the fabric to attract the NPs, etc. [21]. For the abovementioned reasons and adverse impact, a robust solution needs to be implemented in the manufacturing, consumer use, and disposal of nanotextiles so that applications of NPs bring changes thinking about mother nature.

1.3 An Economy for Solution

With the negative consequences that the "extract, manufacture, consume, and dispose" economy has on the environment, it is required to find solutions that might not replace the economy fully but give a better outcome in a long-term plan of action. Some concepts may readily be thought of as:

- The enlargement of the recycling industry so that the disposed wastes can be shaped into a regularly used product. But focusing entirely on the recycling process will impose a great deal of constraints on the construction of facilities, investment required, customers' appeal for end recycled products, and so on.
- Another possible suggestion is to extend the life cycle of a product from the consumer end. This will increase the time of disposal, and ultimately the wastes will come into contact with the environment a day before or after.
- An inspection on the production line may be of great help as wastes are generated also at the time of manufacturing. So, less energy consumption, reuse of scarce resources, and a more efficient production process can help in building a sustainable economy that works for betterment.

None of the above-mentioned suggestions can become a good idea to address the problem of the economy. This is where the researchers brought out the concept of "circular economy", a closed-loop cycle that ensures the best possible use of scarce resources along with less waste generation.

1.3.1 Circular Economy

The closed-loop flow of material where the raw materials are converted into finished goods and the finished goods are again converted to resources is the concept behind circular economy (CE). This concept finds a balance in the economic growth, environmental sustainability, and a future prospectus to work on [23]. The most promising aspect of this CE is the wide acceptance of the concept by the business community, environmental groups, and private and public sectors [24]. The economy works on the reuse of resources, regeneration, and restore energies for future use, leading to a fine line between "reduce–reuse–recycle–restore" [8, 23, 24]. The businesses are required to find the contact between their products and nature because within the upcoming days people will be more concerned about their health and environment more than anything. The governments are inching toward promoting a better world along with a recent activity mentioned by Anne and Phil in their work saying the $25 Trillion business forecasting in 2030 can be increased with the adaptation of the CE concept [24]. CE is expected to reduce the consumption of materials, lowering of costs as wastage are kept in minimal value and price volatility. Anne and Phil revealed a three-tier classification as a diversity of CE under:

- Minimal changes requirements in production process to adopt the minimal wastage planning and promote closed-loop planning of the production

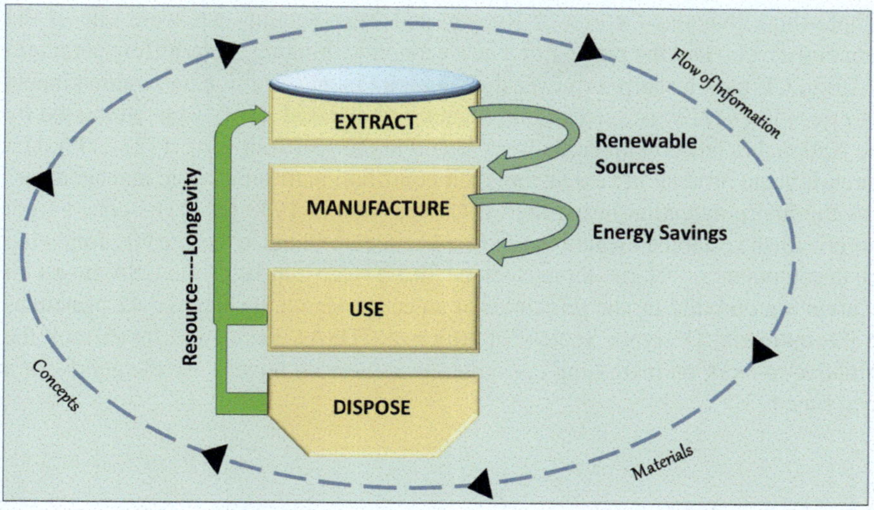

Fig. 7 Representation of circular-based economy

- Building a pattern in reusing the landfills waste and seeing the waste as resources, the concept was ignored earlier. Waste is termed like this because humans did not explore the potential benefits of them, and a gold resource was in hibernation so far.
- Three C model, consumption, collaboration, and coordination, is what is expected to transform the linear economy into a round-shaped circular economy (Fig. 7) [24].

2 Principles of Circular Economy

2.1 Historical Development of CE

Not finding a single inventor of the concept, CE was industrially introduced in Germany in 1990s with the view of managing raw materials use and resources. U.S Professor John Lyle was among the first contributors working on this concept. The eco-city concept developed in Japan had created an impression on building a strategy on reuse, reduce, and recycle (3R) [23]. Our nature, the environment itself is the biggest example of circularity in the ecosystem, the world reverses the energy, process, and materials but if we continue to pollute the world heavily, this will take so long to engulf the tragedy [25].

The wave of questioning the mainstream culture of economy prevailed right after the world wars, mostly in the sort of 1970s era. People started thinking about nature when they were shocked after the news of ozone layer depletion, a hole that made people think. Nicolas Georgescu-Roegen and Fritz Schumacher were one of the economists who laid the pioneer of a new economic thinking. The shifting paradigm of LE to CE took a sharp turn over the unemployment condition and unimaginable oil prices in 1973. Waste reduction policies and disposal issues were addressed by the National Science Foundation in 1978 and in the year following it [25]. "Product durability and product life extension, their contribution to solid waste management" was the first publication in the field if CE reported in 1982 [25]. The concept got larger with the establishment of the eco-park model in China in the late 90s, following the development of "Harmonious Society" in 2000 [23]. In 1992, the USA gave a lot of attention on building the principles of an eco-design and proposed 12 principles in the Industrial Designer Society of America (IDSA). Korea and Japan took the initiative to work on increasing consumer awareness on how to use the goods for a long time [23].

2.2 *Industrial Implementation*

The worldwide development of CE has become possible due to the industry level application was adopted in industries like manufacturing, automotive, construction, aerospace, and many more. The manufacturing industries were based on the concept of remanufacturing after World War II for focusing more on saving the resources and utilizing them for so long [25]. Though the eco-industrial parks were introduced in the 1960s in Denmark, the tradition got more advanced in the Chinese territory with the bottom up approach. This is a self-organized system that controls the material flow, emission of CO_2, and waste management in the area [24]. A promising sector in believing the linear economic flow is the construction industry, 30% responsible for extraction of natural resources in 2020 [26]. The circular economy model developed in this industry based on the life cycle stages such as project designing, manufacturing, construction, operation, and termination. Modular building designing, adaptability of existing buildings, alternative use of raw materials, off site construction, waste reduction, management of demolition waste, etc., are some of the factors considered while planning for a circular economy [26]. The aerospace industry has always been concerned with building lightweight structures so that aviation becomes faster and a growing opportunity for world people. The sustainability concern has led to the promotion of "eco-composite" use rather than fiber-reinforced polymers [27]. The work of Bachmaan et al. reported of "Eco-COMPASS" project which was conducted by EU/China collaboration, and the vision of this project was to give a clean sky to the world. The aviation industry had caused 2% of the greenhouse gas emissions that worked as a motivation for the initiators [27]. The automotive industry has recently been familiar with "Eco-Innovation" which is a quantitative methodology to forecast the green economy [28]. Even the tourism industry worldwide is adopting the changes required for ensuring a sustainable business model as the increased number of tourist destinations face numerous challenges regarding the GHG emissions, plastic wastes, and maritime pollution control management [29].

The application of CE in different value chains has been explained in the work of Winans et. al. with focusing on the resources like wood and paper, plastics, metals, chemicals, agricultural products and waste, water, and land [23]. The process engineering industry, like paper, consumes a considerable amount of energy and the solid wastes generated needs to be reused in the processing to accept CE. Among the pressure from the environmentalists in the past century, the road to sustainable use of wood has been successful due to the collection of selected used materials as sources and invention of processing technologies enhanced the preparation of recycled white paper [23, 30]. The plastic industry is in an alarming condition due to the excessive application of plastic materials, and it was assumed that 80% of the total produced plastics, accounting to 6.3 billion tons, has been disposed of into the environment [31]. EU has continuously been in the promotion of CE availability in this industry, but economic viability of technical solutions of wastes, micro-plastic pollution in river, sewage and sea water, investment in research and development, separation of additives from base plastics, and implementation of policies are vital matters to look at. Some of the solutions could be business to consumer (B2C) reuse model developed by EU, reverse vending schemes, innovative product designs with less amount of additives used, designing out the wastes, and many more [23, 31]. Similarly, steel and iron industries face issues regarding manufacturing mostly from blast oxygen furnace (steel preparation), long energy-consuming processing for iron and ore (natural rock) mining, excessive waste production, structural deficiencies of materials, transportation issues, quality or grading in reused materials, desulphurization techniques, etc. [30].

2.2.1 Problems of Linear Economy in Clothing and Fashion

Readymade garments and textile industry are one of the promising industries that evolved during the last 50–60 years in many developing countries and low-income countries. With the rise of the industrial revolution, the fashion strategy of people has changed all over the world, and the rise of fast fashion concepts required more production capacity in this sector. But the expenses of this exchange or transition are heavier into the field of environment. The following terms are related to each other in understanding the problems, and these are:

- Fast fashion
- Fiber production
- Extensive water usage and contamination.

The concept of fast fashion is the demand of garments in every changed season at a cheaper price and higher in quality. The seasonal trend actually diminishes the optimistic lifetime of the garments, and this resulted in disposal of the clothes in landfills. The renowned brands like H&M, Zara, exist in the business by making branded products available to the middle class society and influencing their mindset of buying season-based clothing. The higher production rate and requirements from the brands somehow demoralized the environment-friendly concept from the industry. The companies may think about the production process optimization, but the destination of the disposed goods is of real concern [32].

The core of garments manufacturing is the production of fiber and spinning them into yarns. Natural fibers like cotton, linen, wool, silk, etc., dominated just before the arrival of the synthetic man-made fibers. The chemical production of these fibers along with incorporation of desirable characteristics has made the world in quest of excessive manufacturing of synthetic materials. The main concern started with the exposure of microplastics into the sea in an alarming way. Not only the disposal exposed plastic pollution into the water but also the excessive rigorous washing of clothes have released microparticles into the sewage or drainage water. Besides waterways, CO_2 emissions from the manufacturing process are becoming a major issue in raising the global warming issue [32]. Most of the processes in traditional garments manufacturing involve the consumption of a wholesome of water, a t-shirt randomly uses water equivalent to consumption of drinking water of an adult in 3 years. The energy consumption of the synthetics is 2–3 times higher than in producing the natural ones, an issue that needs to be fixed in relation to the circular economy.

2.2.2 Tools and Principles Implemented

This portion has explained several principles and models applied in regard to the implementation of circular economy in textile, clothing, and fashion industries. Specific works have been described broadly for the better utilization of any fixed content. The published literature of Saha et al. enabled a systematic correlation between sustainability implementation via circular economy concept by studying 114 textile clothing industries, those are located in Bangladesh, India, and Vietnam [33]. Some of the questions regarding CE implementation strategies, techniques, and action plans have been discussed in their work. The hypotheses and models used in their work are based on the economic, environmental, and social performances for sustainability calculation and take, make, distribute, use, and recover model for the analysis of CE. The proposed strategies included an environmental idea collaboration between suppliers, manufacturers, and buyers on cost, quality, and quantity. A robust material selection process to value the supply chain in building CE environments is required to control cost, quality, and environmental factors. Retailers need to share information regarding sustainability with the factories so that they can ensure a better processing or production system. CE-based training work, low energy consumption studies, lean approach, eco-based design, renewable raw material sources, and expertise in waste management have been given utmost importance in creating sustainable action plans [33].

The studies of Chen et al. prescribed a DEMATEL framework to understand the obstacles in the implementation of CE in the textile industry [34]. Twelve core obstacles have been identified by applying the fuzzy Delphi method along with the Decision-Making Trial and Evaluation Laboratory (DEMATEL) model. The evaluation resulted in three primary and key barriers such as consumer awareness, lack of business models and frameworks, and lack of information exchange system. This concept is somewhat similar to the findings of Saha et al. who also asked about the communication gap as a great barrier in implementing CE [35]. Interpretive structural modeling (ISM) is another method of interpretation, but DEMATEL worked with higher efficiency in that regard. According to this model, textile sectors could not implement the CE model due to high global competitiveness, sustainable production process demand, and increasing environmental problems [36].

Reviewing the Dutch textile industries, Fischer et al. explained three conceptual models on building the institutional initiatives toward CE [35]. The initial point was described as a framework to exhibit the transition toward CE by an institutional approach, mid-focal point was to find out the accelerators and practitioners in the collaboration process, and finally, the team focused on finding the key organizational elements in this regard. There are four strategies explained in case of businesses that want to achieve CE, and those are: circular design strategies, optimal use strategies, value recovery strategies, and network organization strategies. One of the major concern was the lack of clarity in case of quality proposition for the textiles prepared from post-consumer materials (Fig. 8) [35].

The CE exercise can be divided into three specific categories as micro, meso, and macro-levels, where product as a service model, circular supply model, product life extension model, etc., could affect the process positively a lot. The meso-level mainly focused on the closed-loop economy building, and government bodies with policymakers are kept in the macro-level of this thinking [34]. But thinking about the building blocks in establishing a circular economy, fiber is the core element and critical for the value chain. Some of the technological innovations like virtualization, dematerialization, feedback, etc., are making the concept more feasible to implement [36]. The product service system (PSS) model is explained to have a good transitional

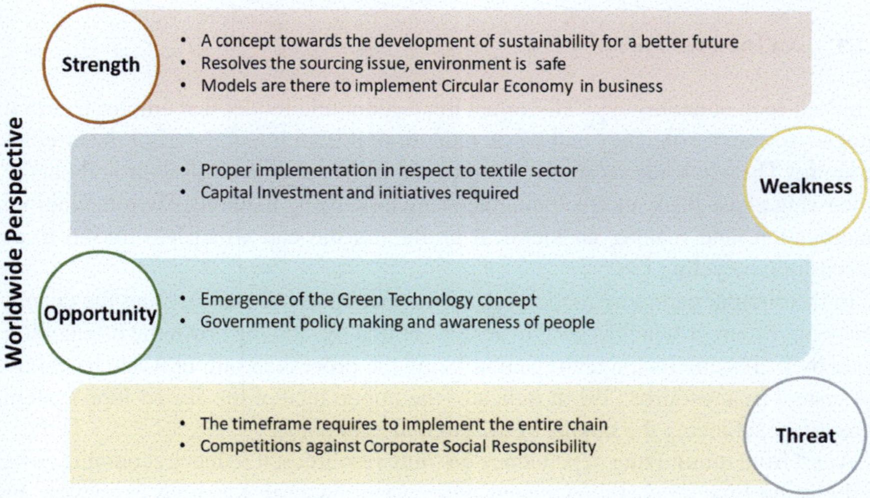

Fig. 8 SWOT analysis on the understanding of the implementation of circular economy in textiles

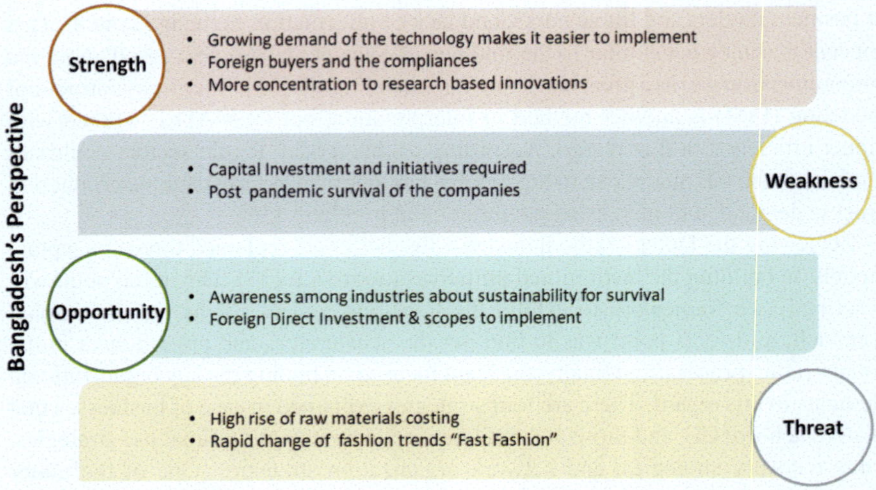

Fig. 9 SWOT analysis on the understanding of the implementation of circular economy in textiles in Bangladesh

model to make a connection between the market demand and flow of information without distortion from retailers to manufacturers and suppliers. The creation of a bullwhip effect is one of the reasons that stopped the circulation of new designs along with a great unused inventory of mountains to be left out in landfills (Fig. 9) [36, 37].

2.3 Closed-Loop Material Flow

2.3.1 An Integral Part of CE

A closed-loop economy typically refers to an economic model that emphasizes two things—resource efficiency and waste reduction. It is an integral part of the circular economy. The whole idea is about reincorporating the generated waste throughout the processing phase back into the manufacturing module by methods like remediation, modification, and making an identical product at the end which is referred to as closed-loop recycling [38].

In the circular economy model, one of the main purposes is lowering the dependency on resources which is tried to achieve at first by slowing down the product life cycle by increasing its longevity and utilizing the process or output waste back into the system as a resource [39]. It indicates the importance of the closed-loop system in terms of achieving the targets of the circular economy.

Apart from minimizing dependency on finite resources, it reduces economic costs at the same time as costs associated with waste disposal, resource extraction, or purchase steps are ended. It promotes the design of products and processes with a focus on reuse, recycling, or repurposing, which is like the economic idea of capital preservation. A prime example can be the British car and aerospace company Rolls Royce. As they transitioned their processes toward waste reduction and kept resources in circulation, in terms of costs, time, and operational efficiencies, they achieved significant savings [39].

From businesses, if we switch the focus toward the national systems, it is said that the changes made in the system through closed-loop thinking have certain positive implications for the developed countries unlike reduced vulnerability to resource scarcity, supply risks, price change shocks, and increase social and environmental wellbeing [40].

A closed-loop economy came up with the larger objective of circular economy. They make sure that items and resources preserve their worth by staying in circulation and serving several life cycles. This interrelation promotes a more sustainable and regenerative economic arrangement that addresses both environmental concerns and economic opportunities.

2.3.2 Recycling of Nano-finished Materials

In recent years, there has been a growing consensus that changes in environmental policies and practices, including recycling techniques, are desirable and, in many cases, economically beneficial for environmental and legislative reasons. Globalization and the decrease in the life cycle of products, technologies, and services have increased competition among recycling businesses. Recycling discarded industrial and municipal waste is beneficial in creating a closed-loop economy. However, businesses are struggling to respond to changing market demands and recycling efforts due to outmoded organizational structures [41].

Nano-finishing is comparatively a newer approach in terms of enhancing the functional properties of consumer-grade textiles. As nanotechnology is an emerging field, The Nanotechnology Consumer Product Inventory (CPI) was developed in 2005 by the Woodrow Wilson International Center for Scholars and the Project on Emerging Nanotechnology for transparent information regarding its usage. The CPI groups entries into eight generally accepted consumer goods categories based on publicly available consumer product classification systems where the clothing subcategory archived 78 products in between 2012 and 2014 [42]. The growth of its usage is getting higher day by day due to its better outcomes and wider arena of improvement which also opens a new stream of waste increasing its load (nano-finished textile waste). So, integrating the recycling approach within it globally at the earliest time is a necessity in terms of sustaining this technology, reducing dependency on limited sources, and protecting the environment as it ultimately contributes to achieving a closed-loop economy.

Many areas are still untouched in terms of nano-finish material recycling. NM recycling processes from waste sources are investigated by providing an understanding of the chemical composition and structural properties of the NMs and discussing methodologies for their efficient separation from the matrix using magnetic separation, antisolvents, aqueous dispersion, colloidal solvents, centrifugation/solvent evaporation, and other techniques [43]. Most R&D has taken place in terms of extracting nanoparticles from e-waste and recycling it, but other fields are still untapped. Both chemical and mechanical recycling options are there to explore the potential of recycling nanomaterials extracted from discarded nano-finished textiles and reusing them in the process. Although there is a gap or unavailability of specific nanotextile recycling-based research work, certain chemical methods have been proven so far to retrieve nanoparticles and reuse them. A brief idea is tried to provide in the following points:

2.3.3 Ionic Liquids

Ionic liquids (ILs) are substances that are entirely made up of ions and have a melting point below 100 °C [44]. High polarity, minor volatility, high thermal stability, high ionic conductivity, low melting point, and structural designability are some of the physicochemical features of ILs. Over the last two decades, ILs have been regarded as promising solvents with distinct properties in organic synthesis, catalysis, electrochemistry, metal separation, gas separation, biomass processing, pharmaceuticals, tribology, and energy storage devices such as batteries, supercapacitors, and fuel cells, among others [45].

Ionic liquids have shown prominent effects in terms of deconstructing polymers and recycling nanoparticles individually. Polymers such as poly(ethylene terephthalate) (PET), poly(3-hydroxybutyrate) (PHB), poly(bisphenol A carbonate), chitin, poly(lactic acid), nylon-6, and polyethylene can be deconstructed using ILs [46]. This ultimately ends up converting the polymer into small molecules. Many subsequent factors are found to be associated with the efficiency of this dissociation. Acidic and basic ILs work for different types of polymers; acidic works better on PET, whereas basic for PHB as it influences the mechanism of deconstruction. Another factor is the steric effect or size/shape of ILs molecule. When PET using amino acid-based ionic liquids is broken down (ILs), glycine yielded 43% of small molecules; meanwhile, lysine and alanine showed lesser deconstructing although all three of them had the same acid dissociation constants and were treated under the same condition [46]. Furthermore, different experimental parameters such as temperature, polymer concentration, and the presence of impurities can impact the effectiveness of ILs in terms of polymer deconstruction. It was seen that glycolysis of PET at even a small fraction of water loading conditions reduced the conversion mechanism. On the other hand, a faster reaction in PET dissociation can take place when a higher temperature is introduced [46]. All these indicate the effectivity of ILs in terms of polymers.

In the modern economy, synthetic polymer is becoming the base of textile materials like polyester, nylon, rayon, acrylic, etc., which are mostly comprised of the polymers discussed so far. To recycle nanomaterials from nanoparticle textile materials, it is necessary to first extract the particle which includes the deconstruction of the textile materials, and ILs show potential in this field. Also exploring whether it is viable for natural polymer deconstruction is another potential area. But here a

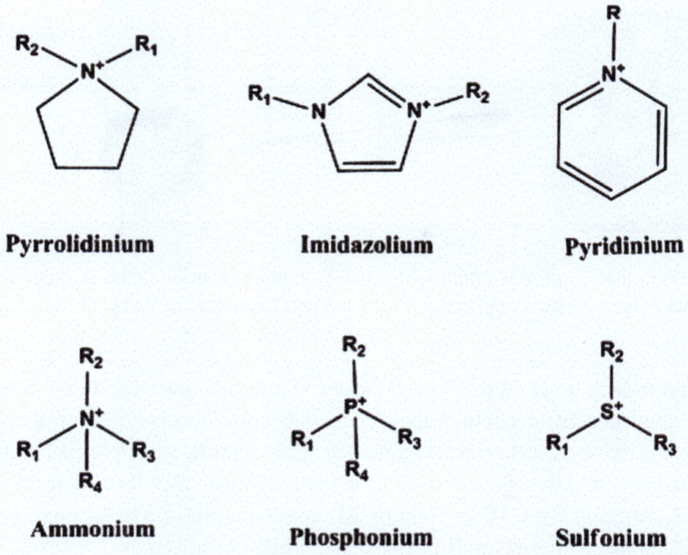

Fig. 10 Chemical structure of different types of ionic liquids [48]

question is raised, although this solution deconstructs polymers, can nanomaterials be separated from the solution or not?

Various types of ILs, including imidazolium, pyridinium, ammonium, piperidinium, and phosphonium, have been found to play a pivotal role in both the synthesis and recycling of nanomaterials [47]. Some of them are shown in Fig. 10 [48].

Ionic liquids (ILs) offer a promising option for the recycling of nanomaterials. These environmentally acceptable alternatives to standard organic solvents have little vapor pressure, are nonflammable, and have low melting points. Besides, ILs can be utilized to successfully dissolve nanomaterials. This technique is frequently reversible, allowing for the recovery and recycling of nanomaterials. The nanomaterials can be isolated from the ILs by altering ambient variables such as temperature and pressure [47]. The potential to separate nanomaterials from IL solvent is somewhat cleared out from this research work. It still needs to be identified whether there is any effect of this solution on the properties of nanomaterials. But overall, it can be a potential pathway for recycling nano-finish particles seen from the analysis.

2.3.4 Microemulsion

Textiles go through numerous wash cycles in both the home and industrial phases. The possibility of nanoparticle shedding, such as the most common silver, when washing nano-treated textiles has already been investigated [49]. This can be a potential route to extract nanomaterials and recycle them to connect them back to the route.

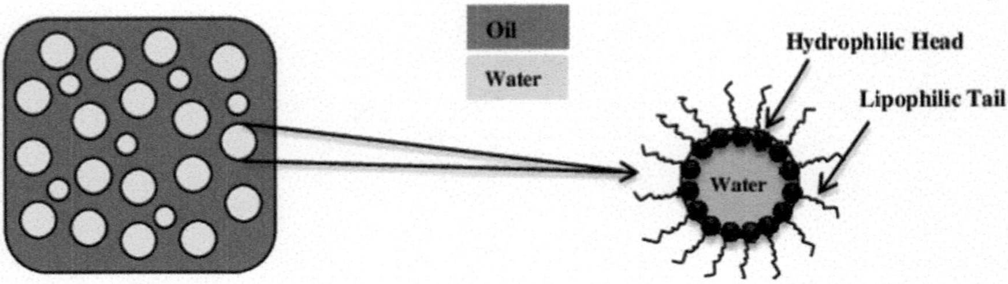

Fig. 11 W/O emulsion and emulsifiers on the dispersed phase's surface; reused under the Creative Commons Attribution-NonCommercial 4.0 International License (Fig. 11) [52]

Significant research works have been conducted in terms of extracting nanomaterials from water samples among which usage of microemulsion is a potential way.

A microemulsion is a thermodynamically stable fluid, as opposed to kinetically stable emulsions, which separate over time into oil and water. Microemulsions have particle sizes ranging from 10 to 300 nm. Microemulsions appear as clear or translucent solutions due to their small particle size [50]. Amongst a few types of emulsions, water-in-oil (W/O) emulsions, also known as inverse emulsions, are created by dispersing water droplets in the oil phase. W/O emulsions can encapsulate and transport hydrophilic functional components [51] which is graphically shown here [52]:

It was found that CdS and ZnS nanoparticles can be isolated from a solution induced within water-in-decane microemulsions [53]. Temperature plays a crucial role in triggering phase transition by lowering or increasing it. At the critical temperature, the liquid–liquid phase transition finally takes place where both nanoparticles get separated. For example, CdS NPs may migrate to the upper phase, while ZnS NPs may move to the lower phase. Afterward, both phases are extracted physically. The mixed nanoparticles are then separated from the distinguished phases.

Although this is a viable option, it still has significant feasibility concerns in terms of large-scale implementation because it includes a complex procedure, several phases, and additional stages. In the current day, the nanofiltration process can play an important role in making this separation and extraction procedure easier. However, complications can occur in terms of filter durability, identification, and separation from other particles, making it feasible in the end. Although nanofiltration is becoming more popular, there is a dearth of information on how to use it to filter nanoparticles from home and industrial textile effluent.

2.3.5 Other Pathways

Aside from the possibilities outlined, a few other recycling processes have been investigated. Cloud point extraction is a well-known method that can extract nanoparticles [54]. Because it is a solvent-based approach, it has the potential to recover nanoparticles from nanotextile waste, but there is no concrete evidence of this yet. Another possibility is to directly produce CNTs from textile waste fabrics [55]. That is classified as upcycling since a higher value-added product is produced, but ultimately because it is made from textile waste and may be used to improve the functionality of textile materials by using inside it, the closed-loop cycle is fulfilled in some way. Recycling of nano-finish parties and integration of closed-loop economy in nanotextile finishing still have a wide range of possibilities to study. Successful implementation can ultimately contribute not only to environmental conservation but also to economic gains. The effectiveness of recycled nanomaterials is an area that needs to be prioritized to achieve a sustainable closed-loop economy.

2.4 Regenerative Practices

2.4.1 The Concept of Regeneration

The third principle of the circular economy is to regenerate nature [56]. Some researchers even suggest that "regenerative, bio-based production" is one of the seven pathways that circular economy strategies must follow to contribute to achieving sustainable development goals [57]. However, speculation raised about why and how this concept of regeneration emerged.

Conventional take–make–waste comes with various inherent issues, particularly in transitioning the balance of nature. Usage of non-renewable resources, abundant generation of waste but environmentally harmful disposal, toxic substances for resource extraction, non-innovative approach, significant reliance on energy sources, etc., turn the linear economy unsustainable in terms of ecological conservation, as all the mentioned aftereffects pose substantial negative influence on the natural systems. The production and usage of nanoparticles in a linear approach.

have already shown adverse impacts on nature which are shown in Fig. 12.

At (1) [58], t is evident that the dependency of major nanomaterials is on the finite elements of earth like titanium and zinc ore, which cannot be replaced naturally over a human-relevant period. The extraction process even involves deforestation, destruction of habitat, pollution of environmental components, etc. Overall, it indicates the unsustainable approach associated with this route. Afterward, (2) [58] showcases two major reducing agents (sometimes used as stabilizing agent too), widely used for the commercial level synthesis of AuNP / Ag nanoparticles that are comparably highly toxic for natural systems than bio-based reducing agents from soybean seeds and sugar beet pulp. Hydrazine itself is persistent in water which can produce harmful byproducts while reacting with other pollutants and eventually harm aquatic life. Lastly shown in (3) [58], incineration is a popular technique to manage waste globally, but in terms of handling nano waste, it comes up with a significant implication which has been mentioned. Accumulation of nanoparticles with air can increase the levels of particulate matter and aerosols which eventually degrades the quality of air and interrupts the natural atmospheric behavior.

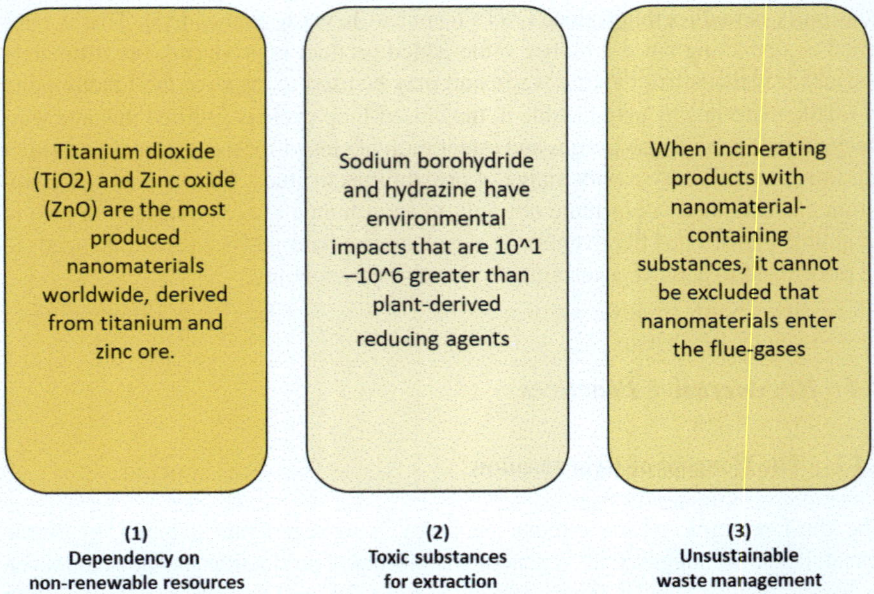

Fig. 12 Adverse impacts of linear economy on natural systems considering nanomaterials

The whole scenario is trying to indicate how the widely adapted approach, the linear economy has been depleting the ecosystem since the start of the industrial era. To restore and revitalize natural systems that have been transitioned for ages, a major principle of circular economy regenerative practices emerged.

Regeneration is essentially the stimulation of natural systems' self-renewal potential to reactivate ecological processes that have been destroyed or over-exploited by human intervention [59]. Regenerative activities emphasize the repair and renewal of the ecosystem. These methods stand out for their dedication to enhancing the resilience and health of ecosystems. Regenerative approaches go beyond the conventional approach of sustainability, which tries to protect current resources, by actively refilling and strengthening the natural capital upon which they rely.

Natural systems are a priority in regenerative approaches because we understand that they are the basis of our society and economies. Regenerative methods foster a more harmonious and balanced relationship between human activities and the environment by actively trying to enhance and restore ecosystems. By concentrating on nature restoration, these methods actively enhance the health and vitality of our natural environment, leaving a positive legacy for future generations. So, it is necessary in terms of the production of any material to oversee what materials/ steps during production can imbalance the environment and find another alternative to reduce that impact also implications upon its disposal on the ecosystem. This responsible way can help in achieving the regeneration principle.

2.4.2 Necessity and Adaption in Terms of Nano-finishing

As nano-finishing comes to improve the quality of textile products and add functionality to them, several types of nanomaterials play an important role in achieving that like polyacetylene, polypyrrole, polyaniline, Au, Ag, Pd, Si, CuO, ZnO, carbon nanotube (CNT), TiO2, chitosan, MXenes, and graphene oxide (GO) nanoparticles with potential applications in wound healing, air purification, drug delivery, cosmetics, renewable energy generation, and electronic applications such as fabrication of on-body diodes, transistors, and circuitry [60]. Most of them are engineered and come up with their own set of problems as described previously. But how they enter the chain of the ecosystem is an important part.

Pre-production, production, and post-production all the three steps come up with their own set of impacts on transitioning the natural ecosystem. Both pre-production (dependency on non-renewable resources) and production (toxic substances for extraction) have been previously discussed. In terms of the post-production phase, which indicates that the usage and disposal part of any nano-finished textile can contribute to the accumulation of nanoparticles in the environmental components in certain ways. Long-term usage of nano-treated materials ultimately causes the degradation and release of nanomaterials. Also, washing and cleaning of products can be another route. According to studies, silver-treated socks can leak up to 650 μg of silver into 500 mL of water [61].

Another probable route can be the landfilling of nano-treated materials. Landfilling is a complex process where the behavior of nanoparticles may undergo alterations upon disposal. This complexity arises from the diverse materials found in landfills, including organic matter that degrades slowly and other substances that change their properties over time. These variables can have a substantial impact on how nanoparticles interact and adapt in the waste environment. When these types of releases occur, they eventually reach the aquatic environment and soil, where they change form by agglomerating or decomposing, and are then transported to living creatures (Fig. 13) [62].

If a summary is drawn at this point with the help of the lifecycle of an ENP used in textile material to show its potential of altering the natural system throughout different phases, it would be clear why it is necessary to adopt regenerative practices in this arena (Fig. 14).

From the different production phases particularly in the last two [63, 64], it is clear how the whole life cycle can have a significant negative effect on the environment, its components, and living animals of different spheres ultimately hampering the overall chain of the ecosystem. As an alternative to primarily stop this burden and help the natural system to conserve back, nano-finishing route can adopt two types of regenerative practices. To tackle the impacts production phase green synthesis can be a potential track. Meanwhile transitioning to bio-based nanomaterials can resolve issues deriving from the rest two.

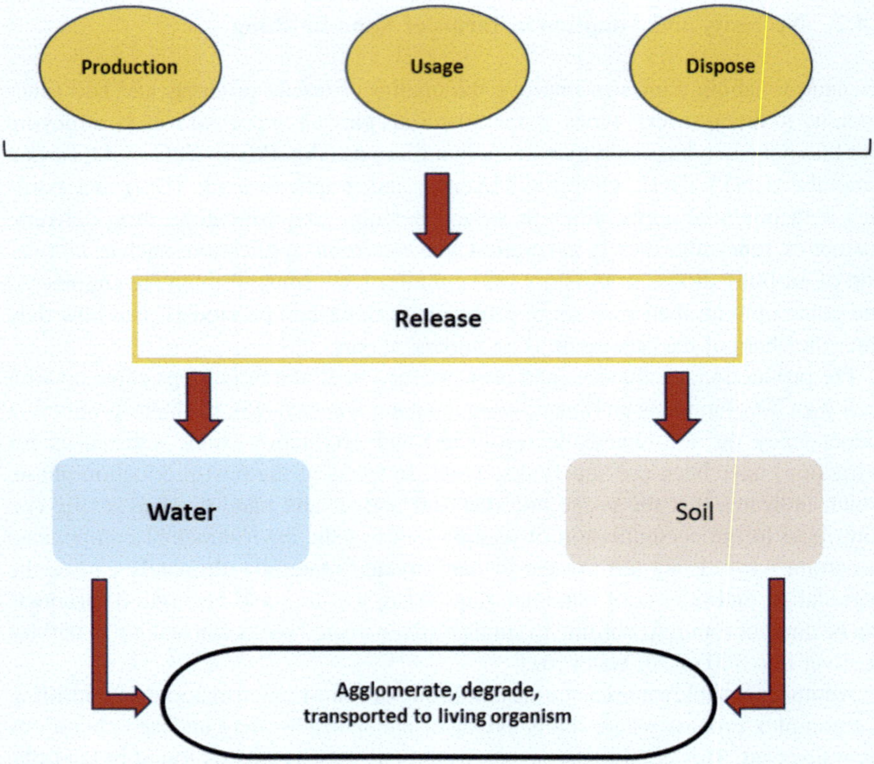

Fig. 13 Behavior of manufactured nanoparticles in the environment

2.4.3 Green Synthesis

Green synthesis is a revolutionary method to produce nanomaterials that are sustainable and possess significantly positive effects on nature. It is an alternative to the conventional and toxic synthesis method of nanomaterials. Whereas the toxic technique is dependent on hazardous chemical substances and extensive environmentally harmful parameters, the green synthesis relies on ingredients or substances derived from nature, economically feasible to use, and clean for human health.

The process of green synthesis can be used to create a wide range of nanomaterials which makes it a flexible way of synthesis. Bacteria, yeast, fungi, algae species, and certain plants serve as substrates for the green synthesis of nanomaterials [65]. Metal and metal oxides are two widely used nanoparticle types in textile nanofinishing. Many studies have previously discovered ways to synthesize them in an environmentally responsible manner, such as aloe vera leaf extract can be used to generate Ag nanoparticle, which acts as a reducing and stabilizing agent [66]. Another method for extracting metal nanoparticles, Cu nanoparticles, is to use neam leaf extract, which acts as a reducing agent [67]. When it comes to metal oxides, ZnO

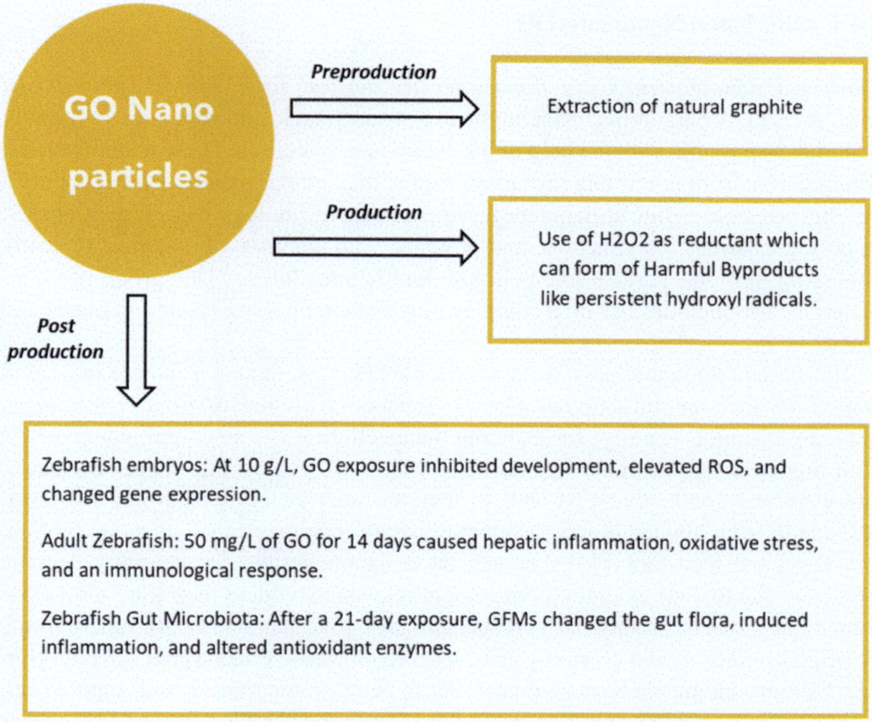

Fig. 14 Effect of different stages of nanoparticles production on overall environment

has the potential to be extracted using orange peels [68]. These are now comparable to both top down and bottom up conventional metal and metal oxide production processes that encourage the use of reducing agents such as sodium borohydride [69, 70] that can help in identifying the more environmentally viable alternate. Reducing agents employed in green synthesis are all directly sourced from nature, and they are biodegradable and non-persistent elements, whereas NaHB4 creates H2 gas when it meets water, which can provide a flammability threat and affect the quality of the water. For a variety of compelling reasons, green synthesis is the superior choice.

Apart from the usage of sustainable chemicals, the fact that less energy consumption and usage of resources is associated with this process compared to the typical one makes green synthesis cost-efficient. It can be promoted as an ideal process as the process incorporates organic materials and methods which ultimately contribute to preserving as well as enhancing biodiversity. These approaches help green synthesis to align with the principles of regeneration and restoration.

2.4.4 Bio-based Nanomaterials

Bio-based nanomaterials are nanomaterials derived from natural, plant-based sources. [71]. A paradigm change in the use of sustainable and ecologically friendly materials is possible with the help of bio-based nanomaterials. These nanomaterials, which derive from renewable resources, follow the circular materials bioeconomy's guiding principles while utilizing the benefits of nanotechnology. They largely consist of polysaccharides, polyphenols, and proteins, with plant-based materials typically containing polysaccharides like cellulose and hemicellulose. This group of nanomaterials also includes lignin, a complex polyphenolic polymer found in plants and wood.

Bio-based nanomaterials have a noticeable effect on the environment in several ways. First, their natural abundance lessens reliance on limited fossil-based resources, reducing resource scarcity; for example, nanocellulose is a potential nanomaterial with numerous applications such as self-cleaning, UV blocking, flame resistance, and electrical conductivity, as well as features such as lightweight and low cost. It is made from abundant and sustainable natural resources such as wood, cotton, and vegetable feedstock [72]. Second, these nanomaterials can naturally degrade over time due to their inherent biodegradability, which reduces their long-term environmental impacts. Additionally, recyclable bio-based nanomaterials adhere to the principles of the circular economy and cut down on waste. Banana peel, coconut coir, eggshell, groundnut shell, mango peel, onion peels, pomegranate peel, sapota peel, rice husks, watermelon rind, orange peel, and tamarind shell are common biowastes utilized in the manufacturing of nanoparticles [73]. They are useful for a variety of applications thanks to their biocompatibility, without having negative impacts on living things. These materials are accessible and economically viable due to their low production costs. Some of them also have the potential to be versatile in terms of application. For instance, chitosan, a cationic biopolymer obtained by the whole or partial deacetylation of chitin, is the primary component of crustacean shells such as shrimp, squid, crab, and lobster, the exoskeleton of terrestrial insects such as honeybees and silkworm, and cell walls of fungi such as molds, yeast, and ray fungi [74]. Chitosan nanoparticles have antifungal, antibacterial, and antiviral activities and have been shown to be efficient against diseases present in seeds and soil. [74]. It is also effective in UV protection [75], enhancing dyeing properties [76] etc.

These explorations are recent and show huge potential in terms of replacing engineered nanomaterials and their toxic consequences on the environment. By introducing bio-based nanomaterials into goods and procedures, we reduce environmental impact while supporting the principle of regeneration which ultimately will be helpful in terms of achieving a circular economy in the finishing process of textiles.

2.4.5 Challenges and Prospects

The scalability of green nanoparticles is restricted by the limited handiness of feedstock materials, like, only in Cyprus and southern Turkey, *Lithodorahispidula* leaf extract is available to produce Pd nanoparticle. This regional dependence complicates the expansion of industry. The problem gets worse with seasonal availability, such as cotton leaves or *Sargassumfusiforme* [77]. Certain green high-tech method requires higher temperature and extended period of reaction times. This raises issues like high energy consumption, negative environmental impact, etc.

Moreover, changes in nanoparticle size and quality have an impact on large-scale production and applications that demand consistency. The quality of nanoparticles is another concerning issue. The enormous variation of nanoparticle sizes seen in green synthesis, such as NZVI NPs (varying from 63 to 381 nm) or Ag NPs (ranging from 5 to 100 nm), highlights quality control difficulties [77]. Bio-based nanomaterials come up with similar sets of uncertainty. However, these difficulties also offer chances for environmentally advantageous regenerative techniques. We can aid in the restoration of the environment by funding continuous research and development to produce adaptable nanomaterials like chitosan or by improving green synthesis techniques. Continuous usage of traditional artificial nanoparticles may cause environmental deterioration, but concentrating on the creation of sustainable substitutes and methods will assist us in moving toward a more regenerative strategy, where nature can heal and flourish.

3 Conclusion

Take–make–dispose, because of this linear economy's methodology, natural systems are currently in a terrible situation. Due to the amount of waste handling and reliance on scarce and expensive natural resources for production, it has been demonstrated to be inefficient economically as well as environmentally. This strategy comes at a bad moment for the economy, the environment, and public health. The circular economy will soon play a significant role in this regard globally as the globe is quickly moving toward sustainable alternatives. To reach the goal of a sustainable future, industries may use circular economy concepts like reuse, recycle, and regenerate. Even though incorporating this paradigm into traditional products can be challenging, it could be a game-changer for new technology. There is no disagreement on the fact that nanotechnology will become standard practice in the future. It has a wide range of applications, particularly in textile finishing. When the linear technique is used, it is clear how this can hurt all aspects of life. Every single particle has its own set of issues, whether they are caused by the harmful chemicals used to create it or their release during the washing or disposal of textiles. Here, several of the principles of the circular economy produce significant effects, such as green synthesis, recycling using ionic liquids, and nanoparticles made from biowaste. However, because the discipline is still in its infancy, there are not many well-defined guidelines or pieces of study that can tell us enough to know whether adopting these strategies on a bigger scale is feasible. To increase the functionality of textiles without harming the environment or the economy, the field still has a lot of areas to investigate and finally implement the nano-finishing concept.

References

1. Exbrayat J-M, Moudilou EN, Lapied E (2015) Harmful effects of nanoparticles on animals. J Nanotechnol 2015:1–10. https://doi.org/10.1155/2015/861092
2. Nobre AG, Martínez JAE, Florêncio O (2021) Mineral nanotechnology in circular economy. In: Iano Y, Saotome O, Kemper G, Mendes De Seixas AC, Gomes De Oliveira G (eds) Proceedings of the 6th Brazilian technology symposium (BTSym'20). Smart Innovation, Systems and Technologies, vol 233. Springer International Publishing, Cham, pp 220–226. https://doi.org/10.1007/978-3-030-75680-2_26
3. Gottardo S et al (2021) Towards safe and sustainable innovation in nanotechnology: state-of-play for smart nanomaterials. NanoImpact 21:100297. https://doi.org/10.1016/j.impact.2021.100297
4. Taghavi SM, Momenpour M, Azarian M (2013) Effects of nanoparticles on the environment and outdoor workplaces. Electron Physician 706–712. https://doi.org/10.14661/2013.706-712
5. Distaso M (2020) Potential contribution of nanotechnolgy to the circular economy of plastic materials. Acta Innov 37:57–66. https://doi.org/10.32933/ActaInnovations.37.5
6. Soares EV, Soares HMVM (2021) Harmful effects of metal(loid) oxide nanoparticles. Appl Microbiol Biotechnol 105(4):1379–1394. https://doi.org/10.1007/s00253-021-11124-1
7. Can E et al (2011) Nanotechnological applications in aquaculture-seafood industries and adverse effects of nanoparticles on environment. J Mater Sci Eng 5:605–609
8. Ciasullo MV, Castellani P, Rossato C, Troisi O (2019) Sustainable business model innovation. Sinergie Ital J Manage 37(1)
9. Gubeladze D, Pavliashvili S (2020) Linear economy and circular economy—current state assessment and future vision. Int J Innov Technol Econ 5(32):0–4. https://doi.org/10.31435/rsglobal_ijite/30122020/7286
10. Morseletto P (2023) Sometimes linear, sometimes circular: states of the economy and transitions to the future. J Clean Prod 390:136138. https://doi.org/10.1016/j.jclepro.2023.136138
11. Niinimäki K, Peters G, Dahlbo H, Perry P, Rissanen T, Gwilt A (2020) The environmental price of fast fashion. Nat Rev Earth Environ 1(4):189–200. https://doi.org/10.1038/s43017-020-0039-9
12. Didenko NI, Klochkov YS, Skripnuk DF (2018) Ecological criteria for comparing linear and circular economies. Resources 7(3). https://doi.org/10.3390/resources7030048
13. Sakamoto M, Ahmed T, Begum S, Huq H (2019) Water pollution and the textile industry in Bangladesh: flawed corporate practices or restrictive opportunities? Sustainability 11(7). https://doi.org/10.3390/su11071951
14. Napper IE, Barrett AC, Thompson RC (2020) The efficiency of devices intended to reduce microfibre release during clothes washing. Sci Total Environ 738. https://doi.org/10.1016/j.scitotenv.2020.140412
15. Nowack B, Bucheli TD (2007) Occurrence, behavior and effects of nanoparticles in the environment. Environ Pollut 150(1):5–22. https://doi.org/10.1016/j.envpol.2007.06.006
16. Tarannum N, Divya, Gautam YK (2019) Facile green synthesis and applications of silver nanoparticles: a state-of-the-art review. RSC Adv 9(60):34926–34948. https://doi.org/10.1039/c9ra04164h
17. Malakar A, Snow DD (2020) Nanoparticles as sources of inorganic water pollutants. INC
18. Stone V, Johnston H, Clift MJD (2007) Air pollution, ultrafine and nanoparticle toxicology: cellular and molecular interactions. IEEE Trans Nanobioscience 6(4):331–340. https://doi.org/10.1109/TNB.2007.909005
19. Ehsanifar M, Banihashemian SS, Farokhmanesh F (2021) Exposure to urban air pollution nanoparticles and CNS disease, pp 520–526. https://doi.org/10.32474/OJNBD.2021.05.000223
20. Lorenz C et al (2012) Characterization of silver release from commercially available functional (nano)textiles. Chemosphere 89(7):817–824. https://doi.org/10.1016/j.chemosphere.2012.04.063

21. Almeida L, Ramos D (2017) Health and safety concerns of textiles with nanomaterials. IOP Conf Ser Mater Sci Eng 254(10). https://doi.org/10.1088/1757-899X/254/10/102002
22. McGillicuddy E et al (2017) Silver nanoparticles in the environment: sources, detection and ecotoxicology. Sci Total Environ 575:231–246. https://doi.org/10.1016/j.scitotenv.2016.10.041
23. Winans K, Kendall A, Deng H (2015) The history and current applications of the circular economy concept. Renew Sustain Energy Rev 68:825–833. https://doi.org/10.1016/j.rser.2016.09.123
24. Velenturf APM, Purnell P (2021) Principles for a sustainable circular economy. Sustain Prod Consum 27:1437–1457. https://doi.org/10.1016/j.spc.2021.02.018
25. Lieder M, Rashid A (2016) Towards circular economy implementation: a comprehensive review in context of manufacturing industry. J Clean Prod 115:36–51. https://doi.org/10.1016/j.jclepro.2015.12.042
26. Benachio GLF, do Freitas MCD, Tavares SF (2020) Circular economy in the construction industry: a systematic literature review. J Clean Prod 260:121046. https://doi.org/10.1016/j.jclepro.2020.121046
27. Bachmann J et al (2021) Eco-composites for interior and secondary structures
28. Maldonado-Guzmán G, Garza-Reyes JA, Pinzón-Castro Y (2021) Eco-innovation and the circular economy in the automotive industry. Benchmarking 28(2):621–635. https://doi.org/10.1108/BIJ-06-2020-0317
29. Zorpas AA, Navarro-Pedreño J, Panagiotakis I, Dermatas D (2021) Steps forward to adopt a circular economy strategy by the tourism industry. Waste Manag Res 39(7):889–891. https://doi.org/10.1177/0734242X211029087
30. Reh L (2013) Process engineering in circular economy. Particuology 11(2):119–133. https://doi.org/10.1016/j.partic.2012.11.001
31. Bucknall DG (2020) Plastics as a materials system in a circular economy: plastics in the circular economy. Philos Trans R Soc 378(2176)
32. Chen X, Memon HA, Wang Y, Marriam I, Tebyetekerwa M (2021) Circular economy and sustainability of the clothing and textile industry. Mater Circ Econ 3(1):1–9. https://doi.org/10.1007/s42824-021-00026-2
33. Saha K, Dey PK, Papagiannaki E (2021) Implementing circular economy in the textile and clothing industry. Bus Strateg Environ 30(4):1497–1530. https://doi.org/10.1002/bse.2670
34. Chen WK, Nalluri V, Hung HC, Chang MC, Lin CT (2021) Apply dematel to analyzing key barriers to implementing the circular economy: an application for the textile sector. Appl Sci 11(8). https://doi.org/10.3390/app11083335
35. Fischer A, Pascucci S (2017) Institutional incentives in circular economy transition: the case of material use in the Dutch textile industry. J Clean Prod 155:17–32. https://doi.org/10.1016/j.jclepro.2016.12.038
36. Kumar PS, Suganya S (2018) Systems and models for circular economy. Elsevier, Amsterdam
37. Coscieme L, Manshoven S, Gillabel J, Grossi F, Mortensen LF (2022) A framework of circular business models for fashion and textiles: the role of business-model, technical, and social innovation. Sustain Sci Pract Policy 18(1):451–462. https://doi.org/10.1080/15487733.2022.2083792
38. Ribul M, Lanot A, Tommencioni Pisapia C, Purnell P, McQueen-Mason SJ, Baurley S (2021) Mechanical, chemical, biological: Moving towards closed-loop bio-based recycling in a circular economy of sustainable textiles. J Clean Prod 326:129325. https://doi.org/10.1016/j.jclepro.2021.129325
39. Camilleri MA (2018) Closing the loop for resource efficiency, sustainable consumption and production: a critical review of the circular economy. Int J Sustain Dev 21(1–4):1–17. https://doi.org/10.1504/IJSD.2018.100802
40. Sandberg B, Kaartemo V (2022) Circular economy in global firm: a system change perspective to closed loop thinking. Supervisors 2003–2005
41. Bányai T (2004) Recycling and networking. Acta Polytech 44(2). https://doi.org/10.14311/554
42. Vance ME, Kuiken T, Vejerano EP, McGinnis SP, Hochella MF, Hull DR (2015) Nanotechnology in the real world: redeveloping the nanomaterial consumer products inventory. Beilstein J Nanotechnol 6(1):1769–1780. https://doi.org/10.3762/bjnano.6.181

43. Dericiler K, Berktas I, Dogan S, Menceloglu YZ, Saner Okan B (2022) General techniques for recovery of nanomaterials from wastes. In: Nanomaterials recycling, Elsevier, 2022, pp 147–174. https://doi.org/10.1016/B978-0-323-90982-2.00008-1
44. Lei Z, Chen B, Koo Y-M, MacFarlane DR (2017) Introduction: ionic liquids. Chem Rev 117(10):6633–6635. https://doi.org/10.1021/acs.chemrev.7b00246
45. Shah FU, An R, Muhammad N (2020) Properties and applications of ionic liquids in energy and environmental science. Front Chem 8. https://doi.org/10.3389/fchem.2020.627213
46. Christoff-Tempesta T, Epps TH (2023) Ionic-liquid-mediated deconstruction of polymers for advanced recycling and upcycling. ACS Macro Lett 12(8):1058–1070. https://doi.org/10.1021/acsmacrolett.3c00276
47. Abdelhamid HN (2022) Ionic liquids for nanomaterials recycling. In: Nanomaterials recycling. Elsevier, Amsterdam, pp 269–287. https://doi.org/10.1016/B978-0-323-90982-2.00024-X
48. Saleem A, Rahim A, Muhammad N, Abbas F (2020) Ionic liquids modified sensors and biosensors for detection of environmental contaminants, pp 259–273. https://doi.org/10.1007/978-3-030-45116-5_9
49. Mitrano DM, Rimmele E, Wichser A, Erni R, Height M, Nowack B (2014) Presence of nanoparticles in wash water from conventional silver and nano-silver textiles. ACS Nano 8(7):7208–7219. https://doi.org/10.1021/nn502228w
50. Fink JK (20212)Dispersions, Emulsions, and Foams," in *Petroleum Engineer's Guide to Oil Field Chemicals and Fluids*, Elsevier, 2012, pp. 663–694. https://doi.org/10.1016/B978-0-12-383844-5.00021-0.
51. Pan Y et al (2021) Stability and rheological properties of water-in-oil (W/O) emulsions prepared with a soyasaponin-PGPR system. Futur Foods 4:100096. https://doi.org/10.1016/j.fufo.2021.100096
52. Akbari S, Nour AH (2018) Emulsion types, stability mechanisms and rheology: a review. Int J Innov Res Sci Stud 1(1):11–17. https://doi.org/10.53894/ijirss.v1i1.4
53. Myakonkaya O, Guibert C, Eastoe J, Grillo I (2010) Recovery of Nanoparticles made easy. Langmuir 26(6):3794–3797. https://doi.org/10.1021/la100111b
54. Hartmann G, Schuster M (2013) Species selective preconcentration and quantification of gold nanoparticles using cloud point extraction and electrothermal atomic absorption spectrometry. Anal Chim Acta 761:27–33. https://doi.org/10.1016/j.aca.2012.11.050
55. Yousef S, Kalpokaitė-Dičkuvienė R, Baltušnikas A, Pitak I, Lukošiūtė SI (2021) A new strategy for functionalization of char derived from pyrolysis of textile waste and its application as hybrid fillers (CNTs/char and graphene/char) in cement industry. J Clean Prod 314:128058. https://doi.org/10.1016/j.jclepro.2021.128058
56. "Regenerate nature," Ellen MacArthur Foundation. https://ellenmacarthurfoundation.org/regenerate-nature
57. Garcia-Saravia Ortiz-de-Montellano C, Samani P, van der Meer Y (2023) How can the circular economy support the advancement of the sustainable development goals (SDGs)? A comprehensive analysis. Sustain Prod Consum 40:352–362. https://doi.org/10.1016/j.spc.2023.07.003
58. Pati P, McGinnis S, Vikesland PJ (2014) Life cycle assessment of 'green' nanoparticle synthesis methods. Environ Eng Sci 31(7):410–420. https://doi.org/10.1089/ees.2013.0444
59. Siddiqi KS, Husen A, Rao RAO (2018) A review on biosynthesis of silver nanoparticles and their biocidal properties. J Nanobiotechnol 16(1). https://doi.org/10.1186/s12951-018-0334-5
60. Shah MA, Pirzada BM, Price G, Shibiru AL, Qurashi A (2022) Applications of nanotechnology in smart textile industry: a critical review. J Adv Res 38:55–75. https://doi.org/10.1016/J.JARE.2022.01.008
61. Benn TM, Westerhoff P (2008) Nanoparticle silver released into water from commercially available sock fabrics. Environ Sci Technol 42(11):4133–4139. https://doi.org/10.1021/es7032718
62. Hendricks N, Olatunji O, Ofori I, Gumbi BP (2023) Occurrence, fate, and impact of engineered metal/carbonaceous nanomaterials in the environment, detection, and quantitation methods. Int J Environ Sci Technol 20(11):12937–12954. https://doi.org/10.1007/s13762-023-04977-8

63. Chen LH et al (2020) Photocatalytic properties of graphene/gold and graphene oxide/gold nanocomposites synthesized by pulsed laser induced photolysis. Nanomaterials 10(10):1–12. https://doi.org/10.3390/nano10101985
64. Malhotra N et al (2020) Toxicity studies on graphene-based nanomaterials in aquatic organisms: current understanding. Molecules 25(16):1–22. https://doi.org/10.3390/molecules25163618
65. Huston M, DeBella M, DiBella M, Gupta A (2021) Green synthesis of nanomaterials. Nanomaterials 11(8):2130. https://doi.org/10.3390/nano11082130
66. Kumar DA, Palanichamy V, Roopan SM (2014) "Green synthesis of silver nanoparticles using *Alternanthera dentata* leaf extract at room temperature and their antimicrobial activity. Spectrochim Acta Part A Mol Biomol Spectrosc 127:168–171. https://doi.org/10.1016/j.saa.2014.02.058
67. Gurudevi D Sirisha C (2023) Green Synthesis Of Copper Nanoparticles Using Neem (*Azadirachta indica*) leaf extract and their antimicrobial activity. (2):383–389
68. Gonçalves RA, Toledo RP, Joshi N, Berengue OM (2021) Green synthesis and applications of ZnO and TiO2 nanostructures. Molecules 26(8):2236. https://doi.org/10.3390/molecules26082236
69. Mehr F, Khanjani M, Vatani P (2015) Synthesis of nano-Ag particles using sodium borohydride. Orient J Chem 31(3):1831–1833. https://doi.org/10.13005/ojc/310367
70. Shahid R, Soliman HMA, Fathy M, Muhammed M (2012) Novel low temperature route for large scale synthesis of ZnO. Int J Sci. 1:153–161. ISSN 2305-3925
71. Bunekar N, Tsai TY (2022) Bio-based nanomaterials for properties and applications. In: Bio-based nanomaterials. Elsevier, Amsterdam, pp 67–72. https://doi.org/10.1016/B978-0-323-85148-0.00001-4
72. Salah SM (2013) Application of nano-cellulose in textile. J Text Sci Eng 03(04). https://doi.org/10.4172/2165-8064.1000142
73. Aswathi VP, Meera S, Maria CGA, Nidhin M (2023) Green synthesis of nanoparticles from biodegradable waste extracts and their applications: a critical review. Nanotechnol Environ Eng 8(2):377–397. https://doi.org/10.1007/s41204-022-00276-8
74. Dutta P, Kumari A, Mahanta M (2023) Chitosan nanoparticle: synthesis, characterization, and use as plant health materials. https://doi.org/10.5772/intechopen.106502
75. Marković D et al (2023) Antimicrobial and UV protective chitosan/lignin multilayer nanocoating with immobilized silver nanoparticles. J. Appl. Polym Sci 140(19). https://doi.org/10.1002/app.53823
76. Mosaad RM, Alhalafi MH, Emam E-AM, Ibrahim MA, Ibrahim H (2022) Enhancement of antimicrobial and dyeing properties of cellulosic fabrics via chitosan nanoparticles. Polymers (Basel) 14(19):4211. https://doi.org/10.3390/polym14194211
77. Noor Javed M et al (2023) Potential and challenges in green synthesis of nanoparticles: a review. **19** [Online]. Available: http://xisdxjxsu.asia

SPRINGER NATURE

GPSR Compliance

The European Union's (EU) General Product Safety Regulation (GPSR) is a set of rules that requires consumer products to be safe and our obligations to ensure this.

If you have any concerns about our products, you can contact us on ProductSafety@springernature.com

In case Publisher is established outside the EU, the EU authorized representative is:

Springer Nature Customer Service Center GmbH
Europaplatz 3
69115 Heidelberg, Germany

The manufacturer's authorised representative in the EU is Springer Nature Customer Service Centre GmbH, Europaplatz 3, 69115 Heidelberg, Germany. If you have any concerns regarding our products, please contact ProductSafety@springernature.com

Printed and bound by CPI Group (UK) Ltd, Croydon, CR0 4YY
30/09/2026
02225253-0018